W0116005

PRESTRESSED CONCRETE

Dr. G.S. PANDIT
Formerly, Senior Professor of Structural Engg.,
Malaviya Regional Engg. College,
JAIPUR (Rajasthan)

Dr. S.P. GUPTA
Department of Structural Engg.,
Malaviya Regional Engg. College,
JAIPUR (Rajasthan)

CBSPD

CBS Publishers & Distributors Pvt Ltd

New Delhi • Bengaluru • Chennai • Kochi • Kolkata • Lucknow • Mumbai
Hyderabad • Jharkhand • Nagpur • Patna • Pune • Uttarakhand

Prestressed Concrete

Copyright © Authors Publisher

ISBN: 978-81-239-0153-4

First Edition: 1993
Reprint: 1995, 2008, 2009, 2010, 2011, 2012, 2014, 2015, 2016, 2018, 2019, 2023

All rights reserved. No part of this publication may be reproduced, stored in a retrieval system, or transmitted, in any form or by any means, electronic, mechanical, photocopying, or otherwise, without the prior permission of the publisher.

Published by **Satish Kumar Jain** and produced by **Varun Jain** for
CBS Publishers & Distributors Pvt Ltd
4819/XI Prahlad Street, 24 Ansari Road, Daryaganj, New Delhi 110 002, India
Ph: 011-23289259, 23266861   Website: www.cbspd.com
                                     e-mail: delhi@cbspd.com

Corporate Office: 204 FIE, Industrial Area, Patparganj, Delhi 110 092, India
Ph: 011-4934 4934   Fax: 011-4934 4935   e-mail: publishing@cbspd.com;
                                              publicity@cbspd.com

Branches

- **Bengaluru:** Seema House 2975, 17th Cross, KR Road, Banasankari 2nd Stage, Bengaluru 560 070, Karnataka, India
  Ph: +91-80-26771678/79   e-mail: bangalore@cbspd.com
- **Chennai:** 7, Subbaraya Street, Shenoy Nagar, Chennai 600 030, Tamil Nadu, India
  Ph: +91-44-26680620, 26681266   Fax: +91-44-42032115   e-mail: chennai@cbspd.com
- **Kochi:** 42/1325, 1326, Power House Road, Opp KSEB, Power House, Ernakulum Kochi 682 018, Kerala, India
  Ph: +91-484-4059061-65,67   Fax: +91-484-4059065   e-mail: kochi@cbspd.com
- **Kolkata:** 147, Hind Ceramics Compound, 1st Floor, Nilgunj Road, Belghoria, Kolkata 700056, West Bengal, India
  Ph: +033-25633055, 033-25633056   e-mail: kolkata@cbspd.com
- **Lucknow:** Basement, Khushnuma Complex, 7 Meerabai Marg (Behind Jawahar Bhawan), Lucknow-226001, UP, India
  Ph: +91-522-4000032   e-mail: tiwari.lucknow@cbspd.com
- **Mumbai:** PWD Shed, Gala no 25/26, Ramchandra Bhatt Marg, Next to JJ Hospital Gate no. 2, Opp. Union Bank of India Noorbaug, Mumbai-400009, Maharashtra, India
  Ph: 022-66661880/89   e-mail: mumbai@cbspd.com

Representatives

• Hyderabad	0-9885175004	• Jharkhand	0-9811541605	• Nagpur	0-9421945513
• Patna	0-9334159340	• Pune	0-9923910676	• Uttarakhand	0-9716462459

Printed at Neekunj Print Process, Haryana, India

Preface

Ever since the introduction of high tensile steel by Freyssinet and the successful development of techniques of mechanical prestressing in the decade of thirties, prestressed concrete occupies a dominant position in the field of concrete construction. The reason is not far to seek. In the words of Guyon, there is probably no structural problem to which prestressing cannot provide a solution and often a revolutionary one. Prestressed concrete offers several advantages over the other two main structural materials, viz. reinforced concrete and structural steel. It is, therefore, not surprising that prestressed concrete is being used for almost every type of structural concrete system. However, there are certain areas in which prestressed concrete excels. A good example is the domain of long span highway and railway bridges in which prestressed concrete often competes favourably with other structural materials. There are also some areas in which prestressed concrete holds almost an absolute monopoly. The field of supersize containment structures is an excellent example. Without the use of prestress, the construction of containers with multi-million litre capacity would have remained only a dream.

In view of the pre-eminent position of prestressed concrete in civil engineering construction, all prominent codes of practice contain specifications for prestressed concrete design. While the American code ACI: 318-1989 contains a separate chapter on prestressed concrete, the European CEB-FIP Model Code devotes a significant portion of its provisions to prestressed concrete. The British code BS: 8110; Part I: 1985 contains two sections devoted exclusively to prestressed concrete. The Bureau of Indian Standards has brought out a separate code IS: 1343-1980 for prestressed concrete. The importance of prestressed concrete in the discipline of civil engineering is reflected by the fact that it is taught as a full course at several universities and technical institutes in India. While it is being offered as an elective at undergraduate level, it is taught as one of the several courses at post-graduate level.

There are several good books on prestressed concrete written by authors from leading industrialised countries. However, these books are based on codes of practice of the respective countries. A few of them are still adopting the Imperial System of units. The few indigenous works on the subject suffer from lack of clarity, inadequate coverage of basic material and inconsistent or illogical distribution of topics. They are not based on the latest versions of the codes of practice. It is disturbing to note that some of the errors in our code have been carried over in these books. For instance, the equation for T_{c1} in Section 22.5.4.1 of IS: 1343-1980 has not been adjusted properly during conversion to SI units. The error has been overlooked by several authors leading to erroneous computations. The illustrative examples are generally found lacking in that they do not bring out several finer aspects of analysis and design.

The main features of the present book are

(i) lucidity of presentation

(ii) critical discussion and adoption of Indian code together with the latest versions of British, American and European codes

(iii) use of SI units in text, illustrative examples and codes of practice

(iv) preview of the subject matter in the first chapter in simple and lucid form

(v) thorough coverage of the basic information regarding materials, systems of prestressing and losses of prestress in three separate chapters

(vi) Comprehensive discussion of flexural members in two chapters and that of tension and compression members in separate chapters

(vii) treatment of the latest developments in composite and partially prestressed beams in two separate chapters

(viii) simple but comprehensive treatment of statically indeterminate prestressed elements such as continuous beams, ring beams, cylindrical walls, pressure vessels and slabs

(ix) separate chapter on precast prestressed elements

(x) 197 carefully selected fully worked out illustrative examples

(xi) 208 diagrams to facilitate easy reading

(xii) 201 unsolved problems with answers

(xiii) list of selected references with each chapter.

The entire treatment is marked by simplicity and ease of comprehension, free from distraction and mathematical jargon. The load balancing approach, generalized in the form of free body approach by the authors, has been used extensively in the analysis and design, particularly for statically indeterminate prestressed elements.

G.S. PANDIT
S.P. GUPTA

Glossary of Terms

ABUTMENT—block or device to hold the prestressing force temporarily until transfer in pretensioning method

ANCHORAGE—device to anchor the tendon to the concrete member in post-tensioning or device to anchor tendon during hardening of concrete in pretensioning

ANCHORAGE STRESSES—stresses caused by anchorage

ANCHORAGE ZONE—zone in which anchorage stresses are created

BALANCE LOAD—part of the external load counteracted by prestress

BENDING CONCORDANCE—concordance in respect of bending moments only

BONDED POST-TENSIONING—post-tensioned construction in which tendons are bonded to concrete through grouting of the ducts

BONDED MEMBER—prestressed concrete member in which the tendons are bonded to the concrete either directly or through grouting

BULKHEAD—same as abutment

BURSTING FORCE—force tending to cause bursting of the duct due to tendon force

BUTTON HEAD—anchoring device in Prescon system

CABLE—coaxial assembly of prestressing wires

CABLE LINE—Line along which total resultant prestressing force acts

CABLE PROFILE—profile or shape of the cable line

CAMBER—negative (upward) deflection

CENTRE OF COMPRESSION—centroid of axial stress or point of action of resultant axial compression at any cross-section of a member

CHARACTERISTIC LOAD—load which has 95 per cent probability of not being exceeded during the life of the structure

CHARACTERISTIC STRENGTH—strength of material below which not more than 5 per cent of the test results are expected to fall

CONCORDANCE—absence of secondary forces due to prestressing

CONCORDANT CABLE PROFILE—cable profile which produces concordance

CREEP IN CONCRETE—increase in strain of concrete with time due to sustained stress

CREEP COEFFICIENT—the ratio of creep strain to elastic strain in concrete

DEGREE OF PRESTRESSING—ratio of decompression moment to working moment

DUCT—tunnel like hole formed in concrete member for housing the prestressing tendon

ECCENTRICITY—distance from centroid of the cross-section

END BLOCK—portion near end of member with appreciable anchorage stresses

FINAL PRESTRESS—prestress which exists after all substantial losses have occurred

FINAL TENSION—tensile force in prestressing tendons after all substantial losses

GROUT—cement or cement-sand slurry pumped into the duct for bonding prestressing tendon with concrete

GUNITING—technique of placing concrete by spraying under pressure

HOSE—solid or hollow cylindrical piece for forming the duct in a concrete member

INITIAL PRESTRESS—prestress in the concrete at transfer

INITIAL TENSION—maximum stress induced in the prestressing tendon at ther time of the stressing operation

INTERNAL FORCES—stress resultants, i.e. tension or compression, shea force, bending moment and twisting moment

JACKING END—end of the member where jack is inserted for pulling the prestressing tendon

KERN POINTS—extreme limits of centre of compression for safe stress or no tension condition

LENGTH EFFECT—wave or wobble effect in computing the frictional loss

LIMIT STATE—the state at which a limiting value of strength or serviceability is attained

LINE OF THRUST—the line along which the resultant compression or thrust on the section acts

LINEAR PRESTRESSING—technique of prestressing by separate tendons as opposed to continuous tendons in circular prestressing

LINEAR TRANSFORMATION—shifting the cable profile without changing the intrinsic shape or end eccentricities in a continuous beam

LONG TERM—after a long interval of time or at infinite time

NEUTRAL AXIS—axis of zero stress

PARTIAL PRESTRESSING—prestressing in which only part of the service load is carried by prestressing steel and the remaining part by reinforcing steel

PARTIAL PRESTRESSING RATIO (PPR)—ratio of ultimate moment resisted by prestressing steel to total ultimate moment

POSTCRACKING RANGE—range of loading from cracking load to collapse load

POST-TENSIONING—tensioning of prestressing tendons after concreting

PRECRACKING RANGE—range of loading from cracking load to cracking load

PRESSURE GROUTING—grouting the duct under pressure

PRESSURE LINE—line of thrust or line along which resultant compression acts

PRESTRESSING — technique of creating permanent stresses to counteract stresses caused by service loads

PRESTRESSING BED — bed or platform used for prestressing concrete elements by pretensioning method

PRESTRESSED CONCRETE — concrete in which permanent internal stresses are deliberately introduced, usually by tensioned steel, to counteract to the desired degree the stresses caused in the member in service

PRESTRESSING INDEX — ratio of yield load of prestressing steel to yield load of total steel

PRESTRESSING JACK — hydraulic or mechanical device for pulling prestressing tendon

PRESTRESSING MOMENT — bending moment due to prestress

PRETENSIONING — tensioning prestressing steel before concreting

PRIMARY PRESTRESSING MOMENT — product of prestressing force and its eccentricity

RENINFORCING STEEL — unprestressed reinforcement

RESULTANT MOMENT — algebraic sum of bending moments due to prestress and external loads

SAG — vertical distance between the lowest point and the line joining the ends of the tendon

SECONDARY FORCES — internal forces caused by prestress in an unloaded continuous beam

SECONDARY PRESTRESSING MOMENT — bending moment due to support reactions caused by prestress in an unloaded continuous beam

STRESSES AT TRANSFER — initial stresses in concrete and prestressing tendons at transfer of prestress

TENDON — wire, cable, bar or strand used to impart prestress to concrete

TOTAL CONCORDANCE — concordance with respect to bending moments and also axial compression

TRANSFER — act of transferring stress in prestressing tendons from the jacks or prestressing bed to the concrete member

TRANSMISSION LENGTH — the distance required at the end of a pretensioned tendon for developing the maximum tendon stress by bond

UNBALANCED LOAD — part of the external load not counteracted

WAVE EFFECT — effect of friction caused by waves in concrete surface inside the duct

WOBBLE EFFECT — same as wave effect

About the Authors

G.S. Pandit, formerly Senior Professor of Structural Engineering at Malaviya Regional Engineering College, Jaipur, earned his M.Sc. and Ph.D. degrees from the University of Alberta, Canada. He has vast teaching experience and has authored over 100 research papers and 150 other publications in leading technical journals. He is the co-author of a book 'Structural Analysis – a matrix approach' and a monograph on 'Torsion in Concrete Structures' written with the financial support of University Grants Commission. Dr. Pandit received the Malaviya Regional Engineering College award for outstanding research and was a co-winner of the Certificate of merit for his paper in the Journal of the Institution of Engineers (India). For several years he served on the Committee 438 (Torsion) of the American Concrete Institute for coordination and liaison of research activities in India.

S.P. Gupta, Reader in the Department of Structural Engineering at Malaviya Regional Engineering College, Jaipur obtained his B.E. and M.Sc. degrees from Delhi University and Ph.D. from the University of Rajasthan. He has teaching experience of over 25 years. Dr. Gupta is the author of over twenty five research papers and is the co-author of a book 'Structural Analysis – a matrix approach', a monograph on 'Torsion in Concrete Structures' and 'Civil Engineering through Objective Type Questions'.

Contents

1

Basic Concepts

1.1 INTRODUCTION

Every structure is designed to serve a specific purpose or to perform a function or a set of functions. In doing so, the structure receives loads, known as service loads, at certain points and transmits them safely and efficiently to some other points. For instance a building structure receives its service loads in the form of the occupancy loads on its floors and transmits them through its system of beams and columns to the foundations. Similarly a bridge structure receives its service loads in the form of wheel loads on its deck and transmits them to the foundations through the piers and abutments. In the process of transferring the service loads from one point to another, the structure develops internal stresses known as service stresses.

The general principle of prestressing is to create prior stresses whose nature or sign is opposite to that of the service stresses. The process of prestressing may be accomplished through a variety of devices or techniques known as the prestressing systems. According to Freyssinet, popularly known as the father of prestressing, prestressing is more than a technique. It is a general philosophy or principle of inducing a favourable stress condition opposed to the stress conditions caused by the service loads. According to Guyon, there is probably no structural problem to which prestressing cannot provide a solution and often a revolutionary one. The immediate result of prestressing is that, in general the magnitude of the net stresses due to the combination of the prestress and the service loads is smaller than that of the service stresses. The prestress may also eliminate completely certain undesirable stresses such as the tensile stresses in concrete structures. Thus the technique of prestressing produces a superior structure capable of performing its function more effectively. As the net stresses for a given set of loads are reduced on account of prestressing, the load carrying capacity of the prestressed structure is higher than that of the corresponding non-prestressed structure. Thus the prestressed structures are able to support heavier loads over longer spans as compared to the non-prestressed structures. It is this principal feature of prestressed structures which has revolutionised the construction industry over the past fifty years. Today prestressed structures are able to compete favourably with other types of structures in the domain of long-span construction. Prestressed concrete holds a complete monopoly in the area of super-size containment structures. Prestressed structures may be expected to enter new avenues and push forward the frontiers of civil engineering construction in the years ahead.

Although the general principle of prestressing may be used for variety of construction materials, only prestressed concrete has been discussed herein. Prestressed concrete structures constitute the bulk of prestressed structures being built today. Due to the inherent weakness of cement concrete in tension, the main object of prestressing a concrete structure may be the elimination of the tensile stresses. The result is a crack free structure with its associated benefits. A detailed comparison of prestressed concrete and reinforced concrete is given in Sec. 1.14. The other sections of this chapter are devoted to the basic concepts and principles of this interesting material-prestressed concrete.

1.2 DEVELOPMENT OF PRESTRESSED CONCRETE

The basic principle of prestressing has been known from olden days. For instance, the old practice of heating the steel rim before slipping it on the wooden wheel of a bullock cart is an example of the application of prestress. The wire-wound gun barrels constitute an example of circular prestressing. The tightening of ropes or steel bands around wooden staves of a barrel and tensioning of bicycle spokes are other examples of prestressing techniques. The first attempt to induce prestress in concrete was made by Jackson of United States in 1886 when he obtained a patent for construction of artificial stone and concrete pavements in which prestress was introduced by tensioning the reinforcing rods set in sleeves. In 1888 Doehring of West Germany used tensioned wires to prestress slabs and small beams. These early attempts were not successful because the prestress was totally lost within a few months. An attempt was made to compensate for the loss of prestress by retensioning the reinforcing bars. Even this technique had only a very limited success. Besides, the technique was not found to be economical. The real break-through in the technique of prestressing came in the year 1928 when Freyssinet made it clear that worthwhile prestress cannot be created by tensioning ordinary reinforcing bars. Instead of the conventional reinforcing bars, he proposed the use of high tensile steel having an ultimate strength several times greater than that of mild steel. The conventional mild steel reinforcing bars can be prestressed to a stress of only 140 MPa producing a strain of $140/(2 \times 10^5) = 0.0007$ which is smaller than the loss of strain due to creep and shrinkage of concrete and other factors responsible for the loss of prestress. This explains why the early attempts to produce prestress by tensioning the reinforcing bars did not succeed. Due to its superior ultimate strength the initial prestress in high tensile steel can be much higher. For instance, if the initial stress in high tensile steel is 1200 MPa, the initial strain is $1200/(2 \times 10^5) = 0.006$. If the loss of strain due to various causes is as high as 0.001, the total loss of prestress is $(0.001/0.006) \times 100 = 16.7$ per cent. Thus the residual or effective prestress is as high as 83.3 per cent. These simple calculations clearly show why high tensile steel must be used instead of ordinary reinforcing steel to obtain a substantial effective prestress. Since the introduction of high tensile steel by Freyssinet, prestressed concrete made rapid strides in the construction industry. In the last few decades prestressed concrete has been used for the construction of industrial structures, shell roofs, folded plates, stadia, airport hangers, long span bridges, curved skew flyovers, containment structures, marine structures, nuclear pressure vessels, transmission poles, railway sleepers and variety of other structures.

In Europe the lead in the development of prestressed concrete was taken by France and Belgium. However, It was quickly followed by other European countries such as United Kingdom, Germany, Switzerland, Italy, Soviet Union and Holland. In 1939 the French engineer Freyssinet prefected his anchoring device in the form of male and female cones and his double acting jack. A year later (1940) Professor Magnel of Belgium developed his well known prestressing system in which the prestressing wires are anchored by means of sand-wich plates and flat wedges. Europe took the lead in the domain of bridges. The shortage of reinforcing steel in Europe during the years following world war II gave impetus to prestressed concrete bridges which required lesser quantities of steel. By 1951 as many as 175 bridges were constructed together with approximately 50 buildings using prestressing technique. As many as 350 out of 500 bridges constructed in Germany during the years 1949 to 1953 were of prestressed concrete. Since 1965 approximately half the bridges constructed in Germany are of prestrssed concrete. Hoyer of Germany developed the technique known as long line system (Hoyer System) which is particularly suitable for mass production of small precast prestressed units such as electric poles and railway sleepers in a central plant. Eriksson and Ulf Bjnggren of Sweden developed the technique for the production of large size units for incorporation in the framework of industrial structures. In 1978 Soviet Union produced 25 million cubic metres of prestressed concrete most of which was in the form of precast pretensioned elements for buildings.

The United States of America took the lead in the domain of prestressed concrete containment structures. The technique of circular prestressing for the construction of super-size water tanks was developed by Preload Company which constructed as many as 1000 prestressed concrete tanks in USA and other parts of the world during the years 1935 to 1963. The first application of linear prestressing in United States was in the year 1949 when the construction of Philadelphia Walnut Lane Bridge was started. A survey by Bureau of Public Roads showed that during the years 1957-60, more than 2000 prestressed concrete bridges were authorized for construction. By 1961 United States had more than 200 plants producing 1.5 million cubic metres of precast prestressed elements annually for use in buildings and bridges. By the year 1975, the number of plants increased to 500. The consumption of prestressing steel increased threefold during the years 1965-74. The economy of post-tensioning methods for the construction of flat slabs is reflected from the fact that 1.9 million square metres of post-tensioned flat slabs were constructed during the year 1974. The rapid development of prestressed concrete construction in United States is indicated by the formation of Prestressed Concrete Institute in 1954 and Post-tensioning Institute in 1976 and several important publications such as Proceedings of the World Conference on Prestressed Concrete in San Francisco in 1957, Proceedings of Western Conferences on Prestressed Concrete Buildings in California in 1960, Prestressed Concrete

Handbook in 1971 and Post-tensioning Manual. The applications of prestressed concrete for the construction of offshore structures, oceangoing barges and large containment structures for nuclear power stations were reviewed at joint sessions of American Concrete Institute (ACI), Prestressed Concrete Institute (PCI) and International Federation for, prestressing (FIP) in Philadelphia in 1975.

The Latin American countries were not far behind United States in prestressed construction. Several long span bridges have been constructed in Brazil. Prestressed concrete has been used in the construction of arch bridges in Venezuela. Several long span bridges with spans upto 100m have been constructed in Cuba. Mass production of precast prestressed elements in centralised plants has been undertaken by Argentina.

In India, prestressed concrete was used in 1940 for the construction of army tank garages in Meerut city. Initially the major applications of prestressed concrete were in the domain of long span bridges on national highways crossing the mighty rivers. Of the 222 bridges built by 1975, as many as 50 were of prestressed concrete. The early prestressed concrete bridges were for those for Assam Rail Links and across Palar river in Tamil Nadu. In the last three decades prestressed concrete has been extensively used in almost every field of construction. India has also entered in a big way in the mass production of precast prestressed elements. The country has a very large number of factories producing transmission poles and other small precast elements. The Indian Railway has established numerous plants for the manufacture of prestressed concrete sleepers. It is difficult to take account of innumerable prestressed concrete structures built in India. However some of the prominent ones are enumerated below.

(A) Buildings

    (i) Rajasthan Atomic Power Project, Kota⁻
       Prestressed dome, height = 49.17 m

    (ii) Workshop, Madras Port
       North light roof truss, span = 18.3 m

    (iii) Wet dock transit sheds, Madras Port
       Precast post-tensioned girders, span = 30.5 m

    (iv) Jetty work, Madras Port
       Prestressed girders spanning 18.3 m to carry pipelines from the tanker berth and to serve as a motorable way over the jetty

    (v) Fertilizer factory at Namrup
       Precast hollow blocks over prestresesd beams to reduce dead weight in seismic prone area

    (vi) Dum-Dum airport hanger at Calcutta
       Five cylindrical shells of 48 × 10.2 m chord

    (vii) Boeing hanger at Santa Cruz airport, Bombay
       Ten cylindrical shells of 45 × 9 m chord

(viii) DCM factory at Ghaziabad
Twin bow-string girder, hollow prestressed concrete box girder stringer of 45 m span

(B) Bridges

(i) Ganga Bridge at Patna, Bihar
46 spans of 121 m each

(ii) Bassein Creek Bridge, Bombay
Length = 555.32 m, maximum span = 114.6 m

(iii) Thana Creek Bridge, Bombay
Length = 1850 m, maximum span = 48.8 m

(iv) Jalangi Bridge, West Bengal
Length = 222.5 m, maximum span = 39.955 m

(v) Bhagirath Bridge, Murshidabad, West Bengal
Length = 280.61 m, maximum span = 78.39 m

(vi) Ajoy Bridge, West Bengal
Length = 524 m, maximum span = 45.14 m

(vii) Sone Bridge at Chopan, Bihar
Length = 1006.5 m, maximum span = 45.75 m

(viii) Coleroon Bridge at Anaikaranchattram
14 spans of 48.5 m

(ix) Wardha Bridge, Maharashtra
Length = 219.14 m, span = 40.56 m

(x) Kunwary River Bridge at Dimmi
2 spans of 39.67 m each and one span of 40.56 m

(C) Containment Structures

(i) Storage silo at Neyveli Lignite Project
261 m long and 28.7 m wide at springing level

(ii) Storage silo at Nangal, Punjab
312.9 m long and 34.4 m wide at springing level

(iii) Kilpauk Water Tower, Tamil Nadu
Capacity = 6820 m³, internal diameter = 31.72 m, depth of water = 8.54 m

(iv) Solan Reservoir, Himachal Pradesh
Capacity = 8183 m³, internal diameter = 41.48 m, depth of water = 6.71 m

(v) CIR Emergency Water Storage Tank, Trombay
Capacity = 3864.1 m³, internal diameter = 24.4 m, depth of water = 6.71 m

(vi) Madras Atomic Power Project

World's first double walled, containment structure with inner wall in prestressed concrete and the outer shielding wall in reinforced masonry, diameter of prestressed inner vessel = 39.62 m

(vii) Narora Atomic Power Project

Double wall reactor building with an outer wall in reinforced concrete and the inner containment wall in prestressed concrete

(D) Other Structures

(i) Wardha aqueduct, Karad, Maharashtra

Longitudinally prestressed box section 3.5 × 4.37 m, capacity = 30 m^3, also carrying one lane traffic on top slab over 20 spans of 47 m each

(ii) Bhima aqueduct, Maharashtra

Truncated circular cross section of 4.8 m diameter with a capacity of 42.5 m^3, also carrying one lane traffic on top slab over 22 spans of 41.5 m and 2 cantilever end spans of 17 m each, the circular part prestressed transversely by cables anchored along edges of the top slab.

(iii) Kunu Syphon, Gwalior, Madhya Pradesh

Internal diameter = 6.1 m,

total length = 1788 m

(iv) Bhandardara Dam, Ahmednagar, Maharashtra

Prestressing used as an emergency measure for strengthening against horizontal crack along the upstream face and going down to the downstream toe in an inclined fashion

(v) Koyna Dam, Maharashtra

Strengthened by cables each having 64 wires of 8 mm diameter in vertical holes drilled through the body of the dam delivering a prestress force of 2700 kN per cable.

1.3 NATURE OF PRESTRESSED CONCRETE

Due to the intrinsic weakness of concrete in tension, the main object of prestressing a concrete structure is to overcome the tensile stresses by inducing compressive stresses in the form of prestress. Although it is technically possible to induce compressive stresses in concrete without the use of steel, the most common and effective way is to use high tensile steel for inducing prestress. The prestressing tendons in the form of wires, bars, multi-wire strands and cables placed at appropriate locations in the cross-section of the member, are stretched by means of jacks. If the jacks rest against abutments, bulkheads or steel moulds during the stretching of tendons and concrete is placed around the stretched steel, the technique is called pretensioning. In this case, the high tensile steel is released from the bulkheads after concrete has gained sufficient strength and the prestress is transferred to concrete through bond. On the other hand, if the jacks rest against the hardened concrete of the member itself, which has been precast a few weeks in advance, the technique is called post-tensioning. In this case, the jacks can be released only after the prestress of the high tensile steel is transferred as a compressive force in concrete by means of end anchorages.

In all prestressed members, whether pretensioned or post-tensioned, the two main elements are the high tensile steel which is initially in tension and concrete which carries an equal and opposite compressive force. Thus the two main elements, viz. the prestressing high tensile steel and the concrete together constitute a self equilibrating system. As the two elements induce or develop prestress by straining each other, a prestressed concrete member may be visualised as a case of self straining. While the resultant forces in the two elements are equal in magnitude the manner in which they carry these forces are quite different. The high tensile wires, bars, multiwire strands and cables have negligible stiffness againt shear, bending and torsional deformations. Hence high tensile prestressing steel essentially carries only axial tension. The total tensile force in all tendons may be assumed to act along the axis of an imaginary cable which coincides with the line, called the cable line, connecting the centroids of the tensile forces in individual tendons at all cross-sections over the entire length of the member. The other element, i.e. concrete, has appreciable stiffness against the deformations caused by the four types of internal forces, viz. the axial and shear forces and the bending and twisting moments. Consequently, it is capable of resisting all four types of internal forces. The stress distribution in concrete must be such that it provides stress resultants equal and opposite to the four internal forces at every cross-section of the member.

As the prestressing steel is in a state of simple tension, free from shear forces and bending and twisting moments, it is easy to draw the free body diagram by considering the equilibrium of the free body of prestressing steel. Thus all forces acting on the cable can be readily computed if the axial tension is known. The forces imposed on concrete by the prestressing steel are equal in magnitude and opposite in sign to those acting on prestressing steel. The net forces on concrete can be calculated by combining these forces, which are caused only due to the effect of prestress, with the external forces. Once these net forces are known, the structure may be analysed for stresses like any ordinary non-prestressed structure. The effect of prestress on the stresses and deformation of the structure is fully taken care of by considering the forces imposed by the prestressing tendons on the surrounding concrete. This approach, known as the free body approach, is very useful in the analysis of internal forces, particularly in statically indeterminate structures.

1.4 SIGN CONVENTION

The sign convention adopted throughout this book conforms to those most widely utilised in structural analysis and have, therefore, received almost universal acceptance. The tensile forces and the corresponding tensile stresses and strains are taken to be positive while the compressive forces, stresses and strains are treated as negative quantities. As a prestressed concrete member represents a self-straining system, the prestressing force exerts a tensile force on prestressing steel and an equal compressive (negative) force on concrete. Hence taking P to represent the numerical (unsigned) value of the prestressing force, the force exerted by prestressing on concrete is $(-P)$ while it is $(+P)$ on prestressing steel. The corresponding stresses and strains in the two materials are treated likewise. The sign convention for beams is summarized in Table 1.4.1. Transverse loads, deflections, eccentricity of prestress e_p, eccentricity of centre of compression e_c and the distance of any point from the centroidal axis y are taken to be positive when they are downward. For frames, the common frame convention is adopted in which clockwise couples and rotations are taken to be positive.

Table 1.4.1 Sign Convention

S. No.	Item	Positive	Negative
1.	Axial forces, stresses and strains	tensile	compressive
2.	Distances y, e, and e_c	below centroidal axis	above centroidal axis
3.	Transverse loads and forces	downward	upward
4.	Shear force	downward to right	upward to right
5.	Bending moment	sagging	hogging
6.	Rotation or slope	clockwise	anticlockwise
7.	Deflection	downward	upward

1.5 MEMBER PROFILE

The members of prestrssed concrete skeletal structures are frequently non-prismatic. For instance, the cross-sectional area of a prestressed concrete continuous girder is larger at the supports as compared to that at midspan. The member profile is defined by the shape of the centroidal axis of the member and the shape and size of its cross-sections. The centroidal axis of a member is the line joining the centroids of all cross-sections of the member. The centroidal axis of straight prismatic member is a straight line. On the other hand, the centroidal axis of a non-prismatic member may be a straight line or a curve. For instance, if the top face of non-prismatic girder is straight and the bottom face is curved, the centroidal axis is a curved line. The member profile defines the geometrical properties of the member such as the position and shape of the centroidal axis, and cross-sectional properties of the member.

The location of the centroid of a prestreseed concrete member varies slightly depending upon whether the gross area, the net area of concrete or the transformed area is considered in the computation. As the variation in the position of the centroid is very small, the centroidal axis is generally taken as the line connecting the centroids of the gross cross-sectional areas of the member at all points along its length.

The cross-sectional properties of a prestressed concrete uncracked section may be determined on the basis of the following three types of sections:

(i) *Gross section*

The presence of steel is ignored and the cross-sectional properties are determined on the basis of overall section.

(ii) *Transformed or equivalent section*

The transformed or equivalent section is obtained by replacing the steel area by its equivalent concrete area taken equal to m times the area of steel where m is the modular ratio.

(iii) *Net concrete section*

The area of concrete lost due to presence of steel in pretensioned members or the cable ducts in post-tensioned members is deducted from the gross section to compute the net concrete area.

Consider for instance a rectangular section of breadth b and overall depth D. On the basis of gross-section, the cross-sectional area A, depth of the centroidal axis from top face y_T and moment of intertia I are given by the equations,

$$A = bD \qquad (1.5.1a)$$

$$y_T = \frac{D}{2} \qquad (1.5.1b)$$

$$I = \frac{1}{12} bD^3 \qquad (1.5.1c)$$

Assuming the total area of prestressing steel A_p to be located at its centroid at an effective depth d_p from the top face, the cross-sectional properties on the basis of transformed section may be expressed as

$$A = bD + (m - 1) A_p \qquad (1.5.2a)$$

$$I = \frac{1}{3} by_T^3 + \frac{1}{3} b (D-y_T)^3 + (m-1) A_p (d_p-y_T)^2 \qquad (1.5.2b)$$

in which y_T may be determined by taking moments about centroidal axis,

$$\frac{by_T^2}{2} = b \frac{(D-y_T)^2}{2} + (m-1) A_p (d_p - T) \qquad (1.5.2c)$$

For a pretensioned member, the sectional properties on the basis of net area of concrete are given by

$$A = bD - A_p \qquad (1.5.3a)$$

$$I = \frac{1}{3} b y_T^3 + \frac{1}{3} b (D-y_T)^3 - A_p (d_p-y_T)^2 \qquad (1.5.3b)$$

in which y_T may be determined from,

$$\frac{by_T^2}{2} = b \frac{(D-y_T)^2}{2} - A_p (d_p-y_T) \qquad (1.5.3c)$$

For a post-tensioned member having total area of ducts A_d located at its centroid at depth d_d from the top face, the cross-sectional properties on the basis of net concrete section may be determined from Eq. (1.5.3) in which A_p and d_p are replaced by A_d and d_d respectively. The correctness of using the three sets of sectional properties in the determination of stresses due to various causes may now be examined. In determining the stresses caused by prestressing only, it is evident that the properties based on net concrete section are appropriate because prestressing steel transfers its force to the net concrete section. The stresses caused by superimposed loads are resisted jointly by concrete and steel in a pretensioned member, and in a post-tensioned member after effective grouting (Fig. 1.5.1a). Hence in these cases, the use of sectional properties based on transformed section are appropriate. In the case of unbonded members or in bonded members before grouting (Fig. 1.5.1b), the sectional properties based on net concrete section are appropriate because the stresses in the members due to superimposed loads are resisted essentially by the net concrete section. Similar recommendations are made in Sec. 18.3.1 of IS: 1343-1980. It may, however, be pointed out that the three sets of sectional properties differ only slightly from each other. Hence the cross-sectional properties based on gross-section may be used for the sake of simplicity.

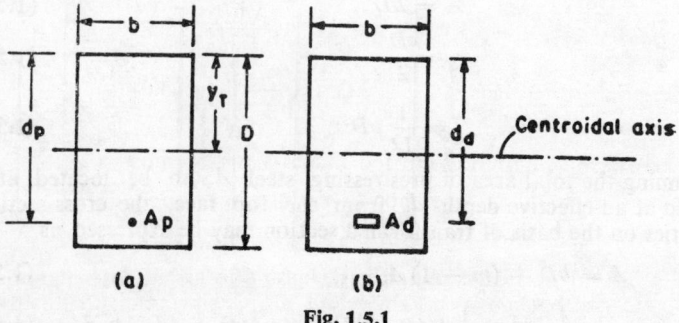

(a)   (b)

Fig. 1.5.1

EXAMPLE 1.5.1

Figure 1.5.2 (a) shows the profile of a pretensioned roof beam of an industrial shed. The beam has a rectangular cross-section. At a cross-section 2.5 m from left end, determine the cross-sectional area A, position of centroid and moment of inertia I about centroidal axis based on (i) gross area, (ii) transformed area and (iii) net area of concrete. Take modular ratio, $m = 7.5$.

Solution

The cross-section at quarter point is shown in Fig. 1.5.2(d). Area of prestressing steel,

$$A_p = 18 \times \frac{\pi}{4} \times 5^2 = 353.6 \text{ mm}^2$$

The depth of the centroid of prestressing steel d_p is determined by taking moments about the top face.

$$d_p = \frac{3 \times 0.05 + 5 \times 0.25 + 5 \times 0.30 + 5 \times 0.35}{18}$$

$$= 0.2583 \text{ m}$$

(i) *Based on gross area*

$$A = 0.20 \times 0.45 = 0.090 \text{ m}^2$$

$$y_T = \frac{D}{2} = \frac{0.45}{2} = 0.2250 \text{ m}$$

$$I = \frac{1}{12} \times 0.20 \times 0.45^3 = 0.0015187 \text{ m}^4$$

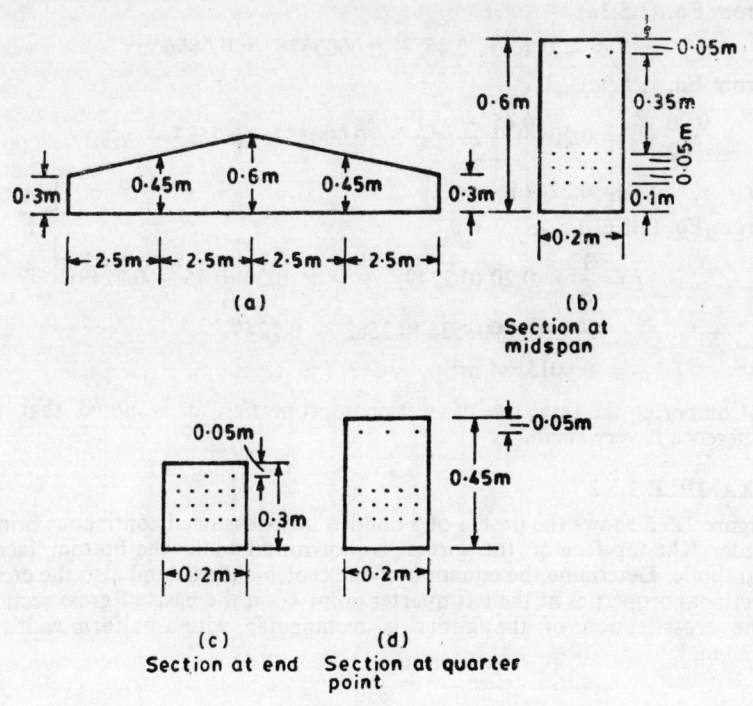

(a)

(b)
Section at midspan

(c)
Section at end

(d)
Section at quarter point

Fig. 1.5.2

(ii) *Based on transformed area*

Using Eq. (1.5.2a)

$$A = 0.20 \times 0.45 + (7.5 - 1) \times 0.0003536 = 0.0923 \text{ m}^2$$

From Eq. (1.5.2c),

$$\frac{0.20\, y_T^2}{2} = 0.20 \frac{(0.45 - y_T)^2}{2}$$

$$+ \mathbf{I}(7.5 - 1) \times 0.0003536\,(0.2583 - y_T)$$

or $\qquad y_T = 0.2258$ m

Using Eq. (1.5.2b),

$$I = \frac{1}{3} \times 0.20 \times 0.2258^3 + \frac{1}{3} \times 0.20\,(0.45 - y_T)^3$$

$$+ (7.5 - 1) \times 0.0003536\,(0.2583 - 0.2258)^2$$

$$= 0.0015211\ \mathrm{m}^4$$

(iii) *Based on net area*

From Eq. (1.5.3a),

$$A = 0.20 \times 0.45 - 0.0003536 = 0.0896\ \mathrm{m}^2$$

From Eq. (1.5.3c),

$$\frac{0.20\, v_T^2}{2} = 0.20 \frac{(0.45 - y_T)^2}{2} - 0.0003536\,(0.2583 - y_T)$$

or $\qquad y_T = 0.2249$ m

From Eq. (1.5.3b),

$$I = \frac{1}{3} \times 0.20\,(0.2249)^3 + \frac{1}{3} \times 0.20\,(0.45 - 0.2249)^2$$

$$- 0.0003536\,(0.2583 - 0.2249)^2$$

$$= 0.0015184\ \mathrm{m}^4$$

Comparing the three sets of sectional properties, it is noted that the difference is very small.

EXAMPLE 1.5.2

Figure 1.5.3 shows the profile of a bonded post-tensioned continuous bridge girder. The top face of the girder is horizontal while the bottom face is parabolic. Determine the equation of the centroidal axis and also the cross-sectional properties at the left quarter point D on the basis of gross section. The cross-section of the girder is rectangular with a uniform width of 0.30 m.

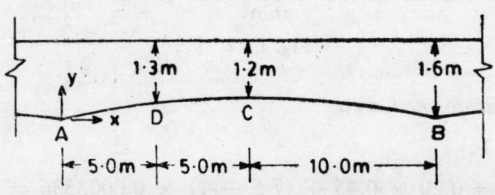

Fig. 1.5.3

Solution

Let y be the ordinate of the centoridal axis at any point.

$$y_A = y_B = 0.8 \text{ m}$$

$$y_C = 0.4 + \frac{1.2}{2} = 1.0 \text{ m}$$

Hence the equation of the centroidal axis may be written as

$$y = 0.8 + \frac{4(1.0 - 0.8) \times (20 - x)}{20^2}$$

$$= 0.8 + 0.002 \times (20 - x)$$

At the quarter point D,

$$y_D = 0.8 + 0.002 \times 5(20 - 5) = 0.95 \text{ m}$$

The sectional properties at D are,

$$A = 0.30 \times 1.30 = 0.39 \text{ m}^2$$

$$y_T = \frac{1.3}{2} = 0.65 \text{ m}$$

$$I = \frac{1}{12} \times 0.3 \times 1.3^3 = 0.05493 \text{ m}^4$$

1.6 CABLE LINE OR CABLE PROFILE

The cable line or cable profile in a prestressed concrete member may be defined as the line along which the total resultant prestressing force acts. Let a prestressed concrete member have n cables carrying prestressing forces $P_1, P_2, \ldots, P_n$ respectively. The total resultant prestressing force P is the sum of forces in individual cables.

$$P = P_1 + P_2 + \ldots + P_n \tag{1.6.1}$$

The distance of the resultant prestressing force P from the centroid of the cross-section,

$$e_p = \frac{P_1 e_{p1} + P_2 e_{p2} + \ldots + P_n e_{pn}}{P_1 + P_2 \ldots \ldots + P_n} \tag{1.6.2}$$

where, $e_{p1}, e_{p2}, \ldots, e_{pn}$ = eccentricities of forces $P_1, P_2, \ldots P_n$ respectively. If the tensile stress f_p in all the cables is the same,

$$P_1 = A_{p1} f_p$$
$$P_2 = A_{p2} f_p \tag{1.6.3}$$
$$P_n = A_{pn} f_p$$

in which $A_{p1}, A_{p2}, \ldots, A_{pn}$ are the cross-sectional areas of cables 1, 2, ..., n. It follows from Eqs. (1.6.2) and (1.6.3) that if the tensile stress in all the cables is the same, the resultant force P acts at the centroid of the prestressing steel. In this case, the cable line joins the centroids of the cross-sectional areas of all cables at different cross-sections of the member. Alternatively, the cable line may be visualised as the axis of a single imaginary cable of cross-sectional area $A_p = A_{p1} + A_{p2} + \ldots + A_{pn}$ located along the centroid of all cables. This concept of the cable line is extremely useful in the analysis of a prestressed concrete member. Although the member has several cables, they may be replaced by a single imaginary cable carrying the total prestressing force P along the centroid of all cables for the sake of simplicity of analysis. The axis of the equivalent single imaginary cable is called the cable line. The eccentricity of this cable is the eccentricity of prestress e_p measured from the centroid of the cross-section.

Alternatively, the position of the cable line or the centroid of the pre-stressing force at any cross-section may be determined by taking moments of the tendon forces about the top face. If $A_{p1}, A_{p2}, \ldots, A_{pn}$ are the cross-sectional areas; $f_{p1}, f_{p2}, \ldots, f_{pn}$ are the stresses and $d_{p1}, d_{p2}, \ldots, d_{pn}$ are the distances from top face for the prestressing tendons $1, 2, \ldots, n$, the distance of the centroid of prestressing force from the top face d_p is given by

$$d_p = \frac{A_{p1} f_{p1} d_{p1} + A_{p2} f_{p2} d_{p2} + \ldots + A_{pn} f_{pn} d_{pn}}{A_{p1} f_{p1} + A_{p2} f_{p2} + \ldots + A_{pn} f_{pn}} \qquad (1.6.4)$$

If all tendons have the same stress,

$$f_{p1} = f_{p2} = \ldots = f_{pn} \qquad (1.6.5)$$

then the distance of the centroid of prestressing force from top is given by

$$d_p = \frac{A_{p1} d_{p1} + A_{p2} d_{p2} + \ldots + A_{pn} d_{pn}}{A_{p1} + A_{p2} + \ldots + A_{pn}} \qquad (1.6.6)$$

= distance of centroid of prestressing steel from top face

If y_T is the distance of the centroid of the effective uncracked section from the top face, the eccentricity of prestress e_p may be expressed as

$$e_p = d_p - y_T \qquad (1.6.7)$$

If $d_p = y_T$, or $e_p = 0$, the member is said to be concentrically prestressed. In this case the compressive stress caused by prestressing force is uniform over the entire cross-section. On the other hand, if $e_p \neq 0$, the member is eccentrically prestressed. If d_p is lesser than y_T so that the prestressing force acts above the centroid of the section, the prestressing force imposes a sagging bending moment in addition to axial compression at the cross-section under consideration. On the other hand, if d_p is larger than y_T, the section is subjected to a hogging (negative) bending moment in addition to axial compression due to prestress. In the case of eccentric prestress, the intensity of prestress across the section is represented by a trapezoidal stress diagram.

EXAMPLE 1.6.1

Determine the position of the cable line at the left quarter point in the beam shown in Fig. 1.5.2. The stresses in the wires in the bottom, second, third and top layers are 950, 1000, 1050 and 1100 MPa respectively.

Solution

The cable line represents the position of the centroid of the prestressing force. Taking moments about the top face, the distance d_p of the cable line from top is given by

$$d_p = \frac{\frac{\pi}{4} \times 5^2 \; [3 \times 1100 \times 0.05 + 5 \times 1050 \times 0.25 + 5 \times 1000 \times 0.30 + 5 \times 950 \times 0.35]}{\frac{\pi}{4} \times 5^2 \; [3 \times 1100 + 5 \times 1050 + 5 \times 1000 + 5 \times 950]}$$

$$= 0.2536 \text{ m}$$

As the centroid of the cross-sectional area of prestressing steel is located at 0.2583 m from top, it follows that the cable line is located at a distance of $(0.2583 - 0.2536) = 0.0047$ m above the centroid of prestressing steel.

EXAMPLE 1.6.2

Determine the equation for the cable line in the beam shown in Fig. 1.6.1. Cable 1 is straight while cables 2 and 3 have parabolic profiles. All cables carry the same force. Determine the position of the cable line at left quarter point D.

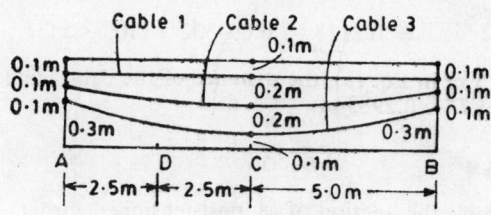

Fig. 1.6.1

Solution

The distances of the centroid of the prestressing force from the top face are 0.20 m and 0.30 m at end section and midspan section respectively. Hence the parabolic cable line is represented by the equation,

$$d_p = 0.20 + \frac{4 \, (0.30 - 0.20) \, x \, (10 - x)}{10^2}$$

$$= 0.2 + 0.004 \, x \, (10 - x) \tag{a}$$

It may be noted that the cable line coincides with the axis of cable 2. Putting $x = 2.5$ m in Eq. (a), the distance of the cable line from top face at quarter point D is given by

$$d_p = 0.275 \text{ m}$$

EXAMPLE 1.6.3

If the prestressing forces in cables 1, 2 and 3 in Example 1.6.2 are 160, 200 and 240 kN respectively, determine the equation of the cable line and its position at quarter point.

Solution

Distances of the cable line from the top face at the end sections A and B are given by

$$(d_p)_A = (d_p)_B = \frac{160 \times 0.10 + 200 \times 0.20 + 240 \times 0.3}{160 + 200 + 240}$$

$$= 0.2133 \text{ m}$$

Similarly the distance of the cable from the top face at C is given by

$$(d_p)_C = \frac{160 \times 0.10 + 200 \times 0.30 + 240 \times 0.50}{160 + 200 + 240}$$

$$= 0.3266 \text{ m}$$

Hence the parabolic cable line is represented by the equation

$$d_p = 0.2133 + \frac{4 (0.3266 - 0.2133) x (10 - x)}{10^2}$$

$$= 0.2133 + 0.004532 \, x \, (10 - x) \tag{a}$$

Putting $x = 2.5$ m in Eq. (a), the distance of the cable line from top face at quarter point D is 0.2983 m.

EXAMPLE 1.6.4

Figure 1.6.2 shows the section of a post-tensioned girder. Determine the eccentricity of the cable line on the basis of gross area. The effective prestressing forces in cables 1, 2, 3, 4 and 5 are 440, 460, 480, 500 and 520 kN respectively.

Solution

The cross-sectional area may be divided into nine parts as indicated in the figure. The areas of these parts are

$$A_1 = A_2 = 0.0700 \text{ m}^2$$
$$A_3 = A_4 = 0.0250 \text{ m}^2$$
$$A_5 = A_6 = 0.0175 \text{ m}^2$$
$$A_7 = A_8 = 0.00625 \text{ m}^2$$
$$A_9 = 0.2400 \text{ m}^2$$

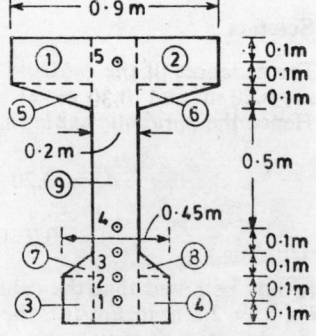

Fig. 1.6.2

Hence total cross-sectional area, $A = 0.4775 \text{ m}^2$

The distance of the centroid of the cross-section from top,

$$y_T = \frac{2 \times 0.0700 \times 0.10 + 2 \times 0.025 \times 1.10 + 2 \times 0.0175 \times 0.2333 \\ + 2 \times 0.00625 \times 0.9667 + 0.2400 \times 0.60}{0.4775}$$

$$= 0.4885 \text{ m}$$

Distance of resultant prestressing force from top,

$$d_p = \frac{440 \times 1.10 + 460 \times 1.00 + 480 \times 0.90 + 500 \times 0.80 + 520 \times 0.10}{440 + 460 + 480 + 500 + 520}$$

$$= 0.7617 \text{ m}$$

Hence the eccentricity of the cable line,

$$e_p = d_p - y_T$$
$$= 0.7617 - 0.4885$$
$$= 0.2732 \text{ m}$$

1.7 CENTRE OF COMPRESSION AND LINE OF THRUST

A compression member is said to be compressed axially when the resultant compressive force C at every cross-section of the member acts at the centroid of the cross-section. The line along which the resultant compressive force C acts is called the line of thrust or pressure line. The point at which C acts on any cross-section is called centre of compression. Thus every cross-section of a compressed member has its centre of compression at which the resultant of the compressive stress over the whole cross-section acts. The line joining the centres of compression of all cross-sections of the member is, therefore, the line of thrust. In an axially or concentrically compressed member, the line of thrust coincides with the centroidal axis. If the two lines do not concide, the member is eccentrically compressed. The distance e_c between the centre of compression and the centroid of the section is the eccentricity of the centre of compression or that of the resultant compressive force C. The resultant bending moment at any section M_r due to eccentric compressive force C is the product of $(-C)$ and e_c.

$$M_r = - Ce_c \qquad (1.7.1)$$

Alternatively, an axial force C and a bending moment M are together equivalent to an eccentric force C acting at an eccentricity e_c given by

$$e_c = - \frac{M_r}{C} \qquad (1.7.2)$$

It follows that the effect of bending moment M_r is to shift the centre of compression from the centroid through a distance e_c given by Eq. (1.7.2). The shift is upwards if the applied bending moment M_r is sagging because a sagging bending moment produces a compressive stress near the top fibres. In a similar manner, it may be explained why a hogging moment tends to shift the centre of compression downwards.

In the case of transversely loaded prestressed concrete beam having total prestressing force P with eccentricity e_p, the resultant bending moment M_r may be expressed as

$$M_r = M_p + M = - Pe_p + M \qquad (1.7.3)$$

in which M is the external bending moment. As the only axial force at any cross-section of the beam is the prestressing force, the resultant compression $C = P$. Hence from Eq. (1.7.2), the eccentricity of the centre of compression in a prestressed concrete beam may be expressed as

$$e_c = - \frac{M_r}{C} = - \frac{M_r}{P} = e_p - \frac{M}{P} \qquad (1.7.4)$$

In an externally unloaded beam having $M = 0$,

$$e_c = e_p \qquad (1.7.5)$$

Equation (1.7.5) shows that in an unloaded prestressed concrete beam, the line of thrust concides with the cable line. As the external moment M is applied, the line of thrust shifts from the cable line. The vertical intercept between the line of thrust and the cable line at any cross section is equal to (M/P). When M becomes numerically equal to M_p but is opposite in sign, i.e. for $M = - M_p$, the eccentricity of line of thrust $e_c = 0$. Hecne for $M = - M_p$, the line of thrust coincides with the centroidal axis of the member.

The position of the centre of compression or the line of thrust is important because the stresses at any cross-section depend on it. In a prestressed concrete beam with total compression $C = P$, the stress f at any point at a distance y from the centroidal axis may be determined from

$$= - \frac{P}{A} + \frac{(P\,e_c)\,y}{I} \qquad (1.7.6)$$

in which A and I are the cross-sectional area and its moment of inertia respectively. The first term on the right of Eq. (1.7.6) represents the direct stress due to P. The second term represents the bending stress caused by the resultant bending moment $M_r = M_p + M = - P\,e_c$ in accordance with Eq. (1.7.1). The value of e_c at any section of the beam can be determined readily from Eq. (1.7.4).

EXAMPLE 1.7.1

Determine the position of the cable line at midspan on the basis of gross area in the beam of Example 1.5.1 assuming that the prestress in all wires is 1100 MPa. Hence determine the position of centre of compression if the external sagging bending moment acting at the section is 0, 20, 32.40 and 40 kN.m respectively. Comment on the position of centre of compression for different values of bending moment.

Solution

Prestressing force in each wire $= \dfrac{\pi}{4} \times 5^2 \times 1100 \times 10^{-3}$

$$= 21.61 \text{ kN}$$

Hence total prestressing force,

$$P = 18 \times 21.61 = 388.98 \text{ kN}$$

The distance of the resultant prestressing force or cable line from top face,

$$d_p = \frac{3 \times 0.05 + 5 \times 0.40 + 5 \times 0.45 + 5 \times 0.50}{18}$$

$$= 0.3833 \text{ m}$$

Hence the eccentricity of prestressing force or cable line at midspan,

$$e_p = (d_p - y_T) = 0.3833 - 0.30 = 0.0833 \text{ m}$$

The eccentricity e_c of the centre of compression or line of thrust at midspan for different values of applied moment M may be determined from Eq. (1.7.4).

For

$$M = 0, \ e_c = e_p = 0.0833 \text{ m}$$

$$M = 20 \text{ kN.m}, \ e_c = 0.0833 - \frac{20}{388.98} = 0.0319 \text{ m}$$

$$M = 32.40 \text{ kN.m}, \ e_c = 0.0833 - \frac{32.40}{388.98} = 0$$

$$M = 40 \text{ kN.m}, \ e_c = 0.0833 - \frac{40}{388.98} = -0.0195 \text{ m}$$

As the applied sagging bending moment increases, the centre of compression or the line of thrust moves upwards progressively. For $M=0$, it coincides with the cable line. For $M = 20$ kN.m, it is located above the cable line and below the centroid of the section. For $M = 32.40$ kN.m, i.e. $M = -M_p$, it coincides with the centroid of the section. Finally for $M = 40$ kN.m, it is located above the centroid of the section. The same results may be obtained by determining the resultant bending moment, M_r and using Eq. (1.7.2).

EXAMPLE 1.7.2

Determine the position of the centre of compression in the cross-section of Example 1.6.4 when the external bending moment M acting at the section is 140.88 (hogging) and 1202.40 (sagging) kN.m respectively.

Solution

Total prestressing force,

$$P = 440 + 460 + 480 + 500 + 520$$
$$= 2400 \text{ kN}$$

From Example 1.6.4, eccentricity of prestress,

$$e_p = 0.2732 \text{ m}$$

Hence bending moment due to prestress,

$$M_p = -Pe_p = (-2400)(0.2732) = -655.68 \text{ kN·m}$$

For $\qquad M = -140.88$ kN.m,

$$M_r = M + M_p = -796.56 \text{ kN m}$$

Hence from Eq. (1.7.2),

$$e_c = -\frac{(-796.56)}{2400} = 0.3319 \text{ m}$$

The centre of compression or line of thrust is located at 0.3319 m below the centroid of the section.

For $\qquad M = 1202.40$ kN.m,

$$M_r = M + M_p = 1202.40 - 655.68 = 546.72 \text{ kN.m}$$

Hence from Eq. (1.7.2),

$$e_c = -\frac{546.72}{2400} = -0.2278 \text{ m}$$

The centre of compression or line of thrust is located at 0.2278 m above the centroid of the section.

1.8 INTERNAL FORCES DUE TO PRESTRESSING

When the cable line coincides with the centroidal axis, the member is said to be concentrically prestressed. In this case, the only internal force induced by prestressing is axial compression. It is, however, possible to induce two or more of the four types of internal forces, viz. axial compression, shear force, bending moment and twisting moment by means of eccentric prestressing, in which the cable line does not coincide with the centroidal axis. The magnitude and nature of the internal forces induced by prestressing depend upon the total prestressing force as well as the cable profile.

The three types of internal forces, which are commonly induced by prestressing in a structural member are axial compression, bending moment and shear force. Consider, for instance, the simply supported beam loaded uniformly having a curved cable line as shown in Fig. 1.8.1 (a). To determine the internal forces due to prestress, consider the equilibrium of the free body of a small element of the beam enclosed by sections 1 and 2 at distance dx apart as shown in Fig. 1.8.1 (b). Let θ be the inclination of the cable line in the element to the centrodial axis which is taken to be

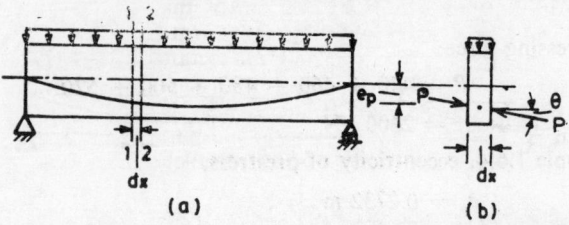

(a)   (b)

Fig. 1.8.1

horizontal. The inclined force P exerted by the prestress at sections 1 and 2 can be resolved into horizontal component ($P \cos \theta$) and vertical component ($P \sin \theta$). As the angle θ is generally small, $\cos \theta$ and $\sin \theta$ may be taken equal to 1 and θ respectively. The horizontal component P is eccentric and is, therefore, equivalent to an axial compressive force P and a bending moment ($-Pe_p$) in which e_p is the eccentricity of the cable line in the element considered. The bending moment M_p induced by prestress is hogging because the eccentric force P acts below the centroid. The vertical component ($P \sin \theta$) induces shear force V_p which may be expressed as

$$V_p = - P \sin \theta = - P \theta \qquad (1.8.1)$$

The minus sign in Eq. (1.8.1) emphasizes that the shear force V_p due to prestress is opposite in sign as compared to shear force V due to external loading. Consequently, the net or resultant shear force V_r may be expressed as

$$V_r = V + V_p = V - P \sin \theta = V - P\theta \qquad (1.8.2)$$

Similarly, the net or resultant bending moment may be expressed as

$$M_r = M + M_p = M - Pe_p \qquad (1.8.3)$$

in which M is the sagging bending moment due to external loading. The foregoing analysis shows that the internal forces induced by prestress are axial compression P, bending moment M_p of magnitude (Pe_p) and shear force V_p of magnitude ($P \sin \theta$). While the axial force is always compressive, the signs of M_p and V_p depend upon the position and inclination of cable line relative to the centoridal axis. The bending moment M_p is sagging or hogging depending upon whether the cable line is above or below the centroidal axis. In an inclined or vertical member, the sign of M_p is such that it produces compressive stress at the face towards which the cable line is placed relative to centroidal axis. The sign of shear force V_p depends upon the rate of change of e_p or the inclination of the cable line. In order to derive the advantages of prestress, the cable profile should be chosen in such a way that the signs of V_p and M_p are opposite to those of V and M caused by service loads. In this manner, the shear force and bending moment due to service loads are counteracted by means of prestress.

It is possible to induce a torque T_p by prestressing to counteract the torque T caused by service loads. To achieve this objective, the prestressing tendons should be placed in such a manner that the force in them have transverse components with a definite lever arm about the centroidal axis. Consider, for instance, the member of rectangular cross-section shown in Fig. 1.8.2 prestressed by four straight tendons each carrying a prestressing force P_1. Each of the four tendons may be visualized as a part of a rectangular spiral. Tendons 1 and 3 are laid parallel to the front and rear faces while tendons 2 and 4 are placed parallel to top and bottom faces. The prestressing force in tendon 1 or 3 has a longitudinal component $P_1 \cos \theta_1$ and a transverse component $P_1 \sin \theta_1$ with a lever arm ($b_1/2$) about the centroidal axis. Similarly, the prestressing force in tendon 2 or 4 has a longitudinal component $P_1 \cos \theta_2$ and a transverse component $P_1 \sin \theta_2$

with a lever arm equal to $(d_1/2)$ about the centroidal axis. Consequently, the effect of prestress is to induce an axial compressive force,

$$P = 2P_1 (\cos \theta_1 + \cos \theta_2) \qquad (1.8.4)$$

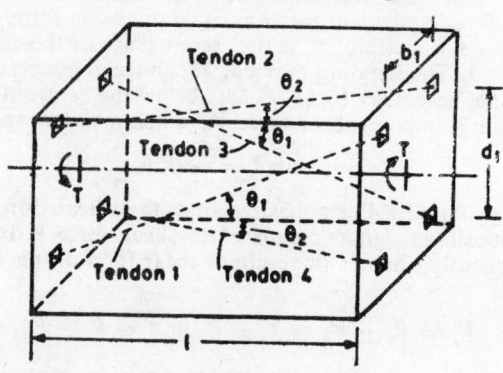

Fig. 1.8.2

and a twisting moment,

$$T_p = 2P_1 \left(\frac{b_1}{2} \sin \theta_1 + \frac{d_1}{2} \sin \theta_2 \right) \qquad (1.8.5)$$

As the inclinations θ_1 and θ_2 of the tendons to the centroidal axis are small, the following approximations may be used:

$$\cos \theta_1 = \cos \theta_2 = 1$$
$$\sin \theta_1 = \theta_1$$
$$\sin \theta_2 = \theta_2 \qquad (1.8.6)$$

With these approximations, the axial compression P and the torque T_p induced by prestress may be expressed as

$$P = 4P_1$$
$$T_p = \frac{2P_1 b_1 d_1}{1} \qquad (1.8.7)$$

EXAMPLE 1.8.1

The cable line of a simply supported post-tensioned beam is parabolic with zero eccentricities at the ends and a sag of 0.40 m at midspan. The total prestressing force is 1000 kN. Determine the internal forces induced by prestressing at the left quarter point.

Solution

Referring to Fig. 1.8.3, the parabolic cable profile is represented by the equation,

$$e_p = \frac{4S}{l^2} x(l - x)$$

The slope of the cable line at any point,

$$\tan \theta = \frac{d\,e_p}{dx} = \frac{4S}{l^2}\,(l - 2x)$$

Consequently, the internal forces caused by prestress may be expressed as

$$\text{Axial force} = P \cos \theta \approx P$$

$$V_p = -P \sin \theta \approx -P \tan \theta$$

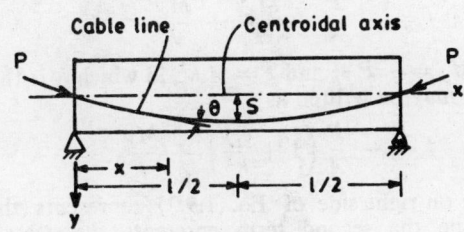

Fig. 1.8.3

or

$$V_p = \frac{-4\,PS}{l^2}\,(l - 2x)$$

$$M_p = -P\,e_p = \frac{-4\,PS}{l^2}\,x\,(l - x)$$

Putting $P \doteq 1000$ kN, $l = 20$ m, $x = 5$ m and $S = 0.40$ m,

$$\text{Axial force} = 1000 \text{ kN (compressive)}$$

$$V_p = \frac{-4 \times 1000 \times 0.4}{20^2}\,(20 - 2 \times 5) = -40 \text{ kN}$$

$$M_p = \frac{-4 \times 1000 \times 0.4}{20^2} \times 5\,(20 - 5) = -300 \text{ kN.m}$$

EXAMPLE 1.8.2

Referring to Fig. 1.8.2 determine the internal forces induced by prestressing given that the prestressing force in each cable, $P_1 = 250$ kN, $b_1 = 0.40$ m, $d_1 = 0.80$ m and $l = 20$ m.

Solution

Using Eq. (1.8.7),

$$\text{Axial force} = 4 \times 250 = 1000 \text{ kN (compressive)}$$

$$T_p = \frac{2 \times 250 \times 0.40 \times 0.80}{20} \doteq 8 \text{ kN.m}$$

1.9   NORMAL STRESSES IN PRESTRESSED CONCRETE MEMBERS

As shown in Sec. 1.8, prestressing imposes an axial compressive (negative) force P and a bending moment M_p at any section. Consequently, prestressing induces a direct stress $(-P/A)$ and a bending stress $(M_p\,y/I)$ at any point distant y from the centroidal axis of the section. If the external bending moment is M, the bending stress caused by it is $(M\,y/I)$. Hence the net normal stress at any point in an uncracked prestressed concrete section may be expressed as

$$f = - \frac{P}{A} + \frac{M_p y}{I} + \frac{My}{I} \tag{1.9.1}$$

Noting that $M_p = - P\,e_p$ and $I = A\,k^2$, in which k is the radius of gyration, Eq. (1.9.1) may be written as

$$f = - \frac{P}{A}\left(1 + \frac{e_p y}{k^2}\right) + \frac{My}{I} \tag{1.9.2}$$

The first term on right side of Eq. (1.9.2) represents the stress due to prestressing while the second term represents the stress due to external moment M. To derive an alternative expression for normal stress, note that

$$M_p + M = - P\,e_p + M = - P\left(e_p - \frac{M}{P}\right) \tag{1.9.3}$$

Hence using Eq. (1.7.4),

$$M_p + M = - P\,e_c \tag{1.9.4}$$

in which e_c is the eccentricity of the centre of compression or line of thrust. Hence Eq. (1.9.1) may be written as

$$f = - \frac{P}{A} - \frac{(Pe_c)}{I}\,y \tag{1.9.5}$$

$$= - \frac{P}{A}\left(1 + \frac{e_c\,y}{k^2}\right) \tag{1.9.6}$$

Equation (1.9.5) shows that normal stress is the sum of the direct stress $(-P/A)$ and the bending stress caused by the resultant moment M_r,

$$M_r = M_p + M = - P\,e_c \tag{1.9.7}$$

EXAMPLE 1.9.1

Determine the stresses due to prestressing at the section at left quarter point of the pretensioned beam of Example 1.5.1 based on gross area. The prestress in each wire is 1200 MPa.

Solution

Using the data of Example 1.5.1,

$$P = 353.6 \times 1200 = 0.424320 \text{ MN}$$
$$e_p = (d_p - y_T) = (0.2583 - 0.2250) = 0.0333 \text{ m}$$

Hence bending moment due to prestressing,

$$M_p = - P\,e_p = - 0.424320 \times 0.0333$$

$$= - 0.0141425 \text{ MN.m}$$

Using Eq (1.9.1), the stress at the top fibre f_T due to prestressing may be obtained by putting $M_p = - 0.0141425$ MN.m, $M = 0$ and $y = -0.225$ m.

$$f_T = - \frac{0.424320}{0.09} + \frac{(-0.0141425)\,(-0.225)}{0.0015187}$$

$$= - 2.62 \text{ MN/m}^2 \text{ or MPa}$$

Similarly, the stress at the bottom fibre f_B is obtained by taking $y = 0.225$ m.

$$f_B = - \frac{0.424320}{0.09} + \frac{(-0.0141425)\,(0.225)}{0.0015187}$$

$$= - 6.81 \text{ MN/m}^2 \text{ or MPa}$$

The entire section is in compression and the stress varies linearly from $- 2.62$ MN/m^2 at top to $- 6.81$ MN/m^2 at bottom.

EXAMPLE 1.9.2

Determine the stresses in the section of Example 1.6.4 on the basis of gross cross-sectional area when it carries an external bending moment equal to (i) 1.2024 MN.m and (ii) $- 0.14088$ MN.m.

Solution

Using the data of Example 1.6.4,

$$P = 440 + 460 + 480 + 500 + 520 = 2400 \text{ kN} = 2.4 \text{ MN}$$
$$M_p = - Pe_p = - 2.4\,(0.2732) = - 0.655680 \text{ MN.m}$$

Moment of inertia about the centroidal axis,

$$I = 0.07741 \text{ m}^4$$

(i) For external bending moment, $M = 1.2024$ MN.m (sagging)
Using Eq. (1.9.1), the stress at the top fibre f_T may be obtained by putting $y = - 0.4885$ m,

$$f_T = - \frac{2.40}{0.4775} + \frac{(-0.655680)\,(-0.4885)}{0.07741}$$

$$+ \frac{1.2024\,(-0.4885)}{0.07741}$$

$$= - 8.48 \text{ MN/m}^2 \text{ or MPa}$$

The stress at the bottom fibre is obtained by taking $y = 0.7115$ m,

$$f_B = -\frac{2.40}{0.4775} + \frac{(-0.655680)(0.7115)}{0.07741}$$

$$+ \frac{1.2024(0.7115)}{0.07741} = 0$$

(ii) For external bending moment, $M = 0.14088$ MN.m (Hogging), the stress at top fibre,

$$f_T = -\frac{2.40}{0.4775} + \frac{(-0.655680)(-0.4885)}{0.07741}$$

$$+ \frac{(-0.14088)(-0.4885)}{0.07741} = 0$$

The stress at the bottom fibre,

$$f_B = -\frac{2.40}{0.4775} + \frac{(-0.655680)(0.7115)}{0.07741}$$

$$+ \frac{(-0.14088)(0.7115)}{0.07741}$$

$$= -12.35 \text{ MN/m}^2 \quad \text{or MPa}$$

1.10 ACTIVE AND PASSIVE FORCES

In the case of non-prestressed structures, the external loads are active forces and the reactive forces due to the resistances of the materials are the passive forces. In the context of prestressed concrete structures, a clear distinction should be drawn between the forces in prestressing steel and non-prestressed reinforcing steel. The prestressing force created by jacking of prestressing steel is an active force. It may, therefore, be visualized as a kind of external loading. Its magnitude and eccentricity, however, are completely in the hands of the designer. He is required to determine them so as to create the most favourable conditions of stress and deformation. The active forces in the perstressing cables remain practically constant in the entire working range. The slight increase in stress due to external loads, which may be considered as a passive part of the force in the cables, is small compared to the active part and may be ignored. It follows that in a prestressed concrete beam the total tension in steel and the total compression in concrete may be taken equal to a constant effective prestressing force P. In a prestressed concrete section shown in Fig. 1.10.1 (a), the centre of compression coincides with the cable line when the external bending moment M is zero as discussed in Sec. 1.7. When a sagging bending moment M is applied which produces stresses in concrete represented by the stress diagram of Fig. 1.10.1 (b), the centre of compression rises through a distance (jd_p) which becomes the lever arm between the equal and opposite forces P and C. As P and C do not change, the lever-arm (jd_p) increases proportionately with applied bending moment M.

$$jd_p = \frac{M}{P} = \frac{M}{C} \tag{1.10.1}$$

It is interesting to note the contrast in the structural responses of prestressed and reinforced concrete beams. In a prestressed concrete beam, C remains constant while jd_p increases proportionately with M. On the contrary, in a reinforced concrete section shown in Fig. 1.10.2 (a), the internal lever arm (jd_p) remains practically constant because the position of the neutral axis and the centre of compression located at the centroid of the stress diagram shown in Fig. 1.10.2 (b) remain practically unchanged during the entire working range. Consequently, the total compression in concrete C and total tension in reinforcing steel T increase proportionately with the applied bending moment M. The stress in the steel in a reinforced concrete beam varies from zero to the full working stress as the load increases from zero to the full working or service load. On the other hand, the steel in a prestressed concrete beam carries practically constant stress when the load increases from zero to the full working load. It is, therefore, evident that prestressing steel is less prone to fatigue failure as compared to reinforcing steel. An interesting aspect of the active and passive roles of prestressing steel and reinforcing bars is worth mentioning. In a reinforced concrete beam, an increase in reinforcing steel can do no harm unless it leads to congestion creating difficulty in placement and compaction of concrete. On the other hand, even a small increase in the prestressing steel and the consequent increase in active prestressing force may create unacceptable stresses and deflections. Hence prestressing steel has to be designed more carefully than reinforcing steel.

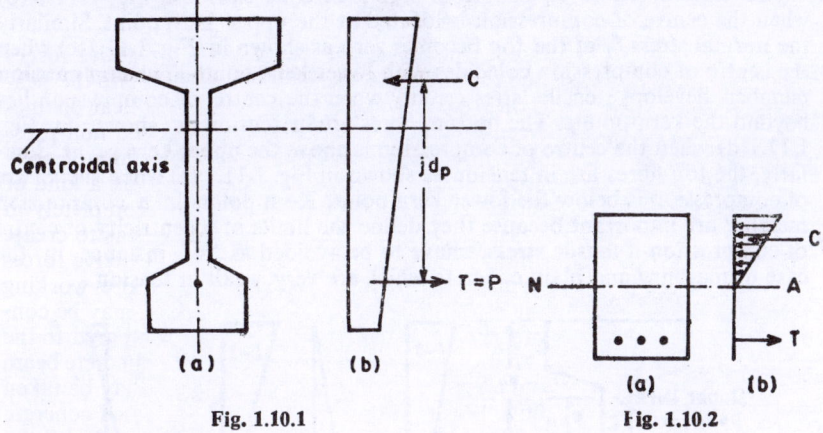

Fig. 1.10.1   Fig. 1.10.2

1.11 KERN POINTS

The kern points in the cross-section of a compression member are the points representing extreme positions of the centre of compression without producing tensile stress at the extreme top or bottom fibres. When the centre of compression moves to the upper extreme limit, called the upper kern point, the compressive stress at the bottom fibre f_B reduces to zero. Similarly, when the centre of compression moves to the lower extreme limit, called the lower kern point, the compressive stress at the extreme top fibre f_T reduces to zero. These conditions, defining the kern points, may be used to determine their positions. Thus referring to Fig. 1.11.1 (a), and noting that the axial compressive force C in a prestressed concrete member is equal to the prestressing force P, the eccentricity of the upper kern point e_{k_u} may be determined from the condition,

$$f_B = - \frac{P}{A} + \frac{(- P e_c) y_B}{I} = 0 \qquad (1.11.1)$$

Putting eccentricity of centre of compression, $e_c = - e_{k_u}$ and $I = Ak^2$,

$$e_{k_u} = \frac{k^2}{y_B} \qquad (1.11.2)$$

Similarly, the eccentricity of the lower kern point e_{kl} may be determined from the condition,

$$f_T = - \frac{P}{A} + \frac{(- Pe_c)(- y_T)}{I} = 0 \qquad (1.11.3)$$

Putting $e_c = e_{kl}$ and $I = Ak^2$, the following expression for the eccentricity of the lower kern point is obtained,

$$e_{kl} = \frac{k^2}{y_T} \qquad (1.11.4)$$

The normal stress f_B at bottom fibre is zero as shown in Fig. 1.11.1 (b) when the centre of compression is located at the upper kern point. Similarly the normal stress f_T at the top becomes zero as shown in Fig. 1.11.1(c) when the centre of compression coincides with lower kern point. The compression member develops tensile stresses only when the centre of compression lies beyond the kern points. The bottom fibres are in tension as shown in Fig. 1.11.1(d) when the centre of compression is above the upper kern point. Similarly, the top fibres are in tension as shown in Fig. 1.11.1 (e) when the centre of compression is below the lower kern point. Kern points in a compression member are important because they define the limits of eccentricity of centre of compression if tensile stresses have to be avoided as for instance in the case of masonry and plain concrete which are very weak in tension.

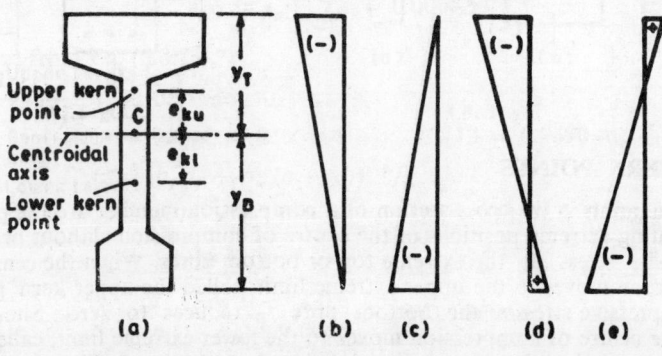

Fig. 1.11.1

In the case of a symmetrical cross-section of depth D, $y_T = y_B = D/2$. Hence in a symmetrical cross-section, the kern points are also symmetrically located with eccentricity e_k given by

$$e_{ku} = e_{kl} = e_k = \frac{k^2}{(D/2)} \qquad (1.11.5)$$

In the case of a rectangular section having $A = bD$, $I = bD^3/12$ and $k^2 = D^2/12$

$$e_{ku} = e_{kl} = D/6 \qquad (1.11.6)$$

Equation (1.11.6) shows that in order to eliminate tensile stresses, the centre of compression must be located within the middle third of the cross-section of the member. This is the well known middle third rule according to which the line of thrust in a compression member having a rectangular cross-section must lie within its middle third in order to avoid tensile stress in the longitudinal direction.

EXAMPLE 1.11.1

Determine the positions of the kern points in the cross-section of the beam of Example 1.6.4 on the basis of gross cross-sectional area. Determine the bending moment required to shift the centre of compression from the cable line to (i) lower kern point and (ii) upper kern point. Determine the extreme fibre stresses in each case.

Solution

Using the data of Example 1.6.4,

Moment of inertia, $I = 0.07741$ m^4

$$k^2 = \frac{I}{A} = 0.16211 \text{ m}^2$$

Distance of the upper kern point from centroidal axis,

$$e_{ku} = \frac{k^2}{y_B} = \frac{0.16211}{0.7115} = 0.2278 \text{ m}$$

Distance of lower kern point from centroidal axis,

$$e_{kl} = \frac{k^2}{y_T} = \frac{0.16211}{0.4885} = 0.3319 \text{ m}$$

The positions of centroid of cross-section, the centroid of prestressing steel, upper kern point and lower kern point are represented by points C, D, U and L respectively in Fig. 1.11.2.

(i) The centre of compression coincides with the centroid of the prestressing force in the unloaded condition. Hence, for $M = 0$, the centre of compression is located at D. The bending moment required to depress it to the point L,

$$M = P (e_p - e_{kl}) = 2400 (0.2732 - 0.3319)$$
$$= -140.88 \text{ kN.m}$$

(ii) The bending moment required to elevate the centre of compression to the point U,

$$M = P(e_p + e_{ku})$$
$$= 2400(0.2732 + 0.2278) = 1202.40 \text{ kN.m}$$

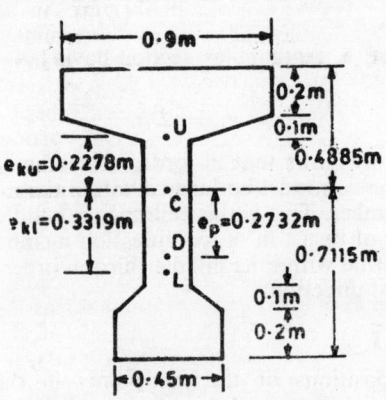

Fig. 1.11.2

The extreme fibre stresses for external bending moment, $M = -0.14088$ MN.m and $M = 1.2024$ MN.m have already been calculated in Example 1.9.2. The same results may be obtained from Eq. (1.9.5) by noting that the centre of compression is located at the lower kern point when $M = -0.14088$ MN.m and it is located at the upper kern point when $M = 1.2024$ MN.m.

(i) $M = -0.14088$ MN.m and $e_c = e_{kl} = 0.3319$ m

$$f_T = -\frac{2.40}{0.4775} - \frac{2.40(0.3319)(-0.4885)}{0.07741} = 0$$

$$f_B = -\frac{2.40}{0.4775} - \frac{2.40(0.3319)(0.7115)}{0.07741}$$

$$= -12.35 \text{ MN/m}^2 \quad \text{or} \quad \text{MPa}$$

(ii) $M = 1.2024$ MN.m and $e_c = -e_{ku} = -0.2278$ m

$$f_T = -\frac{2.40}{0.4775} - \frac{2.40(-0.2278)(-0.4885)}{0.07741}$$

$$= -8.48 \text{ MN/m}^2 \quad \text{or} \quad \text{MPa}$$

$$f_B = -\frac{2.40}{0.4771} - \frac{2.40(-0.2278)(0.7115)}{0.07741} = 0$$

1.12 STATIC INDETERMINACY

The effect of prestress is to induce an axial compression P and a bending moment M_p at every cross section of a prestressed concrete member. These forces may be visualized as a kind of internal forces which produce their own displacements similar to those caused by external loads. If the member is kept in equilibrium by a statically determinate system of support reactions, the displacements of the member due to prestress can occur freely without restraint from the supports. In this case, the act of prestressing does not induce any support reactions. Hence in an externally unloaded prestressed member having a statically determinate support system, the support reactions are zero. The same is not true in the case of a prestressed member with a statically indeterminate support system. In this case, the supports restrict the displacements of the member caused by prestress. In doing so, support reactions are set up which produce secondary bending moment M_{ps} at every cross-section of the member. The net bending moment due to prestressing is the sum of the primary moment M_{pp} and the secondary moment M_{ps}.

$$M_p = M_{pp} + M_{ps} = -Pe_p + M_{ps} \qquad (1.12.1)$$

The support reactions due to prestress and the secondary moments M_{ps} arise only in statically indeterminate structures whereas they are absent in statically determinate structures. As the magnitude of secondary moments may be comparable with that of the primary moments, they cannot be ignored.

Consider, for instance, the simply supported unloaded beam with a straight cable line as shown in Fig. 1.12.1 (a). As the cable line is below

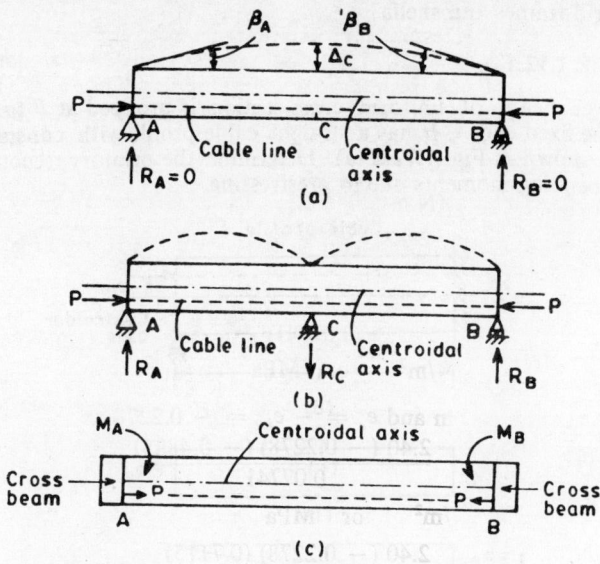

Fig. 1.12.1

the centroidal axis of the beam, the bending moment due to prestress is hogging. The beam deflects upwards causing a deflection Δ_c at centre of span C and slopes β_A and β_B at the ends A and B. In a simply supported beam, these displacements occur freely without restraint from the supports. Hence the support reactions $R_A = R_B = 0$ in the case of a simply supported beam. If the beam has a simple support at C as shown in Fig. 1.12.1 (b) making the beam continuous with a statically indeterminate support system, it is evident that the deflection Δ_c at C cannot occur. In order to make the deflection at C equal to zero, a downward reaction R_C is set up by support C. As the sum of R_A, R_B and R_C must be zero for equilibrium, R_A and R_B act upwards. These secondary forces give rise to secondary moments M_{ps} which are sagging (positive) in the present case. If the beam frames rigidly to cross beams at the ends A and B as shown in Fig. 1.12.1 (c), the end slopes β_A and β_B cannot occur freely as they must be equal to the torsional rotations of the respective cross beams at intersection points for the sake of compatibility of deformation. Due to their torsional stiffness, the cross beams impose sagging bending couples at the points A and B. Consequently sagging couples M_A and M_B are set up at the ends A and B which produce secondary moments M_{ps} at every cross-section of the beam.

The development of secondary forces due to prestressing may be viewed as an additional complexity in the analysis of statically indeterminate structures. However, the complexity can be eliminated to a considerable extent by determining forces imposed by the prestressing steel on the concrete member using the free body approach. When these forces are combind with the external forces, the structure can be analysed as an ordinary non-prestressed structure. This powerful approach is useful in the analysis of statically indeterminate structures such as continuous beams, slabs, rigid frames and grid frames and shells

EXAMPLE 1.12.1

A cantilever beam of uniform cross-section is propped at B to the same level as the fixed end A. It has a straight cable profile with constant eccentricity as shown in Fig. 1.12.2 (a). Determine the primary, secondary and resultant bending moments due to prestressing.

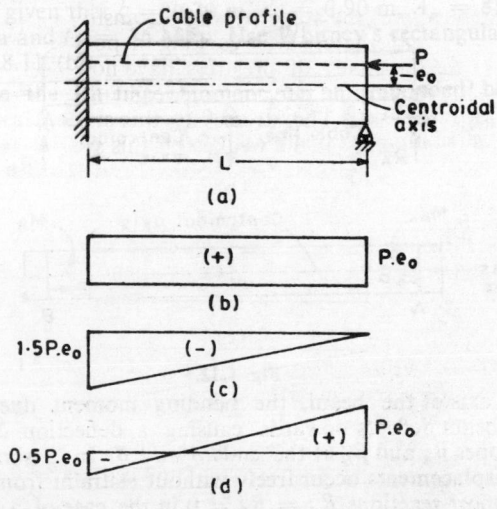

Fig. 1.12.2

Solution

As the eccentricity of prestress is constant, the primary bending moment is also constant and may be expressed as

$$M_{pp} = - P e_p = (- P)(- e_o) = P e_o$$

It is represented by the primary moment diagram of Fig. 1.12.2 (b). If the beam were not propped, it would deflect upward at B under the action of primary moments which are sagging in nature. Consequently, the prop reaction R_B acts downwards. It may be determined from the condition of zero deflection at the point B.

$$\frac{R_B L^3}{3EI} - \frac{(P e_0) L^2}{2EI} = 0$$

$$R_B = \frac{3}{2} \frac{P e_0}{L}$$

The secondary moment diagram caused by the secondary force in the from of prop reaction is shown in Fig. 1.12.2 (c). The resultant moment diagram due to prestressing shown in Fig. 1.12.2 (d) is obtained by combining the primary and secondary moment diagrams of Figs. 1.12.2 (b) and (c) respectively.

1.13 LOADING STAGES

The complexity of prestressed concrete design arises mainly from the multiplicity of loading stages in prestressed concrete structures. The following stages of loading are commonly considered in the design of prestressed concrete structures.

(i) *Initial stage*

The initial stage denotes the stage at stress transfer in pretensioned members and the stage immediately after jacking in post-tensioned members. At this stage, most of the losses of prestress have not occurred. The external loading at this stage is often due to self weight although other dead loads may also act in some cases. The stresses at this stage are often the most severe. Ironically, concrete is the weakest at this stage.

(ii) *Final stage*

It is the stage at which the structure is pressed into service. At this stage, most of the losses of prestress have occurred and the structure carries its full working or service loads. The final stage is important in elastic design because it represents a frequent, if not permanent, loading condition.

(iii) *Stages of lifting, transporting, hoisting and launching*

These stages arise in precast prestressed members at the respective operations during construction. The stresses induced during these operations are often quite diverse in magnitude and nature. These stresses are often large enough to demand careful planning of prestressing operations.

(iv) *Stage of decompression*

It represents the stage of loading at which the prestress at the level of centroid of prestressing steel or at the extreme fibre in flexural tension zone is just overcome by external moment. As flexural tensile stresses start developing when the loading exceeds the decompression stage, it is useful in the classification of beams and in the formulation of design philosophy.

(v) *Cracking stage*

As very often one of the primary objects of prestressing is to produce a crack free structure, the importance of cracking stage is self evident. The cracking stage marks a transition point in the load response characteristics of a prestressed concrete beam. It helps to distinguish partially prestrssed beams which develop flexural cracks at the working load level. If cracking has to be prevented at the working load level, the prestress must produce an adequate margin of safety against cracking.

(vi) *Stage of limiting internal forces*

The prestressing force has to be designed to produce acceptable margins of safety against the loading stages corresponding to the ultimate or limit state in compression or tension, shear, flexure and torsion. For any prestressed concrete member, each of these limit states corresponds to a level of loading which must not be lower than the level of working loads multiplied by the respective load factors recommended by codes of practice.

(vii) *Stage of limiting serviceability*

The requirements of adequate functioning and serviceability of a structural member may impose certain limits on deflection, crack width, vibrations and other structural responses. To ensure a satisfactory performance of a prestressed concrete structure, the provisionsof all relevant codes of practice have to be satisfied by appropriate planning of prestressing operations.

In addition to the foregoing common stages of loading, the designer may have to consider certain other loading stages in particular constructions. For instance, several other loading stages may have to be considered in composite construction depending upon the materials used and the sequence of constructional operations. Another interesting example is the construction of a bridge deck by cantilever method which entails numerous loading stages calling for the application of prestressing force in several phases to ensure proper functioning and safety at every loading stage. It may become necessary to use temporary prestressing cables which may be removed later at the appropriate stage of construction. The challenge of prestressed concrete design lies in judicious selection of the quantum of prestressing force and the cable layout as well as planned sequence and phasing of stressing operations in order to ensure best performance at all stages of loading.

1.14 COMPARISON WITH REINFORCED CONCRETE

As pointed out in Sec. 1.2, the first attempt at prestressing was directed towards improving the ordinary reinforced concrete. Those members in which the flexural tensile stresses due to the working load are only partly overcome by prestressing may actually be visualized as improved high strength reinforced concrete. These members described as Type 3 by Indian code IS: 1343-1980, are only partially prestressed. These members, which develop flexural cracks at the working load level, derive their properties partly from prestressed concrete. On the other hand, the members in which the flexural tensile stresses at the working load level are totally eliminated or reduced to small value much lower than the cracking stress, have properties quite different from those of the conventional reinforced concrete. The performance of these members, described as Types 1 and 2 in IS: 1343 —1980, is superior to that of reinforced concrete in many ways. The main advantages of prestressed concrete over reinforced concrete are as follows.

(i) *Efficient use of concrete*

In reinforced concrete, the concrete located in the flexural tension zone is ineffective due to the low tensile strength of concrete. On the other hand, the entire section is effective in the case of prestressed concrete because the whole cross-section is under compression under the working or service load. This fact justifies the use of high strength concrete in prestressed concrete in contrast to reinforced concrete in which about 60 per cent of concrete in the flexural tension zone is ineffective.

(ii) *Efficient use of steel*

High tensile steel in the form of wires, bars, multi-wire strands and cables are used in prestressed concrete unlike the conventional reinforcing bars in reinforced concrete. The ultimate strength of high tensile steel is 4 to 6 times grater than that of reinforcing steel. Consequently, the working stress in high tensile steel in prestressed concrete is as high as 800 to 1500 MPa as compared to a working stress of only 130 to 230 MPa in the case of reinforcing bars. It follows that in order to develop a given tensile force, the quantity of steel required in prestressed concrete is much smaller as compared to that in reinforced concrete. Thus the congestion of steel, often witnessed in reinforced concrete, is eliminated in the case of prestressed concrete.

(iii) *Reduction in shear*

Whenever the cable line is not parallel to the centroidal axis of the member, the prestressing force generally produces a shear force opposite in direction to that due to the service loads. Hence the net shear force in prestressed concrete is smaller than in reinforced concrete. Consequently, the shear resistance of a prestressed concrete member is higher than that of the corresponding reinforced concrete member. The diagonal tension is also greatly reduced on account of the compressive stress in concrete induced due to prestressing. These facts explain why it is possible to use thin webs in flanged prestressed concrete section as compared to stocky webs of reinforced concrete section.

(iv) Reduction in deflection

The prestressing force produces a negative (upward) deflection or camber in prestressed concrete beams. Hence for a given span and service load, the net deflection in prestressed concrete beam is smaller than in a reinforced concrete beam. This fact explains the superiority of prestressed concrete beams in carrying heavy loads over long spans. The technique of prestressing opens many possibilities towards improving the deformation characteristics of structures.

(v) Reduction in self weight

Due to the more efficient use of high grade concrete and the reductions in shear force and deflection, it is possible to employ slender flanged sections with thin flanges and webs in prestressed concrete as against the stocky rectangular sections common in reinforced concrete. The use of slender prestressed concrete members permits a substantial reduction in the self weight and the consequent reduction in the bending moment and shear force due to self weight. The reduction of self weight is a major consideration in the choice of the type of structure and the material in the case of long span structures, as in these structures, the self weight becomes a dominant part of the total load.

(vi) Freedom from cracks

Reinforced concrete structures have the undesirable property of developing flexural cracks at the working load level. Reinforced concrete members subjected to direct tension also develop cracks under the working load. The prestressing force prevents these cracks by inducing compressive stress at the appropriate locations. The result is a crack free structure with its several attendant advantages. The freedom from cracks has far reaching effects on the basic properties of concrete. It leads to a significant improvement in the structural response, deformation characteristics, resistance to ingress of moisture and other corrosive elements, resistance to freezing and thawing and resistance to the ill effects of thermal changes. It is, perhaps, not too much to say that the freedom from cracks due to prestressing produces an entirely new material superior in so many ways as compared to ordinary concrete.

(vii) Leak proof structure

The prevention of cracking due to prestressing greatly increases the resisance to the ingress of water and other fluids leading to an almost impervious structure. This property is very important in the case of containment structures such as water tanks as it produces a leak proof structure. It eliminates the need for the water proofing compounds commonly used with reinforced concrete tanks, thereby effecting substantial economy.

(viii) *High reliability*

A prestressed concrete member in its entire useful life often undergoes the most severe stress condition at the time of prestressing. The stress in the high tensile steel reduces with time due to loss of prestress and never reaches the initial stress even when the structure is fully loaded. Similarly, the concrete is subjected to the most severe stresses at a young age of a few days or weeks at the time of presressing. The concrete becomes stronger with time. Hence. if the prestrssed concrete member survives the most adverse conditions at the time of prestressing without showing any undesirable effects. it is unlikely to do so during its entire useful life. In fact, the operation of prestressing constitutes the most critical acceptance test for the structure. If it passes the test, it may very well be expected to perform its service satisfactorily ever after. This ensures a high reliability and an automatic quality control in prestressed concrete structures.

(ix) *High fatigue resistance*

The stress in steel in reinforced concrete varies from zero to working stress as the load increases from zero to full service load. On the contrary, the stress in prestressing steel remains practically constant from the unloaded to loaded conditions. The variation in steel stress due to the changes in loading is generally smaller than the stress variation of approximately 15 to 20 percent due to loss of prestress. It follows that the prestressing steel is subjected to a much smaller variation in stress as compared to reinforcing steel. Hence, the prestressing steel is not as prone to a fatigue failure as the reinforcing steel. Due to its high strength and freedom from cracks, the prestressed concrete exhibits a higher fatigue resistance in bending and shear as compared to reinforced concrete. The repetitive cyclic tests in torsion have also established the superiority of prestressed concrete in resisting a fatigue failure.

(x) *High durability*

Prestressed concrete as a material of construction is now about half a century old. During this period, it has been clearly established that the durability of prestressed concrete is better than that of reinforced concrete. This is particularly true when the structure is located in a hostile environment. The higher durability of prestressed concrete is mainly due to the absence of cracks in prestressed concrete. The prestressed concrete with its higher imperviousness and resistance to the entry of water and offensive fumes and gases, offers a better protection to prestressing steel. Thus prestressed concrete may be expected to have a longer life in chemical plants, marine and offshore structures and under extreme climatic conditions. The high grade dense concrete used in prestresed structures also contributes significantly to a high life expectancy. The superior fatigue resistance of prestressed concrete also leads to a longer useful life as compared to reinforced concrete.

(xi) *Dead load compensation*

An interesting feature of prestressed concrete, which contributes significantly to its economy, arises from the fact that the dead load including the self weight, effective at the time of prestressing, may be counteracted by merely increasing the eccentricity of prestressing force without any increase in the area of the prestressing steel. Thus a prestressed concrete girder may carry the dead load without any extra cost. As the dead loads dominate over the live loads with increasing span, the prestressed concrete construction shows spectacular economy in the field of long span structures.

(xii) Economy

Due to more efficient use of concrete and reductions in shear force and deflection, the prestressed concrete members are much slender as compared to reinforced concrete members. The net saving of concrete may be of the order of 50 to 60 per cent. The saving in the volume of steel due to the use of high tensile steel may be as high as 80 per cent. Thus there is a significant saving in both concrete and steel. Although some economy due to prestressing is possible in almost every field of concrete construction, a spectacular economy is achieved in the domain of long span construction. In bridge construction with spans ranging from 30 m to 100 m, prestressed concrete may compare favourably with reinforced concrete and structural steel. In the field of large containment structures, prestressed concrete may provide the most economical design. In the case of supersized tanks, prestressed concrete is not only the most economical but perhaps the only feasible option.

When prestressed concrete was first introduced successfully, it was thought that it will make reinforced concrete obsolete. It did not do so. Nor is it likely to happen in the future. Actually reinforced concrete and prestressed concrete play complementary roles. There are many areas in which reinforced concrete is more appropriate as a material of construction. The main disadvantages of prestressed concrete are as follows:

(a) Complexity of design

The complexity of prestressed concrete design arises mainly from the multiplicity of the loading stages for which the design must be checked. Besides, the prestressing force being an active force unlike the passive force offered by the reinforcing bars, an excess of prestressing force may be as unsafe as the lack of it. Hence the prestressing force has to be critically designed.

(b) Skilled supervision and control

Prestressed concrete calls for a much higher skilled supervision and control as compared to reinforced concrete. The need for high grade concrete, the higher dimensional accuracy and the smaller tolerances permitted in prestressed concrete construction demand very close supervision. The operations associated with the prestressing require experience and expertise. The need for a high quality control also demands a high degree of supervision.

(c) Costly equipment

Considerable investment on tools and plant is necessary in prestressed concrete construction. The equipment depends upon the system of prestressing used. Although the cost of the equipment varies substantially for different systems of prestressing, it is still quite substantial to be beyond the reach of ordinary contractors. Besides, the investment on the equipment may not be justified on economic ground unless the equipment is used continuously.

(d) *Economy*

Prestressed concrete requires the use of high grade concrete and high tensile steel, both of which cost more than their counterparts in reinforced concrete, viz. normal concrete and conventional reinforcing steel. Although the extra expense on high grade concrete is only marginal, the high tensile steel may cost 2 to 3 times as much as reinforcing steel. Thus the saving in cost due to the reduced quantities of concrete and steel is offset partly by the higher cost of the two materials. As significant reduction in the quantity of the two materials due to prestressing is possible only in certain areas such as long span structures, prestressed concrete can compete favourably with reinforced concrete only in these areas. The investment on costly prestressing equipment has also to be kept in veiw in comparing the relative economies of reinforced and prestressed concretes. Both reinforced concrete and prestressed concrete have their areas of application on economic basis. Sometimes the combination of the two materials, known as composite construction, may be found to be the most economical.

REFERENCES

1.1. Guyon, Y., *Prestressed Concrete*, C.R. Books Limited, London, Fourth Edition 1960.

1.2. IABSE Preliminary Publications, *Seminar on Problems of Prestressing*, Madras, January 30 to February 2, 1970.

1.3. IABSE Publication, *Prestressed Concrete Works in India*, 1973.

1.4. IABSE Preliminary Publication, *Seminar on Prestressed Concrete Structures*, Bombay, January 18-21, 1975.

1.5. IABSE Preliminary Publication, *Seminar on Tall Structures and Use of Prestressed Concrete in Hydraulic Structures*, Srinagar, May 24-26, 1984.

1.6. Lin, T.Y. and Kulka, F., *Fifty year advancement in concrete bridge construction*, Journal of Construction Division, ASCE, September 1975, pp 491-510.

1.7. Pandit, G.S. and Gupta, S.P., *Free-body approach for analysis of prestressed concrete beams*, The Indian Concrete Journal, Vol. 54, No. 6, June 1980, pp. 154-61.

1.8. Pandit, G.S. and Gupta, S.P., *Numerical analysis and design of prestressed concrete continuous beams*, The Indian Concrete Journal, Vol. 55, No. 4, April 1981, pp. 99-105.

1.9. PCI Design Handbook *Precast Prestressed Concrete*, Prestressed Concrete Institute, Chicago, Illinois, 1978.

1.10. *Post-Tensioning Manual*, Post-Tensioning Institute, Phoenix, Arizona, 1976.

1.11. Symposium, *T.Y. Lin Symposium on Prestressed Concrete*, Special Commemorative Issue, Journal of the Prestressed Concrete Institute, Vol. 21, No. 5, September-October 1976.

PROBLEMS

1.1.   Define centre of compression and line of thrust in a compression member.

      A member carries an axial compression of 1000 kN. Determine the eccentricity of centre of compression at a section at which the bending moment is (i) 400 kN.m, (ii) zero, and (iii) −600 kN.m.

1.2.   A column carries a compression of 500 kN. Determine the bending moment if the eccentricity of centre of compression is (i) −0.4 m, (ii) zero, and (iii) 0.6 m.

1.3.   Show that in the absence of external bending moment, the line of thrust in a prestressed concrete member coincides with the cable line.

1.4.   At a section of a prestressed concrete beam there are three cables located at 0.4, 0.5 and 0.6 m from top face. Each cable transmits a prestressing force of 200 kN. Determine the eccentricity of prestress and prestressing moment if the centroid of the section is located at 0.3 m from top face.   ʹ

1.5.   At a section of a prestressed concrete beam, prestressing forces of 310, 300, 290 and 150 kN are transmitted by cables located at 0.1, 0.2, 0.3 and 0.9 m from the bottom face. The centroid of the section is located at 0.5 m from the bottom face. Determine the eccentricity of prestress and prestressing moment.

1.6.   The eccentricities of six cables each carrying a force of 300 kN at a section are 0.5, 0.4, 0.3, 0.2, 0.1 and −0.3 m. Determine the eccentricity of prestress and the prestressing moment.

1.7.   A simply supported beam carries a uniform load of 20 kN/m inclusive of self weight over a span of 20 m. The beam has a prestressing force of 2000 kN at an eccentricity of 0.3 m. Determine the position of centre of compression at mid-span section.

1.8.   A cantilever beam having an effective span of 8 m carries a uniform load of 15 kN/m inclusive of own weight. Determine the position of centre of compression at the free and fixed ends if the prestressing force of 960 kN is applied at constant eccentricity of −0.2 m.

1.9.   A prestressed concrete beam carries a central concentrated load of 300 kN over a span of 10 m. The cable profile comprises two straight segments giving zero eccentricity at the ends and an eccentricity of 0.5 m at mid-span. The prestressing force is 800 kN. Determine the internal forces due to prestressing and also the resultant internal forces at left quarter point. Ignore self weight of the beam.

1.10.   A prestressed concrete beam has a parabolic cable profile with an eccentricity of 0.3 m at mid-span and no eccentricity at the ends. The prestresing force is 720 kN. The beam carries a uniform load of 20 kN/m inclusive of own weight over a span of 12 m. Determine the internal forces due to prestressing and also the resultant internal forces at the section at a distance of 2 m from left end.

1.11.   Determine the normal stresses due to prestressing and the resultant stresses in the beam of problem 1.9. The beam has a rectangular cross-section with overall dimensions 0.2 × 0.8 m.

1.12. Determine the normal stresses due to prestressing and the resultant stresses in the beam of problem 1.10. The beam has a rectangular cross-section with overall dimensions 0.2 × 0.6 m.

1.13. Show that the theoretical lower and upper limits of eccentricity of kern points in a symmetrical I-section are $D/6$ and $D/2$ respectively in which D is the overall depth of the section.

1.14. Determine the eccentricities of the upper and lower kern points of T-section shown in Fig. P.1.14.

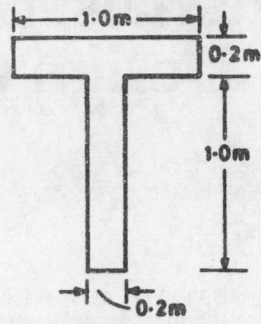

Fig. P 1.14

1.15. The T-section of Prob. 1.14 carries a prestressing force of 2400 kN at an eccentricity of 0.6 m. Determine the external bending moment required to shift the centre of compression to (i) lower kern point and (ii) upper kern point.

1.16. A beam restrained against end rotations has a parabolic cable profile with zero end eccentricities and sag S at midspan. Determine primary, secondary and resultant bending moments due to prestressing at the midspan section. The prestressing force is P and the span of the beam is l.

1.17. If the fixed beam of problem 1.16 carries an external uniform load of w per unit length, determine the resultant bending moment at the midspan section.

2

Materials, Loads and Design Concepts

2.1 INTRODUCTION

Among the variety of materials used in prestressed concrete construction, the two main constituents are the high grade concrete and the high tensile steel. They constitute the bulk of a prestressed concrete structure. In addition, a variety of minor items of materials are also required in a prestressed concrete structure. Plain or deformed reinforcing bars are required for shear reinforcement and as longitudinal bars, particularly in partially prestressed members. Non-prestressed reinforcing steel is needed also in the end blocks of prestressed concrete girders. Grout usually in the form of sand-cement slurry is used to fill the space between the cable and the duct in post-tensioned members. Metal alloys and plastics are commonly used for the sheathing which prevents the bond between cable and concrete in post-tensioned members.

2.2 HIGH GRADE CONCRETE

For a variety of reasons including economic considerations, the use of high grade concrete in prestressed concrete construction is considered not only desirable but essential. The main reasons for the use of high grade concrete are as follows.

(i) *Whole section effective*

In prestressed concrete construction, the entire cross-section of the member is effective in resisting the applied loads unlike reinforced concrete in which concrete located in the flexural tension zone is ineffective. As the entire concrete is utilized effectively in the load resisting mechanism, the use of high grade concrete is more justified in prestressed concrete as compared to reinforced concrete.

(ii) *High shear stress*

Due to the slender webs of prestressed concrete flanged sections common in practice, the shear stresses are high. These shear stresses have to be borne by concrete if diagonal cracking has to be prevented. For this reason, high grade concrete with relatively higher tensile strength has to be used.

(iii) *High bearing stress*

The end anchorages of post-tensioned members bear against concrete at the ends of the prestressed members. They transfer the tensile force in the cable as a bearing pressure on concrete, thereby producing high intensities of bearing stress. Concrete must be able to resist high bearing stresses if the end anchorages have to be small in size for the sake of economy. The need for high grade concrete is, therefore, evident.

(iv) *Loss of prestress*

The loss of prestress due to elastic shortening and creep of concrete decreases as the strength of concrete increases. The modulus of elasticity of concrete increases with the grade of concrete and is believed to be proportional to the square-root of the compressive strength of concrete. Thus for a given compressive stress, the high grade concrete suffers a relatively smaller strain, thereby decreasing the loss of prestress. Hence, in order to reduce loss of prestress, high grade concrete is desirable.

(v) *Reduction of self weight*

The desirability of a smaller self weight is evident because it means saving of concrete. But tne more important reason is the reduction in the dead load moment caused by a reduction in the self weight of the structure. The reduction in the dead load moment becomes crucial in the case of long-span structures in which the dead loads become more dominant than the live loads. In order to achieve economy through saving in the volume of concrete and the additional economy due to reduction in the dead load moment, the use of high strength concrete is recommended.

Most codes of practice prescribe a certain minimum strength for prestressed concrete. A higher compressive strength is generally prescribed for pretensioned members because the stress transfer takes place at a younger age in pretensioning. It is also justified because pretensioning is often carried out under highly controlled conditions in a well supervised central plant. As concrete with a cube strength of 50 MPa may cost only 15 per cent more as compared to concrete with cube strength of 25 MPa, the desirability of using higher grade concrete is evident. However, special problems of technique, supervision and control may arise when concrete with a cube strength greater than 50 MPa is prescribed. The Indian code IS: 1343-1980 prescribes a minimum characteristic cube strength of 40 MPa for pretensioning, and range of 30 to 60 MPa for post-tensioning.

2.2.1 Quality Control

The production of high grade concrete calls for good quality materials, proper mix design, skilled supervision and adequate quality control through well planned tests. The quality control of concrete is a specialised field which has been dealt with exhaustively in the treatises on the subject. The main points in the production of high grade concrete are briefly summarized below.

(i) *Quality cement*

Cement being the only binding material in concrete is mainly responsible for the gain of strength. Hence, cement should be of good quality, fresh and sound. It should satisfy the code requirements in respect of fineness, strength and other physical and chemical properties. Ordinary portland cement, portland slag cement (slag content not exceeding 50%), rapid hardening portland cement and high strength ordinary portland cement are commonly used. The relevant Indian codes for these "cements are IS: 269-1976, IS: 455-1976, IS: 8041-1978 and IS: 8112-1976 respectively.

(ii) *Coarse sand*

Clean coarse sand generally satisfying the requirements of IS: 383-1970 is required for high grade concrete. Sand should be free from silt, clay and organic material. To satisfy this requirement, sand should be washed thoroughly until the water draining out is visibly clean. Coarse sand with a fineness modulus not less than 2 is commonly recommended. Coarse sand requires a smaller quantity of cement paste to coat it due to its relatively smaller surface area.

(iii) *Well graded coarse aggregates*

In order to produce good dense concrete, coarse aggregate particles of varying sizes blended on the principle of either gap grading or continuous grading and generally satisfying the requirements of IS: 383-1970 should be used. The coarse aggregates should be free form flaky and sharp angular pieces. Coarse aggregate should be clean, free from weeds and other organic materials, hard and inert, not susceptible to attack by chemicals and weathering agencies.

(iv) *Clean water*

Water used for mixing and curing of concrete should meet the requirements of IS: 456—1978. It should be clean, free from salts and other soluble compounds and with a PH value generally not less than 6. It is generally believed that water fit for drinking is acceptable for mixing and curing of concrete.

(v) *Proper mix design*

The object of mix design is to determine the proportions of the constituent materials, viz. cement, sand, coarse aggregate, water and admixtures, if any, in the most economical mix which produces good dence concrete with a compressive strength equal to the specified target strength. As cement generally accounts for the major portion of the cost of the mix, the most economical mix uses the minimum quantity of cement. Hence the mix design may be directed towards the selection of the proportions of the. materials with minimum cement consistent with the target strength of concrete. The strength of concrete increases with a decrease in the water— cement ratio. But a lower water-cement ratio leads to a reduction in, the workability of concrete. As the strength is affected adversely if the concrete is not workable, the lowest water-cement ratio consistent with workability for the type of job and the equipment used for compaction has to be determined. The workability is sometimes increased by the addition. of admixtures. All these factors have to be taken into account in a proper mix design.

Several methods of mix design have been proposed and treated exhaustively in literature. Among them the following methods which are widely used for mix design may be {mentioned.

(a) ACI mix design method

The American Concrete Institute gives mix design procedure for no-slump concrete (0 to 25 mm slump) according to ACI: 211-65 and for normal weight and heavy weight concrete in accordance: with ACI: 211.1-77 with maximum 21 day cylinder compressive strengths of 47.5 MPa and 45 MPa respectively.

(b) U S B R mix design practice

The US Bureau of Reclamation Concrete Manual 1975 gives mix design. method for air entrained concrete when water reducing and set controlling admixtures are used for maximum 28 day cylinder compressive strength of 45.5 MPa.

(c) British mix design method

The British method gives design procedure for normal concrete mixes without air entrainment for maximum 28 days cube compressive strength of 75. MPa. The procedure recommended in 1975 supersedes the earlier procedure. (Road Note No. 4 method, 1950).

(d) IS mix design method

The Bureau of Indian Standards recommends design procedure for normal concrete mixes without air entrainment both for medium and high strength concrete in accordance with IS: 10262-1982. The method is similar to USBR method as well as the method specified in IRC: 44-1976 for pavements.

A common feature of the four methods is that the water-cement ratio is chosen for the target mean strength from empirical strength versus water-cement ratio relationships. The water content is chosen for the required workability. The methods, however, differ slightly in respect of the volume of aggregates in the concrete mix. The latest British method does not consider the combined aggregate grading curves unlike that in Road Note No. 4. It implies that aggregates of any grading may be used as long as they are within the grading limits prescribed by appropriate specifications.

(vi) *Thorough mixing*

A thorough mixing of the ingredients is essential for the production of high grade concrete. Thorough mixing can be achieved by a well designed concrete mixer. The fine and coarse aggregates may first be mixed thoroughly, then the cement and finally water may be added and mixed thoroughly. The time taken for mixing should be reasonably short so that the concrete can be transported and placed before the initial setting starts.

(vii) *Efficient transportation and placement*

In prestressed concrete construction, concrete may have to be transported over considerable distance before placement in its final position. The fresh concrete should be transported and placed without causing segregation. The time required for transportation and placement should be short enough keeping in view the initial setting time of cement used. A variety of equipment including conveyor belts, aerial ropeways and overhead cranes may facilitate transportation and placement.

(viii) *Thorough compaction*

Concrete can never attain its desired strength unless it is compacted thoroughly. Full compaction is possible only when concrete has a workable consistency. A variety of high frequency vibrators such as the internal needle type or external form vibrators are used for achieving full compaction. Vibration as a means of compaction has to be applied critically so that it is neither insufficient nor excessive. Excessive vibration leads to segregation and bleeding The air pockets or voids due to insufficient compaction reduce the strength of concrete drastically. They also reduce the impermeability and consequently the durability of concrete, particularly under adverse environmental conditions. In order to produce high grade concrete essential for prestressing, a stiff mix with a low water-cement ratio (0.4 to 0.5) is used. The stiff mixes can not be compacted adequately without powerful vibrators.

(ix) *Adequate curing*

For the development of high strength it is necessary to ensure that the mixing water does not leave concrete during the period in which the chemical reactions between cement and water take place and concrete gains most of its strength. If concrete is permitted to dry during the first few weeks, the chemical reactions responsible for the development of strength may be hindered. Curing of concrete may be effected either by soaking the concrete wet or by maintaining 100 per cent relative humidity around concrete. Sometimes water-proof curing compounds, which prevent the exit of mixing water from concrete, are used, particularly when the water for curing is scarce or at those points of the structure which cannot be kept wet. Steam curing is often used in factories engaged in the mass production of precast pretensioned elements such as transmission poles, railway sleepers, fencing posts and small beams. Steam curing hastens the development of strength by accelerating the chemical processes.

(x) *Good formwork*

Dimensional accuracy, strength and water tightness are the main attributes of a good formwork. The joints of the formwork should be water-tight so that the cement paste, which is mainly responsible for the development of strength does not leak out. Leaky joints of formwork must be strong to resist the deformation due to powerful vibrators. In the individual mould system of pretensioning, the jacks bear against the moulds which must, therefore, be strong enough to transmit the jacking force. Strong formwork is also necessary for achieving full compaction of concrete because only then it can offer adequate lateral support to fresh concrete during the process of compaction.

(xi) *Skilled supervision and control*

High grade concrete cannot be produced without skilled supervision and control. Skilled supervision should ensure that all the points enumerated earlier are attended to. The tests for the quality control of concrete may be divided into three types.

(a) Tests on ingredients

Adequate tests should be carried out on cement, sand, coarse aggregate and water to ensure that they are of the approved quality.

(b) Tests on fresh concrete

The tests should ensure that the materials have been mixed thoroughly and that the mix is workable. Visual inspection by trained and experienced concrete supervisors as well as field and laboratory tests are needed for this purpose.

(c) Tests on hardened concrete

There are many tests to ensure that the concrete has the minimum specified density, strength, impermeability and durability. However, the most important property which is generally considered as the index of quality, is the compressive strength of concrete. Hence, the most crucial test in the quality control of concrete is the compression test on standard 150 mm cubes or standard cylinders 150 mm in diameter and 300 mm in height. Other tests may be needed depending upon the service requirements of the structure.

2.2.2 Physical Properties

The physical properties of concrete depend upon the proportions and properties of its ingredients as well as on the method of placement, compaction and curing. Concrete which is a heterogeneous material neither obeys Hooke's law nor is fully elastic. The stress-strain characteristics are complicated further due to its properties of creep and shrinkage. Most of the physical properties of concrete, however, are related directly or indirectly with the compressive strength of concrete.

2.2.2.1 Compressive Strength

The compressive strength of concrete f_c is the most important parameter representing the quality of concrete because several other properties such as shear and tensile strength, modulus of elasticity, bond, impact, abrasion resistance and durability are related directly or indirectly to compressive strength. It, therefore, appears logical that most codes of practice classify the grades of concrete on the basis of its characteristic compressive strength. In European and several other countries the characteristic compressive strength is based on tests on 150 mm standard cubes, whereas in North America it is based on standard cylinders of 150 mm diameter and 300 mm height at the age of 28 days.

The most important single parameter affecting the compressive strength of concrete is the water-cement ratio. The compressive strength of concrete at a given age and under normal temperature decreases with increase in water-cement ratio and was first expressed by Abrams by the relationship,

$$f_{cy} = \frac{k_1}{k_2^{(w/c)}}$$

where, f_{cy} = Cylinder compressive strength of concrete

w/c = Water-cement ratio by weight

k_1, k_2 = Constants to be determined emprically for the particular type of mix

The compressive strength of concrete is also influenced by the chracteristics of cement, the characteristics and proportions of aggregates, degree of compaction, efficiency of curing, temperature during curing period, conditions of tests and the age at the time of tesing.

The compressive strength determined from tests depends upon the shape and size of the test specimens. The cube strength f_{cu} is greater than the cylinder strength f_{cy} mainly on account of the larger slenderness ratio of stan dard cylinder as compared to standard cube. The tests indicate that the ratio (f_{cy}/f_{cu}) may be taken equal to 0.8. The tests also indicate that a smaller specimen records a higher compressive strength mainly due to a lower statistical probability of a weaker zone or flaw in a smaller specimen. The characteristic compressive strength f_{ck}, which is used to classify concretes according to its grade in the Indian codes, is based on compression tests on standard cubes and is defined as the compressive strength below which not more than 5% of the cube tests are likely to fall. The characteristic compressive strength f_c' based on tests on standard cylinders may be defined in a similar way. The ACI code, however, defines f_c' as the characteristic compressive strength below which no more than 10% of the cylinder tests are likely to fall. The compressive strength of concrete increases with its age although at a progressively slower rate. Tests on concrete having proportions of ordinary portland cement, sand and coarse aggregate of $1:2:4$ showed compressive strength of approximately 5, 14, 21, 30 and 35 MPa at the ages of 1, 3, 7, 28 and 90 days respectively. When rapid hardening portland cement was used, the compressive strengths at the same ages were approximately 7, 18, 25, 33, and 38 MPa respectively. Tests have shown t hat the   gain of strength with time is quicker in high strength mixes with low water-cement ratio. For a water-cement ratio of 0.4, the ratio of strength after 1, 3 and 7 days to the strength at 28 days was found to be 0.19, 0.53 and 0.78 respectively. The values of the same ratio for water-cement ratio of 0.6 were found to be 0.12, 0.42 and 0.68 respectively. Tests on concrete made with British ordinary portland cement have shown that the ratio of strength at 7 days to that at 28 days lies between 0.60 and 0.77 with majority of results falling above 0.67. British code CP: 114-69 permits the strength at 7 days not less than (2/3) rd of the strength at 28 days. Tests on concrete made with American type I cements indicated a slightly higher ratio of 7 and 28 days strengths. The development of strength with age is quicker in hot climate as compared to that in cooler weather.

The development of higher strength in a shorter period is of particular economic importance in respect of the pretensioning technique. A high early strength can effect substantial saving in cost per unit by reducing the period of one cycle between two successive castings. A high early strength permits earlier stress transfer and quicker reuse of the moulds and other equipment used in pretensioning method. The rate of strength gain with time can be accelerated by the use of

(a) Rapid hardening cement and
(b) Steam curing

Rapid hardening cement, also known as high early strength cement, contains a higher proportion of tricalcium silicate (sometimes as high as 70 percent) and has a higher specific surface (325 m^2/kg) due to finer grinding of the clinker. These properties are mainly responsible for accelerated hydration and a quicker gain of strength. Steam curing is generally effected by circulating steam at atmospheric pressure to maintain a temperature of approximately 70 °C inside the steam jacket. High pressure steam curing, also known as autoclaving, is generally not recommended for structural applications because it reduces the bond with steel to approximately half the normal value obtained by ordinary curing. A typical steam curing cycle may be divided into the following four phases:

(i) *Presteaming period*

A presteaming period of 1 to 3 hours is allowed because an early rise of temperature at the time of setting of concrete may be detrimental to concrete. The rise of temperature due to steam curing creates air pressure in the pores of fresh concrete which may interfere with the setting process.

(ii) *Temperature rise period*

The rate of initial temperature rise after the presteaming period is of the order of 10 to 20 °C/hour reaching a maximum temperature not greater than 85-90 °C. A high rate of heating has to be avoided because it may create thermal shocks. The adverse effect of an early rise of temperature is more severe for a higher water-cement ratio. Consequently, steam curing is more appropriate for high grade concrete with low water-cement ratio.

(iii) *Maximum temperature period*

During this period of approximately 4 hours, the maximum temperature in the range of 70 °C to 90 °C is held constant. Temperature higher than 90 °C is not recommonded because it does not increase the strength of concrete. Temperature higher than 70 °C tends to create dilational tendencies due to expansion of concrete.

(iv) *Cooling period*

In thi period lasting for about 5 hours, the steam is cut off and concrete is allowed to soak in the residual heat and moisture of the steam chamber.

In order to achieve optimum results, the steam curing cycle has to be planned keeping in view the type of product, the type of plant and the manufacturing technique. For a concrete of given proportions and period of curing, there is a definite curing temperature which produces the maximum compressive strength at the end of the curing cycle. With proper steam curing a compressive strength equal to 70 per cent of the 28 days strength can be achieved in a period of 16 to 24 hours.

2.2.2.2 Tensile Strength

The tensile strength of concrete is not of much relevance in design of reinforced concrete beams as it is ignored in the computation of flexural strength. In the case of prestressed concrete beams, however, the tensile strength of concrete assumes special importance becuse it can be utilized in computing the flexural capacity in an elastic design at the working load level. The beams in which a limited amount of tensile stress is permitted under working loads have a more economical section as compared to beams in which no tension is allowed under working loads.

The determination of the true tensile strength of concrete is not easy because the measured strength depends upon the type and conditions of test, the shape and size of the test specimen, the stress gradient on the failure surface and unintentional eccentricity of loading. Three types of tests, viz. the direct tension test, the split (Brazilian) test and modulus of rupture test are commonly used for the determination of tensile strength of concrete. An extensive study by Wright has shown that the tensile strength determined from split test is approximately 1.5 times that determined from direct tension test and (2/3)rd of that determined from modulus of rupture test. The split test gives more consistent results as compared to the other two types of tests and is believed to give a value closest to the true tensile strength of concrete f_t. The lower value of tensile strength determined from direct tension test is mainly due to unintentional eccentricity. While the higher value determined from the modulus of rupture test is due to the stress gradient and consequent redistribution of stresses. In the modulus of rupture test the measured tensile strength is greater for centre point loading as compared to third point loading because in the former case the plane of failure is predetermined and has, therefore, a lower statistical probability of having a weak zone or flaw in the failure plane.

The tensile strength of concerete f_t increases with the compressive strength of concrete f_c although at a progressively smaller rate. It may be expressed b the relationship

$$f_t = c \, (f_c)^k \qquad (2.2.1)$$

in which c and k are constants which depend upon the properties of concrete mix and the type of test. Tests indicate that the constant k varies between 0.50 and 0.75 although the former value is widely accepted. Assuming that $k = 0.50$ and $f_c = f_c'$, the constant c depends upon the type of test. Tests indicate that k lies in the range 0.25 to 0.33 in direct tension test, 0 33 to 0.50 in split tension test and 0.50 to 0.75 in modulus of rupture test. Extensive tests at Laboratories of Portland Cement Association (PCA), USA on concrete with compressive cylinder strength ranging from 5 to 65 MPa showed that the ratio of direct tensile strength to compressive strength varied from 0.11 to 0.07 and the ratio of modulus of rupture (third point loading) to compressive strength varied from 0.23 to 0.11. The European Concrete Committee (CEB) assumes that the direct tensile strength may be determined by taking $c = 0.3$, $k = 2/3$ and $f_c = f_c'$ in Eq. (2.2.1). Tests at the Building Research Station, U.K., on 100 mm cubes and 100 × 100 × 400 mm centrally loaded beam showed that $k = 0.5$ and c varied between 0.52 to 0.87 with an average value of 0.70. According to PCA, the flexural strength or modulus of rupture may be determined from Eq. (2.2.1) by taking $c = 0.68$, $k = 0.5$ and $f_c = f_{cu}$. The ACI code recommends that the flexural strength may be determined from Eq. (2.2.1) by taking $c = 0.63$, $k = 0.5$ and $f_c = f_c'$. The Indian code IS: 456-1978 permits the determination of the flexural strength from Eq. (2.2.1) by taking $c = 0.7$, $k = 0.5$ and $f_c = f_{ck}$. In general, it may be mentioned that the intrinsic strength of concrete varies between 10 to 15 per cent of the compressive strength of concrete.

2.2.2.3 Types of Strains

The strains in concrete can be divided into 4 types.

(i) *Elastic strains*

The term elastic strain is generally used to denote the short term strain in the direction of the applied stress which occurs immediately on load application. The term elastic may appear to be somewhat ambiguous because the stress-strain curve for concrete is neither linear nor the strain is fully recoverable.

(ii) *Lateral strains*

Like other structural materials concrete undergoes lateral strains whose sign is opposite to that of the strain in the direction of applied stress. The Poisson's ratio taken as the ratio of lateral and longitudinal strains varies from 0.15 to 0.22 depending upon the type of concrete. The average value of 0.17 is commonly adopted as a representative value of Poisson's ratio of concrete.

(iii) *Creep strain*

Creep strain of concrete is the deferred time dependant strain under the action of sustained loads which occurs over a long period extending over several years although the rate of strain decreases continuously with time. Creep strain is important because it can be several times greater than the instantaneous or short term elastic strain of concrete. Creep strains affect the entire strain distribution, deformation characteristics and consequently the distribution of internal forces in concerete structures. Creep strain occurs essentially due to the deformation of cement paste and the resulting adjustment of aggregate particles. Creep and shrinkage strains interact with each other. Hence in exact terms, it is not possible to delink them or use the principle of superposition. However, in practical terns, the creep strain may be taken as the difference of the deferred strain in specimen under sustained loading and the deferred strain in a companion unloaded specimen kept under identical conditions of temperature and humidity. The main factors affecting the creep strain are the level of applied sustained stress, strength of concrete, type of coarse aggregate, water-cement and aggregate cement ratios. Due to its decisive influence on the deformation characteristics of concrete structures, the phenomenon of creep has been investigated extensively. The following are some of the salient features of creep phenomenon.

(a) The rate of creep strain decreases progressively with time. Tests at the University of California showed that of the total creep strain recorded in twenty years, 18-35% (average 26%), 40 to 70% (average 55%) and 60 to 83% (average 76%) occurred in the first two weeks, 3 months and 1 year respectively. The creep after a period of 2, 5, 10, 20, and 30 years was 14, 20, 26, 33 and 36 per cent higher as compared to creep after one year. The total creep after infinite time may be taken approximately equal to (4/3) times the creep strain after a period of one year. Of the total creep strain occurring after infinite time, 25%, 50% and 75% may be assumed to occur in the first 2 weeks, 3 months and 1 year respectively.

(b) At relatively low levels of sustained stress, the total creep strain over a long period is proportional to the instantaneous or short term elastic strain. Hence the ratio of total creep strain to the elastic strain, known as the ereep coefficient, may be taken to be constant upto working load level. For intensity of sustained stress higher than ($f'/3$) approximately, the creep coefficient increases progressively with higher intensity of sustained stress.

(c) The creep strain is inversely proportional to the strength of concrete. As the strength is inversely proportional to the water-cement ratio, the creep strain increases with increase in water-cement ratio.

(d) Although the aggregates themselves do not creep, they control creep by exerting a restraining effect. Hence aggregates having higher strength and modulus of elasticity exert a greater restraint to the deformation of the cement paste, thereby reducing the creep strain. Creep strain is progressively lesser for sandstone, basalt, gravel, granite, quartz and limestone. Concrete with sandstone aggregate may have a creep strain greater than two times that in the concrete with limestone aggregate.

(e) Concrete loaded at a younger age creeps more because the cement gel is relatively weak at a younger age. The creep strain of specimens loaded at the age of 90 days was approximately 10% lower than that of specimens loaded at 28 days.

(f) Tests showed that the creep strain was appreciably greater for low heat cement as compared to normal portland cement. Low heat cement of higher fineness gave larger creep but the opposite trend was noticed in the case of normal portland cement.

(g) Smaller specimens creep more than larger ones.

(h) The creep strain decreases with increase in relative humidity. Tests showed that the creep strain at 50% relative humidity was 40% and 20% greater as compared to those at 70% and 100% relative humidity.

(i) Depending upon the conditions of tests, the creep strain after a period of 20 years ranged from 1 to 5 times (average value 3 times) the instantaneous elastic strain. When the effect of shrinkage was also included, the total strain due to shrinkage and creep after 20 years ranged from 1 to 11 times the instantaneous elastic strain.

(j) Part of the creep strain is recoverable. It takes longer to recover a given creep strain than the period over which it occurs. In general 80 to 90% of the creep is recovered in the same period in which it occurs.

(iv) Shrinkage strain

Shrinkage like creep is a time-dependent phenomenon as it continues over a long period of time although at a progressively slower rate. However, creep occurs under sustained load while shrinkage occurs without the application of load. Shrinkage of concrete is caused mainly due to the exist of moisture from the body of concrete. Hence any factor which reduces the outward movement of moisture, tends to reduce the rate of shrinkage strain. The factors affecting shrinkage are mostly the same as those for creep except that ambient temperature and relative humidity play a dominant role in the shrinkage phenomenon. The shrinkage strain varies widely between 0 to 0.001 depending upon the type of mix and conditions of temperature and humidity. For certain types of cements and aggregntes under high humidity, it may be even negative (swelling).

The shrinkage of concrete may be divided into following four types.

(a) Plastic shrinkage

Plastic shrinkage occurs due to evaporation of water from the concrete surface while it is still in the plastic stage. It may result into plastic shrinkage cracks on the surface as well as inside the body of concrete if mixing water is allowed to evaporate at a very fast rate. Plastic shrinkage can be reduced by covering the surface with polythene sheeting immediately on finishing operation by monomolecular coatings, by fog spray or by working at night.

(b) Drying shrinkage

The shrinkage of hardened concrete does not occur due to loss of free water. It is the exit of adsorbed water in the gel pores which is responsible for drying shrinkage. Drying shrinkage represents the major part of the total shrinkage. It is one of the most undesirable properties of concrete as it is difficult to control and results in unsightly permanent cracks in slabs, pavements and other concrete structures.

(c) Autogeneous shrinkage

The shrinkage of concrete which occurs at constant temperature without movement is called autogeneous shrinkage. As autogeneous shrinkage strain is generally as small as 0.0001, it is of little practical importance.

(d) Carbonation shrinkage

This type of shrinkage occurs due to atmospheric carbon dioxide reacting with calcium hydroxide producing calcium carbonate. Carbonation shrinkage occurs because the volume of calcium carbonate thus formed is smaller than the volume of calcium hydroxide. As carbonation shrinkage is only a small fraction of drying shrinkage, it is of little practical significance.

Among the foregoing types of shrinkage, the drying shrinkage representing the major part of the total shrinkage is the most important. The following are the salient features in respect of drying shrinkage.

(1) Most of the shrinkage occurs in the first 2 or 3 months. Thus 14 to 34, 40 to 80 and 66 to 85 per cent of 20 year shrinkage may be expected in the first 2 weeks, 3 months and 1 year respectively.

(2) Shrinkage strain decreases with increase in relative humidity. Thus the shrinkage at 50 per cent relative humidity may be 35 to 40 per cent greater than that at 70 per cent relative humidity.

(3) Shrinkage strain decreases with increase in aggregate-cement ratio and decrease in water-cement ratio. Tests on concrete prisms having 125 mm square cross-section and a water-cement ratio of 0.5 stored at 21 °C and 50 percent relative humidity showed that the shrinkage strain over a period of 6 months decreased from 0.0012 to 0.0003 as aggregate-cement ratio was increased from 3 to 7. Under similar conditions of tests, the shrinkage strain decreased from 0.00085 to 0.00040 as the water- cement ratio decreased from 0.7 to 0.4 for aggregate-cement ratio of 5.

(4) As harder and denser aggregates exert a greater restraint on the shrinkage of cement paste, the shrinkage strain decreases progressively for sandstone, gravel, basalt, granite, limestone and quartz aggregate. Thus the shrinkage in concrete using sandstone aggregate may be more than twice that for quartz aggregate. Lightweight concretes have a larger shrinkage due to smaller density, hardness and modulus of elasticity and larger absorption of lightweight aggregates.

(5) The shrinkage strain is smaller for cements having larger proportion of tricalcium silicate and smaller proportion of alkalies and oxides of sodium and potassium.

(6) The shrinkage strain is greater for smaller specimens which have relatively larger surface area per unit volume. However, the scale effect vanishes after certain limit.

(7) A part of shrinkage is recoverable by restoration of lost water.

(8) The shrinkage strain varies widely under different conditions and may range from 0 to 0.001. Under normal or average conditions, the shrinkage strain varies between 0.0002 to 0.0006 in prestressed concrete constructions.

2.2.2.4 Stress-strain Relationship

The so called elastic short term strain discussed in Sec. 2.2.2.3 has a non-linear relationship with the applied uniaxial compressive stress f as shown in Fig. 2.2.1. While the exact shape of the stress-strain curve depends upon the type of mix and the physical properties of ingredients, the rising part of the curve is believed to be close to a parabola. The descending part of the curve is generally very close to a straight line. The maximum or peak stress f_m is attained at a strain ϵ_0 which is approximately equal to 0.002. The ultimate or crushing strain ϵ_u of unconfined concrete in uniaxial compression generally lies between 0.003 and 0.004 with 0.0035 as a mean value.

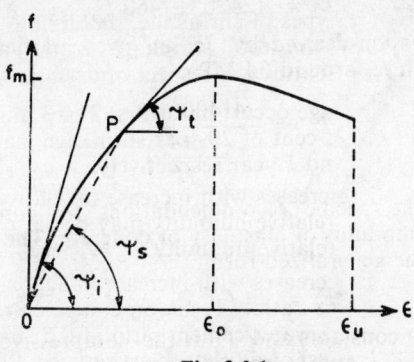

Fig. 2.2.1

2.2.2.5 Modulus of Elasticity

Due to the non-linearity of the stress-strain curve, the slope of the curve shown in Fig. 2.2.1 changes continuously.

The initial tangent modulus E_{ci} is defined as the slope of the tangent at origin.

$$E_{ci} = \tan \psi_i \qquad (2.2.2)$$

The tangent modulus E_{ct} at any point P is defined as the slope of the tangent at P.

$$E_{ct} = \tan \psi_t = \frac{df}{d\epsilon} \qquad (2.2.3)$$

The secant modulus E_{cs} at any point P is defined as the slope of the chord OP joining P to the origin O.

$$E_{cs} = \tan \psi_s \qquad (2.2.4)$$

The secant modulus is always smaller than initial tangent modulus but larger than tangent modulus

$$E_{ci} > E_{cs} > E_{ct} \qquad (2.2.5)$$

The tangent modulus decreases continuously with increasing stress in the ascending part of the curve, becomes zero at peak stress and becomes negative in the descending part of the curve.

The curvature of the initial portion of the ascending part of the curve corresponding to relatively low level of stress is small. Hence it is often approximated by a straight line joining the origin to a point at which the stress f is equal to $0.33 f_{ck}$. The modulus of elasticity of concrete E_c used in common practice is taken equal to the slope of this straight line. Hence the modulus of elasticity is the secant modulus at a point at which the stress is $0.33 f_{ck}$ or $0.40 f_c'$.

Hognested proposed the following empirical expression for modulus of elasticity of normal weight concrete.

$$E_c = 12420 + 460 f_{cy} \qquad (2.2.6)$$

The empirical expression derived by Jensen gives particularly good results for concrete strength f_{cy} around 34 MPa.

$$E_c = \frac{41400}{\left(1 + \dfrac{14}{f_c'}\right)} \qquad (2.2.7)$$

All major codes make specific recommendations for computing the value of short term static modulus of elasticity of concrete. The code expressions may be written in the generalized form,

$$E_c = c \, (f_c)^k \qquad (2.2.8)$$

in which c and k are constants and f_c is the compressive strength of concrete. The values of c, k and f_c according to some major codes are shown in Table 2.2.1 in which the unit weight of concrete w_c is in kg/m^3.

Table 2.2.1

S. No.	Code	c	k	f_c	Remarks
1.	ACI : 318	$0.043 \, w_c^{1.5}$	0.5	f_{cy}	General
2.	ACI : 318	4730	0.5	f_{cy}	For normal weight concrete
3.	CEB	$0.0016 \, w_c^2$	0.33	$(f_c' + 8)$	General
4.	CEB	9500	0.33	$(f_c' + 8)$	For normal weight concrete
5.	CP : 110	$0.0017 \, w_c^2$	0.33	f_{cu}	General
6.	CP : 110	9100	0.33	f_{cu}	For normal weight concrete
7.	IS : 1343	5700	0.5	f_{ck}	For normal weight concrete

2.2.2.6 Modular ratio

The modular ratio m is defined as the ratio of moduli of elasticity of any two materials. In the context of prestressed concrete, the modular ratio is defined as the ratio of modulus of elasticity of prestressing steel E_s and the modulus of elasticity of concrete E_c discussed in Sec. 2.2.2.5.

$$m = \frac{E_s}{E_c} \qquad (2.2.9)$$

The modulus of elasticity of prestressing steel generally ranges from 1.95×10^5 to 2.10×10^5 MPa. The values of modulus of elasticity of concrete E_c for the grades of concrete used in prestressed concrete in accordance with *IS* code are shown in Table 2.2.2. The last two lines of the table give the values of modular ratio for $E_s = 1.95 \times 10^5$ and 2.10×10^5 MPa respectively. It is seen that m varies from 4.42 to 6.73 in accordance with the provisions of IS code.

Table 2.2.2

f_{ck}, MPa	30	35	40	45	50	55	60
E_c, MPa	31220	33722	36050	38237	40305	42272	44152
m ($E_s = 1.95 \times 10^5$ MPa)	6.25	5.78	5.41	5.10	4.84	4.61	4.42
m ($E_s = 2.10 \times 10^5$ mPa)	6.73	6.23	5.83	5.49	5 21	4.97	4.76

2.3 HIGH TENSILE STEEL

High tensile steel is a high carbon steel with ultimate tensile strength not less than 980 MPa as per IS : 2090-1983. It is seen in Sec. 1.2 that for prestressing it is necessary to use high tensile steel which can be tensioned to much higher stress so that even after the losses, discussed in Chapter 4, the effective prestress is a major part of initial prestress. The dramatic increase in tensile strength is essentially due to an increase in carbon content. The carbon content ranges from 0.60 to 0.85 per cent in high tensile steel as against 0.20 to 0.30 per cent in reinforcing steel. It shows that although carbon is a minor constituent, it plays a dominant role in controlling the physical properties of steel. The increase in tensile strength due to higher carbon content is at the cost of ductility or energy absorbing capacity. This is evident from the fact that the ultimate elongation ranges from 2.5 to 6.0 per cent for high tensile steel as against 10 to 30 per cent for different grades of reinforcing steel. The higher carbon content also results in greater hardness and brittleness. Thus high tensile steel cracks when bent to sharp curvature.

A typical stress-strain curve for high tensile steel wire is shown in Fig. 2.3 1. It is straight upto the limit of proportionality and becomes non-linear thereafter. Comparing it with the stress-strain curve for mild steel which is also shown on the same figure, it is noted that the stress-strain curve for high tensile steel is marked by the absence of distinct yield point, plastic range and strain-hardening range. As high tensile steel does not have a distinct yield point, it is taken arbitrarily equal to the stress corresponding

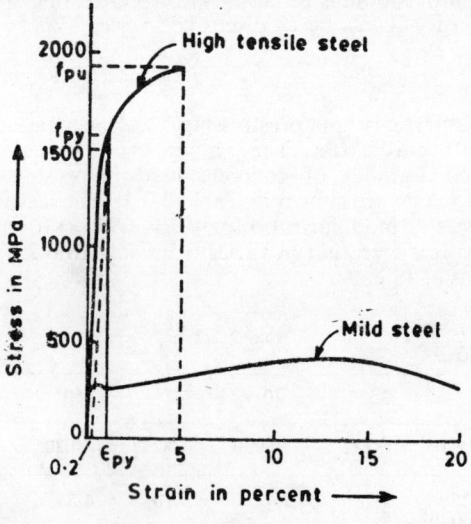

Fig. 2.3.1

to 0.1 or 0.2 per cent permanent set. Most codes of practice including IS : 1343-1980 take 0.2 per cent, also known as 0.2 per cent proof stress as the yield point. As an alternative, the stress corresponding to 0.7 or 1.0 per cent strain is sometimes taken as the yield point. The minimum ultimate strength f_{pu}, yield stress f_{py} and the ultimate elongation ϵ_{pu} are usually specified by the manufacturer. Prestressing tendons are often given stress relieving treatment to increase the proportional limit and also to reduce creep and stress concentration. Typical stress-strain curves for prestressing steel as per IS : 1343-1980 are shown in Fig. 2.3.2.

High tensile tendons used for prestressing are available commercially in the form of wires, bars, strands and cables. Basically there are only two types, viz. wires ranging from 2 to 8 mm diameter and bars above 10 mm diameter. Strands and cables are fabricated by grouping together several wires in order to reduce the number of units to be handled during the stressing operations. Strands are fabricated by twisting together two or more high tensile wires. Among the strands, the 7-ply strand fabricated using seven high tensile wires is most common. It is made by using a single central wire which forms the core with six wires laid around it in a single layer, all the helices having the same pitch and direction. Due to their spiral configuration, strands are superior in respect of bond. Cables are fabricated by arranging several high tensile wires or strands parallel to each other with the help of spacer plates and helical springs. The number and diameter of wires in a cable depend upon the system of post-tensioning adopted. For precast pretensioned concrete, individual wires are used for relatively small members and strands for relatively large members. Strands, bars and cables are used for post-tensioned members.

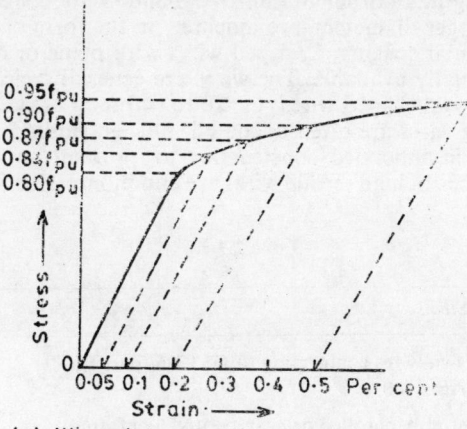

(a) Wires (stress relieved), strands and bars

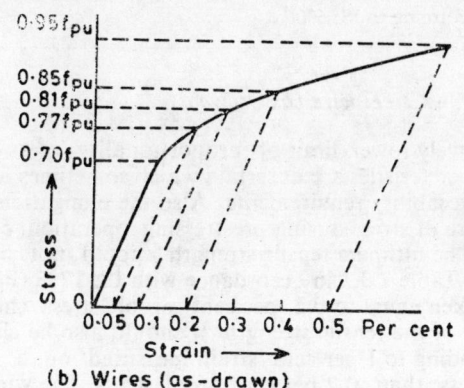

(b) Wires (as-drawn)

Fig. 2.3.2

Although high tensile steel differs largely from reinforcing steel in respect of strength and ductility, its modulus of elasticity is practically the same as that of reinforcing steel. The Indian code IS: 1343-1980 recommends that the modulus of elasticity of prestressing steel may be determined from test or manufacturers certificate. In the absence of such information, the code recommends the values given in Table 2.3.1.

2.3.1 High Tensile Steel Wires

High tensile steel wires are cold drawn from steel ingots made by the open hearth, electric duplex, acid bassemer, basic oxygen process or a combination of process. The realignment of steel crystals on account of successive drawing through a series of dies increases the tensile strength although the ductility is slightly reduced. Thus smaller the diameter of wire, the greater is its tensile strength. In order to improve bond with concrete, the wires, particularly of larger diameters, are indented in the form of small depressions at close regular spacing. Crimped wires with plane or heilcal crimping are also commercially available. The wires are generally circular in cross-section although oval shaped wires of 20 to 40 mm² area have also been used. Prestressing wires are often uncoated but galvanised wires are also used particularly in unbonded construction to prevent possible corrosion. The following types of high tensile wires are commonly used for prestressing.

Table 2.3.1

Type of Steel	E_s, kN/mm²
(i) Plain cold drawn wires conforming to IS: 1785 (part I), 1785 (part II) and 6003.	210
(ii) High tensile steel bars rolled or heat treated conforming to IS: 2090	200
(iii) Strands conforming to IS: 6006	195

(i) Plain hard drawn steel wire (as drawn)

Due to the relatively lower limit of proportionality of as drawn wire, the deformation characteristics are uncertain which sometimes leads to problems linked with serviceability requirements. Also the elongation, which is often used as a measure of stress during prestressing operation, cannot be predicted accurately. The ultimate tensile strength should not be less than the values shown in Table 2.3.2 in accordance with IS: 1785 (Part-II)-1983. The yield strength taken equal to 0.2 per cent proof stress should not be less than 75 per cent of the tensile strength. It should also be checked that the stress corresponding to 1 per cent strain measured on a gauge length of 200 mm is not less than 0.2 per cent proof stress. The wire should satisfy the reverse bend test according to which it should not show cracks when bent three times in reverse directions through 90° on a radius shown in the table.

(ii) Plain hard drawn steel wire (stress relieved)

The limit of proportionality of as drawn wires is increased by means of several types of stress relieving treatments. The treatments reduce the creep and streess concentration. The techniques commonly used for stress relieving are as follows:

(a) Time-stress treatment. In this treatment wire is prestretched to a stress higher than the initial prestress to be adopted in the construction. The technique can be used to increase the proportional limit to approximately 60 to 70 per cent of the ultimate strength which itself remains unaffected by the treatment.

(b) **Time-temperature treatment:** In this technique the wire is heated to a temperature of 400-430°C for the period of 30 to 40 seconds either by passing it through a bath of molten lead or a hot tunnel such as a heated ceramic tube.

The ultimate tensile strength should not be less than the values shown in Table 2.3.2 in accordance with IS: 1785 (Part-1)—1983. Wires of 5, 7 and 8 mm diameter may be manufactured for higher strengths having a minimum guaranteed tensile strength indicated by the numbers given in parentheses. The yield strength taken equal to 0.2 per cent proof stress should not be less than 86 per cent of the tensile strength. It should also be checked that the stress corresponding to 1 per cent strain measured on a gauge length of 240 mm is not less than 0.2 per cent proof stress. The wire should satisfy the reverse bend test according to which it should not show cracks when bent three times in reverse directions through 90° on a radius shown in the table. The wire should not show a relaxation of stress greater than 5 per cent of initial stress in 1000 hours and 3.5 per cent of initial stress in 100 hours.

Table 2.3.2   Minimum requirements of high tensile wires

Type	Nominal diameter, mm	Tolerance, mm	Tensile Strength, MPa	Ultimate elongation, per cent	Radius for bend test, mm
Wire—as drawn	3.0	± 0.02	1765	—	10.0
	4.0	± 0.03	1715	—	12.5
	5.0	± 0.03	1570	—	15.0
Wire—stress	2.5	± 0.025	2010	2.5	7.5
relieved	3.0	± 0.04	1865	2.5	10.0
	4.0	± 0.05	1715	3.0	12.5
	5.0	± 0.05	1570 (1715)	4.0	15.0
	7.0	± 0.05	1470 (1570)	4.0	20.0
	8.0	± 0.05	1375 (1470)	4.0	25.0
Indented wire	3.0	± 0.05	1865	2.5	10.0
	4.0	± 0.05	1715	3.0	12.5
	5.0	± 0.05	1570	4.0	15.0

(iii) *Indented wire*

Indented wires are generally preferred for pretensioning due to their superior bond with concrete. The indentations which are in the form of small depressions of uniform depths at regular pitch may have a circular or elliptic shape. They are placed in two lines diametrically opposite and are staggered so that no two indentations are opposite to each other. The ultimate tensile strength should not be less than the values shown in Table

2.3.2 in accordance with IS: 6003-1983. The yield strength taken equal to 0.2 per cent proof stress should not be less than 85 per cent of the tensile strength. It should also be checked that the stress corresponding to 1 per cent strain measured on a gauge length of 200 mm is not less than 0.2 per cent proof stress. The wire should satisfy the reverse bend test according to which it should not show cracks when bent three times in reverse directions through 90° on a radius shown in the table. Wires having indentations deeper than 3 per cent of nominal diameter should withstand two reverse bends without fracture. The wire should not show a relaxation of stress greater than 5 per cent of initial stress in 1000 hours and 3.5 per cent of initial stress in 100 hours.

High tensile wire is generally supplied by the manufacturer in the form of coils. The wires are cut to the required length and assembled at the plant or the construction site. While grease and loose rust must be removed before use, the presence of thin adherent rust is often considered desirable in improving the bond.

2.3.2 High Tensile Steel Bars

High tensile steel is manufactured by the open hearth, electric duplex, acid bassemer, basic oxygen (LD) process or a combination of these processes with the addition of necessary alloying elements. The steel is hot rolled into bars and then processed to give the required physical properties. The nominal diameters of bars as per IS: 2090-1983 are 10, 12, 16, 20, 25, 28 and 32 mm with a tolerance of ± 0.5 mm for bars upto 25 mm diameter and ± 0.6 mm for bars above 25 mm. The code requires that the tensile strength, 0.2 per cent proof stress and ultimate elongation should not be less than 980 MPa, 784 MPa and 10 per cent respectively. The relaxation of stress under constant strain over a period of 1000 hours should not be greater than 49 MPa. The threading at the ends of the bar, if required, is either rolled or cut.

2.3.3 High Tensile Steel Strands

Strands are fabricated by twisting together high tensile steel wires in such a manner that they do not unravel when cut and do not fly out of position when cut without seizing. The minimum physical requirements for strands manufactured from 2, 3 and 7 wires as per IS: 6006-1983 are shown in Table 2.3.3. The seven-wire strand has a central wire, at least 1.5 percent greater than the surrounding wires, enclosed tightly by six helically placed outer wires. The length of lay, which is defined as the distance measured along a straight line parallel to the strand forming one complete spiral of a wire around the strand. should not be less than 12 times nor greater than 16 times the nominal diameter of strand. The length of lay for two-wire and three-wire strands should range from 24 to 36 times the diameter

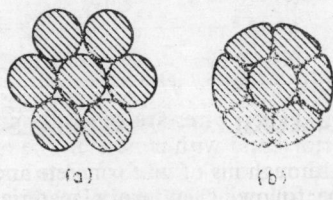

(a)   (b)

Fig. 2.3.3

of the element wire. The strand may be supplied as-spun (Fig. 2.3.3. a) or after it is made to pass through a die (Fig. 2.3.3 b). After the stranding process, the strands are stress relieved by time-temperature treatment. Thereafter, the strands are wound into coils or reels of sufficiently large diameter (not smaller than 600 mm). The elongation measured on a gauge length of not less than 600 mm should not be less than 3.5 per cent immediately prior to the fracture of any of the component wire. The wire should not show a relaxation of stress greater than 5 percent of initial stress in 1000 hours and 3.5 per cent of initial stress in 100 hours.

Table 2.3.3   Minimum requirements of high tensile strands

S. No.	Type and designation	Nominal diameter, mm		Tolerance, mm	Nominal cross— sectional ares of strand, mm²	0.2 per cent proof load, kN	Breaking load, kN
		Wire	Strand				
1.	Two wire strand						
	(a) 2—ply 2mm	2	—	± 0.03	6.28	10.84	12.75
	(b) 2—ply 3mm	3	—	± 0.03	14.14	21.67	25.50
2.	Three wire strand						
	3—ply 3mm	3	—	± 0.03	21.21	32.46	38.25
3.	Seven wire strand (Class I)						
	(a) 6.3mm 7—ply	—	6.3	± 0.4	25.1	37.81	44.48
	(b) 7.9 mm 7—ply	—	7.9	± 0.4	37.4	58.60	68.95
	(c) 9.5mm 7—ply	—	9.5	± 0.4	51.6	§79.40	93.41
	(d) 11.1 mm 7—ply	—	11.1	± 0.4	70.3	105.86	124.54
	(e) 12.5 mm 7—ply	—	12.5	± 0.4	92.9	139.90	164.58
	(f)15.2 mm 7—ply	—	15.2	± 0.4	138.7	192.83	226.86
4.	Seven wire strand (class II)						
	(a) 9.5 mm 7—ply	—	9.5	± 0.4	54.8	86.96	102.31
	(b) 11.1 mm 7—ply	—	11.1	± 0.4	74.2	117.21	137.89
	(c) 12.5 mm 7—ply	—	12.5	± 0.4	98.7	156.15	183.71
	(d)15.2 mm 7—ply	—	15.2	± 0.4	140.0	222.23	261.44

2.4 AUXILIARY MATERIALS

Although high grade concrete and high tensile steel are the main constituents, the following auxiliary materials are frequently required in prestressed concrete construction.

2.4.1 Reinforcing Steel

Prestressed concrete beams require untensioned reinforcing steel in the form of web reinforcement to resist shear and torsion. Even when concrete by itself is capable of resisting the design shear force and torque, a certain amount of nominal reinforcement is generally provided to satisfy the codal requirements. Other linear members such as transmission poles, railway sleepers and fencing posts have reinforcing steel in the form of ties. In most prestressed members, reinforcing steel is used for control on crack width and for other serviceability requirements. In partially prestressed members, the design moment is carried jointly by prestressed high tensile steel and untensioned reinforcing steel. In actual practice, there is hardly any prestressed concrete member which does not contain a certain amount of reinforcing steel.

Reinforcing steel is marketed in various forms and grades. It can be broadly classified into mild steel, medium tensile steel and high yield strength steel. All codes of practice on reinforced concrete contain provisions for acceptable forms of reinforcing steel. The minimum requirements of reinforcing steel for different grades of steel are specified in IS: 432 (Part I), 1139, 1786 and 1566.

2.4.2 Extractable Cores and Sheathing

In the case of post-tensioned members it is necessary to prevent bond between prestressing tendons and concrete so that the tendons can be stressed against hardened concrete when it has developed adequate strength. This objective can be achieved either by means of extractable cores or by covering the tendons with sheathing. Extractable cores are placed in the mould, before concreting. They are withdrawn after the concrete has hardeneds thereby forming ducts at the desired positions. Thereafter, the prestressing tendons are threaded through the preformed ducts. The acceptable rubber core may be either solid or hollow. The solid core is sometimes stiffened by an axial steel rod. The core is often easily withdrawn by pulling from one end because the cross-sectional area decreases due to Poisson's ratio effect. To facilitate withdrawal, the hollow core may be inflated before concreting and deflated during extraction. The withdrawal becomes difficult in the case of long ducts. In such cases it is better to use sheathing or conduit which remains permanently embeded in concrete. The main advantage of extractable core is that it can be used several times with accompanying benefits in cost. On the other hand, sheathing saves the cost incurred in extracting operations. The sheathing or conduit, which is generally made from galvanised ferrous metal, light alloys or plastic, should have the following properties:

(i) They should be sufficiently water tight to prevent concrete laitance penetrating in them during concreting.

(ii) They should have bores sufficiently large to allow being easily threaded on to the cable or bar in long lengths.

(iii) They should be of sufficient strength as not to be dented or deformed during handling or concreting. Generally they have spirally wound corrugations to provide transverse rigidity and longitudinal flexibility.

A maximum tolerance of 5 mm is generally permitted on the alignment of extractable cores and sheathing which must be secured adequately against possible movement during placement and compaction of concrete. For this purpose the cores and conduits are supported by transverse steel rods placed under and over them at a spacing of approximately 1 m.

2.4.3. Grout

The prestressing tendons in post-tensioned members are housed in preformed ducts. In order to prevent corrosion of prestressing tendons and also to prevent collapse of the member in the event of failure of end anchorages, the ducts are subsequently filled by cement grout in what is described as bonded construction. The desirable properties of grout are fluidity and low sedimentation in the plastic state as well as low shrinkage and high durability in the hardened state. Hence the grout is generally in the form of neat cement with water-cement ratio of 0.5 approximately but not exceeding 0.55. The code IS: 1343-1980 specifies a 7-day compressive strength of 17 MPa for 100 mm cubes. In the case of ducts, fine sand passing 150 micron IS sieve may be added. Sometimes admixtures are also added to improve the performance of the grout. The grout is injected into the duct under pressure by pumps. The intensity of pressure commonly used for grouting ranges from 0.5 MPa to 1.7 MPa. The grout is injected through holes in the anchorage heads and cones or pipes buried in the concrete members. Grout is injected from one end of the cable duct until it is forced out from the other end. Sometimes grouting may be done from both ends until it is ejected from the central vent. Before the commencement of grouting, all unwanted openings should be sealed Wherever possible the grout must proceed uphill from the lowest point so that air and water which are lighter than the grout are pushed ahead instead of being entrapped. Air vents, which are provided at all crests and at a spacing of 20 m to 30 m, should be closed successively in the direction of flow when grout flowing out of them has the same consistency as that at the injection end. The remote end of the duct should be treated similarly. A grouting pressure of 0.7 MPa is developed before closing the injection end. In case where grouting has to be done in vertical ducts of considerable length, special additives should be used to prevent segregation of water from the grout. This precaution eliminates the possibility of a portion of the cable near the top remaining exposed due to lack of grout resulting in its serious corrosion. The grouting should not be done in extremely cold weather to prevent the possibility of ice formation and consequent voids in the ducts. A minimum temperature of 1.7°C has been recommended for grouting by Prestressed Concrete Institute specifications.

2.5 LIMIT STATES AND DESIGN APPROACHES

According to CEB-FIP model code, a structure is considered unfit for use when it exceeds a particular state, called a limit state, beyond which it infringes one of the criteria governing its performance or use. With this generalized definition of a limit state, each one of the stages of loading enumerated in Sec. 1.13 constitutes a limit state. The limit states can be broadly classified into two main categories:

(a) Limit state of ultimate strength or collapse

A structure is said to have reached the limit state of ultimate strength or collapse when it has reached its maximum load carrying capacity due to

(i) loss of equilibrium of a part or the whole of the structure considered as a rigid body

(ii) rupture of critical sections of the structure in axial tension or compression, shear, flexure, torsion or any combination of these internal forces

(iii) transformation of the structure into a mechanism

(iv) buckling due to elastic or plastic instability

In the case of some special structures, fatigue and excessive deformation may also be the causes for the collapse of the structure.

(b) Limit states of serviceability

A structure may not be able to satisfy the serviceability requirements due to

(i) excessive deformation which may adversely affect the appearance or efficiency of the structure

(ii) local damage which may adversely affect the appearance or efficiency of the structure or lead to excessive maintenance or corrosion and

(iii) excessive vibration which may cause discomfort, alarm or impairment of proper functioning of the structure.

In addition to the limit states of serviceability due to the three causes cited above, there may be other limit states of serviceability depending upon the type of structure and its intended functions. In the context of prestressed concrete, the stages of loading such as initial or final stages and those during transportation, lifting, hoisting and launching may be regarded as limit states of serviceability.

A satisfactory design based on the limit state concept must satisfy the requirements for all possible limit states. The limit state design is sometimes confused with the ultimate strength design. It must be recognised that the limit state of ultimate strength or collapse is just one of the several limits to be considered in limit state design although it is usually the most important and dominant limit state. It may be emphasized that the limit state design has a much wider scope as compared to ultimate strength design or working stress design which may be viewed as two of the several limit states covered in a comprehensive limit state concept. Some engineers feel that the method of working stress design should be dispensed with because the linear elastic theory used in this method is incapable of giving a rational analysis for a cracked structure. While this argument may appear justifiable in the case of reinforced concrete which cracks under working loads, it may not have the same weight in the context of prestressed concrete structures which generally remain uncracked under normal service loads. Even in the case of reinforced concrete structures, the model used for flexural analysis in working stress method is essentially the same as that in the ultimate strength method. It is in respect of shear and torsion that the working stress method gives quite unrealistic results. To sum up, it may be stated that the limit state design method which takes care of all possible limit states is most suitable particularly in the context of reinforced and prestressed concrete structures.

Although there are several limit states which have to be considered in a comprehensive limit state design, there are generally only two or three critical limit states which control the design. The structure may be designed for these critical limit states and checked for all other relevant limit states. The modifications required, if any, as a result of these checks are usually very minor. In the context of prestressed concrete, the common practice is to design the structure for the initial stage at stress transfer before losses and the final stage after losses using the linear elastic theory for uncracked section. The cross-sectional properties and the prestressing force are chosen so that the stresses under the action of relevant working or service loads at the two stage do not exceed the permissible stresses. The design is then checked for all other limit states. particularly the limit state of collapse. In keeping with this common practice, the elastic design for initial and final stages under service loads is given the importance it deserves.

Codes of practice specify permissible stresses in materials to ensure adequate performance at the limit state of serviceability both at the initial stage as well as the final stage. The permissible stresses at the initial stage are chosen carefully to ensure that the structure does not show any undesirable signs such as cracking of concrete or shapping of prestressing tendons at jacking or stress transfer. The permissible stresses at the final stage are selected in such a manner that the materials can withstand them over infinitely long period without showing any signs of distress. In order to ensure these requirements, the permissible stresses must incorporate sufficient margins of safety over the characteristic strength of the respective materials. The choice of permissible stresses reflects experience gained through years of practice and extensive research on properties of materials.

Due to the social and economic consequences of structural failures, the limit state of collapse clearly dominates structural design. To minimize, if not eliminate, the possibility of a disastrous collapse, it is necessary to ensure that the limit state of collapse is sufficiently higher than the limit state of serviceability pertaining to the final stage. In order to achieve this objective, the codes of practice specify partial safety factors or strength reduction factors and load factors. For instance, the Indian code on prestressed concrete. IS: 1843-1980 specifies partial safety factors on materials and loads. These factors reflect the uncertainities associated with material properties and variations in load intensities. The American code. ACI: 318-1989 specifies strength reduction factors reflecting possible variations in resistances and load factors representing possible overloads. The design ultimate loads, called the factored design loads are obtained by multiplying the service loads by the respective load factors. In order to arrive at appropriate values of partial safety factors, strength reduction factors and load factors, due consideration is given to the nature of materials and the characteristics of service loads.

2.6 PERMISSIBLE STRESSES

The permissible stresses are relevant to the initial stage. the final stage and other serviceability limit states. At these stages, a prestressed concrete section is designed by the elastic theory utilizing the permissible stresses in constituent materials, viz. high grade concrete, prestressing steel and reinforcing steel.

2.6.1 Permissible Stresses in Concrete

The permissible stresses in concrete are related to its characteristic compressive strength. They depend upon the nature of stress. Thus the codes of practice specify permissible stresses in concrete in respct of flexural and direct compression, tension, shear, bond and bearing. These are discussed in appropriate sections as shown in Table 2.6.1.

Table 2.6.1 Permissible stresses in concrete

S. No.	Type of stress	Section
1.	Compressive stress	6.7
2.	Tension	
	(a) Full prestressing (Type 1)	6.7
	(b) Limited prestressing (Type 2)	6.7
	(c) Partial prestressing (Type 3)	10.10
3.	Shear stress	7.2.3
4.	Bond stress	7.5.3
5.	Bearing stress	7.6.2

2.6.2. Permissible Stresses in Prestressing Steel

Prestressing tendons generally carry their maximum tensile stress at the time of jacking. Due to loss of prestress on account of several factors, the stress in tendons decreases and it seldom reaches its initial or jacking stress even when the structure carries its full working load. Only in the case of overloads, the tendons may carry a stress greater than the initial stress. Consequently the codes of practice specify an upper limit on maximum permissble initial stress in prestressing steel. The Indian code IS: 1343–1980 specifies that at the time of initial tensioning, the maximum tensile stress f_{pi} immediately behind the anchorages shall not exceed 80 per cent of the ultimate tensile strength f_{pu} of prestressing tendons. The minimum requirements in respect of the ultimate strength of the prestressing steel are given in Sec. 2.3. The ACI code specifies a maximum jacking stress of $0.80\ f_{pu}$ or $0.94\ f_{py}$, whichever is smaller. It also specifies that the stress in pretensioned tendons immediately after transfer of prestress should not be greater than $0.82 f_{py}$ or $0.74\ f_{pu}$, whichever is smaller. In the case of post-tensioned tendors after anchorage the stress should not be greater than $0.70 f_{pu}$. The CEB-FIP code specifies the maximum tensile stress in the tendons, after releasing the jacks and so bringing the anchorages into operation, equal to 0.75 times the characteristic tensile strength or 0.85 times the 0.10 per cent characteristic proof stress, whichever is smaller.

The upper limit upto which prestressing tendons are permitted to be stressed depends upon social consequences of rupture, system of prestressing, degree of prestressing, quality of steel, type of tendon, time interval before grouting and practicability of replacing of a broken tendon, care should be exercised during jacking, if the jacking stress is greater than the proportional limit.

2.6.3 Permissible Stresses in Reinforcing Steel

The permissible stresses in reinforcing steel depend upon the grade of steel, the diameter of bar and the type of stress, i.e. tensile or compressive. Codes of practice on reinforced concrete specify the permissible stresses in reinforcing steel. The Indian code IS: 456-1978 gives the permissible stresses in reinforcing steel in Sec. 44.2. Table 16.

2.7 PARTIAL SAFETY FACTORS

The partial safety factors are mainly relevant to the limit state of collapse. The values of partial safety factors are chosen to ensure a reasonably low probability of collapse of the structure. As the materials as well as the loads are given to uncertainities and variations, the partial safety factors are applied to materials strengths and load intensities. The partial safety factors for materials γ_m take account of the nature of materials, manufacturing process and the probable scatter in characteristic strength of the material. The partial safety factors for loads γ_f depend upon the type of load and also the manner in which the loads combine with each other to form the design load. While specifying the partial safety factors under the action of various combinations of loads the guiding principle is that the acceptable margin or safety against collapse is maintained for all possible or reasonable load combinations. For instance, a value smaller than 1.0 is assigned to the partial safety factor for dead load if a smaller dead load creates a more adverse condition. Such a contingency often arises in respect of stability against overturning and stress reversal. Partial safety factors for wind and seismic forces are also specified but it is generally assumed that wind and seismic forces do not act simultaneously. The criterion for design based on the limit state of collapse is that the reduced material strengths and the enhanced loads due to application of partial safety factors should just take the structure to its limit state of collapse.

The Indian code IS: 1343-1980 specifies the following partial safety factors for materials.

$$\gamma_m = 1.5 \text{ for concrete}$$
$$= 1.15 \text{ for steel}$$

These factors should be used in design based on limit state of collapse. The code recommends values of partial safety factors for loads to be used in the design based on limit states of collapse as well as the limit states of serviceability. The recommended values of partial safety factors for loads are shown in Table 2.7.1 in which DL, LL and WL represent dead, live and wind loads respectively. The partial safety factors for loads according to British Code BS: 8110–1985 are similar except for minor changes as indicated by numbers in parentheses in the table. As wind and earthquake forces are assumed not to act simultaneously, the wind force WL should be replaced by earthquake load EL whenever the earthquake load acts. While considering the combination comprising dead load and wind load, the lower value of 0.9 should be used for dead load if it gives a more critical condition in the consideration of stability against overturning and stress reversal. The γ_f values for limit state of serviceability given in Table 2.7.1 should be used in computing the short term effects. In computing the long term effects due to creep, the dead load and that part of the live load which is of permanent nature should be considered.

Table 2.7.1 Values of partial safety factors γ_f for loads

S. No.	Load combination	Limit state of collapse			Limit state of serviceability		
		DL	LL	WL	DL	LL	WL
1.	DL + LL	1.5 (1.4)	1.5 (1.6)	—	1.0	1.0	—
2.	DL + WL	1.5 (1.4) or 0.9	—	1.5 (1.4)	1.0	—	1.0
3.	DL + LL + WL	1.2	1.2	1.2	1.0	0.8	0.8

The CEB-FIP code contains comprehensive provisions in respect of partial safety factors. The partial safety factors for materials take into account the following effects:

(a) the probability of unfavourable deviations of the strength of materials from the specified charactristic values

(b) the probable difference in the strengths of materials in the actual structure and the control specimens

(c) the probability of local weaknesses in materials or elements

(d) the probable inaccuracies in the computations of strengths of structural elements and

(e) the probable variations in dimensions and geometrical properties of the structure.

The partial safety factors for loads take into account the following effects:

(a) the probability of unfavourable variations in characteristic loads

(b) the probability of possible combinations of various types of loads such as dead, live, wind and earthquake loads and

(c) the probable inaccuracies in the assessment of characteristic load.

While the partial safety factors are relevant to all limit states, they are primarily relevant to the limit state of collapse bacause it is the safety aspect which is being considered. The characteristic strength of materials should be divided by γ_m to determine the design strengths. Similarly, the characteristic load should be multiplied by γ_f to determine the design loads.

2.8 LOAD FACTORS AND STRENGTH REDUCTION FACTORS

These factors are relevant to the limit state of collapse. The strength reduction factors are similar to the partial safety factors for materials. The basic difference is that whereas the partial safety factors for materials are applied to the characteristic strengths of materials, the strength reduction factors are applied to the computed capacities of structural elements. The load factors are similar to the partial safety factors for loads except that they are relevant only to the limit state of collapse. The computed ultimate strengths of structural elements are multiplied by the strength reduction factors to take account of possible undesirable variations in constituent materials. The service loads are multiplied by the load factors to obtain the factored or design ultimate load. It is then checked that the reduced ultimate strengths of the members at all critical sections are not smaller than the internal forces produced by the most unfavourable combination of the design ultimate loads.

The foregoing concepts of ultimate strength design are incorporated in the ACI code which specifies the following strength reduction factors:

(a) flexure with or without axial forces   0.90

(b) axial compression with or without flexure

    (ii) spiral reinforcement   0.75

    (ii) other reinforcement   0.70

(c) axial tension without flexure   0.90

(d) shear   0.85

(e) torsion   0.85

(f) bearing on concrete   0.70

The ACI code specifies load factors equal to 1.4 and 1.7 for dead load and live load respectively. When the effect of wind has also to be included, the three load combinations shown in Table 2.8.1 should de considered and the capacities of all critical Sections should be based on the most unfavourable combination. As the wind and earthquake forces are assumed not to act simultaneously, the wind forces should be replaced by earthquake forces in the analysis for earthquake forces. The load factors for earthquake forces are obtained by multiplying those for wind load by 1.10. When the effect of lateral earth pressure has also to be included in design, the load combination comprising dead load, live load and lateral earth pressure should be considered with load factors equal to 1.4, 1.7 and 1.7 respectively. It should also be checked that load factors equal to 0.9,0 and 1.7 for the three types of loads do not produce a more unfavourable combination. The foregoing criterion for lateral pressure due to earth also applies to that due to liquid except that the load factor for lateral pressure due to liquid should be taken equal to 1.4. Also the vertical pressure due to the liquid should be considered as the dead load with due regard to the variations in liquid depths. The impact effect, if any, should be included in the load factor for live load. When the effects of differential settlement, shrinkage, creep and thermal changes causing a load TL have to be included in design, the load factors for these effects, dead load and live load should be taken equal to 1.05, 1.05 and 1.275 respectively. However it should be checked that the ultimate strength is not less than 1.4 (DL + TL).

Table 2.8.1   Values of load factors for limit state of collapse

S. No.	Load combination	DL	LL	WL
1.	DL + LL	1.4	1.7	—
2.	DL + WL	0.9	—	1.3
3.	DL + LL + WL	1.05	1.275	1.275

REFERENCES

2.1. Davis, R.E. and Troxell, G.E, *Modulus of elasticity and Poisson's ratio for concrete and influence of age and other factors upon these values*, Proceedings, ASTM, 1929.

2.2. Davis, R.E. and Troxell, G.E., *Properties of concrete and their influence on prestressed design*, Journal of the American Concrete Institute, Proceedings Vol. 50, January 1954, pp. 381-91.

2.3. Handbook on Concrete Mixes, SP 23 (S&T), Bureau of Indian Standards, 1982.

2.4. Meyer, A., *Uber den E:nfluss des Wasserzementwertes auf die Fruhfestikeit von Beton*, Betonstein Zeitung, No. 8, 1963, pp. 391-41.

2.5. Neville, A.M., *Creep of Concrete: plain, reinforced and prestressed*, Amsterdam, North Holland, 1970.

2.6. Neville, A.M., *Properties of Concrete*, Pitman Publishing Limited, Third Edition, 1981.

2.7. Orchard, D.F., *Concrete Technology, Vol. I*, Applied Science Publishers Limited (London), Third Edition, 1973.

2.8 Pandit, G.S., *Strength of concrete under multi axial stress*, The Indian Concrete Journal, Vol. 47, No. 7 July 1973, pp. 266-9.

2.9 Polivka, M., *Grouts for post-tensioned prestressed concrete members*, Journal of the Prestressed Concrete Institute, 1961.

2.10. *Post-Tensioning Manual*, Post-Tensioning Institute, Phoenix, Arizona, 1976, pp. 143-8.

2.11. PCI *Committee, Recommended practice for grouting of post-tensioned prestressed concrete*, PCI Committee on post-tensioning, Journal of the Prestressed Concrete Institute, Vol. 17, No, 6, November-December, 1972.

2.12. Troxell, G.E.; Raphel, J.M. and Davis, R.E., *Long time creep and shrinkage tests of plain and reinforced concrete*, Proceedings ASTM, Vol. 58, 1958, pp. 1101-20.

2.13. Wright, P.J.F., *Comments on an indirect tensile test on concrete cylinders*, Magazine of Concrete Research, Vol. 7, No. 20, July 1955, pp. 87-95.

3

Systems of Prestressing

3.1 CLASSIFICATION OF PRESTRESSED CONCRETE

Prestressed concrete can be classified in various ways as illustrated by the chart of Fig.3.1.1. As the first major classification, prestressed concrete may be divided into the following types:

(i) precast
(ii) cast-in-situ and
(iii) composite

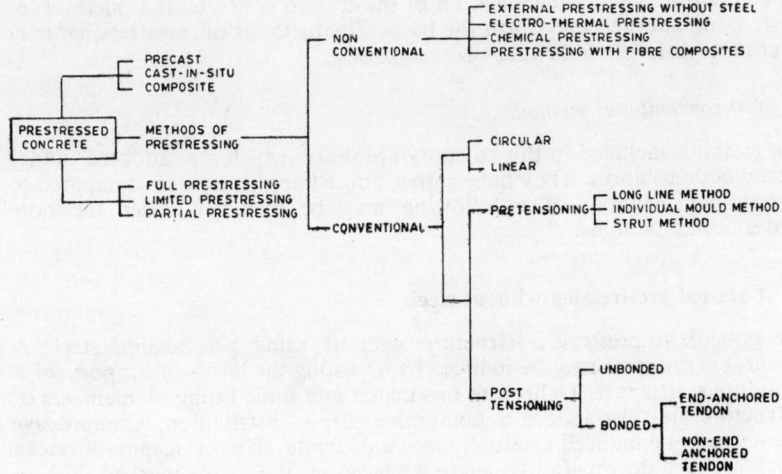

Fig. 3.1.1

Precast prestressed members are made in a central plant or at a place near the work site to reduce transportation. Cast-in-situ prestressed concrete is placed in its final position and does not require transportation, lifting, hoisting and launching. The third type is the composite concrete made by

---

judicious combination of precast and cast-in-situ concretes to derive the benefits of both types. Often the beam or rib is of precast prestressed concrete and the top slab is of cast-in-situ reinforced concrete. Cast-in-situ concrete firmly connects the precast concrete and imparts rigidity to the composite construction.

As a second major classification, prestressed concrete members may be divided into three types:

(i) *Full prestressed members (Type 1)*

These members do not have any tensile stress under the action of working or service ;loads.

(ii) *Members with limited prestress (Type2)*

These members develop some tensile stress under working loads but it is smaller than the tensile strength of concrete. Hence these members like fully prestressed members remain uncracked under working loads.

(iii) *Partially prestressed members (Type 3)*

Like reinforced concrete members, these members crack under working loads. They contain both prestressing steel and reinforcing steel which resist the working loads jointly. Partial prestressing combines the merits of both types of materials, viz. prestressed concrete and reinforced concrete.

For the third major classification of prestressed concrete, the method of prestressing adopted may form the basis. The methods of prestressing may be grouped into the following two categories:

(i) *Non-conventional methods*

The methods included in this category are those which are adopted under special circumstances. They have rather limited application as compared to conventional methods. The following may be included under the non-conventional methods:

1. External prestressing without steel

It is possible to prestress a structure without using prestressing steel. A favourable prestress may be induced by adjusting the levels of supports of a continuous structure. Deliberate mis-match and force fitting of members of a structure may also induce a favourable stress distribution. Compressive stresses may be induced externally into a concrete slab by means of jacks which are withdrawn after concrete is placed in the gaps marked 'G' in Fig. 3.1.2 and allowed to develop adequate strength. Similar technique may be adopted for other structures wherein the jacks are used temporarily for creating a desired deformation of the structure which is prevented from returning to its normal unstressed condition by the concrete placed in apropriately located gaps between the structure itself and its unyielding supports. Although the technique may appear to be attractive, it has only limited application because a major part of the externally applied prestress is often lost on account of shrinkage and creep of concrete. The technique can, however, be used for compensation of loss of prestress due to these factors.

In the applications of external prestressing, large forces are frequently required which are conveniently delivered by means of flat jacks. A flat jack, which is essentially an inflatable steel vessel, looks like a dish on both faces as shown in Fig.3.1.3 (a). It comprises a pair of moveable membranes in the form of thin circular steel plates. Fig. 3.1.3 (b) shows the end view of the jack before inflation. When fluid is pumped in, the membranes move out as shown in the inflated jack of Fig. 3.1.3 (c), thereby exerting a large thrust on any structural member bearing against the steel plates. The fluid may be oil or water for temporary inflation and cement grout or solidifying resin for permanent inflation. Flat jacks have relatively small travel ranging generally from 15 mm to 35 mm but they can deliver a large force. If larger travels are required, a series of jacks, generally not exceeding three, may be placed one over the other like a sandwich. The flat jacks manufactured in India by Freyssinet Prestressed Concrete Company have overall diameter, capacity and travel ranging from 70 to 920 mm, 30 to 9100 kN and 15 to 35 mm respectively.

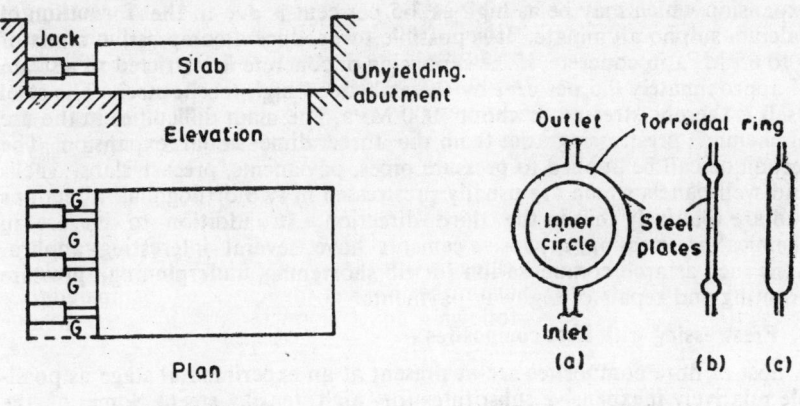

Fig. 3.1.2   Fig. 3.1.3

2. Electro-thermal prestressing

The electro-thermal technique of prestressing has been widely used in Soviet Union and East European countries for the manufacture of small precast prestressed elements. The technique can be used both for pretensioned and post-tensioned members. In the case of pretensioned members, the prestressing tendons are heated to a temperature of 300 to 400°C by passing a low voltage (30 V) and high amperage (300 to 1100 A) electric current for a few minutes in order to produce a substantial elongation (0.4 to 0.5 per cent). The tendons are gripped by end anchors in their heated condition and allowed to cool to a temperature of at least 90°C before concreting. In this manner, it is possible to induce in tendons a stress of 500 − 600 MPa. In the case of post-tensioned members, the tendons coated with thermoplastic material such as sulphur or low melting alloys are buried in concrete. After the concrete has gained sufficient strength, an electric current of low voltage and high amperage is passed through the tendons. As the bar is heated, the thermo-plastic material melts permitting the elongation of the tendon. The tendons are in the form of high tensile steel bars with threaded ends protruding from the ends of the concrete members. To keep the bar in its elongated position, nuts are tightened before cooling due to stoppage of electric current. As a result of cooling, the thermo-plastic material solidifies restoring the bond with surrounding concrete. The main advantage of electro-thermal technique is that it dispenses with the costly equipment needed in the conventional mechanical prestressing together with the maintenance problems associated with it.

3. Chemical prestressing

The technique of chemical prestressing which utilizes the phenomenon of self-straining due to expansive cement was developed in France by Lossier in 1944. Expansive cement usually comprises 75 per cent ordinary portland cement, 15 per cent high alumina cement and 10 per cent gypsum. The expansion which may be as high as 3-5 per cent is due to the formation of calcium sulpho aluminate. It is possible to produce a compressive stress of 4 to 65 MPa in concrete, if the expansion of concrete is restricted to a strain of approximately 0.5 per cent by the embedded high tensile steel. The steel itself is thereby stressed to about 1000 MPa. The main difficulties in the use of chemical prestressing arise from the three dimensional expansion. The technique can be applied to pressure pipes, pavements, precast slabs, shells and well panels which are usually prestressed in two orthogonal directions and are relatively thin in the third direction. In addition to its use in chemical prestressing, expansive cements have several interesting applications such as arch compensation for rib shortening, underpinning, pressure grouting and repair of highway payments.

4. Prestressing with fibre composites

A host of fibre composites are at present at an experimental stage as possible relatively inexpensive substitutes for high tensile steel. Some of the composites being investigated are carbon fibre reinforced plastic (Grafill A), Aromatic polyamide bundle (paraffil Type F) and glass fibred plastic. These composites have a lower modulus of elasticity ranging from 30 to 110 MPa resulting in smaller loss of prestress. They have very low coefficient of thermal expansion and a tensile strength as high as 1900 MPa. The main problems in the use of these composites arise from their excessive elongation, low bond strength and poor machinability with the consequent anchorage difficulties. The durability and fatigue resistance of these composites are also yet to be established.

(ii) Conventional methods

In the conventional methods of prestressing, concrete is precompressed by stretching high tensile steel mechanically and transferring its force to concrete by means of bond or end anchorages. On the basis of the type of prestressed strcutures, the conventional methods may be classified into two types.

1. Circular prestressing

Circular prestressing is generally used for pressure vessels and containment structures such as pressure pipes, pressure vessels of nuclear power stations, water tanks, bunkers and silos. In this case prestressing steel is in the form of continuous circles or endless loops spiralling around the concrete structures.

2. Linear prestressing

Linear prestressing is commonly adopted in buildings and bridges. In this case the prestressing tendons may be straight or curved but they do not go round the structures in the form of endless loops as in the case of circular prestressing.

As an alternative broad classification, the prestressing techniques can be divided into two categories:

1. Pretensioning

In pretensioning, the tendons are stretched temporarily against some external device before placing concrete. The concrete is then placed around the elongated tendons and allowed to develop sufficient strength before the tendons are released from the external device thereby transferring the prestress to concrete. The prestressing techniques may be subdivided further into three types on the basis of the the external device used for stretching the tendons.

(A) Long line method

This method also known as the Hoyer method utilizes a pair of strong abutments or bulkheads against which the prestressing tendons are streched temporarily until the transfer of stress ro concrete.

(B) Individual mould method

In this method the prestressing tendons are stretched temporarily against the mould until the concrete placed in the moulds becomes strong enough for the transfer of prestress by bond.

(C) Strut method

In this method the external medium required to hold the presiressing tendons temporarily in their elongated condition is a strut which is often in the form of an internal steel tube against which the tendons are stretched. The steel tube is withdrawn after the release of tendons and the consequent stress transfer to concrete.

2. Post-tensioning

Unlike pretensioning, the post-tensioning techniques do not require any external support for keeping the prestressing tendons in their stretched position. Instead, the concrete member to be prestressed itself acts as the reaction element against which the *prestressing tendons are stretched by means of jacks. The method is described as post-tensioning because the prestressing tendons are stretched only after the concrete gains sufficient strength to sustain the force in the tendons. There are several post-tensioning methods which differ from each other in respect of the type of jack, anchorage device and the form of prestressing tendons used. The post-tensioned members or tendons are of two types:

(A) Unbonded

In this case the prestressing tendons are not bonded to the surrounding concrete. In order to protect the tendons from possible corrosion, they are either galvanized or covered with a protective coating or grease.

(B) Bonded

In this case the prestressing tendons are bonded to the surrounding concrete through the medium of grout injected into the cable duct. The bonded tendons may be further classified into two types:

(a) End-anchored tendons

In this case the prestress is transferred to concrete jointly by bond and end anchorage. In the event of failure of bond, the end anchorage serves as a second line of defence.

(b) Non-end-anchored tendons

As there are no end anchorages, the transfer of prestress occurs solely through bond. It is evident that the member will collapse should a bond failure occurs.

3.2 PRETENSIONING TECHNIQUES

The technique of pretensioning is generally used for the mass production of small precast concrete elements in a central plant or factory. Thus pretensioning may be used for the mass production of fencing posts, electric and telephone poles, floor and wall panels, small and medium size beams, foundation piles, railway sleepers and a variety of other precast prestressed elements. After manufacture in a central plant, these elements may be transported to the site. In the technique of pretensioning, the high tensile prestressing steel is first tensioned against external supports such as abutments, moulds or struts. Next, concrete is placed around the prestressing steel. The prestressing steel is released from the external supports as soon as the concrete develops sufficient strength and bond with steel. The tensile force in the prestressing steel is thus transferred from the external supports to the concrete member as a compressive force. The prestressing tendons are generally straight between the two external supports. However, it is possible to drape tendons to achieve a crooked cable profile. For instance, the prestressing tendons may be depressed at mid-span in order to produce a larger hogging bending moment due to prestress at the mid-span section. The tendons may be draped by means of a hydraulic jack which deflects the tendon vertically after they are stressed. Alternatively, the tendon may pass through a hook-bolt secured to the floor of the prestressing bed. In this case the loss of prestress due to friction at the hook-bolt should be taken into account. Several prestressing plants utilize these techniques for draping the prestressing tendons in pretensioning system. The systems of prestressing may be classified according to the type of external supports used for holding the prestressing steel before transfer of stress to concrete members.

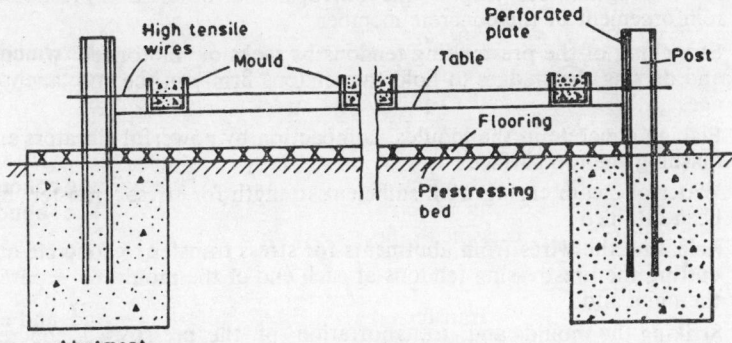

Fig.3.2.1

3.2.1 Long Line System (Hoyer System)

In this system of pretensioning, the prestressing steel which is generally in the form of high tensile wires or multi-wire strands is stretched across two external supports known as abutments or bulkheads. The abutments are placed at the ends of a striplike slightly elevated platform known as the prestressing bed. Figure 3.2.1 shows the prestressing bed together with the moulds for casting the precast prestressed concrete members. The moulds are placed on the prestressing bed in a line longitudinally like the coaches in a train. As the prestressing bed may be as long as 100 m or even more, several moulds can be placed in a single line. The required number of prestressing wires or strands are threaded through all the moulds in a line at proper location and spacing. Hence, the tensioning operation for all the members in a single line is completed by stretching the wires only once. Thus the labour on the stressing operation is reduced considerably. Several prestressing beds are often laid out parallel to each other in a large industrial shed of a central plant. An overhead crane catering to all prestressing beds in the shed is an added facility for the easy handling of the moulds and the prestressed elements. The prestressing beds may have a width of about 1-2 m. The consecutive beds may be separated from each other by a space of about 1.5 m width for the movement of hand trolleys and the workers. The moulds placed over the prestressing bed may be covered by steel jackets to facilitate circulation of steam around the moulds for steam curing of concrete. After the concrete gains sufficient strength, the prestressing wires are cut, generally by gas cutting, one by one to separate the prestressed members from each other. The moulds are then struck off and the prestressed elements are shipped to the curing yard.

The operations which constitute one complete cycle in the production of precast prestressed concrete members by the long line system of pretensioning are as follows:

1. Preparation, oiling and placement of the moulds on prestressing bed
2. Threading of high tensile steel wires through the end plates of the moulds and the anchorage blocks in the abutments or bulkheads
3. Positioning and securing of the stirrups and other non-prestressed reinforcement of the concrete member
4. Stretching of the prestressing tendons by jacks or motorised winches and driving the wedges to hold the tendons firmly in the end anchorages
5. Placing concrete in the moulds, compaction by powerful vibrators and finishing the top surface
6. Water or steam curing until sufficient strength for stress transfer has been attained
7. Releasing the wires from abutments for stress transfer to concrete and cutting the prestressing tendons at each end of the mould to separate the units and
8. Striking the moulds and transportation of the presstresed concrete members to the curing yard

One complete cycle may take 3 to 5 days if ordinary portland cement and water curing are used. The period of one cycle can be reduced to 24 hours through the u e of rapid hardening cement and steam curing. The effect may be to increase the production and the monthly turn over of the plant several times. The optimization of equipment and labour thus achieved leads to significant financial benefits.

3.2.2 Individual Mould System

In this system the moulds for the concrete members are used as external supports for holding the tensioned prestressing steel. In the process of tensioning the steel, the jacks bear against the moulds. Before the steel is released to transfer the force to concrete, the moulds have to carry the force exerted by the jacks. The moulds are, therefore, made from structural steel strong enough to carry the jacking force. The moulds are often mounted on trolleys which can be driven to the steam chamber after concreting As discussed in Sec. 2.2.2.1, steam curing accelerates the development of compressive and bond strengths thereby permitting an earlier transfer of stress to concrete. The period for one cycle, i.e. the time required from one concreting to the next concreting is reduced considerably leading to much higher production.

3.2.3 Strut System (Shorer Chalos System)

The method was developed in United States and has now become practically obsolete. In this method a strut, often in the form of a central steel tube, is used as the reaction element against which high tensile wires are stretched. While concrete is allowed to develop bond with prestressing wires, the bond with steel tube is prevented. After concrete has gained sufficient strength, the wires are released and the tube is withdrawn for reuse. The hollow space created on withdrawal of the tube is subsequently filled up by grout. The problems associated with stretching the wires against the tube, placement of concrete and withdrawal of the tube make the method unpopular.

3.3 POST-TENSIONING TECHNIQUES

The basic difference between the pretensioning and post-tensioning techniques is that whereas in pretensioning techinques the prestressing steel is stressed against external supports before placing concrete, the stressing in post-tensioning techniques is carried out after the concrete to be prestressed has gained sufficient strength. As the prestressing jacks bear against the member itself, the need for external supports is eliminated. Very often the concrete member to be prestressed is cast in its final position. Hence the need for transportation of the member is also eliminated. It is for this reason that post-tensioning techniques are widely used for large concrete members which can not be transported easily. Whereas the pretensioning techniques are found suitable for the mass production of small units in a central plant, the post-tensioning techniques are appropriate for the prestressing of large structures at the construction site. There is a large number of techniques of post-tensioning. They differ from each other in respect of the jacks for tensioning the cables, the formation of the cable itself and the end anchorages Although the techniques of post-tensioning differ in respect of the equipment used, the basic operations in all techniques are the same and may be described as follows:

1. Erection of formwork
   Generally the formwork is erected for the permanent and final position of the post-tensioned member. Sometimes the member is cast and prestressed on the ground [and then lifted and launched to its final position by means of cranes.

2. Placement of extractable cores
   For the formation of ducts for housing the prestressing cables at the appropriate locations, sometimes the prestressing cables themselves are placed instead of the cores. The cables are covered by sheathing to prevent bond with concrete.

3. Placement of stirrups and other non-prestressed reinforcement
   Conduits for forming the grout vents are also placed at proper locations.

4. Placement of concrete, compaction by powerful vibrators and finisshing the top surface

5. Curing for the specified period of about four weeks until the concrete attains the specified strength

6. Withdrawal of extractable cores, threading the cable through the ducts and tensioning the cables in accordance with the particular system of post-tensioning adopted

   The cables and tensioned in the specified sequence

7. Grouting the ducts by means of a grout pump as discussed in Sec. 2.4.3

8. Striking the formwork in appropriate sequence

A large number of systems of post-tensioning are currently being used. The post-tensioning systems differ from each other in respect of the type of anchorage, the type of prestressing tendons, the type of jacks, the size and shape of duct and the equipment required for prestressing. The end anchorages work on the principles of mechanical friction through wedging, looping and direct bearing through thread and nut or buttonheads or a combination of these principles. Some prominent systems are listed in Table 3.3.1 along with the main features which characterise the systems. The systems which are most widely used and known internationally are discussed below.

Table 3.3.1 Prominent Post-tensioned Prestressing Systems

S. No.	System	Country	Tendon	Duct	Method of tensioning	Anchorage
1.	Freyssinet	France	wires and strands	circular	simultaneously	male-female cone
2.	Magnel-Blaton	Belgium	wires	rectangular	two wires	flat wedges
3.	Gifford-Uddal CCL	U.K.	wires	rectangular	one wire	split-circular wedges
4.	Lee Mc Call	U.K.	bars	circular	one bar	thread and nut
5.	Prescon	U.S.A.	wires	circular	one wire	button-heads
6.	Texas PI	U.S.A.	wires	circular	one wire	button-heads
7.	B B R V	Switzerland	wires	rectangular	simultaneously	button-heads
8.	Baur-Leonhardt	Germany	wires and strands	—	simultaneously	loops
9.	Billner	Germany	wires	—	simultaneously	loops
10.	Leoba	Germany	wires	rectangular	simultaneously	loops
11.	Roebling	USA	strands	circular	one strand	thread and nut
12.	Dywidag	Germany	threadbar	circular	one bar	thread and nut

3.3.1 Freyssinet System

The Freyssinet system is indisputably the most popular system of post-tensioning. The system was developed by Freyssinet of France in 1939. The popularity of the system lies in its unique features.

The Freyssinet system uses cables having 8, 10, 12, 14, 16, 18, or 24 wires of 5,7 or 8 mm diameter. In addition to these cables, multi-strand cables have also been introduced. The common multi-strand cables designated as 6 T 13 and 12 T 13 comprise six and twelve 7-ply strands of 13 mm nominal diameter. The properties of the prestressing cables being manufactured in India are shown in Table 3.3.2.

Table 3.3.2 Properties of Freyssinet Prestressing Cables

S. No.	Designation	Nominal area, mm²	Nominal ultimate strength, kN	Minimum sheath diameter. mm
	Multi-wire cables :			
1.	12 φ 5	235	376	33
2.	12 φ 7	462	691	39
3.	12 φ 8	603	844	48
4.	24 φ 7	922	1382	57
5.	24 φ 8	1206	1688	60
	Multi-strand cables:			
6.	6 T 13	557	1002	48
7.	12 T 13	1115	2004	66

An anchorage unit at each end of a cable comprises a hollow cylinder and a conical plug. The hollow cylinder, called a female cone, which is embedded in concrete, is shown in Fig. 3.3.1 (a). The outer surface of the female cone is corrugated for the sake of better bond. The dimensions of the longitudinal conical hole match those of the conical plug, also called the male cone, as shown in Fig. 3.3.1 (b). The female cone is made of high grade concrete or mortar and is heavily reinforced with closely spaced spiral reinforcement. The male cone which is also made of high grade concrete, has as many longitudinal grooves on its surface as the number of wires or strands in the multi-wire cable or multi-strand cable so that each groove provides seating for one wire or strand. The plug has a central steel tube which provides access to the cable duct for grouting. The complete assembly of a Freyssinet anchorage unit is shown in Fig. 3.3.1 (c).

The double acting Fryessi jack which operates in two stages is illustrated schematically in Fig. 3.3.1 (c). The jack essentially comprises three parts, viz. grooved foot A, outer casing B and inner piston C. The immobile part A is in the form of a hollow grooved foot resting against the female cone embedded in concrete. The high tensile steel wires and strands of the Fryessi cable emerging from the conical hole in the female cone run along the longitudinal grooves on the periphery of the immobile foot and are anchored by means of flat wedges to part B which is in the form of outer casing. The longitudinal grooves placed symmetrically on the outer surface of the foot provide room for the wires to emerge from the female cone. As the wires are located concentrically with the jack on its outer surface in a symmetrical arrangement, the jack is self centring. In the first stage, oil is pumped into chamber 1 which pushes the outer casing towards left. In this way wires may be elongated to the desired extent. In the second stage oil is pumped into chamber 2 thereby pushing the piston C towards right. As the piston rests against the male cone, it is driven firmly into the female cone, thereby anchoring the wires effectively. The oil pressure in the two chambers is then

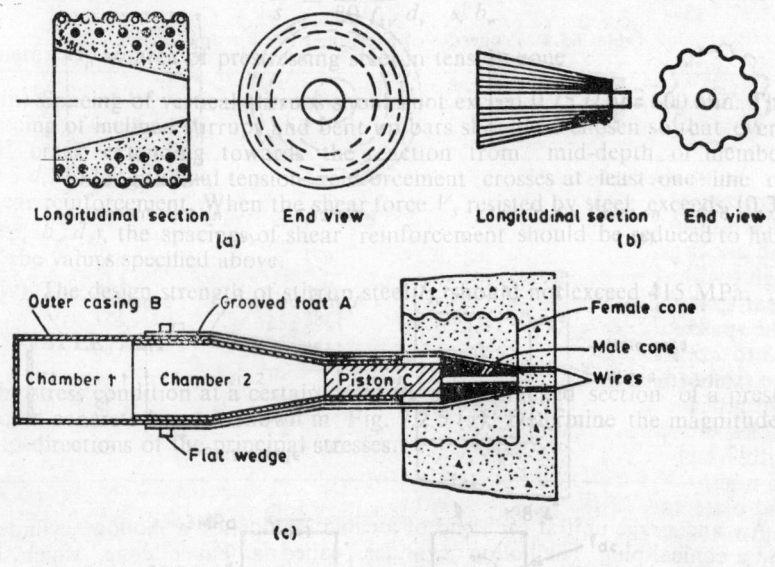

Longitudinal section   End view   Longitudinal section   End view
(a)   (b)

(c)

Fig. 3.3.1

relased slowly and the flat wedges on the outer casing are withdrawn to release the jack from the cable. The jack is provided with springs which bring the casing and the piston back to their normal positions as soon as the oil pressure is released. After removal of the jack, the cable duct is grouted through the tube in the male cone. The main features of some of the models of Fryessi jack which are manufactured in India are given in Table 3.3.3.

3.3.2 Magnel-Blaton System

The system developed by Magnel and Blaton of Belgium is next in popularity only to Freyssinet system. It has been widely used for a variety of prestressed concrete structures all over the world. The prestressing cables used in this system have 16, 24, 32, 40, 48, 56 or 64 high tensile steel wires of 5 or 7 mm diameter. An anchorage unit which anchors 8 wires comprises a steel sandwich plate and four steel flat wedges. The sandwich plate has

Table 3.3.3 Particulars of Freyssinet Prestressing Jacks

S. No.	Designation	Travel, mm	Rated capacity, kN	Extended overall length, mm	Overall diameter, mm	Weight, N	Suitable for
1.	U–3	200	780	1120	215	800	Multi-wire cables
2.	U–5	300	780	1120	215	800	Multi-wire cables
3.	S–6	300	2000	1323	300	1880	Multi-wire cables
4.	S–7	300	900	1230	250	1300	Multi-wire cables
5.	S–8	300	1000	1270	280	1480	cables

four tapering grooves as shown in Fig. 3.3.2 (a). The tapered flat wedge shown in Fig. 3.3.2. (b) fits into one of the grooves in the sandwich plate and anchors two wires. The assembly of one anchorage unit together with high tensile steel wires is shown in Fig. 3.3.2 (c). The vertical and horizontal clear space between neighbouring wires in the cable is generally 4.75 mm. The spacing is maintained by means of spacer plates located at a spacing of 1.3 to 2.5 m. The ducts are formed by means of extractable cores which are stiffened by central steel rod. For cables comprising 5 mm wires, the width of the duct is 55 mm and the depth ranges from 55 to 200 mm for cables with minimum (16) and maximum (64) wires. Similarly, for cables with 7 mm wires, the duct has a width of 65 mm and a depth ranging from 65 mm to 220 mm. Depending upon the number of wires, the size of the end plate varies from 130 × 130 mm to 320 × 210 mm for cables with 5 mm wires and 150 × 150 mm to 390 × 300 mm for cables with 7 mm wires. After the stretching operations, the entire assembly of sandwich plate

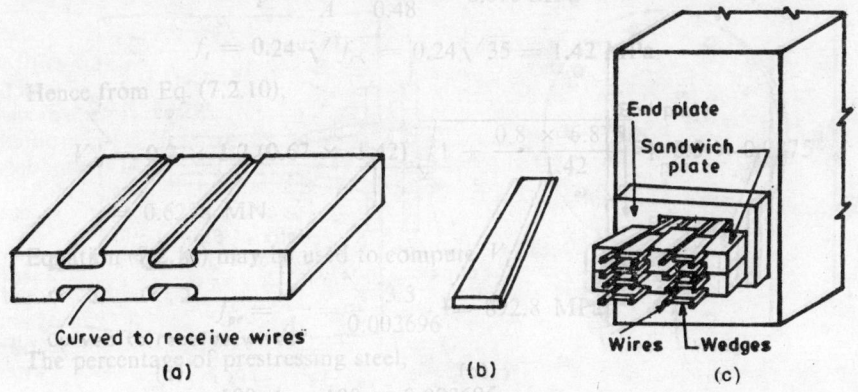

Curved to receive wires
(a)

(b)

End plate
Sandwich plate

Wires Wedges
(c)

Fig. 3.3.2 General View of Anchorage

and flat wedges is encased in concrete for protection and prevention from corrosion. The hydraulic jack used in this system stresses only two wires at a time and is, therefore, lighter and less expensive than Freyssi jack.

3.3.3 Gifford-Udall CCL System

The system developed in Britain is suitable for small and medium jobs and has been widely used in many countries including India. The prestressing tendons are in the form of cables having 8 to 12 wires of 5 or 7 mm diameter. The parallelism of the wires is maintained by spacer plates (Fig. 3.3.3 (a)) located at regular spacing with first and last spacers being 500 mm from the ends. The diameter of the spacer plate is 39 mm for cable of 8 wires and 51 mm for cable of 12 wires. There are two types of anchorages used in this system.

(i) Plate Anchorage

The plate anchorage comprises anchor grips, bearing plate, thrust ring and steel helix which are arranged as shown in Fig. 3.3.3 (b). The wires which have a circular formation in the cable are flared out to a rectangular formation near the end of the beam. For this purpose, the cable duct is enlarged by means of an extractable rubber end former passing through the thrust ring and the helix. The overall dimensions of the bearing plate and thrust ring for a cable with 12 wires are 190 × 150 mm as shown in Fig. 3.3.3 (c) and (d) respectively. Each wire of the cable has its own anchor grips bearing against the steel bearing plate. The anchor grip comprises a barrel with a conical hole (Fig. 3.3.3. (e)) and a pair of conical split wedges (Fig. 3.3.4. (f)) which fit inside the conical hole in the barrel. The inner faces of the split wedges are serrated for better grips on wires.

(ii) Tube Anchorage

The tube anchorage comprises conical wedges, the bearing plate and tube assembly as shown in Fig. 3.3.3 (g). The tube assembly along with its

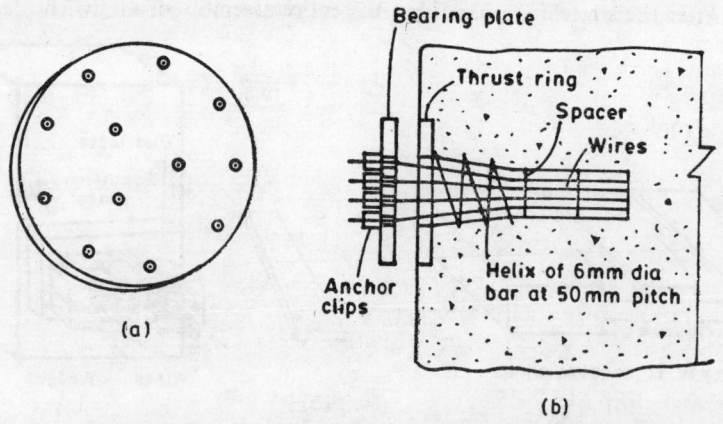

(a)

(b)

Fig. 9.3.3 (Contd)

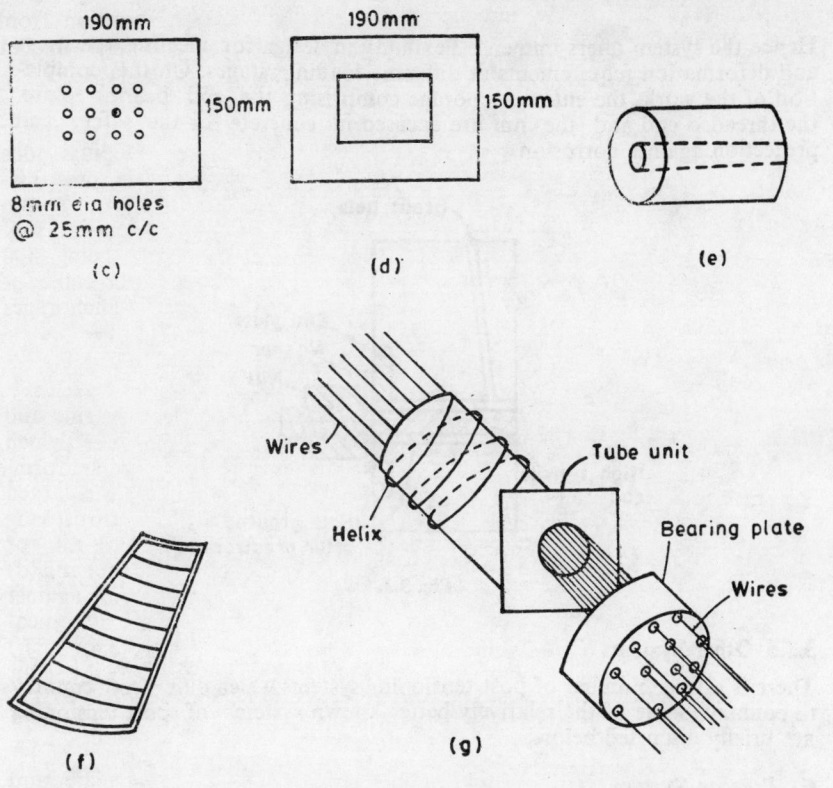

Fig. 3.3.3

helix is embedded in concrete at appropriate locations near the end of the beam and permits the wires to flare out into the tapered holes in the bearing plate. The circular wedges are driven directly into the tapered holes of the bearing plate for gripping the wires.

The hydraulic jack used in this system pulls only one wire at a time and is equipped for the measurement of elongations and the force exerted. The jacks with extensions of 380, 510 and 640 mm have overall lengths of approximately 890, 1020 and 1140 mm respectively.

3.3.4 Lee Mc Call System

The popular Lee Mc Call system developed in Britain is known for its simple anchorage device. The prestressing tendons used in the system are in the form of high tensile steel bars with threaded ends. The bars are elongated, one at a time, by means of a hydraulic jack and a nut is tightened against the end bearing plate as shown in Fig. 3.3.4. The main advantage of the system is that prestressing force can be increased or decreased at any stage to suit the requirements for the respective stages of loading.

Hence the system offers immense flexibility in design for meeting the stress and deformation requirements at different loading stages. On the completion of the work, the entire anchorage comprising the end bearing plate, the threaded end and the nut are encased in concrete for the safety and protection against corrosion.

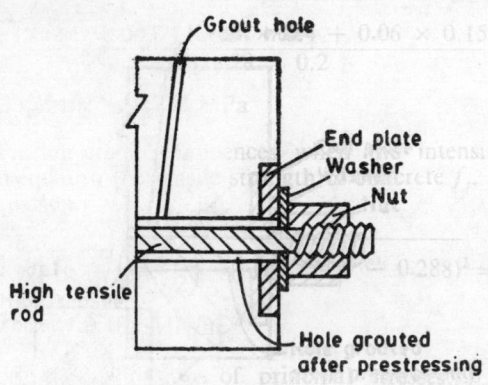

Fig. 3.3.4

3.3.5 Other Systems

There is a large number of post-tensioning systems which differ from country to country. Some of the relatively better known systems of post-tensioning are briefly discussed below:

(i) Prescon System

The Prescon system of United States depends for anchorage on direct bearing of button heads which are cold formed at the ends of the high tensile steel wires of the prestressing cables. The wires are threaded through a barrel-like washer which has threads on its cylinder surface as shown in Fig. 3.3.5. The washer is screwed onto the anchor element of a hydraulic jack and pulled

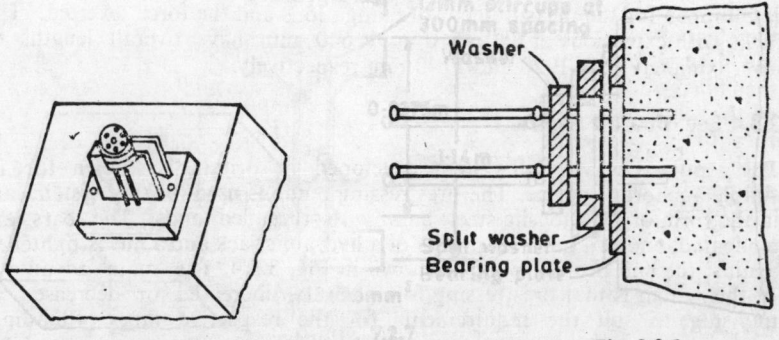

Fig. 3.3.5

Fig. 3.3.6

to produce the required elongation of the wire. Before releasing the jack, a steel shim of appropriate length is inserted to keep the wires in the stressed condition. The cable may have upto 16 wires arranged in circular formation. After the stressing operation the cable duct is grouted through the hole in the washer.

(ii) *Texas PI System*

This system is similar to the Prescon system except that the prestressing wire has two button heads at the jacking end as shown in Fig. 3.3.6. The outer button head is used for pulling and the inner button head for anchorage. A split circular shim is inserted between the washer and the bearing plate after elongating the wire. In this manner, the length of shim and the length of wire to be encased in concrete are considerably reduced.

(iii) *BBRV System*

The system developed by Birkenmeier, Brandestini, Ros and Vogt of Switzerland also utilizes cold formed button heads bearing against a machined anchorage fixture. The tendons, shop fabricated to the required lengths, comprise several wires of 6 mm diameter which are pulled simultaneously. Whereas the tendons used in building construction have upto 52 wires, those used in bridge construction and pressure vessels of nuclear power stations may have as many as 170 wires.

(iv) *Baur-Leonhardt System*

This system, developed in Germany, is noted for delivering very high prestressing force in a single stressing operation. In this system, the structure to be prestressed is separated from its semicircular end blocks by gaps which accommodate the jacks abutting against them as shown in Fig. 3.3.7. The prestressing tendons in the separated form of wires or strands are taken continuously through the structure and round the semi-circular blocks as similar to that in circular prestressing. The high capacity jacks are then operated to push the end blocks away from the structure thereby tensioning the tendons to the desired stress. The gaps are then filled with concrete and allowed to gain adequate strength before the jacks are released and

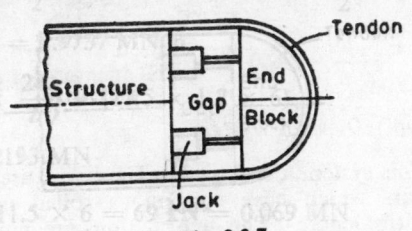

Fig. 3.3.7

withdrawn. Sometimes built-in jacks are made of reinforced concrete and are permanently buried in concrete to become a part of the structure.

(v) *Billner System*

In this syssem, developed in United States, the structure is separated into two parts at midspan by a comblike partition. The prestressing tendons passing through the openings in the partitions are not bonded to concrete except at the two ends where they are anchored by means of loops. The tendons are stressed and the structure prestressed by operating the jacks placed on either side of the partition. In contrast to Baur-Leonhardt system, the Billner system is more appropriate for relatively small structures.

(vi) *Leoba System*

This system, developed in Germany, also utilizes end loops for anchorage. At the non-jacking end, the wires are hooked and concreted. At the jacking end, the wires are connected to a short rod of high tensile steel which is anchored by nuts and washers. A short Tee is formed at one end of the rod around which the wires are looped. The prestressing cable generally comprises 12 wires arranged in two horizontal layers inside a rectangular corrugated metal sheath.

(vii) *Roebling System*

In the Roebling system, in which strands are used as prestressing tendons, the wires are flared out at the end in the form of bushing and buried in zinc poured into a conical funnel of cast steel tube having internal as well as external threading. The inner threading is utilized for pulling the strand. After the desired elongation, a nut is tightened on the outer threading against the bearing plate as shown in Fig. 3.3.8.

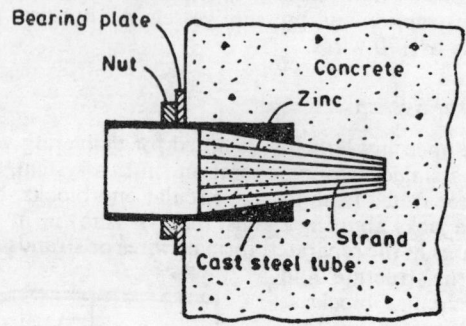

Fig. 3.3.8

(viii) *Dywidag System*

This system, developed in Germany, uses proof stressed alloy steel threadbars which have continuous rolled in threadlike deformations over their entire lengths. The deformations are more durable than machined threads and allow anchorages and couplers to thread on to the threadbar at any point. Consequently, the threadbar can be cut to any desired length or extended by means of couplers. The other features of the system are similar to those of Lee Mc Call system.

3.4 COMPARISON OF PRESTRESSING SYSTEMS

The discussion in the preceding sections shows that there are several prestressing systems and techniques. Each one of them has its own special features which may be most appropriate to particular structural requirements and job conditions. Sometimes more than one prestressing system or technique may be used to obtain optimum results in a prestressed concrete construction. In large structural systems, the best results are often achieved by a suitable combination of reinforced concrete with either precast pretensioned concrete or with cast-in-place post-tensioned concrete. The recent trends point towards increasing use of partial prestressing in which the degree of prestressing is chosen to produce the most favourable stress and deformation conditions. While the non-conventional methods are used only in special structural applications, an overwhelming majority of prestressed concrete structures adopt conventional mechanical methods of prestressing by means of high tensile steel. The pretensioning techniques are eminently suitable for the mass production of relatively small precast elements in a central well-equipped plant capable of producing high quality products. These plants often use steam curing to reduce the period of one cycle to a single day in order to boost the annual turnover. Steam curing is often found to be more economical in plants using individual mould system in which the trolley mounted moulds are conveniently driven to the steam chamber for curing. Although steam or hot water curing can also be used in the long-line system, air curing is often preferred. The cost of labour is lesser in long-line system in which a single stressing operation is required for several units cast in a single line on the prestressing bed. Part of the economy achieved in mass production of precast prestressed units by pretensioning in a central plant is lost on account of the cost involved in transportation, lifting, hoisting and launching. The damage and breakage during transportation also adds to the debit side. Hence large structures are invariably post-tensioned. While post-tensioning saves the cost of transportation, it entails higher cost on formwork and anchorages. Table 3.4.1 gives a comparison of the two most important types of prestressing systems, viz. pretensioning and post-tensioning systems. The structures, which are post-tensioned, are mostly of

Table 3.4.1 Comparison of Pretensioning and Post-tensioning Systems

S. No.	Aspect	Pretensioning	Post-tensioning
1.	Suitability	Suitable for mass production of small and medium units	Suitable for large in-situ construction
2.	Transportation	Considerable	Little or none
3.	End anchorages	Not required	Required
4.	Grouting	Not required	Required
5.	Loss of prestress	Approximately 20 per cent	Approximately 15 percent
6.	Dead load counter-action	Only self weight can be counteracted	Self weight as well as other dead loads can be counter-acted
7.	Cable profile	Only straight or crooked profile	Any curved profile feasible
8.	Formwork or moulds	Reusable several times	Forms generally used only once
9.	Capital investment	High	Modest

the bonded type although they entail additional cost for grouting. Bonded members have better deformation characteristics, more uniform distribution of cracks under overload condition and higher ultimate strength as compared to unbonded members. The possibility of corrosion of prestressing steel also does not exist in the case of bonded members. They may be considered to be more durable and reliable as the loss of prestress due to slipping of cables is prevented both by grout and the end anchorages.

REFERENCES

3.1. Bilner, K.P. and Carlson, R.W., *Electrical prestressing of reinforcing steel*, Journal of the American Concrete Institute, Proceedings Vol. 39, June 1943, pp. 585-92.

3.2. Dobell, C., *Patents and code relating to prestressed concrete*, Journal of the American Concrete Institute, Proceedings Vol. 46, May 1950, pp. 713-24.

3.3 Lin, T.Y. and Klein, A., *Chemical prestressing of concrete elements using expanding cements*, Journal of the American Concrete Institute, Proceedings Vol. 60, No. 9, September 1963, pp. 1187-1218.

3.4. Lossier, H., *Cements with controlled expansion and their application to prestressed concrete*, The Structural Engineer (London), Vol. 24, No. 10, October 1946, pp. 505-34.

3.5. Madhava Rao, A.G.; Parameswaran, V.S. and Ramchandra Murthy, D.S., *Precast concrete channel flooring units using high strength deformed bars prestressed by an electro-thermal method*, The Indian Concrete Journal, December 1973.

3.6. Mikhaillov, V.V., "*Recent developments in automatic manufacture of prestressed members in the USSR*", Journal of the Prestressed Concrete Institute, September 1961.

3.7. Nawy, E.G. and Newerth, G.E., *Fiberglass reinforced concrete slabs and beams*, Journal of the Structural Division, ASCE, Vol. 103, No. ST 2, February 1977, pp. 421-40.

3.8. Rubinsky, I.A. and Rubinsky, A., *A preliminary investigation of the use of fibreglass for prestressed concrete*, Magazine of concrete Research, September 1954, pp. 71-8.

3.9 Shorer, H., *Prestressed cnocrete design principles and reinforcing units*, Journal of the American Concrete Institute, Proceedings Vol. 39, June 1943, pp. 493.521.

3.10. Skramtaev, B.G., *Electro-thermic method of pretensioning bar reinforcement of precast reinforced concrete*, Journal of the Prestressed Concrete Institute, Vol. 6, No. 3, 1961, pp. 57-71.

3.11. Slater, W.M., *Stage post-tensioning-a versatile and economic construction technique*", Journal of the Prestressed Concrete Institute, Vol. 20, No. 1, January-February 1975, pp. 14-27.

4

Loss of Prestress

4.1 INTRODUCTION

The initial tensile stress in prestressing steel, i.e. the stress at transfer in pretensioned members and the jacking stress in post-tensioned members, undergoes a reduction due to a number of factors. Consequently, the prestressing force decreases. The drop in prestressing force is called the loss of prestress. Any factor which tends to reduce the initial prestrain in prestressing tendons makes its own contribution towards eventual loss of prestressing force. If ΔP is the total reduction in the prestressing force, the loss of prestress due to all factors may be expressed as

$$\Delta P = P_i - P \tag{4.1.1}$$

where P_i = initial prestressing force, i.e. the prestressing force at stress transfer in pretensioned members or the jacking force in post-tensioned members

P = effective prestressing force after losses of prestress

The loss of prestress occurs due to reduction in tensile stress and strain in tendons and may, therefore, be expressed as

$$\Delta P = A_p (f_{pi} - f_p) = A_p E_s (\epsilon_{pi} - \epsilon_p) \tag{4.1.2}$$

where A_p = cross-sectional area of prestressing steel

f_{pi}, ϵ_{pi} = initial stress and strain in prestressing steel before losses

f_p, ϵ_p = effective stress and strain in prestressing steel after losses

E_s = modulus of elasticity of prestressing steel (Sec. 2.3)

The loss of prestress δP due to a single factor may be expressed as

$$\delta P = A_p. \delta f_p = A_p. E_s. \delta \epsilon_p \tag{4.1.3}$$

where $\delta f_p, \delta \epsilon_p$ = changes in stress and strain in prestressing steel due to the factor considered

The factors which contribute to loss of prestress are the elastic shortening, creep and shrinkage of concrete, relaxation of prestressing steel, slip and deformation of anchorages, friction between tendon and duct, and stressing of tendons in stages. All these factors tend to reduce the tensile prestrain in steel and therefore make their own contribution towards loss of prestress. The quantum of loss due to each factor depends upon a variety of contributions such as the technique of prestressing adopted, the physical properties of concrete and steel, and the environmental factors, particularly temperature and relative humidity. Several of these conditions often combine to affect the magnitude of loss of prestress. The loss of prestress due to creep and shrinkage of concrete is progressive and generally extends over several months. Most of the losses, however, occur in the first few days after stress transfer or jacking. Due to complex interaction of several factors responsible for loss of prestress, the exact determination of the loss of prestress is difficult. However, in common practice, the loss of prestress due to all factors is approximately 15 to 20 per cent.

4.2   ELASTIC SHORTENING

The strain in concrete due to application of stress is discussed in Sec. 2.2.2.3. As the stress in concrete due to prestressing is compressive, it produces elastic shortening in concrete and a consequent reduction in the initial tensile strain in prestressing steel. In a pretensioned member, the loss due to elastic shortening occurs immediately at transfer of prestressing force from the external support to the member. As the jacking is against the member itself in the case of post-tensioning, the loss due to elastic shortening can be eliminated unless the post-tensioning operation is carried out in stages.

Consider a tendon located at a distance y from the centroidal axis. The elastic compressive stress in concrete at the level of the tendon due to prestressing force P_i may be expressed as

$$f_e = \frac{P_i}{A} - \frac{(-P_i\, e_p)\, y}{I} = \frac{P_i}{A}\left(1 + \frac{e_p\, y}{k^2}\right) \qquad (4.2.1)$$

Assuming perfect bond between the tendon and the surrounding concrete, the tendon undergoes the same elastic shortening as surrounding concrete. Hence,

$$\delta\, \epsilon_p = \frac{f_e}{E_c} \qquad (4.2.2)$$

The reduction in the tensile stress in the tendon may, therefore, be expressed as

$$\delta f_P = E_s\, \delta\epsilon_p = \frac{E_s}{E_c}\, f_e = m f_e \qquad (4.2.3)$$

The average loss of stress in all tendons may be computed from Eq. (4.2.3) by taking,

$f_e = f_{ep}$ = compressive stress in concrete at the level of centroid of prestressing force

$$\delta f_p \ \text{(average)} = m f_{ep} \qquad (4.2.4)$$

where $f_{ep} = \dfrac{P_i}{A} - \dfrac{(-P_i\, e_p)}{I} e_p = \dfrac{P_i}{A}\left[1 + \dfrac{e_p^2}{k^2}\right]$

The loss of prestressing force due to elastic shortening may, therefore, be expressed as

$$\delta P = A_p \, \delta f_p = m \, A_p \, f_{ep} \qquad (4.2.5)$$

The loss due to elastic shortening decreases with increase in strength of concrete because E_c is greater and m is smaller for stronger concrete. As the loss due to elastic shortening can be predicted with reasonable accuracy, it may be compensated by increasing the initial stress in prestressing steel by appropriate amount. In the case of post-tensioned members, the tendons are stressed successively one by one. Consequently, whenever a tendon is stressed, the tendons which have been tensioned and anchored earlier lose part of their prestress due to elastic shortening. The loss due to elastic shortening in post-tensioned members on account of stressing in stages is discussed in Sec. 4.8.

In the foregoing analysis, the loss of prestress due to elastic shortening caused by prestressing force only has been taken into account. However, the effect of external loads acting on the member can be included in the expression for the loss of prestress given by Eq. (4.2.4) provided the stress f_{ep} is taken as the stress caused by prestressing force as well as external loads as given by the following equation:

$$f_p = \frac{C}{A} - \frac{(M_p + M) \, e_p}{I} \qquad (4.2.6)$$

Where C = axial compressive force due to prestress and external loads

   M = bending moment due to external loads

In the preceding discussion, the attention was focussed at a particular cross-section of the member. The average loss of prestress over the entire span may be determined by replacing f_e in Eq. (4.2.3) and f_p in Eq. (4.2.4) by the respective average values. Hence, the average loss of stress in a particular tendon over its entire span l may be detetermined from

$$\delta f_p \text{ (average for one tendon)} = \frac{m}{l} \int_0^l f_e \, dx \qquad (4.2.7)$$

Similarly, the average loss of prestress for all tendons over the entire span,

$$\delta f_p \text{ (average for all tendons)} = \frac{m}{l} \int_0^l f_{ep} \, dx \qquad (4.2.8)$$

It follows that the average loss of prestress over the entire span depends upon the shape of the cable line and the distribution of external bending moment, M.

EXAMPLE 4.2.1

Determines los of prestress due to elastic shortening in the pretensioned beam whose cross-section is shown in Fig. 4.2.1. All wires are straight and parallel to centroidal axis of the beam. The beam has 20 wires of 5 mm diameter carrying an initial stress of 1100 MPa. Take modular ratio, $m = 7.5$.

Solution

$$A = 0.15 \times 0.40 = 0.06 \ \text{m}^2$$

$$I = \frac{1}{12} \times 0.15 \times 0.40^3$$

$$= 0.0008 \ \text{m}^4$$

$$k^2 = \frac{I}{A} = \frac{1}{75}$$

$$P_i = 20 \times \frac{\pi}{4} \times 5^2 \times 1100$$

$$= 0.432143 \ \text{MN}$$

$$\frac{Pi}{A} = 7.2 \ \text{MPa}$$

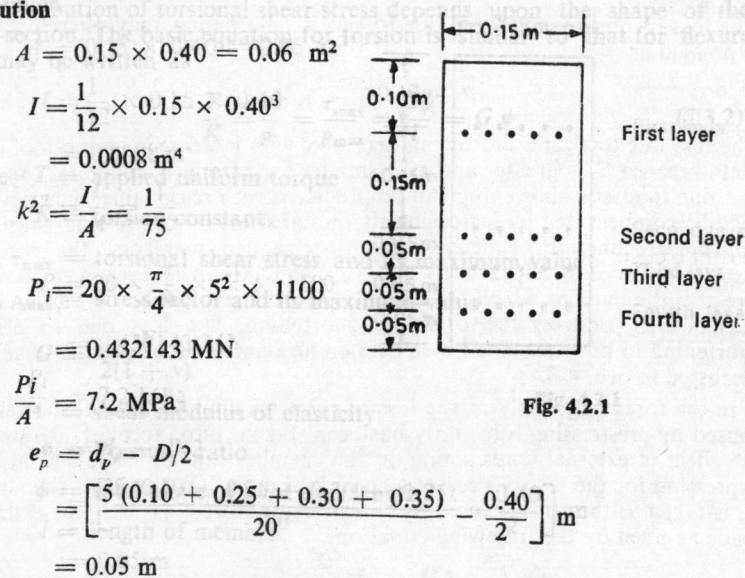

Fig. 4.2.1

$$e_p = d_v - D/2$$

$$= \left[\frac{5(0.10 + 0.25 + 0.30 + 0.35)}{20} - \frac{0.40}{2} \right] \text{m}$$

$$= 0.05 \ \text{m}$$

Elastic compressive stress at top,

$$f_{eT} = \frac{P_i}{A} \left[1 + \frac{e_p \, y}{k^2} \right]$$

$$= 7.2 \left[1 + \frac{(0.05)(-0.2)}{(1/75)} \right] = 1.8 \ \text{MPa}$$

Elastic compressive stress at bottom,

$$f_{eB} = 7.2 \left[1 + \frac{(0.05)(0.2)}{\left(\dfrac{1}{75} \right)} \right] = 12.6 \ \text{MPa}$$

The compressive stress varies linearly from 1.8 MPa at top to 12.6 MPa at bottom. Hence the compressive stresses at the level of first, second, third and fourth layers are 4.50, 8.55, 9.20 and 11.25 MPa respectively. Using Eq. (4.2.3), the loss of prestress in the wires of first, second, third and fourth layers are 33.75, 64.13, 74.25 and 84.38 MPa respectively. Hence,

$$\text{Average loss of Prestress} = \tfrac{1}{4} [33.75 + 64.13 + 74.25 + 84.38]$$

$$= 64.13 \ \text{MPa}$$

Alternatively, the average loss of prestress may be determined by using Eq. (4.2.4). The centroid of the prestressing force is located at 0.25 m from top. Hence the compressive stress at the level of the centroid, $f_{ep} = 8.55$ MPa. The average loss of prestress is, therefore, equal to $7.5 \times 8.55 = 64.13$ MPa.

EXAMPLE 4.2.2

A pretensioned beam 0.15×0.40 m has a draped cable line as shown in Fig. 4.2.2. Determine the average loss of prestress in all wires due to elastic shortening. Which wire has the maximum average loss and what is its magnitude. The beam has 20 wires of 5 mm diameter carrying an initial stress of 1100 MPa. Take $m = 7.5$.

Solution

From Example 4.2.1, eccentricity of prestress at midspan is equal to 0.05 m. Similary, the eccentricity of prestress at ends is found to be (-0.025) m. The cable line comprises straight lines as shown in Fig. 4.2.2. Due to symmetry, only half the beam need be considered. The equation of the cable line in the left half of the beam with A as origin may be written as

$$e_p = -0.025 + 0.075 \left(\frac{x}{l/2} \right)$$

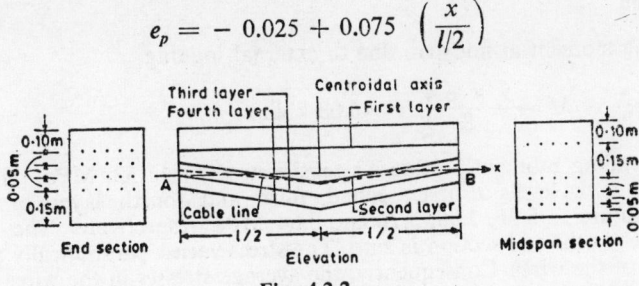

Fig. 4.2.2

Hence the compressive stress in concrete surrounding the cable line may be expressed as

$$f_{ep} = \frac{P_i}{A} + \frac{(P_i \cdot e_p) e_p}{I} = \frac{P_i}{A} \left[1 + \frac{e_p^2}{k^2} \right]$$

Hence, the average compressive stress in concrete,

$$f_{ep} \text{ (average)} = \frac{P_i}{A} \cdot \frac{1}{(l/2)} \int_0^{l/2} \left(1 + \frac{e_p^2}{k^2} \right) dx$$

Substituting for e_p and integrating,

$$f_{ep} \text{ (average)} = 1.047 \left(\frac{P_i}{A} \right) = 7.54 \text{ MPa}$$

Hence using Eq. (4.2.4),

$$\delta f_p \text{ (average for all wires)} = 7.5 \times 7.54 = 56.55 \text{ MPa}$$

The wires in the fourth layer have the maximum average loss of prestress. The equation for the centre line of any wire in the fourth layer in the left half of the beam may be written as

$$y = 0.05 + 0.1 \left(\frac{x}{l/2} \right)$$

Hence using Eq. (4.2.4), the average loss of prestress in the wires of fourth layer,

$$\delta f_p = m \left[\frac{P_i}{A} \cdot \frac{1}{(l/2)} \int_0^{l/2} \left(1 + \frac{e_p.y}{k^2} \right) dx \right]$$

$$= 7.5 \times 7.2 \times 1.141 = 61.61 \text{ MPa}$$

EXAMPLE 4.2.3

If the beam of Example 4.2.1 carries a uniform load of 9 kN/m inclusive of self weight over a span of 6.4 M, determine the loss of prestress due to elastic shortening under the combined effect of prestressing force and external moment.

Solution

Bending moment at midspan due to external loading,

$$M = \frac{9 \times 6.4^2}{8} = 46.08 \text{ kN.m}$$

This bending moment produces a tensile stress of 11.52 MPa. Hence the stresses at the levels of first, second, third and fourth layers of wires at midspan are − 5.76, 2.88, 5.76 and 8.64 MPa respectively. The stress in all wires at the end section is zero. The stress varies parabolically along the length of the wires. Consequently, the average stresses in the wires are equal to 2/3rd of the maximum stresses at midspan. Therefore, the average stresses in the wires of first, second, third and fourth layers are (− 3.84), 1.92, 3.84 and 5.76 MPa respectively. Multiplying these stresses by modular ratio, $m = 7.5$, the corresponding changes in the stresses in the wires are (−28.8), 14.4 28.8 and 43.2 MPa. The wires of the first layer lose some of their prestress while the wires in other layers have a gain of prestress. Hence the net losses of prestress in the wires of the four layers due to combined effect of prestressing force and external moment are

First layer   = 33.75 + 28.80 = 62.55 MPa

Second layer = 64.13 − 14.40 = 49.73 MPa

Third layer   = 74.25 − 28.80 = 45.45 MPa

Fourth layer = 84.38 − 43.20 = 41.18 MPa

As the initial stress in all wires is 1100 MPa, the net stress in the wires of first, second, third, and fourth layers after elastic shortening are 1037.45, 1050.27, 1054.55 and 1058.82 MPa respectiveiy.

EXAMPLE 4.2.4

The post-tensioned beam shown in Fig. 4.2.3 has a parabolic cable profile. Determine the avearge increase of prestress due to a uniform load w per unit length.

Solution

The equation of the cable with origin at A is

$$e_p = \frac{4\,S}{L^2}\, x\,(L - x)$$

The bending moment at a distance x from A,

$$M = \frac{w}{2}\, x\,(L - x)$$

Hence the tensile stress at any section at a distance x from A in concrete surrounding the cable line,

$$f_e = \frac{M \cdot e_p}{I} = \frac{2\,w\,S}{L^2\,I}\, x^2\,(L - x)^2$$

Using Eq. (4.2.7), the average increase in prestress,

$$\delta f_p \text{ (average)} = \frac{mw\,SL^2}{15\,I} \qquad (4.2.9)$$

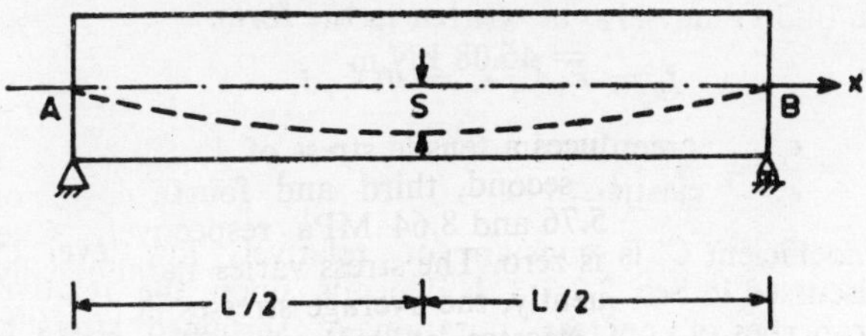

Fig. 4.2.3

The increase of stress due to loading is maximum at midspan. It may be expressed as

$$(\delta f_p)\text{ max } = m f_e = m\,\frac{\left(\dfrac{wL^2}{8}\right)S}{I} \qquad (4.2.10)$$

Dividing Eq. (4.2.9) by (4.2.10), it is observed that the ratio of average increase of stress over the entire span to the maximum increase of stress at midspan is $(8/15)$.

4.3 CREEP OF CONCRETE

As discussed in Sec. 2.2.2.3, the deferred strain in concrete under sustained stress called creep strain continues for several months although at a progressively decreasing rate. Consequenely, the loss of prestress due to creep of concrete increases at a progressively decreasing rate with time. As the creep strain is caused by sustained stress, only permanent loads are responsible for creep strain. The Indian Code IS: 1343-1980 recommends that only permanent loads including prestressing force should be considered in the computation of creep strain. The creep due to live load stresses, erection stresses and other stresses of short duration may be ignored.

The loss of prestress in a tendon at any cross-section of a prestressed concrete member may be expressed as

$$\delta f_p = E_s \, \epsilon_c = E_s \, \epsilon_{cc} f_e \qquad (4.3.1)$$

where ϵ_c = total creep strain in concrete surrounding the tendon under consideration

ϵ_{cc} = creep strain per unit stress in MPa

f_e = elastic compressive stress in concrete surrounding the tendon

The value of unit creep srrain ϵ_{cc} may be taken as 48×10^{-6} for pre-tensioning and 36×10^{-6} for post-tensioning in accordance with the British code. If the cube strength of concrete at transfer f_{ci} is less than 40 MPa, these values may be increased by the ratio $(40/f_{ci})$. If the maximum stress anywhere in the cross-section exceeds one third of the cube strength of concrete, the values of ϵ_{cc} should be increased linearly becoming 25 percent greater when the maximum stress is equal to half the cube strength.

Equation (4.3.1) may also be written in the form,

$$f_p = E_s \, C_c \, \epsilon_e = m \, C_c \, f_e \qquad (4.3.2)$$

where $C_c = \dfrac{\epsilon_c}{\epsilon_e} = \dfrac{\text{creep strain}}{\text{elastic strain}}$

The creep coefficient C_c is constant for relatively low level of sustained stress as discussed in Sec. 2.2.2.3. It depends upon the relative humidity, physical properties of concrete, duration of sustained stress and age of concrete at loading. The Indian Code IS: 1343—1980 recommends that in the absence of experimental data and detailed information on the effect of the variables, the creep coefficient C_c may be taken as 2.2, 1.6 and 1.1 when the age of concrete at loading is 7 days, 28 days and 1 year respectively. For the determination of loss due to creep at different stages it may be assumed that approximately half and three-fourth of the ultimate total creep strain occurs in one and six months after loading.

The average loss of prestress in all tendons at the cross-section under consideration may be obtained by replacing f_e in Eq. (4.3.2) by the elastic stress at the centroid of the prestressing force f_{ep}.

$$\delta f_p \text{ (average)} = m.C_c. f_{ep} \qquad (4.3.3)$$

in which f_{ep} may be taken as the elastic compressive stress due to all permanent forces including prestressing force. It may be determined from Eq. (4.2.6) in which C and M should include only permanent forces. The average loss of prestress in a particular tendon and in all tendons over the entire span due to creep of concrete may be determined by the following equations which may be derived in the same way as Eq. (4.2.7) and (4.2.8).

$$\delta f_p \text{ (average for one tendon)} = m. \, C_c. \frac{1}{l} \int_0^l f_e \, dx \qquad (4.3.4)$$

$$\delta f_p \text{ (average for all tendons)} = m \cdot C_c. \frac{1}{l} \int_0^l f_{ep} \, dx \qquad (4.3.5)$$

EXAMPLE 4.3.1

Determine the loss due to creep of concrete for the beam of Example 4.2.1. The strength of concrete at transfer of stress, $f_{ci} = 30$ MPa. Take $E_s = 0.2 \times 10^6$ MPa.

Solution

Using the British Code, the unit creep strain, $\epsilon_{rc} = 48 \times 10^{-6} \times (40/30) = 64 \times 10^{-6}$. From Example 4.2.1, the compressive stresses at the levels of first, second, third and fourth layers are 4.50, 8.55, 9.90 and 11.25 MPa respectively. Hence, using Eq. (4.3.1), the respective losses due to creep of concrete are 57.60, 109.44, 126.72 and 144.0 MPa.

EXAMPLE 4.3.2

Determine the loss of prestress in wires of the beam of Example 4.2.1. Take creep coefficient, $C_c = 1.5$.

Solution

Putting $m = 7.5$, $C_c = 1.5$ and values of f_e as in Example 4.3.1 into Eq. (4.3.2), the loss due to creep of concrete in the first, second third and fourth layers are found to be 50.63, 96.19, 111.38 and 126.56 MPa respectively.

EXAMPLE 4.3.3

If the beam of Example 4.2.1 carries a permanent uniform load of 9 kN/m inclusive of own weight over a span of 6.4 m, determine the loss due to creep of concrete. Take $C_c = 1.5$.

Solution

In Example 4.2.3, the loss of prestress due to elastic shortening under the combined effect of prestress and external load of 9 kN/m is computed. The loss due to creep of concrete is C_c times greater than the loss due to elastic shortening. Hence the loss due to creep in the wires of first, second, third and fourth layers are 93.83, 74.59, 68.18, and 61.76 MPa respectively.

EXAMPLE 4.3.4

Derive an expression for the losses of prestress due to creep of concrete at section C of the post-tensioned beam shown in (a) Fig. 4.3.1 and (b) Fig. 4.3.2 if it is (i) bonded and (ii) unbonded.

Solution

(a) For the cable line shown in Fig. 4.3.1.

(i) *Bonded Beam*

The compressive stress at section C,

$$f_e = \frac{P_i}{A}\left[1 + \frac{e^2_{pC}}{k^2}\right]$$

Hence using Eq. (4.3.2), the loss of prestress at section C,

$$\delta f_p = m \cdot \mathscr{C}_c \cdot \frac{P_i}{A} \left[1 + \frac{e_{pC}^2}{k^2} \right] \qquad (4.3.6)$$

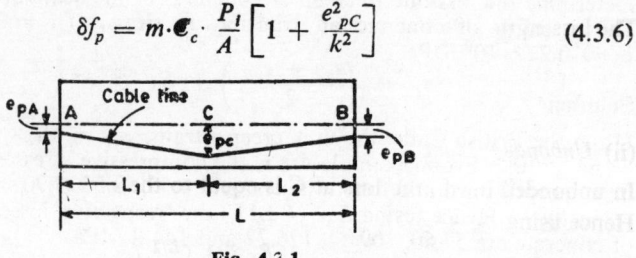

Fig. 4.3.1

(ii) Unbonded Beam

As the tendons are free inside the duct in the case of unbonded beam, the loss of prestress is the same at all sections and is equal to the average loss. In segment AC, the equation for the cable line with origin at A may be written as

$$e_p = e_{pA} + \frac{e_{pC} - e_{pA}}{L_1}$$

Hence using Eq. (4.3.2), the average loss of prestress in segment AC,

$$\delta f_p = m \cdot C_c \cdot \frac{P_i}{A} \cdot \frac{1}{L_1} \int_0^{L_1} \left(1 + \frac{e_p^2}{k^2} \right) dx$$

$$= m \cdot C_c \cdot \frac{P_i}{A} \left[1 + \frac{e_{pA}^2 + e_{pC}^2 + e_{pA} \cdot e_{pC}}{3k^2} \right]$$

Similarly, the average loss of prestress in segment BC,

$$\delta f_p = m \cdot C_e \cdot \frac{P_i}{A} \left[1 + \frac{e_{pB}^2 + e_{pC}^2 + e_{pB} \cdot e_{pC}}{3k^2} \right]$$

Hence the average loss of prestress over the entire length of the beam,

$$\delta f_p = m \cdot C_c \cdot \frac{P_i}{A} \cdot \frac{1}{L} \left[L_1 \left(1 + \frac{e_{pA}^2 + e_{pC}^2 + e_{pA} \cdot e_{pC}}{3k^2} \right) \right.$$

$$\left. + L_2 \left(1 + \frac{e_{pB}^2 + e_{pC}^2 + e_{pB} \cdot e_{pC}}{3k^2} \right) \right] \qquad (4.3.7)$$

(b) For the cable line shown in Fig. 4.3.2

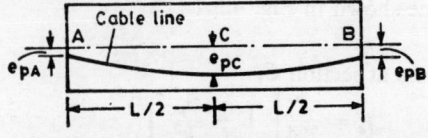

Fig. 4.3.2

(i) *Bonded Beam*

The loss of prestress at section C is given by Eq. (4.3.6) in which

$$e_{PC} = \left(\frac{e_{pA} + e_{pB}}{2} \right) + S$$

(ii) *Unbonded Beam*

In unbonded beam, the loss at C is equal to the average loss.
Hence using Eq. (4.3.5),

$$\delta f_p = m \cdot C_c \cdot \frac{P_i}{A} \cdot \frac{1}{L} \int_0^{L/2} \left(1 + \frac{e_p^2}{k^2} \right) dx \qquad (a)$$

where e_p is represented by the equation,

$$e_p = e_{pA} + (e_{pB} - e_{pA}) \frac{x}{L} + \frac{4S \, x \, (L - x)}{L^2}$$

in which x is the horizontal distance from A. Substituting for e_p in Eq. (a),

$$f_p \text{ (average)} = m \, C_c \frac{P_i}{A} \Big[1 + \frac{1}{3k^2} (e_{pA}^2 + e_{pB}^2 + e_{pA} \, e_{pB} - 4e_{pA}S$$
$$+ 2e_{pB}S + 1.6 \, S^2) \Big] \qquad (c)$$

In the particular case of symmetry in which $e_{pA} = e_{pB} = 0$,

$$f_p \text{ (average)} = mC_c \frac{P_i}{A} \left[1 + \frac{8}{15} \frac{S^2}{k^2} \right] \qquad (4.3.8)$$

EXAMPLE 4.3.5

Derive an expression for the average loss of prestress due to creep of concrete in a post-tensioned beam carrying a permanent uniform load w per unit length inclusive of own weight. The cable line is parabolic without any eccentricity at ends and sag S at mid-span. Hence determine the average loss of prestress, given that $w = 27$ kN/m, $L = 12$ m, $P_i = 1.62$ MN and $S = 0.15$ m. The cross section of beam is rectangular, 0.3×0.9 m. Take $m = 7.5$ and $C_c = 1.5$.

Solution

Combining the results of Examples 4.2.4 and 4.3.4, Eq. (4.2.9) and (4.3.8), the net average loss of prestress due to the combined effect of prestressing force P_i and the permanent external load w may be expressed as

$$\delta f_p = m \, C_c \frac{P_i}{A} \left[1 + \frac{8}{15} \frac{S_2}{k_2} \right] - \frac{mC_c \, w \, SL^2}{15I}$$

Inserting the given values,

$$\delta f_p = 79.5 - 24 = 55.5 \text{ MPa}$$

4.4 SHRINKAGE OF CONCRETE

The complex phenomenon of shrinkage of concrete is discussed in Sec. 2.2.2.3 Assuming perfect bond between prestressing tendons and concrete, the loss of prestress due to shrinkage of concrete may be determined from

$$f_p = E_s \cdot \epsilon_{sh} \qquad (4.4.1)$$

in which ϵ_{sh} is the shrinkage strain in concrete surrounding the tendons. As discussed in Sec. 2.2.2.3, the shrinkage strain depends upon a large number of factors, particularly, the relative humidity and the composition of concrete. Although the shrinkage of concrete begins as soon as it starts drying, only that part of shrinkage which occurs after stress transfer is responsible for loss of prestress.

The Indian Code IS: 1343-1980 recommends that in the absence of test data, the shrinkage strain may be taken equal to 0.0003 in pretensioning and it may be determined from the following equation in the case of post-tensioning,

$$\epsilon_{sh} = \frac{0.0002}{\log_{10}(t + 2)} \qquad (4.4.2)$$

in which t is age of concrete at transfer in days. This value of shrinkage strain may be increased by 50 per cent in dry atmospheric conditions subject to a maximum value of 0.0003. For the determination of loss due to shrinkage at different stages, it may be assumed that approximately half and three-fourth of the ultimate shrinkage strain occur in one month and six months after the commencement of drying.

The British Code specifies that for pretensioning with stress transfer between 3 to 5 days after concreting, the shrinkage strain may be taken equal to 0.0003 and 0.0001 for relative humidity equal to 70% (indoor exposure) and 90% (outdoor exposure) respectively. In the case of post-tensioning with stress transfer between 7 and 14 days after concreting, the shrinkage strain may be taken equal to 0.0002 and 0.00007 for relative humidity equal to 70% and 90% respectively.

EXAMPLE 4.4.1

Determine the loss of prestress due to shrinkage of concrete given by the equation

$$\epsilon_{sh} = 0.00055 \left(1 - \frac{0.0024\ v}{s}\right)(1.5 - 0.015\ RH) \qquad (4.4.3)$$

in which v and s are the volume and surface area of the member in mm units. The relative humidity, $RH = 60$ per cent. The member has a symmetrical I-section of 1.20 m overall depth, 0.60 m wide flanges and thickness of flanges and web equal to 0.20 m. Take $E_s = 2 \times 10^5$ MPa.

Solution

Considering 1 mm length of the member,

$$v = 0.4 \times 10^9 \text{ mm}^3$$
$$s = 4.4 \times 10^6 \text{ mm}^2$$

Hence using Eq. (4.4.3),

$$\epsilon_{sh} = 0.00055 \left(1 - \frac{0.0024 \times 0.4 \times 10^9}{4.4 \times 10^6} \right) (1.5 - 0.015 \times 60)$$

$$= 0.000168$$

Using Eq. (4.4.1), the loss of prestress,

$$\delta f_p = 2 \times 10^5 \times 0.000168 = 33.6 \text{ MPa}$$

4.5 RELAXATION OF STEEL

The relaxation of prestressing steel, defined as the reduction of stress under constant strain, is due to creep. While both concrete and steel creep under sustained stress the basic difference is that whereas concrete creeps at relatively low stress, prestressing steel creeps only at relatively high stress. Consequently, the prestressing steel does not show any loss of prestress due to relaxation unless the stress is higher than approximately half the ultimate strength. The loss due to relaxation depends mainly on the type of steel, the initial prestress and the temperature. As discussed in Sec. 2.3, the creep of prestressing steel can be reduced by appropriate treatment. The creep of steel continues for a long time although the rate of creep decreases continuously. According to the CEB-FIP code,

$$\log \left(\frac{\delta f_p}{f_{pi}} \right) = k_1 + k_2 \log t \qquad (4.5.1)$$

in which δf_p is the loss of prestress due to relaxation after time t and f_{pi} is the initial prestress. For stabilized steel, which has been given appropriate treatment for reduction of creep, the constant k_1 has a more marked influence than k_2.

The Indian code IS: 1343-1980 recommends that the loss due to relaxation may be determined from appropriate tests. In the absence of test data, the loss due to relaxation at 27°C for 1000 hours may be taken equal to 0, 35, 70 and 90 MPa corresponding to the initial prestress f_{pi} equal to 0.5, 0.6, 0.7 and $0.8 f_{pu}$ respectively. These values may have to be increased in the case of higher temperature and large lateral loads. No reduction in the loss due to relaxation should be permitted for a tendon with a load equal to or greater than the relevant jacking force that has been applied for a short time prior to the anchoring of the tendon.

EXAMPLE 4.5.1

Determine the loss of prestress due to relaxation of steel as per provisions of the I.S. Code if the initial stress $f_{pi} = 1100$ MPa and the ultimate strength $f_{pu} = 1650$ MPa.

Solution

$$\frac{f_{pi}}{f_{pu}} = \frac{1100}{1650} = 0.667$$

As per IS code, $\delta f_p = 35$ MPa for $f_{pi} = 0.6 f_{pu}$

and $\delta f_p = 70$ MPa for $f_{pi} = 0.7 f_{pu}$

Hence using linear interpolation, the loss of prestress, $\delta f_p = 58.35$ MPa

4.6   SLIP AND DEFORMATION OF ANCHORAGES

The end anchorages, which are responsible for holding the prestressing tendons in their elongated condition, undergo a certain amount of slip and deformation before they become fully effective. As a result the tendons suffer a loss of prestress due to the consequent reduction in their strain. The loss of prestress due to slip and deformation of anchorages depends mainly on the technique of prestressing, the type of anchorage and the length of tendon between the end anchorges. In the case of pretensioning, in which the stress transfer to concrete takes place through bond, the loss of prestress occurs at the time of anchoring the tendons against the external device. In long line system, the length of the tendons is generally very large compared to the slip of anchorages. Consequently, it may be ignored. The loss may become appreciable in individual mould and strut systems of pretensioning. In post-tensioning, the loss due to slip and deformation of anchorages occurs at the stress transfer and is generally considered to be of greater relevance than in pretensioning. In anchorages which utilize friction wedges, the loss is essentially due to the slip of wedges before they are housed firmly in the tapering cone or groove. The amount of slip of the wedges depends upon the manner in which they are driven before the jack is released. For instance, in the case of Freyssinet system in which the male cone is pushed hard into the female cone by the large force delivered by the Freyssi jack, the slip is negligible. The loss is insignificant also in the systems which utilize thread and nut for anchorage as in Lee Mc Call system. In a similar way, the loss is very small in systems which use the principles of direct bearing and looping for anchorage.

As the slip and deformation of anchorage is independent of the length of the tendon, the loss is inversely proportional to the length of the tendon between the end anchorages. If a tendon of length L undergoes reduction of elongation equal to δL due to slip and deformation of anchorages at its ends, the loss of prestress may be expressed as

$$\delta f_p = \frac{\delta L}{L} \cdot E_s \qquad (4.6.1)$$

In determining δL, the slip and deformation of both anchorages have to be taken into account. The slip at the dead or non-jacking end, which is generally smaller than that at the jacking end, may be eliminated completely by stressing and releasing the tendon once before stressing it for anchorage. Although theoretically some loss occurs due to deformation of anchorages, it is generally negligible in comparison to the loss due to slip. The loss due to slip can be anticipated and may be compensated by slight overstressing the tendon as long as it does not exceed the permissible limits. The British Code permits a temporary overstress of 10 per cent at the time of jacking to compensate for the loss.

EXAMPLE 4.6.1

Determine the loss of prestress due to a net displacement of 3 mm caused by slip and deformation of anchorages (i) in a prestressing bed with anchorages 100 m apart, (ii) in individual mould system using 8 m long moulds and (iii) in a post-tensioned railway sleeper of 2.75 m length. Take $E_s = 2 \times 10^5$ MPa.

Solution

Using Eq. (4.6.1),

(i) $\delta f_p = \dfrac{0.003}{100} \times 2 \times 10^5 = 6$ MPa

(ii) $\delta f_p = \dfrac{0.003}{8} \times 2 \times 10^5 = 75$ MPa

(iii) $\delta f_p = \dfrac{0.003}{2.75} \times 2 \times 10^5 = 218.2$ MPa

4.7 FRICTION IN TENDON AND DUCT

In post-tensioning, in which a prestressing tendon is housed in a preformed duct or sheathing, the effective prestressing force at any point in the tendon away from the jacking end is smaller than the jacking force on account of friction between the tendon and duct or sheathing. The loss due to friction occurs due to the following effects:

(i) *Curvature Effect*

The loss due to curvature of tendon depends upon its cumulative change of direction. It is zero for straight tendons. In a tendon with reverse curvature in the form of two or more waves, the cumulative change of direction is the arithmetic sum of the angles between the tangents at each end of the waves. For example, the cumulative change of direction between the jacking end A and any point P in Fig. 4.7.1 may be expressed as

$$\theta = \theta_1 + \theta_2 + \theta_3 + \theta_4$$

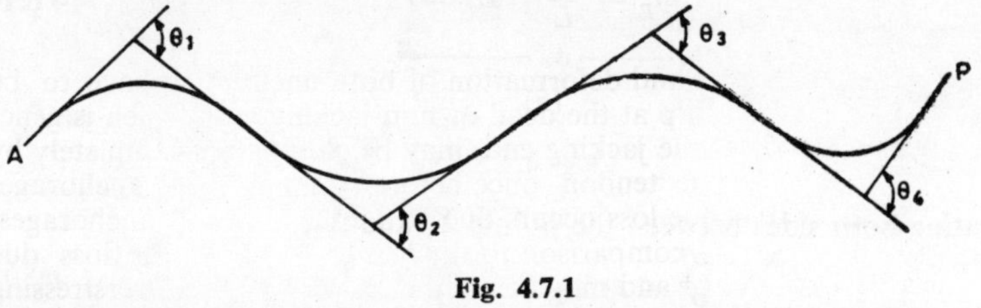

Fig. 4.7.1

The friction loss due to curvature effect increases with θ in accordance with an exponential law.

(ii) *Wave Effect*

The wave effect, also known as wobble or length effect, is also responsible for loss due to friction. The actual duct or sheathing differs from it stheoretical alignment due to the presence of unintentional waves or wobbles along its inner surface. Consequently, the tendon has additional points of contact with the duct or sheathing resulting in frictional loss. This type of loss depends upon the manner in which the duct is formed or sheathing is supported during concreting. The frictional loss increases with the length of the tendon in accordance with an exponential law.

To determine the frictional loss between any two points X_1 and X_2 shown in Fig. 4.7.2 (a), consider a small element of the cable of length dx as shown in Fig. 4.7.2. (b). The element subtends an angle $\delta\theta$ at its centre of curvature. The normal pressure exerted by the duct or sheathing is evidently equal to $P\,\delta\theta$. Consequently, the frictional force acting on the element is $(\mu\,P\,\delta\theta)$ in which μ is the coefficient of friction between the tendon and the duct or sheathing. As the cable line is very flat, its slope at any point is very small. Hence, the equilibrium of all forces along x-axis gives,

$$P = (P - \delta P) + \mu \cdot P \cdot \delta\theta$$

or
$$\frac{\delta P}{P} = \mu \cdot \delta\theta$$

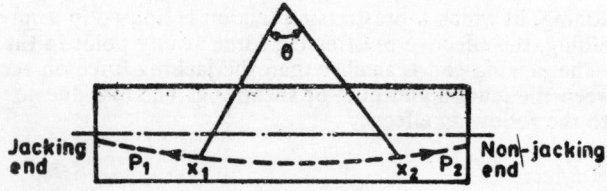

(a)

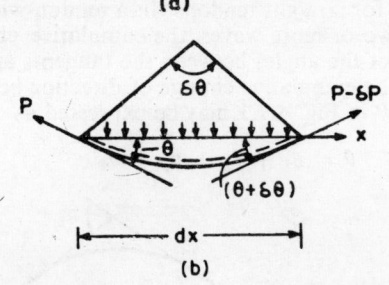

(b)

Fig. 4.7.2

Integrating both sides between the points X_1 and X_2,

$$\frac{P_1}{P_2} = e^{\mu\theta}$$

or
$$\frac{P_2}{P_1} = e^{-\mu\theta} \qquad (4.7.1)$$

A similar analysis may be carried out for the wave effect. Combining the curvature and wave effects,

$$\frac{P_2}{P_1} = e^{-(\mu\theta + kx)} \tag{4.7.2}$$

where k = coefficient for wave effect

x = distance between X_1 and X_2

Hence the total loss of prestress due to curvature and wave effects may be expressed as

$$\delta P = P_1 - P_2 = P_1 \left[1 - e^{-(\mu\theta + kx)} \right] \tag{4.7.3}$$

If the term $(\mu\theta + kx)$ is small (less than approximately 0.2), Eq. (4.7.2) and (4.7.3) may be approximated as

$$P_2 = P_1 (1 - \mu\theta - kx) \tag{4.7.4}$$

$$\delta P = P_1 (\mu\theta + kx) \tag{4.7.5}$$

The value of coefficient of friction depends upon the type of contact surfaces. The Indian Code IS: 1343-1980 specifies μ equal to 0.55 for steel moving on smooth concrete, 0.30 for steel moving on steel fixed to duct and 0.25 for steel moving on lead. In the case of circular prestressing, the code specifies μ equal to 0.45 for steel moving on smooth concrete, 0.25 for steel moving on steel bearers fixed to the concrete and 0.10 for steel moving on steel rollers. The code also specifies that the effect of reverse friction should be taken into consideration in such cases where the initial tension applied to prestressing tendon is partially released. The coefficient of friction may be reduced considerably by using a variety of materials such as greases, oils, oil and graphite mixtures and paraffin. According to the Indian code, the value of k ranges from 0.0015 to 0.0050 per metre length of the tendon. According to British code, k is equal to 0.0033 for normal condition and 0.0017 for rigid sheath or closely supported duct formers. For greased strands running in plastic sleeves, the value of k may be taken as 0.0025

EXAMPLE 4.7.1

For the post-tensioned beam with a parabolic cable line shown in Fig. 4.7.3, determine the difference of prestressing forces at any two points M and N distant x_1 apart. Hence determine the net prestressing forces at C and B if prestressing force P_i is appplied at jacking end A.

Take $P_A = 1000$ kN, $l = 20$ m, $S = 0.5$ m, $\mu = 0.30$ and

$$k = 0.0033 \text{ per metre}$$

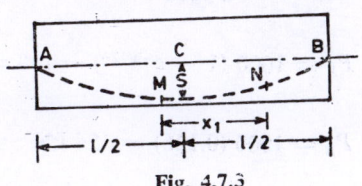

Fig. 4.7.3

Solution

The equation of cable line with origin at A is

$$e_p = \frac{4\,Sx\,(l - x)}{l^2}$$

Hence the slope at any point M at a distance x from A,

$$\theta_M = \tan\theta_M = \frac{de_p}{dx} = \frac{4\,S}{l^2}(l - 2\,x)$$

Similarly, the slope at point N,

$$\theta_N = \frac{4\,S}{l^2}[l - 2\,(x + x_1)]$$

Hence the change of direction of the cable line between the points M and N,

$$\theta = \theta_M - \theta_N = \frac{8\,Sx_1}{l^2}$$

Using the exact Eq. (4.7.3), the loss of prestress between the points M and N,

$$\delta P = P_M - P_N = P_M\left[1 - e^{-\left(\frac{8\mu\,Sx_1}{l^2} + kx_1\right)}\right] \qquad (4.7.6)$$

Using the approximate Eq. (4.7.5),

$$\delta P = P_M - P_N = P_M\left(\frac{8\mu\,Sx_1}{l^2} + kx_1\right) \qquad (4.7.7)$$

To determine the prestressing forces P_C at midspan and P_B at non-jacking end, the point M may be made to coincide with A.

For the segment AC,

$$P_A = 1000 \text{ kN} \qquad \mu = 0.3 \qquad S = 0.5 \text{ m} \qquad x_1 = 10 \text{ m}$$
$$l = 20 \text{ m} \qquad k = 0.0033$$

Using Eq. (4.7.6),

$$P_A - P_C = 1000\,(1 - e^{-0.063}) = 61.1 \text{ kN}$$

Using Eq. (4.7.7),

$$P_A - P_C = 1000\,(0.063) = 63 \text{ kN}$$

For the segment AB,

$$P_A = 1000 \text{ kN} \qquad \mu = 0.3 \qquad S = 0.5 \text{ m} \qquad x_1 = 20 \text{ m}$$
$$l = 20 \text{ m} \qquad k = 0.0033$$

Using Eq. (4.7.6),

$$P_A - P_B = 1000\,(1 - e^{-0.126}) = 118.4 \text{ kN}$$

Using Eq. (4.7.7)

$$P_A - P_B = 1000\,(0.126) = 126 \text{ kN}$$

EXAMPLE 4.7.2

Figure 4.7.4 shows the profile of one of the cables in a continuous beam. Determine the force in the cable at the non-jacking end D if the prestressing force in the cable immediately behind the anchorage at the jacking end A is 100 kN. Take $\mu = 0.25$ and $k = 0.0017$.

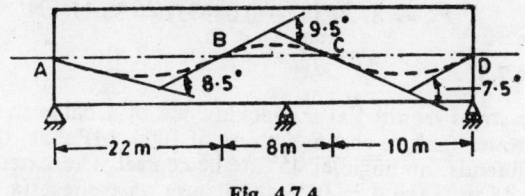

Fig. 4.7.4

Solution

(i) Using the exact expression

$$\theta = 8.5° + 9.5° + 7.5° = 25.5° = 0.445 \text{ radian}$$

$$x = 22 + 8 + 10 = 30 \text{ m}$$
$$\mu\theta + kx = 0.17925$$

Using Eq. (4.7.2), the force at D,

$$P_D = 100 \, e^{-0.17925} = 83.59 \text{ kN}$$

(ii) Using the approximate expression

$$\mu\theta + kx = 0.17925$$

Using Eq. (4.7.4), the force at D,

$$P_D = 100 \, (1 - 0.17925) = 82.08 \text{ kN}$$

The value obtained by using approximate expression may be improved by dividing the cable into segments AB, BC and CD and computing forces in each segment.

For the segment AB,

$$\theta = 8.5° = 0.148 \text{ radian} \qquad x = 22 \text{ m}$$
$$\mu\theta + kx = 0.0744$$

Hence using Eq. (4.7.4),

$$P_B = 100 \, (1 - 0.0744) = 92.56 \text{ kN}$$

For segment BC,

$$\theta = 9.5° = 0.166 \text{ radian} \qquad x = 8 \text{ m}$$
$$\mu\theta + kx = 0.0551$$

Hence using Eq. (4.7.4),

$$P_C = 92.56 \, (1 - 0.0551) = 87.46 \text{ kN}$$

For segment CD,

$$\theta = 7.5° = 0.131 \text{ radian} \qquad x = 10 \text{ m}$$

$$\mu\theta + kx = 0.04975$$

Hence using Eq. (4.7.4),

$$P_D = 87.46 (1 - 0.04975) = 83.11 \text{ kN}$$

EXAMPLE 4.7.3

Determine the stress required at the jacking end of a cable in the cylindrical wall of a water tank to ensure a stress of 1000 MPa at the dead end. The cable subtends an angle of 45° at the centre. The external diameter of the tank is 45 m. Take $\mu = 0.25$ and ignore wave effect ($k = 0$).

Solution

As the change of direction is equal to the angle subtended by the cable at centre,

$$\theta = 45° = 0.786 \text{ radian}$$

Taking f_{p1} and f_{p2} as the stresses in the cable at the jacking and dead ends and using Eq. (4.7.1),

$$f_{p1} = f_{p2} \, e^{\mu\theta}$$

$$= 1000 \, e^{0.197}$$

$$= 1218 \text{ MPa}$$

Alternatively, using the approximate Eq. (4.7.4),

$$f_{p2} = f_{p1} (1 - \mu\theta)$$

$$1000 = f_{p1} (1 - 0.197)$$

or $\qquad\qquad f_{p1} = 1245 \text{ MPa}$

4.8   STRESSING IN STAGES

In post-tensioning, the tendons are stressed successively one at a time. Consequently there is progressive loss of prestress due to elastic shortening. Whenever a particular tendon is stressed, a loss of prestress occurs in all tendons which have been stressed earlier. Consider, for instance, a post-tensioned member having tendons with jacking forces equal to P_1, P_2, ... P_n. Let the eccentricities of the tendons at a particular section under consideration be e_{p1}, e_{p2}, ... e_{pn}. The elastic compressive stress in concrete at the level of jth tendon due to the stressing of the kth tendon may be expressed as

$$f_e = \frac{P_k}{A} - \frac{(- P_k \, e_{pk}) \, e_{pj}}{I}$$

Consequently, using Eq. (4.2.3), the loss of prestress in j th tendon due to the stressing of k th tendon may be expressed as

$$(\delta f_{pj})_k = m\left[\frac{P_k}{A} - \frac{(-P_k e_{pk})e_{pj}}{I}\right] \qquad (4.8.1)$$

Hence the loss of prestressing force in j th tendon due to stressing of k th tendon,

$$(\delta P_j)_k = m\, A_{pj}\left[\frac{P_k}{A} - \frac{(-P_k e_{pk})e_{pj}}{I}\right] \qquad (4.8.2)$$

in which A_{pj} is the cross-sectional area of jth tendon. Similar losses occur in jth tendon due to the stressing of all tendons from $(j+1)$ th to n th tendon. Hence the total loss in j th tendon due to elastic shortening when all tendons have been stressed may be expressed as

$$\delta P_j = m\, A_{pj}\left[\frac{1}{A}\sum_{k=j+1}^{n}P_k + \frac{e_{pj}}{I}\sum_{k=j+1}^{n}P_k\, e_{pk}\right] \qquad (4.8.3)$$

Assigning different values to j ranging from 1 to n, the loss due to elastic shortening on account of stressing in stages in any tendon may be determined. It is evident that whereas the maximum loss ocurs in the first tendon, there is no loss in the last tendon. Hence as an approximation, the average loss in the initial prestressing force P_i due to stressing in stages may be expressed as

$$\delta P_i = \tfrac{1}{2}\, m f_{ep} \qquad (4.8.4)$$

in which f_{ep} is the elastic compressive stress at the centroid of the prestressing force.

The Indian Code IS: 1343-1980 specifies that the loss in tendons due to stressing in stages may be taken equal to half the product of the stress in concrete adjacent to the tendons averaged along their lengths and the modular ratio. Alternatively, the loss of prestress may be computed more precisely based on the sequence of tensioning.

EXAMPLE 4.8.1

Figure 4.8.1 shows the midspan section of a post-tensioned beam. The cables are stressed one by one in the sequence indicated in the figure. The jacking force for each cable is 432 kN. Determine the loss of prestress in the third cable due to elastic shortening when the fifth cable is stressed. Also determine the net loss of stress in all cables at the end of the stressing operation. Take $m = 7.5$.

Solution

$$A = 0.3 \times 1.2 = 0.36 \text{ m}^2$$

$$I = \frac{1}{12} \times 0.3 \times 1.2^3 = 0.0432 \text{ m}^4$$

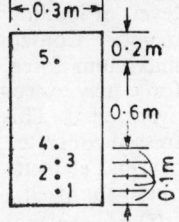

Fig. 4.8.1

Hence using Eq. (4.8.1), loss of prestress in third cable due to stressing of fifth cable,

$$\delta f_p = 7.5 \left[\frac{0.432}{0.36} - \frac{(-0.432)(-0.4)(0.3)}{0.0432} \right] = 0$$

Using Eq. (4.8.3), loss of prestress in first cable due to stressing of all subsequent cables,

$$\delta f_p = 7.5 \left[\frac{0.432 \times 4}{0.36} + \frac{0.5}{0.0432} \{0.432 \times 0.4 + 0.432 \right.$$

$$\left. \times 0.3 + 0.432 \times 0.2 + 0.432 \,(-0.4)\} \right]$$

$$= 54.75 \text{ MPa}$$

Similarly loss of prestress in second, third and fourth cables are found to be 30.75, 10.5 and -6.0 MPa. It may be pointed out that the prestress in fourth cable increases as indicated by negative sign. No loss of prestress occurs in the last (fifth) cable.

4.9 TOTAL AMOUNT OF LOSS OF PRESTRESS

The foregoing discussion shows that the loss of prestress or the reduction in the prestressing force in high tensile steel depends mainly on the following factors:

(i) Physical properties of concrete
(ii) Physical properties of steel
(iii) Environmental conditions, particularly the temperature and relative humidity
(iv) System of prestressing
(v) Manner and sequence of stressing
(vi) Shape of the cable profile

The prestressing force starts decreasing immediately after the stressing operation. Although the drop in prestressing force continues for several months, the major part of the loss of prestress occurs in the first few weeks. The rate of loss of prestress decreases continuously with time until it becomes negligible.

The stress in the prestressing steel starts increasing when the prestressed member is loaded. At a certain level of load, the prestressing force may reach the same value which it had at the time of stressing. However, this level of loading is often higher than the working load of the prestressed member. Consequently, the initial prestressing force before losses is often the maximum force during the entire useful life of the member. The prestressing force may exceed the initial prestressing force only when the member is overloaded. This fact represents one of the several advantages of prestressed concrete. The high tensile steel, which constitutes the main load bearing element, is subjected to an acceptance test during the stressing operation itself.

The complexity of the time-dependent phenomena, viz. creep and shrinkage of concrete and relaxation of steel, is enhanced further due to their interde-

pendence. The net loss of prestress due to combined effect of the three phenomena is likely to be overestimated if the losses are computed separately and
added. For a more precise assessment, the interaction method may be used
in which creep of concrete and relaxation of steel are expressed as functions
of time and stress level whereas shrinkage of concrete and loads acting on
the structure are expressed as functions of time. The useful life of the
structure is divided into suitable number of steps on logarithmic time scale
and the computations for loss of prestress are programmed for solution by
a digital computer.

As the loss of prestress depends upon a large number of complex and
interdependent factors, it is difficult to determine it precisely. It is, however,
possible to indicate the range of probable loss under ordinary conditions. It
is generally believed that the total loss of prestress is approximately 20 to
25 percent in pretensioning and 15 to 20 per cent in the case of post-tensioning. Under special or abnormal conditions, each type of loss should be
analysed critically for a better estimate of the loss of prestress. The loss of
prestress can be reduced by using higher grade of concrete, low relaxation
steel, temporary overstressing and restressing of tendons and also by stressing the tendons from both ends.

A reasonably accurate assessment of the loss of prestress is important for
the prediction of the behaviour of the structure at working load level. An
underestimate of the loss leads to premature cracking and undesirable deflections, particularly for large span/depth ratios. On the other hand, an
overestimate of the loss may lead to undesirable camber. The behaviour at
the limit state of collapse is, however, unaffected by errors in the computation of loss of prestress.

If the total initial prestress force at the time of transfer of prestress from
steel to concrete is denoted by P_i and the effective prestressing force after all
losses is denoted by P, then the loss of prestress $(P_i - P)$ may be expressed
by introducing the reduction factor for effective prestress or loss factor,

$$\Delta P = P_i - P = (1 - \eta) P_i \qquad (4.9.1)$$

Although the processes responsible for the loss of prestress may start even
before the transfer of prestress and may continue even after the final stage
when the structure is pressed into service, it is generally assumed for the
sake of design that the losses start from the time of the transfer of prestress
and terminate at the final stage. As pointed out earlier, the exact computations of losses of prestress is difficult. However, for the sake of design, η
may be taken to lie between 0.75 and 0.80 for pretensioned members and
between 0.80 and 0.85 for post-tensioned members in the absence of precise
computations.

EXAMPLE 4.9.1

Determine the precentage of temporary overstressing required to compensate
for loss of prestress due to elastic shortening of concrete and a total slip
of 3 mm in the anchorages for a transmission pole manufactured by individual mould system. The length of pole is 8 m. The design initial prestress
in concrete and prestressing steel are 6 and 1100 MPa respectively. Take
$E_s = 2 \times 10^5$ MPa and $m = 7.5$.

Solution

Loss due to elastic shortening

$$\delta f_p = m f_e = 7.5 \times 6 = 45 \text{ MPa}$$

Loss due to slip of anchorages,

$$\delta f_p = \frac{0.003}{8} \times 2 \times 10^5 = 75 \text{ MPa}$$

Hence total loss to be compensated $= 45 + 75 = 120$ MPa

Percentage of temporary overstressing $= \dfrac{120}{1100} \times 100 = 10.9$

EXAMPLE 4.9.2

Figure 4.9.1 shows the profiles of cables in a post-tensioned beam of rectangular cross-section 0.3×1.0 m. The beam is symmetrical about the midspan section. The cables are stressed from the left end one by

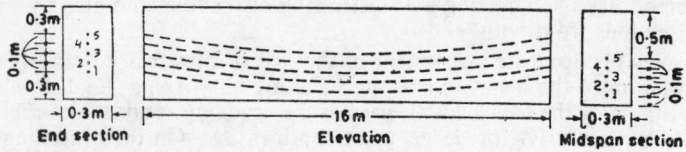

Fig. 4.9.1

one in sequence starting from lowermost cable. Determine the jacking stress required for each cable so that a net stress of 1000 MPa and a force of 360 kN are transmitted at midspan. The net displacement due to slip and deformation of anchorages is 3.5 mm. Take $\mu = 0.3$, $k = 0.0033/\text{m}$, $m = 8.5$ and $E_s = 2 \times 10^5$ MPa.

Solution

Loss of prestress due to slip,

$$\delta f_p = \frac{0.0035}{16} \times 2 \times 10^5 = 43.8 \text{ MPa}$$

As all the cables are parallel to each other, the loss of prestress due to friction is the same in all cables and is given by

$$\delta f_p = 1000 \, (e^{\,\mu\theta + kx} - 1)$$
$$= 1000 \, (e^{\,0.0414} - 1)$$
$$= 42.3 \text{ MPa}$$

As the cables are stressed one by one, they suffer loss of prestress due to elastic shortening as follows:

First cable, $\delta f_p = 7.5 \left[\dfrac{0.36 \times 4}{0.3} + \dfrac{0.3}{0.025} \{0.36\,(0.3) + 0.36\,(0.2) \right.$

$$\left. + 0.36\,(0.1) + 0.36\,(0) \} \right] = 61.9 \text{ MPa}$$

Second cable, $\delta f_p = 7.5 \left[\dfrac{0.36 \times 3}{0.3} + \dfrac{0.3}{0.025} \{0.36\,(0.2) + 0.36\,(0.1) \right.$

$$\left. + 0.36\,(0)\} \right] = 36.7 \text{ MPa}$$

Third cable, $\delta f_p = 7.5 \left[\dfrac{0.36 \times 2}{0.3} + \dfrac{0.2}{0.025} \{0.36\,(0.1) + 0.36\,(0)\} \right]$

$$= 20.2 \text{ MPa}$$

Fourth cable, $\delta f_p = 7.5 \left[\dfrac{0.36}{0.3} + \dfrac{0.1}{0.025} \times 0.36\,(0) \right] = 9 \text{ MPa}$

Fifth cable, $\delta f_p = 0$

The net loss of prestress in cables 1, 2, 3, 4 and 5 due to slip, friction and stressing in stages are 148.0, 122.8, 106.3, 95.1 and 86.1 MPa respectively. Hence the jacking stresses should be 1148.0, 1122.8, 1106.3, 1095.1 and 1086.1 MPa respectively.

REFERENCES

4.1. Cooley, E.H , *Friction in post-tensioned prestressing systems,* Research Report No. 1, Cement and Concrete Association (London) 1953.

4.2. Glodowski, R.J. and Lorenzetti, J.J., *An interaction method for prestress losses in a prestressed concrete structure,* Journal of the Prestressed Concrete Institute, March-April 1972.

4.3. Hernandez, H.D. and Gamble, W.L., *Time-dependent losses in prestressed concrete construction,* Structural Research Series No. 417, University of Illinois, Urbana, May 1975.

4.4. Lin, T.Y., *Cable friction in post-tensioning,* Journal of the Structural Division, ASCE., November 1956.

4.5. Magura, D.D.; Sozen, M.A. and Siess, C.P., *A study of stress relaxation in prestressing reinforcement,* Journal of the Prestressed Concrete Institute, Vol 9, No. 2, April 1964, pp. 13-57.

4.6. Marks, J.D. and Keifer, O., *Long-term field study of stresses in a post-tensioned concrete girder bridge,* First International Symposium on Concrete Bridge Design, ACI Publication SP-23, 1969, pp. 631-54.

4.7. PCI Report, *Recommendations for estimating prestress loss,* Report of PCI Committee on Prestress Losses, Journal of the Prestressed Concrete Institute, Vol. 20, No. 4, July-August 1975, pp. 43-75.

4.8. Sinno, R. and Furr, H.L., *Computer program of predicting prestress loss and camber,* Journal of the Prestressed Concrete Institute, Vol. 17, No. 5, September-October 1972, pp. 27-38.

4.9. Zia, P.; Preston, H.K.; Scott, N.L. and Workman, E.B., *Estimating prestress losses,* ACI-ASCE Committee on Prestressed Concrete Recommended Procedure, Concrete International, Vol. i, No. 6, June 1979, pp. 32-8.

PROBLEMS

4.1. A pretensioned transmission pole has a rectangular cross-section 0.1×0.1 m at top and 0.1×0.2 m at bottom. It is prestressed concentrically with an initial force of 90 kN. Determine the average loss of prestress due to elastic shortening Take $m = 7.5$.

4.2. A pretensioned beam of rectangular cross-section 0.2×0.3 m has a straight cable line parallel to centroidal axis with constant eccentricity of 0.05 m. The initial prestressing force is 360 kN. Determine the average loss of prestress due to elastic shortening. Take $m = 7.5$.

4.3. A post-tensioned beam of rectangular section 0.3×0.6 m has a parabolic cable line with zero eccentricity at ends and 0.15 m eccentricity at midspan. Determine the average increase in prestress due to uniform load of 18 kN/m over its entire span of 10 m. Take $m = 7.5$

4.4. Determine the loss of prestress due to creep of concrete in the pole of Prob. 4.1 if the strength of concrete at stress transfer, $f_{ci} = 5$ MPa. Take $m = 7.5$ and $E_s = 200$ kN/mm^2.

4.5. Determine the loss of prestress due to creep of concrete in the pole of Prob. 4.1 taking creep coefficient $C_c = 1.5$ and m = 7.5.

4.6. Determine the loss of prestress due to creep of concrete in the beam of Prob. 4.2 taking $C_c = 1.2$.

4.7. A post-tensioned beam of rectangular section 0.2×0.6 m has a parabolic cable line with an eccentricity of 0.1 m at midspan and zero eccentricity at ends. The initial prestressing force is 720 kN. Determine the loss of prestress at midspan section due to creep of concrete for (i) bonded beam and (ii) unbonded beam. Take $C_c = 1.5$ and m = 7.5

4.8. If the beam of Prob. 4.7 carries a permanent uniform load of 15 kN/m over a span of 10 m, determine the average loss of prestress due to creep of concrete.

4.9. Using Eq. (4.4.3.), determine the loss of prestress due to shrinkage of concrete in a post-tensioned girder having an overall depth of 1.15 m. The girder has a T-section with flange 0.6×0.15 m and web 1.0×0.2 m. The relative humidity is 40%. Take $E_s = 195$ kN/mm^2.

4.10. Determine the loss of prestress due to shrinkage of concrete in a post-tensioned beam in which the stress transfer takes place 13 days after concreting. Take $E_s = 210$ kN/mm^2.

4.11. Discuss the loss of prestress due to relaxation of steel. Describe the treatment given to prestressing steel for reducing this loss.

4.12. Determine the loss of prestress due to relaxation of steel as per IS code, if initial stress $f_{pi} = 880$ MPa and ultimate strength $f_{pu} = 1600$ MPa.

4.13. If the slip of wedges at the jacking end and dead end are 3 mm and 2 mm respectively, determine the loss of prestress. The distance between the anchorages is 25 m. Take $E_s = 200$ kN/mm^2.

4.14. The profiles of three cables in a post-tensioned beam are shown in Fig. P. 4.14. Determine the prestressing force in the cables at midspan if the jacking force in each cable is 250 kN. Take $\mu = 0.55$ and $k = 0.003/m$

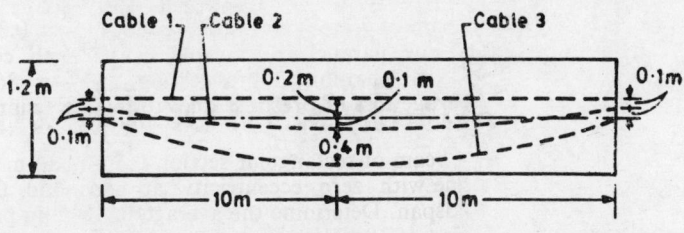

Fig. P 4.14

4.15. A pretensioned member with rectangular section 0.4×0.75 m has a straight cable line with a constant eccentricity of 0.125 m. The initial prestressing force is 1.8 MN. Determine the loss of prestress due to elastic shortening, creep and shrinkage of concrete and relaxation of steel taking $m = 7.5$, $C_c = 1.5$, $\epsilon_{sh} = 0.0003$ and $E_s = 2 \times 10^5$ MPa. The relaxation of steel is 3%. Hence determine the stress in steel after losses if the initial stress is 1150 MPa.

5

Analysis for Flexure

5.1 INTRODUCTION

Flexure produces normal stresses of opposite sign on the two sides of the neutral axis. A favourable distribution of prestress in a flexural member can be achieved by placing the centre of gravity of prestressing steel at an appropriate location in the flexural tension zone. Hence the technique of prestressing finds a ready application to flexural members. The majority of members in a skeletal structure may be described as flexural members. Flexure also occurs in non-skeletal structures such as folded plates and shells. Consequently, it is possible to derive advantage from the technique of prestressing in an overwhelming majority of concrete structures. The analysis of stresses in an uncracked flexural member may be carried out on the basis of elastic theory using any one of the following three approaches:

(i) *Theory of Flexure Approach*

In this approach, the flexural stresses due to the bending moments caused by prestressing force, dead load and live load are computed by simple theory of bending and added to the direct stresses caused by prestress.

(ii) *Line of Thrust Approach*

In this approach, the normal stresses at any cross-section are evaluated by locating the position of the centre of compression or the line of thrust and treating the cross section as an eccentrically loaded column. This approach is similar to one used in reinforced concrete.

(iii) *Load Balancing or Freebody Approach*

In this approach, a prestressed concrete beam is visualised as a self straining system in which the forces acting on the free bodies of prestressing steel and concrete are equal in magnitude and opposite in sign. The freebody diagram of prestressing steel is considered first. As the forces acting ont he freebody of concrete are equal and opposite to those acting on steel, the freebody of concrete may be considerd next and the stresses in concrete evaluated therefrom. The approach is of special interest in the analysis of statically indeterminate structures.

The analysis of flexural stresses using the three approaches enumerated above is discussed in the following sections.

5.2 FLEXURE THEORY

In simply supported beams subjected to sagging bending moment, the centroid of the prestressing tendons in the major portion of the beam is placed below the centroid of the cross section so that the prestressing force produces direct compression as well as hogging bending moment in the prestressed beam. The stresses in prestressing steel are maximum at the time of transfer of prestress. They decrease continuously with time as the losses of prestress occur. The flexural member is also subjected to bending moments due to self weight, dead loads, live loads and other loads at different stages of loading during its useful life. While there are many stages which may have to be taken into account depending upon the system of prestressing, the method of construction and the service conditions, the two stages which need to be considered invariably in the working stress design are the initial and final stages. In the initial stage, the prestressing force P_i is at its maximum and the self weight together with some of the permanent loads may also act on the member. If the gravity loads at the initial stage are described as dead loads, the initial stage is characterised by the prestressing force P_i together with the dead loads. The final stage is, therefore, characterised by the effective prestress P after the losses of prestress in combination with the entire dead and live loads known as service loads.

(i) *Initial Stage*

Figure 5.2.1 shows the cross-section of a flexural member subjected to the prestressing force with an eccentricity e_p. The cross-section is symmetrical

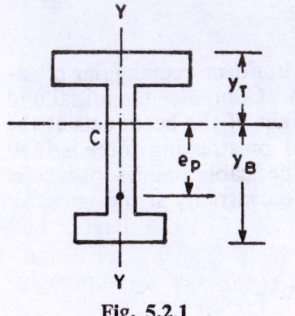

Fig. 5.2.1

with respect to the vertical axis yy passing through the centroid C. Treating the cross-section as an eccentrically loaded column subjected to an eccentric load P_i, the stresses at the top fibre f_T and bottom fibre f_B due to prestress are given by the following equations in accordance with the sign convention discussed in Sec. 1.4.

$$f_T = \frac{-P_i}{A} + \frac{(-P_i e_p)(-y_T)}{I}$$

$$= \frac{-P_i}{A}\left[1 - \frac{e_p y_T}{k^2}\right] \qquad (5.2.1)$$

$$f_B = \frac{-P_i}{A} + \frac{(-P_i e_p)(y_B)}{I}$$

$$= \frac{-P_i}{A}\left[1 + \frac{e_p y_B}{k^2}\right] \qquad (5.2.2)$$

As the direct and the bending stresses due to prestress are of the same sign at the bottom fibre and they are of opposite sign at the top fibre, the eccentric prestress produces a high compressive stress at the bottom fibre and a low compressive or tensile stress at the top fibre. Combining the stresses due to prestress with those caused by the dead load moment M_D, the stresses at top and bottom fibres at the initial stage are given by the equations,

$$f_T = \frac{-P_i}{A} + \frac{P_i e_p y_T}{I} - \frac{M_D y_T}{I} \qquad (5.2.3)$$

$$f_B = \frac{-P_i}{A} - \frac{P_i e_p y_B}{I} + \frac{M_D y_B}{I} \qquad (5.2.4)$$

(ii) *Final Stage*

The final stage is characterised by the effective prestressing force P after losses and the full working or service load which is the sum of the dead and live loads. Hence the stresses at the top and bottom fibres at the final stage may be expressed as

$$f_T = \frac{-P}{A} + \frac{P e_p y_T}{I} - \frac{(M_D + M_L) y_T}{I} \qquad (5.2.5)$$

$$f_B = \frac{-P}{A} - \frac{P e_p y_B}{I} + \frac{(M_D + M_L) y_B}{I} \qquad (5.2.6)$$

where M_L = bending moment due to live loads

At the initial stage, the compressive stress at the bottom fibre is high and is approximately equal to the working stress in compression. The stress at the top is close to zero or the working stress in tension. At the final stage the compressive stress at the top is close to the working stress in compression and the stress at the bottom fibre is close to zero or the working stress in tension.

EXAMPLE 5.2.1

A post-tensioned prestressed concrete beam of uniform rectangular cross-section 0.4×0.6 m has an effective span of 8 m. Compute the initial and final stresses assuming that only the self weight of the beam acts at the initial stage. The live load is 30 kN/m. The total prestressing force is 1680 kN before losses and 1440 kN after losses. The cable line is parabolic with an eccentricity of 0.1 m at midspan and no eccentricity at the ends.

Solution

Cross-sectional area, $A = 0.4 \times 0.6 = 0.24$ m^2

Moment of inertia, $I = \frac{1}{12} \times 0.4 \times 0.6^3 = 0.0072$ m^4

Taking unit weight of concrete as 25 kN/m^3,

Self weight of beam $= 0.4 \times 0.6 \times 25 = 6$ kN/m

Dead load moment, $M_D = \frac{6 \times 8^2}{8} = 48$ kN. m

Live load moment, $M_L = \frac{30 \times 8^2}{8} = 240$ kN. m

(i) Initial Stage

Using Eq. (5.2.3) and (5.2.4),

$$f_T = \frac{-1.680}{0.24} + \frac{1.680 \times 0.1 \times 0.3}{0.0072} - \frac{0.048 \times 0.3}{0.0072} = -2 \text{ MPa}$$

$$f_B = \frac{-1.680}{0.24} - \frac{1.680 \times 0.1 \times 0.3}{0.0072} + \frac{0.048 \times 0.3}{0.0072} = -12 \text{ MPa}$$

(ii) Final Stage

Using Eq. (5.2.5) and (5.2.6),

$$f_T = \frac{-1.440}{0.24} + \frac{1.440 \times 0.1 \times 0.3}{0.0072} - \frac{0.288 \times 0.3}{0.0072} = -12 \text{ MPa}$$

$$f_B = \frac{-1.440}{0.24} - \frac{1.440 \times 0.1 \times 0.3}{0.0072} + \frac{0.288 \times 0.3}{0.0072} = 0$$

The distribution of stresses at the initial and final stages are shown in Fig. 5.2.2.

EXAMPLE 5.2.2

A prestressed concrete beam having a cross-section shown in Fig. 5.2.3 carries a uniform load of 12 kN/m due to its own weight at the initial stage over a span of 16 m. Determine the prestressing force and its eccentricity to produce net stresses equal to 0 and -12 MPa at the top and bottom fibres respectively.

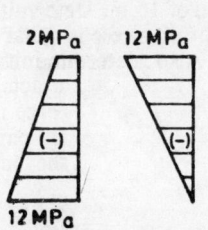

2MPa   12MPa

(–)   (–)

12MPa

(a) Inital stage (b) Final stage

Fig. 5.2.2

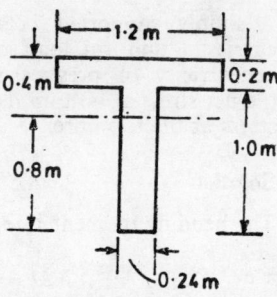

Fig. 5.2.3

Solution

$$A = 1.2 \times 0.2 + 0.24 \times 1 = 0.48 \text{ m}^2$$

$$y_T = \frac{1.2 \times 0.2 \times 0.1 + 0.24 \times 1 (0.2 + 0.5)}{1.2 \times 0.2 + 0.24 \times 1}$$

$$= 0.4 \text{ m}$$

$$y_B = 0.8 \text{ m}$$

Moment of inertia about centroidal axis,

$$I = \frac{1}{12} \times 1.2 \times 0.2^3 + 1.2 \times 0.2 \,(0.4 - 0.1)^2$$

$$+ \frac{1}{12} \times 0.24 \times 1^3 + 0.24 \times 1 \,(0.8 - 0.5)^2$$

$$= 0.064 \text{ m}^4$$

$$k^2 = \frac{I}{A} = 0.1333 \text{ m}^2$$

Bending moment at midspan due to self weight,

$$M_D = \frac{0.012 \times 16^2}{8} = 0.384 \text{ MN. m}$$

Hence using Eq. (5.2.3) and (5.2.4),

$$f_T = \frac{-P_i}{0.48} \left[1 - \frac{e_p \times 0.4}{0.1333} \right] - \frac{0.384 \times 0.4}{0.064} = 0 \qquad \text{(a)}$$

$$f_B = \frac{-P_i}{0.48} \left[1 + \frac{e_p \times 0.8}{0.1333} \right] + \frac{0.384 \times 0.8}{0.064} = -12 \text{ MPa (b)}$$

Solving Eq. (a) and (b),

$$P_i = 1.92 \text{ MN} \qquad e_p = 0.5333 \text{ m}$$

EXAMPLE 5.2.3

A simply supported beam having a cross-section shown in Fig. 5.2.3 carries a uniform load of 30 kN/m over a span of 16 m. Determine the eccentricity of prestress if the average intensity of prestress is 4 MPa and the net stress at bottom fibre is 2 MPa (tensile). Also determine the net stress at the top fibre.

Solution

The bending moment due to external loads,

$$M = \frac{0.030 \times 16^2}{8} = 0.960 \text{ MN·m}$$

The net stress at bottom fibre,

$$f_B = -4 \left[1 + \frac{e_p \times 0.8}{0.1333} \right] + \frac{0.96 \times 0.8}{0.064} = 2 \text{ MPa}$$

$$e_p = 0.25 \text{ m}$$

The stress at the top fibre,

$$f_T = -4 \left[1 - \frac{0.25 \times 0.4}{0.1333} \right] - \frac{0.96 \times 0.4}{0.064} = -7 \text{ MPa}$$

5.3 LINE OF THRUST APPROACH

The centre of compression and the line of thrust are discussed in section 1.7. The position of centre of compression in an uncracked section at any stage may be determined from Eq. (1.7.4).

(i) Initial Stage

The dead load at the initial stage produces sagging bending moment M_D in the simply supported beam. Hence using Eq. (1.7.4), the eccentricity of centre of compression at the initial stage may be expressed as

$$e_c = e_p - \frac{M_D}{P_i} \qquad (5.3.1)$$

The dead load moment M_D shifts the line of thrust upwards from the cable line through a vertical distance jd_p given by

$$jd_p = \frac{M_D}{P_i} \qquad (5.3.2)$$

The shift of line of thrust jd_p represents the internal lever arm between the resultant compressive and tensile forces in the cross-section of the beam. Knowing the eccentricity of line of thrust, the extreme fibre stresses may be determined from the equations,

$$f_T = \frac{-P_i}{A} + \frac{(-P_i e_c)(-y_T)}{I}$$

$$= \frac{-P_i}{A}\left[1 - \frac{e_c y_T}{k^2}\right] \qquad (5.3.3)$$

$$f_B = \frac{-P_i}{A} + \frac{(-P_i e_c) y_B}{I}$$

$$= \frac{-P_i}{A}\left[1 + \frac{e_c y_B}{k^2}\right] \qquad (5.3.4)$$

(ii) Final Stage

The dead and live loads at the final stage produce sagging bending moment $(M_D + M_L)$ in a simply supported beam. Hence using Eq. (1.7.4), the eccentricity of centre of compression at the final stage may be expressed as

$$e_c = e_p - \frac{(M_D + M_L)}{P} \qquad (5.3.5)$$

The moment $(M_D + M_L)$ shifts the line of thrust upwards from the cable line through a vertical distance jd_p given by

$$jd_p = \frac{M_D + M_L}{P} \qquad (5.3.6)$$

The shift of line of thrust jd_p represents the internal lever arm between the resultant compressive and tensile forces in the cross-section of the beam. Knowing the eccentricity of line of thrust, the extreme fibre stresses may be determined from the equations,

$$f_T = \frac{-P}{A} + \frac{(-Pe_c)(-y_T)}{I}$$

$$= \frac{-P}{A}\left[1 - \frac{e_c\,y_T}{k^2}\right] \qquad (5.3.7)$$

$$f_B = \frac{-P}{A} + \frac{(-Pe_c)(y_B)}{I}$$

$$= \frac{-P}{A}\left[1 + \frac{e_c\,y_B}{k^2}\right] \qquad (5.3.8)$$

EXAMPLE 5.3.1

Check the initial and final stresses in the post-tensioned beam of Ex. 5.2.1.

Solution

(i) *Initial Stage*

$$M_D = 0.048 \text{ MN}\cdot\text{m}$$

$$k^2 = \frac{0.0072}{0.24} = 0.03 \text{ m}^2$$

Using Eq. (5.3.1),

$$e_c = 0.1 - \frac{0.048}{1.68} = 0.0714 \text{ m}$$

Using Eq. (5.3.3) and (5.3.4),

$$f_T = \frac{-1.68}{0.24}\left[1 - \frac{0.0714 \times 0.3}{0.03}\right] = -2 \text{ MPa}$$

$$f_B = \frac{-1.68}{0.24}\left[1 + \frac{0.0714 \times 0.3}{0.03}\right] = -12 \text{ MPa}$$

(ii) *Final Stage*

$$M_D + M_L = 0.288 \text{ MN}\cdot\text{m}$$

Using Eq. (5.3.5),

$$e_c = 0.1 - \frac{0.288}{1.44} = -0.1 \text{ m}$$

Using Eq. (5.3.7) and (5.3.8),

$$f_T = \frac{-1.44}{0.24}\left[1 - \frac{(-0.1)(0.3)}{0.03}\right] = -12 \text{ MPa}$$

$$f_B = \frac{-1.44}{0.24}\left[1 + \frac{(-0.1)(0.3)}{0.03}\right] = 0$$

EXAMPLE 5.3.2

Solve the problem of Example 5.2.2 using the line of thrust approach.

Solution

Using Eq. (5.3.3) and (5.3.4),

$$f_T = \frac{-P_i}{0.48}\left[1 - \frac{e_c \times 0.4}{0.1333}\right] = 0 \qquad \text{(a)}$$

$$f_B = \frac{-P_i}{0.48}\left[1 + \frac{e_c \times 0.8}{0.1333}\right] = -12 \text{ MPa} \qquad \text{(b)}$$

Solving Eq. (a) and (b),

$$P_i = 1.92 \text{ MN}$$
$$e_c = 0.3333 \text{ m}$$

Using Eq. (5.3.1),

$$0.3333 = e_p - \frac{0.384}{1.92}$$

or

$$e_p = 0.5333 \text{ m}$$

EXAMPLE 5.3.3

Solve the problem of Example 5.2.3 using the line of thrust approach.

Solution

The stress at the bottom fibre,

$$f_B = -4\left[1 + \frac{e_c \times 0.8}{0.1333}\right] = 2 \text{ MPa}$$

$$e_c = -0.25 \text{ m}$$

$$\frac{P}{A} = 4 \quad \text{or} \quad P = 4 \times 0.48 = 1.92 \text{ MN}$$

Using Eq. (5.3.1),

$$-0.25 = e_p - \frac{0.96}{1.92}$$

or

$$e_p = 0.25 \text{ m}$$

5.4 LOAD BALANCING OR FREEBODY APPROACH

In simply supported post-tensioned beams subjected to sagging bending moments, the prestressing cables are usually curved with concavity upwards. As the cables are under tension, they tend to become straight and are, therefore, subjected to downward transverse pressure by the surrounding concrete. Hence, the concrete beam is subjected to an upward pressure which balances the entire or part of the transverse load acting on the beam. This concept, known as the load balancing approach, was first given by Lin.

A more general approach which may be described as the freebody approach, visualises the prestressed concrete beam as self-strained structure. In the unloaded condition, the forces acting on the prestressing steel and the concrete beam are equal and opposite. Hence to compute the stresses in a prestressed concrete beam, the freebody of the prestressing steel may be considered first. As the shear and flexural stiffnesses of the prestressing cables are negligible, they can carry only axial (tensile) forces. The freebody diagram for the prestressing steel can, therefore, be drawn readily. The following simplifying assumptions are generally made:

(i) The prestressing force P is uniform throughout the span of the beam or the length of the cable.

(ii) The slope of the cable line is small so that
$$\sin \theta = \tan \theta = \theta; \qquad \cos \theta = 1 \qquad (5.4.1)$$

(iii) The prestressing force P remains unaffected on account of the transverse load acting on the beam.

Having drawn the freebody diagram of the prestressing steel, the freebody of the concrete beam may be considered next, in order to evaluate the stresses in concrete.

Figure 5.4.1 (a) shows a simply supported prestressed concrete beam carrying any arbitrary transverse loading of intensity w over an effective span l. The freebody diagram of the prestressing steel is shown in Fig. 5.4.1 (b). If the sag of the cable line S is small compared to the span l, the slope of the cable line is small so that the horizontal and vertical components of

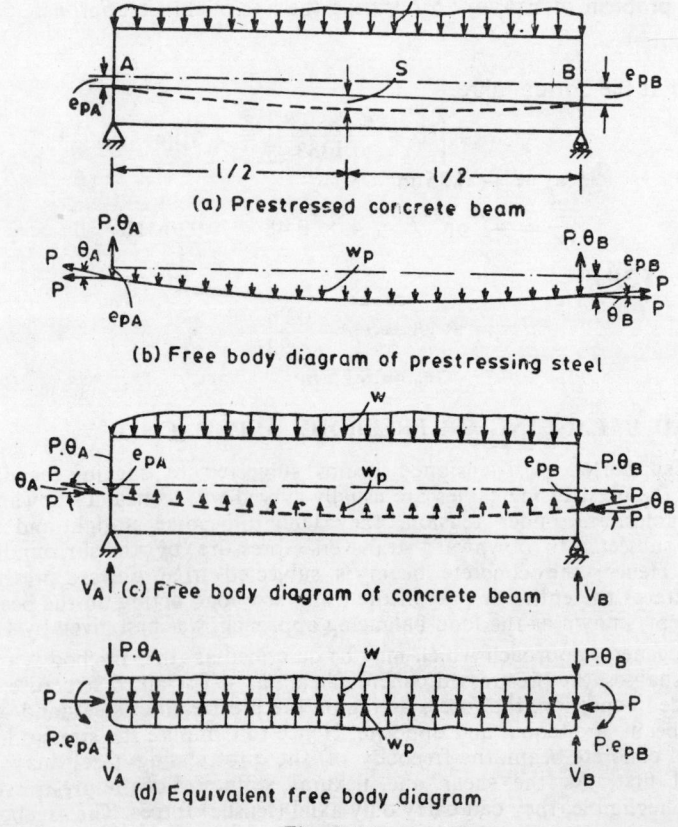

(a) Prestressed concrete beam

(b) Free body diagram of prestressing steel

(c) Free body diagram of concrete beam

(d) Equivalent free body diagram

Fig. 5.4.1

the prestressing force P at the end A are $P \cos \theta_A = P$ and $P \sin \theta_A = P$ θ_A. Similarly the components of the prestressing force P at the end B are P and $P \theta_B$ in the horizontal and vertical directions respectively. For the equilibrium of vertical forces, the resultant of the transverse pressure w_p must be equal to $P (\theta_A + \theta_B)$.

$$P (\theta_A + \theta_B) = \int_0^l w_p \, dx \qquad (5.4.2)$$

The transverse pressure w_p exerted by concrete on the prestressing steel depends upon the shape of the cable line. The freebody of the concrete beam is shown in Fig. 5.4.1(c) in which the forces are equal and opposite to those acting on the prestressing steel. In addition, the external load w and the support ractions V_A and V_B also act on the beam. It may be noted that the support reactions are not affected by prestressing as the forces due to pre-stress constitute a self-equilibrating system of forces. The equivalent freebody diagram which may be used for computation of stresses is shown in Fig. 5.4.1(d). The approximations of Eq. (5.4.1) have been used in Fig. 5.4.1 (b), (c) and (d).

To derive an expression for the transverse pressure w_p, consider a small element of the cable of length dx as shown in Fig. 5.4.2. For the equilibrium of vertical forces,

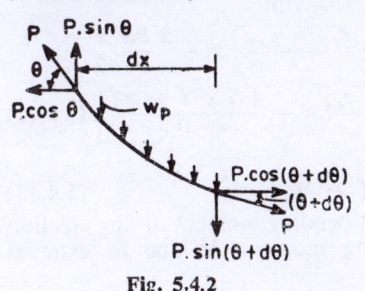

Fig. 5.4.2

$$w_p \, dx + P \sin (\theta + d\theta) = P \sin \theta$$

Using the approximations of Eq (5.4.1),

$$w_p = - P \frac{d\theta}{dx} \qquad (5.4.3)$$

As the rate of change of slope $\frac{d\theta}{dx}$ represents the curvature of the cable line,

$$w_p = - P \frac{d^2 e_p}{dx^2} \qquad (5.4.4)$$

The minus sign on the right sides of Eq. (5.4.3) and (5.4.4) arises because the rate of change of slope $\left(\frac{d\theta}{dx} \right)$ or curvature is negative.

If the unsymmetrical cable line shown in Fig. 5.4.1 (a) is parabolic, the equation of the cable line with origin at A may be written as

$$e_p = e_{pA} + (e_{pB} - e_{pA}) \frac{x}{l} + \frac{4S}{l^2} x (l - x) \qquad (5.4.5)$$

Hence the slope of the cable line at any point,

$$\frac{de_p}{dx} = \frac{e_{pB} - e_{pA}}{l} + \frac{4S}{l^2} (l - 2x) \qquad (5.4.6)$$

The slopes at A and B are

$$\left(\frac{de_p}{dx} \right)_{x=0} = \frac{4S}{l} + \frac{e_{pB} - e_{pA}}{l} = \theta_A \qquad (5.4.7)$$

$$\left(\frac{de_p}{dx} \right)_{x=l} = \frac{- 4S}{l} + \frac{e_{pB} - e_{pA}}{l} = \theta_B \qquad (5.4.8)$$

The curvature of the cable line,

$$\frac{d\theta}{dx} = \frac{d^2 e_p}{dx^2} = \frac{-8\ S}{l^2} \qquad (5.4.9)$$

Hence it follows from Eq. (5.4 4.) that the upward pressure exerted the cable on the concrete beam is unifrom and is given by

$$w_p = \frac{8\ PS}{l^2} \qquad (5.4.10)$$

Using Eq. (5.4.7), (5.4.8) and (5.4.10), it may be verified readily that the forces caused by prestressing shown in Fig. 5.4.1 (d) satisfy the three equations of static equilibrium. Consequently, these forces constitute a self-equilibrating system of forces.

Referring to the freebody diagram of Fig. 5.4.1 (d), the resultant bending moment at any section X at a distances x from A may be written as

$$M_r = \left[V_A\ x - \overline{W}_{AX}\ (x - \bar{x}_{AX}) \right] - P\ e_{pA} - P\theta_A\ x + w_p\ x\ \frac{x}{2} \qquad (5.4.11)$$

in which $\overline{W}_{AX}$ is the resultant of external load on the portion AX and $\bar{x}_{AX}$ is the distance of $\overline{W}_{AX}$ from A. The quantity inside the square brackets represents the bending moment M caused by external loads. Hence substituting for θ_A and w_p from Eq. (5.4.7) and (5.4.10),

$$M_r = M - P\ e_{pA} - P\left[\frac{4\ S}{l} + \frac{e_{pB} - e_{pA}}{l}\right] x + \frac{8\ PS}{l^2}\ \frac{x^2}{2}$$

$$= M - P\left[e_{pA} + \frac{e_{pB} - e_{pA}}{l}\ x + \frac{4\ S\ x\ (l - x)}{l^2}\right] \qquad (5.4.12)$$

Hence using Eq. (5.4.5),

$$M_r = M - P e_p = M + M_p \qquad (5.4.13)$$

Equation (5.4.13) shows that the resultant bending moment at any section is equal to the algebraic sum of the bending moment M due to external loading and M_p due to prestressing.

If the cable line is symmetrical about midspan, i.e. in the case $e_{pA} = e_{pB}$, the Eq. (5.4.5) of the cable line becomes

$$e_\eta = e_{pA} + \frac{4\ S\ x\ (l - x)}{l^2} \qquad (5.4.14)$$

Hence,

$$\theta_A = \theta_B = \frac{4\ S}{l} \qquad (5.4.15)$$

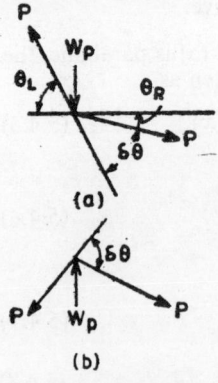

(a)

(b)

Fig. 5.4.3

As discussed in Sec. 3.2, the straight tendons in pretensioning technique are sometimes depressed at appropriate points in order to obtain a more favourable cable line. In this case the tendon has a sudden change of direction or a kink at certain points. Consider the freebody of a small element of the prestressing tendon carrying a force P in the vicinity of the kink as shown in Fig. 5.4.3. For the equilibrium of vertical forces,

$$P\ \sin\ \theta_R + W_p - P\ \sin\ \theta_L = 0 \qquad (5.4.16)$$

Hence using the approximations of Eq. (5.4.1),

$$W_p = P(\theta_L - \theta_R) = P \, \delta\theta \qquad (5.4.17)$$

in which $\delta\theta$ is the sudden change in the direction of the tendon, $(\theta_L - \theta_R)$. The concentrated vertical force W_p exerted by concrete on the tendon at the kink is downward if the kink points in the downward direction as shown in Fig. 5.4.3 (a). In this case the concentrated force W_p exerted on concrete by the cable is upward. On the other hand, if the kink points upward as shown in Fig. 5.4.3 b), the force W_p acting on the cable is upward and, therefore, that acting on concrete is downward.

EXAMPLE 5.4.1

Draw the shear force and bending moment diagrams due to prestressing for the simply supported beam shown in Fig. 5.4.4 (a). The prestressing force is 1000 kN.

Solution

The free body of the cable is shown in Fig. 5.4.4 (b). The slopes of the four straight segments of the cable are $\theta_{AD} = [0.2 - (-0.1)]/5 = 0.06$, $\theta_{DC} = 0.04$, $\theta_{CE} = 0.04$ and $\theta_{EB} = 0.02$. These slopes at the points D, C and E are shown in Fig 5.4.4 (b). Hence using Eq. (5.4.17), the concentrated vertical forces at D, C and E are 20, 80 and 20 kN respectively. The freebody diagram of the concrete beam is shown in Fig. 5.4.4. (c) in which the forces are equal and opposite to those shown in Fig. 5.4.4 (b). In addition, the end couples due to the eccentricity of prestress also act at the ends. The shear force and bending moment diagrams derived from the freebody of Fig. 5.4.4 (c) are shown in Fig. 5.4.4 (d) and (e) respectively.

EXAMPLE 5.4.2

Analyse the stresses at midspan in the beam of Example 5.2.1 using the freebody approach.

Solution

(i) *Initial Stage*

The freebody of the cable at the initial stage is shown in Fig. 5.4.5 (a). The horizontal component of the prestressing force at the end A or B is

$$P_i \cos \theta_A \approx P_i = 1680 \text{ kN} = 1.680 \text{ MN}$$

The vertical component of the prestressing force at the end A or B is

$$P_i \sin \theta_A \approx P_i \theta_A = P_i \frac{4S}{l} = \frac{1.680 \times 4 \times 0.1}{8} = 0.084 \text{ MN}$$

The uniform load on the cable,

$$w_p = \frac{8\,P_i\,S}{l^2} = \frac{8 \times 1.680 \times 0.1}{8^2} = 0.021 \text{ MN/m}$$

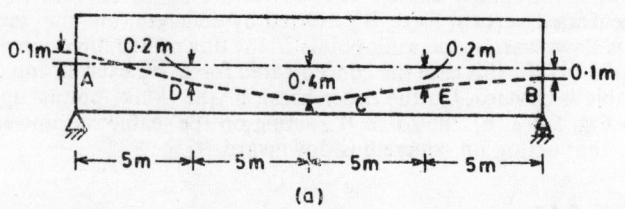

(a)

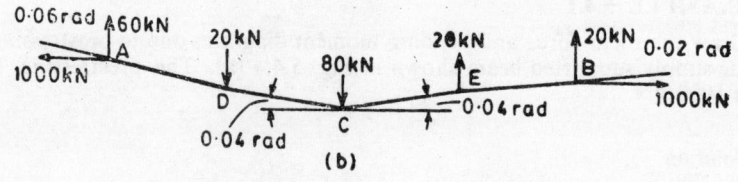

(b)

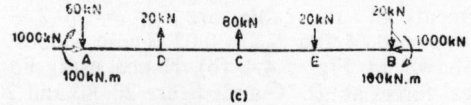

(c)

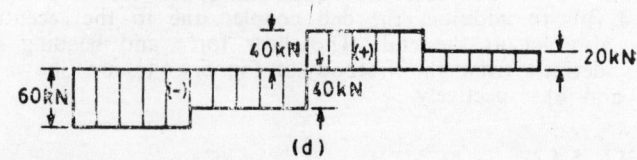

(d)

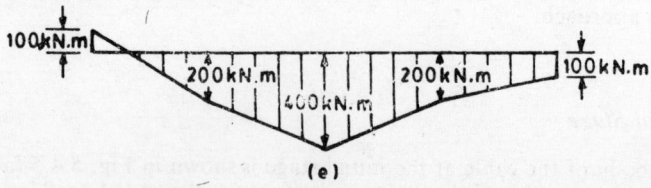

(e)

Fig. 5.4.4

The freebody of the concrete beam is shown in Fig. 5.4.5 (b). The self weight acting downward is 0.006 MN/m and w_p acting upward is 0.021 MN/m. Hence the beam carries a net upward uniform load of 0.015 MN/m at the initial stage. The hogging bending moment at midspan,

$$M = \frac{0.015 \times 8^2}{8} = 0.120 \text{ MN.m}$$

The bending stress at top,

$$f_T = \frac{(-0.120)(-0.3)}{0.0072} = 5 \text{ MPa}$$

The bending stress at bottom,

$$f_B = \frac{(-0.120)(0.3)}{0.0072} = -5 \text{ MPa}$$

$$\text{Direct stress} = \frac{-P_i}{A} = -\frac{1.680}{0.24} = -7 \text{ MPa}$$

Net stress at top, $f_{Ti} = 5 - 7 = -2$ MPa

Net stress at bottom, $f_{Bi} = -5 - 7 = -12$ MPa

(ii) *Final Stage*

The freebody diagrams of the cable and the concrete beam at the final stage are shown in Fig. 5.4.5 (c) and (d) respectively. At the final stage, the net uniform load on the beam is 0.018 MN/m downward. The final stresses at top and bottom are −12 MPa and zero respectively. The results coincide with those obtained in Example 5.2.1.

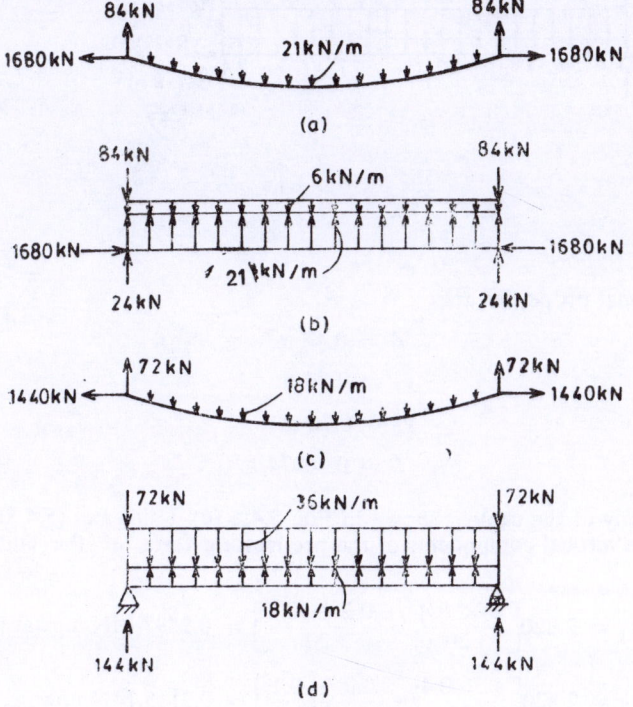

Fig. 5.4.5

EXAMPLE 5.4.3

Analyse the stresses at midspan in the simply suppoited beam shown in Fig. 5.4.6 (a). The beam has a parabolic cable profile. The prestressing force is 2.820 MN. The beam carries a uniform load of 0.04675 MN/m inclusive of self weight.

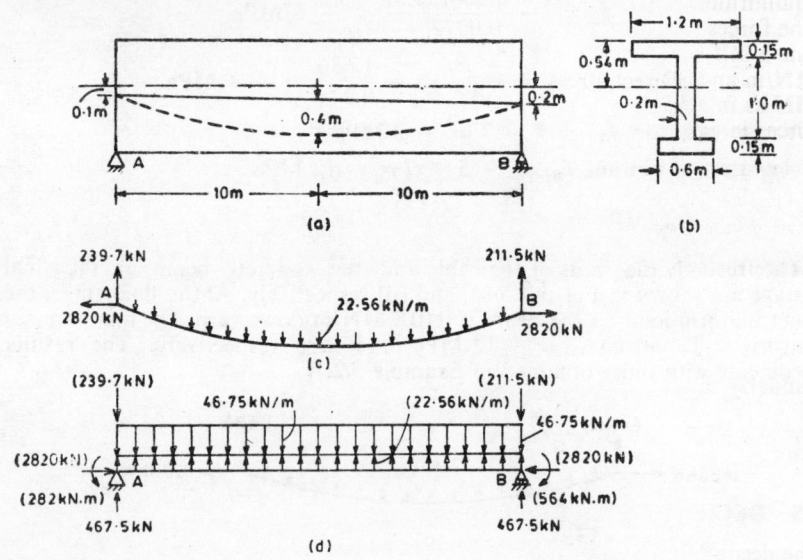

Fig. 5.4.6

Solution

The sectional properties are

$$A = 0.47 \text{ m}^2$$
$$y_T = 0.54 \text{ m}$$
$$y_B = 0.76 \text{ m}$$
$$I = 0.10074 \text{ m}^4$$

The freebody of the cable is shown in Fig. 5.4.6 (c). Using Eq. (5.4.7) and (5.4.8), the vertical components of the prestressing force at the ends are upward.

$$P\,\theta_A = 2.820\left[\frac{4 \times 0.4}{20} + \frac{0.2 - 0.1}{20}\right] = 0.2397 \text{ MN upward}$$

$$P\,\theta_B = 2.820\left[\frac{4 \times 0.4}{20} - \frac{0.2 - 0.1}{20}\right] = 0.2115 \text{ MN upward}$$

Using Eq. (5.4.10), the uniform load on the cable due to prestress,

$$w_p = \frac{8 \times 2.820 \times 0.4}{20^2} = 0.02256 \text{ MN/m}$$

It may be checked that the forces acting on the freebody of the cable are in equilibrium. The freebody of the concrete beam is shown in Fig. 5.4.6 (d). The forces shown in parentheses are caused due to prestress. They constitute a self equilibrating system of forces. The external load of 0.04675 MN/m and the support reactions at A and B each of magnitude 0.4675 MN as in an ordinary non-prestressed beam also constitute a system of forces in equilibrium. The bending moment at midspan,

$$M = -0.282 + (0.4675 - 0.2397) \times 10 - (0.04675 - 0.02256)$$
$$\times 10 \times 5 = 0.7865 \text{ MN·m}$$

Hence the stress at top,

$$f_T = \frac{-2.820}{0.47} + \frac{0.7865 \,(-0.54)}{0.10074} = -10.216 \text{ MPa}$$

Similarly the stress at bottom,

$$f_B = \frac{-2.820}{0.47} + \frac{0.7865 \,.0.76)}{0.10074} = -0.066 \text{ MPa}$$

5.5 DECOMPRESSION MOMENT

The decompression moment M_{ac} is the bending moment which just overcomes the maximum compressive stress due to prestress at the extreme fibre. In the context of a simply supported beam, the decompression moment is the sagging bending moment which reduces the compressive stress at the bottom fibre to zero. As discussed in Sec 1.11, the stress at bottom fibre is zero when the centre of compression is located at the upper kern point. Hence the shift of the centre of compression due to the decompression moment or the lever arm jd_p must be equal to the sum of e_p and e_{ku}.

$$jd_p = \frac{M_{dc}}{P} = e_p + e_{ku} \tag{5.5.1}$$

Hence the decompression moment is given by the equation,

$$M_{dc} = P\,(e_p + e_{ku}) \tag{5.5.2}$$

Alternatively, the decompression moment may be determined using its basic definition. As the stress at the bottom fibre is zero under the action of the decompression moment, Eq. (1.9.1) gives,

$$\frac{M_{dc}\, y_B}{I} - \frac{P}{A} + \frac{(-Pe_p)\, y_B}{I} = 0 \tag{5.5.3}$$

or

$$\frac{P y_B}{A k^2} \left[\frac{M_{dc}}{P} - e_p \right] = \frac{P}{A}$$

Putting $k^2/y_B = e_{ku}$ according to Eq. (1.11.2) and solving for the decompression moment,

$$M_{dc} = P (e_p + e_{ku})$$

which coincides with Eq. (5.5.2). The decompression moment may be computed from Eq. (5.5.3) or directly from Eq. (5.5.2).

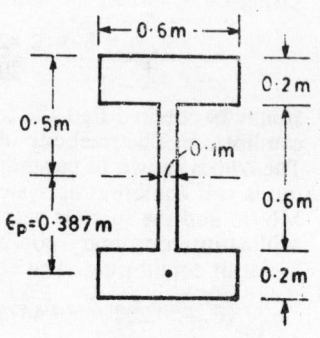

EXAMPLE 5.5.1

Compute the decompression moment for the symmetrical I-section shown in Fig. 5.5.1. The prestressing force is 1.800 MN.

Fig. 5.5.1

Solution

$$y_T = y_B = 0.5$$

$$A = 2 \times 0.6 \times 0.2 + 0.1 \times 0.6 = 0.30 \text{ m}^2$$

$$I = \frac{1}{12} \times 0.6 \times 1^3 - \frac{1}{12} \times 0.5 \times 0.6^3 = 0.041 \text{ m}^4$$

$$k^2 = I/A = 0.1367 \text{ m}^2$$

$$e_{ku} = e_{kl} = e_k = \frac{k^2}{y_T} = 0.273 \text{ m}$$

Using Eq. (5.5.2)

$$M_{dc} = 1.800 (0.387 + 0.273) = 1.188 \text{ MN·m}$$

Alternatively, from Eq. (5.5.3),

$$\frac{M_{dc} \times 0.5}{0.41} - \frac{1.800}{0.30} + \frac{(-1.800 \times 0.387) \times 0.5}{0.041} = 0$$

which gives, $M_{dc} = 1.188$ MN·m

5.6 CRACKING MOMENT

A prestressed concrete beam is just on the point of flexural cracking when the tensile stress at the extreme fibre is equal to the tensile strength of concrete in bending. The bending moment, which creates this condition in the beam, is known as the cracking moment M_{cr}. It may, therefore, be computed from the equation,

$$\frac{M_{cr} \, y_B}{I} - \frac{P}{A} + \frac{(-Pe_p) \, y_B}{I} = f_r \tag{5.6.1}$$

where f_r = modulus of rupture or the tensile strength of concrete in bending

In Eq. (5.6.1) putting,

$$f_r = \frac{M_{cr, o} \, y_B}{I} \tag{5.6.2}$$

where $M_{cr,\,o}$ = cracking moment of an ordinary non-prestressed beam

$$\frac{(M_{cr} - M_{cr,\,o})\,y_B}{I} - \frac{P}{A} + \frac{(-Pe_p)\,y_B}{I} = 0 \qquad (5.6.3)$$

Equation (5.6.3) is similar to Eq. (5.5.3) except that M_{d_c} has been replaced by $(M_{cr} - M_{cr,\,o})$. Proceeding in the same manner as for the derivation of Eq. (5.5.2), the following expression for the cracking moment is obtained:

$$M_{cr} = P\,(e_p + e_{ku}) + M_{cr,\,o} \qquad (5.6.4)$$

Substituting from Eq. (5.5.2),

$$M_{cr} = M_{dc} + M_{cr,\,o} \qquad (5.6.5)$$

Equation (5.6.5) shows that the cracking moment of a prestressed concrete beam is the sum of the moment required to cause decompression at the bottom fibre and the cracking moment of an ordinary non-prestressed beam. This is evident because the cracking moment has not only to create decompression but to produce tensile stress equal to the modulus of rupture f_r at the bottom fibre. The cracking moment marks an important point in the flexural response of a prestressed concrete beam. If the applied moment is smaller than the cracking moment, the entire section is effective. But if the applied moment is larger than the cracking moment, part of the section becomes ineffective due to flexural cracking.

EXAMPLE 5.6.1

Compute the cracking moment of the prestressed concrete beam of Example 5.5.1. The modulus of rupture of concrete $f_r = 3$ MPa.

Solution

From Eq. (5.6.1),

$$\frac{M_{cr} \times 0.5}{0.041} - \frac{1.800}{0.300} + \frac{(-1.800 \times 0.387)\,(0.5)}{0.041} = 3$$

which gives $M_{cr} = 1.434$ MN·m
or, from Eq. (5.6.4),

$$M_{cr} = 1.800\,(0.387 + 0.273) + \frac{3 \times 0.041}{0.5}$$

$$= 1.434 \text{ MN·m}$$

5.7   STRESS IN PRESTRESSING STEEL

The inital stress in prestressing tendons decreases due to the factors responsible for loss of prestress. As discussed in Chapter 4, some of these losses occur immediately at stress transfer while others continue for a considerable period of time. Hence the stress in prestressing steel in a member which does not carry any external load depends upon the initial stress and the cumulative loss of prestress at any time due to all factors causing loss of prestress. In actual practice, the member receives external load while the loss of prestress is still continuing. As the effect of external load is generally to increase the stress in prestressing steel, two opposite effects overlap each other, i.e. decrease of stress due to loss of prestress and increase of stress due to external loads. The first external load on a prestressed concrete beam is often due to its own weight. Generally the hogging moment imposed by prestressing on a simply supported beam is large enough to cause upward deflection causing the beam soffit to break away from the formwork, thereby bringing the self weight of the beam into action. Hence the self weight of the beam is often the essential part of the dead load at the initial stage. When the structure is pressed into service, other dead and live loads act on the beam. The stress in prestressing steel at any stage f_p may be expressed as

$$f_p = f_{pi} - \delta f_p + f_{pM} = f_{pe} + f_{pM} \qquad (5.7.1)$$

where f_{pi} = initial stress in prestressing steel

δf_p = loss of prestress at the stage considered

f_{pM} = increase in stress due to bending moment M acting at the stage considered

f_{pe} = effective prestress at commencement of loading

In the precracking range, the increase in stress in prestressing steel at any section in a prestressed concrete bonded beam due to moment M is

$$f_{pM} = m f_e = \frac{M e_p}{I} \qquad (5.7.2)$$

in which m is the modular ratio. The elastic stress f_e at the level of the centroid of prestressing steel at the section of maximum bending moment M under the action of full service load does not generally exceed 14 MPa. Hence for modular ratio, $m = 6$, the increase in stress f_{pM} due to loading does not exceed 84 MPa. As the cumulative loss of prestress δf_p due to all factors generally ranges from 150 to 300 MPa, it is evident that in accordance with Eq. (5.7.1), the net stress f_p in prestressing steel, even when the member carries its full working load, does not exceed the initial stress f_{pi}. Hence the initial stress f_{pi} is generally the maximum stress experienced by prestressing steel during the entire useful life of the structure unless it is exposed to overload conditions.

In an unbonded beam, the tendon is free inside the duct. Hence it has the same stress over the entire length if friction between the cable and the duct is ignored. Consequently, the increase of stress in prestressing steel f_{pM} due to bending moment M corresponds to the average value of elastic stress f_e along the entire span.

$$f_{pM} = \frac{m}{l} \int_0^l f_e \, dx \qquad (5.7.3)$$

As the average value of elastic stress f_e over the whole span is smaller than value at the critical section of maximum moment, it follows that the increase of stress at the critical section is smaller in unbonded beams as compared to bonded beams. For instance, referring to Example 4.2.4, the increase in stress in an unbonded beam is $(8/15)$ times that in the corresponding bonded beam if the load is uniform and cable line is parabolic with no eccentricity at ends.

As the load on the beam is increased progressively, the bending moment increases to the decompression moment M_{dc} at which the stress at the bottom fibre becomes zero. The centre of compression rises progressively with increasing load and reaches the upper kern point at decompression moment M_{dc}. With further increase in load a stage is reached at which the stress at the bottom fibre becomes equal to the tensile strength of concrete in bending. This stage characterised by incipient flexural cracking under bending moment M_{cr}, called cracking moment, marks the end of elastic action with the whole section behaving like an ideal homogeneous and isotropic material. When the load is increased beyond the cracking stage, the flexural cracks emanate from the bottom fibre and propagate upwards. The centre of compression and neutral axis continue to move upwards although at a progressively slower rate. On the other hand, the stress in prestressing steel increases at much faster rate after cracking in almost direct proportion with the increase in load or bending moment. In the precracking range, the internal moment of resistance keeps pace with increasing external bending moment by increasing its internal lever arm while the stress in prestressing steel remains practically constant. On the other hand, in the post-cracking range, the lever arm remains practically constant while the increasing moment is resisted by rising tensile stress in prestressing steel. In beams having low proportion of steel, the rise of tensile stress in steel may lead to its sudden breaking before crushing of concrete causing disastrous complete collapse. In this type of undesirable failure, the stress in prestressing steel is equal to the ultimate stress of the material, $f_p = f_{pu}$ In most practical beams, however, the eventual failure occurs by the crushing of concrete while the stress in prestressing steel remains smaller than f_{pu}. The stress in steel at collapse of the beam depends upon the shape of the cross-section and the stress-strain curves for steel and concrete. In a bonded beam, the strain in steel at collapse can be determined from compatibility of strains in accordance with Navier's hypothesis. The corresponding stress is then obtained from the stress-strain curve for prestressing steel.

In an unbonded beam, the stress in prestressing steel in the post-cracking range is greater than that in the corresponding bonded beam at any given load. The tendon being free inside the duct. the tendon stress is same everywhere and is equal to the integration of strains at the level of the tendon over its entire length. To determine the elongation of the tendon and the consequent increase in stress due to any given load, the strain at the level of the tendon may be computed at several sections using Navier's hypothesis and strain compatibility. The average value of these strains multiplied by E_s gives the increase in stress in the tendon due to the applied load. As the average value of the strain is smaller than the strain at the section of maximum moment. it is evident that the stress in tendons of unbonded beam increases at a slower rate in comparison to bonded beams. The unbonded beam may have the ultimate moment at collapse as much as 20 per cent smaller than that of bonded beam because the total tension in tendons is smaller in the unbonded beams. Another undesirable feature of unbonded beams is that they develop wider cracks at larger spacing as the tensile strains concentrate at these wide cracks. Due to these reasons. the unbonded beams are not preferred in practice. The possibility of corrosion of tendons and loss of prestress in the event of failure of end anchorages also discourage the use of unbonded beams. The distribution of tensile strains can be made more uniform resulting in finer cracks at closer spacing by the provision of non-prestressed reinforcing bars.

EXAMPLE 5.7.1

A tendon in a post-tensioned beam has a parabolic shape with zero eccentricity at ends and a sag of 0.25 m at mid-span. The beam has a rectangular section 0.20 x 0.60 m and a span of 10 m. Determine the increase in stress in the tendon due to uniform load of 14.4 kN/m when the tendon is (i) bonded (ii) unbonded. Take m = 6.

Solution

Bending moment at mid-span,

$$M = \frac{0.0144 \times 10^2}{8} = 0.18 \text{ MN.m}$$

$$I = \frac{1}{12} \times 0.20 \times 0.60^3 = 0.0036 \text{ m}^4$$

(i) Bonded Beam

Elastic stress at the level of prestressing tendon,

$$f_e = \frac{0.18}{0.0036} \times 0.25 = 11.5 \text{ MPa}$$

Hence the increase of stress in the tendon due to external load,

$$f_{pM} = m f_e = 6 \times 12.5 = 75 \text{ MPa}$$

(ii) Unbonded Beam

Referring to Example 4.2.4, the uniform increase in stress in the tendon in the unbonded beam is (8/15) times the increase in stress in bonded beam at the midspon section. Hence,

$$f_{pM} = \frac{8}{15} \times 75 = 40 \text{ MPa}$$

5.8 ULTIMATE MOMENT

The ultimate flexural strength of a prestressed concrete beam is an important property because it gives the margin of safety against failure or collapse. The codes of practice specify load factors for dead, live and other kinds of loads. The ultimate moment must be equal to or greater than the bending moment due to the working loads multiplied by the respective load factors to ensure adequate margin of safety as per the provisions of the codes. The basic assumptions of the ultimate strength analysis of prestressed concrete beams are similar to the assumptions made in the working stress analysis of reinforced concrete beames. These basic assumptions are:

(i) Concrete being very weak in tension, it is incapable of taking any tensile stress. All tension is assumed to be carried by steel.

(ii) Perfect bond exists between concrete and steel except in unbonded beams.

(iii) Navier's hypothesis is valid, i.e. plane sections remain plane during bending. It follows that the strain distribution along the depth of cross-section is linear and the strains are proportional to the distance from the neutral axis upto failure.

The basic difference in the assumptions in the working stress analysis and the ultimate strength analysis is that in the latter analysis, the Hooke's law is given up and instead the nonlinear stress-strain curves for concrete and steel are utilized. Referring to Fig. 5.8.1, the ultimate moment M_u of a prestressed concrete section is equal to the tensile force in steel T multiplied by the internal lever arm (jd_p).

$$M_u = T\,(jd_p) = A_p f_p\,(jd_p) \tag{5.8.1}$$

where A_p = area of prestressing steel

  f_p = stress in prestressing steel at failure

  jd_p = internal level arm which is the vertical distance between the centre of compression and the centroid of prestressing steel

From Eq. (5.8.1) it follows that the ultimate moment depends upon the following sectional and material properties:

(i) Size and Shape of the Cross-section

The internal lever arm depends upon the size and shape of cross-section. It increases with the effective depth of the section d_p. It also increases with the breadth b in the compression zone (shown shaded) because for any particular value of area in the compression zone, the depth of neutral axis x_u decreases as b increases and consequently lever arm (jd_p) increases.

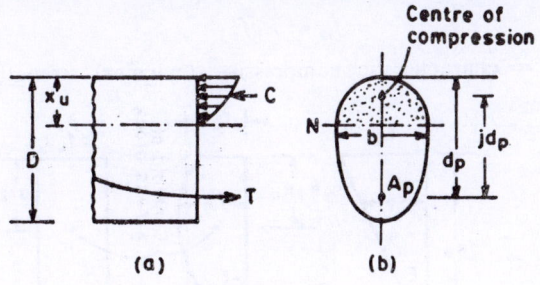

(a)   **(b)**

Fig. 5.8.1

(ii) Proportion of Prestressing Steel

The ultimate moment increases with the total tension in prestressing steel which is the product of A_p and f_p. With higher proportion of prestressing steel, it is able to develop a larger tensile force.

(iii) *Grade of Concrete*

With stronger concrete, the area in compression zone decreases. Consequently, x_u decreases and jd_p increases. It follows that the ultimate moment M_u increases for higher grade concrete.

(iv) *Type of Section*

In a bonded beam the stress in prestressing steel f_p at ultimate moment at the critical section is greater than that in an unbonded beam as discussed in Sec. 5.7. Consequently, the ultimate moment in a bonded beam is larger than that in an unbonded beam.

(v) *Physical Properties of Concrete and Steel*

The ultimate moment is affected by the stress-strain curves for concrete and prestressing steel. It is particularly influenced by the ultimate or crushing strain of concrete ϵ_u and the ultimate stress of prestressing steel f_{pu}.

5.8.1 Stress Block

As the strain in concrete varies linearly from zero at the neutral axis to the ultimate crushing strain ϵ_u at the top extreme fibre, the stress block which shows the variation of the compressive stress in concrete, is the same as the stress-strain curve for concrete. The stress-strain relationship for concrete is discussed in Sec. 2 2.2.4. As the load on a simply supported beam is increased progressively, the flexural cracks which commence at the bottom fibre propagate upwards and the neutral axis shifts upwards until the limit state of collapse is reached. Figure 5 8.2 (a) shows the strain diagram for concrete in which the strain varies linearly from zero at the level of the neutral axis to the ultimate crushing strain ϵ_u at the top fibre. The general stress block is shown in Fig. 5.8.2 (b) in which the stress f at any strain ϵ depends upon the stress-strain curve of Fig. 2.2.1. The general stress block may be characterised by the constants k_1, k_2 and k_3. The maximum stress f_m corresponding to a strain ϵ_o may be expressed as

$$f_m = k_3 f_c' \tag{5.8.2}$$

where $f_c' = $ characteristic compressive (cylinder) strength of concrete

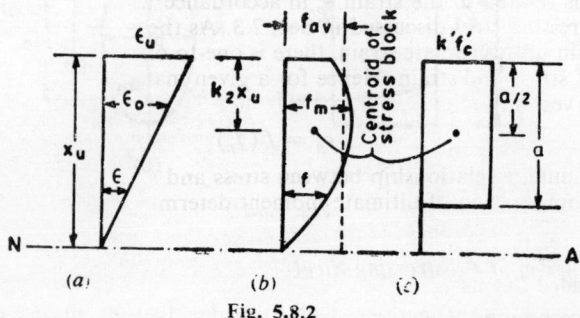

Fig. 5.8.2

The average ordinate of the stress block f_{av} may be expressed as

$$f_{av} = k_1 f_m = k_1 k_3 f'_c \qquad (5.8.3)$$

The distance $\bar{x}$ of the centroid of the stress block from the top face,

$$\bar{x} = k_2 x_u \qquad (5.8.4)$$

where x_u = depth of neutral axis at ultimate stage

For simplicity in the computation of ultimate flexural strength, Whiteney replaced the general stress block of Fig. 5.8.2 (b) by the rectangular stress block shown in Fig. 5.8.2 (c) so that the area and the distance of the centroid from the top face are the same for the two stress blocks. Hence,

$$a = 2 k_2 x_u \qquad (5.8.5)$$

Equating the areas of two stress blocks,

$$k' = \frac{k_1 k_3}{2 k_2} \qquad (5.8.6)$$

If the constants k_1, k_2 and k_3, which depend upon the shape of the stress-strain curve, are known, the ordinate and depth of the equivalent rectangular stress block can be determined. For instance, if $k_1 = k_3 = 0.85$ and $k_2 = 0.425$, the ordinate and depth of the equivalent rectangular stress block are equal to $0.85 \, f'_c$ and $0.85 \, x_u$ respectively. Due to its simplicity and also due to the fact that the shape of the stress block has only a marginal effect on the ultimate moment, the rectangular stress block has been adopted by several codes of practice.

5.8.2   Stress-Strain Curve for Prestressing Steel

Equation (5.8.1) shows that the stress in prestressing steel at collapse of the beam is the key to the problem of determination of the ultimate strength in flexure. If the steel stress is known, the position of neutral axis and the internal lever arm can be determined by using Navier's hypothesis and compatibility of strains. As discussed in Sec. 5.7, the stress in prestressing steel f_p is equal to the ultimate stress of the material f_{pu} only for low proportions of prestressing steel. In most practical beams, the steel stress at collapse is smaller than f_{pu} as failure occurs by crushing of concrete. The stress f_p is related to the strain ϵ_p in accordance with the stress-strain curve for prestressing steel discussed in Sec. 2.3. As the stress rises continuously with strain upto ultimate strain, there is one-to-one correspondence between values of stress and strain. Hence for a given material, f_p is a function of ϵ_p and vice versa,

$$\epsilon_p = f(f_p) \qquad (5.8.7)$$

It is this unique relationship between stress and strain which is the basis for the computation of ultimate moment determined from strain compatibility.

5.8 3   Bonded Beams

In a bonded beam, the maximum stress in prestressing steel occurs at the section of maximum bending moment. As the transverse load on the beam is increased progressively beyond the cracking load, the flexural cracks propagate upwards thereby reducing the area in the compression zone. At the same time the total compression C increases, because it is equal to the total tension T for equilibrium at any stage of loading.

$$C = T = A_p f_p \qquad (5.8.8)$$

It follows that the intensity of stress in concrete in the compression zone increases progressively until the concrete crushes, unless the prestressing tendons reach the breaking point earlier. If the depth of the neutral axis at the section of maximum moment is x_u, the internal lever arm at incipient collapse of the beam is

$$jd_p = (d_p - k_2 x_u) \qquad (5.8.9)$$

Hence the ultimate moment may be expressed as

$$M_u = A_p f_p (d_p - k_2 x_u) \qquad (5.8.10)$$

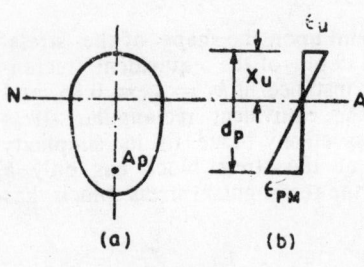

(a)   (b)

Fig. 5.8.3

If the failure occurs due to breaking of prestressing tendons, $f_p = f_{pu}$. In this case, the depth of the neutral axis in an arbitrary section shown in Fig. 5.8.3 (a) may be determined from the equation,

$$T = A_p f_{pu} = C = \int_0^{x_u} f b \, dx \qquad (5.8.11)$$

in which b is the variable breadth of the section in the compression zone and f is compressive stress in concrete in accordance with the general stress block.

When the failure occurs due to crushing of concrete before the tendons reach their breaking stress, $f_p < f_{pu}$. In this case f_{pu} may be determined from the compatibility of strains. The net strain ϵ_{pM} caused by the applied bending moment M_u may be expressed as

$$\epsilon_{pM} = \epsilon_p - \epsilon_{pe} \qquad (5.8.12)$$

in which ϵ_{pe} is the initial strain in prestressing steel before the commencement of loading. Hence, referring to Fig. 5.8.3 (b), the condition of compatibility of strains gives,

$$\frac{\epsilon_u}{x_u} = \frac{\epsilon_{pM}}{(d_p - x_u)} = \frac{\epsilon_p - \epsilon_{pe}}{d_p - x_u} \qquad (5.8.13)$$

Substituting from Eq. (5.8.7), the strain compatibility condition may be expressed as

$$\frac{\epsilon_u}{x_u} = \frac{f(f_p) - \epsilon_{pc}}{d_p - x_u} \qquad (5.8.14)$$

The depth of neutral axis x_u and the stress in prestressing steel f_p may be determined from solving Eq. (5.8.11) and (5.8.14) simultaneously. Due to the non-linearity of the stress strain curves for concrete and steel, it is more convenient to adopt the iterative procedure described by the following steps:

(i) Assume a reasonable value of stress in prestressing steel f_p at collapse of the beam.

(ii) Determine the strain ϵ_p using the stress-strain curve for prestressing steel.

(iii) Determine the net strain ϵ_{pM} due to bending moment M_u from Eq. (5.8.12).

(iv) Determine x_u from Eq. (5.8.13) using the specified or assumed value of ϵ_u.

(v) Determine the total compression C from Eq. (5.8.11).

(vi) As $C = T$. determine the stress in prestressing steel, $f_p = T/A_p$.

(vii) Compare f_p determined in step (vi) with the stress assumed in step (i). If the difference is substantial, iterate until f_p is determined to the required accuracy.

(viii) Knowing f_p, determine ultimate moment M_u from Eq. (5.8.10).

The foregoing general analysis can be used for any arbitrary cross-section. In common practice, prestressed concrete beams have rectangular, flanged or box sections.

(A) Rectangular Section

Consider a rectangular section shown in Fig. 5.8.4. If the failure occurs by breaking of prestressing steel, $f_p = f_{pu}$. In this case the equality of total tension and total compression gives,

$$A_p f_{pu} = b x_u k_1 k_3 f'_c \qquad (5.8.15)$$

Hence,
$$x_u = \frac{A_p f_{pu}}{b k_1 k_3 f'_c} \qquad (5.8.16)$$

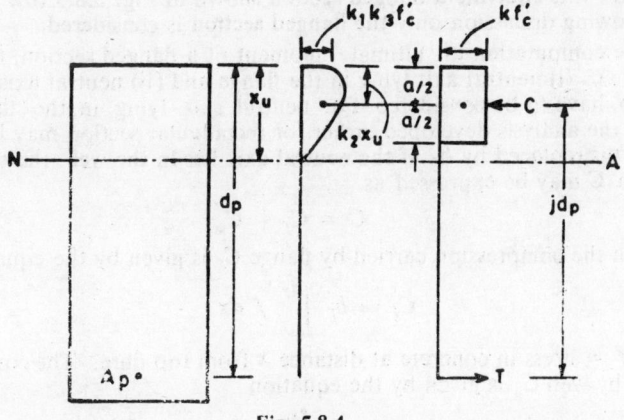

Fig. 5.8.4

The ultimate moment is given by

$$M_u = A_p f_{pu} (d_p - k_2 x_u)$$ (5.8.17)

in which x_u may be determined from Eq. (5.8.16).

If Whitney's rectangular stress block is adopted, the depth of the stress block,

$$a = \frac{A_p f_{pu}}{b (k' f_c')}$$ (5.8.18)

Hence the ultimate moment may be computed from

$$M_u = A_p f_{pu} \left(d_p - \frac{a}{2} \right)$$ (5.8.19)

If the failure occurs by crushing of concrete without breaking of steel $f_p < f_{pu}$. The equality of compressive and tensile forces gives

$$A_p f_p = b x_u (k_1 k_3 f_c')$$ (5.8.20)

This equation contains two unknowns, viz. f_p and x_u. A second equation relating the two unknowns is obtained from the strain compatibility Eq. (5.8.14). The unknowns may be determined by solving the two equations simultaneously or by adopting the iterative procedure discussed earlier. The ultimate moment may then be determined from Eq. (5.8.10). If Whitney's stress block is used, Eq. (5.8.20) takes the form

$$A_p f_p = b a (k' f_c')$$ (5.8.21)

and the ultimate moment may be expressed as

$$M_u = A_p f_p \left(d_p - \frac{a}{2} \right)$$ (5.8.22)

(B) Flanged and Box Sections

In the analysis for flexure, the box section shown in Fig. 5.8.5 (a) may be replaced by its equivalent flanged section shown in Fig. 5.8.5 (b). Hence in the following discussion only the flanged section is considered.

In the computation of ultimate moment of a flanged section, two possibilities, viz. (i) neutral axis lying in the flange and (ii) neutral axis lying in the web, have to be considered. For neutral axis lying in the flange, i.e. $x_u < d_f$, the analysis developed earlier for rectangular section may be used in which b is replaced by b_f. If the neutral axis lies in the web, the total compression C may be expressed as

$$C = C_f + C_w$$ (5.8.23)

in which the compression carried by flange C_f is given by the equation,

$$C_f = b_f \int_0^{d_f} f \, dx$$ (5.8.24)

where, $f = $ stress in concrete at distance x from top fibre. The compression carried by web C_w is given by the equation

$$C_w = b_w \int_{d_f}^{x_u} f \, dx$$ (5.8.25)

The centroids of the compressive forces C_f and C_w from the top face are

$$\bar{x}_f = \frac{\int_0^{d_f} fx\,dx}{\int_0^{d_f} f\,dx} \qquad (5.8.26)$$

and

$$\bar{x}_w = \frac{\int_{d_f}^{x_u} fx\,dx}{\int_{d_f}^{x_u} f\,dx} \qquad (5.8.27)$$

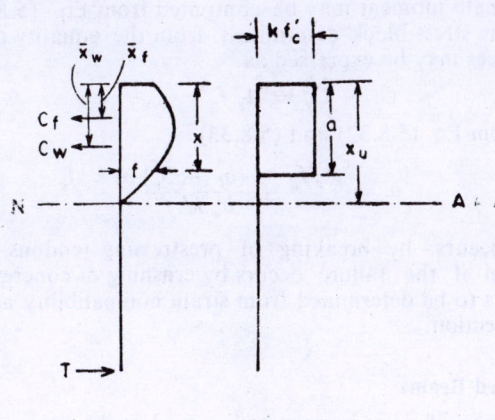

Fig. 5..5

In Eq. (5.8.24) to (5.8.27), the compressive stress in concrete f should be expressed as function of x measured from the top face. The depth of the neutral axis may be determined from

$$A_p f_p = C = C_f + C_w \qquad (5.8.28)$$

The ultimate moment may be determined from

$$M_u = C_f (d_p - \bar{x}_f) + C_w (d_p - \bar{x}_w) \qquad (5.8.29)$$

in which C_f and C_w are given by Eq. (5.8.24) and (5.8.25) respectively.

The use of Whitney's rectangular stress block simplifies the computations considerably. For neutral axis located in the flange, the equality of compressive and tensile forces gives

$$a = \frac{A_p f_p}{b_f k' f_c'} \qquad (5.8.30)$$

and the ultimate moment may be expressed as

$$M_u = A_p f_p \left(d_p - \frac{a}{2} \right) \qquad (5.8.31)$$

For neutral axis lying in the web as shown in Fig. 5.8.5,

$$C_f = b_f d_f k' f_c' \qquad (5.8.32)$$

$$C_w = b_w (a - d_f) k' f_c' \qquad (5.8.33)$$

$$\bar{x}_f = \frac{d_f}{2} \qquad (5.8.34)$$

$$\bar{x}_w = d_f + \frac{(a - d_f)}{2} = \frac{a + d_f}{2} \qquad (5.8.35)$$

Hence the ultimate moment may be computed from Eq. (5.8.29) in which the depth of the stress block determined from the equality of compressive and tensile forces may be expressed as

$$C_f + C_w = A_p f_p \qquad (5.8.36)$$

Substituting from Eq. (5.8.32) and (5.8.33),

$$a = \frac{A_p f_p - (b_f - b_w) d_f k' f_c'}{b_w k' f_c'} \qquad (5.8.37)$$

If the failure occurs by breaking of prestressing tendons, $f_p = f_{xu}$. On the other hand if the failure occurs by crushing of concrete, f_p which is less than f_{pu}, has to be determined from strain compatibility as discussed for a rectangular section.

5.8.4 Unbonded Beams

The determination of stress in prestressing steel at the section of maximum bending moment is far more difficult than in the corresponding bonded beam. In a bonded beam, the strain in prestressing steel can be determined from strain compatibility and Navier's hypothesis. In an unbonded beam, the strain in prestressing steel at the section of maximum moment is not the same as in surrounding concrete. Instead, it is equal to the strain in concrete at the level of prestressing steel integrated over the entire span of the beam. Before cracking of the beam, the average strain can be evaluated by integration using elastic theory as explained in Sec. 5.7. But in the post-cracking range, the elastic theory is no longer applicable. The complexity of the problem increases because the extent of cracking and the position of neutral axis vary from section to section. The strain in concrete at the level of prestressing steel depends not only on the variation of bending moment and shape of the cable line but also on the extent of cracking of the beam. It is, however, evident that the stress in prestrsssing steel at the section of maximum moment in an unbonded beam at collapse is smaller than that in the corresponding bonded beam. The lower stress in prestressing steel in an unbonded beam at failure is primarily responsible for its lower ultimate moment which may be 10 to 30 per cent smaller than that of the corresponding bonded beam. The tests have shown that the increase in stress in prestressing steel in unbonded beams ranged from 70 to 550 MPa due to the application of ultimate load. When concrete reached its crushing strain, the stress f_p was observed to be far below the ultimate stress f_{pu}.

In the post-cracking range, the structural action of an unbonded beam with straight tendons is very similar to that of a tied arch. Referring to Fig. 5.8.6 (a), the shaded portion in the compression zone acts like the arch rib and the tendons behave like the tie. The compressive force in the rib and the equal tensile force in the tie constitute the internal couple which opposes the external bending moment. Just as the tie prevents the spreading of the arch rib, the tendons anchored at their ends restrict the spreading of the arch rib formed by the concrete in the compression zone.

In an unbonded beam with substantially curved tendons shown in Fig. 5.8.6(b), the structural response of the tendons is similar to that of a sus-

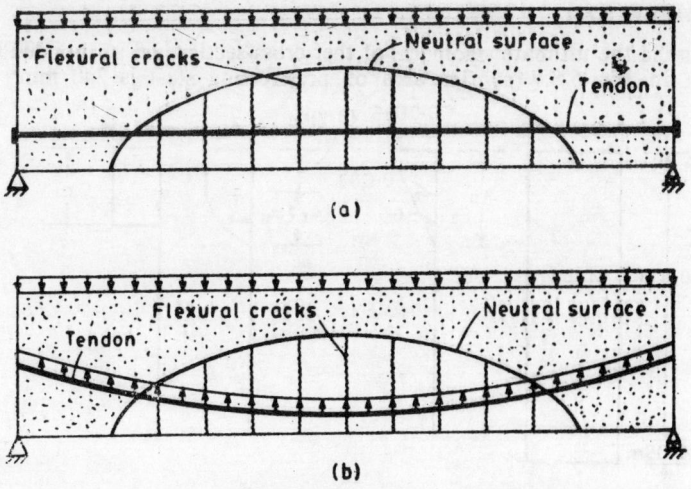

(a)

(b)

Fig. 5.8.6

pension cable. The tendons anchored at their ends tend to straighten out, thereby pushing the concrete beam upwards as conceived in the load balancing approach. As the sag of the cable is small compared to the span, the stress in the tendon is practically the same over the entire length. As the stage of collapse is approached, the structural response of an unbonded beam with curved tendons too approaches that of a tied arch. In either case, the failure occurs due to crushing of the concrete arch rib in the compression zone.

The analysis for ultimate strength for bonded beams discussed in Sec. 5.8.3 may be used for unbonded beams if the stress in tendons at collapse of the beam is known. Due to the difficulty of predicting the tendon stress by theoretical approach, the following empirical expression based on tests at several research centres has been recommended:

$$f_p = f_{pe} + 69 + \frac{1.4 f'_c}{100 \, \rho_p} \qquad (5.8.38)$$

where f_p = stress in prestressing steel at ultimate moment

ρ_p = proportion of prestressing steel

= $\dfrac{A_p}{b d_p}$ for rectangular sections

= $\dfrac{A_p}{b_f \, d_p}$ for flanged sections

f_{pe} = effective stress in prestressing steel at commencement of loading

b = width of compression zone

d_p = effective depth

EXAMPLE 5.8.1

Determine the ultimate moment of the cross-section of a bonded beam shown in Fig. 5.8.7 (a). The area of prestressing steel is 240 mm². The

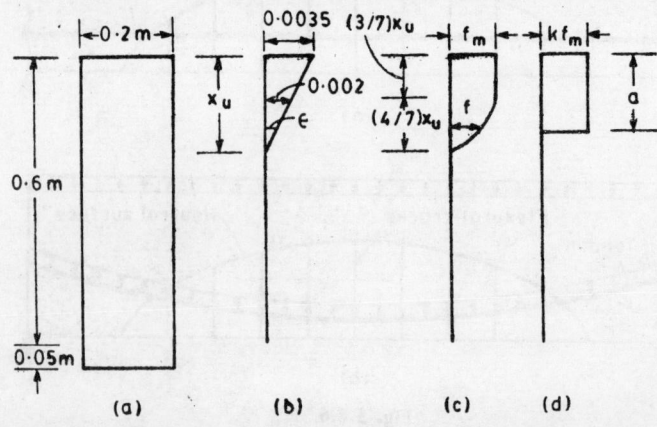

Fig. 5.8.7

failure occurs by breaking of prestressing tendons. The ultimate stress of prestressing steel, $f_{pu} = 1600$ MPa. Use the idealized stress block or its equivalent rectangular stress block shown in Fig. 5.8.7 (c) and (d) corresponding to the strain diagram of Fig. 5.8.7 (b). Take $f_m = 35$ MPa.

Solution

As the failure occurs by breaking of tendons,

$$T = A_p f_{pu} = 0.000240 \times 1600 = 0.384 \text{ MN}$$

$$C = T = 0.384 \text{ MN}$$

(i) Using Idealized Stress Block

$$\text{Area of stress block} = f_m \left(\frac{3}{7} x_u\right) + f_m \times \frac{2}{3} \left(\frac{4}{7} x_u\right) \tag{a}$$

$$= 0.81 f_m x_u$$

Hence average ordinate of stress block,

$$f_{av} = 0.81 f_m$$

Distance of centroid of stress block from top,

$$k_2 x_u = \frac{f_m \left(\frac{3}{7} x_u\right)\left(\frac{3}{14} x_u\right) + f_m \left(\frac{2}{3} \times \frac{4}{7} x_u\right)\left(\frac{3}{7} x_u + \frac{3}{8} \times \frac{4}{7} x_u\right)}{0.81 f_m x_u}$$

$$= 0.416 x_u \tag{b}$$

Hence depth of neutral axis,

$$x_u = \frac{C}{bf_{av}} = \frac{0.384}{0.20 \times 0.81 \times 35} = 0.0677 \text{ m}$$

The ultimate moment of the section,

$$M_u = T jd_p = T(d_p - k_2 x_u)$$

$$= 0.384 (0.60 - 0.416 \times 0.0677)$$

$$= 0.2196 \text{ MN·m}$$

(ii) Using Equivalent Rectangular Stress Block

The centroid of the rectangular stress block will be at the same depth as that of the idealized stress block if

$$a = 2k_2 x_u = 2 \times 0.416 \times 0.0677 = 0.0564 \text{ m}$$

Equating the areas of the two stress blocks, the ordinate of the rectangular stress block,

$$k f_m = \frac{0.81 \times 0.0677 \times f_m}{0.0564} = 0.972 f_m$$

Hence the internal lever arm,

$$jd_p = \left(d_p - \frac{a}{2}\right) = 0.60 - 0.0282 = 0.5718 \text{ m}$$

and $M_u = Cjd_p = 0.20 \times 0.0564 \times 0.972 \times 35 \times 0.5718$

$$= 0.2194 \text{ MN.m}$$

EXAMPLE 5.8.2

Determine the ultimate moment of the rectangular cross-section of a bonded beam shown in Fig. 5.8.7 (a), given that the area of prestressing steel, $A_p = 960$ mm². Use the stress block of Fig. 5.8.7 (c) and the idealized stress-strain curve for prestressing steel shown in Fig. 5.8.8. Failure occurs by crushing of concrete without breaking of tendons. The effective prestress in steel at the commencement of loading, $f_{pe} = 1100$ MPa. Take $f_m = 35$ MPa.

Solution

The stress in steel at collapse may be determined from strain compatibility and the ultimate moment may be computed by iteration.

To start the first cycle of iteration, assume $f_p = 1350$ MPa. The corresponding strain from stress-strain curve of Fig. 5.8.8 is

$$\epsilon_p = 0.006 + \frac{(0.0125 - 0.006)(1350 - 1200)}{(1500 - 1200)} = 0.00925$$

As $f_{pe} = 1100$ MPa, the effective prestrain at the commencement of loading,

$$\epsilon_{pe} = \frac{1100}{2 \times 10^5} = 0.0055$$

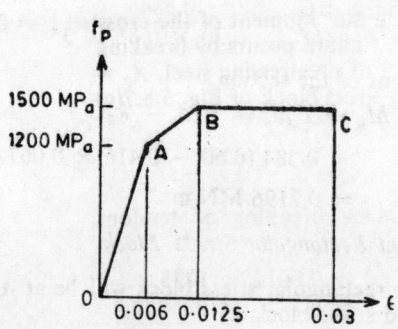

Fig. 5.8.8

Hence the net strain due to bending moment M,

$$\epsilon_{pm} = \epsilon_p - \epsilon_{pe} = 0.00925 - 0.0055 = 0.00375$$

Taking crushing strain of concrete, $\epsilon_u = 0.0035$ and using Eq. (5.8.13),

$$x_u = 0.2897 \text{ m}$$

As the average ordinate of the stress block from Eq. (a) of Example 5.8.1 is $0.81 f_m$, the total compression,

$$C = 0.20 \times 0.2897 \times 0.81 \times 35 = 1.6426 \text{ MN}$$

As $C = T$, the stress in prestressing steel,

$$f_p = \frac{T}{A_p} = \frac{C}{A_p} = \frac{1.6426}{0.000960} = 1711 \text{ MPa}$$

For the next iteration, f_p may be taken equal to 1400 MPa which gives f_p equal to 1488 MPa at the end of the cycle. In the third cycle f_p may be taken equal to 1420 MPa which gives $f_p = 1414$ MPa at the end of the cycle. Further iterations indicate that $f_p = 1418.66$ MPa approximately and $x_p = 0.2403$ m.

Hence from Eq. (a) of Example 5.8.1,

$$C = 0.20 \times 0.2403 \times 0.81 \times 35 = 1.3625 \text{ MN}$$

From Eq. (b) of Example 5.8.1,

$$k_2 x_u = 0.416 \times 0.2403 = 0.10 \text{ m}$$

Hence the ultimate moment,

$$M_u = C (d_p - k_2 x_u) = 1.3625 (0.60 - 0.10)$$
$$= 0.68125 \text{ MN·m}$$

It may be verified that the same ultimate moment is obtained from the equivalent rectangular stress block of Fig. 5.8.7 (d).

EXAMPLE 5.8.3

Determine the ultimate moment of the cross-section of a bonded beam shown in Fig. 5.8.9. Failure occurs by breaking of tendons at a stress of 1725 MPa· The area of prestressing steel, $A_p = 1200$ mm². Use the stress block of Fig. 5.8.7(c). Take $f_m = 40$ MPa.

Solution

As the failure occurs by breaking of tendons, $f_p = f_{pu}$. Hence,

$$C = T = A_p f_{pu} = 0.0012 \times 1725$$
$$= 2.07 \text{ MPa}$$

Assuming that the neutral axis lies in the flange and noting from Example 5.8.1 that the average stress, $f_{av} = 0.81 f_m$,

$$x_u = \frac{C}{b_f f_{av}} = \frac{2.07}{0.75 \times 0.81 \times 40}$$
$$= 0.0852 \text{ m}$$

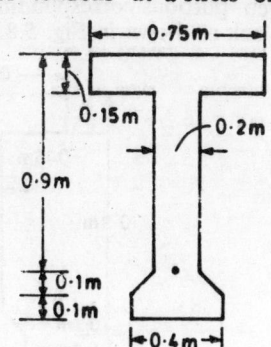

Fig. 5.8..9

Hence the neutral axis lies in the flange as assumed. From Example 5.8.1,

$$k_2 \, x_u = 0.416 \, x_u = 0.0354 \text{ m}$$

Hence the ultimate moment of the section,

$$M_u = T \, (d_p - k_2 \, x_u)$$
$$= 2.07 \, (0.90 - 0.0354)$$
$$= 1.7897 \text{ MN·m}$$

It may be verified that the same value of ultimate moment is obtained if the equivalent rectangular stress block of Fig. 5.8.7 (d) is used.

EXAMPLE 5.8.4

Compute the ultimate moment of the beam of Example 5.8.3 if the area of prestressing steel, $A_p = 3300 \text{ mm}^2$. Use the stress block of Fig. 5.8.7 (c).

Solution

$$C = T = A_p f_{pu} = 0.0033 \times 1725 = 5.6925 \text{ MN}$$

Let it be assumed that the neutral axis lies in the web and that the parabolic part of the stress block is located entirely in the web. This assumption is valid only if,

$$d_f \leqslant \left(\frac{3}{7} \, x_u\right) \quad \text{or} \quad x_u \geqslant 0.35 \text{ m}$$

The total compression in concrete may be expressed as

$$C = C_f + C_{w1} + C_{w2} \tag{a}$$

in which C_f, C_{w1} and C_{w2} are the compressive forces in the flange and the web portions corresponding to rectangular and parabolic parts of the stress block as shown in Fig. 5.8.10.

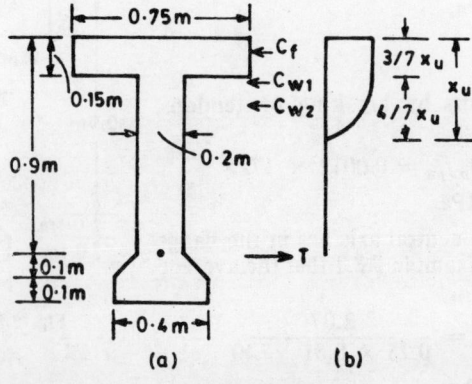

Fig. 5.8.10

$$C_f = 0.75 \times 0.15 \times 40 = 4.5 \text{ MN}$$

$$C_{w1} = 0.2 \left(\frac{3}{7} x_u - 0.15 \right) \times 40 = (3.429\, x_u - 1.2) \text{ MN}$$

$$C_{w2} = 0.2 \times \frac{4}{7} x_u \times 40 \times \frac{2}{3} = 3.048\, x_u \text{ MN}$$

Substituting into Eq. (a),

$$5.6925 = 4.5 + 3.429\, x_u - 1.2 + 3.048\, x_u$$

or

$$x_u = 0.3694 \text{ m}$$

As $x_u > 0.35$ m, the parabolic part of the stress block lies entirely in the web as assumed earlier. In case $x_u < 0.35$ m, the parabolic portion of the stress block is truncated and the corresponding forces will have to be determined by integration.

Putting $x_u = 0.3694$ m,

$$C_f = 4.5 \text{ MN} \qquad C_{w1} = 0.0666 \text{ MN} \qquad C_{w2} = 1.1259 \text{ MN}$$

The distances of the centroids of forces C_f, C_{w1} and C_{w2} from the top are

$$\bar{x}_f = 0.075 \text{ m}$$

$$\bar{x}_{w1} = 0.15 + \left(\frac{3}{7} x_u - 0.15 \right) \times \frac{1}{2} = 0.1542 \text{ m}$$

$$\bar{x}_{w2} = \frac{3}{7} x_u + \frac{3}{8} \times \frac{4}{7} x_u = 0.2375 \text{ m}$$

Hence the ultimate moment of the section,

$$M_u = C_f (d_p - \bar{x}_f) + C_{w1} (d_p - \bar{x}_{w1}) + C_{w2} (d_p - \bar{x}_{w2})$$
$$= 4.5 (0.90 - 0.075) + 0.0666 (0.90 - 0.1542)$$
$$+ 1.1259 (0.90 - 0.2375)$$
$$= 4.50\text{?}1 \text{ MN·m}$$

EXAMPLE 5.8.5

Using the equivalent rectangular stress block of Fig. 5.8.7 (d), determine the ultimate moment of the beam of Example 5.8.4.

Solution

$$C = T = 5.6925 \text{ MN as in Example 5.8.4}$$

Let it be assumed that the depth of rectangular stress block a is greater than d_f. Hence the total compressive force,

$$C = [b_f\, d_f + b_w (a - d_f)]\, k f_m \qquad (a)$$

From Example 5.8.1, the average ordinate of stress block,

$$k f_m = 0.972 f_m = 38.88 \text{ MPa}$$

on substitution in Eq. (a),

$$a = 0.3196 \text{ m}$$

Hence the ultimate moment of the section,

$$M_u = b_f \, d_f \, kf_m \, (d_p - 0.5 \, d_f) + b_w \, (a - d_f) \, kf_m \left(d_p - d_f - \frac{a - d_f}{2} \right)$$

$$= 4.4858 \; \text{MN·m}$$

This value is slightly lower than that obtained in Example 5.8.4. The difference arises because the ordinate of equivalent rectangular stress block is computed for a rectangular section. Consequently, a small error is introduced when the rectangular stress block is used for a flanged section.

EXAMPLE 5.8.6

Determine the ultimate moment of the cross section of a bonded beam shown in Fig. 5.8.9 if the failure occurs due to crushing of concrete. The area of prestressing steel, $A_p = 3600$ mm^2 and the effective prestress before commencement of loading, $f_{pe} = 960$ MPa. Use the equivalent rectangular stress block of Fig. 5.8.7 (d) and the idealized stress-strain curve for prestressing steel shown in Fig. 5.8.8. Take $f_m = 40$ MPa.

Solution

As the failure occurs due to crushing of concrete, the stress in prestressing steel at collapse may be determined by iterative procedure. Assume $f_p = 1450$ MPa. The strain in prestressing steel according to stress-strain curve of Fig. 5.8.8,

$$\epsilon_p = 0.0060 + \frac{(0.0125 - 0.0060)}{(1500 - 1200)} (1450 - 1200) = 0.01142$$

The strain ϵ_{pe} corresponding to stress f_{pe},

$$\epsilon_{pe} = \frac{960}{2 \times 10^5} = 0.00480$$

Hence the net strain caused by applied moment,

$$\epsilon_{pM} = \epsilon_p - \epsilon_{pe} = 0.00662$$

Hence from Eq. (5.8.13),

$$x_u = 0.3113 \; \text{m}$$

$$a = 2k_2 \, x_u = 2 \times 0.416 \times 0.3113 = 0.2590 \; \text{m}$$

$$C = [b_f \, d_f + b_w \, (a - d_f)] \, kf_m$$

$$= [0.75 \times 0.15 + 0.2 \, (0.2590 - 0.15)] \times 0.972 \times 40$$

$$= 5.2216 \; \text{MN}$$

$$f_p = \frac{T}{A_p} = \frac{C}{A_p} = \frac{5.2216}{0.0036} = 1450.4 \; \text{MPa}$$

As this value of f_p is very close to the value assumed in the beginning, no further iteration is required.

$$M_u = T \left(d_p - \frac{a}{2} \right) = 5.2216 \left(0.90 - \frac{0.2590}{2} \right)$$
$$= 4.0232 \text{ MN} \cdot \text{m}$$

EXAMPLE 5.8.7

Determine the ultimate moment of the cross section of an unbonded beam shown in Fig. 5.8.11 (a) using Whitney's rectangular stress block of Fig. 5.8.11(b). The area of prestressing steel $A_p = 240$ mm^2 and $d_p = 0.6$ m. The compressive (cylinder) strength of concrete $f'_c = 35$ MPa. The effective stress at the commencement of loading $f_{pe} = 1100$ MPa.

Solution

$$\rho_p = \frac{A_p}{bd_p} = \frac{0.00024}{0.2 \times 0.6} = 0.002$$

Using Eq. (5.8.38), the stress in prestressing steel at collapse,

$$f_p = 1100 + 69 + \frac{1.4 \times 35}{100 \times 0.002}$$
$$= 1414 \text{ MPa}$$

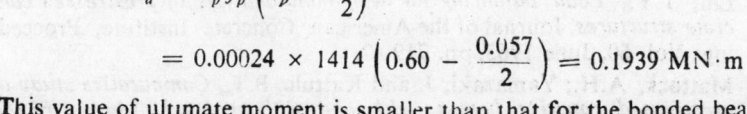

Depth of rectangular stress block,

$$a = \frac{A_p f_p}{b (0.85 f'_c)} = \frac{0.00024 \times 1414}{0.2 (0.85 \times 35)}$$
$$= 0.0570 \text{ m}$$

(a)   **(b)**

Fig. 5.8.11

$$M_u = A_p f_p \left(d_p - \frac{a}{2} \right)$$
$$= 0.00024 \times 1414 \left(0.60 - \frac{0.057}{2} \right) = 0.1939 \text{ MN} \cdot \text{m}$$

This value of ultimate moment is smaller than that for the bonded beam of Example 5.8.1. Direct comparison, however, is not possible because different stress blocks have been used in the two examples.

EXAMPLE 5.8.8

Determine the ultimate moment of the cross section of an unbonded beam shown in Fig. 5.8.9. Given that $A_p = 2700$ mm^2, $f_{pe} = 1100$ MPa and $f'_c = 35$ MPa. Use the rectangular stress block of Fig. 5.8.11 (b).

Solution

$$\rho_p = \frac{A_p}{b_f d_p} = \frac{0.0027}{0.75 \times 0.90} = 0.004$$

Using Eq. (5.8.36),

$$f_p = 1100 + 69 + \frac{1.4 \times 35}{100 \times 0.004} = 1291.5 \text{ MPa}$$

Equating total tension and total compression,

$$A_p f_p = [b_f d_f + b_w (a - d_f)] (0.85 f_c')$$

or $0.0027 \times 1291.5 = \|0.75 \times 0.15 + 0.2 (a - 0.15)] (0.85 \times 35)$

or $a = 0.1736$ m

$$M_u = b_f d_f (0.85 f_c') \left(d_p - \frac{d_f}{2}\right)$$

$$+ b_w (a - d_f) (0.85 f_c') \left(d_p - d_f - \frac{a - d_f}{2}\right)$$

$$= 2.8648 \text{ MN m}$$

REFERENCES

5.1 Billet, D.F. and Appleton, J.H., *Flexural strength of prestressed concrete beams*, Journal of the American Concrete Institute, Proceedings Vol. 50, June 1954, pp. 837-54.

5.2 Evans, R.H., *Extensibility and modulus of rupture of concrete*, The Structural Engineer (London), Vol. 24, No. 12, December 1946, pp. 636-59.

5.3 Hognestad E.; Hanson, H.W. and McHenry, D., *Concrete stress distribution in ultimate strength design*, Journal of the American Concrete Institute, Proceedings Vol. 52, December 1955, pp. 455-79.

5.4 Janney, J.R; Hognestad, E. and Mc Henry, D., *Ultimate flexural strength of prestressed and conventionally reinforced concrete beams*, Journal of the American Concrete Institute, Proceedings Vol. 52, February 1956, pp. 601-20.

5.5 Lin, T.Y., *Load balancing for design and analysis of prestressed concrete structures*, Journal of the American Concrete Institute, Proceedings Vol. 60, June 1963, pp. 719-42.

5.6 Mattock, A.H.; Yamazaki, J. and Kattule, B.T., *Comparative study of prestressed concrete beams, with and without bond*, Journal of the American concrete Institute, Proceedings Vol. 68, February 1971, pp. 116-25.

5.7 Pandit, G.S. and Gupta, S.P., *Free body approach for analysis of prestressed concrete beams*, The Indian Concrete Journal, Vol. 54, No. 6, June 1980, pp. 154-61.

5.8 Rogers, G.L., *Validity of certain assumptions in the mechanics of prestressed concrete*, Journal of the American Concrete Institute, Proceedings Vol. 49, December 1953, pp. 317-30.

5.9 Rusch, H., *Researches towards a general flexural theory for structural concrete*, Journal of the American Concrete Institute, Proceedings Vol. 57, July 1960.

5.10 *Tentative recommendations for prestressed concrete*, Journal of the American Concrete Institue, Proceedings Vol. 54, January 1958, pp. 545-78.

5.11 Warwaruk, J.; Sozen, M.A. and Siess, C.P., *Strength and behaviour in flexure of prestressed concrete beams*, Engineering Experiment Station Bulletin No. 464, University of Illinois, 1962.

5.12 Yamazaki, J., Kattula, B.T. and Mattock, A.H., *A comparison of the behaviour of post-tensioned prestressed concrete beams with and without bond*, Report SM 69-3, University of Washington, College of Engineering, Structures and Mechanics December 1969, 94 p.

PROBLEMS

5.1 A pretensioned beam of rectangular cross-section 0.30 × 0.90 m has straight prestressing tendons at an effective eccentricity of 0.35 m. At the initial stage the prestressing force is 1.89 MN and sagging bending moment is 0.378 MN·m. At the final stage the prestressing force reduces to 1.485 MN and the total sagging bending moment is 0.7425 MN·m. Determine the stress at the initial and final stages.

5.2 Verify the stresses of Prob. 5.1 by locating the position of line of thrust.

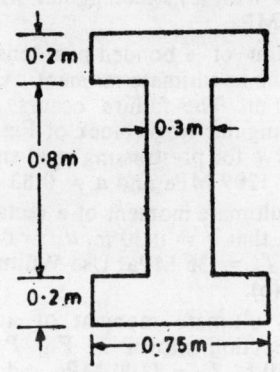

Fig. P 5.4

5.3 Verify the stresses of Prob. 5.1 using the free body approach.

5.4 A post-tensioned girder of 20 m span has a cross-section shown in Fig. P. 5.4. It has a parabolic cable line with eccentricities equal to 0.5 m, − 0.2 m and − 0.2 m at midspan and the two ends. At the initial stage the prestressing force is 3.78 MN and the dead load is 17.08 kN/m inclusive of self weight. At the final stage, the prestressing force reduces to 3.24 MN and the the total load is 50.16 kN/m. Determine the extreme fibre stresses at midspan section at the initial and final stages.

5.5 Verify the stresses of Prob. 5.4 by locating the position of line of thrust.

5.6 Verify the stresses of Prob. 5.4 using the free body approach.

5.7 Determine the decompression moment at the final stage for the beam of Prob. 5.1.

5.8 Determine the intensity of uniform load required to cause decompression at the final stage at midspan section of the girder of Prob. 5.4.

5.9 Determine the bending moment required to cause flexural cracking in the beam of Prob. 5.1. Given that the modulus of rupture $f_r = 4$ MPa.

5.10 Compute the intensity of loading required to cause flexural cracking at midspan in the girder of Prob. 5.4. Take $f_r = 4.5$ MPa.

5.11 Determine the ultimate moment of a rectangular cross-section 0.36×0.80 m of a pretensioned beam. Given that $A_p = 810$ mm² and $d_p = 0.75$ m. The failure occurs by breaking of tendons at a stress of 1700 MPa. Use the stress block of Fig. 5.8.7 in which $f_m = 32$ MPa.

5.12 Detemine the ultimate moment of the beam of Prob. 5.11 if $A_p = 1620$ mm². The failure occurs ue to crushing of concrete. The effective stress in prestressing steel at commencement of loading is 971 MPa. Use the stress block of Fig. 5.8.7 (c), and the idealised stress-strain curve for prestressing steel shown in Fig. 5.8.8.

5.13 Determine the ultimate moment of a bonded beam whose cross-section is shown in Fig. P 5.4. Given that $A_p = 1650$ mm² and $d_p = 1.10$ m. The failure occurs due to breaking of tendons at ultimate stress of 1620 MPa. Use Whitney's rectangular stress block of Fig. 5.8.11 (b). Take $f'_c = 32$ MPa.

5.14 The cross-section of a bonded post-tensioned beam is shown in Fig. P 5.4. Determine its ultimate moment. Given that $A_p = 3300$ mm² and $d_p = 1.10$ m. The failure occurs by crushing of concrete. Use Whitney's rectangular stress block of Fig. 5.8.11 (b) and the idealised stress-strain curve for prestressing steel shown in Fig. 5.8.8. Take $f'_c = 28$ MPa, $f_{pe} = 1209$ MPa and $a = 0.83\, x_u$.

5.15 Determine the ultimate moment of a rectangular section of an unbonded beam, given that $b = 0.30$ m, $d_p = 0.90$ m, $A_p = 810$ mm², $f_{pe} = 1050$ MPa and $f'_c = 36$ MPa. Use Whitney's rectangular stress block of Fig. 5.8.11 (b).

5.16 Determine the ultimate moment of an unbonded beam having a symmetrical l-section shown in Fig. P 5.4. Given that $A_p = 4125$ mm², $d_p = 1.10$ m, $f_p = 1100$ MPa and $f'_c = 36$ MPa. Use Whitney's rectangular stress block of Fig. 5.8.11 (b).

Design for Flexure

6.1 INTRODUCTION

The flexural moment is the most dominant internal force in the majority of prestressed concrete members. Although flexure is generally associated with shear and axial force and even torsion in some cases, it is flexure which largely controls the overall cross-sectional dimensions, the numbers and position of the prestressing tendons and the magnitude of the prestressing force. The decisive influence of flexure on the design parameters underlines the importance of flexure in the design process.

6.2 CLASSIFICATION OF BEAMS

For the purpose of design, the Indian Code IS: 1343-1980 classifies flexural members into following three types:

(i) Type 1 (Full Prestress)

Fully prestressed members are proportioned so that the maximum bending moment under any possible combination of working loads M_w is smaller than the decompression moment, $M_w < M_{dc}$. Consequently, the members are free from flexural axial tension in the entire working load range. In particular, the members do not have flexural tension at the two crucial stages, viz. the initial and the final stages.

(ii) Type 2 (Limited Prestress)

Members with limited prestress are proportioned so that the maximum bending moment under any possible combination of working loads M_w is larger than the decompression moment but smaller than the cracking moment, $M_{dc} < M_w < M_{cr}$. Consequently, these members develop flexural tension under full working load but it is not large enough to cause flexural cracking. The members, therefore, remain uncracked during the entire working range.

(iii) Type 3 (Partial Prestress)

Partially prestressed members are proportioned so that the maximum bending moment M_w under any possible combination of working loads is larger than the cracking moment $M_w > M_{cr}$. Consequently, these members undergo flexural cracking under full working load. In these members, the working moment M_w is resisted partly by prestressing tendons and partly by non-prestressed reinforcing bars.

The discussion in this chapter is restricted to beams of Types 1 and 2 which remain flexurally uncracked during the entire working range. Partially prestressed beams (Type 3) are discussed in a separate chapter (CHAPTER 10).

6.3 DESIGN PRINCIPLES

As pointed out in Sec. 1.13, the complexity of prestressed concrete design arises primarily from the multiplicity of loading stages. A satisfactory design must ensure acceptable performance under all limit states of strength and serviceability at different stages of loading. For the design of flexural members of Types 1 and 2, which remain flexurally uncracked in the entire working range, it is a common practice to proportion the members for the two critical working load stages, viz. the initial stage and the final stage, using the elastic or working load theory based on entire uncracked section. The chosen proportions are then checked for all other limit states related to strength and serviceability. In particular it should be checked that the ultimate strength of the member provides adequate safety against collapse as per the provisions of relevant codes of practice.

As pointed out in Sec. 5.2, the initial and final stages are the two critical stages in the working range. If σ_{cbci} and σ_{cti} are the maximum permissible compressive and tensile stresses at the initial stage and σ_{cbc} and σ_{ct} are the respective stresses at the final stage, the conditions to be satisfied for an acceptable design may be written as follows.

(i) *At initial stage*

$$f_T \leqslant \sigma_{cti} \qquad (6.3.1)$$

$$f_B \leqslant \sigma_{cbci} \qquad (6.3.2)$$

(ii) *At final stage*

$$f_T \leqslant \sigma_{cbc} \qquad (6.3.3)$$

$$f_B \leqslant \sigma_{ct} \qquad (6.3.4)$$

These conditions are sufficient for the derivation of expressions for the minimum values of section moduli to ensure acceptable stress conditions at the two critical stages.

6.4 EXPRESSIONS FOR SECTION MODULI

Consider a prestressed concrete member having an arbitrary section shown in Fig. 6.4.1 (a). At the initial stage, the tensile stress at top fibre due to prestress alone,

$$f_T \text{ (tensile)} = \frac{P_i e_p y_T}{I} - \frac{P_i}{A} = \frac{P_i}{A}\left[\frac{e_p y_T}{k^2} - 1\right] \qquad (6.4.1)$$

and the compressive stress at bottom fibre due to prestress alone,

$$f_B \text{ (compressive)} = \frac{P_i e_p y_B}{I} + \frac{P_i}{A} = \frac{P_i}{A}\left[\frac{e_p y_B}{k^2} + 1\right] \quad (6.4.2)$$

Similarly the tensile stress at top fibre and the compressive stress at bottom fibre due to prestress alone at the final stage may be expressed as

$$f_T \text{ (tensile)} = \frac{P e_p y_T}{I} - \frac{P}{A} = \frac{P}{A}\left[\frac{e_p y_T}{k^2} - 1\right] \quad (6.4.3)$$

$$f_B \text{ (compressive)} = \frac{P \epsilon_p y_B}{I} + \frac{P}{A} = \frac{P}{A}\left[\frac{e_p y_B}{k^2} + 1\right] \quad (6.4.4)$$

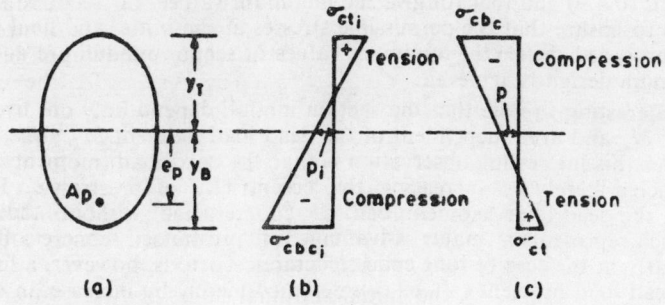

Fig. 6.4.1

The limiting or critical stress diagrams reflecting the conditions expressed by Eq. (6.3.1) to (6.3.4) are shown in Fig. 6.4.1 (b) and (c) for the initial and final stages respectively. Using Eq. (6.4.1) to (6.4.4), the four conditions of stress expressed by Eq. (6.3.1) to (6.3.4) may be written as

$$\frac{P_i}{A}\left[\frac{e_p y_T}{k^2} - 1\right] - \frac{M_D}{Z_T} \leqslant \sigma_{cti} \quad (6.4.5)$$

$$\frac{P_i}{A}\left[\frac{e_p y_B}{k^2} + 1\right] - \frac{M_D}{Z_B} \leqslant \sigma_{cbci} \quad (6.4.6)$$

$$-\frac{P}{A}\left[\frac{e_p y_T}{k^2} + 1\right] + \frac{M_D}{Z_T} + \frac{M_L}{Z_T} \leqslant \sigma_{cbc} \quad (6.4.7)$$

$$-\frac{P}{A}\left[\frac{e_p y_B}{k^2} + 1\right] + \frac{M_D}{Z_B} + \frac{M_L}{Z_B} \leqslant \sigma_{ct} \quad (6.4.8)$$

where
$$M_D = \text{dead load moment at initial stage}$$

$$M_L = \text{live load moment at final stage}$$

$$Z_T = \frac{I}{y_T} = \text{section modulus with respect to top fibre}$$

$$Z_B = \frac{I}{y_B} = \text{section modulus with respect to bottom fibre}$$

Adding Eq. (6.4.5) and (6.4.7) and solving for Z_T,

$$Z_T \geqslant \frac{M_L}{(\sigma_{cbc} + \sigma_{cti}) - \left(\dfrac{P_i - P}{A}\right)\left(\dfrac{e_p\, y_T}{k^2} - 1\right)} \tag{6.4.9}$$

Similarly adding Eq. (6.4.6) and (6.4.8) and solving for Z_B,

$$Z_B \geqslant \frac{M_L}{(\sigma_{cbci} + \sigma_{ct}) - \left(\dfrac{P_i - P}{A}\right)\left(\dfrac{e_p\, y_B}{k^2} + 1\right)} \tag{6.4.10}$$

Equations (6.4.9) and (6.4.10) give the minimum values of section moduli required to ensure that the permissible stresses at the initial and final stages are not exceeded. When the minimum values of section moduli are adopted, an optimum design is achieved.

It is interesting to note that the section moduli depend only on live load moment M_L and are independent of the dead load moment M_D. The explanation for this interesting observation is that the dead load moment can be counteracted merely by increasing the eccentricity of prestress e_p. In this manner, the dead load moment can be counteracted without additional cost which represents a major advantage of prestressed concrete beams, particularly in the case of long span structures. There is, however, a limit to which dead load moment can be counteracted merely by increase in e_p and is given by Eq. (6.5.5).

Equations (6.4.9) and (6.4.10) are general equations which are applicable to both types of beams whether pretensioned or post-tensioned. Some particular cases may now be considered. In the case of beams with symmetrical cross-section, $y_B = y_T = (d/2)$ and $Z_B = Z_T$. The comparison of Eq. (6.4.9) and (6.4.10), therefore, shows that Z_B is larger. Consequently, Eq. (6.4.10) for the section modulus Z_B with respect to bottom fibres controls the size of the symmetrical section. The term $(P_i - P)$, representing loss of prestress between the initial and final stages, may be taken equal to 20 to 25 per cent for pretensioning and 15 to 20 per cent for post-tensioning. The terms $(e_p\, y_T/k^2)$ and $(e_p\, y_B/k^2)$ generally vary from 1 to 3. However, their exact values are not needed for a tentative choice of the section in a preliminary design in which the loss of prestress $(P_i - P)$ may be ignored. In this case Eq. (6.4.9) and (6.4.10) reduce to

$$Z_T \geqslant \frac{M_L}{(\sigma_{cbc} + \sigma_{cti})} \tag{6.4.11}$$

$$Z_B \geqslant \frac{M_L}{(\sigma_{cbci} + \sigma_{ct})} \tag{6.4.12}$$

The values of section moduli obtained from Eq. (6.4.11) and (6.4.12) may be increased by approximately 20 per cent to account for the loss due to prestress.

6.5   SAFE ZONE FOR LINE OF THRUST

The safe zone for line of thrust or pressure line is the zone within which the line of thrust must be confined to ensure that the extreme fibre stresses at the initial and final stages do not exceed the maximum permissible stresses. Consequently, the lower limit of eccentricity of centre of compression e_{cl} is determined from the conditions that the compressive stress at bottom fibre and the tensile stress at top fibre at the initial stage do not exceed σ_{cbci} and σ_{cti} respectively.

$$\frac{P_i}{A}\left(\frac{e_{cl}\, y_B}{k^2} + 1\right) = \sigma_{cbc\,i}$$

or

$$e_{cl} = \left(\frac{\sigma_{cbc\,i}\, A}{P_i} - 1\right)\frac{k^2}{y_B} \qquad (6.5.1a)$$

and

$$\frac{P_i}{A}\left(\frac{e.\, y_T}{k^2} - 1\right) = \sigma_{cti}$$

or

$$e_{cl} = \left(\frac{\sigma_{cti}\, A}{P_i} + 1\right)\frac{k^2}{y_T} \qquad (6.5.1b)$$

The smaller of the two values gives the limiting or lowest position of the centre of compression.

Similarly, the upper limit of centre of compression is determined from the condition that the compressive stress at the top fibre and tensile stress at the bottom fibre at the final stage do not exceed σ_{cbc} and σ_{ct} respectively.

$$\frac{P}{A}\left(\frac{e_{cu}\, y_T}{k^2} + 1\right) = \sigma_{cbc}$$

or

$$e_{cu} = \left(\frac{\sigma_{cbc}\, A}{P} - 1\right)\frac{k^2}{y_T} \qquad (6.5.2a)$$

$$\frac{P}{A}\left(\frac{e_{cu}\, y_B}{k^2} - 1\right) = \sigma_{ct}$$

or

$$e_{cu} = \left(\frac{\sigma_{ct}\, A}{P} + 1\right)\frac{k^2}{y_B} \qquad (6.5.2b)$$

The smaller of the two values given by Eq. (6.5.2) determines the limiting value or highest position of the line of thrust. The limiting lower and upper eccentricities of the centre of compression e_{cl} and e_{cu} may be determined at all cross-sections of a beam. The lines joining these points give the lower and upper limits for the line of thrust or pressure line. In the case of a straight prismatic beam shown in Fig. 6.5.1 (a), the lower and upper limits of line of thrust are represented by lines $A_2\, B_2$ and $A_3\, B_3$ which are respectively at distances e_{cl} and e_{cu} from the centroidal axis $A_1\, B_1$. The shaded area between the two limits represents the safe zone for the line of thrust.

The limits imposed by Eq. (6.5.1a) and (6.5.2a) relate to permissible compressive stresses. If compressive stresses are not critical, the limits of safe zone are controlled by Eq. (6.5.1b) and (6.5.2b) relating to permissible tensile stresses. For Type 1 beams, the permissible tensile stresses are zero, $\sigma_{cti} = \sigma_{ct} = 0$. Consequently, Eq. (6.5.1b) and (6.5.2b) reduce to

$$e_{cl} = \frac{k^2}{y_T} = e_{kl} \qquad (6.5.3)$$

and

$$e_{cu} = \frac{k^2}{y_B} = e_{ku} \qquad (6.5.4)$$

ws that if compressive stresses are not critical, the limits of the safe zone for the line of thrust are the same as the upper and lower kern lines for Type 1 beams (full prestress) as shown in Fig. 6.5.1 (b).

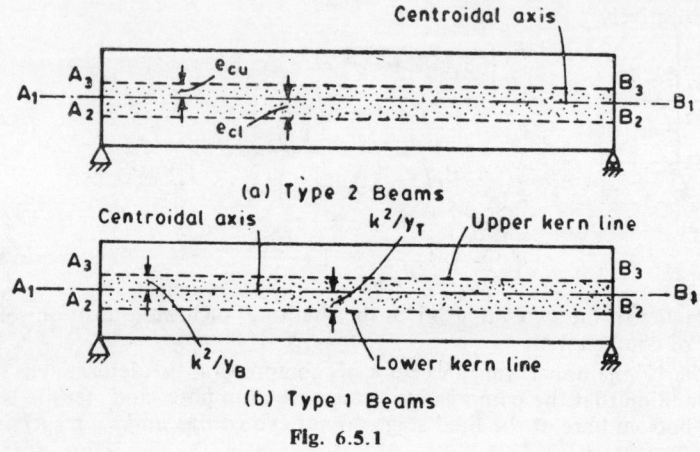

(a) Type 2 Beams

(b) Type 1 Beams

Fig. 6.5.1

In an unloaded beam, the line of thrust coincides with the cable line. Hence in the absence of external bending moment, the maximum permissible eccentricity of prestress is e_{cl}. Allowing minimum concrete cover c as practical requirement, the maximum practical value of eccentricity of prestress is $(y_B - c)$. It follows that the maximum permissible increase in eccentricity of prestress from its value under no load condition is $(y_B - c - e_{cl})$. It is evident, therefore, that the maximum sagging moment in the form of dead load moment at the initial stage which can be neutralised by merely increasing the eccentricity of prestress,

$$M_D \text{ (max)} = P_i (y_B - c - e_{cl}) \tag{6.5.5}$$

6.6   SAFE ZONE FOR CABLE LINE

A safe zone for cable line is the zone within which the cable line should be confined to ensure safe stresses at initial and final stages. If M_D is the sagging bending moment at the initial stage, the cable line should not lie at a distance greater than (M_D/P_i) below the lower limit of the line of thrust $A_1 C_1 B_1$ because the effect of M_D is to raise the line of thrust from the cable line through a distance (M_D/P_i). Hence referring to Fig. 6.6.1, the lower limit of the cable line $A_2 C_4 B_2$ is obtained by depressing the lower limit of the line of thrust $A_2 C_2 B_2$ through a distance (M_D/P_i). At the final stage, the total service load moment $(M_D + M_L)$ due to dead and live loads raises the line of thrust from the cable line through a vertical distance equal to $(M_D + M_L)/P$. Consequently, the upper limit $A_3 C_5 B_3$ of the cable line is obtained by depressing the upper limit of the line of thrust $A_3 C_3 B_3$ through a distance equal to $(M_D + M_L)/P$. The shaded area located between the lower limit $A_2 C_4 B_2$ and upper limit $A_3 C_5 B_3$ is the safe zone for the cable line. The stresses at the initial and final stages are safe as long as the cable line lies entirely within its safe zone.

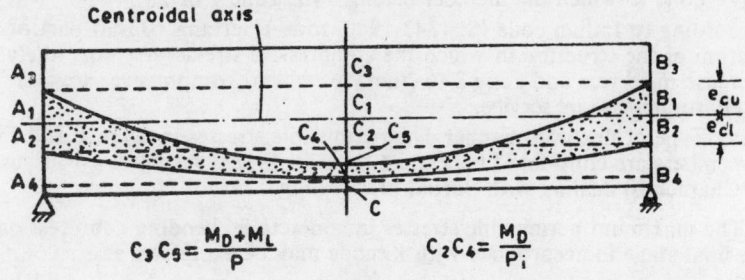

Fig. 6.6.1

Some interesting conclusions can be drawn from the positions of lower and upper limits of the cable line.

(i) If the lower limit goes outside the section or does not leave sufficient cover at its lowest point, complete neutralization of dead load moment M_D is not possible merely by increase in e_p. In this case the section moduli depend not only upon live load moment but also upon the dead load mo-ment. For complete neutralization of dead load moment, it is necessary to increase either the size of the section which means more concrete or the prestressing force which means more steel. If the lower limit touches the line A_4 B_4 corresponding to the minimum concrete cover c, the entire dead load moment can just be neutralized in accordance with Eq. (6.4.11) or (6.4.13).

(ii) If the upper limit goes outside the section or does not leave adequate concrete cover at its lowest point, the stresses at the final stage are unsafe unless the depth of the beam or prestressing force is increased.

(iii) If the lower and upper limits intersect each other, there is no safe zone in the region located between the two points of intersection. In this case an under-design is indicated which may be corrected by increasing the size of the cross-section.

(iv) If the two limits are wide apart at their lowest points, an over-design is indicated which may be corrected by reducing the size of the cross-section.

(v) If the two limits touch each other and also leave the minimum neces-sary cover, an optimum design is achieved.

6.7 PERMISSIBLE STRESSES IN CONCRETE

The codes specify maximum permissible stresses or working stresses to ensure adequate serviceability in the working range. The maximum permis-sible stresses in concrete are discussed in this section while those in prestressing steel are given in Sec. 2.3. The maximum permissible stresses in concrete depend upon:

(i) Grade of concrete. Larger stresses are permitted in stronger or higher grade concrete.

(ii) Technique of prestressing, viz. pretensioning or post-tensioning

(iii) Stage considered, viz. initial or transfer stage and final stage

(iv) Zone to which the member belongs, viz. zone 1 or 2.

According to Indian code IS: 1343-1980, zone 1 pertains to that part of the section of the structure in which the compressive stresses are not likely to increase in service and zone 2 to that in which compressive stresses are likely to increase in service.

(v) Type of flexural member The permissible stresses in Type 1 and Type 2 members are considered here while those in Type 3 members are discussed in Chapter 10 dealing with partial prestressing.

The maximum permissible stresses in concrete in bending compression at the final stage in accordance with IS code may be expressed as

$$\sigma_{cbc} = c_1 f_{ck} \tag{6.7.1}$$

in which coefficient c_1 varies linearly from 0.41 for M-30 concrete to 0.35 for M-60 concrete for zone 1. For zone 2, c_1 varies linearly from 0.34 for M-30 concrete to 0.27 for M-60 concrete. Values of c_1 for different grades of concrete are shown in Table 6.7.1. The values of coefficient c_1 given in the Table are applicable both for pretensioning as well as post-tensioning.

Table 6.7.1   Values of Coefficients c_1 and c_2

f_{ck}, MPa	c_1		c_2	
	Zone 1	Zone 2	Pretensioning	Post-tensioning
30	0.4100	0.3400	—	0.5400
35	0.3929	0.3200	—	0.5116
40	0.3800	0.3050	0.5100	0.4833
45	0.3700	0.2933	0.4925	0.4550
50	0.3620	0.2840	0.4750	0.4267
55	0.3555	0.2764	0.4575	0.3983
60	0.3500	0.2700	0.4400	0.3700

The maximum permissible stresses in concrete in bending compression at the initial stage as per IS code,

$$\sigma_{cbci} = c_2 f_{ci} \tag{6.7.2}$$

where f_{ci} = compressive (cube) strength of concrete at initial stage or at transfer

The coefficient c_2 varies linearly with f_{ck}. Values of c_2 for different grades of concrete are shown in Table 6.7.1. The code specifies that strength of concrete at transfer f_{ci} should not be less than $0.5 f_{ck}$. It should be determined by tests at the time of stress transfer for all major structures.

Equations (6.7.1) and (6.7.2) give maximum permissible stresses in concrete in bending compression. The maximum permissible stresses in direct compression are smaller due to lack of stress gradient and consequent redistribution of stresses. The direct compressive stress at final stage,

$$\sigma_{cc} = 0.8 \ \sigma_{cbc} \qquad (6.7.3)$$

and the direct compressive stress at initial stage,

$$\sigma_{cci} = 0.8 \ \sigma_{cbci} \qquad (6.7.4)$$

The permissible ensile stress in concrete depends upon the method of construction and the type of member. In the case of precast prestressed members with mortar or concrete joints, no tension is allowed at the joints at any stage of loading. If the member has no joint, the permissible tensile stress depends upon the type of member. No tension is permitted anywhere in Type 1 members (full prestress) in the entire working range. In Type 2 members (limited prestress) the maximum permissible tensile stress is 3 MPa. However, it may be increased to 4.5 MPa if a part of the service load is of temporary nature provided that the stress remains compressive under the action of permanent part of the service load. The permissible tensile stress in Type 3 members (partial prestress) is discussed in Chapter 10.

Similar working stresses are specified by other codes. The British Code BS: 8110-1985 specifies that the design tensile stress in Type 2 members should not exceed f_{cr} for pretensioned members and $0.8 \ f_{cr}$ for post-tensioned members in which f_{cr} is the design flexural tensile strength of concrete which may be taken equal to $0.45\sqrt{f_{ck}}$. These limiting design tensile stresses may be increased by upto 1.7 MPa where tests show that the increased stress does not exceed 0.75 times the stress at the appearance of the first crack and the compressive stress in concrete due to prestress after all losses is not less than 3 MPa. The limiting design tensile stresses may be further increased by upto 1.7 MPa where the design service load is exceptionally high and is of temporary nature provided that the stress under normal service conditions is compressive. Whenever the limiting design flexural tensile stresses are increased for either of the foregoing reasons, the pretensioned tendons should be well distributed in the flexural tension zone and the post-tensioned tendons should be supplemented, whenever necessary by means of bonded reinforcement near the tension face. The British Code also specifies that the design tensile stress at transfer should not exceed 1 MPa for Type 1 members. For Type 2 members, it should not exceed $0.45\sqrt{f_{ci}}$ for pretensioned members and $0.36\sqrt{f_{ci}}$ for post-tensioned members in which f_{ci} is the cube strength of concrete at stress transfer. The ACI code permits a maximum compressive stress equal to $0.45f'_c$ at final stage and $0.60 f_{ci}$ at the initial stage. It permits a tensile stress equal to $0.50\sqrt{f'_c}$ at the final stage. However, a tensile stress equal to $\sqrt{f'_c}$ is permitted when the analysis, based on a cracked section and bilinear load-deflection relationship, shows that the immediate and long time deflections are within the specified limits. At the initial stage a maximum tensile stress equal to $0.25\sqrt{f_{ci}}$ is permissible except that it may be increased to $0.50\sqrt{f_{ci}}$ at the ends of simply supported beams.

6.8 PRELIMINARY DESIGN

The object of preliminary design is to arrive at tentative proportions which form the basis for final design and working drawings. When preliminary design is executed with care, only minor ch nges may be needed to meet all relevant strength, serviceability and practical requirements. The preliminary design may be carried out using working stress method based on uncracked section and linear elastic theory. Although the proportions in a preliminary design are generally chosen to meet the stress requirements at initial and final stages, certain other stages may also require attention in special cases. The process of preliminary design necessarily entails a certain amount of trial-and-error. However, the computations can be reduced by adopting a design procedure described by the following steps:

(1) Select an appropriate overall depth of the beam D. It depends upon its span, intensity of loading, and aesthetic and headroom requirements. An increase in depth results in greater flexural strength and stiffness. It also helps in reducing the prestressing force and costly prestressing steel. Hence the maximum depth consistent with aesthetic and headroom requirements may be selected. An increase in depth may, however, lead to increase of self weight and dead load moment M_D at initial stage which may not be compensated without extra prestressing steel in accordance with Eq. (6.4.11) and (6.4.13). Hence the overall depth should be chosen with due care. In most practical cases, the overall depth varies from (span/15) to (span/25) depending upon the intensity of loading. The overall depth may be increased by approximately 50 per cent for cantilever beams. For fixed beams and continuous beams, the overall depth may be reduced approximately by 50 per cent.

(2) Determine the dead load moment M_D inclusive of that due to self weight at the initial stage. The self weight of the beam has to be assumed tentatively. The aspect ratio of rectangular sections generally varies from 1.5 for small beams to 3.0 for large ones. Hence the cross sectional area of a rectangular section, for the computation of self weight, may be taken to range from $(D^2/1.5)$ to $(D^2/3)$. In the case of flanged sections, the aspect ratio of the equivalent rectangular section having the same depth and cross-sectional area generally ranges from 3 to 4 depending upon the size of the beam. Hence the cross-sectional area may be taken to range from $(D^2/3)$ to $(D^2/4)$. The computations have to be revised if the assumed value of self weight is substantially different from the actual self weight.

Bennett has proposed the following empirical expression for determintion of self weight w_b in MN per metre length of the beam:

$$w_b = \frac{1.6\, c_1\, w_L}{1 - 1.4\, c_1} \tag{6.8.1}$$

where $c_1 = \dfrac{c_2\, c_3\, w_c\, l}{f_{ck}} \dfrac{(l/D)}{(d_p/D)^2}$

c_2 = moment coefficient (e.g. 0.125 for simply supported beam)

w_c = unit weight of concrete in MN/m^3

w_L = live load in MN/m

l = span of the beam

The dimensionless constant c_3 ranges from 6 to 7.5 for rectangular sections and I-section beams of short spans and from 4 to 5 for flanged T or I section beams of long spans. The ratio (d_p/D) usually ranges from 0.85 to 0.95 and may be taken conservatively as 0.85.

(3) Determine the live load moment M_L which acts on the beam at the final stage due to superimposed service loads. The total working moment at the final stage,

$$M_W = M_D + M_L \tag{6.8.2}$$

In the case of bridges and other structures subjected to moving loads, it may be found to be convenient to replace the moving loads by their equivalent uniform load. The effect of impact should also be included in the equivalent static uniform load.

(4) Determine the effective prestressing force from the equation,

$$P = \frac{M_w}{j\,d_p} \tag{6.8.3}$$

As the prestressing tendons must have adequate protective concrete cover c, and the working moment M_w should not shift the centre of compression from the cable line to beyond the safe upper limit of line of thrust, the maximum internal lever-arm shown in Fig. 6.8.1 is

$$J\,d_p = (y_B - c) + e_{cu} \tag{6.8.4}$$

in which e_{cu} is given by Eq. (6.5.2). The maximum available lever arm generally varies from 0.45 D to 0.75 D for Type 1 beams depending upon the shape of the cross-section. It is smallest for rectangular section, medium for symmetrical flanged and box sections and least for T and unsymmetrical I-sections. Hence values of lever arm equal to 0.5 D, 0.6 D and 0.7 D may be tried respectively for the three types of cross-sections. In order to utilize full flexural capacity of the section and to save costly prestressing steel, it is necessary to maximize the internal lever arm by

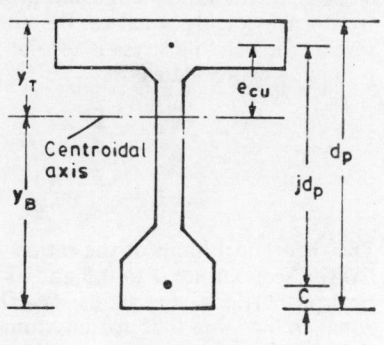

Fig. 6.8.1

(i) adopting the largest eccentricity of prestress consistent with the practical requirement of adequate cover.

(ii) adopting a flanged section to increase the upper limit of pressure line e_{cu} or eccentricity of upper kern point e_{ku}.

(iii) using unsymmetrical I−section with wide top flange and narrow bottom flange in order to increase y_B.

In Type 2 beams in which flexural tension is permissible at the working load level, the width of the safe zone for the line of thrust is larger than that in Type 1 beams. This is evident from Eq. (6.5.1) and (6.5.2) as well as from Fig. 6.5.1. Consequently, the maximum available internal lever arm is larger in Type 2 beams as compared to Type 1 beams. The increase in lever arm is essentially due to upward shift of the upper limit of the safe zone for pressure line when tensile stress is allowed at working load. The increase in lever arm depends upon the cross-sectional properties and the intensity of permissible tensile stress. For Type 2 beams, the maximum available lever arm may be taken equal to 0.6, 0.7 and 0.8 times overall depth D for the three types of sections, viz. rectangular, symmetrical−I and unsymmetrical-I or T section.

A larger lever arm increases the moment capacity at the working and ultimate stages and improves the deformation characteristics of the beam due to increased stiffness. After computing P from Eq. (6.8.3), the initial prestressing force P_i may be determined from

$$P_i = \frac{P}{\eta} \tag{6.8.5}$$

in which the loss factor η may be taken to lie approximately between 0.75 and 0.80 for pretensioning and between 0.80 to 0.85 for post-tensioning.

(5) Determine a suitable cross-sectional area A for the beam. The minimum requirement of cross-sectional area may be determined from the most critical stress diagrams at the initial and final stages shown in Fig. 6.4.1(b) and (c) respectively, in which the extreme fibre stresses are equal to the maximum permissible stresses at the respective stages. The intensities of stress p_i at the initial stage and p at the final stage at the level of the centroid are evidently equal to the intensities of direct stress due to prestressing as the bending stress is zero at the level of the centroid. Hence p_i and p may be expressed as

$$p_i = \frac{P_i}{A} = \frac{y_T}{D} \left(\sigma_{cbci} + \sigma_{cti} \right) - \sigma_{cti} \tag{6.8.6a}$$

$$p = \frac{P}{A} = \frac{y_B}{D} \left(\sigma_{cbc} + \sigma_{ct} \right) - \sigma_{ct} \tag{6.8.6b}$$

The theoretical limits of the ratios (y_T/D) and (y_B/D) for unsymmetrical flanged sections are 0 to 0.5 and 0.5 to 1.0 respectively. However for practical proportions, the ratios (y_T/D) and (y_B/D) may be taken tentatively equal to 0.45 and 0.55 for unsymmetrical I-sections and equal to 0.35 and 0.65 for T-sections.

In the case of symmetrical sections, $y_T = y_B = D/2$,

$$p_i = \frac{P_i}{A} \left(\frac{\sigma_{cbci} - \sigma_{cti}}{2} \right) \tag{6.8.7a}$$

$$p = \frac{P}{A} \left(\frac{\sigma_{cbc} - \sigma_{ct}}{2} \right) \tag{6.8.7b}$$

For unsymmetrical sections of Type 1 beams,

$$p_i = \frac{P_i}{A} = \frac{y_T}{D} \sigma_{cbci} \tag{6.8.8a}$$

$$p = \frac{P}{A} = \frac{y_B}{D} \sigma_{cbc} \tag{6.8.8b}$$

In the case of symmetrical sections of Type 1 beams,

$$p_i = \frac{P_i}{A} = 0.5 \, \sigma_{cbc\,i} \tag{6.8.9a}$$

$$p = \frac{P}{A} = 0.5 \, \sigma_{cbc} \tag{6.8.9b}$$

The minimum requirement of cross-sectional area A is evidently the larger of the two values computed for the initial and final stages. The cross-sectional area actually adopted may be slightly higher than the minimum requirement computed from the above equations. This results in a lower intensity of prestress which may be preferable for several reasons:

(i) It eliminates the ill effects of excessive prestress such as excessive camber and possible cracking at initial stage.

(ii) The creep deformations and the loss of prestress due to creep of concrete are reduced because prestress constitutes a major part of permanent load and is, therefore, mainly responsible for creep deformations and loss of prestress.

(iii) For a given cross sectional area, a smaller intensity of prestress means smaller prestressing force, saving of costly prestressing steel and reduction in labour involved in prestressing operations. Although a low value of intensity of prestress is preferable, it should be realized that a lower value of intensity of prestress leads to a larger area of concrete to develop a given prestressing force and consequently leads to more concrete and increased self weight. The choice of intensity of prestress is guided by serviceability requirements at initial and final stages as per Eq. (6.8.6) and codal restrictions on deformations of the beam.

(6) Determine section moduli Z_T and Z_B using Eq. (6.4.9) and (6.4.10). In the first instance. the factors $(e_p \, y_T/k^2)$ and $(e_p \, y_B/k^2)$ may be taken equal to 3.0 for rectangular sections and 2 for flanged sections. The section modulus Z_B with respect to bottom fibre generally governs the design due to its larger value as compared to Z_T.

(7) Select a cross section which meets the requirements in respect of over-all depth D, cross-sectional area A and section moduli Z_T and Z_B determined in steps 1, 5 and 6. The choice of the shape of the cross-section is influenced by the following observations:

(i) A rectangular section is preferred for its simple formwork and ease of fabrication. It has the drawback of having a lower moment of inertia and section moduli for a given cross-sectional area. Rectangular section is suitable for small pretensioned beams. It is generally considered unsuitable for long span post-tensioned beams in which the reduction of self weight is crucial.

(ii) Symmetrical I-section offers relatively high moment of inertia and section moduli. Its drawbacks are costlier formwork and lower torsional strength and stiffness.

(iii) Unsymmetrical I and T-sections have the same features as symmetrical I-section. However, they have an additional advantage of providing room for larger eccentricity of prestress and internal lever arm. They are, therefore, commonly used for long span post-tensioned beams.

(iv) Box section offers high moment of inertia and section moduli. Its main advantage is its high torsional strength and stiffness. Its main drawback is its complicated and costly formwork. It is commonly used in long span bridges.

The cross-sectional proportions of prestressed concrete members are selected on functional, aesthetic and practical considerations. They also depend on type of structure and theoretical aspects. Consequently, the proportions vary widely particularly in unusual structures. However, for common structures, the usual range for the proportions may be indicated for general guidance. Three types of sections, viz. rectangular, open flanged and box sections deserve consideration due to their common occurrence. The aspect ratio of rectangular section (b/D) varies widely depending upon the type of structure. For instance, the aspect ratio may be as low as 1.5 for small precast pretensioned beams and it may be as high as 10 or even more for the edge beam of cylindrical shell. The minimum width is controlled by lateral and torsional strength and stiffness, flexural and torsional buckling and the practical requirement of adequate concrete cover on either side.

The proportions of component rectangles of thin-walled open sections such as I, T, L and channel sections are chosen so that each component can adequately perform its respective function. The function of the top flange is generally to carry flexural compression. Hence the width and thickness of top flange should be sufficient to carry flexural compression without lateral buckling. In bridge girders, the width of the top flange may be chosen equal to the spacing of the girders so that no formwork is needed for in-situ top slab. The width of the top flange generally ranges from $0.4\,D$ to $0.6\,D$ and the thickness from $0.15\,D$ to $0.25\,D$. The width and thickness of the bottom flange is selected so that it can accommodate the tendons with sufficient concrete cover. The depth and thickness of the web are proportioned to enable it to perform its main function of carrying flexural shear. The ratio (b_w/b_f) commonly ranges from 0.2 to 0.3. In the case of large girders with curved tendons, a minimum thickness of 120 to 150 mm is considered mendatory for sufficient concrete cover on either side of the tendons. The web thickness may be as small as 40 to 60 mm in the case of small pretensioned beams.

The box sections which are commonly used for bridge girders due to their superior torsional properties, have single or multiple cells of rectangular or trapezoidal shape. The aspect ratio of the cells usually ranges from 1 to 3. The minimum wall thickness is controlled by the dual criteria of adequate strength against buckling and practical requirement regarding concrete cover. The wall thickness commonly ranges from 0.1 to 0.4 times the smaller dimension of the cell. For walls thinner than one-tenth of the smaller dimension of the cell, the stiffness and possible buckling of the wall have to be considered. In order to facilitate demoulding, to avoid honey-combing and to reduce stress concentration, adequate fillets are provided at all re-entrant corners.

(8) Determine safe zone for cable line in accordance with the procedure described in Sec. 6.6. If the tentative selection of the cross-section in step 7 is good, the safe zone is quite narrow near mid-span and allows adequate concrete cover. Otherwise the cross-section may be given suitable changes

in accordance with the conclusions drawn in Sec. 6.6 from the shape of the safe zone for cable line. Practical and codal provisions should be kept in view in arriving at a suitable cross-section for the beam.

(9) Locate the cable line inside the safe zone. In the case of pretensioned beams having straight tendons at constant eccentricity, the cable line should preferably coincide with the lower limit for line of thrust, i.e. $e_p = e_{cl}$, unless the tendons are depressed at mid-span. The advantage of draping the tendons is an increase in the internal lever arm and a full or partial compensation of dead load moment at initial stage. In the case of posttensioned beams, the cable line should preferably be placed at the lowest point inside its safe zone at mid-span or section of maximum moment subject, of course, to the practical requirement of adequate concrete cover. At the ends of the beam, the cable can be located anywhere between the lower and upper limits of its safe zone. If the cable line is raised to its upper limit at the ends, the shear force induced by prestress is increased and consequently, a larger part of external shear force is neutralized by prestress. However, the increased curvature of the tendons leads to greater loss of prestress due to friction. On the other hand, if the cable line is placed at the lower limit of its safe zone, the frictional loss as well as the shear force neutralized by prestress are reduced. Often a compromise is struck between the two extremes and the cable line is located at or near the centroidal axis at the ends of the beam. If frictional loss is reduced by restretching or stretching at both ends, it may be advantageous to raise the cable line at ends above the centroidal axis.

(10) Check the preliminary design for initial and final stages and make appropriate changes to satisfy the stress requirements at the two stages. The objective of the guidelines in the preceding steps is served if the preliminary design requires only minor changes to satisfy all requirements of strength and serviceability in accordance with code provisions. While such a design is acceptable for strength and serviceability, it is not necessarily the most economical unless it minimizes the total cost including materials, formwork and labour.

(11) Select appropriate number of prestressing tendons and their positions in order to obtain the desired prestressing force and eccentricity of prestress. Although the cable line must lie entirely within the safe zone, the individual tendons may be placed anywhere in the cross section as long as they satisfy practical requirements of minimum spacing and concrete cover and subject to the condition that the centroid of forces in tendons lies inside the safe zone at all sections. Selecting a suitable value of initial stress in prestressing steel f_{pi}, the cross-sectional area of tendons may be determined from

$$A_p = \frac{P_i}{f_{pi}} \tag{6.8.10}$$

Sufficient tendons should be provided to supply the required area A_p. The profile of each tendon is carefully selected in final design keeping in view all the practical requirements for the particular system of prestressing adopted and other codal provisions. However, in checking strength and serviceability requirements, the chosen cable line may be used.

EXAMPLE 6.8.1

Design a rectangular section for a pretensioned beam for an industrial shed. The effective span is 15 m. The beam carries only its own weight at stress transfer. It has to carry a superimposed load of 4.2 kN/m at the final stage. Concrete is of M-40 grade. The strength of concrete at stress transfer, $f_{ci} = 30$ MPa. The ultimate tensile strength of high tensile steel, $f_{pu} = 1600$ MPa. The beam has to be of Type 1 (full prestress). Take working stresses pertaining to zone 1 as per Indian Code.

Solution

The design is carried out using the steps enumerated in Sec. 6.8.

(1) As the girder carries light loading, overall depth of the beam may be taken equal to (span/25).

$$D = \frac{15}{25} = 0.60 \text{ m}$$

(2) Adopting the aspect ratio, $D/b = 3$, the breadth $b = 0.20$ m. Hence taking the unit weight of concrete, $w_c = 0.024 \text{ MN/m}^3$, the self weight of the beam,

$$w_b = 0.20 \times 0.60 \times 1 \times 0.024 = 0.00288 \text{ MN/m}$$

Alternatively, using Eq. (6.8.1) and taking $c_2 = 0.125$, $c_3 = 6$, $d_p/D = 0.85$,

$$c_1 = \frac{0.125 \times 6 \times 0.024 \times 15 \times 25}{40 \times 0.85^2} = 0.2336$$

$$w_b = \frac{1.6 \times 0.2336 \times 0.0042}{1 - 1.4 \times 0.2336} = 0.00233 \text{ MN/m}$$

Adopting the larger value of w_b, the dead load moment at the initial stage,

$$M_D = \frac{0.00288 \times 15^2}{8} = 0.0810 \text{ MN.m}$$

(3) The live load moment at the final stage,

$$M_L = \frac{0.0042 \times 15^2}{8} = 0.1181 \text{ MN.m}$$

$$M_w = M_D + M_L = 0.0810 + 0.1181 = 0.1991 \text{ MN.m}$$

(4) Taking the internal lever arm $= jd_p = 0.5$ D,

$$P = \frac{M_w}{jd_p} = \frac{0.1991}{0.5 \times 0.60} = 0.6637 \text{ MN}$$

Adopt $P = 0.700$ MN. Taking loss of prestress equal to 20 per cent,

$$P_i = \frac{0.700}{0.8} = 0.875 \text{ MN}$$

(5) From Table 6.7.1

$$\sigma_{cbc} = 0.380 f_{ck} = 0.380 \times 40 = 15.20 \text{ MPa}$$
$$\sigma_{cbci} = 0.51 f_{ci} = 0.51 \times 30 = 15.30 \text{ MPa}$$

Also for Type 1 beams (full prestress).

$$\sigma_{ct} = \sigma_{cti} = 0$$

Hence using Eq. (6.8.9),

$$p_i = \frac{1}{2} (15.30 - 0) = 7.65 \text{ MPa}$$

$$p = \frac{1}{2} (15.20 - 0) = 7.60 \text{ MPa}$$

Hence,

$$A = \frac{P_i}{p_i} = \frac{0.875}{7.65} = 0.1144 \text{ m}^2$$

or
$$A = \frac{P}{p} = \frac{0.700}{7.60} = 0.0921 \text{ m}^2$$

The requirement of area is greater at initial stage.

(6) Taking $e_p \, y_T/k^2 = e_p \, y_B/k^2 = 3$ for rectangular section, $A = .120$ m² as in step 2, $P_i = 0.875$ MN and $P = 0.700$ MN as in step 4, and using Eq. (6.4.9) and (6.4.10),

$$Z_T > \cfrac{0.1181}{(15.20 + 0) - \left(\cfrac{0.875 - 0.700}{0.12} \right) (3-1)}$$

$$> 0.0096 \text{ m}^3$$

$$Z_B > \cfrac{0.1181}{(15.30 + 0) - \left(\cfrac{0.875 - 0.700}{0.12} \right) (3 +1)}$$

$$> 0.0125 \text{ m}^3$$

As the minimum required value of Z_B is larger, the design is controlled by the section modulus with respect to the bottom fibre.

(7) Adopting a rectangular section 0.20×0.60 m,

$$A = 0.20 \times 0.60 = 0.12 \text{ m}^2$$

$$Z_T = Z_B = \frac{1}{6} \times 0.20 \times (0.60)^2 = 0.012 \text{ m}^3$$

$$k^2 = \frac{I}{A} = \frac{\left(\frac{1}{12} \right) \times 0.20 \times (0.60)^3}{0.20 \times 0.60} = 0.030 \text{m}^2$$

The chosen section satisfies the requirements of overall depth (step 1) and cross-sectional area (step 5). It also satisfies the requirements of section moduli (step 6) approximately.

(8) The safe zone for cable line may now be determined. Using Eq. (6.5.1), the lower limit of line of thrust,

$$e_{cl} = \left(\frac{15.30 \times 0.12}{0.8750} - 1 \right) \left(\frac{0.03}{0.30} \right) = 0.1098 \text{ m}$$

or

$$e_{cl} = (0 + 1) \left(\frac{0.03}{0.30} \right) = 0.10 \text{ m}$$

Hence referring to Fig. 6.8.2 (a), line $A_2 C_2 B_2$ at a distance of 0.10 m below the centroidal axis $A_1 C_1 B_1$ represents the safe lower limit of the line of thrust. Using Eq. (6.5.2), the upper limit of line of thrust,

$$e_{cu} = \left(\frac{15.20 \times 0.12}{0.7000} - 1 \right) \left(\frac{0 \cdot 03}{0.30} \right) = 0.1606 \text{ m}$$

or

$$e_{cu} = (0 + 1) \left(\frac{0.03}{0.30} \right) = 0.10 \text{ m}$$

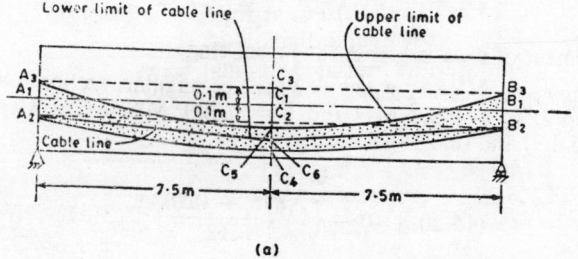

(a)

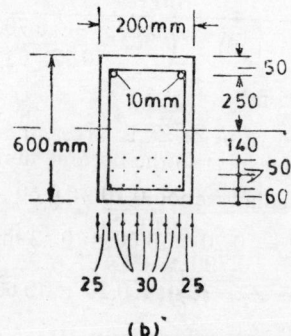

(b)

Fig. 6.8.2

Hence line $A_3 C_3 B_3$ at a distance of 0.10 m above the centroidal axis $A_1 C_1 B_1$ represents the safe upper limit of the line of thrust. The parabola A_2

$C_4 B_2$ represents the lower limit of the cable line in which the midspan ordinate,

$$C_2 C_4 = \frac{M_D}{P_i} = \frac{0.081}{0.875} = 0.0926 \text{ m}$$

The parabola $A_3 C_5 B_3$ represents the upper limit of the cable line in which the midspan ordinate

$$C_3 C_5 = \frac{M_w}{P} = \frac{0.1991}{0.700} = 0.2844 \text{ m}$$

The shaded area between the lower and upper parabolas represents the safe zone for cable line.

(9) The minimum permissible eccentricity of cable line at midspan,

$$C_1 C_5 = 0.2844 - 0.10 = 0.1844 \text{ m}$$

The maximum permissible eccentricity of cable line at midspan,

$$C_1 C_4 = 0.0926 + 0.10 = 0.1926 \text{ m}$$

Hence an eccentricity of 0.19 m at midspan may be adopted. The eccentricity may be reduced linearly to 0.10 m at the two ends. The line $A_2 C_6 B_2$, therefore, represents the chosen cable line.

(10) Check the stresses at initial and final stages using Eq. (5.2.3) to (5.2.6).

(a) *Initial Stage*

$$P_i = 0.875 \text{ MN}$$
$$e_p = 0.19 \text{ m}$$
$$f_T = \frac{-0.875}{0.12} + \frac{0.875 \times 0.19}{0.012} - \frac{0.081}{0.012} = -0.19 \text{ MPa}$$
$$f_B = \frac{-0.875}{0.12} - \frac{0.875 \times 0.19}{0.012} + \frac{0.081}{0.012} = -14.40 \text{ MPa}$$

(b) *Final Stage*

$$P = 0.700 \text{ MN}$$
$$e_p = 0.19 \text{ m}$$
$$f_T = \frac{-0.700}{0.12} + \frac{0.700 \times 0.19}{0.012} - \frac{0.1991}{0.012} = -11.34 \text{ MPa}$$
$$f_B = \frac{-0.700}{0.12} - \frac{0.700 \times 0.19}{0.012} + \frac{0.1991}{0.012} = -0.32 \text{ MPa}$$

These stresses are within permissible limits. Hence the chosen section is adequate.

(11) Using 18 high tensile wires, each of 7 mm diameter, the initial stress in each wire,

$$f_{pi} = \frac{875000}{18 \times \frac{\pi}{4} \times 7^2} = 1263 \text{ MPa}$$

Hence the ultimate strength required as per IS code,

$$f_{pu} = \frac{f_{pi}}{0.8} = \frac{1263}{0.8} = 1578 \text{ MPa}$$

which is less than the specified strength.

To obtain the requisite eccentricity of 0.19 m, the wires may be arranged in three layers as shown in Fig. 6.8.2 (b). Two untensioned reinforcing bars, each of 10 mm diameter, may be provided near the top to act as hanger bars as shown in the figure.

An alternative design is shown in Fig. 6.8.3 in which the objective of reducing the eccentricity of prestress towards the ends is achieved by decreasing the depth linearly on both sides of the midspan section. In this case all wires run straight from one end to the other end. This design is preferable where the physical facilities for depressing the wires at midspan are not available. The additional advantage of sloping top face for a roof beam is that the slope necessary for drainage of rain water is automatically provided.

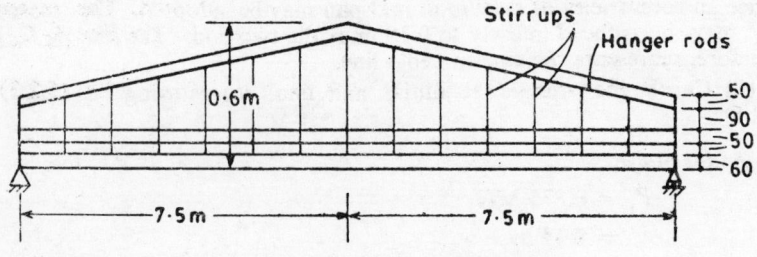

(a) Elevation

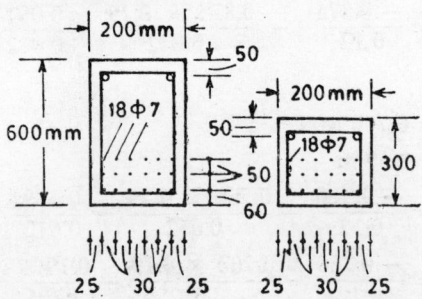

(b) Midspan section   (c) End section

Fig. 6.8.3

EXAMPLE 6.8.2

Design a post-tensioned beam for an auditorium for an effective span of 24 m. The beam carries only its own weight at stress transfer. The superimposed load at the final stage is 21.5 kN/m. Adopt a symmetrical I-section. Concrete is of M-35 grade. Use Freyssinet system of post-tensioning. The beam has to be of Type 1 (full prestress). Take working stresses pertaining to zone 1 as per IS code. The strength of concrete at stress transfer, $f_{ci} = 28$ MPa.

Solution

(1) Overall depth of the beam may be taken equal to (span/20) as the intensity of loading is medium.

$$D = \frac{24}{20} = 1.2 \text{ m}$$

(2) For the computation of self weight, take

$$A = \frac{D^2}{3} = \frac{1.2^2}{3} = 0.48 \text{ m}^2$$

Hence weight per unit length,

$$w_b = 0.48 \times 1 \times 0.024 = 0.0115 \text{ MN/m}$$

Alternatively, using Eq. (6.8.1) and taking $c_2 = 0.125$, $c_3 = 4$ and $d_p/D = 0.85$,

$$c_1 = \frac{0.125 \times 4 \times 0.024 \times 24 \times 20}{35 \, (0.18)^2} = 0.2278$$

$$w_b = \frac{1.6 \times 0.2278 \times 0.0215}{1 - 1.4 \times 0.2278} = 0.0115 \text{ MN/m}$$

$$M_D = \frac{0.0115 \times 24^2}{8} = 0.8280 \text{ MN·m}$$

(3) $$M_L = \frac{0.0215 \times 24^2}{8} = 1.5480 \text{ MN·m}$$

$$M_w = M_D + M_L = 2.3760 \text{ MN·m}$$

(4) Taking the internal lever arm equal to 0.6 D for symmetrical I-section,

$$P = \frac{2.3760}{0.6 \times 1.2} = 3.300 \text{ MN}$$

Taking loss of prestress equal to 15 per cent,

$$P_i = \frac{P}{0.85} = \frac{3.300}{0.85} = 3.8824 \text{ MN}$$

(5) From Table 6.7.1,

$$\sigma_{cbc} = 0.3922 f_{ck} = 0.3929 \times 35$$

$$= 13.750 \text{ MPa}$$

$$\sigma_{cbci} = 0.5116 f_{ci} = 0.5116 \times 28 = 14.325 \text{ MPa}$$

Also for Type 1 beams (full prestress),

$$\sigma_{ct} = \sigma_{cti} = 0$$

Hence using Eq. (6.8.9),

$$p_i = 0.5! \sigma_{cbci} = 7.163 \text{ MPa}$$
$$p = 0.5 \, \sigma_{cbc} = 6.875 \text{ MPa}$$

Hence
$$A = \frac{P_i}{p_i} = 0.5420 \text{ m}^2$$

or
$$A = \frac{P}{p} = 0.48 \text{ m}^2$$

(6) Taking $e_p \, y_T/k^2 = e_p \, y_B/k^2 = 2$ for a flanged section, $A = 0.48 \text{ m}^2$ as in slep 2, $P_i = 3.8824 \text{ MN}$ and $P = 3.300$ as in step 4 and using Eq. (6.4.9) and (6.4.10),

$$Z_T \geqslant \frac{1.5480}{(13.750 + 0) - \left(\dfrac{3.8824 - 3.3000}{0.48} \right)(2 - 1)}$$

$$\geqslant 0.1235 \text{ m}^3$$

$$Z_B \geqslant \frac{1.5480}{(14.325 + 0) - \left(\dfrac{3.8824 - 3.3000}{0.48} \right)(2 + 1)}$$

$$\geqslant 0.1449 \text{ m}^3$$

(7) Keeping in view the requirements in steps 1, 5 and 6, the symmetrical I-section shown in Fig. 6.8.4 (a) may be tried. A fillet of 0.10 m side may be provided at each re-entrant corner.

$$A = 2 \times 0.75 \times 0.20 + 0.80 \times 0.20 + 0.4 \times 0.5 \times 0.10 \times 0.10$$
$$= 0.48 \text{ m}^2$$

$$I = \frac{0.75 \times 1.20^3}{12} - \frac{0.55 \times 0.80^3}{12} + \frac{4 \times 0.10 \times 0.10^3}{36}$$

$$+ 4 \times \tfrac{1}{2} \times 0.10 \times 0.10 \left(0.40 - \frac{0.10}{3} \right)$$

$$= 0.09188 \text{ m}^4$$

$$Z_B = Z_T = \frac{I}{y} = \frac{0.09188}{0.60} = 0.1531 \text{ m}^3$$

The chosen cross-section meets the requirements of steps 1, 5 and 6 reasonably well.

(8) The safe zone for the cable line shown shaded in Fig. 6.8.4 (b) may be obtained by following the procedure similar to that in Example 6.8.1. The lower and upper limits of the line of thrust are located at distances 0.2460 m and 0.3190 m below and above the centroidal axis respectively. The midspan ordinate of the lower parabola,

$$C_2 C_4 = \frac{M_D}{P_i} = \frac{0.8280}{3.8824} = 0.2133 \text{ m}$$

The midspan ordinate of the upper parabola,

$$C_3 C_5 = \frac{M_w}{P} = \frac{2.3760}{3.3000} = 0.720 \text{ m}$$

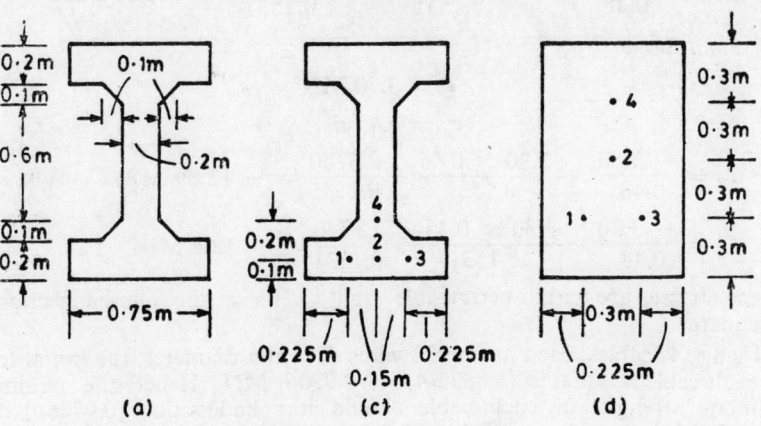

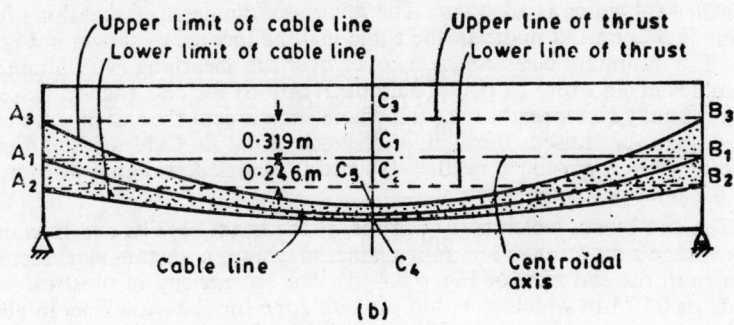

Fig. 6.8.4

(9) The minimum permissible eccentricity of cable line at midspan,

$$C_1 C_5 = 0.7200 - 0.3190 = 0.4010 \text{ m}$$

The maximum permissible eccentricity of the cable line at midspan,

$$C_1 C_4 = 0.2460 + 0.2133 = 0.4593 \text{ m}$$

Hence a parabolic cable line with an eccentricity of 0.45 m at the midspan may be adopted.

(10) Check the stresses at initial and final stages using Eq. (5.2.3) to (5.2.6).

(a) *Initial Stage*

$$P_i = 3.8824 \text{ MN}$$
$$e_p = 0.45 \text{ m}$$

$$f_T = \frac{-3.8824}{0.48} + \frac{3.8824 \times 0.45}{0.1531} - \frac{0.8294}{0.1531} = -2.27 \text{ MPa}$$

$$f_B = \frac{-3.8824}{0.48} - \frac{3.8824 \times 0.45}{0.1531} + \frac{0.8294}{0.1531} = -13.91 \text{ MPa}$$

(b) *Final Stage*

$$P = 3.30 \text{ MN}$$

$$e_p = 0.45 \text{ m}$$

$$f_T = \frac{-3.30}{0.48} + \frac{3.30 \times 0.45}{0.1531} - \frac{2.3760}{0.1531} = -12.69 \text{ MPa}$$

$$f_B = \frac{-3.30}{0.48} - \frac{3.30 \times 0.45}{0.1531} + \frac{2.3760}{0.1531} = -1.06 \text{ MPa}$$

These stresses are within permissible limits. Hence the chosen section is adequate.

Using 4 cables, each having 24 wires of 7 mm diameter, the initial force in each cable is equal to $(3.8824/4) = 0.9706$ MN. Hence the minimum ultimate strength of each cable should not be less than $(0.9706/0.8) = 1.2133$ MN. Referring to Table 3.3.2, the ultimate strength of each cable is 1.3820 MN which is adequate. The minimum diameter of sheathing for the cable is 57 mm. At midspan the cable may be located as shown in Fig. 6.8.4 (c). The minimum clear concrete cover over the sheathing is 71.5 mm which is sufficient. In order to obtain suitable locations for end anchorages, cables 1 and 3 may run straight over the whole span except the end blocks in which they may be raised through a distance of 0.2 m. Cables 2 and 4 may be raised with a parabolic profile with end eccentricities equal to zero and -0.3 mres pectively.

The end block, which is the portion of the beam near its end in which the anchorage zone stresses are appreciable, may have a rectangular section as shown in the end view of Fig. 6.8.4 (d). The eccentricity of prestress at the ends is 0.075 m which is within the safe zone for the cable line. In addition to prestressing cables, untensioned reinforcing bars and stirrups may also be provided in accordance with the code provisions.

EXAMPLE 6.8.3

Design a post-tensioned bridge girder of unsymmetrical I-section for an effective span of 30 m. The girder has to carry its own weight at the stage of stress transfer. The superimposed load due to all effects at the final stage is 39 kN/m. Concrete is of M-35 grade. Adopt Magnel-Blaton system of post-tensioning and working stresses as per IS code for Type 2 (limited prestress) members. The strength of concrete at stress transfer, $f_{ci} = 24$ MPa. Take $f_{pu} = 1540$ MPa.

Solution

(1) As the girder carries heavy intensity of loading, the overall depth may be taken equal to (span/15),

$$D = \frac{30}{15} = 2 \text{ m}$$

(2) Assuming the aspect ratio of the equivalent rectangular section equal to 4, the cross-sectional area,

$$A = \frac{D^2}{4} = 1 \text{ m}^2$$

Taking the unit weight of concrete equal to 0.024 MN/m³, weight per unit length,

$$w_b = 1 \times 1 \times 0.024 = 0.024 \text{ MN/m}$$

Alternatively, using Eq. (6.8.1) and taking $c_2 = 0.125$, $c_3 = 4$ and $d_p/D = 0.85$,

$$c_1 = \frac{0.125 \times 4 \times 0.024 \times 30 \times \left(\frac{30}{2}\right)}{35 \times 0.85^2} = 0.2135$$

$$w_b = \frac{1.6 \times 0.2135 \times 0.039}{1 - 1.4 \times 0.2135} = 0.019 \text{ MN/m}$$

Adopting the larger value of w_b, the dead load moment at the initial stage,

$$M_D = \frac{0.024 \times 30^2}{8} = 2.700 \text{ MN·m}$$

(3) The live load moment at the final stage,

$$M_L = \frac{0.039 \times 30^2}{8} = 4.3875 \text{ MN·m}$$

$$M_w = M_D + M_L = 7.0875 \text{ MN·m}$$

(4) Taking the internal lever arm equal to 0.75 D for the unsymmetrical I-section,

$$P = \frac{M_w}{jd_p} = \frac{7.0875}{0.75 \times 2} = 4.725 \text{ MN}$$

Taking loss of prestress equal to 15 per cent,

$$P_i = \frac{4.725}{0.85} = 5.5588 \text{ MN}$$

(5) From Table 6.7.1,

$$\sigma_{cbc} = 0.3929 f_{ck} = 0.3929 \times 35 = 13.75 \text{ MPa}$$

$$\sigma_{cbci} = 0.5116 f_{ci} = 0.5116 \times 24 = 12.2784 \text{ MPa}$$

Assuming that the girder has no joints, $\sigma_{ct} = \sigma_{cti} = 2$ MPa as per IS code provisions. Taking $y_T/D = 0.45$ and $y_B/D = 0.55$ for unsymmetrical I-section and using Eq. (6.8.6),

$$p_i = 0.45 (12.2784 + 3) - 3 = 3.8753 \text{ MPa}$$

$$p = 0.55 (13.75 + 3) - 3 = 6.2125 \text{ MPa}$$

Hence,
$$A = \frac{P_i}{p_i} = \frac{5.5588}{3.8753} = 1.4344 \text{ m}^2$$

$$A = \frac{P}{p} = \frac{4.7250}{6.2125} = 0.7606 \text{ m}^2$$

(6) Taking $e_p\, y_T/k^2 = e_p\, y_B/k^2 = 2$ for a flanged section, $A = 1 \text{ m}^2$ as in step 2, $P_i = 5.5588$ MN and $P = 4.725$ MN as in step 4 and using Eq. (6.4.9) and (6.4.10),

$$Z_T \geqslant \frac{4.3875}{(13.75 + 3)\left(\dfrac{5.5588 - 4.7250}{1}\right)(2 - 1)}$$

$$\geqslant 0.2757 \text{ m}^3$$

$$Z_B \geqslant \frac{4.3875}{(12.2784 + 3)\left(\dfrac{5.5588 - 4.7250}{1}\right)(2 + 1)}$$

$$\geqslant 0.3434 \text{ m}^3$$

(7) Keeping in view the requirements in steps 1, 5 and 6, the unsymmetrical I-section shown in Fig. 6.8.5 (a) may be tried.

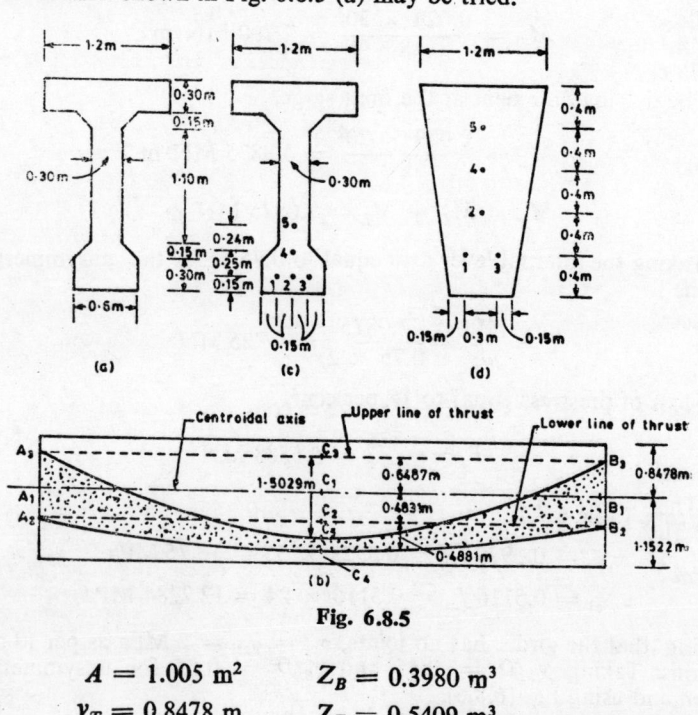

Fig. 6.8.5

$A = 1.005 \text{ m}^2$ $Z_B = 0.3980 \text{ m}^3$

$y_T = 0.8478 \text{ m}$ $Z_T = 0.5409 \text{ m}^3$

$y_B = 1.1522 \text{ m}$ $w_b = 1.005 \times 1 \times 0.024$
$\phantom{y_B = 1.1522 \text{ m}}$ $ = 0.02412 \text{ MN/m}$

$$I = 0.4586 \text{ m}^4 \qquad M_D = \frac{0.02412 \times 30^2}{8} = 2.7135 \text{ MN} \cdot \text{m}$$

$$k^2 = 0.4563 \text{ m}^2 \qquad M_L = 4.3875 \text{ MN} \cdot \text{m}$$

$$k^2/y_B = 0.3960 \text{ m} \qquad P = 4.7250 \text{ MN}$$

$$k^2/y_T = 0.5382 \text{ m} \qquad P_i = 5.5588 \text{ MN}$$

(8) The lower limit of the line of thrust corresponds to the lesser of the two values given by Eq. (6.5.1),

$$e_{cl} = 0.3960 \left(\frac{12.2784 \times 1.005}{5.5588} - 1 \right) = 0.4831 \text{ m}$$

$$e_{cl} = 0.5382 \left(\frac{3 \times 1.005}{5.5588} + 1 \right) = 0.8301 \text{ m}$$

The upper limit of the line of thrust corresponds to the lesser of the two values given by Eq. (6.5 2),

$$e_{cu} = 0.5382 \left(\frac{13.75 \times 1.005}{4.725} - 1 \right) = 1.0358 \text{ m}$$

$$e_{cu} = 0.3960 \left(\frac{3 \times 1.005}{4.725} + 1 \right) = 0.6487 \text{ m}$$

The safe zone for the line of thrust is located between the straight lines A_2 B_2 and A_3 B_3 shown in Fig. 6.8.5 (b). The maximum ordinate of the lower parabola corresponding to the lower limit of the cable line,

$$C_2 C_4 = \frac{M_D}{P_i} = \frac{2.7135}{5.5588} = 0.4881 \text{ m}$$

The maximum ordinate of the upper parabola corresponding to the upper limit of the cable line.

$$C_3 C_5 = \frac{M_w}{P} = \frac{(2.7135 + 4.3875)}{4.725} = 1.5029 \text{ m}$$

Safe zone for cable line is shown by shaded area in the figure.

(9) According to Fig. 6.8.5 (b), the minimum permissible eccentricity of prestress,

$$e_p = 0.8542 \text{ m}$$

This eccentricity may be adopted. This gives concrete cover,

$$c = 1.1522 - 0.8542 = 0.2890 \text{ m}$$

which is adequate.

(10) The stresses at the initial and final stages may now be checked using Eq. (5 2.3) to (5.2.6).

(a) *Initial Stage*

$$f_T = \frac{-5.5588}{1.005} + \frac{5.5588 \times 0.8542 \times 0.8478}{0.4586} - \frac{2.7135 \times 0.8478}{0.4586}$$

$$= -1.77 \text{ MPa}$$

$$f_B = \frac{-5.5588}{1.005} - \frac{5.5588 \times 0.8542 \times 1.1522}{0.4586} + \frac{2.7135 \times 1.1522}{0.4586}$$

$$= -10.64 \text{ MPa}$$

(b) *Final Stage*

$$f_T = \frac{-4.7250}{1.005} + \frac{4.7250 \times 0.8542 \times 0.8478}{0.4586} - \frac{7.101 \times 0.8478}{0.4586}$$

$$= -10.37 \text{ MPa}$$

$$f_B = \frac{-4.7250}{1.005} - \frac{4.7250 \times 0.8542 \times 1.1522}{0.4586} + \frac{7.101 \times 1.1522}{0.4586}$$

$$= 3 \text{ MPa}$$

Comparing these stresses with the permissible stresses computed in step 5, it is noted that the stresses are acceptable. Hence the design is adequate. However. there is a scope for a slight reduction in the cross-section.

(11) Maximum permissible initial stress as per IS code,

$$f_{pi} = 0.8 f_{pu} = 0.8 \times 1540 = 1232 \text{ MPa}$$

Using 5 mm diameter wires,

$$\text{Number of wires} = \frac{5.5588}{1232 \times \frac{\pi}{4} \times 5^2 \times 10^{-6}} = 229.7$$

Hence 5 cables each of 48 wires may be adopted. Initial stress in **each** wire,

$$f_{pi} = \frac{5.5588}{5 \times 48 \times \frac{\pi}{4} \times 5^2 \times 10^{-6}} = 1179 \text{ MPa}$$

The cables may be located at midspan section as shown in Fig. 6.8.5 (c). The centroid of the prestressing steel is located at 0.2980 m from bottom face. Hence,

$$e_p = 1.1522 - 0.2980 = 0.8542 \text{ m}$$

The position of the cables at the ends of the beam is shown in Fig. 6.8.5 (d). Cables 1 and 3 run straight parallel to the centroidal axis except that they are raised gradually through a vertical distance of 0.25 m in the end blocks. Cables 2, 4 and 5 are raised in a parabolic profile to points indicated in the end view. It may be verified readily that the cable line lies entirely within its safe zone.

6.9 LIMIT STATES OF STRENGTH

Code provisions specify requirements to ensure adequate safety against collapse or limit state of strength under the action of four types of internal forces. viz. bending moment, shear force, twisting moment and axial force. The other limit states of strength correspond to (i) static equilibrium of a part or of the whole of the structure considered as a rigid body, (ii) transformation of the structure into a mechanism. (iii) buckling due to elastic or plastic instability, (iv) punching shear and (v) fatigue. As explained in Sec. 2.7, the desired margin of safety is achieved through partial safety factors in respect of materials and loads. Alternatively the same objective is achieved through strength reduction factors and load factors. As flexure is often the most dominant internal action, the ultimate strength in flexure is of great practical importance. The code provisions for determination of ultimate strength in flexure are described in the following subsections.

6.9.1 Indian Code

According to IS: 1343 — 1980, the ultimate moment of a prestressed concrete beam may be expressed as

$$M_u = A_p f_p (d_p - 0.42 x_u) \qquad (6.9.1)$$

in which the stress in prestressing steel f_p and the depth of the neutral axis x_u at limit state of collapse depend mainly upon the effective proportion of prestressing steel defined by

$$\omega_p = \frac{A_p}{b \, d_p} \left(\frac{f_{pu}}{f_{ck}} \right)$$

$$= \rho_p \left(\frac{f_{pu}}{f_{ck}} \right) \text{ for rectangular sections} \qquad (6.9.2 \text{ a})$$

$$\omega_p = \frac{A_p}{b_f d_p} \left(\frac{f_{pu}}{f_{ck}} \right)$$

$$= \rho_p \left(\frac{f_{pu}}{f_{ck}} \right) \text{for flanged sections} \qquad (6.9.2 \text{ b})$$

in which b_f is the width of the compression flange. As shown in Table 6.9.1, the values of f_p and x_u in the case of bonded beams depend upon the technique of prestressing, viz. pretensioning and post-tensioning. In the case of unbonded beams, these values depend upon the span/depth ratio l/D as shown in Table 6.9.2, in which f_{pu}, the effective prestress in steel after losses, should not be less than $0.45 f_{pu}$.

Table 6.9.1   Values of f_p and x_u for bonded beams

ω_p	$f_p/(0.87 f_{pu})$		x_u/d_p	
	Pretensioning	Post-tensioning	Pretensioning	Post-tensioning
0.025	1.0	1.0	0.054	0.054
0.050	1.0	1.0	0.109	0.109
0.10	1.0	1.0	0.217	0.217
0.15	1.0	1.0	0.326	0.316
0.20	1.0	0.95	0.435	0.414
0.25	1.0	0.90	0.542	0.488
0.30	1.0	0.85	0.655	0.558
0.40	0.9	0.75	0.783	0.653

Table 6.9.2   Values of f_p and x_u for unbonded beams

ω_p	(f_p/f_{pe}) for (l/D) equal to			(x_u/d_p) for (l/D) equal to		
	30	20	10	30	20	10
0.025	1.23	1.34	1.45	0.10	0.10	0.10
0.050	1.21	1 32	1.45	0.16	0.16	0.18
0.10	1.18	1.26	1.45	0.30	0.32	0.36
0.15	1.14	1.20	1.36	0.44	0.46	0.52
0.20	1.11	1.16	1.27	0.56	0.58	0.64

Equation (6.9.1) is valid for rectangular sections and also for flanged sections in which the neutral axis lies within the flange. If the neutral axis lies in the web, the analysis discussed in Sec. 5.8 may be used. Equation (6.9.1) is based on the stress block shown in Fig. 5.8.7 (c) recommended by IS: 456—1978. The average ordinate of the stress block is equal to $0.36 f_{ck}$ and the depth of the centre of comprsssion is equal to $0.42 x_u$ from the top face.

6.9.2 British Code

According to BS: 8110-1985, the ultimate moment of resistance of a prestressed concrete beam with bonded or unbonded tendons, all of which are located in the tension zone, may be computed from

$$M_u = A_p f_p (d_p - 0.45 x_u) \tag{6.9.3}$$

Equation (6.9.3) is applicable for a rectangular section and also for a flanged section in which the thickness of flange is not less than $0.9 x_u$. For beams with bonded tendons, the values of f_p and x_u may be determined from Table 6.9.3. For beams with unbonded tendons, the value of f_p in MPa and x_u may be determined from,

$$f_p = f_{pe} + \frac{7000}{(l/d_p)} (1 - 1.7 \omega_p) \tag{6.9.4}$$

$$x_u = 2.47 \left(\omega_p \frac{f_p}{f_{pu}} d_p \right) \tag{6.9.5}$$

Table 6.9.3   Values of f_p and x_u for bonded beams

ω_p	$f_p/0.87 f_{pu}$ for f_{pe}/f_{pu} equal to			x_u/d_p for $\dfrac{f_{pe}}{f_{pu}}$ equal to		
	0.6	0.5	0.4	0.6	0.5	0.4
0.05	1.0	1.0	1.0	0.11	0.11	0.11
0.10	1.0	1.0	1.0	0.22	0.22	0.22
0.15	0.99	0.97	0.95	0.32	0.32	0.31
0.20	0.92	0.90	0.88	0.40	0.39	0.38
0.25	0.88	0.86	0.84	0.48	0.47	0.46
0.30	0.85	0.83	0.80	0.55	0.54	0.52
0.35	0.83	0.80	0.76	0.63	0.60	0.58
0.40	0.81	0.77	0.72	0.70	0.67	0.62
0.45	0.79	0.74	0.68	0.77	0.72	0.66
0.50	0.77	0.71	0.64	0.83	0.77	0.99

The value of f_p should not be taken greater than $0.7\ f_{pu}$. The length l in Eq. (6.9.4) may be taken equal to the length of unbonded tendons between the end anchorages.

6.9.3 American Code

The ultimate moment capacity of a prestressed concrete section may be taken as the total tension in prestressing steel and unstressed reinforcing steel in the tension zone multiplied by the internal lever arm

$$M_u = (A_p f_p + A_s f_s)\ \text{(lever arm)} \tag{6.9.6}$$

where A_s = area of unstressed or auxiliary reinforcing steel in tension zone

f_s = stress in unstressed steel in tension zone

In the absence of untensioned reinorcing steel, the ultimate moment in accordance with American Code ACI:318 — 1989 may be expressed as

$$M_u = \Phi\ A_p f_p \left(d_p - \frac{a}{2} \right)$$
$$= \Phi\ A_p f_p\ (d_p - k_2\ x_u) \tag{6.9.7}$$

where $\bar{\phi}$ = capacity reduction factor which may be taken as 0.9 for flexure.

a = depth of the equivalent rectangular stress block having an ordinate equal to $0.85 f_c'$

k_2 = factor for the depth of the centroid of the equivalent rectangular stress block

The factor k_2 may be determined from,

$$k_2 = 0.425 \ \text{for} \ f_c' \leqslant 28 \ \text{MPa} \tag{6.9.8 a}$$

$$= 0.425 - 0.025 \left(\frac{f_c'}{7} - 4 \right) \text{for} \ f_c' > 28 \ \text{MPa} \tag{6.9.8 b}$$

The stress in prestressing steel may be determined as follows:

(a) *Beams with bonded prestressing tendons*

$$f_p = f_{pu} \left[1 - \frac{\gamma_p}{2k_2} \left\{ \rho_p\ \frac{f_{pu}}{f_c'} + \frac{d}{d_p} (\omega - \omega') \right\} \right] \tag{6.9.9}$$

where γ_p = 0.55 for f_{py}/f_{pu} not less than 0.80

$\quad\quad$ = 0.40 for f_{py}/f_{pu} not less than 0.85

$\quad\quad$ = 0.28 for f_{py}/f_{pu} not less than 0.90

$$\omega = \frac{A_s f_y}{bd f_c'}$$

$$\omega' = \frac{A'_s}{bd}\frac{f_y}{f_c'}$$

A_s, A'_s = areas of untensioned reinforcements in tension and compression zone respectively.

If any untensioned steel in compression zone is taken into account, the term $\left[\rho_p \frac{f_{pu}}{f_c'} + \frac{d}{d_p}(\omega - \omega')\right]$ in equation (6.9.9) should not be taken less than 0.17. Equation (6.9.9) is valid when the effective depth of compression steel A'_s measured from compression face is not greater than 0.15 d_p. The favourable effect of compression steel on ultimate moment reflected by Eq. (6.9.9) may not be available when its effective depth is more than 0.15 d_p because the compression steel may then have a stress much lower than its yield stress.

(b) Beams with unbonded tendons

(i) $\dfrac{\text{Span}}{\text{depth}} < 35$

$$f_p = f_{pe} + 69 + \frac{f_c'}{100\,\rho_p} \tag{6.9.10}$$

The stress f_p should not be taken greater than f_{py} nor ($f_{pe} + 415$).

(ii) $\dfrac{\text{Span}}{\text{depth}} > 35$

$$f_p = f_{pe} + 69 + \frac{f_c'}{300\,\rho_p}$$

The stress f_p should not be taken greater than f_{py} nor ($f_{pe} + 207.5$).

When deformed bars are used as untensioned reinforcement in the tension zone, their contribution to total tension may be determined by assuming them to have yielded. When other types of steel are used, they may be included in the strength computation only when strain compatibility analysis discussed in Sec. 5.8 or 10.5 is used to determine stresses in steel. Equation (6.9.9) which is an approximation to the more accurate method based on strain compatibility and equilibrium may underestimate the ultimate strength in case of beams with high percentages of reinforcement. The foregoing equations are applicable to rectangular sections and flanged sections in which the neutral axis is located within the compression flange. The method of strain compatibility and equilibrium should be used when the neutral axis lies outside the compression flange and also when a part of the prestressing steel is located in the compression zone.

6.9.4 European Code

The CEB--FIP Model Code for Concrete Structures-1978 recommends the use of stress block shown in Fig. 6.9.1 (b) corresponding to the strain dia-

gram of Fig. 6.9.1 (a). The code also permits the use of the simplified rectangular stress block shown Fig. 6.9.1 (c). The stress ck of Fig. 6.9.1 (b) is
similar to that recommended by IS: 456-
1978. The coefficient 0.85 in Fig. 6.9.1
(c) is to be used for sections in which
the compression zone is either constant
or it increases towards the extreme fibre
in compression. The coefficient should
be reduced to 0.80 for sections such as
triangluar or circular sections in which
the width of the compression zone decreases towards the extreme fibre. The
partial safety factor for concrete γ_c is to
be taken equal to 1.5 under normal

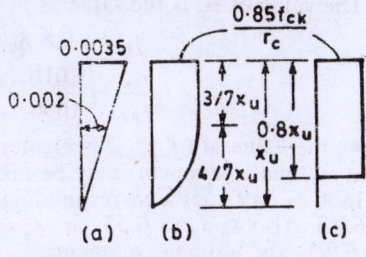

Fig. 6.9.1

conditions of supervision a.. control. It may be increased or decreased
appropriately depending upon the prevailing conditions of supervision and
control.

Example 6.9.1 Check the adequacy of the beam designed in Example
6.8.1 in respect of ultimate bending moment at midspan using (i) Indian
code, (ii) British code and (iii) American code.

Solution The midspan cross-section of the beam in Example 6.8.1 is
rectangular, 0.2×0.6 m having the following properties:

$$A_p = 693 \text{ mm}^2 \qquad\qquad d_p = 0.49 \text{ m}$$
$$f_{pu} = 1600 \text{ MPa} \qquad\qquad f_{ck} = 40 \text{ MPa}$$
Also, $\qquad M_D = 0.081 \text{ MN·m} \qquad\qquad M_L = 0.1181 \text{ MN·m}$
$$M_w = M_D + M_L = 0.1991 \text{ MN·m}$$

(i) *Indian Code*

Effective proportion of prestressing steel,
$$\omega_p = \frac{693 \times 10^{-6} \times 1600}{0.2 \times 0.4 \times 40} = 0.283$$

Referring to Table 6.9.1 and using linear interpolation for x_u,
$$\frac{f_p}{0.87 f_{pu}} = 1 \qquad \frac{x_u}{d_p} = 0.617$$
Hence, $\qquad f_p = 0.87 \times 1600 = 1392 \text{ MPa}$
$$x_u = 0.617 \times 0.49 = 0.3021 \text{ m}$$

Using Eq. (6.9.1), the ultimate moment capacity,
$$M_u = 693 \times 10^{-6} \times 1392 \ (0.49 - 0.42 \times 0.3021)$$
$$= 0.3503 \text{ MN·m}$$

Referring to Table 2.7.1, the partial safety factor for dead load as well as
live load is 1.5. Hence design ultimate moment,
$$M_{ud} = 1.5 \times 0.081 + 1.5 \times 0.1181$$
$$= 0.2987 \text{ MN·m}$$

As M_{ud} is less than M_u, the section is adequate.

(ii) *British Code*

The value of ω_p is the same as in (i)

$$f_{pe} = 0.8\, f_{pi} = 0.8 \times 1263 = 1010 \text{ MPa}$$

$$\frac{f_{pe}}{f_{pu}} = \frac{1010}{1600} = 0.63125$$

As the value of (f_{pe}/f_{pu}) is greater than 0.6, the stress in prestressing steel at ultimate moment may be taken conservatively equal to $0.86 \times 0.87 \times 1600 = 1197$ MPa corresponding to $f_{pe}/f_{pu} = 0.6$ in accordance with Table 6.9.3. Also $x_u/d_p = 0.53$ or $x_u = 0.53 \times 0.49 = 0.2597$ m. Using Eq. (6.9.3), the ultimate moment,

$$M_u = 693 \times 10^{-6} \times 1197 (0.49 - 0.45 \times 0.2597)$$

$$= 0.3095 \text{ MN·m}$$

Referring to Table 2.7.1, the partial safety factors for dead and live loads are 1.4 and 1.6 respectively. Hence design ultimate moment,
$$M_{ud} = 1.4 \times 0.081 + 1.6 \times 0.1181$$

$$= 0.3024 \text{ MN·m}$$

As M_{ud} is less than M_u, the section is adequate.

(iii) *American Code*

$$\frac{f_{py}}{f_{pu}} = \frac{1360}{1600} = 0.85$$

Hence $\qquad \gamma_p = 0.4$

$$\rho_p = \frac{693 \times 10^{-6}}{0.2 \times 0.49} = 0.0070714$$

$$f_c' = 0.8\, f_{ck} = 0.8 \times 40 = 32 \text{ MPa}$$

From Eq. (6.9.8b),

$$k_2 = 0.425 - 0.025 \left(\frac{32}{7} - 4 \right) = 0.4107$$

Hence using Eq. (6.9.9),

$$f_p = 1600 \left[1 - \frac{0.4}{2 \times 0.4107} \times 0.0070714 \times \frac{1600}{32} \right]$$

$$= 1324.5 \text{ MPa}$$

Equating total tension and total compression, the depth of equivalent rectangular stress block,

$$a = \frac{693 \times 10^{-6} \times 1324.5}{0.2 \times 0.85 \times 32} = 0.1687 \text{ m}$$

Hence from Eq. (6.9.7), the ultimate moment

$$M_u = 0.9 \times 693 \times 10^{-6} \times 1324.5 \left(0.49 - \frac{0.1687}{2} \right)$$

$$= 0.3351 \text{ MN·m}$$

As discussed in Sec. 2.8, the load factors for dead and live loads are 1.4 and 1.7 respectively. Hence design ultimate moment,

$$M_{ud} = 1.4 \times 0.081 + 1.7 \times 0.1181 = 0.3142 \text{ MN} \cdot \text{m}$$

As M_{ud} is less than M_u, the section is adequate.

Example 6.9.2 Using IS code, determine the ultimate moment capacity of a post-tensioned bonded beam of T-section having a flange 1.5 m wide and 0.3 m thick. The total area of prestressing steel is 4725 mm^2 and its centroid is 2.1 m below the top face. Concrete is of $M-45$ grade and ultimate tensile strength of prestressing steel is 1500 MPa.

Solution Assuming that the neutral axis lies in the flange, effective proportion of prestressing steel,

$$\omega_p = \frac{4725 \times 10^{-6} \times 1500}{1.5 \times 2.1 \times 45} = 0.05$$

Referring to Table 6.9.1,

$$\frac{f_p}{0.87 f_{pu}} = 1 \qquad \frac{x_u}{d_p} = 0.109$$

Hence,

$$f_p = 0.87 \times 1500 = 1305 \text{ MPa}$$

$$x_u = 0.109 \times 2.1 = 0.2289 \text{ m}$$

As neutral axis lies in the flange, the values given in Table 6.9.1 are applicable. Using Eq. (6.9.1), the ultimate moment capacity,

$$M_u = 4725 \times 10^{-6} \times 1305 (2.10 - 0.42 \times 0.2289)$$

$$= 12.356 \text{ MN} \cdot \text{m}$$

Example 6.9.3 Using the IS code, check the adequacy of the section of the unbonded beam in Example 5.8.7 if the dead and live load moments are 0.07 and 0.05 MN·m respectively. The span/depth ratio, $\frac{l}{D} = 20$.

Solution The beam of rectangular cross-section 0.20×0.65 m has following properties:

$$A_p = 240 \text{ mm}^2 \qquad d_p = 0.60 \text{ m}$$

$$b = 0.20 \text{ m} \qquad f_{pu} = 1600 \text{ MPa}$$

$$f_c' = 35 \text{ MPa} \qquad f_{pe} = 1100 \text{ MPa}$$

Taking $f_{ck} = f_c'/0.8 = 43.75$ MPa, effective proportion of prestressing steel,

$$\omega_p = \frac{240 \times 10^{-6} \times 1600}{0.2 \times 0.6 \times 43.75} = 0.073$$

Referring to Table 6.9.2 and using linear interpolation,

$$\frac{f_p}{f_{pe}} = 1.293 \qquad \frac{x_u}{d_p} = 0.2339$$

Hence, $\qquad f_p = 1.293 \times 1100 = 1421.5 \text{ MPa}$

$$x_u = 0.2339 \times 0.6 = 0.1403 \text{ m}$$

Hence from Eq. (6.9.1),

$$M_u = 240 \times 10^{-6} \times 1421.5 (0.60 - 0.42 \times 0.1403)$$

$$= 0.1846 \text{ MN} \cdot \text{m}$$

Referring to Table 2.7.1, the partial safety factor for dead load as well as live load is 1.5. Hence design ultimate moment,

$$M_{ud} = 1.5 \times 0.07 + 1.5 \times 0.05$$

$$= 0.18 \text{ MN} \cdot \text{m}$$

As M_{ud} is less than M_u, the section is adequate.

6.10 LIMIT STATES OF SERVICEABILITY

In addition to the limit states of strength discussed in Sec. 6.9, the code provisions specify requirements to ensure adequate serviceability at all stages under service conditions. The limit states of serviceability specified by the codes correspond to (i) deflection, (ii) crack width, (iii) vibration and (iv) compression. The limit states of deflection and crack width which relate to the deformation of the member are discussed in Sections 8.6 and 10.10. The limit state of vibration is relevant to structures which are subjected to vibration due to action caused by wind forces, machinery or convoys of vehicles. In such structures it is necessary to take measures to prevent discomfort or alarm on the part of the users. It is also necessary to check that excessive vibrations do not infringe on the normal functioning of the structure. The provisions in respect of the limit state of compression ensure that the compressive stresses at all stages in the working range do not exceed the maximum permissible stresses. The most crucial stages in the working range are often the initial stage or the stage at stress transfer and the final stage. The maximum permissible compressive stresses at these stages are discussed in Sec. 6.7. It is, however, necessary to check that the compressive stresses at all other stages in the working range do not exceed the specified maximum values. The stages to be considered depend upon the technique adopted and the type of structures. For instance, in the case of precast pre-stressed elements manufactured in a central plant. it is necessary to check the stresses during lifting, transportation, hoisting and launching. The stresses during some of these operations may be opposite in nature to those caused by the normal service loads. For instance, a precast segment of a simply supported bridge girder may act like a cantilever during launching. In this case temporary prestressing tendons may be provided which may be removed at a later stage. Alternatively, these tendons may become a part of the overall arrangement of tendons When prestressing is carried out in stages, it is necessary to check the stresses at every stage. Stresses may also have to be checked at various stages of the removal of formwork. The Indian Code IS: 1343 1980 specifies permissible compressive stresses only at the two crucial stages, viz. the initial and the final stages. The permissible stresses at other stages should correspond to the age of concrete at the stage under consideration. The Indian Code relates the maximum permissible tensile stresses with the limit state of crack width as discussed in Sec. 6.7.

REFERENCES

6.1 Bennett, E.W., *A graphical method for the design of prestressed beams*, Concrete and Constructional Engineering, Vol. 53, November 1959, pp. 399-403.

6.2 CEB-FIP Joint Committee International, *Recommendations for the design and construction of concrete structures*, Cement and Concrete Association (London), June 1970.

6.3 FIP-CEB Joint Committee, *Practical recommendations for the design and construction of prestressed concrete structures*, June 1966.

6.4 Lin, T.Y. and Scordelis, A.C., *Selection and design of prestressed concrete beam sections*, Journal of the American Concrete Institute, Proceedings Vol. 49, November 1953, pp. 209-24.

6.5 Ramaswamy, G.S. and Raman, N.V., *Optimum design of prestressed concrete sections for minimum cost by non-linear programming*, Sixth FIP Congress, Prague, June 1970.

6.6 Rowe, R.E.; Cranston, W.B. and Best, B.C., *New concepts in the design of structural concrete*, Structural Engineer (London), Vol. 43, 1965, pp. 399-403.

PROBLEMS

6.1 A rectangular section 0.25×0.60 m has an initial prestressing force of 1.2 MN and final prestressing force of 0.9 MN. Determine the lower and upper limits of the line of thrust if the permissible stresses in compression, $\sigma_{cbci} = \sigma_{cbc} = 16$ MPa and permissible tensile stresses, $\sigma_{cti} = \sigma_{ct} = 0$.

6.2 Determine the lower and upper limits of the line of thrust in the section of Problem 6.1 if $\sigma_{cbci} = \sigma_{cbc} = 16$ MPa and $\sigma_{cti} = \sigma_{ct} = 3$ MPa.

6.3 Determine the minimum and maximum permissible eccentricities of prestress in the section of Problem 6.1 if $M_D = 0.12$ MN·m and $M_L = 0.105$ MN·m.

6.4 Determine the minimum and maximum eccentricities of prestress in the section of Problem 6.1 if $M_D = 0.12$ MN·m, $M_L = 0.105$ MN·m and the permissible stresses $\sigma_{cbci} = \sigma_{cbc} = 16$ MPa and $\sigma_{cti} = \sigma_{ct} = 3$ MPa.

6.5 Determine the maximum dead load moment which can be compensated merely by increasing the eccentricity of prestress without increasing the prestressing force in the section of Problem 6.1. The required value of minimum concrete cover, $c = 0.06$ m.

6.6 Determine the maximum value of M_D which can be compensated by increasing e_p in the section of Problem 6.1 if the permissible stresses, $\sigma_{cbci} = \sigma_{cbc} = 16$ MPa, $\sigma_{cti} = \sigma_{ct} = 3$ MPa and minimum concrete cover, $c = 0.06$ m.

6.7 Design the midspan section of a precast pretensioned beam of a residential building floor. The effective span is 4 m. The beam has to carry only its own weight at the initial stage. The total service load exclusive of self weight at the final stage is 15 kN/m. Use M-40 concrete. Ultimate strength of high tensile steel, $f_{pu} = 1750$ MPa. Use provisions of IS code.

6.8 Design a precast pretensioned beam of rectangular section for an industrial building to carry only its own weight at the initial stage and a total working load of 9 kN/m exclusive of own weight at the final stage over an effective span of 12.5 m. Use M-45 concrete and the provisions of IS code.

6.9 A post-tensioned girder of a warehouse has to carry only its self weight at the initial stage and a total working load of 33 kN/m exclusive of self weight at the final stage over an effective span of 27 m. Design the girder using a symmetrical I-section and determine a suitable cable line. Adopt M-35 concrete, Freyssinet system and provisions of IS code.

6.10 Design a post-tensioned bridge girder of unsymmetrical I-section to carry only its own weight at the initial stage and total working load of 16 kN/m execlusive of self weight at final stage over an effective span of 27 m. Adopt M-40 concrete, Lee Mc Call system and provisions of IS code.

6.11 Check the adequacy of the beam in Problems 6.7 to 6.10 at the limit state of collapse as per Indian, British and American codes.

6.12 Discuss the provisions in respect of partial safety factors for materials and loads in accordance with Indian, British and European codes. Also discuss the provisions for load factors and strength reduction factors as per American coed.

7

Shear, Torsion, Bond and Anchorage Zone

7.1 INTRODUCTION

Although bending moment is generally the most dominant internal force which often governs the size of cross-section of a flexural member, the adequacy of design has also to be checked for other types of internal forces. Due to the common use of flanged sections with thin webs in prestressed concrete structures, shear stresses assume special importance in respect of strength and serviceability of prestressed concrete members. In those structures in which torsion arises either for equilibrium or for compatibility of deformations, it is necessary to verify that the torsional stresses by themselves or in combination with other types of stresses do not violate the strength and serviceability requirements. The nature and role of bond stresses in prestressed concrete beams is different from those in reinforced concrete beams In addition to the conventional role as in reinforced concrete structures, the bond stresses in prestressed concrete serve the primary function of stress transfer in pretensioned members and provide the second line of defence in bonded post-tensioned members in the event of failure of end anchorages. The end blocks of prestressed concrete members are subjected to severe localised stresses, called anchorage zone stresses, which call for careful analysis and design to ensure proper functioning of anchorages. The tendency for bursting, splitting and spalling of concrete due to the application of large concentrated forces delivered by the prestressing tendons must be checked by the provision of adequate reinforcement in the end blocks.

7.2 SHEAR

It is shown in Sec. 1.8 that the net or effective shear force at any section of a prestressed concrete member with a draped cable line is smaller than the external shear force due to the applied loads. The distribution of shear stresses due to the effective shear force depends on the geometry of the cross-section and the extent of cracking. In an uncracked section, the distribution of shear stresses may be determined with reasonable accuracy by the conventional elastic theory. The distribution changes continuously as the extent of cracking progresses and reaches a limiting state at the stage of ultimate strength or collapse of the member.

7.2.1 Uncracked Section

In the precracking range, the distribution of shear stresses may be determined on the basis of elastic theory ignoring the presence of steel and treating concrete as a homogeneous, isotropic and elastic material obeying Hooke's law. Hence the intensity of shear stress v at any point P at a distance y from the centroidal axis of the section shown in Fig. 7.2.1 may be expressed as

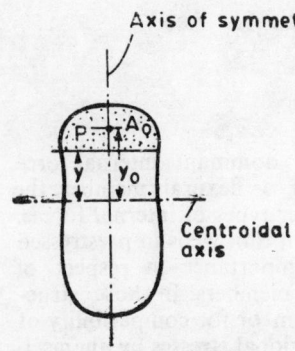

Axis of symmetry

Centroidal axis

Fig. 7.2.1

$$v = \frac{V}{Ib}(A_0\, \bar{y}_0) \qquad (7.2.1)$$

where V_r = effective or net shear force
$\quad = V - P \sin \theta$

b = width of section at the point considered

A_0 = area of the section (shown shaded) between the extreme fibre and the layer through the point under consideration

$\bar{y}_0$ = distance of centroid of area A_0 from centroidal axis

The distribution of shear stresses in sections commonly used for prestressed concrete members are shown in Fig. 7.2.2. The normal stress f at the point considered at distance y from centroidal axis,

$$f = \frac{-P}{A} + \frac{M_p + M}{I}\, y \qquad (7.2.2)$$

The principal stresses f_1, f_2 at the point may be expressed as

$$f_1, f_2 = \frac{f}{2} \pm \sqrt{\left(\frac{f}{2}\right)^2 + v^2} \qquad (7.2.3)$$

As the second term on the right side of Eq. (7.2.3) is numerically greater than the first term, the two principal stresses are always opposite in sign, i.e. one is tensile and the other is compressive. The principal tensile stress, called diagonal tension, may be expressed as

$$f_{dt} = \frac{f}{2} + \sqrt{\left(\frac{f}{2}\right)^2 + v^2} \qquad (7.2.4)$$

The principal compressive stress, called diagonal compression, is given by

$$f_{dc} = \frac{f}{2} - \sqrt{\left(\frac{f}{2}\right)^2 + v^2} \qquad (7.2.5)$$

The inclination of the principal stresses, β is given by the equation,

$$\tan 2\beta = \frac{-2v}{f} \qquad (7.2.6)$$

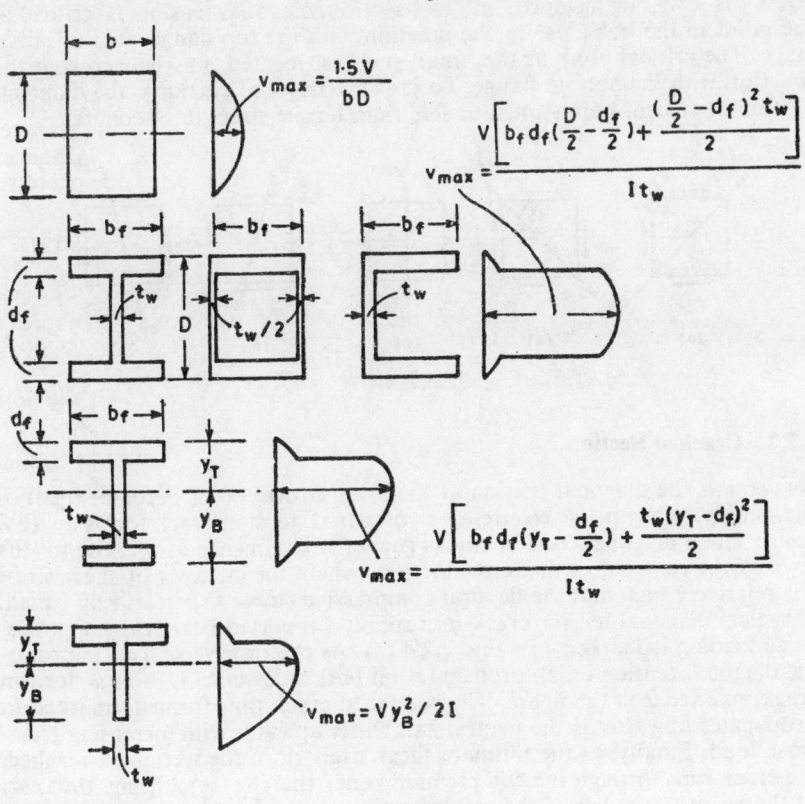

$$v_{max} = \frac{1.5V}{bD}$$

$$v_{max} = \frac{V\left[b_f d_f\left(\frac{D}{2} - \frac{d_f}{2}\right) + \frac{\left(\frac{D}{2} - d_f\right)^2 t_w}{2}\right]}{I t_w}$$

$$v_{max} = \frac{V\left[b_f d_f\left(y_T - \frac{d_f}{2}\right) + \frac{t_w(y_T - d_f)^2}{2}\right]}{I t_w}$$

$$v_{max} = V y_B^2 / 2I$$

ig. 7.2.2

Equation (7.2.6) has two roots differing by 90° which shows that the two principal stresses are orthogonal. Regarding the direction of diagonal tension, it may be noted that the inclination of diagonal tension β is 135° in the case of pure shear in which v is positive and f is zero as shown in Fig. 7.2.3 (a). On the other hand $\beta = 45°$ in the case of pure shear in which v is negative as shown in Fig. 7.2.3 (b). The inclination of diagonal tension, β =90° in the case of simple compression as shown in Fig. 7.2.3 (c). Combining Figures 7.2.3 (a) and (c), it is evident that β lies between 0° and 135° if f is compressive and v is positive as shown in Fig. 7.2.3 (d). On the other hand β lies between 45° and 90° if f is compressive and v is negative as shown in Fig. 7.2.3 (e).

Equation (7.2.4) shows that it is not possible to eliminate diagonal tension by uniaxial prestressing however high the intensity of prestress. Only biaxial prestress can do so. However, biaxial prestressing is not practicable in flexural members such as beams due to the problems associated with the anchorage of transverse tendons. Equation (7.2.4) shows that diagonal tension increases with (v/f) ratio if the normal stress f is compressive. Thus diagonal tension is critical at points at which v is high and the compressive stress f is low. For instance, in an I-section, diagonal tension is critical at the point in the web close to the junction with the top flange at the initial stage. The critical point at the final stage is located in the web near the junction with the bottom flange. To prevent diagonal cracking, the diagonal tension at critical points must be less than tensile strength of concrete.

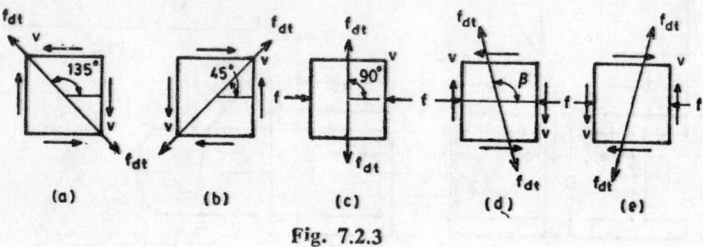

Fig. 7.2.3

7.2.2 Cracked Section

As soon as the diagonal tension at the most critical point becomes equal to the tensile strength of concrete, a diagonal tension crack forms at that point. These diagonal tension cracks generally commence at or close to the level of the centroid of the cross-section where the intensity of shear stress v is relatively high and the flexural compressive stress f is relatively small. A typical diagonal tension crack in a simply supported beam carrying transverse loading is marked 1 in Fig. 7.2.4 (a). As the transverse load increases the diagonal tension crack propagates on both sides until it joins a flexural crack, marked 2 in the figure. The combined crack thus formed, marked 3, propagates upwards as the neutral axis shifts upwards with increasing transverse load. Finally, as the ultimate shear strength of the section is reached, the crack runs through the compression zone, thereby separating the part of the beam to the left of the crack from the rest of the beam. The portion to the left of the crack shown in Fig. 7.2.4 (b) slides upwards along the failure surface which separates it from the right portion. The mechanism of shear failure is complex and has been the subject of extensive research since the decade of fifties. Several internal forces enter the complex shear failure mechanism. The distribution of these internal forces is governed by several parameters related to the geometry of the cross-section, type of loading and the physical properties of concrete and steel. The main internal forces which oppose the shear failure are the shear resistance F_1 due to concrete in the compression zone, the tensile forces F_2 in the vertical legs of the stirrups intersected by the failure crack, the forces F_3 due to interlocking of coarse aggregate particles and the dowel forces F_4 in reinforcing bars and F_5 in prestressing tendons. However, for the sake of simplicity and a certain degree of conservatism in respect of sudden and disastrous shear failure, only the forces F_1 and F_2 are generally relied upon in the computation of shear strength. Adopting this simplified approach for design purposes, the ultimate shear force or shear capacity V may be expressed as

$$V = V_c + V_s \qquad (7.2.7)$$

in which V_c is the ultimate shear force resisted by concrete in flexural compression zone and V_s is the ultimate shear force resisted by shear reinforcement, i.e. the vertical legs of stirrups intersected by the failure crack. The failure crack starts with a vertical orientation from the bottom face but it changes its direction continuously with progressively decreasing inclination as it moves upwards. Taking the average inclination of the failure crack as 45°, the number of stirrups intersected by the crack is approximately equal to (d_v/s) in which d_v is the effective depth of the section in shear and s is spacing of stirrups. Hence the shear force resisted by stirrups V_s may be expressed as

$$V_s = \Omega_v \, \frac{A_v f_{sy} d_v}{s} \qquad (7.2.8)$$

where A_v = area of stirrups (both legs) for resisting shear

f_{sv} = yield stress of stirrups reinforcement

Ω_v = factor reflecting effectiveness of stirrup reinforcement in shear

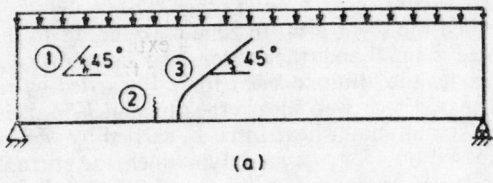

(a)

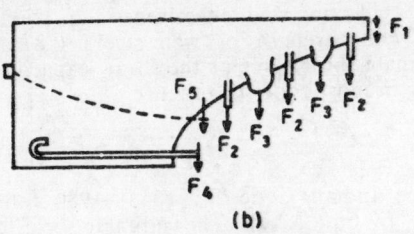

(b)

Fig. 7.2.4

Combining Equations (7.2.7) and (7.2.8), the ultimate shear strength of a prestressed concrete beam with stirrup reinforcement may be expressed as

$$V = V_c + \Omega_v \, \frac{A_v f_{sy} d_v}{s} \qquad (7.2.9)$$

Equation (7.2.9) forms the basis for the shear provisions of prominent codes. They, however, differ in respect of shear force V_c resisted by concrete.

7.2.3 Code Provisions

The earlier versions of code provisions were based on the elastic theory for shear. As shear reinforcement remains dormant in the entire precracking range in which elastic theory can be justified, it was soon realised that meaningful and realistic provisions for shear should be based on ultimate shear strength at the limit state of collapse. The current code provisions on shear are based on extensive tests on reinforced and prestressed concrete beams. In respect of shear failure of a simply supported beam carrying transverse loading, four zones may be distinguished extending from left support to the midspan section. Zone 1 extending a short distance (generally believed to be approximately equal to $D/2$) from the edge of the left support is practically free of cracks, partly due to the compressive stress in vertical direction caused by support reaction. Zone 2 extending to the right of zone 1 is subjected to high shear force and a bending moment which is relatively low to cause flexural cracks. Zone 2 develops web shear cracks near the neutral axis due to diagonal tension at a shear force V_{co} in a flexurally uncracked section. Zone 3 extending to right of zone 2 is subjected to significant shear force and a bending moment which is relatively high to cause flexural cracking. Zone 3 develops inclined flexures-hear cracks at a shear force V_{cr} in a flxurally cracked section. The shear force V_{cr} at the formation of inclined flexure-shear crack under the combined action of flexure and shear is controlled, among other factors, by the moment-shear (M/V) ratio, called the shear arm. In zone 4 extending to the right of zone 3, the shear force is small and, therefore, shear failure is ruled out. In several prominent codes the ultimate shear force V_c carried by concrete is taken to be the smaller of V_{co} at web shear cracking and V_{cr} at inclined flexure-shear cracking. The ultimate shear force V_s carried by web reinforcement in most codes is based on 45° truss model in which the stirrups serve as vertical ties. The code expressions for V_s have the general form given by Eq. (7.2.8). The factor Ω_r reflecting the effectiveness of stirrups in resisting shear and probable understrength of web steel is 0.87 according to IS : 1343-1980. The ultimate shear force or the shear capacity V is taken as the sum of V_c and V_s in most codes of practice.

7.2.3.1 Indian Code

The provisions of the Indian Code IS : 1343—1980 follow the general approach represented by Eq. (7.2.7). The ultimate shear force resisted by concrete V_c is taken to be the smaller of the ultimate shear strength of the concrete for the section uncracked in flexure V_{co} and the ultimate shear strength of concrete for section cracked in flexure V_{cr} given by Eq. (7.2.10) and (7.2.11).

$$V_{co} = bD\,(0.67\,f_t)\ \sqrt{1 + \frac{0.8\,p}{f_t}} + P\sin\theta \qquad (7.2.10)$$

where b, D = breadth and overall depth of cross-section

$f_t = 0.24\,\sqrt{f_{ck}}$

= ultimate tensile strength of concrete

f_{ck} = characteristic compressive (cube) strength of concrete

$$p = \frac{P}{A}$$

= average effective intensity of prestress

θ = inclination of cable line at the section under consideration

In the case of flanged sections, the breadth of the section b should be replaced by the breadth (thickness) of the web b_w. If the centroid of the uncracked section lies in the flange, the average intensity of prestress p should be replaced by the intensity of prestress at the junction of the web and the flange.

The ultimate shear strength of section cracked in flexure V_{cr} is expressed as

$$V_{cr} = \left(1 - \frac{0.55 f_{pe}}{f_{pu}} \right) b\, d_p\, v_c + \frac{M_0}{(M/V)} \qquad (7.2.11)$$

where

f_{pe} = effective intensity of prestress after losses which should not be taken greater than $0.6 f_{pu}$

f_{pu} = characteristic ultimate strength of prestressing steel

d_p = effective depth of prestressing steel

v_c = ultimate shear stress resisted by concrete in accordance with Table 7.2.1

M/V = moment-shear ratio (shear arm) at ultimate load

M_0 = bending moment causing decompression at the level of centroid of prestressing steel given by the equation

$$M_0 = (0.8 f_{cp}) \frac{I}{(d_p - y_T)} \qquad (7.2.12)$$

Table 7.2.1 Ultimate shear stress resisted by concrete, v_c

$\dfrac{100 A_p}{b\, d_p}$	f_{ck} in MPa		
	30	35	40 and above
0.25	0.37	0.37	0.38
0.50	0.50	0.50	0.51
0.75	0.59	0.59	0.60
1.00	0.66	0.67	0.68
1.25	0.71	0.73	0.74
1.50	0.76	0.78	0.79
1.75	0.80	0.82	0.84
2.00	0.84	0.86	0.88
2.25	0.88	0.90	0.92
2.50	0.91	0.93	0.95
2.75	0.94	0.96	0.98
3.00	0.96	0.99	1.01

in which y_T is the distance of centroid of uncracked section from top and f_{cp} is the compressive stress due to prestress alone at the level of centroid of prestressing steel given by the equation,

$$f_{cp} = \frac{P}{A} + \frac{Pe_p^2}{I} \tag{7.2.13}$$

The code specifies that V_{cr} should not be taken less than $(0.1 \ \sqrt{f_{ck}}) \ (bd_p)$ and the vertical component of prestressing force $(P \sin \theta)$ should be ignored in computing V_{cr}. The value of V_{cr} calculated at a particular section may be assumed to be constant for a distance equal to $0.5 \ d_p$ in the direction of increasing moment.

When the ultimate shear force V exceeds the shear force V_c resisted by concrete, the area of shear reinforcement in the form of stirrups with vertical legs should be computed from the eqution,

$$\frac{A_v}{s} = \frac{V - V_c}{(0.87 \ f_{sy}) \ d_v} \tag{7.2.14}$$

where f_{sy} = characteristic yield strength of stirrup steel

d_v = effective depth in shear (larger of d_p and d_s)

d_s = effective depth of non-prestressed reinforcing steel

The code also contains the following provisions in respect of shear reinforcement:

(i) Shear reinforcement need not be provided in members of minor importance and in structures such as slabs, walls and pile caps in which V is grenerally less than $0.5 \ V_c$.

(ii) When V is greater than $0.5 \ V_c$ but less than V_c, only minimum nominal reinforcement should be provided in accordance with the following equation,

$$\frac{A_v}{bs} = \frac{0.4}{0.87 \ f_{sy}} \tag{7.2.15}$$

in which f_{sy} should not be taken greater than 415 MPa.

(iii) In rectangular beams, the strirrups should pass around a reinforcing bar, a tendon or a group of tendons at both corners in the tension zone.

(iv) The spacing of stirrups s should not exceed $0.75 \ d_v$ nor 4 times the web thickness b_w in the case of flanged beams. When V exceeds $1.8 \ V_c$, the spacing should not exceed $0.5 \ d_v$.

(v) The transverse spacing between individual vertical legs of the stirrups should not exceed $0.75 \ d_v$.

(vi) In order to prevent excessive shear reinforcement and possible crushing of concrete due to diagonal compression, it is specified that the ultimate shear force V should not exceed $(bd \ v_{max})$ in which v_{max} is the maximum permissible ultimate shear stress as per Table 7.2.2.

Table 7.2.2 Maximum permissible ultimate shear stress, v_{max}

f_{ck}, MPa	30	35	40	45	50	55 and above
v_{max}	3.5	3.7	4.0	4.3	4.6	4.8

7.2.3.2 British Code

The provisions of British Code BS: 8110-1985 are similar to those of the Indian Code except for the following points:

(a) The effective stress in pressing tendons after all losses f_{pe} in the case of beams containing untensioned reinforcing steel A_s in the tension zone may be determined from

$$f_{pe} = \frac{P}{A_p + \dfrac{A_s f_y}{f_{pu}}}$$

(b) The ultimate shear stress resisted by concrete v_c may be obtained from Table 7.2.3.

Table 7.2.3 Values of v_c, the design concrete shear stress, in MPa

$\dfrac{100 (A_p + A_s)}{bd}$	Effective depth in mm							
	125	150	175	200	225	250	300	400
$\leqslant 0.15$	0.45	0.43	0.41	0.40	0.39	0.38	0.36	0.34
0.25	0.53	0.51	0.49	0.47	0.46	0.45	0.43	0.40
0.50	0.67	0.64	0.62	0.60	0.58	0.56	0.54	0.50
0.75	0.77	0.73	0.71	0.68	0.66	0.65	0.62	0.57
1.00	0.84	0.81	0.78	0.75	0.73	0.71	0.68	0.63
1.50	0.97	0.92	0.89	0.86	0.83	0.81	0.78	0.72
2.00	1.06	1.02	0.98	0.95	0.92	0.89	0.86	0.80
3.00	1.22	1.16	1.12	1.08	1.05	1.02	0.98	0.91

Note: 1. The effective depth d is measured from the compression face to the centroid of tension steel $(A_s + A_p)$.

2. For grades of concrete higher than M-25, the v_c values given in the table should be multiplied by $\left(\dfrac{f_{ck}}{25}\right)^{1/3}$. For grades of concrete higher than M-40 f_{ck} should be taken equal to 40 MPa.

(c) The maximum premissible ultimate shear stress v_{max} may be taken as $0.8 \times \sqrt{f_{ck}}$ but not greater than 5 MPa.

(d) When V does not exceed $(V_c + 0.4 \; bd)$, minimum shear reinforcement should be provided in accordance with Eq. (7.2.15).

(e) When V exceeds $(V_c + 0.4 \; bd)$, shear reinforcement should be provided in accordance with Eq. (7.2.14).

7.2.3.3 American Code

According to American Code ACI: 318-1989, the factored ultimate shear force,

$$V_u \leqslant 0.85\,(V_c + V_s) \qquad (7.2.16)$$

For beams having effective prestressing force not less than 40 per cent of the tensile strength of flexural reinforcement, the ultimate shear force resisted by concerete V_c may be taken as

$$V_c = \left(0.05\ \sqrt{f_c'} + 4.9\ \frac{V_u\,d_p}{M_u}\right)b_w\,d_p \qquad (7.2.17)$$

where f_c' = characteristic compressive (cylinder) strength of concrete

M_u = factored ultimate moment occurring simultaneously with V_u at the section considered

The value of V_c should not be taken less than $(0.17\ \sqrt{f_c'}\,b_w\,d_p)$ nor greater than $(0.42\ \sqrt{f_c'}\,b_w\,d_p)$.

For a more precise computation, shear force resisted by concrete V_c should be taken equal to V_{cw} or V_{ci}, whichever is smaller, in accordance with Eq. (7.2.18) and (7.2.19). Note that V_{cw} and V_{ci} are similar to V_{co} and V_{cr} in Indian and British codes. The nominal shear strength V_{cw}, provided by concrete when diagonal cracking results from excessive principal tensile stress in web, should be determined from

$$V_{cw} = (0.29\ \sqrt{f_c'} + 0.3\,p)\,b_w\,d_v + P\sin\theta \qquad (7.2.18)$$

where p = compressive stress at the centroid of cross-section or at junction of flange and web if centroid of section lies in flange

d_v = effective depth in shear

$\quad\ = d_p$ or $0.8\,D$, whichever is greater

Alternatively, V_{cw} may be taken equal to the shear force which creates a principal tensile stress equal to $0.33\ \sqrt{f_c'}$ at the centroid of the section or at the junction of flange and web when centroidal axis lies in the flange. In pretensioned members in which the section at a distance equal to $(D/2)$ from face of support is closer to the end of the member than the transfer length of the prestressing tendons (50 times diameter of strand 100 times diameter for single wire), the reduced prestress should be considered in computing V_{cw} assuming a linear variation of presteress over the transfer length. The value of V_{cw} determined in this manner should be treated as the upper limit of V_c computed from Eq. (7.2.17).

The nominal shear strength V_{ci} provided by concrete, when diagonal cracking results from combined shear and moment, may be computed from

$$V_{ci} = 0.05\ \sqrt{f_c'}\,b_w\,d_v + V_D + V_L\,\frac{M_{cr}}{M_{max}} \qquad (7.2\ 19)$$

but V_{ci} need not be taken less than $0.14\ \sqrt{f_c'}\,b_w\,d_v$

where V_D = shear force at the section due to unfactored (working) dead load

$\quad\quad V_L$ = shear force at the section due to factored (ultimate) externally applied loads occurring simultaneously with M_{max}

$\quad\quad M_{\text{max}}$ = maximum   factored (ultimate) moment at section due to externally applied loads

$\quad\quad M_{cr}$ = momenl causing flexural cracking at the section due to exter-ₗ nally applied loads as per Eq. (7.2.20)

$$M_{cr} = \frac{I}{y_B} (0.5 \sqrt{f_c'} + p_B - f_D) \tag{7.2.20}$$

where y_B = distance of centroid of uncracked section from bottom

$\quad\quad p_B$ = effective intensity of prestress at bottom fibres

$\quad\quad f_D$ = stress at bottom (tension) face due to unfactored dead load

Values of V_L ond M_{max} should be computed from the load combination causing maximum moment to occur at the section.

When factored (ultimate) shear force V_u exceeds $0.85\ V_c$, shear reinforcement should be provided in accordance with the following equations. In the case of vertical stirrups,

$$\frac{A_v}{s} = \frac{V_u - 0.85\ V_c}{0.85\ f_{sy}\ d_v} \tag{7.2.21}$$

In the case of inclined stirrups having an inclination α to horizontal,

$$\frac{A_v}{s} = \frac{V_u - 0.85 V_c}{0.85\ f_{sy}\ d_v\ (\sin\alpha + \cos\alpha)} \tag{7.2.22}$$

The shear force resisted by a single or a group of bent up reinforcing bars should be taken equal to $(A_b f_{sy} \sin\alpha)$ in which A_b is the cross-sectional area and f_y is the yield stress of bent up bars inclined at an angle α to the horizontal. The shear force resisted by bent up bars should not be taken greater than $(0.25 \sqrt{f_c'}\ b_w\ d_v)$. When the shear reinforcement comprises series of single or group of parallel bent up bars at different distances from the support, Eq. (7.2.22) may be used. Only the central 3/4th part of a bent up bar should be considered as effective in resisting shear. If more than one type of shear reinforcement is used to reinforce the same part of a beam, shear resisted should be taken equal to the sum of the shears resisted by different types of reinforcements. In order to prevent excessive shear reinforcement and possible crushing of concrete due to diagonal compression, the code specifies that the shear force resisted by steel should not exceed $(0.67 \sqrt{f_c'}\ b_w\ d_v)$.

The code also contains the following provisions:

(i) No shear reinforcement need be provided in slabs, footings and beams having total depth not greater than 250 mm, 2.5 times thickness of flange or half the width of web, whichever is greater.

(ii) A minimum nominal shear reinforcement should be provided in all beams in which $0.85 (0.5\ V_c) < V_u < 0.85\ V_c$ in accordance with the following equation:

$$\frac{A_v}{b_w\ s} = \frac{0.34}{f_{sy}} \qquad (7.2.23)$$

In beams in which the effective prestressing force is not less than 40 per cent of the tensile strength of the flexural reinforcement, the minimum shear reinforcement may be determined from

$$\frac{A_v}{s} = \frac{A_p}{80}\frac{f_{pu}}{f_{sy}}\frac{1}{d_r}\sqrt{\frac{d_v}{b_w}} \qquad (7.2.24)$$

where A_p = area of prestressing steel in tension zone

(iii) Spacing of vertical stirrups should not exceed $0.75\ D$ nor 600 mm. The spacing of inclined stirrups and bent up bars should be chosen so that every $45°$ crack extending towards the reaction from mid-depth of member $(0.5\ d_r)$ to longitudinal tension reinforcement crosses at least one line of shear reinforcement. When the shear force V_s resisted by steel exceeds $(0.33\ \sqrt{f_c'}\ b_w\ d_v)$, the spacings of shear reinforcement should be reduced to half of the values specified above.

(iv) The design strength of stirrup steel f_{sy} should not exceed 415 MPa.

EXAMPLE 7.2.1

The stress condition at a certain point in an uncracked section of a prestressed concrete beam is shown in Fig. 7.2.5 (a). Determine the magnitudes and directions of the principal stresses.

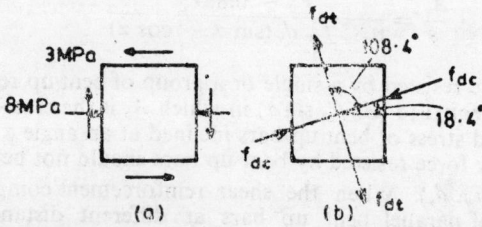

(a)

(b)

Fig. 7.2.5

Solution

$$f_x = -8 \text{ MPa}$$

$$v = 3 \text{ MPa}$$

Hence, using Eq. (7.2.4), diagonal tension,

$$f_{dt} = -\frac{8}{2} + \sqrt{\left(\frac{-8}{2}\right)^2 + 3^2} = 1 \text{ MPa}$$

and using Eq. (7.2.5), diagonal compression,

$$f_{dc} = -\frac{8}{2} - \sqrt{\left(\frac{-8}{2}\right)^2 + 3^2} = -9 \text{ MPa}$$

Directions of principal stresses may be determined from Eq. (7.2.6),

$$\tan 2\beta = \frac{-2 \times 3}{-8} = 0.75$$

Hence, $2\beta = 36.87°$ or $216.87°$

or $\beta = 18.4°$ or $108.4°$

As v is positive, the inclination of diagonal tension β lies between 90° and
135°. Consequently, the inclination of diagonal compression and diagonal
tension are 18.4° and 108.4° respectively as shown in Fig. 7.2.5 (b).

EXAMPLE 7.2.2

A prestressed concrete beam has a parabolic cable line with an eccentricity
of 0.35 m at midspan and 0.15 m at the ends. The beam has a symmetrical
I-section as shown in Fig. 7.2.6. It carries a dead load of 9 kN/m inclusive
of self weight at the initial stage and a live load of 15 kN/m at final stage
over a span of 20 m. Determine the magnitude and
direction of diagonal tension in the web at point A at
the initial stage and at point B at the final stage at an
uncracked section 5 m from the left support. Initial
and final prestressing forces are 2.54 MN and 2.16
MN respectively.

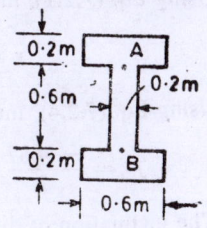

Solution

Area of cross-section, $A = 0.36 \text{ m}^2$

Moment of Inertia, $I = 0.428 \text{ m}^4$

Fig. 7.2.6

The sag of cable line at midspan,

$$S = 0.35 - 0.15 = 0.2 \text{ m}$$

Taking the origin at left support, the cable line may be represented by the
equation,

$$e_p = 0.15 + \frac{4Sx\,(l - x)}{l^2} = 0.15 + \frac{20x - x^2}{500}$$

Hence the eccentricity of prestress at left quarter point,

$$e_p = 0.15 + \frac{20 \times 5 - 5^2}{500} = 0.3 \text{ m}$$

The slope of cable line at left quarter point,

$$\tan \theta = \frac{de_p}{dx} = \frac{20 - 2x}{500} = 0.02$$

(i) *Initial Stage*

The bending moment due to prestress at initial stage,

$$M_{pi} = -2.54 \times 0.3 = -0.762 \text{ MN·m}$$

The bending moment due to dead load,

$$M_D = \frac{0.009 \times 20^2}{8} = 0.45 \text{ MN·m}$$

Hence the net bending moment,

$$M_r = M_D + M_{pi} = -0.312 \text{ MN.m}$$

The normal stress at point A,

$$f = \frac{-2.54}{0.36} + \frac{(-0.312)(-0.3)}{0.0428} = -4.87 \text{ MPa}$$

As external shear force, $V = 0.045$ MN, the net shear force V_r, at left quarter point,

$$V_r = V - P_i \sin \theta = 0.045 - 2.54 \times 0.02$$

$$= -0.0058 \text{ MN}$$

Using Eq. (7.2.1), intensity of shear stress at A,

$$v = \frac{(-0.0058) \times 0.12 \times 0.4}{0.0428 \times 0.2} = -0.033 \text{ MPa}$$

Using Eq. (7.2.4), intensity of diagonal tension at A,

$$f_{dt} = \frac{-4.87}{2} + \sqrt{\left(\frac{-4.87}{2}\right)^2 + (-0.033)^2} = 0.00022 \text{ MPa}$$

The inclination of diagonal tension from Eq. (7.2.6),

$$\beta = 89.6°$$

(ii) *Final Stage*

The bending moment due to prestress at the final stage,

$$M_p = -2.16 \times 0.3 = -0.648 \text{ MN.m}$$

The bending moment due to total load,

$$M = \frac{0.024 \times 20^2}{8} = 1.2 \text{ MN.m}$$

Hence net bending moment,

$$M_r = M + M_p = 0.552 \text{ MN.m}$$

The normal stress at point B,

$$f = \frac{-2.16}{0.36} + \frac{0.552 \times 0.3}{0.0428} = -2.13 \text{ MPa}$$

As external shear force, $V = 0.12$ MN, the net shear force at left quarter point,

$$V_r = V - P \sin \theta = 0.12 - 2.16 \times 0.02 = 0.0768 \text{ MN}$$

Using Eq. (7.2.1), the intensity of shear stress at B,

$$v = \frac{0.0768 \times 0.12 \times 0.4}{0.0428 \times 0.2} = 0.43 \text{ MPa}$$

Using (Eq. (7.2.4), the intensity of diagonal tension at B,

$$f_{dt} = \left(\frac{-2.13}{2}\right) + \sqrt{\left(\frac{-2.13}{2}\right)^2 + (0.43)^2} = 0.084 \text{ MPa}$$

The inclination of diagonal tension from Eq. (7.2.6),

$$\beta = 101°$$

EXAMPLE 7.2.3

For the beam at its final stage in Ex. 7.2.2, determine the intensity of uniform load required to cause at the left quarter point (i) flexural cracking and (ii) diagonal tension cracking at the centroid. Modulus of rupture of concrete, $f_{cr} = 5.4$ MPa and direct tensile strength, $f_t = 3.5$ MPa. Also determine the inclination of web shear crack at centroid at the left quarter point.

Solution

(i) *Flexural Cracking*

Using the data of Ex. 7.2.2, the normal stress at bottom fibre at left quarter point,

$$f = -\frac{2.16}{0.36} + \frac{(-0.648)(0.5)}{0.0428} = -13.57 \text{ MPa}$$

Flexural cracking commences at the left quarter point when the net stress at the bottom fiber under the combined action of external load and prestress becomes equal to modulus of rupture f_{cr}. Hence, if w is the intensity of uniform load required to cause flexural cracking,

$$\frac{w \times 20^2}{8} \times \frac{0.5}{0.0428} - 13.57 = 5.40$$

or $$w = 0.0325 \text{ MN/m}$$

(ii) *Diagonal Tension Cracking*

Normal stress at centroid,

$$f = -\frac{2.16}{0.36} = -6 \text{ MPa}$$

Intensity of unifrom force due to prestress,

$$w_p = \frac{8 PS}{l^2} = \frac{8 \times 2.16 \times 0.2}{20^2} = 0.00864 \text{ MN/m (upward)}$$

If w is the intensity of downward load required to cause diagonal tension crack, the net downward load is $(w - 0.00864)$ MN/m. Hence the net shear force at left quarter point,

$$V_r = 5\,w - 0.0432 \text{ MN}$$

Hence from Eq. (7.2.1), the intensity of shear stress,

$$v = \frac{(5\,w - 0.0432)\,(0.12 \times 0.4 + 0.06 \times 0.15)}{0.0428 \times 0.2}$$

$$= 33.294\,w - 0.288 \text{ MPa}$$

The diagonal tension crack commences when the intensity of diagonal tension becoms equal to the tensile strength of concrete f_t.
Hence from equation,

$$\left(-\frac{6}{2}\right) + \sqrt{\left(-\frac{6}{2}\right)^2 + (33.294\,w - 0.288)^2} = 3.5$$

or $\qquad\qquad w = 0.182 \text{ MN/m}$

Using Eq. (7.2.6), the inclinations of principal stresses, $\beta = 31.3°$ and $121.3°$. Hence diagonal tension crack forms at the level of centroid of the section with an inclination of $31.3°$ with the horizontal at an intensity of load equal to 0.182 MN/m.

EXAMPLE 7.2.4

A prestressed concrete beam carries uniform load over a span of 24 m. It has an effective prestressing force of 3.3 MN. Determine the ultimate shear strength at the left quarter point at which the beam has the cross-section shown in Fig. 7.2.7 using the provisions of (i) Indian Code and (ii) British Code. The cable line is parabolic with eccentricities equal to zero at ends and 0.45 m at midspan. Concrete is of M-35 grade. The ultimate stress of prestressing steel, $f_{pu} = 1600$ MPa and yield stress of stirrups, $f_{sy} = 250$ MPa.

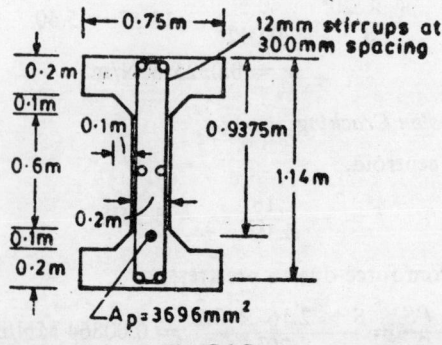

Fig. 7.2.7

Solution

(i) *Indian Code*

Cross-sectional properties are as follows:

$$A = 0.48 \text{ m}^2$$
$$y_T = y_B = 0.6 \text{ m}$$
$$I = 0.09188 \text{ m}^4$$
$$A_p = 3696 \text{ mm}^2 = 0.003696 \text{ m}^2$$
$$d_s = 1.14 \text{ m}$$

Hence,
$$d_v = 1.14 \text{ m}$$

Equation (7.2.10) may be used to compute V_{co}. The parabolic cable line may be represented by the equation,

$$e_p = \frac{4\,S \times (l - x)}{l^2} = \frac{4 \times 0.45\, x\, (24 - x)}{24^2}$$

$$= 0.003125\, x\, (24 - x)$$

Hence at the left quarter point,

$$e_p = 0.003125 \times 6\, (24 - 6) = 0.3375 \text{ m}$$
$$d_p = 0.6 + 0.3375 = 0.9375 \text{ m}$$

$$\tan \theta \approx \sin \theta = \frac{d\, e_p}{dx} = 0.003125\, (24 - 2x)$$

$$= 0.003125\, (24 - 2 \times 6)$$

$$= 0.0375$$

The effective intensity of prestress,

$$p = \frac{P}{A} = \frac{3.3}{0.48} = 6.875 \text{ MPa}$$

$$f_t = 0.24 \sqrt{f_{ck}} = 0.24\sqrt{35} = 1.42 \text{ MPa}$$

Hence from Eq. (7.2.10),

$$V_{co} = 0.2 \times 1.2\, (0.67 \times 1.42) \sqrt{1 + \frac{0.8 \times 6.875}{1.42}} + 3.3 \times 0.0375$$

$$= 0.6278 \text{ MN}$$

Equation (7.2.11) may be used to compute V_{cr}.

$$f_{pe} = \frac{P}{A_p} = \frac{3.3}{0.003696} = 892.8 \text{ MPa}$$

The percentage of prestressing steel,

$$\frac{100\, A_p}{b_w\, d_p} = \frac{100 \times 0.003696}{0.2 \times 0.9375} = 1.9712$$

Hence from Table 7.2.1, the shear stress resisted by concrete of grade M-35,

$$v_c = 0.8554 \text{ MPa}$$

Using Eq. (7.2.13),

$$f_{ep} = \frac{3.3}{0.48} + \frac{3.3 \times 0.3375^2}{0.09188} = 10.966 \text{ MPa}$$

Hence from Eq. (7.2.12),

$$M_o = 0.8 \times 10.966 \left(\frac{0.09188}{0.3375}\right) = 2.3883 \text{ MN.m}$$

In the case of uniform load of intensity w, the bending moment at quarter point,

$$M = \frac{3 \, w \, L^2}{32}$$

and the shear force at the quarter point,

$$V = \frac{w \, L}{4}$$

Hence the shear arm at the quarter point,

$$\frac{M}{V} = \frac{3 \, L}{8} = 9 \text{ m}$$

Using Eq. (7.2.11),

$$V_{cr} = \left(1 - \frac{0.55 \times 892.8}{1600}\right) \times 0.2 \times 0.9375 \times 0.8554 + \frac{2.3883}{9}$$
$$= 0.3765 \text{ MN}$$

As V_c is lesser of V_{co} and V_{cr},

$$V_c = 0.3765 \text{ MN}$$

The ultimate shear strength of the section V may be computed from Eq. (7.2.14).

$$A_v = 2 \times \frac{\pi}{4} \times 12^2 = 226.3 \text{ mm}^2 = 0.0002263 \text{ m}^2$$

Hence,

$$\frac{0.0002263}{0.3} = \frac{V - 0.3765}{0.87 \times 250 \times 1.14}$$

or

$$V = 0.5635 \text{ MN}$$

(ii) *British Code*

For $d_v = d_p = 0.9375$ m and $100 \, A_p/(b_w \, d_p) = 1.9712$, the shear stress resisted by concrete in accordance with Table 7.2.3,

$$v_c = 0.80 \text{ MPa}$$

While $V_{co} = 0.6278$ MN as in (i) above, the value of V_{cr} in accordance with Eq. (7.2.11),

$$V_{cr} = \left(1 - \frac{0.55 \times 892.8}{1600}\right) \times 0.2 \times 0.9375 \times 0.80 + \frac{2.3883}{9}$$

$$= 0.3693 \text{ MN}$$

As $V_{cr} > V_{co}$, the shear force resisted by concrete,

$$V_c = V_{cr} = 0.3693 \text{ MN}$$

Using Eq. (7.2.14),

$$\frac{0.0002263}{0.3} = \frac{V - 0.3693}{0.87 \times 250 \times 1.14}$$

or

$$V = 0.5564 \text{ MN}$$

EXAMPLE 7.2.5

Design the shear reinforcement for the post-tensioned beam of Ex. 7.2.4 at the left quarter point using (i) Indian Code and (ii) American Code. The beam carries a dead load of 11.5 kN/m and a live load of 21.5 kN/m at the working load level.

Solution

(ii) *Indian Code*

The load factors for dead load and live load according to Indian Code are 1.5 and 1.5. Hence the ultimate design shear force,

$$V = \left[\frac{(11.5 + 21.5) \times 2.4}{2} - (11.5 + 21.5)\,6\right] \times 1.5$$

$$= 297 \text{ kN or } 0.297 \text{ MN}$$

From the computations of Ex. 7.2.4, ultimate shear force resisted by concrete,

$$V_c = 0.3765 \text{ MN}$$

As the ultimate design shear force is smaller than V_c but greater than $0.5 \times V_c$, only nominal or minimum shear reinforcement is necessary.

Using 10 mm stirrups, the spacing in accordance with Eq. (7.2.15),

$$s = \frac{A_r \times 0.87\, f_{sy}}{0.4 \times b} = \frac{2 \times \frac{\pi}{4} \times 10^2 \times 0.87 \times 250}{0.4 \times 200}$$

$$= 427 \text{ mm}$$

Hence 10 mm two legged stirrups at 425 *mm* spacing may be provided.

(ii) *American Code*

The load factors for dead and live loads according to ACI code are 1.4 and 1.7 respectively. Hence the ultimate design shear force or the factored ultimate shear force,

$$V_u = \frac{(1.4 \times 11.5 + 1.7 \times 21.5) \times 24}{2}$$

$$- (1.4 \times 11.5 + 1.7 \times 21.5) \times 6$$

$$= 315.9 \text{ kN or } 0.3159 \text{ MN}$$

The ACI Code provides two methods for design of shear reinforcement

(a) *Simplified Method*

The ultimate shear force resisted by concrete may be determined from Eq. (7.2.17). The ultimate factored bending moment at quarter point,

$$M_v = (1.4 \times 11.5 + 1.7 \times 21.5) \times \frac{24}{2} \times 6$$

$$- (1.4 \times 11.5 + 1.7 \times 21.5) \times 6 \times \frac{6}{2}$$

$$= 2843 \text{ kN.m} = 2.843 \text{ MN.m}$$

Taking $f'_c = 0.8 f_{ck} = 28$ MPa, Eq. (7.2.17) gives,

$$V_c = \left(0.05 \sqrt{28} + 49 \times \frac{0.3159 \times 0.9375}{2.843} \right) \times 0.2 \times 0.9375$$

$$= 0.1453 \text{ MN}$$

The code specifies that V_c should not be taken less than $0.17 \sqrt{f'_c} b_w d_p = 0.1687$ MN. Using 10 mm stirrups, the spacing in accordance with Eq. (7.2.21),

$$s = \frac{A_v (0.85 f_{sy} d_v)}{V_u - 0.85 V_c}$$

$$= \frac{\left(2 \times \frac{\pi}{4} \times 10^2 \times 10^{-6} \right) \times 0.85 \times 250 \times 0.96}{0.3159 - 0.85 \times 0.1687}$$

$$= 183 \text{ mm}$$

(b) *Precise Method*

The effective depth in shear d_v is the greater of d_p and 0.8 D. Hence,

$$d_v = 0.8 \times 1.2 = 0.96 \text{ m}$$

Hence using Eq. (7.2.8),

$$V_{cw} = (0.29 \sqrt{28} + 0.3 \times 6.875) \times 0.2 \times 0.96 + 3.3 \times 0.0375$$
$$= 0.8144 \text{ m}$$

Alternatively, taking V_{cw} equal to shear stress creating a principal tensile stress equal to $0.33\sqrt{f_c'}$ at centroid,

$$\left(- \frac{6.875}{2}\right) + \sqrt{\left(\frac{6.875}{2}\right)^2 + v_{cw}^2} = 0.33 \sqrt{28}$$

or

$$v_{cw} = 3.879 \text{ MPa}$$

Hence,

$$V_{cw} = v_{cw} \, b_w \, d_p = 3.879 \times 0.2 \times 0.9375$$
$$= 0.7273 \text{ MN}$$

Next V_{ci} may be computed from Eq. (7.2.19).

$$p_D = \frac{3.3}{0.48} + \frac{3.3 \times 0.3375}{0.09188} \times 0.6 = 14.148 \text{ MPa}$$

$$f_D = \frac{\left(\dfrac{0.0115}{2} \times 24 \times 6 - 0.0115 \times 6 \times \dfrac{6}{2}\right) \times 0.6}{0.09188}$$

$$= 4.055 \text{ MPa}$$

From Eq. (7.2.20),

$$M_{cr} = \frac{0.09188}{0.6} (0.5\sqrt{28} + 14.148 - 4.055)$$

$$= 1.9507 \text{ MN.m}$$

$$M_{max} = 21.5 \times 1.7 \times \frac{24}{2} \times 6 - 21.5 \times 1.7 \times 6 \times \frac{6}{2}$$

$$= 1973.7 \text{ kN.m} = 1.9737 \text{ MN.m}$$

$$V_L = \left(21.5 \times 1.7 \times \frac{24}{2}\right) - (21.5 \times 1.7 \times 6)$$

$$= 219.3 \text{ kN} = 0.2193 \text{ MN}$$

$$V_D = \frac{11.5 \times 24}{2} - 11.5 \times 6 = 69 \text{ kN} = 0.069 \text{ MN}$$

Hence from Eq. (7.2.19),

$$V_{ci} = 0.05\sqrt{28} \times 0.2 \times 0.96 + 0.069 + \frac{0.2193 \times 1.9507}{1.9737}$$

$$= 0.3365 \text{ MN}$$

As V_c is the lesser of V_{cw} and V_{ci},

$$V_c = V_{ci} = 0.3365 \text{ MN}$$

Using 10 mm stirrups, the spacing in accordance with Eq. (7.2.21)

$$s = \frac{\left(2 \times \frac{\pi}{4} 10^2 \times 10^{-6} \right) \times 0.85 \times 250 \times 0.96}{0.3159 - 0.85 \times 0.3365} = 1.072 \text{ m}$$

However, the maximum spacing of stirrups to satisfy the requirement of minimum reinforcement in accordance with Eq. (7.2.23),

$$s = \frac{A_v f_{sy}}{0.34 b_w} = \frac{2 \times \frac{\pi}{4} \times 10^2 \times 250}{0.34 \times 200} = 577 \text{ mm}$$

Hence 10 mm stirrups at 575 mm spacing may be adopted.

EXAMPLE 7.2.6

A pretensioned beam of uniform rectangular cross-section 100 mm × 500 mm is prestressed with an effective force of 0.32 MN at constant eccentricity of 180 mm. Design the shear reinforcement at a section uncracked in flexure to resist an ultimate shear force of 0.16 MN using the provisions of Indian Code. The concrete is of M-40 grade. Use 6 mm stirrups with yield stress of 310 MPa.

Solution

As the section is uncracked in flexure, $V_c = V_{co}$

Also, $\theta = 0$ in the case of straight cable line.

$$p = \frac{P}{A} = \frac{0.32}{0.1 \times 0.5} = 6.4 \text{ MPa}$$

$$f_t = 0.24 \sqrt{40} = 1.518 \text{ MPa}$$

Hence using Eq. (7.2.10),

$$V_c = 0.1 \times 0.5 \times 0.67 \times 1.518 \sqrt{1 + \frac{0.8 \times 6.4}{1.518}}$$

$$= 0.1063 \text{ MN}$$

Using Eq. (7.2.14), the spacing of stirrups,

$$s = \frac{\left(2 \times \frac{\pi}{4} \times 6^2 \times 10^{-6} \right) \times 0.87 \times 310 \times 0.430}{0.1600 - 0.1063}$$

$$= 0.122 \text{ m} = 122 \text{ mm}$$

The maximum spacing of stirrups to satisfy minimum requirement in accordance with Eq. (7.2.15),

$$s = \frac{2 \times \frac{\pi}{4} \times 6^2 \times 0.87 \times 310}{100 \times 0.4} = 381 \text{ mm}$$

Code also provides that the spacing should not exceed $0.75 \, d_x = 0.75 \times 430 = 322.5$ mm. Hence the spacing of 122 mm is suitable.

7.3 TORSION

The phenomena of shear and torsion have striking similarities although there are basic dissimilarities too. Both shear and torsion produce shearing stresses which in turn give rise to diagonal tension. However, the distributions of flexural shear stress v and torsional shear stress τ are entirely different. The distribution of torsional shear stress undergoes a total change after cracking. In the precracking range, the torsional stresses, strains and the resulting twist may be computed with reasonable accuracy on the basis of elastic theory ignoring the presence of steel. In the post-cracking range, the torsional behaviour of a prestressed concrete member is governed largely by the amount and disposition of prestressing and reinforcing steel. A prestressed concrete member with a high intensity of prestress fails abruptly in torsion with explosive violence. The violence at failure can be mitigated and a certain degree of ductility can be imparted by the provision of shear reinforcement in the form of closed rectangular stirrups. The mechanism of failure in torsion has been the subject of extensive theoretical and experimental research since the decade of sixties. These studies have led to several theories of torsional failure.

7.3.1 Uncracked Section

The distribution of torsional stresses in an uncracked section of a prestressed concrete member may be determined with reasonable accuracy by using linear theory of elasticity ignoring the presence of steel and treating concrete as a homogeneous, isotropic and elastic material. Using a semi-inverse approach, Saint Venant obtained the correct solution for uniform torsion of a prismatic member having non-circular section of any arbitrary shape. Assuming that in uniform torsion, the cross-section does not distort in its own plane but it suffers warping due to longitudinal displacements which are identical at all cross-sections, he showed that all equations of theory of elasticity are satisfied. Referring to Fig. 7.3.1, the three normal stress components f_x, f_y, f_z and the shear stress component τ_{xy} at any point in prismatic member under uniform torque T are zero. The torsional shear strees τ at any point is the vector sum of the shear stress components τ_{zx} and τ_{zy}.

$$\tau = \sqrt{\tau_{zx} + \tau_{zy}} \qquad (7.3.1)$$

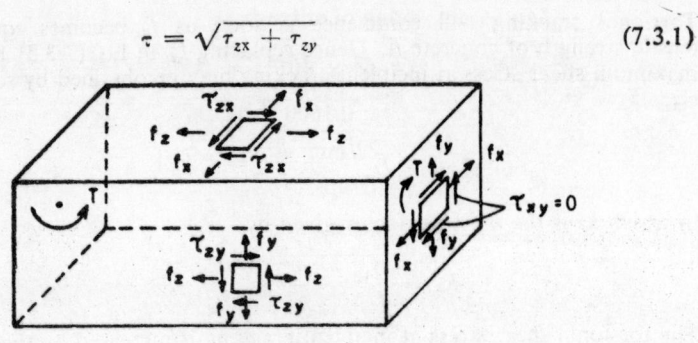

Fig. 7.3.1

The distribution of torsional shear stress depends upon the shape of the cross-section. The basic equation for torsion is similar to that for flexure and may be written as

$$\frac{T}{K} = \frac{\tau}{\rho} = \frac{\tau_{max}}{\rho_{max}} = \frac{G\psi}{l} = G\phi \qquad (7.3.2)$$

where T = applied uniform torque

   K = torsion constant

   τ, τ_{max} = torsional shear stress and its maximum value

   ρ, ρ_{max} = stress factor and its maximum value

   $G = \dfrac{E}{2(1 + v)}$

   = shear modulus of elasticity

   v = Poisson's ratio

   ψ = total angle of twist over length l

   l = length of member

   $\phi = \dfrac{\psi}{l}$ = angle of twist per unit length

Equation (7.3.2) may be used to determine the maximum torsional shear stress τ_{max} and the angle of twist ψ for any torque T in the precracking range. Table 7.3.1 gives values of torsion constant K and maximum stress factor ρ_{max} for common cross-sectional shapes.

In a concentrically prestressed member with intensity of prestress $p = P/A$, the intensity of diagonal tension f_{dt} at the point of maximum torsional shear stress τ_{max} is

$$f_{dt} = \sqrt{\left(\frac{p}{2}\right)^2 + \tau^2_{max}} - \frac{p}{2} \qquad (7.3.3)$$

Torsional cracking will commence as soon as f_{dt} becomes equal to the tensile strength of concrete f_t. Hence replacing f_{dt} in Eq. (7.3.3) by f_t, the maximum shear stress at incipient cracking may be obtained by solving for τ_{max}.

$$\tau_{max} = f_t \sqrt{1 + \frac{p}{f_t}} = f_t k_p \qquad (7.3.4)$$

in which k_p is the prestress factor given by,

$$k_p = \sqrt{1 + \frac{p}{f_t}} \qquad (7.3.5)$$

The torsional shear stress at incipient cracking is increased by the factor k_p on account of prestress. If the limited plasticity of concrete is ignored, a concentrically prestressed member may be assumed to fail in torsion at the

Table 7.3.1

S. No.	Section		K	ρ_{max}	Remarks
1.	Solid circular section		$\dfrac{\pi R^4}{2}$	R	
2.	Hollow circular section		$\dfrac{(R_e^4 - R_i^4)}{2}$	R_e	
3.	Thin circular section		$2\pi R_\theta^3 t$	R_θ	
4.	Rectangular section		$k_1 b^3 d$	$k_2 b$	Constants k_1 and k_2 depend upon aspect ratio, D/b as shown in Table 7.3.2.
5.	Thin rectangular section		$\dfrac{1}{3} t^3 b$	t	
6.	Thin walled open section (flanged section)		$\sum \dfrac{1}{9} t^3 b$	t_{max}	
7.	Thin walled closed cell of varying thickness		$\dfrac{4A_\theta^2}{\displaystyle\int \frac{ds}{t}}$	$\dfrac{2A_\theta}{t_{min} \displaystyle\int \frac{ds}{t}}$	A_θ = area enclosed by mean perimeter
8.	Thin walled closed cell of uniform thickness		$\dfrac{4A_\theta^2 t}{s_\theta}$	$\dfrac{2A_0}{s_0}$	s_θ = mean perimeter
9.	Thin walled single cell with outstanding flanges		$\dfrac{4A_o^2 t}{\displaystyle\int \frac{ds}{t}} + \sum \dfrac{1}{3} t^3 b$	—	$\dfrac{4A_o^2}{\displaystyle\int \frac{ds}{t}} = K_{cell}$ $\sum \dfrac{1}{3} t^3 b = K_{flanges}$

torque producing incipient cracking. It follows that the ultimate torque of a concentrically prestressed member is k_p times greater than that of the corresponding plain concrete member without prestress, T_{up}.

$$T_u = k_p \, T_{up} = T_{up} \cdot \sqrt{1 + \frac{p}{f_t}} \qquad (7.3.6)$$

in which the ultimate torque T_{up} of a non-prestressed plain concrete member determined from Eq. (7.3.2) may be expressed as

$$T_{up} = \frac{K}{\rho_{max}} \, \tau_{max} = \frac{K}{\rho_{max}} \, f_t \qquad (7.3.7)$$

In the absence of prestress, $p = 0$, diagonal tension f_{dt} equals τ_{max} from Eq. (7.3.3). Hence cracking and failure occur as soon as τ_{max} becomes equal to f_t. The expressions for ultimate torque given by Eq. (7.3.6) and (7.3.7) are based on the assumption that failure occurs by diagonal tension. The failure can, however, occur due to diagonal compression if the intensity of prestress p is very high. The change from diagonal tension failure to diagonal compression failure occurs at the transformation point which is believed to correspond to a value of p lying between $0.6 f_c'$ to $0.7 f_c'$. As most codes restrict the intensity of prestress to $0.45 f_c'$, the diagonal compression failure is only of academic interest.

In the case of eccentrically prestressed member, the diagonal tension at any point may be determined from Eq. (7.3.3) by replacing p by normal compressive stress f due to combined effect of prestress and external bending moment M, if any. The diagonal tension is max:mum at the point at which the ratio (τ/f) is maximum. The tension crack due to diagonal tension may be expected to start from this critical point. The critical point is not necessarily the same as the point at which τ is maximum. In the case of concentric prestress, the torsion crack commences at the point at which τ is maximum.

Table 7.3.2   Values of coefficients k_1 and k_2

Aspect Ratio, $\dfrac{D}{b}$	k_1	k_2
1.0	0.1406	0.675
1.2	0.166	0.759
1.5	0.196	0.848
2.0	0.229	0.930
2.5	0.249	0.968
3.0	0.263	0.985
4.0	0.281	0.997
5.0	0.291	0.999
10.0	0.312	1.0
∞	0.333	1.0

7.3.2   Cracked Section

The post-cracking range is characterised by a dramatic change in stress distribution and an abrupt reduction in torsional stiffness to a value ranging from 5 to 10 per cent of the precracking stiffness. Although the cracking torque is increased substantially due to prestress, the ultimate torque may not be increased to the same extent unless the member is reinforced with substantial and effective torsion reinforcement. Consequently, the effect of prestress may be to reduce the post-cracking range, resulting in reduced toughness and ductility. The provision of substantial and effective torsion reinforcement has just the opposite effect as it enhances the post-cracking range resulting in superior ductility and toughness. It also mitigates the abruptness and violence of torsional failure. Hence the provision of a certain minimum amount of torsion reinforcement is very desirable for prestressed concrete members subjected to torsion.

The intensive research supported by extensive experimental work in the past three decades has led to a number of theories for the post-cracking behaviour and ultimate strength of prestressed concrete members in torsion. Prominent among these are the skew bending theory, space truss theory and the compression field theory. These theories may be used to predict the torque-twist response in the entire post-cracking range. The failure mechanisms envisaged in these theories may be used to compute the ultimate torsional strength. Besides, several empirical expressions, which may be expressed in the following general form, have been recommended for the determination of ultimate torque.

$$T = T_c + T_s \qquad (7.3.8)$$

in which T_c is the part of the ultimate torque resisted by concrete which may be expressed as

$$T_c = c_1 \, T_{up} \, k_p \qquad (7.3.9)$$

where c_1 = a dimensionless constant equal to or less than 1

        k_p = prestress factor given by Eq. (7.3.5)

and T_{up} = the torsional strength of corresponding plain concrete section given by Eq. (7.3.7)

The part of the ultimate torque T_s resisted by closed rectangular stirrups may be expressed as

$$T_s = \frac{\Omega_t \, A_t \, f_{sy} \, b' \, d'}{s} \qquad (7.3.10)$$

where Ω_t = factor reflecting the effectiveness of stirrup reinforcement in torsion

       A_t = area of stirrup (one leg) for resisting torsion

       f_{sy} = yield stress of stirrup steel

   b', d' = smaller and larger dimensions of the stirrup measured on centre lines of legs

        s = spacing of stirrups

Equation (7.3.8) is based on the assumptions that concrete resistance T_c is controlled by diagonal tension and all stirrups intersecting the failure surface attain their yield stress. The similarity of Eq. (7.2.7) and (7.3.8) underlines the similarity of approaches for shear and torsion design. The expressions for V_s and T_s are also strikingly similar because the stirrups are responsible for resisting diagonal tension caused either by shear or by torsion.

7.3.3 Code Provisions

The torsion provisions of several prominent codes are based either on skew bending theory or the space truss theory. The Russian, Australian and Indian Codes are based on skew bending theory. According to this theory, failure in combined bending, torsion and shear occurs by skew bending in which the inclined compression zone, called the compression hinge, develops adjacent to either top face (Mode 1), bottom face (Mode 2) or the side face (Mode 3). The failure is assumed to occur by rotation about the oblique compression hinge on a warped failure surface bounded by the compression hinge on one of the four faces and a spiral crack on the remaining three faces. The code clauses are aimed to ensure the provision of sufficient longitudinal bars and closed stirrups to prevent failure in any of the three possible modes of failure. The CEB—FIP Model code and the codes of several West European countries are based on space truss theory of torsion. According to this theory, the ultimate torque is resisted by a space truss comprising the concrete shell (which supplies the compression diagonals), the longitudinal bars (forming the stringers) and the closed stirrups (forming the ties). The theory assumes that the torsional strength of a solid section is the same as that of the equivalent hollow section having a wall thickness which according to CEB—FIP code is equal to one-sixth of the diameter of the largest circle which can be inscribed inside the lines joining the corner longitudinal bars. The code clauses for torsion are framed so as to prevent crushing of compression diagonals and yielding of longitudinal stringers and transverse ties. It is interesting to note that the provisions of the codes based on the two prominent theories can be reconciled as both theories ultimately lead to very similar results.

While all major codes contain torsion provisions for reinforced concrete members, torsion provisions specifically for prestressed concrete members are made in only some of the codes. Whereas the Indian Code IS: 1343—1980 and Australian Code AS: 1481—1978 make distinct provisions for prestressed concrete members, the British Code CP: 110-1972 permits torsion provisions of reinforced concrete members to be used also for prestressed concrete members. The code provisions reflect the general form of the torsional strength represented by Eq. (7.3.8). Due to the prevailing uncertainty, the codes differ in respect of the torque T_c carried by concrete. The value of the constant Ω_t in Eq. (7.3.10), reflecting the effectiveness of stirrups in resisting torsion, also differs in different codes.

7.3.3.1 Indian Code

As torsion rarely occurs alone, the Indian Code IS: 1343-1980 gives provisions for torsion in combination with flexure and shear. These provisions are discussed in Sec. 7.4.3.1.

7.3.3.2   British Code

The British Code specifies that in general where the torsional resistance or stiffness of members has not been taken into account in the analysis of the structure, no specific calculations for torsion are necessary. Adequate control of any torsional cracking is provided by the required nominal shear reinforcement. When it is considered that torsional resistance or stiffness of members at the limit state should be taken into account in the analysis and design of a structure, the following provisions in respect of reinforced concrete may also be used for prestressed concrete members.

To determine the distribution of internal forces in the members of a structure, the torsional rigidity GK may be computed by taking modulus of rigidity G equal to 0.4 times modulus of elasticity of concrete and the torsion constant K equal to half the Saint Venant value calculated for plain concrete section. The torsional shear stress should be computed using plastic theory for torsion. In the case of a rectangular section,

$$\tau = \frac{2\,T}{b^2 \left(D - \dfrac{b}{3} \right)} \tag{7.3.11}$$

in which b and D are the smaller and larger overall dimensions of the rectangular section. The flanged sections such as T, L or I may be treated by dividing them into component rectangles so that $\sum b^3 D$ is maximized. The component rectangles may be assumed to carry torques in proportion to their respective values of $(b^3 D)$. Box sections with wall thickness greater than one-fourth the overall dimension of the box in the direction of measurement may be treated as solid sections. Torsional reinforcement should be provided if τ computed from Eq. (7.3.11) exceeds the shear stress carried by concrete τ_c as per Table 7.3.3. In small beams having the larger dimension of stirrup $y' < 550$ mm, the torsional shear stress should not exceed $(y'/550)\,\tau_c$, max. The torsional reinforcement in the form of closed rectangular stirrups which should be in addition to that required for bending and shear may be determined from

$$\frac{A_t}{s} \geqslant \frac{T}{1.6\,b'd'\,(0.87\,f_{sy})} \tag{7.3.12}$$

and the longitudinal reinforcement A_{sl} for resisting torsion from

$$A_{sl} \geqslant \frac{2\,A_t}{s} \left(\frac{f_{sy}}{f_y} \right) (b' + d') \tag{7.3.13}$$

Table 7.3.3

f_{ck}, MPa	20	25	30	40 or more
τ_c, MPa	0.30	0 33	0. 37	0.42
τ_c, max, MPa	3.35	3.75	4.10	4.75

The yield stresses f_y for longitudinal steel and f_{sy} for stirrups should not be taken greater than 425 MPa. The code also contains the following provisions for detailing of torsional reinforcement:

(i) The spacing of closed rectangular stirrups should not exceed b', 0.5 d' nor 200 mm.

(ii) The longitudinal bars required for torsion should be evenly distributed around the inside perimeter of the stirrup with clear spacing not exceeding 300 mm and with a bar in each one of the four corners.

(iii) The torsion reinforcement should be extended over a distance equal to the larger overall dimension of the section beyond the theoretical cutoff point.

(iv) In flanged sections such as T, L or I, the reinforcement cages should tie the component rectangles together. No torsion reinforcement need be provided in a minor component rectangle in which the torsional shear stress is less than τ_c.

7.3.3.3 European Code

The European Code CEB-FIP-1978 based on space truss theory assumes that the ultimate torsional strength is controlled either by the compressive stress in the concrete diagonals or the tensile stress in reinforcement. The design ultimate torque T should not be greater than the ultimate torques T_d, T_l and T_t defined as follows.

(i) Ultimate torque T_d controlled by compression in walls,

$$T_d = 0.5\, A_o\, t_d\, f_{cd} \sin 2\alpha \qquad (7.3.14)$$

where A_o = area enclosed by shear flow which may be taken approximately equal to the area enclosed by corner longitudinal bars or tendons.

t_d = thickness of compression diagonals

f_{cd} = design compressive strength of concrete obtained by multiplying f_{ck} with appropriate partial safety factors

α = inclination of compression diagonals

(ii) Ultimate torque T_l controlled by longitudinal reinforcement,

$$T_l = \frac{2\, A_o\, A_{sl}\, f_y \tan \alpha}{u_o} \qquad (7.3.15)$$

where A_{sl} = area of longitudinal steel required for torsion

u_o = perimeter of shear flow which may be taken approximately equal to perimeter of polygon connecting centres of corner longitudinal bars or tendons

(iii) Ultimate torque T_t controlled by space truss (concrete diagonals and reinforcement),

$$T_t = T_c + \frac{2\, A_o\, A_t\, f_{sy} \cot \alpha}{s} \qquad (7.3.16)$$

where $T_c = C_1\, \tau_c\, A_o\, t_d$

= contribution of concrete in resisting torsion

τ_c = torsional shear stress resisted by concrete as per Table 7.3.4

Table 7.3.4

f_{cr}, MPa	30	35	40	45	50
τ_c, MPa	0.34	0.38	0.42	0.46	0.50

The constant c_1 should be taken equal to 5 if

$$T < 5 \, \tau_c \, A_o \, t_d$$

and c_1 is equal to zero if

$$T \geqslant 15 \, \tau_c \, A_o \, t_d$$

For intermediate values of T, c_1 should be linearly interpolated.

In order to check the crack width for adequate serviceability, the inclination of compression diagonals determined from space truss theory should be within the range,

$$(0.6 \leqslant \cot \alpha \leqslant 1.667).$$

The additional area of longitudinal steel due to prestressing tendons should be taken equal to the lesser of the following,

$$A_{sl} = \frac{A_p f_{py}}{f_y} \qquad (7.3.17)$$

$$A_{sl} = \frac{(f_{pi} + f_y)}{f_y} \qquad (7.3.18)$$

where A_p = area of prestressing tendons

f_{py} = characteristic yield strength of prestressing tendons

f_{pi} = initial or permanent stress in prestressing steel

EXAMPLE 7.3.1

Determine the ultimate torque of a prestressed concrete beam of rectangular section 0.2×0.4 m having no web reinforcement. The beam has an effective concentric prestressing force of 560 kN. The tensile strength of concrete in direct tension, $f_t = 3.33$ MPa. Also determine the angle of twist per unit length and maximum shear stress when the applied torque is 9.5 kN.m. Take $G_c = 14.5$ GPa.

Solution

Using elastic theory and assuming that the ultimate torque is reached as soon as the principal tensile stress becomes equal to the tensile strength of concrete, Equations (7.3.6) and (7.3.7) give,

$$T_u = T_{up} \, k_p = \frac{K f_t}{\rho_{max}} k_p$$

Using Tables 7.3.1 and 7.3.2,

$$K = k_1 \, b^3 \, d = 0.229 \times 0.2^3 \times 0.4 = 0.0007328 \text{ m}^4$$

$$\rho_{max} = k_2 \, b = 0.930 \times 0.2 = 0.186 \text{ m}$$

Using Eq. (7.3.5),

$$k_p = \sqrt{1 + \frac{0.560}{0.2 \times 0.4} \times \frac{1}{3.33}} = 1.761$$

Hence,

$$T_u = \frac{0.0007328 \times 3.33 \times 1.761}{0.186} = 0.0231 \text{ MN·m}$$

Using Eq. (7.3.2), the angle of twist per unit length,

$$\theta = \frac{T}{GK} = \frac{0.0095}{14500 \times 0.0007328} = 0.000894 \text{ rad/m}$$

Also, maximum shear stress at midpoint of the longer side,

$$\tau_{max} = \frac{T \, \rho_{max}}{K} = \frac{0.0095 \times 0.186}{0.0007328} = 2.411 \text{ MPa}$$

EXAMPLE 7.3.2

The flanged section shown in Fig. 7.3.2 has a uniform intensity of prestress of 6.5 MPa. Determine the cracking torque and the angle of twist per unit length at cracking. Take $f_t = 2.9$ MPa and $G_c = 13.5$ GPa.

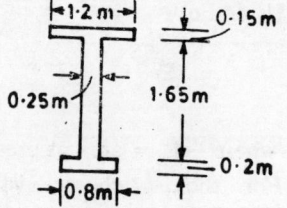

Fig. 7.3.2

Solution

According to elastic theory, the maximum torsional shear stress occurs in the thickest component rectangle, viz. the web. Hence, using Eq. (7.3.4), the shear stress in web at cracking,

$$\tau_w = f_t \, k_p = 2.9 \sqrt{1 + \frac{6.5}{2.9}} = 5.22 \text{ MPa}$$

The maximum shear stresses in component rectangles of thin walled open sections are proportional to their respective thickness. Hence the maximum shear stresses in top and bottom flanges are,

$$\tau_{tf} = 5.22 \times \frac{0.15}{0.25} = 3.13 \text{ MPa}$$

$$\tau_{bf} = 5.22 \times \frac{0.20}{0.25} = 4.18 \text{ MPa}$$

Using Eq. (7.3.7), Table 7.3.1 and taking the component rectangles to be thin,

$$T_{up} = \frac{1}{3} t^2 b \, \tau_{max}$$

Hence the twisting moments at incipient cracking resisted by top flange, bottom flange and web are,

$$T_{tf} = \frac{1}{3} \times 0.15^2 \times 1.2 \times 3.13 = 0.0282 \text{ MN.m}$$

$$T_{bf} = \frac{1}{3} \times 0.20^2 \times 0.8 \times 4.18 = 0.0446 \text{ MN.m}$$

$$T_w = \frac{1}{3} \times 0.25^2 \times 1.65 \times 5.22 = 0.1795 \text{ MN.m}$$

Hence the torque resisted by the whole section at incipient cracking,

$$T = T_{tf} + T_{bf} + T_w = 0.2523 \text{ MN.m}$$

Using Table 7.3.1, the torsion constants for top flange, bottom flange and web are

$$K_{tf} = \frac{1}{3} \times 0.15^3 \times 1.2 = 0.00135 \text{ m}^4$$

$$K_{bf} = \frac{1}{3} \times 0.20^3 \times 0.8 = 0.00213 \text{ m}^4$$

$$K_w = \frac{1}{3} \times 0.25^3 \times 1.65 = 0.00859 \text{ m}^4$$

The torsion constant for the whole section,

$$K = K_{tf} + K_{bf} + K_w = 0.01207 \text{ m}^4$$

Using Eq. (7.3.2), the angle of twist per unit length,

$$\theta = \frac{T}{G_c K} = \frac{0.2523}{13500 \times 0.01207} = 0.00155 \text{ rad/m}$$

EXAMPLE 7.3.3

For the prestressed concrete section shown in Fig. 7.3.3, calculate the maximum torque if the maximum shear stress is limited to 2 MPa. Also determine the maximum intensity of diagonal tension if the section has a uniform prestress of 6.5 MPa.

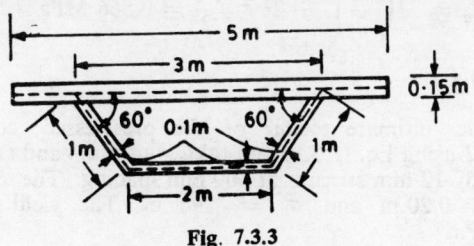

Fig. 7.3.3

Solution

The section is a combination of a closed cell and two outstanding flanges. Using Table 7.3.1, the torsional constant of the cell,

$$K_{cell} = \frac{4 \left(\frac{3+2}{2} \times \frac{\sqrt{3}}{2} \right)^2}{\frac{1}{0.10} + \frac{2}{0.10} + \frac{1}{0.10} + \frac{3}{0.15}} = 0.3125 \text{ m}^4$$

and the stress factor for the cell,

$$\rho_{cell} = \frac{2\left(\dfrac{3+2}{2}\right) \times \dfrac{\sqrt{3}}{2}}{0.1\left[\dfrac{1}{0.10} + \dfrac{2}{0.10} + \dfrac{1}{0.10} + \dfrac{3}{0.15}\right]} = 0.7217 \text{ m}$$

Hence using Eq. (7.3.2), the torque carried by the cell,

$$T_{cell} = \frac{0.3125 \times 2}{0.7217} = 0.866 \text{ MN·m}$$

Using Eq. (7.3.2), the angle of twist per unit length,

$$\phi = \frac{0.866}{0.3125\, G} = \frac{2.7712}{G}$$

Using Table 7.3.1, the torsion constant of the flanges,

$$k_{flanges} = 2 \times \frac{1}{3} \times 0.15^3 \times 1 = 0.00225 \text{ m}^4$$

As the cell and the outstanding flanges must twist through the same angle, the torque resisted by outstanding flanges, according to Eq. (7.3.2),

$$T_{flanges} = G \times \frac{2.7712}{G} \times 0.0025 = 0.0062 \text{ MN·m}$$

Hence the total torque resisted by the section,

$$T = 0.866 + 0.0062 = 0.8722 \text{ MN·m}$$

The maximum diagonal tension occurs in the thinnest part of the cell where the torsional shear stress is maximum (2 MPa). Using Eq. (7.3.3), the the intensity of diagonal tension,

$$f_{dt} = \sqrt{\left(\frac{6.5}{2}\right)^2 + 2^2} - \frac{6.5}{2} = 0.566 \text{ MPa}$$

EXAMPLE 7.3.4

Determine the ultimate torque of the prestressed concrete beam of Example 7.3.2 using Eq. (7.3.8) and taking $c_1 = 0.4$ and $\Omega_t = 1.2$. The web is reinforced by 12 mm stirrups at 300 mm spacing. The dimensions of the stirrups, $b' = 0.20$ m and $d' = 1.90$ m. The yield stress of stirrups, $f_{sy} = 250$ MPa.

Solution

Taking the cracking torque equal to the ultimate strength of the corresponding plain concrete section, the data of Example 7.3.2 gives,

$$T_{up}\, k_p = 0.2523 \text{ MN·m}$$

Hence the torque resisted by concrete in accordance with Eq. (7.3.9),

$$T_c = 0.4 \times 0.2523 = 0.1009 \text{ MN·m}$$

Using Eq. (7.3.10), the torque resisted by steel,

$$T_s = \frac{1.2 \times \frac{\pi}{4} \times 0.012^2 \times 250 \times 0.20 \times 1.90}{0.30} = 0.0430 \text{ MN} \cdot \text{m}$$

Hence the utimate torque,

$$T = T_c + T_s = 0.1009 + 0.0430 = 0.1439 \text{ MN} \cdot \text{m}$$

EXAMPLE 7.3.5

Design the torsion reinforcement for the beam of Example 7.2.4 to develop an ultimate torque of 12.5 kN·m. Take $f_y = f_{sy} = 250$ MPa and $f_{py} = 1280$ MPa.

Solution

(i) *British Code*

Ignoring the fillets and dividing the section into three component rectangles, viz. top flange, web and bottom flange, the shear stress according to plastic theory of torsion (Eq. 7.3.11),

$$\tau = \frac{2 \times 0.0125}{0.20^2 \left(0.75 - \frac{0.20}{3}\right) + 0.20^2 \left(0.80 - \frac{0.20}{3}\right) + 0.20^2 \left(0.75 - \frac{0.20}{3}\right)}$$

$$= 0.2976 \text{ MPa}$$

Using Table 7.3.3, linear interpolation gives,

$$\tau_c = 0.39 \text{ MPa for } M\text{-}35 \text{ concrete}$$

As $\tau < \tau_c$, torsion reinforcement is not required.

(ii) *European code*

For the cross-section of Example 7.2.4, the following cross-sectional properties may be used. It is assumed that the torsional reinforcement comprises two legged stirrups and longitudinal bars in the web.

$$b_l = 0.13 \text{ m}$$
$$d_l = 1.08 \text{ m}$$
$$A_o = b_l d_l = 0.1404 \text{ m}^2$$
$$u_o = 2(b_l + d_l) = 2.42 \text{ m}$$

The torsion reinforcement may be computed from Eq. (7.3.16). The thickness of compression diagonals.

$$t_d = \frac{b_l}{6} = 0.02166 \text{ m}$$

Using Table 7.3.4,

$$\tau_c = 0.38 \text{ MPa for } M\text{-35 concrete}$$

$$5 \, \tau_c \, A_o \, t_d = 5 \times 0.38 \times 0.1404 \times 0.02166 = 0.005778 \text{ MN·m}$$

$$15 \, \tau_c \, A_o \, t_d = 0.017334 \text{ MN·m}$$

As the applied torque of 0.0125 MN·m lies between 0.005778 MN·m and 0.017334 MN·m, the constant $c_1 = 2.093$ by linear interpolation. Hence using Eq. (7.3.16),

$$T_c = 2.093 \times 0.308 \times 0.1404 \times 0.02166 = 0.00242 \text{ MN·m}$$

Taking $\alpha = 45°$ for pure torsion,

$$0.0125 = 0.00242 + 2 \times 0.1404 \times 250 \cot 45° \times \frac{A_t}{s}$$

or
$$\frac{A_t}{s} = 0.1435 \text{ mm}$$

Using 10 mm stirrups, the spacing of stirrups,

$$s = \frac{\frac{\pi}{4} \times 10^2}{0.1435} = 547.5 \text{ mm}$$

An equal volume of longitudinal bars may be provided.

$$A_{sl} = \frac{A_t}{s} u_o = 347.27 \text{ mm}^2$$

Hence six longitudinal bars of 10 mm diameter, one at each corner and one at the midpoint of the longer side may by provided.

It should be checked that the torques controlled by the crushing of compression diagonals T_d and yielding of the longitudinal steel T_l are not less than the applied torque. Using Eq. (7.3.14) and taking $f_{cd} = f_{ck}/1.5 = 23.33$ MPa,

$$T_d = 0.5 \times 0.1404 \times 0.02166 \times 23.33 = 0.03547 \text{ MN·m}$$

$$= 35.47 \text{ kN·m}$$

The additional area of longitudinal steel due to prestressing tendons is the lesser of the areas given by Eq. (7.3.17) and (7.3.18).

$$A_{sl} = \frac{3696}{250} \times 1280 = 18924 \text{ mm}^2$$

$$A_{sl} = \frac{3696}{250} \left(\frac{3.3 \times 10^6}{3696} + 250 \right) = 16896 \text{ mm}^2$$

Adding the contribution of prestressing tendons to the area of longitudinal bars, the net area

$$A_{sl} = 6 \times \frac{\pi}{4} \times 10^2 + 16896 = 17367 \text{ mm}^2 = 0.017367 \text{ m}^2$$

Hence using Eq. (7.3.15),

$$T_l = 2 \times \frac{0.1404}{2.42} \times 0.017367 \times 250 = 0.5038 \text{ MN·m}$$

$$= 503.8 \text{ kN·m}$$

A_s both T_d and T_l are greater than the applied torque, the section is safe against crushing of concrete and yielding of longitudinal steel.

7.4 COMBINED SHEAR AND TORSION

Torsion is frequently accompanied by other types of internal actions, viz. flexure, shear and axial tension or compression. Two or more internal actions interact with each other, i.e. the behaviour under any internal action is affected due to the presence of other internal actions. Among these interactions, the one between shear and torsion is generally the strongest. Due to this reason and also due to the similarity of the shear and torsion phenomena, the provisions for shear and torsion are integrated with each other in most codes of practice.

7.4.1 Uncracked Section

In an uncracked section, the stresses due to two or more internal actions can be superimposed on each other in accordance with the principle of superposition. The flexural shear stress computed from Eq. (7.2.1) and the torsional shear stress computed from Eq. (7.3.2) may be combined vectorially with the normal stress f at any point due to prestress, bending moment and axial force, if any, to obtain the resultant stress. The flexural shear stress v and the torsional shear stress τ are additive on one side of the section and subtractive on the other side as shown in Fig. 7.4.1. The point at which v and τ are additive can be the most critical point for the commencement of diagonal tension crack, particularly if the normal compressive stress f at that point is relatively small. The intensity of diagonal tension f_{dt} at the point under combined stresses v, τ and f may be expressed as

$$f_{dt} = \frac{f}{2} + \sqrt{\left(\frac{f}{2}\right)^2 + (v + \tau)^2} \tag{7.4.1}$$

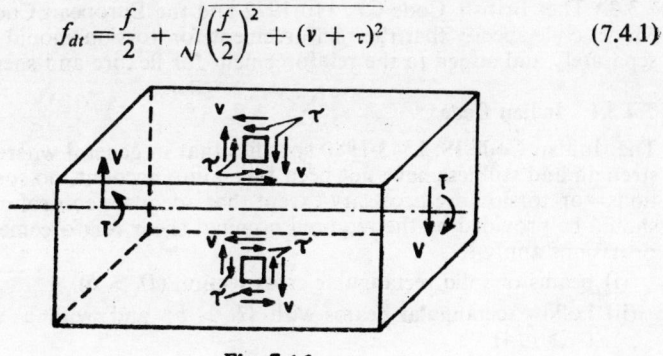

Fig. 7.4.1

The condition of incipient diagonal cracking is attained when f_{dt} becomes equal to the tensile strength of concrete, f_t. In the absence of web reinforcement, the cracking load may be taken conservatively equal to the ultimate load of the member.

7.4.2   Cracked Section

At the formation of the diagonal tension crack, the stress is transferred progressively from concrete to web reinforcement. Both flexural shear and torsional shear demand provision of web reinforcement. Hence the additive principle may be used for computing the area of web steel under combined shear and torsion. According to additive principle, the areas of web steel required for shear and torsion are determined separately, assuming that shear and torsion act by themselves. The steel areas thus computed are then added.

$$A_{vt} = A_v + 2A_t \tag{7.4.2}$$

where A_{vt} = area of stirrup (both legs) for resisting combined shear and torsion

A_v = area of stirrup (both legs) for resisting shear

A_t = area of stirrup (one leg) for resisting torsion

The additive principle forms the basis for the provisions of several prominent codes for dealing with combined shear and torsion.

The shear force V_c and torque T_c resisted by concrete also interact with each other. Due to the complexity of the problem, it is difficult to derive an interaction relationship between V_c and T_c on theoretical basis. Most codes, therefore, adopt interaction relationship based on test results.

7.4.3   Code Provisions

The Indian Code IS: 1343-1980, which is similar to Australian code AS: 1481-1978 presumes that torsion is invariably accompanied by flexure and shear. Consequently, the Indian Code makes recommendations for design of prestressed concrete members subjected to combined torsion, bending and shear. The provisions are based on three possible modes of failure (Modes 1, 2 and 3) in accordance with the skew bending theory discussed in Sec. 7.3.3. The British Code CP: 110-1972 and the European Code CEB—FIP Model code specify that the reinforcement for torsion should be computed separately and added to the reinforcement for flexure and shear.

7.4.3.1   Indian Code

The Indian Code IS: 1343-1980 specifies that in general where the torsional strength and stiffness have not been taken into account, no specific calculations for torsion are necessary except that adequate control on crack width should be provided by the required nominal shear reinforcement. The code provisions apply to:

(i) beams of solid rectangular cross-section $(D > b)$
(ii) hollow rectangular beams with $(D > b)$ and with a wall thickness $(t \geqslant b/4)$
(iii) T-beams and I-beams

The maximum value of the average intensity of prestress in concrete is restricted to $0.3 \, f_{ck}$. The longitudinal and transverse reinforcement for torsion should be designed in accordance with the following provisions:

(a) Longitudinal Reinforcement

The longitudinal reinforcement should be designed to resist the equivalent ultimate bending moment M_{e1} computed from

$$M_{e1} = M + M_t \tag{7.4.3}$$

where M = applied ultimate bending moment at the cross-section acting in combination with torque, T

and M_t is the moment contributed by the applied torque T and is given by the equation

$$M_t = T \sqrt{1 + \frac{2D}{b}} \tag{7.4.4}$$

The sign of M_t should be taken to be the same as that of M. If the numerical value of M is smaller than that of M_t, the member should also be designed to resist the equivalent bending moment M_{e2} computed from

$$M_{e2} = M_t - M \tag{7.4.5}$$

The moment M_{e2} should be taken to be of opposite in sign to that of M. If the numerical value of M is less than or equal to that of M_t, the member should also be designed to resist an equivalent transverse bending moment M_{e3} in the horizontal plane determined from

$$M_{e3} = M_t \left(1 + \frac{b'}{2e}\right) \left[\frac{1 + \frac{2b}{D}}{1 + \frac{2D}{b}} \right] \tag{7.4.6}$$

where b' = shorter centre line dimension of closed rectangular stirrup resisting torsion

$e = \dfrac{T}{V}$

= torque-shear ratio at the section at ultimate stage

The equivalent moment M_{e2} should not be taken acting simultaneously with M_{e1}.

(b) Transverse Reinforcement

The area of closed rectangular stirrup (both legs) resisting combined shear and torsion should be taken as the larger of the values given by Eq. (7.4.7) and (7.4.8).

$$A_{vt} = \frac{M_{ts}}{1.5 \, b_1 \, d_1 \, f_{sy}} \tag{7.4.7}$$

and

$$A_{vt} = A_v + 2A_t \tag{7.4.8}$$

in which A_v is the area of stirrup (both legs) resisting shear computed from

$$A_v = \frac{(V - V_{c1})\,s}{0.87\,f_{sy}\,d_l} \tag{7.4.9}$$

and A_t is the area of stirup (one leg) resisting torsion computed from

$$A_t = \frac{(T - T_{c1})\,s}{0.87\,f_{sy}\,b_l\,d_l} \tag{7.4.10}$$

where b_l, d_l = centre to centre distance between corner bars in the direction of the width and the depth respectively

$$V_{c1} = V_c \left(\frac{e}{e + e_c} \right)$$

= reduced shear force carried by centre

$$T_{c1} = T_c \left(\frac{e}{e + e_c} \right)$$

= reduced torque carried by concrete

$$e_c = \frac{T_c}{V_c}$$

$$T_c^* = 0.125\,b^2\,D \left(1 - \frac{b}{3D} \right) k_p \sqrt{f_{ck}}$$

$$k_p = \sqrt{1 + \frac{12p}{f_{ck}}}$$

= prestress factor

V_c = shear force carried by concrete as given in Sec. 7.2.3.1

*The expression for T_c appearing in IS: 1343-1980 appears to be erroneous. The coefficient 1.5 is reasonable in Imperial units and should be replaced by 0.125 if SI units are used. Also the reduction factor $(1 - b/30)$ should read $1 - \frac{1}{3}\left[\frac{b}{D}\right]$.

In the above analysis b is the breadth of rectangular section in metre and should be replaced by b_w in the case of flanged sections.

The code also contains the following provisions:

(i) The area of web reinforcement A_{vt} should not be less than that given by

$$\frac{A_{vt}}{s} = \frac{0.4b}{0.87 f_{sy}} \tag{7.4.11}$$

(ii) At least one longitudinal bar of diameter not less than 12 mm should be provided in each corner of the stirrups.

(iii) The web reinforcement should be in the form of closed rectangular stirrups normal to axis of the member.

(iv) The spacing of the stirrups should not be greater than 200 mm nor $(b' + d')/4$ where b' and d'' are the shorter and longer centre line dimensions of the stirrup.

(v) Torsional reinforcement should be continued to a distance $(D + b)$ beyond the theoretical cut off point.

7.4.3.2 British Code

The British Code specifies that the reinforcement for torsion should be in addition to that required for flexure and shear. The longitudinal reinforcement required for torsion at the level of tension and compression reinforcement may be provided by using larger bars than those required for bending alone. The code also specifies that in no case should the shear stress $(v + \tau)$ due to combined action of shear and torsion exceed $\tau_{c, max}$ as per Table 7.3.3.

7.4.3.3 European Code

The CEB-FIP code provides that in the case of combined torsion, flexure and/or axial forces, the longitudinal reinforcement in the flexural tension zone should be determined separately for torsion and for axial force and added. On the other hand in the flexural compression zone, the area of longitudinal steel may be reduced on account of flexural compression. The principal stress in the flexural compression zone should be checked for high moment-torque ratio, particularly in the case of box sections. The principal compressive stress should be based on the mean flexural compressive stress and torsional shear stress determined from

$$\tau = \frac{T}{2A_o\, t_d} \tag{7.4.12}$$

The principal compressive stress should not exceed $0.85 f_{cd}$. In the case of combined torsion and shear the code provides a linear interaction relationship.

EXAMPLE 7.4.1

A pretensioned beam of rectangular section 0.20×0.45 m has uniform prestress of 6.4 MPa. It is subjected to a shear force of 72 kN and a torque of 3.6 kN.m. Determine the maximum intensity of shear stress and diagonal tension.

Solution

The maximum intensities of flexural shear stress and torsional shear stress occur at the midpoint of the longer side. The maximum intensity of flexural shear stress is 1.5 times the average shear stress. Hence

$$v_{max} = 1.5 \times \frac{0.072}{0.20 \times 0.45} = 1.2 \text{ MPa}$$

Using Table 7.3.1 and Eq. (7.3.2), maximum torsional shear stress,

$$\tau_{max} = \frac{T}{K}\, \rho_{max} = \frac{T k_2 b}{k_1 b^3 d} = \frac{k_2}{k_1} \frac{T}{b^2 d}$$

$$= \frac{0.947}{0.239} \times \frac{0.0036}{0.20^2 \times 0.45}$$

$$= 0.792 \text{ MPa}$$

At the midpoint of one of the two vertical sides, v_{max} and τ_{max} are additive. Hence the maximum intensity of shear stress,

$$v_{max} + \tau_{max} = 1.2 + 0.792 = 1.992 \text{ MPa}$$

Using Eq. (7.4.1),

$$f_{dt} = \frac{(-6.4)}{2} + \sqrt{\left(\frac{-6.4}{2}\right)^2 + (1.992)^2}$$

$$= 0.569 \text{ MPa}$$

EXAMPLE 7.4.2

For the beam of Example 7.2.4, design the web reinforcement at the left quarter point at which there is a shear force of 0.297 MN combined with a bending moment of 2.673 MN.m and a twisting moment of 0.0125 MN.m.

Solution

(a) *Longitudinal Reinforcement*

According to Eq. (7.4.3), the longitudinal reinforcement has to resist a bending moment,

$$M_{e1} = M + M_t = M + T \sqrt{1 + \frac{2D}{b_w}}$$

$$= 2.673 + 0.0125 \sqrt{1 + \frac{2 \times 1.20}{0.20}}$$

$$= 2.673 + 0.03307$$

$$= 2.70607 \text{ MN.m}$$

It may be checked that the area of prestressing steel $A_p = 3696 \text{ mm}^2$ is sufficient to develop an ultimate moment equal to M_{e1}. Also as M_t is not greater than M, it is not necessary to check the section for moments M_{e2} and M_{e3}.

(b) *Transverse Reinforcement*

Using Eq. (7.4.7).

$$\frac{A_{vt}}{s} = \frac{0.03307}{1.5 \times 0.13 \times 1.08 \times 250}$$

$$= 0.000628 \text{ m}$$

Next. Eq. (7.4.8) will be used

$$e = \frac{T}{V} = \frac{0.0125}{0.297} = 0.042 \text{ m}$$

$$V_c = 0.3765 \text{ MN from Example 7.2.4}$$

$$T_{c_1} = 0.125 \sqrt{\left(1 + \frac{12 \times 6.875}{35}\right) 35}$$

$$\times \left[0.2^2 \times 0.75 \left(1 - \frac{0.2}{3 \times 0.75}\right) \times 2 + \right.$$

$$\left. 0.2^2 \times 0.8 \left(1 - \frac{0.2}{3 \times 0.8}\right)\right]$$

$$= 0.1138 \text{ MN.m}$$

$$e_c = \frac{T_c}{V_c} = \frac{0.1138}{0.3765} = 0.302 \text{ m}$$

$$V_{c1} = V_c \left(\frac{e}{e + e_c}\right) = 0.3765 \left(\frac{0.042}{0.042 + 0.302}\right) = 0.0460 \text{ MN}$$

$$T_{c1} = T_c \left(\frac{e}{e + e_c}\right) = 0.1238 \left(\frac{0.042}{0.042 + 0.302}\right) = 0.0151 \text{ MN.m}$$

As T_{c1} is greater than applied torque T, $A_t = 0$. Hence Using Eq. (7.4.8),

$$\frac{A_{vt}}{s} = \frac{V - V_{c1}}{0.87 f_{sy} d_t} = \frac{0.297 - 0.046}{0.87 \times 25 \times 1.08}$$

$$= 0.001069 \text{ m} = 1.069 \text{ mm}$$

Hence area of transverse reinforcement is controlled by Eq. (7.4.8) which gives a larger area, $A_{vt}/s = 1.069$ mm. Using 10 mm diameter two legged stirrups, the spacing of stirrups, $s = 147$ mm.

7.5 BOND

The tangential stress at the interface of prestressing tendons or reinforcing bars and the surrounding concrete or grout is known as bond stress. It tends to cause relative movement or slip at the interface between the two materials. The bond strength or bond capacity comprises the tangential forces which oppose this relative movement or slip. The factors contributing to bond strength are adhesion, friction, mechanical anchorage and wedge action. The failure of bond known as bond failure occurs as soon as the bond stress exceeds the bond strength. In the context of prestressed concrete, bond may be classified into two types.

7.5.1 Prestress Transfer Bond

As indicated by its name, the prestress transfer bond is responsible for the transfer of prestress from prestressing tendons to concrete. This type of bond stress occurs near the ends of a prestressing tendon in a pretensioned member. As soon as a prestressing tendon is released from the external supports or detached by cutting at the end A of a pretensioned member shown in Fig. 7.5.1 (a), the stress at the free end of the tendon becomes zero. Due to bond with surrounding concrete, the stress in prestressing tendon increases along its length until it attains its full value at some point B located at distance l_t from the free end A. The length of tendon l_t over which the prestressing tendon attains its full stress is known as transfer length, transmission length or anchor length. A short length of the tendon near the free end A loses its bond with concrete. Consequently, the tendon digs into concrete through a small distance AA'. Due to Poisson's ratio effect, the diameter of the prestressing tendon increases continuously from the point B at which the tendon has its full tensile stress to the point A' at which the tendon has no stress as indicated by the broken lines. The increase in diameter of tendon towards the free end over the transfer length of tendon is responsible for bond due to wedge action. The wedge action due to Poisson's ratio effect was first discovered by Hoyer and is known after him as the Hoyer effect. The increased diameter of tendon causes compressive stress between itself and surrounding concrete which is responsible for enhanced friction against slip. This phenomenon may be contrasted with the behaviour of a tension bar in a reinforced concrete member in which the reduced diameter of the reinforcing bar tends to create tensile stress at the interface in addition to adhesion, friction and wedge action. A contribution to bond strength is also made by mechanical anchorage in the case of deformed tendons, such as indented or crimped wire. Consequently, the transfer length l_t is reduced in the case of deformed tendons as compared to smooth tendons.

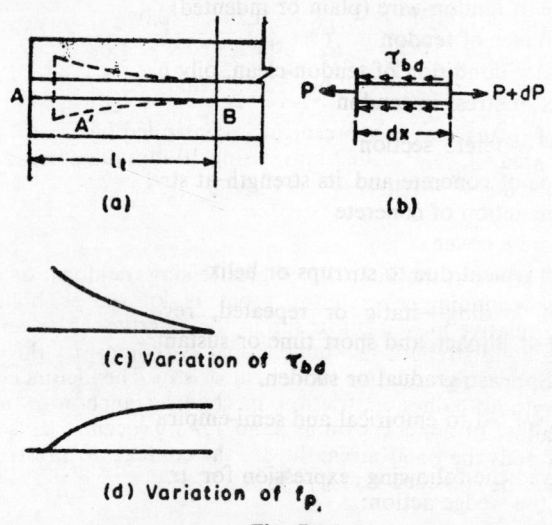

(a)

(b)

(c) Variation of τ_{bd}

(d) Variation of f_p.

Fig. 7.5.1

In order to determine the relationship between the bond stress τ_{bd} and the tensile stress in prestressing tendon f_p, consider the free body of a small element of the tendon in the transfer length shown in Fig. 7.5.1 (b). For the equilibrium in the transfer length of forces along the axis of the tendon,

$$P + \tau_{bd}\, u_t\, d_x = P + dP$$

or

$$\tau_{bd} = \left(\frac{dP}{dx}\right)\frac{1}{u_t} \tag{7.5.1}$$

$$= \frac{A_p}{u_t}\left(\frac{df_p}{dx}\right) \tag{7.5.1 a}$$

where $\quad\quad\quad \tau_{bd}$ = bond stress

$\quad\quad\quad\quad\quad A_p$ = area of prestressing tendon

$\quad\quad\quad\quad\quad f_p$ = stress in prestressing tendon

$\quad\quad\quad\quad\quad u_t$ = perimeter of the tendon

Equation (7.5.1) shows that the bond stress τ_{bd} is proportional to the rate of change of tensile stress in tendon f_p. The variation of τ_{bd} and f_p over the transfer length shown in Fig. 7.5.1 (c) and (d) respectively depends mainly upon the type of prestressing tendon. While the actual variation of f_p is non-linear, certain codes. notably the ACI code, allow the linear variation for the sake of simplicity in design.

Due to practical importance of transfer length and the difficulties in determining it by theoretical methods, several experimental investigations have been carried out for determining l_t. The tests have shown that l_t depends upon several parameters such as:

(1) Condition of tendon:

   (a) type of tendon-wire (plain or indented) or strand

   (b) diameter of tendon

   (c) surface condition of tendon-clean, oily or rusted

   (d) level of stress in tendon

(2) Type of concrete section

   (a) grade of concrete and its strength at stress transfer

   (b) compaction of concrete

   (c) concrete cover

   (d) confinement due to stirrups or helix

(3) Type of loading—static or repeated, reversible or non-reversible, gradual or impact, and short time or sustained.

(4) Type of release-gradual or sudden.

The tests have led to empirical and semi-empirical expressions for length l_t.

Hoyer derived the following expression for transfer length taking into account only the wedge action:

$$l_t = \frac{\phi}{2\mu} \left[(1 + \nu_c) \left(\frac{m}{\nu_s} - \frac{f_{pi}}{E_c} \right) \left(\frac{f_{pe}}{2f_{pi} - f_{pe}} \right) \right] \quad\quad (7.5.2)$$

where $\quad\quad \phi$ = diameter of tendon

$\quad\quad\quad\quad \mu$ = coefficient of friction between concrete and steel

$\quad\quad\quad\quad \nu_c$ = Poisson's ratio for steel

$\quad\quad\quad\quad m$ = modular ratio

$\quad\quad\quad\quad\quad = \dfrac{E_s}{E_c}$

$\quad\quad\quad\quad E_c$ = modulus of elasticity of concrete

$\quad\quad f_{pi}, f_{pe}$ = initial and effective stress in prestressing tendon

The quantity inside the square brackets in Eq. (7.5.2), usually lies between 16 and 18 approximately. Taking it equal to 18, the transfer length may be expressed as

$$l_t = \frac{9\phi}{\mu} \qquad (7.5.3)$$

Taking $\mu = 0.1$, the transfer length,

$$l_t = 90\,\phi \qquad (7.5.4)$$

On the basis of extensive tests, Janney proposed the following expression for transfer length in millimeter,

$$l_t = 1.5\,\frac{f_{pi}}{f'_{ci}}\,\phi - 117 \qquad (7.5.5)$$

where f'_{ci} = cylinder strength of concrete at transfer

For $f_{pi} = 1200$ MPa and $f'_i = 30$ MPa, the transfer length for 5 mm diameter tendon is 183 mm in accordance with Eq. (7.5.5).

Marshall and Krishnamurthy gave the following empirical expression for transfer length which is applicable for plain wires and strands.

$$l_t = 31.6\,(\beta)^{-0.5}\,(f_{ci})^{0.25} \qquad (7.5.6)$$

where f_{ci} = cube strength of concrete at transfer

β = constant which depends upon type of tendon (Table 7.5.1)

Table 7.5.1 Values of β

Tendon	β
Plain Wires	
2 mm	0.144
5 mm	0.0235
7 mm	0.0174
Strands	
10 mm —7 wires	0.144
12.5 mm —7 wire	0.058
18 mm —19 wire	0.0235
19 mm —7 wire	0.0235
Twin twisted wires or	
6.25 mm—7 wire	0.077

7.5.2 Flexural Bond

Flexural bond stress arises due to shear force or variation in bending moment similar to that in a reinforced concrete member. At a section at which there is no shear force and no flexural crack, the flexural bond stress is zero. This type of bond stress occurs in pretensioned as well as post-tensioned bonded members. In pretensioned members, the flexural bond stress acts at the interface of prestressing tendon and surrounding concrete just as in reinforced concrete. In post-tensioned bonded members, the flexural bond stress occurs at the interface of prestressing tendon and grout. The flexural bond stress arises also at the interface of grout and duct in the absence of sheathing. In the case of non-retractable sheathing left permanent in concrete, flexural bond stress occurs at three different interfaces, viz. (i) prestressing tendon and grout, (ii) grout and sheathing and (iii) sheathing and surrounding concrete. In post-tensioned unbonded members, the flexural bond stress is absent. The distribution of flexural bond stress changes abruptly at flexural cracking. Hence flexural bond stress before and after flexural cracking deserves consideration.

(a) Precracking Range

The nature of flexural bond stress at the interface of prestressing tendon and grout is similar to that in a reinforcing bar located in the flexural compression zone. Consider a small element of the prestressing tendon of length dx located at distance y_t from the centroidal axis of the uncracked section. If dM is the change of bending moment over length dx, the change in stress f_p in prestressing tendon may be expressed as

$$df_p = \frac{dM}{I}\, y_t\, m \qquad (7.5.7)$$

in which m is the modular ratio and I is the moment of inertia of the uncracked section about centroidal axis. Also for the equilibrium of forces along the axis of the tendon acting on the free body of the tendon element shown in Fig. 7.5.2,

Fig. 7.5.2

$$dP = A_p\, df_p = \tau_{bd}\, u_t\, dx \qquad (7.5.8)$$

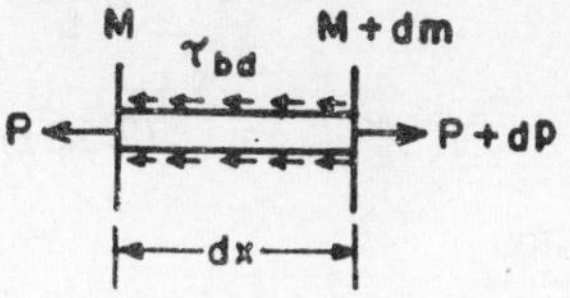

where $u_t =$ perimeter of tendon

Combining Eq. (7.5.7) and (7.5.8),

$$\tau_{bd} = \frac{A_p\, dM\, y_t\, m}{u_t\, dx\, I}$$

$$= \frac{Vm\, A_p\, y_t}{I\, u_t} \qquad (7.5.9a)$$

For a tendon of diameter ϕ,

$$\frac{A_p}{u_t} = \frac{\phi}{4}$$

Hence the expression for bond stress at a flexurally uncracked section may be written as

$$\tau_{bd} = \frac{Vm\, y_t\, \phi}{4I} \qquad (7.5.9b)$$

The flexural bond stress on an uncracked section is generally small and, therefore. seldom creates problem in design. The nature of flexural bond stress between grout and concrete is similar. Proceeding in a manner similar to that used for derivation of Eq. (7.5.9 a), it may be shown that the bond stress at the interface of the grout and concrete may be written as

$$\tau_{bd} = \frac{V \, y_t}{I} \left(\frac{mA_p + m_g \, A_g}{u_g} \right)$$

$$= \frac{Vm \, y_t \, A_p}{I \, u_t} \qquad (7.5.10)$$

where m_g = modular ratio for grout

A_g, u_g = area and perimeter of grout prism or duct

The flexural bond stress at the interfaces between prestressing tendon, grout, sheathing and concrete should be smaller than the respective permissible bond stresses to prevent local bond failure at the section of maximum shear force,

(b) Post-cracking Range

The formation of a flexural crack destroys continuity of the medium, viz. concrete. It is evident that a crack cannot open without a local bond failure at the section of the crack formation. Hence the flexural bond stress is zero at this section. The flexural bond stress increases on either side of the crack until it becomes maximum at some section between any two consecutive cracks. Thereafter it decreases until it becomes zero at the section of next crack. The local bond failure at the sections of flexural cracks do not pose the danger of an overall bond failure as long as the total bond strength along the entire length of the tendon becomes equal to the resultant or cumulative bond stress. The local bond stress at any section in a cracked member may be determined in the same manner as in the reinforcing bars of a reinforced concrete member.

$$\tau_{bd} = \frac{V}{jd_p \left(\sum u_t \right)} \qquad (7.5.11)$$

where V = shear force at the section

jd_p = internal lever arm

$\sum u_t$ = sum of perimeter of prestressing tendons

7.5.3 Code Provisions and Empirical Expressions

The Indian Code IS: 1343-1980 recommends the following values of l_t in the absence of tests subject to the condition that the strength of concrete at transfer is not less than 35 MPa and the tendons are released gradually

(i) plain and indented wires $100 \, \phi$
(ii) crimped wires $65 \, \phi$
(iii) strands $30 \, \phi$

The above values apply for wires not exceeding 5 mm and strands not exceeding 18 mm diameter. The stress in tendon f_p may be assumed to vary parabolically from zero at the free end to its full value at the end of transfer length. The code recommonds a clear overhang equal to $0.5\,l_t$ for simply supported beams and l_t for end restrained beams. The Indian code IS: 1343-1980 does not give permissible values of bond stress specifically. Instead, it controls the bond stress by specifying the minimum permissible values of transfer length. The Indian code on Reinforced Concrete IS: 456-1978 specifies bond stress equal to 1.5, 1.7 and 1.9 MPa respectively for M-30. M-35 and M-40 concretes in the limit state design.

The British Code BS: 8110-1985 permits the use of the following equation for transfer length in pretensioned members in the absence of experimental evidence provided that the initial prestressing force in the tendons is not greater than 75 per cent of the characteristic strength and concrete is fully compacted at the ends.

$$l_t = \frac{K_t\,\phi}{\sqrt{f_{ci}}}$$

where f_{ci} = strength of concrete at transfer in MPa

ϕ = nominal diameter of the tendon

K_t = coefficient which depends upon the type of tendon (Table 7.5.1)

Table 7.5.1   Values of the coefficient, K_t

S. No.	Type of tendon	K_t
1.	Plain or indented wire (including crimped wire with a small wave height)	600
2.	Crimped wire with a total wave height not less than $0.15\,\phi$	400
3.	7-wire standard or super strand	240
4.	7-wire drawn strand	360

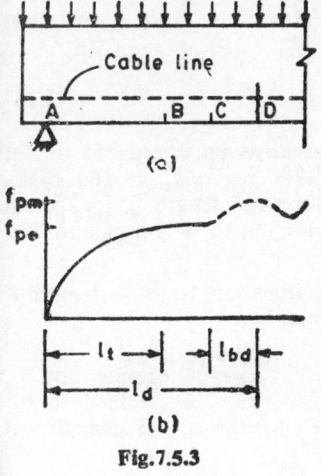

(a)

(b)

Fig.7.5.3

The American Code ACI: 318-1989 recommends that the transfer length may be taken equal to $100\,\phi$ for wires and $50\,\phi$ for strands. A reduced prestressing force in the transfer length should be taken into consideration by assuming a linear variation of stress in the prestresing tendons. Suitable amount of additional reinforcement should be provided within thet ransfer length to guard against flexure, shear and cracking. As the transverse load on a beam is increased, a flexural crack appears at a distance l_d called the development length or embedment length from the support as shown in Fig. 7.5.3 (a). The stress in prestressing tendon increases from zero at A to effective prestress f_{pe} at B over the transfer length l_t as shown in Fig. 7.5.3 (b). The stress remains constant over the length BC and increases to maximum value f_{pm} at the section D at which flexural crack forms. An expression for the length l_{bd}, called the flexural bond length, may be derived by considering the free body of prestressing tendon between points C and D. For the equilibrium of horizontal forces,

$$\frac{\pi}{4} \, \phi^2 \, (f_{pm} - f_{pe}) = l_{bd} \, (\pi \, \phi) \, \tau_{bd}$$

in which ϕ is the diameter of the tendon. Hence,

$$l_{bd} = \frac{(f_{pm} - f_{pe}) \, \phi}{4\tau_{bd}} \tag{7.5.12}$$

As the maximum stress in prestressing tendon f_{pm} cannot exceed its ultimate stress f_{pu}, the maximum value of bond length,

$$l_{bd} \, (\text{max}) = \frac{(f_{pu} - f_{pe}) \, \phi}{4\tau_{bd}} \tag{7.5.13}$$

Taking $\tau_{bd} = 1.725$ MPa, the ACI Code gives the following expression for bond length,

$$l_{bd} = 0.145 \, (f_{pm} - f_{pe}) \, \phi \tag{7.5.14}$$

If the flexural bond length l_{bd} overlaps the transfer length l_t. i.e. if the point C falls to the left of point B, there is a possibility of general bond failure due to bond slip over the entire development length l_d on account of the presence of high bond stress waves in the region. To guard against this type of bond failure, the development length l_d may be expressed as

$$l_d = l_t + l_{bd} \, (\text{max}) \tag{7.5.14}$$

Adopting Eq. (7.5.5) for l_t, Zia and Mustafa proposed the following equation for development length,

$$l_d = 1.5 \left(\frac{f_{pi}}{f'_{ci}} \right) \phi - 117 + 0.181 \, (f_{pu} - f_{pe}) \, \phi \tag{7.5.15}$$

The following equation for development length recommended by ACI code is based on tests carried by Portland Cement Association and Association of American Railroads.

$$l_d = 0.145 \left(f_{pm} - \frac{2}{3} \, f_{pe} \right) \phi \tag{7.5.16}$$

The ACI Code expression is slightly less conservative as compared to Eq. (7.5.15). For instance, taking $f_{pm} = f_{pu} = 1700$ MPa, $f_{pe} = 960$ MPa, $f_{pi} = 1200$ MPa and $f'_{ci} = 28$ MPa, the development length l_d for 5 mm wire according to Eq. (7.5.15) and (7.5.16) works out to 874 mm and 769 mm respectively.

According to CEB-FIP Code, the following values for trans. length l_t are admissible when prestress is transferred gradually.

(i) Indented wires $100 \, \phi$ to $140 \, \phi$

(ii) 7-wire strands $45 \, \phi$ to $90 \, \phi$

These values should be increased by 25 per cent when prestress is transferred by severing the tendons.

EXAMPLE 7.5.1

Determine the transfer length l_t for 5 mm wire and 12.5 mm −7 wire strand. Take $\mu = 0.1, f_{ct} = 35$ MPa, $f'_{ci} = 30$ MPa and $f_{pi} = 1200$ MPa.

Solution

The values of l_t according to code provisions and Eq. (7.5.3), (7.5.5) and (7.5.6) are given in Table 7.5.2.

Table 7.5.2

Code/Equation	Wire	Strand
I.S. Code	500 mm	375 mm
ACI code	500 mm	625 mm
CEB-FIP code	500 mm to 700 mm	562.5 mm to 1125 mm
Hoyer, Eq. (7.5.3)	450 mm	—
Jenney, Eq. (7.5.5)	183 mm	—
Marshall and Krishnamurthy, Eq. (7.5.6)	501 mm	319 mm

EXAMPLE 7.5.2

Determine flexural bond stress for a Freyssinet prestressing cable having 12 wires of 5 mm diameter. The section which has a moment of inertia, $I = 0.09188$ m^4 carries a shear force, $V = 198$ kN. The cable is located 335 mm below centroidal axis. Modular ratio of steel, $m = 7.5$. The diameter of sheathing is 33 mm. The section is uncracked in flexure.

Solution

(i) *Bond between Wire and grout*

Using Eq. (7.5.9 b), the bond stress between wire and grout,

$$\tau_{bd} = \frac{0.198 \times 7.5 \times 0.335 \times 0.005}{0.09188 \times 4} = 0.00677 \text{ MPa}$$

(ii) *Bond between grout and sheathing or sheathing and concrete*

The bond stress between grout and sheathing is approximately the same as that between the sheathing and concrete. Using Eq. (7.5.10),

$$\tau_{bd} = \frac{0.198 \times 7.5 \times 0.335 \times \frac{\pi}{4} \times 0.005^2 \times 12}{0.09188 \times \pi \times 0.033}$$

$$= 0.0367 \text{ MPa}$$

EXAMPLE 7.5.3

A pretensioned beam of rectangular section of 0.30×0.60 m has eight pre-stressing tendons having nominal diameter of 12.5 mm at an effective depth of 0.48 m. Determine the flexural bond stress if the section is subjected to a shear force of 200 kN. The section is cracked in flexure.

Solution

Using Eq. (7.5.11) and taking $jd_p = 0.9 \times 0.48 = 0.432$ m,

$$\tau_{bd} = \frac{0.200}{0.432 \, (8 \times \pi \times 0.0125)}$$

$$= 1.472 \text{ MPa}$$

7.6 BEARING

Bearing stresses of high intensity occur at the end anchorages of post-tensioned members. These stresses do not arise in pretensioned members in which prestress is transferred entirely by bond. In post-tensioned members in which the anchorages are burried in concrete, the transfer of prestress occurs partly by bearing and partly by bond. For instance, in the Freyssinet system, the transfer of prestress occurs partly by bond between the outer lateral surface of the female cone and surrounding concrete and partly by direct bearing of the end of the female cone against concrete. Certain systems of prestressing depend entirely on direct bearing for transfer of prestress. For instance, Prescon system and B.B.R.V. system which utilize buttonheads and Lee Mc Call system depend on direct bearing for the transfer of prestress. It is necessary to keep the bearing stress within permissible limits for the safety of the end block and the anchorages themselves. The end anchorages are designed and fabricated for each particular tendon anchored by it using high expertise and stringent quality control. Hence the chances of failure of end anchorages are extremely remote. The permissible bearing stress in concrete depends upon several factors such as (i) strength of concrete at stress transfer, (ii) amount of reinforcement, (iii) ratio of total or bearing area to punching area, (iv) number and spacing of anchorages, (v) lateral confinement of concrete and (vi) approach for stress analysis. The analysis of stress immediately behind the anchorages is complex due to (a) large number of parameters (b) the three dimensional nature of the problem and (c) the non-linearity, heterogeneity and aniso-tropy of concrete. However, a high value of bearing stress at the anchorages is usually permitted due to the following reasons:

(1) The force in the tendon and, therefore, the bearing stress at the an-chorage which is generally maximum at stress transfer is only temporary because it reduces continuously due to a variety of factors responsible for loss of prestress. Consequently, if the bearing stresses are sustained without trouble during the stressing operation, there is little likelihood of trouble at a subsequent date.

(2) The strength of concrete in the end block keeps increasing conti-nuously with passing of time. Consequently, the bearing strength of con-crete also increases progressively.

(3) Highly skilled supervision is used for placement and compaction of concrete in the end block.

(4) Concrete behind the anchorages has considerable lateral confinement due to the provision of closed stirrups and spiral reinforcement.

Due to the complexity of the problem, several empirical expressions have been proposed for maximum permissible bearing stresses. Most codes of practice also contain clauses related to permissible bearing stresses.

7.6.1 Empirical Expressions

The maximum permissible bearing stress increases with the compressive strength of concrete and the ratio A_{br}/A_{pun}, in which A_{pun} known as the punching area is the area of direct contact between anchorage and concrete and A_{br} known as the bearing area is the largest area of anchorage surface geometrically similar to and concentric with the punching area. The Post-Tensioning Institute Manual 1976 has proposed the following equation for maximum permissible bearing stress f_{br} at the anchorages at stress transfer,

$$f_{br} = 0.8 f'_{ci} \sqrt{\left(\frac{A_{br}}{A_{pun}}\right) - 0.2} \qquad (7.6.1)$$

where f'_{ci} = compressive (cylinder) strength of concrete at stress transfer

The permissible bearing stress f_{br} should not exceed $1.25 f'_{ci}$. Under service loads, the maximum permissible bearing stress may be computed from

$$f_{br} = 0.6 f'_c \sqrt{\frac{A_{br}}{A_{pun}}} \qquad (7.6.2)$$

where f'_c = characteristic compressive (cylinder) strength of concrete. The value of f_{br} should not exceed f'_c.

7.6.2 Code Provisions

The Indian code IS: 1343-1980 recommends that the maximum bearing stress in concrete immediately behind external anchorages after allowing for all losses due to elastic shortening, creep of concrete, relaxation of steel, slip and seating of anchorages etc. should be computed from

$$f_{br} = 0.48 f_{ci} \sqrt{\frac{A_{br}}{A_{pun}}} \qquad (7.6.3)$$

where f_{ci} = compressive (cube) strength of concrere at stress transfer.

The code restricts the upper limit of f_{br} at $0.8 f_{ci}$. The code also specifies that

(i) A 25 per cent increase in permissible bearing stress at the time of tensioning may be permitted provided this temporary stress does not exceed f_{ci}.

(ii) The temporary bearing stress at stress transfer and the permanent bearing stress may be increased suitably if adequate hoop reinforcement is provided in the anchorage zone.

(iii) In the case of embedded anchorages, due allowance should be made for transfer of prestress by friction between the anchorage and the concrete.

(iv) The effective punching area A_{pun} should be taken equal to the contact area of anchorage device which if circular should be replaced by a square of equivalent area. The bearing area A_{br} should be taken as the largest area of that portion of the member which is geometrically similar and concentric to the effective punching area. When a number of anchorages are used, the bearing area A_{br} for each anchorage should be determined in such a way that the bearing areas do not overlap.

(v) If there is a compressive stress on the bearing area as, for instance, in the case of anchorages inside the body of the structure, the total stress should not exceed the limits specified above.

EXAMPLE 7.6.1

Figure 7.6.1 shows the end view of a post-tensioned beam prestressed using Lee Mc Call system. The prestressing tendons comprise four high tensile steel bars each of 16 mm diameter. The rectangular bearing plate common for all tendons is 210 × 300 mm. The diameter of the duct for each bar is 25 mm. Check the bearing stresses given that, $f_{ck} = 24$ MPa. The initial and final stresses in each tendon are 1100 MPa and 930 MPa.

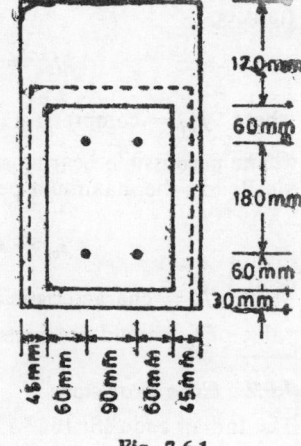

Fig. 7.6.1

Solution

The punching area for the common bearing plate,

$$A_{pun} = 0.210 \times 0.300 - 4 \times \frac{\pi}{4} \times 0.025^2$$

$$= 0.0610 \text{ m}^2$$

The bearing area bounded by broken lines is the largest area geometrically similar to and concentric with the punching area,

$$A_{br} = 0.270 \times 0.360 - 4 \times \frac{\pi}{4} \times 0.025^2$$

$$= 0.0952 \text{ m}^2$$

(i) Indian Code

Using Eq. (7.6.3), the permissible bearing stress,

$$f_{br} = 0.48 \times 30 \sqrt{\frac{0.0952}{0.0610}} = 17.99 \text{ MPa}$$

which is less than $0.8 f_{ci} = 0.8 \times 30 = 24$ MPa. Total prestressing force at stress transfer,

$$P_i = 4 \times \frac{\pi}{4} \times 0.016^2 \times 1100 = 0.885 \text{ MN}$$

Hence the actual bearing stress,

$$f_{br} = \frac{0.885}{0.0610} = 14.51 \text{ MPa}$$

which is safe.

(ii) *American Code*

Using Eq. (7.6.1), the permissible bearing stress at initial stage,

$$f_{br} = 0.8 \times 24 \sqrt{\frac{0.0952}{0.0610}} - 0.2 = 22.40 \text{ MPa}$$

which is less than $1.25 \; f'_{ci} = 1.25 \times 24 = 30$ MPa. The actual bearing stress is 14.51 MPa which is therefore safe. Using Eq. (7.6.2), the permissible bearing stress under service loads,

$$f_{br} = 0.6 \times 28 \sqrt{\frac{0.0952}{0.0610}} = 20.99 \text{ MPa}$$

which is less than 28 MPa. The actual bearing stress,

$$f_{br} = \frac{4 \times \frac{\pi}{4} \times 0.016^2 \times 930}{0.0610} = 12.27 \text{ MPa}$$

which is less than the permissible value.

EXAMPLE 7.6.2

Figure 7.6.2 (a) shows the end view of a post-tensioned girder prestressed using Freyssinet system. There are five prestressing tendons each of which

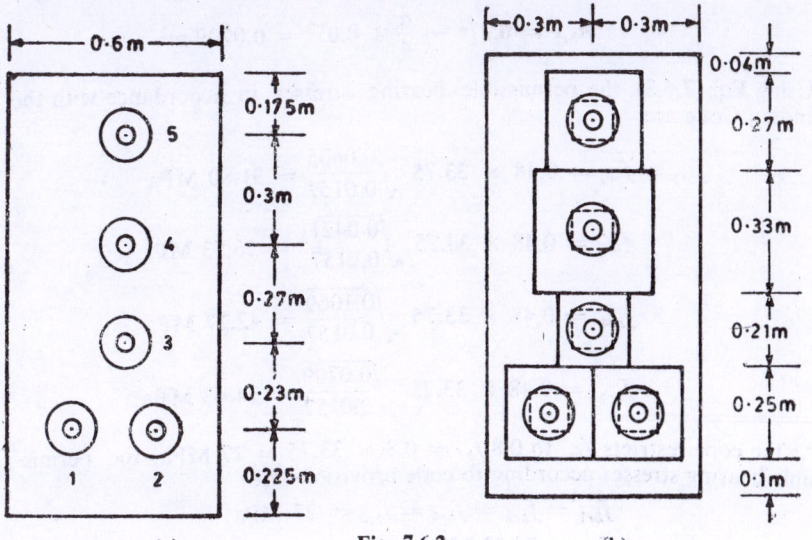

Fig. 7.6.2
(a)   (b)

carries an initial force of 350 kN. The external and internal diameters of the female cone are 150 mm and 50 mm respectively. Check the bearing stresses at the anchorages. Take $f_{ci} = 33.75$ MPa.

Solution

The circular areas of the anchorages may first be replaced by equivalent square areas shown by broken lines in Fig. 7.6.2 (b). The side of the equivalent square,

$$a = \sqrt{\frac{\pi}{4} \times 150^2} = 133 \text{ mm}$$

The punching area for each anchorage,

$$A_{pun} = \frac{\pi}{4} \times 150^2 - \frac{\pi}{4} \times 50^2 = 15714 \text{ mm}^2 = 0.0157 \text{ m}^2$$

The bearing area A_{br} for each anchorage is the largest area which is concentric with and geometrically similar to the equivalent square areas. The bearing areas shown shaded in Fig. 7.6.2 (b) do not overlap and also maximize the total area. The bearing areas for the five anchorages numbered in Fig. 7.6.2 (a) are,

$$A_{br1} = A_{br2} = 0.25^2 - \frac{\pi}{4} \times 0.05^2 = 0.0605 \text{ m}^2$$

$$A_{br3} = 0.21^2 - \frac{\pi}{4} \times 0.05^2 = 0.0421 \text{ m}^2$$

$$A_{br4} = 0.33^2 - \frac{\pi}{4} \times 0.05^2 = 0.1069 \text{ m}^2$$

$$A_{br5} = 0.27^2 - \frac{\pi}{4} \times 0.05^2 = 0.0709 \text{ m}^2$$

Using Eq. (7.6.3), the permissible bearing stresses in accordance with the Indian Code are,

$$f_{br1} = f_{br2} = 0.48 \times 33.75 \sqrt{\frac{0.0605}{0.0157}} = 31.80 \text{ MPa}$$

$$f_{br3} = 0.48 \times 33.75 \sqrt{\frac{0.0421}{0.0157}} = 26.53 \text{ MPa}$$

$$f_{br4} = 0.48 \times 33.75 \sqrt{\frac{0.1069}{0.0157}} = 42.27 \text{ MPa}$$

$$f_{br5} = 0.48 \times 33.75 \sqrt{\frac{0.0709}{0.0157}} = 34.43 \text{ MPa}$$

As the code restricts f_{br} to $0.8 f_{ci} = 0.8 \times 33.75 = 27$ MPa, the permissible bearing stresses according to code provisions are

$$f_{br1} = f_{br2} = f_{br4} = f_{br5} = 27 \text{ MPa}$$
$$f_{br3} = 26.53 \text{ MPa}$$

The actual bearing stress behind the anchorages at the initial stage is

$$f_{br} = \frac{0.350}{0.0157} = 22.29 \text{ MPa which is safe.}$$

EXAMPLE 7.6.3

The overall dimensions of the end block of the post-tensioned girder are 400 × 600 mm. The girder is prestressed by a single cable comprising 64 wires of 5 mm diameter using Magnel Blaton system. The initial stress in each wire is 1200 MPa. The dimensions of the duct located concentrically at the end are 55 × 200 mm. Design the bearing plate given that, $f_{ck} = 35$ MPa and $f_{ci} = 28$ MPa.

Solution

Total prestressing force at initial stage,

$$P_i = 64 \times \frac{\pi}{4} \times 0.005^2 \times 1200 = 1.509 \text{ MN}$$

For a permissible bearing stress, $f_{br} = 0.8\, f_{ci} = 22.4$ MPa, the minimum bearing area required,

$$A_{pnn} = \frac{1.509}{22.4} = 0.0674 \text{ m}^2$$

Adding the area of duct, the area of bearing plate should be equal to 0.0674 + 0.055 × 0.2 = 0.0784 m². Taking the bearing plate to be geometrically similar to the duct, the size of plate should be 216.7 × 361.7 mm. Hence, adopting a bearing plate having dimensions 220 × 365 mm,

$$A_{pun} = 0.220 \times 0.365 - 0.055 \times 0.200 = 0.0693 \text{ m}^2$$

The bearing area, geometrically similar to punching area, is 400 × 545 mm.

$$A_{br} = 0.400 \times 0.545 - 0.055 \times 0.200 = 0.2070 \text{ m}^2$$

Using Eq. (7.6.3), the permissible bearing stress,

$$f_{br}' = 0.48 \times 28 \sqrt{\frac{0.2070}{0.0693}} = 23.23 \text{ MPa}$$

As $0.8\, f_{ci} = 0.8 \times 28 = 22.4$ MPa is smaller, it represents the permissible bearing stress. The actual bearing stress,

$$f_{br} = \frac{1.509}{0.0693} = 21.77 \text{ MPa which is safe}$$

7.7 ANCHORAGE ZONE

In post-tensioned members, which have relatively larger prestressing tendons as compared to pretensioned members, large prestressing forces are delivered over relatively small areas resulting in the formation of zones of high stress concentration with irregular or discontinuous stress distribution near the ends of the member. However, in accordance with Saint Venant's principle, the stress distribution becomes continuous and practically linear at certain distance from the end of the member. According to Saint Venant's principle, if the forces acting in a certain zone of a structure are disturbed by replacing them by their satically equivalent forces, the stress distribution in zones away from the disturbed zone remains practically unchanged. Consequently, the stress distribution at sections away from the end of the member under the action of irregular concentrated anchorage forces is the same as that caused by regular and continuous end forces which are statically equivalent to the actual anchorage forces. Consider, for instance, a single concentric anchorage transmitting prestressing force P along the axis of the member as shown in Fig. 7.7.1 (a). The actual distribution of anchorage force shown in Fig. 7.7.1 (a) may be replaced by its static equivalent in the form of uniform distribution over the entire depth of the member as shown in Fig. 7.7.1 (b). The distribution of longitudinal or horizontal normal stress f_x at sections located at distances 0, $D/4$ and $D/2$ from the end of the member are shown in Fig. 7.7.1 (c), (d) and (e) respectively. In these distributions, the stress f_x is not uniform across the depth of the member. However, the distribution of f_x becomes progressively more uniform at sections located at larger distances from the end of the member until at a section YY the distribution of f_x is practically uniform as shown in Fig. 7.7.1 (f). The uniform distribution of f_x at section YY is caused by the actual anchorage force shown in Fig. 7.7.1 (a) or by its equivalent shown in Fig. 7.7.1 (b) in accordance with Saint Venant's principle. The distance of the section YY from the end of the member is known as the lead-in-length which is generally taken equal to the overall depth of the member D. The portion of the member to the left of YY is known as the anchorage zone or the end block. While the elementary theory may be

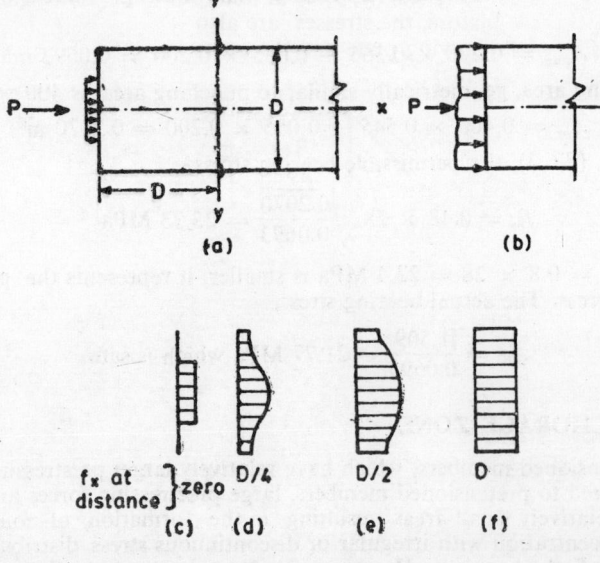

Fig. 7.7.1

used to determine the distribution of stresses outside the end block, it cannot be used for the determination of the complex stress distribution inside the end block. The complexity of the problem of determination of stresses in the anchorage zone is enhanced further due to non-homogeneity, anisotropy and non-linearity of the medium. Due to the practical importance of the problem, several theoretical and experimental investigations have been carried out for the determination of anchorage zone stresses. In this context, the contributions by Magnel, Guyon, Zielinski and Rowe, Iyengar, and Yettram and Robbins may be cited.

7.7.1 Nature of Stresses

The stresses in the end block of a post-tensioned member are truly three dimensional in nature. Generally, there are several anchorages at the end of a member transmitting forces $P_1, P_2 \ldots, P_n$ over their respective bearing areas $A_1, A_2, \ldots, A_n$. Consider the stresses acting on the six faces of a small cubical element located at a point having coordinates (x, y, z) due to prestressing force P_j acting over bearing area A_j of the jth anchorage as shown in Fig. 7.7.2. All the six stress components act on the element, viz. the three normal stresses f_x, f_y and f_z along respective orthogonal axes and the three shear stress components q_{xy}, q_{yz} and q_{zx}. As the magnitudes of all six stress components are comparable with the average intensity of prestress, the stresses are truly three dimensional. The magnitude and nature of the six stress components due to the anchorage force P_j depend upon several parameters such as (i) the magnitude of P_j and its inclination with the axis of the member, (ii) the ratio A_j/A in which A_j is the punching or contact area of jth anchorage and A is the total cross-sectional area, (iii) the ratios (b_j/b) and (D_j/D), (iv) the eccentricities e_y and e_z of P_j with respect to y and z axes and (v) the coordinates (x, y, z) of the point under consideration. In addition to these factors, the stresses are also dependent to some extent on the rigidity of the bearing plate and the amount and disposition of reinforcement in the end block.

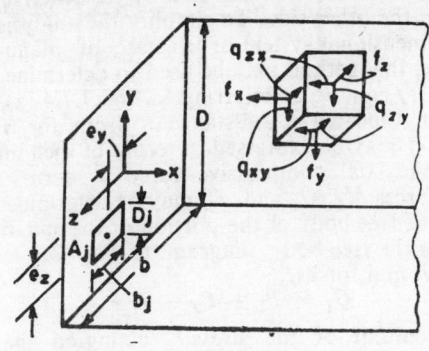

Fig. 7.7.2

Among the three normal stresses, the longitudinal stress f_x is generally compressive in nature while the transverse stresses f_y and f_z may be tensile or compressive depending upon the position of point under consideration.

Due to the high strength of concrete in compression and shear, the normal stress f_x and the shear stresses q_{xy}, q_{yz} and q_{zx} do not create any problem in design. Due to the intrinsic weakness of concrete in tension, the transverse normal stresses f_y and f_z have to be considered carefully because these stresses are tensile in certain zones of the end block. There are two distinct zones known as the bursting zone and the spalling zone which deserve special consideration in design because the transverse stresses f_y and f_z are tensile in these zones. The main objective in the design of end block is to ensure that the tensile forces in these zones are resisted adequately by the provision of suitable reinforcement. The bursting zone due to any particular tendon is located along the axis of the tendon at a distance ranging approximately from 0.1 D to D from the end of the member. The spalling zone is located near the end face of the member on all four sides of the anchorage bearing area. The isobars for the tensile stress f_y, which are the contours of equal tensile stress, in the bursting and spalling zones for some typical cases are shown in Fig. 7.7.3. The number attached to each isobar represents the magnitude of tensile stress f_y as a ratio of the average intensity of prestress P/A. The isobars in the bursting zone and spalling zone are drawn by continuous and broken lines in order to distinguish the two zones. The zone located between the bursting and spalling zones is called the compression zone. The isobars for tensile stress f_z are similar to those for f_y.

7.7.2 Stress Analysis

Several approaches based on deep beam theory, symmetrical prisms and successive resultants, two and three dimensional theories of elasticity and experimental techniques, particularly photoelasticity have been used for the analysis of stresses in the end blocks of post-tensioned members.

7.7.2.1 Magnel's Method

In this method the end block is visualized as a deep beam loaded by anchorage forces on one side and supported by the reactive forces imposed on the end block on the other side. To simplify the analysis, the problem is reduced to a two dimensional system or a state of plane stress in the xy plane. Consequently, the method may be used to determine the three plane, stress components f_x, f_y and q_{xy}. Referring to Fig. 7.7.4 (a), the stresses f_y and q_{xy} at any point A located at a distance x from the free end of the horizontal section 1-1 may be expressed in terms of bending moment M_1, shear force Q_1 and the axial compressive force C_1 across the section 1-1. The three internal forces M_1, Q_1 and C_1 may be determined by considering the equilibrium of the free body of the portion of the end block above the section 1-1 shown in the free body diagram of Fig. 7.7.4 (b). From the equilibrium of horizontal forces,

$$Q_1 = P_I + P_2 - H \tag{7.7.1}$$

in which H is the resultant of the stress f_x acting on the portion of the beam above section 1-1 on the right face of the end block. The normal stress f_x on the right face of the end block may be determined from the elementary theory of bending,

$$f_x = \frac{-P}{A} + \frac{M_p y}{I} \tag{7.7.2}$$

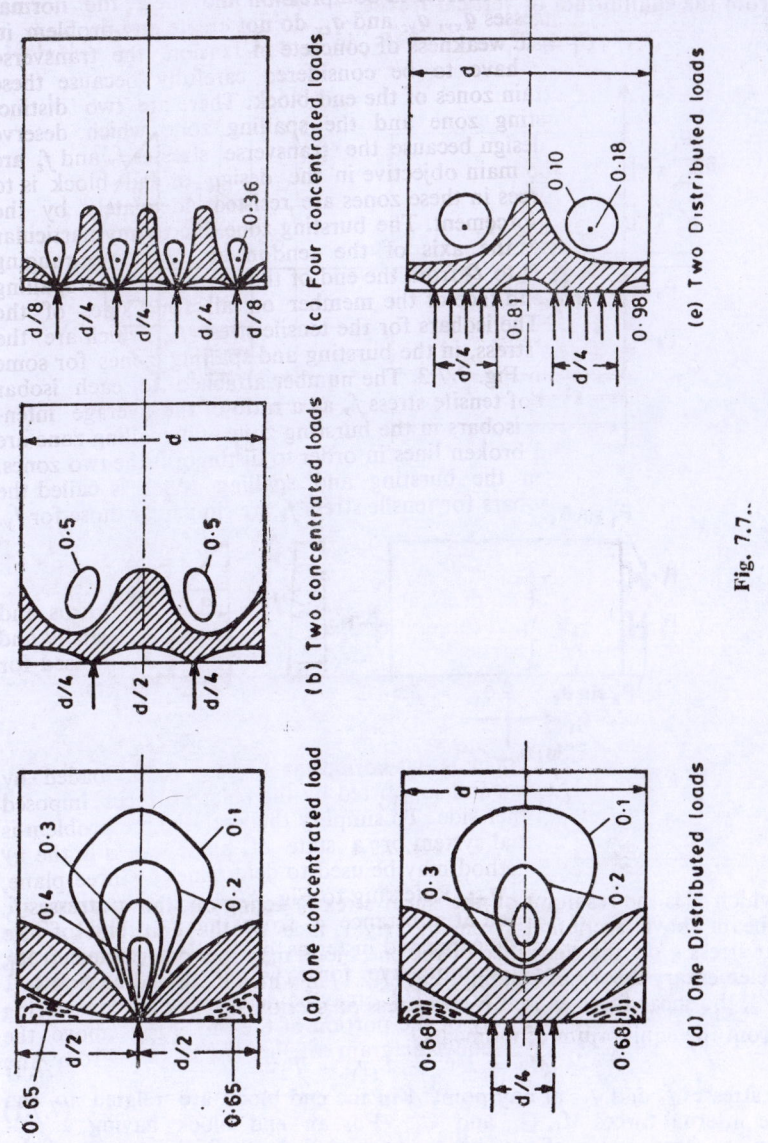

Fig. 7.7.

From the equilibrium of vertical forces,

$$C_1 = P_1 \sin \theta_1 + P_2 \sin \theta_2 - V \qquad (7.7.3)$$

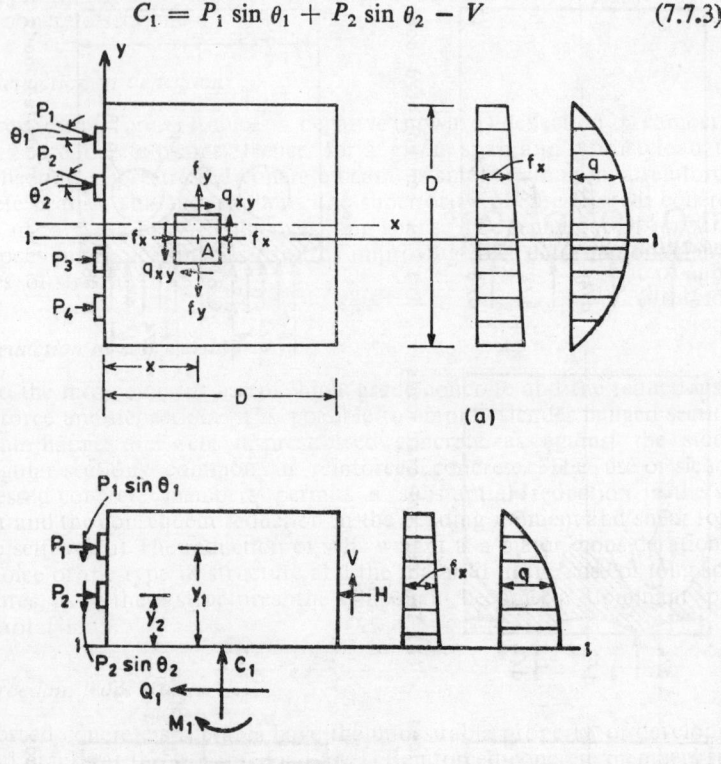

Fig. 7.7.4

In which V is the resultant of the shear stress q acting on the portion of the beam above section 1-1 on the right face of the end block. The shear stress q on the right face of the end block may be determined from the elementary theory of bending, Eq. (7.2.1) in which V_r should be replaced by V_p, the shear force caused by prestress at section 1-1.

From the equilibrium of moments,

$$M_1 = Hy_3 - P_1y_1 - P_2y_2 \qquad (7.7.4)$$

The stresses f_y and q_{xy} at any point A in the end block are related to the three internal forces M_1, Q_1, and C_1. For an end block having a rectangular cross-section of breadth b and overall depth D, these stresses may be expressed as

$$f_y = \frac{K_1 M_1}{bD^2} + \frac{K_2 C_1}{bD} \qquad (7.7.5)$$

$$q_{xy} = \frac{K_3 Q_1}{bD} \qquad (7.7.6)$$

iu which the parameters K_1, K_2 and K_3 depend upon the ratio x/D.

$$K_1 = -20\left(1 - \frac{6x}{D} + \frac{9x^2}{D} - \frac{4x^3}{D^3}\right) \qquad (7.7.7)$$

$$K_2 = 2\left(1 - \frac{12x}{D} + \frac{21x^2}{D^2} - \frac{10x^3}{D^3}\right) \qquad (7.7.8)$$

$$K_3 = -20\left(-\frac{x}{D} + \frac{3x^2}{D^2} - \frac{3x^3}{D^3} + \frac{x^4}{D^4}\right) \qquad (7.7.9)$$

The distribution of normal stress f_x may be determined approximately by assuming 45° dispersion of the anchorage forces. The stress f_x at any point due to each anchorage may force be determined separately and added algebraically to obtain the net value. Referring to Fig. 7.7.5, the stress f_x at point A due to anchorage force P_i at anchorage i may be expressed as

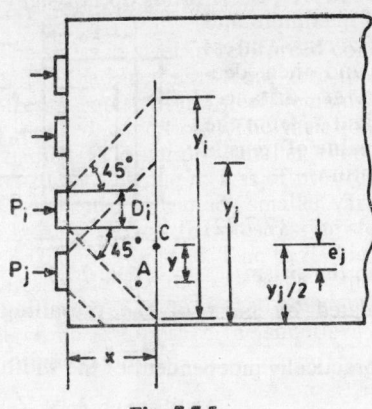

$$f_x = \frac{-P_i}{by_i} \qquad (7.7.10)$$

where $y_i = D_i + 2x$

= depth of the zone on which anchorage force P_i is dispersed

If the anchorage under consideration is located near the top or bottom face one of the 45° dispersion lines cuts the top or bottom face at a distance smaller than x. The anchorage face becomes eccentric with respect to its zone of dispersion. Consider for instance the anchorage force P_j at jth anchorage. The stress f_x at point A in this case may be expressed as

Fig. 7.7.5

$$f_x = \frac{-P_j}{by_j} - \frac{P_j e_j y}{I_j} \qquad (7.7.11)$$

where P_j = anchorage force due to jth tendon

y_j = depth over which the anchorage force P_j is dispersed

e_j = eccentricity of force P_j with respect to the centre C of the depth y_j

y = distance of point A under consideration from the centre C

$I_j = \dfrac{by_j^3}{12}$

7.7.2.2 Zielinski and Rowe's Method

Zielinski and Rowe reported extensive tests on concrete prisms carried out at the Cement and Concrete Association of United Kingdom for the determination of tensile stress f_y and bursting force F_{bst} due to an anchorage force. The prisms were proportioned to simulate the end blocks of post-tensioned members. The strains on the surface of the prism were measured to determine the cracking and ultimate strengths. The parameters investigated included (a) ratio of loaded area to the total cross-sectional area, (b) diameter of cable duct, (c) type of anchorage and (d) type of reinforcement (helical or mats or stirrups). The main conclusions drawn from these tests were as follows:

(i) The stress distribution, cracking strength and ultimate strength are not affected by the type of anchorge (embeded or external), the material of anchorage (concrete or steel) and the manner in which the individual wires are gripped.

(ii) The distribution of stresses is strongly dependent upon the ratios y_{po}/y_o and z_{po}/z_o in which y_o, z_o are the dimensions of the cross section of the prism concentric to and symmetric with the loaded area having dimensions y_{po} and z_{po} as shown in Fig. 7.7.6. However, these ratios do not significantly affect the position of the point of maximum and zero tensile stress in the bursting zone. The tensile stress f_y is zero, increases to a maximum value and then decreases again to zero at points located at distances approximately equal to 0.1 y_o, 0.25 y_o and y_o from the loaded end of the prism. The maximum value of tensile stress f_y may be determined from the equation,

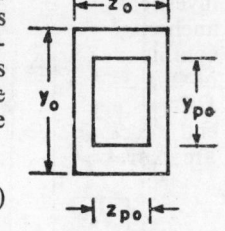

Fig. 7.7.6

$$f_y \text{ (max)} = p \left(0.98 - 0.825 \frac{y_{po}}{y_o} \right) \qquad (7.7.12)$$

where, p = average compressive stress in the prism

The maximum tensile stress is underestimated by most of the prevailing theories.

(ii) The bursting tension F_{bst} which is practically independent of the width of the prism may be determined from

$$F_{bst} = P_k \left(0.48 - 0.40 \frac{y_{po}}{y_o} \right) \qquad (7.7.13)$$

where, P_k is tendon jacking load. Equations (7.7.12) and (7.7.13) are valid for y_{po}/y_o ranging from 0.3 to 0.7.

(iv) The helical reinforcement is more effective than mats and stirrups.

The foregoing conclusions were used to develop a design procedure for the end block of a post-tensioned member having the following salient points:

(1) The reinforcement for resisting the bursting force should be provided in the region extending from 0.1 D to 0.5 D where D is the overall depth of the end block.

(2) As part of the bursting force is resisted by concrete, the reinforcement required to resist bursting force may be computed from the reduced or corrected value of the bursting force given by the equation,

$$F_{bst} \text{ (corrected)} = F_{bst} \left[1 - \left(\frac{f_t}{f_{y \text{ (max)}}} \right)^2 \right] \qquad (7.7.14)$$

where, f_t = permissible tensile strength of concrete

The area of reinforcement may be determined by dividing the corrected bursting force by the permissible stress in steel resisting the bursting force.

(3) For the design of an end block with several anchorages, the end block should be subdivided into as many prisms as the number of anchorages. Each prism should be concentric to and symmetric with the respective loaded area. The areas of the prisms should be as large as possible subject to the condition that they do not overlap. The bursting forces may be calculated separately for each prism and reinforcement should be provided at appropriate locations to resist these bursting forces.

7.7.2.3 Guyon's Method

Guyon carried out extensive theoretical studies based on two dimensional theory of elasticity to determine the distribution of f_x, f_y and q_{xy} in the anchorage zone and verified the results by photoelasticity. The results of the investigations were given in the form of tables and influence lines. The anchorage forces were assumed to be in the form of knife edge loads, discontinuous along the depth but continuous and spread over the entire width of the end block. The eccentricity of anchorage force ranged from 0 to $\pm D/2$ in steps of $D/8$. Referring to Fig. 7.7.7, the stresses at any point in the end block due to the horizontal component P_{jh} of the anchorage force P_j are expressed as

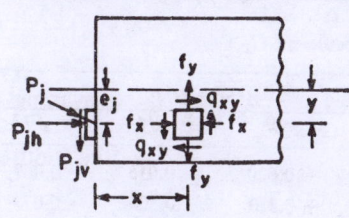

$$f_x = C_1 \frac{P_{jh}}{bD} \tag{7.7.15}$$

$$f_y = C_2 \frac{P_{jh}}{bD} \tag{7.7.16}$$

$$q_{xy} = C_3 \frac{P_{jh}}{bD} \tag{7.7.17}$$

Fig. 7.7.7

in which C_1, C_2 and C_3 depend upon the eccentricity of anchorage force e_j and the location of the point under consideration. In a similar way the stresses due to tangential or vertical component of anchorage force P_{jv} may be expressed as

$$f_x = C_1' \frac{P_{jv}}{bD} \tag{7.7.18}$$

$$f_v = C_2' \frac{P_{jv}}{bD} \tag{7.7.19}$$

$$q_{xy} = C_3' \frac{P_{jv}}{bD} \tag{7.7.20}$$

in which the coefficients C_1', C_2' and C_3' depend upon the eccentricity of the anchorage force e_j and the position of the point under consideration. In order to determine the stresses f_x, f_y and q_{xy} at any point in the anchorage block having several anchorages, each anchorage force should be resolved into its vertical and horizontal components. The stresses at the point under consideration due to each horizontal and vertical component may then be determined by using Eq. (7.7.15) to (7.7.20). The net stresses may be computed by adding the stresses due to component forces algebraically. However, from the point of view of design of the end block, only the transverse stress f_y is of importance. The values of coefficients C_2 and C_2' for f_y related to horizontal and vertical anchorage forces for seven values of eccentricity of applied load are given in Tables 7.7.1. and 7.7.2. The stress f_y generally decreases beyond the point $x = D/2$. To satisfy the boundary conditions, f_y becomes zero at the free surfaces, $y = \pm D/2$. The Tables are useful for determining f_y in the end block. For a concentric anchorage P distributed

Table 7.7.1   Values of Coefficient C_2

Eccentricity	At $y = 0$			At $y = D/4$		
e_p	$x = 0$	$x = D/4$	$x = D/2$	$x = 0$	$x = D/4$	$x = D/2$
$- 3D/8$	+ 1.222	− 0.192	− 0.242	+ 0.368	− 0.036	− 0.128
$- D/4$	+ 0.758	− 0.154	+ 0.024	+ 0.356	− 0.052	- 0.058
$- D/8$	+ 0.566	+ 0.144	+ 0.262	+ 0.386	− 0.122	+ 0.018
0	− ∞, 0.523	+ 0.462	+ 0.314	+ 0.500	− 0.205	+ 0.122
$D/8$	+ 0.566	+ 0.144	+ 0.262	+ 0.670	+ 0.146	+ 0.222
$D/4$	+ 0.758	− 0.154	+ 0.024	− ∞, 1.092	+ 0.588	+ 0.184
$3D/8$	+ 1.222	− 0.192	− 0.242	+ 2.176	− 0.004	− 0.076

Table 7.7.2   Values of Coefficient C_2'

Eccentricity	At $y = 0$			At $y = D/4$		
e_p	$x = 0$	$x = D/4$	$x = D/2$	$x = 0$	$x = D/4$	$x = D/2$
$- 3D/8$	+ 5.656	+ 0.192	− 0.140	− 0.020	− 0.015	− 0.067
$- D/4$	+ 4.000	+ 0.444	− 0.196	+ 0.820	+ 0.304	− 0.077
$- D/8$	+ 5.656	+ 0.192	− 0.144	+ 0.966	+ 0.282	− 0.077
0	+ ∞, − ∞	0	0	+ 1.410	+ 0.060	− 0.059
$D/8$	− 5.656	− 0.192	+ 0.144	+ 3.310	− 0.302	+ 0.011
$D/4$	− 4.000	− 0.444	+ 0.196	+ ∞, − ∞	0.140	+ 0.119
$3D/8$	− 5.656	− 0.192	+ 0.140	− 13.676	− 0.185	+ 0.137

Note: *Sign Convention, Tension-Positive*

uniformly over depth y_{po} as shown in Fig. 7.7.8, Table 7.7.3 gives the maximum value of f_y on the axis of the member and also the position of the points at which f_y is zero and maximum. The maximum tensile stress f_y (max) is given in the non-dimensional form by dividing it by the average intensity of prestress, $p = P/A$. The Table also gives the bursting force F_{bst} expressed in the non-dimensional form by dividing it by the prestressing force P. The Table 7.7.3 shows that as the distribution ratio y_{po}/y_o increases from 0 to 0.9,

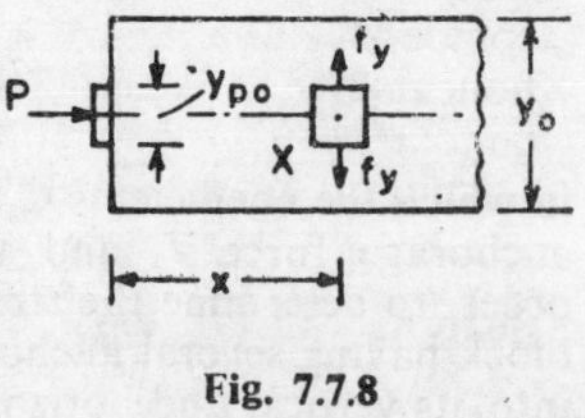

Fig. 7.7.8

Table 7.7.3 Maximum values of f_y and bursting force F_{bst} due to concentric anchorage force

Distribution ratio, y_{po}/y_o	x/D for		f_y (max)/p	F_{bst}/P
	$f_y = 0$	$f_y = f_y$ (max)		
0	0	0.17	0.50	0.31
0.1	0.09	0.24	0.43	0.24
0.2	0.14	0.30	0.36	0.19
0.3	0.16	0.36	0.33	0.15
0.4	0.18	0.39	0.27	0.13
0.5	0.20	0.43	0.23	0.10
0.6	0.22	0.44	0.18	0.08
0.7	0.23	0.45	0.13	0.06
0.8	0.24	0.46	0.09	0.04
0.9	0.27	0.55	0.04	0.02

(i) the maximum tensile stress f_y (max) decreases progressively from $0.5\,p$ to $0.04\,p$. If the anchorage force is distributed uniformly over the entire depth D, the tensile stress f_y becomes zero.

(ii) The points of zero stress and maximum stress f_y (max) continuously shift away from the loaded end reaching maximum values equal to 0 26 D and 0.55 D.

(iii) The bursting force decreases progressively from $0.3\,P$ to $0.02.\,P$. The bursting force vanishes for $y_{po}/y_o = 1$.

Approximate Method

An extensive study of the distribution of stresses in the end blocks undertaken by Guyon revealed that in the case of groups of forces irregularly distributed, the maximum stresses occur on the axes of individual forces, the axes of resultants of groups of forces and on the axis of the total resultant. He, therefore, proposed an approximate empirical approach based on the method of partitioning and concept of symmetric prisms. The approach may be described by the following steps.

(i) Each anchorage force may be resolved into its horizontal and vertical components. Due to the small inclination of the tendons, the vertical components may be ignored. The horizontal components P_1, P_2, ..., P_n may be taken equal to the applied anchorage forces.

(ii) Determine the total resultant P of forces P_1, P_2 . . ., P_n and its eccentricity e_p. Hence determine the intensity of linear distribution of flexural stress at the end of the stress block from the equation,

$$f = \frac{-P}{A} - \frac{(Pe_p)\,y}{I} \qquad (7.7.21)$$

Represent the linear stress distribution by a trapezium at the right end of the end block as shown in Fig. 7.7.9.

(iii) Consider first the total resultant P acting at eccentricity e_p. It may be assumed to act on a concentric symmetric prism of depth $y_o = 2 \times \left(\dfrac{D}{2} - e_p\right)$. Table 7.7.3 may then be used to determine f_y (max) and bursting force F_{bst} in the symmetric prism cut out from the end block. The depth y_{po} of the loaded area on which the resultant force P acts, may be taken equal to the sum of the depths of loaded areas of individual forces. $P_1, P_2, \ldots P_n$.

(iv) The prestressing tendons may be divided suitably into groups depending upon their positions in the end block. For instance, the prestressing tendons may be divided into two groups comprising those located in the upper and lower zones of the end block. Let P_I and P_{II} called partial anchorage forces in groups I and II respectively as shown in Fig. 7.7.10. The end block may then be partitioned into two parts by horizontal section $1 - 1$ in such a manner that the force P_I and the resultant of stress represented by the trapezium CEHG are equal and opposite. It follows that the group or partial resultant P_{II} is equal and opposite to the resultant of the stress represented by trapezium GHFD. The prism symmetric to resultant P_I has a depth $y_{oI} = 2y_I$. In a similar manner consider the prism symmetric to P_{II} having depth $y_{oII} = 2 y_{II}$ as shown in the figure. The tensile stress and the bursting force in each of the two prisms assumed to act as isolated prisms may be computed form Table 7.7.3. The depths of the loaded areas for resultants P_I and P_{II} may be taken equal to the sums of the depths of the loaded areas for the anchorage forces belonging to the respective groups.

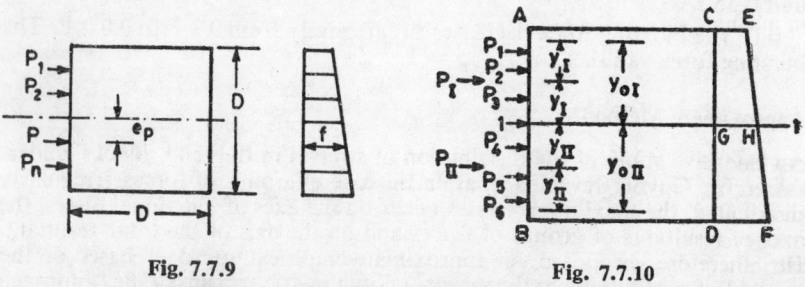

Fig. 7.7.9   Fig. 7.7.10

(v) Finally divide the end block by horizontal partitions into as many parts as the number of anchorage forces. Consider for instance the case of four anchorage forces P_1, P_2, P_3 and P_4 as shown in Fig. 7.7.11. The end block may be partitioned into four parts by horizontal sections 1, 2 and 3 so that the resultants R_1, R_2, R_3 and R_4 of stresses represented by the respective parts of the trapezoidal stress diagram are equal and opposite to the forces P_1, P_2, P_3 and P_4. Prisms symmetric to forces P_1, P_2, P_3 and P_4 confined within the four partitioned parts may then be conceived and the tensile stresses and the bursting force in each prism determined using Table 7.7.3. The distribution of anchorage forces P_1, P_2, P_3 and P_4 is called linear if the anchorage forces are so located that the anchorage forces P_1, P_2, P_3

and P_4 are collinear with the stress resultants R_1, R_2, R_3 and R_4 respectively. Guyon has recommended that the end anchorages should be so located as far as possible so as to achieve linear distribution. The effect of linear distribution is to minimise tensile stresses in the end block, particularly in the spalling zone. If the anchorages are placed in accordance with linear distribution, it may not be necessary to check for the bursting forces due to partial and total resultants.

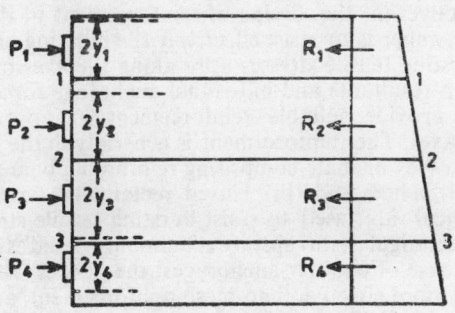

Fig. 7.7.11

7.7.3 Code Provisions

The provisions of the Indian Code IS: 1343-1980, which are similar to the British Code in respect of end blocks of post-tensioned members are as follows:

(1) The bursting tensile force F_{bst} due to a single anchorage force acting on a square loaded area of side y_{po} in an end block of square cross section having side y_o may be computed from the equation,

$$F_{bst} = P_k \left(0.32 - 0.30 \, \frac{y_{po}}{y_o} \right) \qquad (7.7.22)$$

in which P_k should be taken equal to the tendon jacking load in the case of bonded members. For unbonded members, P_k should be taken equal to tendon jacking load or the load in the tendon at the limit state of collapse, whichever is greater. The bursting tensile force should be assumed to be distributed over a length ranging from 0.1 y_o to y_o from the loaded end. The area of the reinforcement, required to resist bursting tensile force, should be computed by dividing the bursting force by 0.87 f_y in which f_y is the yield strength of reinforcement. If the concrete cover to the reinforcement is less than 50 mm, the stress in reinforcement should be limted to that corresponding to a strain of 0.001.

(2) In the case of a rectangular end bloek, the bursting forces in the two principal directions should be determined in accordance with (1) above.

(3) In the case of circular anchorage or loaded area, the actual loaded area should be replaced by an equivalent square area.

(4) In the case of groups of anchorages or end plates, the end blocks should be divided into as many symmetrically loaded prisms as the number

of anchorages or bearing plates. The areas of the prisms should be as large as possible subject to the condition that they do not overlap each other. Each prism may then be treated separately.

(5) Provision of reinforcement should also be made for spalling tension which reaches its maximum value near the loaded face of the member.

7.7.4 Provision of Reinforcement

The main objective in the design of reinforcement in the end block of a post-tensioned member is to resist effectively the bursting and spalling tensile stresses. The bursting tensile stresses arise along the axes of total resultant, partial or group resultants and individual anchorage forces. Consequently, it is logical to provide suitable reinforcement at appropriate locations normal to these axes. The reinforcement is generally in the form of (i) square or rectangular meshes or mats comprising reinforcing bars along two principal directions, (ii) helices and (iii) closed rectangular stirrups. While the meshes and helices are used to resist bursting tensile stresses, the meshes and the closed rectangular stirrups are commonly used to resist spalling tension. In the case of circular anchorages, the helices which are generally considered to be more effective than mesh reinforcement, are often provided around the axes of individual tendons to contain the bursting tensile stresses along these axes. If F_{bst} is the bursting force due to any tendon, the total area of an helix A_s intersected by a section passing through the axis of the tendon may be determined from the equation,

$$A_s = \frac{F_{bst}}{0.87 f_y}$$

in which F_{bst} may be determined from Eq. (7.7.22). Some economy in reinforcement may be achieved by using Eq. (7.7.14) in which the tensile strength of concrete is taken into account. In addition to helix reinforcement along each tendon, mesh reinforcement may be provided to resist tensile stresses normal to the axes of partial and total resultants. The mesh reinforcement resisting tensile stresses normal to the axis of the total resultant may be provided within a distance of D from the loaded end. The meshes for partial resultants may be provided within the lengths of their respective symmetric prisms.

Tensile stresses of relatively high intensity occur in the spalling zone immadiately behind the anchorages, particularly in those cases in which the anchorages are highly eccentric. However, these tensile stresses are confined to small areas. Consequently the total spalling tensile force F_{spl} is not large. In most cases F_{spl} is resisted adequately by providing mesh reinforcement to resist a force of 0.03 P in the two principal directions. This mesh reinforcement should be provided immediately behind the anchorages as close as possible to the end of the member.

EXAMPLE 7.7.1

The end block of a post-tensioned beam has a rectangular cross-section 0.60 × 1.00 m. Determine the position of zero and maximum value of bursting tensile stress and the magnitude of f_y (max) along the axis of the member due to a concentric anchorage force of 2.1 MN distributed uniformly over a depth of 0.20 m.

Solution

(i) *Magnel Blaton method*

Intensity of bearing stress behind the anchorage,

$$f_{br} = \frac{2.1}{0.60 \times 0.20} = 17.5 \text{ MPa}$$

Uniform intensity of stress at the end of the end block,

$$f = \frac{2.1}{0.60 \times 1.0} = 3.5 \text{ MPa}$$

Cut the end block shown in Fig. 7.7.12 (a) along section 1-1 coinciding with the axis of the member. The free body of the portion of the end block above section 1-1 is shown in Fig. 7.7.12 (b). Considering the end block as a deep beam, bending moment as section 1-1

$$M_1 = 3.5 \times 0.60 \times 0.50 \times 0.25 - 17.5 \times 0.60 \times 0.10 \times 0.05$$
$$= 0.21 \text{ MN m}$$

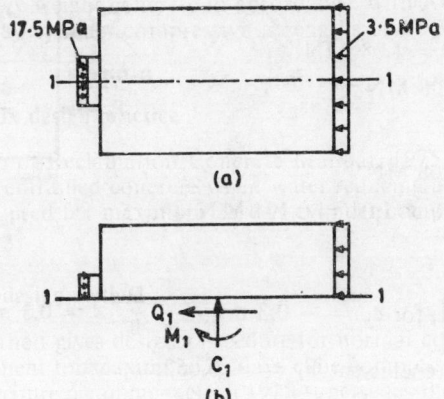

(a)

(b)

Fig. 7.7.12

By symmetry, the shear force Q_1 is zero. Also the axial compression C_1 is zero because the prestressing force does not have a vertical component. Using Eq. (7.7.5) and (7.7.7), the bursting tensile stress,

$$f_y = -20\left(1 - \frac{6x}{D} + \frac{9x^2}{D^2} - \frac{4x^3}{D^3}\right)\frac{M_1}{bd^2} \tag{a}$$

Equation (a) shows that f_y is compressive (negative) for x/D less than 0.25. The bursting tensile stress f_y increases from zero value at $x/D = 0.25$ to its maximum value at $x/D = 0.50$.

$$f_y \text{ (max)} = \frac{5M_1}{bd^2} = \frac{5 \times 0.21}{0.60 \times 1.00^2} = 1.75 \text{ MPa}$$

Also f_y becomes zero at x/D equal to 1.00.

(ii) *Zielinski and Rowe's method*

Using Eq. (7.7.12), maximum bursting tensile stress,

$$f_y \text{ (max)} = 3.5 \,(0.98 - 0.825 \times 0.20) = 2.8525 \text{ MPa}$$

(iii) *Guyon's method*

Using Table 7.7.3, the maximum bursting tensile stress for $D_p/D = 0.20$,

$$f_y \text{ (max)} = 0.36 \, p = 0.36 \times 3.5 = 1.26 \text{ MPa}$$

This maximum stress occurs at $x/D = 0.30$. The Table also shows that $f_y = 0$ at $x/D = 0.14$.

EXAMPLE 7.7.2

Figure 7.7.13 shows the end block of a beam of rectangular cross-section 0.60×1.20 m. The beam has two prestressing tendons, each carrying a force of 1.08 MN. Determine stress f_y at point A.

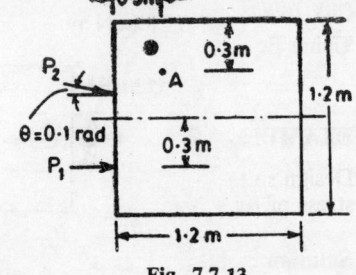

Fig. 7.7.13

Solution

$$P_1 = P_2 = 1.08 \text{ MN}$$
$$P_{1h} = 1.08 \text{ MN}$$
$$P_{1v} = 0$$
$$P_{2h} = 1.08 \text{ MN}$$
$$P_{2v} = 0.1 \times 1.08 = 0.108 \text{ MN}$$

(i) *Effect of P_1*

Using Table 7.7.1, for $e_p = -0.3$ m $= -\dfrac{D}{4}$, $x = 0.3$ m $= \dfrac{D}{4}$ and $y = 0.3$ m $= \dfrac{D}{4}$, the coefficient $C_2 = -0.052$.

(ii) *Effect of P_2*

Using Tables. 7.7.1 and 7.7.2 for $e_p = \dfrac{D}{8}$, $x = \dfrac{D}{4}$ and $y = \dfrac{D}{4}$, the coefficient $C_2 = 0.146$ and $C'_2 = -0.302$.

Combining the effects of anchorage forces P_1 and P_2, the stress f_y in accordance with Eq. (7.7.16) and (7.7.19) may be expressed as

$$f_y = -0.052\, \frac{P_{1h}}{bD} + 0.146\, \frac{P_{2h}}{Db} - 0.302\, \frac{P_{2v}}{bD}$$

$$= -0.052 \times \frac{1.08}{0.60 \times 1.20} - 0.302 \times \frac{0.108}{0.60 \times 1.20} + 0.146 \times$$

$$\frac{1.08}{0.60 \times 1.20}$$

$$= 0.0957 \text{ MPa (tensile)}$$

EXAMPLE 7.7.3

For the end block of Example 7.7.1, determine the bursting force assuming that the total force of 2.1 MN is delivered by a single prestressing tendon.

Solution

(i) *Zielinski and Rowe's method*

Using Eq. (7.7.13), the bursting tensile force,

$$F_{bst} = 2.1 (0.48 - 0.40 \times 0.20) = 0.84 \text{ MN}$$

(ii) *Guyon's method*

Using Table 7.7.3, for $D_{p/D} = 0.20$,

$$F_{bst} = 0.19 P = 0.19 \times 2.1 = 0.399 \text{ MN}$$

(iii) *Indian code*

Using Eq. (7.7.22),

$$F_{bst} = 2.1 (0.32 - 0.30 \times 0.20) = 0.546 \text{ MN}$$

EXAMPLE 7.7.4

Design suitable reinforcement for the end block of Example 7.6.3. The yield stress of reinforcing steel, $f_y = 250$ MPa.

Solution

$$P_k = P_i = 1.509 \text{ MN}$$
$$\frac{y_{po}}{y_o} = \frac{365}{600} = 0.608$$
$$\frac{z_{po}}{z_o} = \frac{220}{400} = 0.550$$

Using Eq. (7.7.22), the bursting force in the vertical direction,

$$F_{bst} = 1.509 (0.32 - 0.30 \times 0.608) = 0.2075 \text{ MN}$$

Similarly, the bursting force in the horizontal direction,

$$F_{bst} = 1.509 (0.32 - 0.30 \times 0.55) = 0.2339 \text{ MN}$$

Using many legged stirrups of 8 mm diameter, the total number of legs required in the vertical and horizontal directions are,

$$n_v = \frac{0.2075}{\frac{\pi}{4} \times 0.008^2 \times (0.87 \times 250)} = 19.0$$

$$n_h = \frac{0.2339}{\frac{\pi}{4} \times 0.008^2 \times (0.87 \times 250)} = 21.4$$

Two types of stirrups, type A and type B, shown in Fig. 7.7.14 may be used. Type A has six vertical legs and type B has six horizontal legs. A mesh comprising one stirrup of type A and another of type B may be provided at

distances of 100, 220, 380 and 600 mm from the loaded end. In addition longitudinal bars of 8 mm diameter may be provided at the corners of stirrups of types A and B to form a reinforcing cage.

As suggested in Sec. 7.7.4, the reinforcement may be designed for spalling tensile force,

$$F_{spl} = 0.03\ P = 0.03\ \eta\ P_i$$

Taking the loss factor $\eta = 0.85$,

$$F_{spl} = 0.03 \times 0.85 \times 1.509$$

$$= 0.0385\ \text{MN}$$

Hence the number of legs of 8 mm diameter required in the vertical and horizontal directions,

$$n_v = n_h = \frac{0.0385}{\dfrac{\pi}{4} \times 0.008^2 \times (0.87 \times 250)}$$

$$= 3.5$$

Hence a mesh formed by combining one stirrup each of type A and type B may be provided immediately behind the anchorages at a distance of 30 mm from the beam end.

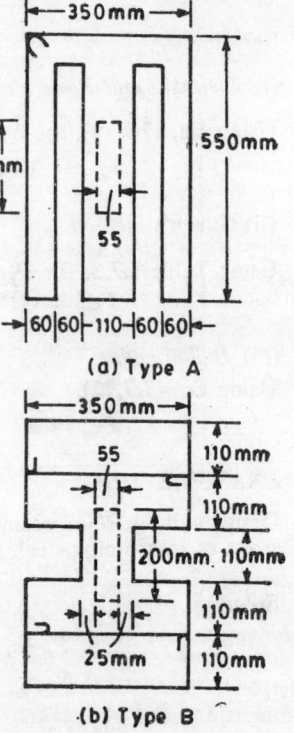

(a) Type A

(b) Type B

Fig. 7.7.14

EXAMPLE 7.7.5

The end block of a beam of rectangular cross-section 0.80 × 1.50 m is prestressed by four Freyssinet cables which are anchored as shown in the end view and elevation of the end block in Fig. 7.7.15 (a) and (b). Each cable has an effective prestressing force of 600 kN. The outer diameter of the female cone is 150 mm. Design appropriate reinforcement for the end block.

Solution

It is proposed to provide spiral reinforcement to resist the bursting force on account of each of the four cables. Additional reinforcement in the form of mats or meshes will be provided to resist the bursting forces due to group and total resultants.

The total prestressing force is 2.4 MN which acts at a distance of 0.675 m from the bottom face. The eccentricity of prestress, $e_p = 0.075$ m. Hence the extreme fibre stresses at the end of the stress block are

$$f_T = \frac{-2.4}{0.80 \times 1.50} + \frac{2.4 \times 0.075 \times 0.75}{\frac{1}{12} \times 0.80 \times 1.50^3} = -1.4 \text{ MPa}$$

$$f_B = \frac{-2.4}{0.80 \times 1.50} - \frac{2.4 \times 0.075 \times 0.75}{\frac{1}{12} \times 0.80 \times 1.50^3} = -2.6 \text{ MPa}$$

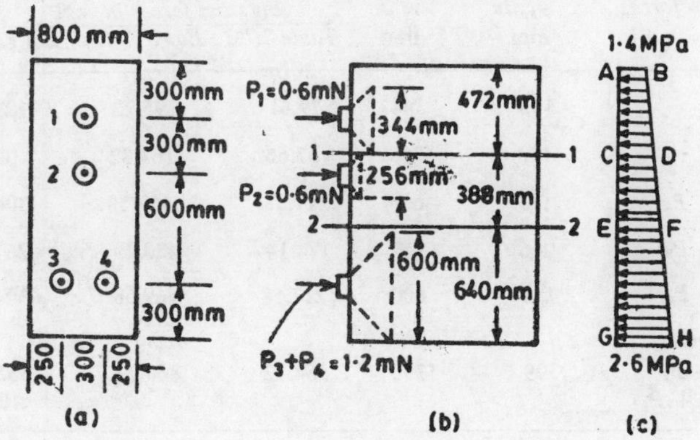

Fig. 7.7.15

The stress distribution at the end of the stress block is shown in Fig. 7.7.15 (c). The end block is first partitioned along sections 1-1 and 2-2 in such a way that the resultants of the compressive stress represented by trapezia ABCD, CDEF and EFGH are equal to 0.6 MN, 0.6 MN and 1.2 MN respectively. Referring to Fig. 7.7.15 (b), it is noted that the depth of the symmetric prism for the top force P_1 is 0.344 m. Similarly, the depths of the symmetric prisms for forces P_2 and $(P_3 + P_4)$ are 0.256 and 0.600 m respectively. The corresponding bursting forces due to force P_1, P_2 and $(P_3 + P_4)$ computed from Table 7.7.3 and Eq. (7.7.13) and (7.7.22) are shown at S. No. 1 to 3 of Table 7.7.4. As the next step, group resultants may be considered. The end block is partitioned by the section 2-2 so that the resultants of the compressive stresses represented by the trapezia ABEF and EFGH are equal to $(P_1 + P_2)$ and $(P_3 + P_4)$ respectively. The depths of the symmetric pirsms for the top and bottom group resultants are 0.820 m and 0.600 m respectively. The corresponding bursting forces are shown at S. No. 4 and 5 of the Table. As the final step, the resultant of all forces may be considered. As this total resultant acts at a distance of 0.675 m from bottom face, the depth of the symmetric prism is equal to 1.350 m. The corresponding bursting force is shown at S. No. 6 of 1.350 m. The corresponding bursting force is shown at S. No. 6 of the table. The Table shows that the bursting forces are the largest according to Zielinski and Rows's Eq. (7.7.13) and smallest according to Guyon's coefficients (Table 7.7.3`. For the sake of conservatism, the reinforcement will be designed for the bursting forces computed from Eq. (7.7.13).

The spiral reinforcement immediately behind each of the four anchorages may be designed to resist half of the bursting force caused by the combined action of P_3 and P_4, i.e, 234.8 kN. This bursting force is larger than those due to P_1 and P_2. Hence the area of spiral reinforcement,

$$A_s = \frac{0.2348}{0.87 \times 250} = 0.00108 \text{ m}^2$$

Table 7.7.4 Bursting Forces

S. No.	Due to forces	y_{po} in mm	y_o in mm	Bursting forces in kN Table 7.7.3	Eq. (7.7.13)	Eq. (7.7.22)
1.	P_1	133	344	79.61	195.22	122.41
2.	P_2	133	256	57.66	163.32	98.49
3.	$(P_3 + P_4)$	133	600	217.58	469.58	304.19
4.	$(P_1 + P_2)$	266	820	174.14	420.29	267.20
5.	$(P_3 + P_4)$	133	600	217.58	469.58	304.19
6.	All forces, $(P_1 + P_2 + P_3 + P_4)$	399	1350	364.32	868.22	555.17

Using 10 mm diameter bar for the spiral, the total number of turns.

$$n = \frac{0.00108}{2 \times \frac{\pi}{4} \times 0.010^2} = 6.87$$

Taking the length of the spiral equal to the depth of the symmetric prism y_o for P_1, the pitch of the spiral,

$$s = \frac{0.344}{6.87} = 0.050 \text{ m}$$

Hence provide 10 mm spiral at a pitch of 50 mm immediatly behind the anchorage. As the diameter of the anchorage is 150 mm, the mean diameter of the spiral may be taken equal to 200 mm.

In addition to the spiral reinforcement for individual anchorages, mat reinforcement may be provided to resist the bursting forces due to group resultants. Using 12 mm diameter bars for the mat, the number of bars required to resist the larger bursting force of 469.58 kN.

$$n = \frac{0.46958}{\frac{\pi}{4} \times 0.012^2 \times (0.87 \times 250)} = 19.08$$

Three mats each having 6 vertical bars at a spacing of 150 mm and an equal number of horizontal bars at a spacing of 280 mm may be provided. As the depths of the symmetric prisms for the two groups ore 0.820 m and 0.600 m, these mats may be provided at distances 400 mm, 550 mm and 700 mm from the loaded end.

Total number of 12 mm diameter bars required to resist the bursting force of 868.22 kN due to total resultant,

$$n = \frac{0.86822}{\frac{\pi}{4} \times 0.012^2 \times (0.87 \times 250)} = 35.27$$

Henee three more mats may be provided at distance 900 mm, 1100 mm and 1350 mm from the loaded end.

The total spalling tension,

$$F_{spl} = 0.03 \times 2.4 = 0.072 \text{ MN}$$

Therefore, the number of 8 mm diameter bars required,

$$n = \frac{0.072}{\frac{\pi}{4} \times 0.008^2 \times (0.87 \times 250)} = 6.56$$

Hence a square mat comprising 8 mm diameter bars at 100 mm spacing in both principal directions may be provided at 50 mm from the loaded end for resisting spalling tension.

REFERENCES

7.1 ACI—ASCE Committee 326, *Shear and diagonal tension*, Proceedings of the American Concrete Institute, Vol. 59, January-February 1962, pp. 1-30 and 277-334.

7.2 Anderson, A.R. and Anderson, R.G., *An assurance criterion for flexural bond in pretensioned hollow core units*, Journal of the American Concrete Institute, Vol. 73, August 1976, pp. 457-64.

7.3 Bennett, E.W. and Balasoorya, B.M.A., *Shear strength of prestressed beams with thin webs failing in inclined compression*, Journal of the American Concrete Institute, Vol. 68, No. 3, March 1971, pp. 204-12.

7.4 Bishara, A., *Prestressed concrete beams under combined torsion, bending and shear*, Journal of the American Concrete Institute, Proceedings Vol. 66, No. 7, July 1969, pp. 525-38.

7.5 Gergely, P. and Sozen, M.A., *Design of anchorage zone reinforcement in prestressed concrete beams*, Journal of the Prestressed Concrete Institute, April 1967, pp. 63-75.

7.6 Guyon, Y., *Contraintes dans les pieces prismatiques soumises a des forces appliquees sur leurs bases, au voisimage daces bases*, Publications of IABSE, Vol. 11, 1951, pp. 165-226.

7.7 Hanson, N.W. and Kaar, P.H., *Flexural bond tests of pretensioned prestressed beams*, Journal of the American Concrete Institute, January 1959, pp. 783-802.

7.8 Hognestad, E. and Janney, J.R. *The ultimate strength of pretensioned prestressed concrete failing in bond*, Magazine of Concrete Research, June 1954.

7.9 Humphreys, R., *Torsional properties of prestressed concrete*, Structural Engineer (London), Vol. 35, No. 6, June 1957, pp. 213-25.

7.10 Iyengar, K.T.S., *Two dimensional theories of anchorage zone stresses in post-tensioned beams*, Journal of the American Concrete Institute, Proceedings Vol. 59, No. 10, 1962, pp. 1443-6.

7.11 Iyengar, K.T.S. and Prabhakara, M.K., *Anchor zone stresses in prestressed concrete beams*, Journal of the Structural Division, ASCE, Vol. 97, No. ST 3, 1971, pp. 807-24.

7.12 Janney, J.R., *Nature of bond in pretensioned prestressed concrete*, Journal of the American Concrete Institute, Proceedings Vol. 50, May 1954, pp. 717-36.

7.13 Karr, P.H. ; Lafraugh, R.W. and Mass, M.A., *Influence of concrete strength on strand transfer length*, Journal of Prestressed Concrete Institute, Vol. 8, No. 5, October 1963, pp. 47-67.

7.14 Kenning, R.W.; Sozen, M.A. and Siess, C.P., *A study of anchorage bond in prestressed concrete*, University of Illinois, Structural Research Series No. 251, June 1962.

7.15 Krishna Murthy, D., *Design of end zone reinforcement to control horizontal cracking in pretensioned concrete members at transfer-1*, The Indian Concrete Journal, Vol. 47, No. 9, September 1973, pp. 346-51.

7.16 Magnel, G., *Design of the ends of prestressed concrete beams*, Concrete and Constructional Engineering, Vol. 44, No. 5, 1949, pp. 141-8.

7.17 Marshall, W.T. and Mattock, A.H., *Control of horizontal cracking in the ends of pretensioned prestressed concrete girders*, Journal of the Prestressed Concrete Institute, Vol. 7, No. 5, October 1962, pp. 56-74.

7.18 Marshall, W.T. and Krishna Murthy, D., *Transmission length of prestressing tendons from concrete cube strengths at transfer*. The Indian Concrete Journal, Vol. 43, No. 7, July 1969, pp. 244-53.

7.19 Mast, P.E. *Short cuts for the shear analysis of standard prestressed concrete members*, Journal of the Prestressed Concrete Institute, Vol. 9, No. 5, October 1964, pp. 15-47.

7.20 Middendorf, K.H., *Anchorage bearing stresses in post-tensioned concrete*, Journal of the American Concrete Institute, November 1960, pp. 580-4.

7.21 Pandit, G.S., *A note on balanced section in combined loading*, Journal of the American Concrete Institute, Vol. 67, No. 5, May 1970, pp. 420-3.

7.22 Pandit, G.S., *Ultimate torque of rectangular reinforced concrete beams*, Journal of the Structural Division, ASCE, Vol. 96, No. ST 9, September 1970, pp. 1987-95.

7.23 Pandit, G.S. and Sharma, A.K., *Design of compression member under biaxial couples*, The Indian Concrete Journal, Vol. 46, No. 1, January 1972, pp. 34-6.

7.24 Pandit, G.S. and Mawal, M.B., *Tests on short columns in torsion*, The Indian Concrete Journal, Vol. 46, No. 11, November 1972, pp. 471-4.

7.25 Pandit, G.S. and Vankappa, V., *Ultimate strength design of reinforced concrete beams under triaxial couples,* The Indian Concrete Journal, Vol. 53, No. 5, May 1979, pp. 132-5.

7.26 Pandit, G.S. and Gupta. S.P., *A comprehensive code format for torsion,* The Indian Concrete Journal, Vol. 58, No. 8, August 1984, pp. 212-8.

7.27 Post-Tensioning Manual, *Post-Tensioning Institute,* Glenview, Illinois, 1976.

7.28 Rowe, R.E., *End block stresses in post-tensioned concrete beams,* Structural Engineer (London), Vol. 41, No. 2. 1963, pp. 54-68.

7.29 Yettram, A. L. and Robbins. K., *Anchorage zone stresses in axially post-tensioned member of uniform rectangular section,* Magazine of Concrete Research (London), Vol. 21, No. 67, 1969, pp. 103-12.

7.30 Zia, P., *Research in torsion of prestressed members,* Journal of the Prestressed Concrete Institute, Vol. 5, 1960, pp. 35-40.

7.31 Zia, P., *Torsional strength of prestressed concrete members,* Journal of the American Concrete Institute, Proceedings Vol. 57, April 1961, pp. 1337-59.

7.32 Zia, P. and McGee, W.D., *Torsion design of prestressed concrete,* Journal of the American Concrete Institute, Vol. 19, No. 2, March-April 1974, pp. 46-65.

7.33 Zia, P. and Mostafa, T., *Development length of prestressed strands,* Journal of the Prestressed Concrete Institute, Vol. 22, No. 5, September-October 1977, pp. 54-65.

7.34 Zielinski, J. and Rowe, R.E., *An investigation of the stress distribution in the anchorage zones of post-tensioned concrete members,* Report No. 9, C.A.C.A. (London), September 1960.

7.35 Zielinski, J. and Rowe. R.E., *The stress distribution associated with groups of anchorages in post-tensioned concrete members,* Research Report No. 13, C.A.C.A. (London), 1962.

PROBLEMS

7.1 Explain why diagonal tension can be reduced but not eliminated by uniaxial prestressing of beams. Derive expressions for diagonal tension and its inclination at any point in an uncracked section of a uniaxially prestressed beam.

7.2 Determine the magnitude of diagonal tension and its inclination at a point in an uncracked section of a uniaxially prestressed beam given that

(i) $f = -4.5$ MPa $\qquad v = 1.5$ MPa

(ii) $f = -4.5$ MPa $\qquad v = -1.5$ MPa

7.3 Discuss the sequence and pattern of cracking of prestressed beams under overload conditions.

7.4 Discuss the forces which contribute to shear resistance of prestressed concrete beams and the relative importance of the parameters affecting them.

7.5 A pretensioned beam of rectangular section 250 × 550 mm has an effective prestressing force of 900 kN at a constant eccentricity of 200 mm. It carries a service load of 25.8 kN/m over an effective span of 11 m. Design the shear reinforcement for the beam. The grade of concrete is M-40.

7.6 A post-tensioned beam has T-section with the flange 1.50 m wide and 0.25 m thick and web 0.30 m thick. The overall depth of the section is 1.8 m. An effective prestressing force of 5.4 MN acts along a parabolic cable line with no eccentricity at ends and an eccentricity of 0.90 m at midspan. It carries a uniform load of 28 kN/m over an effective span of 36 m. Design the shear reinforcement for the beam. The grade of concrete is M-35.

7.7 Discuss the similarities and dis-similarities of the phenomena of shear and torsion. Compare the elastic distributions of flexural and torsional shear stresses in sections commonly adopted for prestressed beams.

7.8 A pretensioned member of rectangular section 250 × 550 mm has a concentric prestressing force of 900 kN. Design the torsion reinforcement to resist a torque of 90 kN.m. The grade of concrete is M-40.

7.9 A post-tensioned member has a T-section with the flange 1.50 m wide and 0.25 m thick and web 0.30 m thick. The overall depth of the member is 1.80 m. It carries a concentric prestressing force of 5.4 MN. Design the torsion reinforcement to resist a torque of 0.45 MN.m The grade of concrete is M-35.

7.10 Discuss the interaction between shear and torsion in prestressed concrete beams.

7.11 A pretensioned beam of rectangular section 250 × 550 mm has an effective prestressing force of 900 kN at a constant eccentricity of 200 mm. It carries a service load of 25.8 kN/m over an effective span of 11 m. The beam also carries a torque of 18 kN.m. Design the shear reinforcement for the beam. The grade of concrete is M—40.

7.12 A post-tensioned beam has a T-section with the flange 1.50 m wide and 0.25 m thick and web 0.30 m thick. The overall depth of the beam is 1.80 m. An effective prestressing force of 5.4 MN acts along a parabolic cable line with no eccentricity at ends and an eccentricity of 0.90 m at midspan. It carries a seismic load of 28 kN/m over an effective span of 36 m. The beam also carries a torque of 0.090 MN.m. Design the shear reinforcement. The grade of concrete is M-35.

*7.13 Discuss the nature of bond stress in prestressed concrete beams and compare it with that in reinforced concrete beams. Distinguish between prestress transfer bond and flexural bond. Derive relevant expressions for each type of bond stress.

7.14 Compare the nature of bond stress in pretensioned beams with that in post-tensioned beams. How is the distribution of bond stress affected by flexural cracking ?

7.15 Discuss the nature of anchorage zone stresses with particular reference to bursting and spalling tension.

7.16 The end block of a post-tensioned beam has a rectangular cross-section 300 × 500 mm. It has a single concentric cable comprising 16 wires of 5 mm diameter. The initial and final stresses in the wires are 1260 MPa and 1010 MPa respectively. The dimensions of the duct are 55 × 55 mm. The grade of concrete is M-40. Design the bearing plate for the Magnel Blaton system of post-tensioning. Also design the reinforcement in the end block.

7.17 A post-tensioned beam is prestressed by a single high tensile bar of 20 mm diameter carrying an initial stress of 1150 MPa and a final stress of 950 MPa. It is anchored at an eccentricity of 50 mm in the end block having a rectangular cross-section 320 × 520 mm. Design a suitable bearing plate for the Lee Mc Call system. Also design the reinforcement for the end block. The grade of concrete is M-35.

7.18 The end block of a post-tensioned beam has a rectangular cross-section 1.20 × 1.20 m. It carries 5 Freyssi cables, each carrying an effective force of 1020 kN. The cables are anchored at distances of 0.30, 0.60, 0.90, 1.50 and 1.80 m from the bottom face. The diameter of the female cone is 150 mm. Design the shear reinforcement for the end block. The grade of concrete is M-35.

8

Deformation and Cracking

8.1 INTRODUCTION

The deformation characteristics of flexural members are closely linked with their serviceability requirements which may be as important as the strength requirements in some cases. The computations for deformation include the determination of curvature, slope or rotation, deflection and crack width. A check on deformations is necessary for the following reasons:

(i) To verify that the code provisions in respect of limit state of deflection for adequate serviceability under working loads is complied with.

(ii) To prevent damage to partitions, claddings, plasters, finishes and other non-structural elements.

(iii) To prevent annoyance and discomfort to users on account of perceptible deflections and unacceptable vibrations.

(iv) To prevent inconvenience in driving on bridges due to excessive deflection or cambre.

(v) To prevent unsightly deflections and psychological feeling of impending distress.

(vi) To prevent distress in connected or adjoining structural elements.

(vii) The computations of deformations are also necessary to establish the eqations of compatibility of deformations for the determination of internal forces in a statically indeterminate structure.

The deformation caused by prestress is generally opposite in sign to that of the deformation due to service loads. For instance, in a simply supported beam, the service loads cause downward (positive) deflection whereas prestress induces upward (negative) deflection called cambre. The net deflection under the combined effect of loads and prestress is evidently the algebraic sum of the two deflections. Excessive deformations create serviceability problems.

A limit on the crack width is also an essential serviceability requirement. A wide crack permits ingress of moisture resulting in corrosion of steel and reduced useful life of the structure. Consequently it is desirable to have the tensile strains in concrete well distributed into several fine cracks along the span instead of a few wide cracks. While cracking is an essential feature of all reinforced concrete beams, cracking occurs only under overload conditions in fully prestressed (type 1) and limited prestress (type 2) beams. Partially prestressed (type 3) beams crack at working load similar to reinforced concrete beams. The width of the cracks can, however, be kept within specified limits by proper proportioning and detailing of prestressed and non-prestressed steel.

A precise determination of the deformation in prestressed concrete beams is difficult due to the non-linearity of the stress-strain curve for concrete and also due to time dependent deformations caused by creep and shrinkage of concrete and relaxation of steel. The complexity of the problem is enhanced further when the beam undergoes flexural cracking. As the bending moment is generally variable, the extent of cracking also varies along the span. The height of flexural cracks, the position of neutral axis and the area in compression zone vary from section to section along the span of the beam. Consequently, the effective cross-sectional area becomes variable after cracking. Hence the determination of slope and deflection by successive integration of curvature becomes difficult. In the computations for displacements due to flexure, it is commonly assumed that the strains are linearly distributed along the depth of a section even after cracking. This assumption, called the Navier's hypothesis, forms the basis for the determination of slope and deflection of beams. The deformation characteristics of a beam are best expressed by moment-curvature and load-deflection relationships. These relationships are practically linear in the precracking range but have progressively increasing non-linearity in the post-cracking range.

8.2 PRECRACKING RANGE

A prestressed concrete member under the action of service loads undergoes two types of deformations.

(i) Short time deformations which occur immediately on the application of prestressing force and service loads. These deformations are elastic in nature and almost completely disappear on the removal of the load.

(ii) Long time deformations which occur mainly due to the combined effect of creep and shrinkage of concrete and relaxation of prestressing steel. A major portion of these long time deformations is non-recoverable.

8.2.1 Short-Time Deformations

Before the formation of flexural cracks, the entire concrete section is effective. The elastic short-time deformations in pre-cracking range may be determined with reasonable accuracy by linear elastic theory using gross concrete section and modulus of elasticity of concrete E_c related to elastic short-time strains discussed in Sec. 2.2.2.3.

A prestressed concrete member carrying service loads is actually subjected to two systems of forces, viz. prestressing force and service loads. The deformations produced by the two systems of forces are generally opposite in sign. In the precracking range, the principle of superposition is valid. Consequently, the deformations due to the combined action of the two systems of forces are equal to the algebraic sum of the deformations caused by the two systems of forces acting separately. The deformations due to the two systems of forces may, therefore, be considered separately.

(a) Prestressing Force

The deformations caused by the prestressing force depend upon:

(i) magnitude of prestressing force

(ii) eccentricity of prestress or shape of cable line

(iii) geometry and size of the cross-sections at all points

(iv) physical properties of constituent materials, particularly concrete and prestressing steel

(v) geometry of the structure and

(vi) type of support system.

The free body approach, discussed in Sec. 5.4, may be used conveniently to determine the deformations caused by the prestressing force. Some simple types of cable lines commonly used in pretensioned and post-tensioned beams may now be considered. Consider for instance, the prestressed concrete beam of Fig. 8.2.1 (a) in which the cable line comprises two straight segments. The free body diagram of concrete beam is shown in Fig. 8.2.1 (b). The change of direction in the cable line at D,

$$\alpha = \theta_A + \theta_B = \frac{e_{pD} - e_{pA}}{l_1} + \frac{e_{pD} - e_{pB}}{l_2} \qquad (8.2.1)$$

Hence the upward force W_p at D,

$$W_p = P\alpha = P\left[\frac{e_{pD} - e_{pA}}{l_1} + \frac{e_{pD} - e_{pB}}{l_2}\right] \qquad (8.2.2)$$

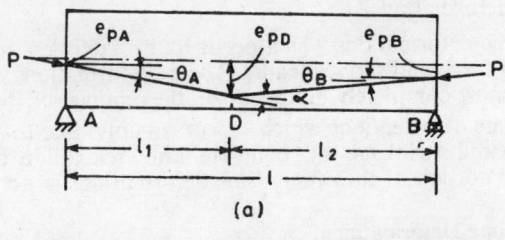

(a)

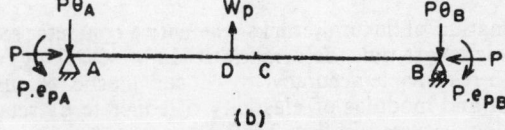

(b)

Fig. 8.2.1

Hence from usual methods of analysis, the slopes β_A and β_B at A and B and deflections Δ_C and Δ_D at mid-span section C and D due to prestress alone may be expressed as

$$\beta_A = \frac{(-Pe_{pA}) \, l}{3E_c \, I} + \frac{(-Pe_{pB}) \, l}{6E_c \, I} - \frac{W_p \, l_1 \, l_2 \, (2l - l_1)}{6lE_c \, I}$$

$$(8.2.3a)$$

$$\beta_B = \frac{(Pe_{pA}) \, l}{6E_c \, I} + \frac{(Pe_{pB}) \, l}{3E_c \, I} + \frac{W_p \, l_1 \, l_2 \, (2l - l_2)}{6l \, E_c \, I}$$

$$(8.2.3\ b)$$

$$\Delta_C = -\frac{W_p \, l_1 \, (3l^2 - 4l_1^2)}{48 \, E_c \, I} - \frac{Pe_{pA} \, l^2}{16E_c \, I} - \frac{Pe_{pB} \, l^2}{16E_c \, I}$$

$$(8.2.3\ c)$$

$$\Delta_D = -\frac{W_p \, l_1^2 \, l_2^2}{3l \, E_c \, I} - \frac{Pe_{pA} \, l_1 \, l_2 \, (l + l_2)}{6l \, E_c \, I} - \frac{Pe_{pB} \, l_1 \, l_z \, (l + l_1)}{6l \, E_c \, I}$$

$$(8.2.3\ d)$$

In these equations, clockwise rotations are taken positive in accordance with the sign convention given in Sec. 1.4.

In the symmetrical case, $l_1 = l_2$ and $e_{pA} = e_{pB}$,

$$\theta_A = \theta_B = \frac{2}{l} (e_{pC} - e_{pA})$$

$$(8.2.4\ a)$$

$$\beta_A = -\beta_B = \frac{-Pe_{pA} \, l}{2E_c \, I} - \frac{W_p \, l^2}{16 \, E_c \, I}$$

$$(8.2.4\ b)$$

$$\Delta_C = \frac{-Pe_{pA} \, l^2}{8E_c \, I} - \frac{W_p \, l^3}{48E_c \, I}$$

$$(8.2.4\ c)$$

If the cable line is straight and has constant eccentricity e_p throughout the span as is common in pretensioned beams,

$$\beta_A = -\beta_B = \frac{-Pe_p \, l}{2E_c \, I}$$

$$(8.2.5\ a)$$

$$\Delta_C = \frac{-Pe_p \, l^2}{8E_c \, I}$$

$$(8.2.5\ b)$$

Next consider the beam with a parabolic cable line shown in Fig. 8.2.2 (a). The free body of the concrete beam is shown in Fig. 8.2.2 (b). From Eq. (5.4.10), the intensity of uniform upward force w_p on account of prestress is

$$w_p = \frac{8 \, PS}{l^2}$$

$$(8.2.6)$$

Hence the short-time elastic slopes β_A and β_B at A and B and the deflection Δ_C at midspan section C,

$$\beta_A = \frac{(-Pe_{pA}) \, l}{3E_c \, I} + \frac{(-Pe_{pB}) \, l}{6E_c \, I} - \frac{w_p \, l^3}{24E_c \, I}$$

$$(8.2.7\ a)$$

$$\beta_B = \frac{(Pe_{pA}) \, l}{6E_c \, I} + \frac{(Pe_{pB}) \, l}{3 \, E_c \, I} + \frac{w_p \, l^3}{24 \, E_c \, I}$$

$$(8.2.7\ b)$$

$$\Delta_C = \frac{(-Pe_{pA}) \, l^2}{16E_c \, I} + \frac{(-Pe_{pB}) \, l^2}{16E_c \, I} - \frac{5 \, w_p \, l^4}{384 \, E_c \, I}$$

$$(8.2.7\ c)$$

In case of symmetry, $e_{pA} = e_{pB} = e_p$,

$$\beta_A = -\beta_B = -\frac{Pe_p\, l}{2E_c\, I} - \frac{w_p\, l^3}{24E_c\, I} \tag{8.2.8 a}$$

$$\Delta_C = -\frac{Pe_p\, l^2}{8\, E_c\, I} - \frac{5w_p\, l^4}{384\, E_c\, I} \tag{8.2.8 b}$$

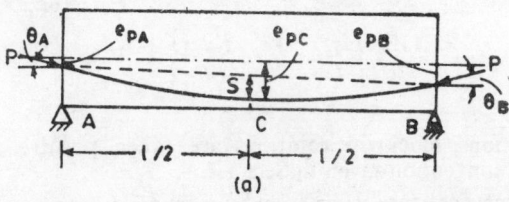

(a)

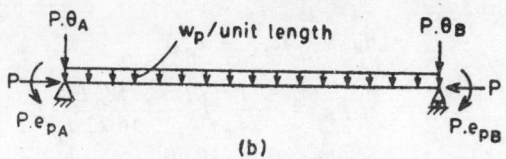

(b)

Fig. 8.2.2

The foregoing expressions for slopes and deflections give the short-time elastic deformation of a beam due to prestress.

(b) Service loads

To determine the deformations due to service loads, the beam may be considered as an ordinary non-prestressed concrete member. Any one of the classical methods of flexural analysis may be used for the computation of deformations. Some of the standard results are given in Table 8.2.1.

(c) Combined loads

The net deformations under the combined action of prestressing force and service loads may be obtained by adding algebraically the deformations due to prestressing force and service loads. Thus the net curvature Φ, the slope or rotation β and the deflection Δ at any section of a beam may be expressed as

$$\Phi = \Phi_p + \Phi_w \tag{8.2.9 a}$$

$$\beta = \beta_p + \beta_w \tag{8.2.9 b}$$

$$\Delta = \Delta_p + \Delta_w \tag{8.2.9 c}$$

It should be noted that the deformations Φ_p, β_p and Δ_p due to prestressing force are generally opposite in sign to the deformations Φ_w, β_w and Δ_w caused by the service loads.

In the foregoing computations for deformations, it is appropriate to consider only the net area of concrete in determining the deformations caused by prestressing force. In the computations for deformations due to service loads, it is appropriate to consider the transformed or equivalent

TABLE 8·2·1

S.No	STRUCTURE	EXPRESSIONS
1		$\alpha_B = \dfrac{ML}{EI}$ $\Delta_B = \dfrac{ML^2}{2EI}$
2		$\alpha_B = \dfrac{WL^2}{2EI}$ $\Delta_B = \dfrac{WL^3}{3EI}$
3		$\alpha_B = \dfrac{wL^3}{6EI}$ $\Delta_B = \dfrac{wL^4}{8EI}$
4		$\alpha_A = \dfrac{ML}{3EI}$ $\qquad \alpha_B = \dfrac{ML}{6EI}$ $\Delta_C = \dfrac{ML^2}{16EI}$
5		$\alpha_A = \alpha_B = \dfrac{ML}{2EI}$ $\Delta_C = \dfrac{ML^2}{8EI}$
6		$\alpha_A = \alpha_B = \dfrac{WL^2}{16EI}$ $\Delta_C = \dfrac{WL^3}{48EI}$

TABLE 8·2·1 (CONTD.)

S.No	STRUCTURE	EXPRESSIONS
7		$\alpha_A = \dfrac{Wab(L+b)}{6LEI}$ $\alpha_B = \dfrac{Wab(L+a)}{6LEI}$ $\Delta_C = \dfrac{Wa(3L^2-4a^2)}{48EI}, \Delta_D = \dfrac{Wa^2b^2}{3LEI}$
8		$\alpha_A = \alpha_B = \dfrac{Wa(L-a)}{2EI}$ $\Delta_D = \Delta_E = \dfrac{Wa^2(L-4a/3)}{2EI}$ $\Delta_C = \dfrac{Wa}{24EI}(3L^2-4a^2)$
9		$\alpha_A = \alpha_B = \dfrac{WL^3}{24EI}$ $\Delta_C = \dfrac{5wL^4}{384EI}$
10		$\Delta_C = \dfrac{WL^3}{192EI}$ $M_A = M_B = \dfrac{WL}{8}$
11		$\Delta_C = \dfrac{Wa^2(3L-4a)}{48EI}, \Delta_D = \dfrac{Wa^3b^3}{3L^3EI}$ $M_A = \dfrac{Wab^2}{L^2}, M_B = \dfrac{Wba^2}{L^2}$
12		$\Delta_C = \dfrac{wL^4}{384EI}$ $M_A = M_B = \dfrac{wL^2}{12}$

area in the case of pretensioned beams and post-tensioned bonded beams. For unbonded post-tensioned beams, the net area of concrete is appropriate for deformations due to prestress as well as service loads. However, for the sake of simplicity, all computations for deformation may be based on the cross-sectional properties of the gross concrete section ignoring the presence of prestressing steel and cable ducts. The simplification appears to be justified because the modulus of elasticity of concrete E_c can not be determined with high accuracy. Besides, it changes continuously with increasing stress. The prestressing force too cannot be determined with high precision due to uncertainities in the loss of prestress due to various factors.

8.2.2 Long-Time Deformations

Here too the principle of superposition may be used, i.e. the effects of pre-stressing force and service loads may be considered separately and then combined to obtain the net long time deformations in the precracking range.

(a) Prestressing Force

As prestressing force is a permanent force, it must be taken into account in the computation of long-time deformations. The long-time deformations due to prestress are influenced by two opposite effects. With the passage of time, the prestressing force decreases thereby causing a reduction in the deformations due to prestress. On the other hand, the strains are enhanced due to the combined effect of creep and shrinkage causing an increase in the deformations. Taking into account the two opposite effects, the curv-ature Φ_{pt} at any section of a prestressed concrete beam after time t may be expressed as

$$\Phi_{pt} = \frac{P_t\, e_p}{E_c\, I} - \frac{(P_i - P_t)\, e_p}{E_c\, I} + \frac{0.5\,(P_i + P_t)\, e_p\, C_{ct}}{E_c\, I}$$

$$= \frac{P_i\, e_p}{E_c\, I}\left[1 + 0.5\, C_{ct}\left(1 + \frac{1}{\eta}\right)\right] \qquad (8.2.10)$$

where P_i = prestressing force at initial stage or at stress transfer

P_t = effective prestressing force after time t

C_{ct} = effective creep coefficient after time t

$\quad = \dfrac{\text{creep strain after time } t}{\text{elastic strain}}$

$\eta_t = \dfrac{P_t}{P_i}$

$\quad$ = reduction factor for effective prestress after time t

After very long time, i.e. for $t = \infty$, the eventual long time curvature may be expressed as

$$\Phi_{pt} = \frac{Pe}{E_c I}\left[1 + 0.5 C_c\left(1 + \frac{1}{\eta}\right)\right] \qquad (8.2.11)$$

Where P = effective prestressing force after all losses have occurred

C_c = eventual long time creep coefficient

$\eta = \dfrac{P}{P_i}$

(b) Service loads

As concrete creeps under the action of only sustained loads, only that portion of the service loads which is of permanent or semi-permanent nature should be considered in the computation of long time deformations. The loads of purely temporary nature such as axle loads on bridges may be ignored in creep computations. On the other hand, the loads of semi-permanent nature such as occupancy loads of library stacks and warehouses should be carefully taken into account. The strains in concrete in the compression zone of a prestressed concrete beam increase due to the combined effect of creep and shrinkage by a factor $(1 + C_{cs})$ in which the coefficient C_{cs} may be defined as

$$C_{cs} = \frac{\text{strain due to combined effect of creep and shrinkage after time } t}{\text{elastic strain}}$$

(8.2.12)

As the curvature Φ_{wt} due to the service loads increases in the same proportion as the strains, it may be expressed as

$$\Phi_{wt} = - \frac{M_w (1 + C_{cs})}{E_c I}$$

(8.2.13)

where C_{cs} = coefficient for the combined effect of creep and shrinkage after time t, Eq. (8.2.12)

M_w = bending moment at the section caused by sustained or permanent portion of service loads

Equation (8.2.13) may also be rewritten as

$$\Phi_{wt} = - \frac{M_w}{E_c' I}$$

(8.2.14)

where $E_c' = \dfrac{E_c}{1 + C_{cs}}$

= reduced modulus of elasticity of concrete.

Equation (8.2.14) is the basis for the approach commonly known as the reduced modulus method for the computation of long-time deformations.

(c) Combined loads

Combining the effects of prestressing force and service loads, the net curvature Φ_t after time t may be expressed as

$$\Phi_t = \Phi_{pt} + \Phi_{wt}$$

(8.2.15)

Substituting from Eq. (8.2.10) and (8.2.14),

$$\Phi_t = \frac{P_t e_p}{E_c I} \left[1 + 0.5 C_{ct} \left(1 + \frac{1}{\eta_t} \right) \right] - \frac{M_w}{E_c' I}$$

(8.2.16)

After computing the curvatures at a number of sections, the slope and deflection at any particular section may be determined by successive integration of curvature utilizing the appropriate boundary conditions for the determination of constants of integration. As direct integration is generally difficult, numerical integration may be adopted.

EXAMPLE 8.2.1

A pretensioned beam of uniform rectangular section 0.15×0.30 m is subjected to an initial prestressing force of 337.5 kN at a constant eccentricity of 0.05 m. Determine the short-time initial camber if only the self weight of the beam is effective at stress transfer. The beam has an effective span of 6 m. Also determine the net central deflection immediately on the application of a concentrated load of 22.5 kN at midspan. Take $E_c = 36050$ MPa. The prestressing force drops to 270 kN at the stage of load application.

Also determine the long-time deflection at midspan of the beam after 3 months and infinite time. The losses of prestress after 3 months and infinite time are 20% and 25% respectively and the corresponding values of C_{cs} are 1.1 and 1.8 respectively. Sixty percent of the transverse load is permanent.

Solution

(a) *Short-time deflections*

$$I = \frac{1}{12} \times 0.15 \times 0.30^3 = 0.0003375 \text{ m}^4$$

The bending moment due to prestress at initial stage and at the stage of loading,

$$M_{pi} = -0.3375 \times 0.05 = -0.016875 \text{ MN.m}$$
$$M_p = -0.270 \times 0.05 = -0.0135 \text{ MN.m}$$

Taking the unit weight of concrete as 24 kN/m^3, the self weight of the beam,

$$w_b = 0.024 \times 0.15 \times 0.3 \times 1 = 0.00108 \text{ MN/m}$$

The free body of the concrete beam at initial stage is shown in Fig. 8.2.3 (a). Hence using Table 8.2.1, the central deflection at initial stage,

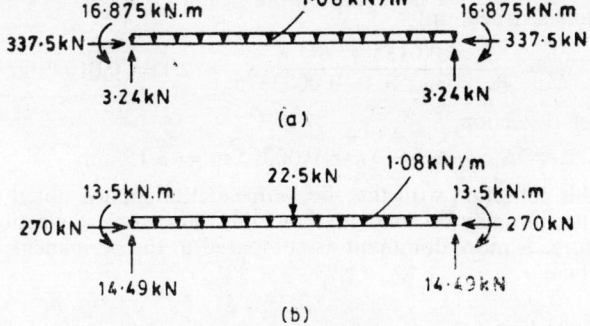

Fig. 8.2.3

$$\Delta_i = \frac{-0.016875 \times 6^2}{8 \times 36050 \times 0.0003375} + \frac{5}{384} \times \frac{0.00108 \times 6^4}{36050 \times 0.0003375}$$

$$= -0.00474 \text{ m} = -4.74 \text{ mm}$$

The minus sign shows that the initial camber (upward deflection) is 4.74 mm. The free body of the concrete beam at the stage of loading is shown in Fig. 8.2.3 (b). Hence, using Table 8.2.1, the central deflection at the stage of loading,

$$\Delta = \frac{-0.0135 \times 6^2}{8 \times 36050 \times 0.0003375} + \frac{5}{384} \times \frac{0.00108 \times 6^4}{36050 \times 0.0003375}$$

$$+ \frac{0.02250 \times 6^3}{48 \times 36050 \times 0.0003375}$$

$$= 0.00483 \text{ m} \quad = 4.83 \text{ mm}$$

(b) *Long-time deflections*

Only permanent loads have to be considered for long-time deflections. For an uncracked beam, the long-time deflections may be computed by multiplying the short-time deflections by the factor $(1 + C_{cs})$.

(i) *After 3 months*

$$1 + C_{cs} = 1 + 1.1 = 2.1$$

The permanent forces are the prestressing force, self weight of the beam and 60 percent of the superimposed load. Hence the upward deflection due to prestress,

$$\Delta_1 = -\frac{0.3375 \times 0.8 \times 0.05 \times 6^2}{8 \times 36050 \times 0.0003375} \times 2.1 = -0.01049 \text{ m}$$

Deflection due to self weight of the beam,

$$\Delta_2 = \frac{5}{384} \times \frac{0.00108 \times 6^4}{36050 \times 0.0003375} \times 2.1 = 0.00315 \text{ m}$$

The permanent part of superimposed load is $0.6 \times 0.0225 = 0.0135$ MN which produces a deflection,

$$\Delta_3 = \frac{0.0135 \times 6^3}{48 \times 36050 \times 0.0003375} \times 2.1 = 0.01049 \text{ m}$$

Hence the net deflection,

$$\Delta = \Delta_1 + \Delta_2 + \Delta_3 = 0.00315 \text{ m} = 3.15 \text{ mm}$$

Comparing this deflection with the short-time deflection, it is noted that the long-time deflection may be smaller than the short-time deflection if the prestressing force is more dominant as compared to the permanent part of the external load.

(ii) *After infinite time*

$$1 + C_{cs} = 1 + 1.8 = 2.8$$

Deflection due to prestress,

$$\Delta_1 = -\frac{(0.3375 \times 0.75 \times 0.05) \times 6^2}{8 \times 36050 \times 0.0003375} \times 2.8 = -0.01311 \text{ m}$$

Deflection due to self weight of the beam,

$$\Delta_2 = \frac{5}{384} \times \frac{0.00108 \times 6^4}{36050 \times 0.0003375} \times 2.8 = 0.00419 \text{ m}$$

Deflection due to permanent portion of superimposed load,

$$\Delta_3 = \frac{0.0135 \times 6^3}{48 \times 36050 \times 0.0003375} \times 2.8 = 0.01398 \text{ m}$$

Hence, the net deflection

$$\Delta = \Delta_1 + \Delta_2 + \Delta_3 = 0.00506 \text{ m} = 5.06 \text{ mm}$$

EXAMPLE 8.2.2

A pretensioned beam has a segmental cable line as shown in Fig. 8.2.4 (a). The beam has a uniform rectangular cross-section 0.2×0.6 m and an effective span of 15 m. It is subjected to an initial prestressing force of 0.84 MN at stress transfer. Determine the short-time initial camber if the beam carries only its own weight at stress transfer. Also determine the net immediate deflection at midspan when two equal loads each equal to 30 kN are applied at quarter points D and E. The prestressing force decreases to 0.672 MN at the stage of load application. Take $E_c = 36050$ MPa.

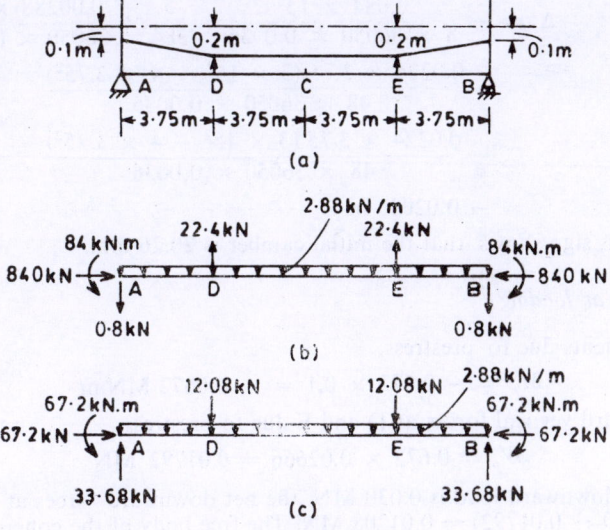

(a)

(b)

(c)

Fig. 8.2.4

Also determine the eventual long-time deflection at midspan, assuming that after infinite time the loss of prestress is 25% and $C_{cs} = 2.1$. Eighty percent of the transverse load is of permanent nature.

Solution

$$I = \frac{1}{12} \times 0.2 \times 0.6^3 = 0.0036 \text{ m}^4$$

Taking unit weight of concrete as 24 kN/m^3, the self weight of the beam,

$$w_b = 0.024 \times 0.2 \times 0.6 \times 1 = 0.00288 \text{ MN/m}$$

(a) *Short-time deflection*

(i) *Initial stage*

End moments due to prestress,

$$M_{pi} = -0.84 \times 0.1 = -0.084 \text{ MN . m}$$

Change of angle of cable line at D and E,

$$\delta\theta = \frac{0.1}{3.75} = 0.02666 \text{ rad}$$

The upward vertical forces at D and E due to prestress,

$$W_p = 0.84 \times 0.0266 = 0.0224 \text{ MN}$$

The free body of the concrete beam at initial stage is shown in Fig. 8.2.4 (b). Hence using Table 8.2.1, the midspan deflection,

$$\Delta_i = -\frac{0.084 \times 15^2}{8 \times 36050 \times 0.0036} + \frac{5}{384} \times \frac{0.00288 \times 15^4}{36050 \times 0.0036}$$
$$- \frac{0.0224 \times 3.75 (3 \times 15^2 - 4 \times 3.75^2)}{48 \times 36050 \times 0.0036}$$
$$- \frac{0.0224 \times 3.75 (3 \times 15^2 - 4 \times 3.75^2)}{48 \times 36050 \times 0.0036}$$
$$= -0.02026 \text{ m}$$

The minus sign shows that the initial camber is 20.26 mm.

(ii) *Stage at loading*

End moments due to prestress,

$$M_p = -0.672 \times 0.1 = -0.0672 \text{ MN.m}$$

The upward vertical forces at D and E due to prestress,

$$W_p = 0.672 \times 0.02666 = 0.01792 \text{ MN}$$

As each downward load is 0.030 MN, the net downward forces at D and E are (0.030 − 0.01792) = 0.01208 MN. The free body of the concrete beam at the stage of loading is shown in Fig. 8.2.4 (c).

Hence, using Table 8.2.1, the midspan deflection,

$$\Delta = -\frac{0.0672 \times 15^2}{8 \times 36050 \times 0.0036} + \frac{5}{384} \times \frac{0.00288 \times 15^4}{36050 \times 0.0036}$$

$$+\frac{0.01208 \times 3.75 \, (3 \times 15^2 - 4 \times 3.75^2)}{48 \times 36050 \times 0.0036}$$

$$+\frac{0.01208 \times 3.75 \, (3 \times 15^2 - 4 \times 3.75^2)}{48 \times 36050 \times 0.0036}$$

$$= 0.00906 \text{ m}$$

(b) Long-time deflection

The permanent part of each of the two concentrated loads is 0.8×0.030 $=0.024$ MN. The long time deflection is $(1 + C_{cs}) = 3.1$ times greater than the short time deflection due to the effects of creep and shrinkage of concrete. Hence, using Table 8.2.1, the midspan deflection,

$$\Delta = 3.1 \left[-\frac{0.0672 \times 15^2}{8 \times 36050 \times 0.0036} + \frac{5}{384} \times \frac{0.00288 \times 15^4}{36050 \times 0.0036} \right.$$

$$+\frac{(0.024 - 0.01792) \times 3.75 \, (3 \times 15^2 - 4 \times 3.75^2)}{48 \times 36050 \times 0.0036}$$

$$\left. +\frac{(0.024 - 0.01792) \times 3.75 \, (3 \times 15^2 - 4 \times 3.75^2)}{48 \times 36050 \times 0.0036} \right]$$

$$= 0.01423 \text{ m}$$

$$= 14.23 \text{ mm.}$$

EXAMPLE 8.2.3

A post-tensioned bonded beam having a uniform section shown in Fig. 8.2.5 carries an effective prestressing force of 2.16 MN. The cable line is para-bolic with zero eccentricity at ends and 300 mm at mid span. The effective span is 24 m. Calculate the short-time end slope and the central deflection when the beam carries a uni-form transverse load which causes decompre-sion at mid-span section. Take $E_c = 33720$ MPa.

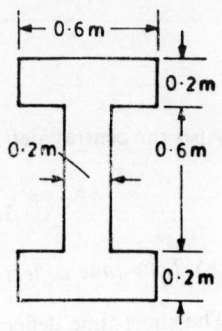

Fig. 8.2.5

Also determine the long-time deflection at midspan, given that $C_{cs} = 2.0$.

Solution

$$A = 0.36 \text{ m}^2$$

$$I = \frac{1}{12} \left(0.6 \times 1^3 - 0.4 \times 0.6^3 \right)$$

$$= 0.0428 \text{ m}^4$$

(a) Short-time deflection

Intensity of stress at bottom fibre at midspan section due to prestress,

$$f_B = \frac{(-\,2.16)}{0.36} + \frac{(-\,2.16 \times 0.3) \times 0.5}{0.0428} = -\,13.57 \text{ MPa}$$

Hence the intensity of uniform load w inclusive of self weight which causes decompression at extreme fibre at the midspan section should cause a tensile stress equal to 13.57 MPa.

$$\frac{\dfrac{w \times 24^2}{8} \times 0.5}{0.0428} = 13.57$$

or

$$w = 0.01613 \text{ MN/m}$$

The intensity of upward force caused by prestress,

$$w_p = \frac{8PS}{l^2} = \frac{8 \times 2.16 \times 0.3}{24^2} = 0.009 \text{ MN/m}$$

Hence the net intensity of downward load,

$$w - w_p = 0.01613 - 0.009 = 0.00713 \text{ MN/m}$$

The free body of the concrete beam is shown in Fig. 8.2.6. Using Table 8.2.1, the end slopes,

$$\alpha_A = \alpha_B = \frac{0.00713 \times 24^3}{24 \times 33720 \times 0.0428} = 0.00285 \text{ rad}$$

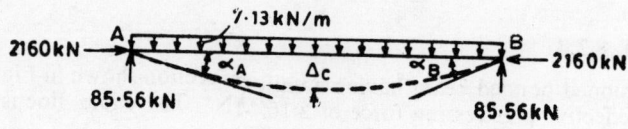

Fig.8.2.6

Fig. 8.2.6

Also the central deflection,

$$\Delta_C = \frac{5}{384} \times \frac{0.00713 \times 24^4}{33720 \times 0.0428} = 0.01713 \text{ m}$$

(b) Long-time deflection

The short-time deflection is increased by the factor $(1 + C_{cs}) = 1 + 2 = 3$ due to creep and shrinkage of concrete. Hence long-time deflection at midspan,

$$\Delta_C = 0.01713 \times 3 = 0.05139 \text{ m}$$

EXAMPLE 8.2.4

A post-tensioned unbonded beam is prestressed by a single prestressing tendon comprising 48 wires of 5 mm diameter. The effective stress in each wire is 1020 MPa. The beam has a uniform cross-section shown in Fig. 8.2.7 (a). It has an effective span of 14.4 m. The cable has a parabolic profile as shown in the figure. Determine short-time end slope and central deflection when the beam carries a load equal to 90 per cent of the load causing incipient cracking at midspan section.

Take $E_c = 33720$ MPa and $f_{,r} = 4.14$ MPa

Also determine the long-time deflection at midspan, given that $C_{cs} = 1.9$.

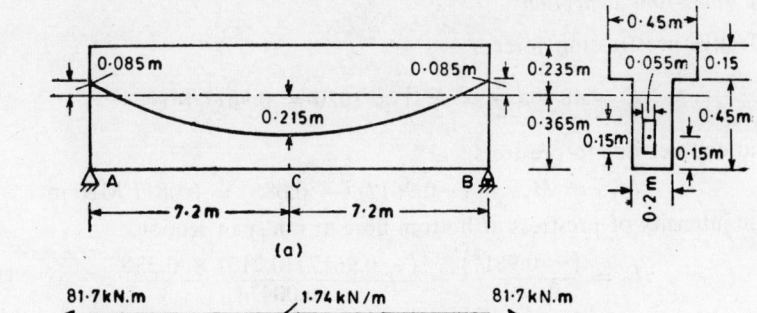

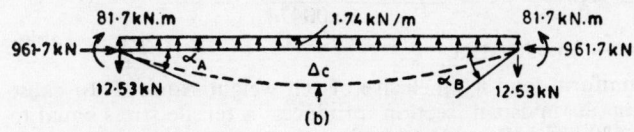

Fig. 8.2.7

Solution

As the beam is unbonded, the cross-sectional area A and moment of inertia I may be based on the net area of concrete.

$$A = 0.45 \times 0.15 + 0.2 \times 0.45 - 0.055 \times 0.15$$
$$= 0.14925 \text{ m}^2$$

Distance of centroid from top face,

$$\bar{x} = \frac{1}{0.14925}(0.45 \times 0.15 \times 0.075 + 0.2 \times 0.45$$
$$\times 0.375 - 0.055 \times 0.15 \times 0.45)$$
$$= 0.235 \text{ m}$$

Hence the eccentricity of cable line at ends,

$$e_{pA} = e_{pB} = -(0.235 - 0.150) = -0.085 \text{ m}$$

and the eccentricity of cable line at midspan,

$$e_{pC} = 0.6 - 0.235 - 0.15 = 0.215 \text{ m}$$

Moment of inertia I about the centroidal axis,

$$I = \frac{1}{12} \times 0.45 \times 0.15^3 + 0.45 \times 0.15 \, (0.235 - 0.075)^2$$

$$+ \frac{1}{12} \times 0.20 \times 0.45^3 + 0.20 \times 0.45 \, (0.365 - 0.225)^2$$

$$- \frac{1}{12} \times 0.055 \times 0.15^3 - 0.055 \times 0.15 \times 0.215^2$$

$$= 0.00474 \text{ m}^4$$

(a) *Short-time deflection*

Effective prestressing force,

$$P = 48 \times \left(\frac{\pi}{4} \times 5^2 \right) \times 1020 = 0.9617 \text{ MN}$$

End couples due to prestress,

$$M_{pA} = M_{pB} = (-0.9617)(-0.085) = 0.0817 \text{ MN·m}$$

The intensity of prestress at bottom fibre at midspan section,

$$f_B = \frac{(-0.9617)}{0.14925} + \frac{(-0.9617)(0.215) \times 0.235}{0.00474}$$

$$= -16.695 \text{ MPa}$$

Hence the total uniform load w_{cr} inclusive of self weight required to cause incipient cracking at midspan section produces a tensile stress equal to $(16.695 + f_{cr}) = 20.835$ MPa at bottom fibre.

$$\frac{w_{cr} \times 14.4^2 \times 0.365}{8 \times 0.00474} = 20.835$$

or
$$w_{cr} = 0.01044 \text{ MN/m}$$

Hence the applied transverse load,

$$w = 0.9 \, w_{cr} = 0.00939 \text{ MN/m}$$

The intensity of upward uniform force due to prestress,

$$w_p = \frac{8PS}{l^2} = \frac{8 \times 0.9617 \,(0.085 + 0.215)}{14.4^2}$$

$$= 0.01113 \text{ MN/m}$$

Hence the net intensity of transverse load,

$$w - w_p = 0.00939 - 0.01113$$

$$= -0.00174 \text{ MN/m}$$

The free body of the concrete beam is shown in Fig. 8.2.7 (b). Using Table 8.2.1, the end slopes,

$$\alpha_A = \alpha_B = \frac{0.0817 \times 14.4}{2 \times 33720 \times 0.00474} - \frac{0.00174 \times 14.4^3}{24 \times 33720 \times 0.00474}$$

$$= 0.03545 \text{ rad.}$$

Also the central defleetion,

$$\Delta_C = \frac{0.0817 \times 14.4^2}{8 \times 33720 \times 0.00474} + \frac{5}{384} \times \frac{0.00174 \times 14.4^4}{33720 \times 0.00474}$$

$$= 0.00715 \text{ m}$$

(b) *Long-time deflection*

$$1 + C_{cs} = 1 + 1.9 = 2.9$$

Hence long-time deflection at midspan,

$$\Delta_C = 0.00715 \times 2.9 = 0.02074 \text{ m}$$

8.3 POST-CRACKING RANGE

Beams with full prestress (Type 1) and limited prestress (Type 2) crack only under overload conditions whereas beams with partial prestress (Type 3) develop flexural cracks even under the working load. As the bending moment is generally variable along the length of the beam, the cracking occurs only in certain regions adjacent to the section of maximum bending moment. The extent of cracking increases as the cracked regions spread out under increasing load. Due to the variation of bending moment from section to section, the heighet of flexural cracks and the depth of the neutral axis also vary along the length of the beam. Consequently, the effective cross-sectional area and the moment of inertia change continuously not only from section to section but also with each load increment. The relatively higher stress in the flexural compression zone in post-cracking range brings in a greater degree of non-linearity of the stress-strain relationship for concrete and a progressively decreasing value of E_c. Further complications are introduced if long-time deformations have to be determined due to the time dependent phenomena such as creep and shrinkage of concrete, relaxation of prestressing steel and variable value of creep coefficient under relatively high compressive stresses.

The principle of superposition used in Sec. 8.2.1 to determine the net deformations under the combined action of prestressing force and service loads is no longer valid in the post-cracking range. Hence both types of forces have to be considered simultaneously in the computation of short-time and long-time deformations of a prestressed concrete cracked member.

8.3.1 Short-time Deformations

The short-time or immediate deformations are practically elastic as most of the deformations are recoverable. The cracks close on the removal of load but they reappear as soon as the load is increased to the level of decompression. There are two main approaches for the determination of short-time deformations of prestressed concrete beams in the post-cracking range.

8.3.1.1 Strain Compatibility Method

In this approach, the compatibility of strains in concrete and prestressing steel based on Navier's hypothesis is utilized to determine the curvatures at a number of sections along the length of the beam. The slope and deflection at any section are then determined by integration of the curvatures.

The height of flexural crack and the position of neutral axis at any section of a transversely loaded beam depends upon the geometry of cross-section, material properties and external bending moment at the section under consideration. In the analysis of flexural curvature at the section, the maximum compressive strain in concrete ϵ_{max} at the extreme top fibre and the depth of the neutral axis x may be treated as the two unknown quantities which have to be determined from the two conditions regarding the equality of (i) tensile and compressive forces and (ii) internal and external moments. Consider, for instance, the prestressed concrete section shown in Fig. 8.3.1 (a). Assuming a linear variation of strain as shown in Fig. 8.3.1 (b), the strain in concrete ϵ at depth x' from top,

$$\epsilon = \frac{x - x'}{x} \epsilon_{max} \tag{8.3.1}$$

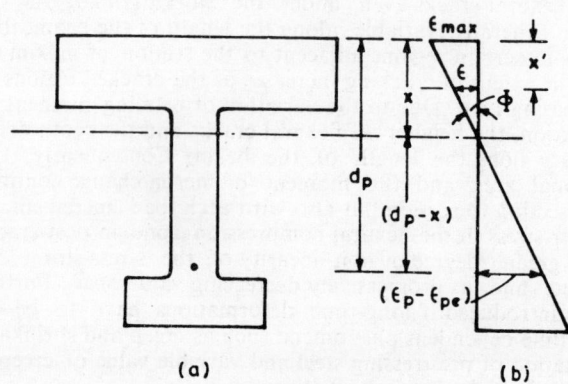

(a)   (b)

Fig. 8.3.1

The corresponding stress in concrete f at depth x' from top is obtained from the stress-strain curve for concrete. The total compression C is obtained by integration.

$$C = \int_o^x f \, dA \tag{8.3.1}$$

in which dA is an element of area in the compression zone at depth x' from top. The tensile strain in prestressing steel is computed from the compatibility of strains.

$$\frac{\epsilon_{max}}{x} = \frac{\epsilon_p - \epsilon_{pe}}{d_p - x} \tag{8.3.3}$$

where ϵ_p = strain in prestressing steel

ϵ_{pe} = effective strain in prestressing steel before the commencement of loading

The net strain ($\epsilon_p - \epsilon_{pe}$) is caused by the applied bending moment M. The stress in prestressing steel f_p is related to the strain ϵ_p through the stress-strain curve for prestressing steel. In this manner, the first equation of equilibrium is obtained.

Total compression, C = Total tension, T

or

$$\int_o^x f \, dA = A_p \, f_p \tag{8.3.4}$$

The distance of the centroid of total compression C from the top,

$$\bar{x}_c = \frac{\displaystyle\int_o^x fx \, dA}{\displaystyle\int_o^x f \, dA} \tag{8.3.5}$$

and the internal lever arm

$$jd_p = d_p - \bar{x}_c \tag{8.3.6}$$

The equality of internal and external moments gives the following equilibrium equation,

$$M = A_p f_p \, (jd_p) \tag{8.3.7}$$

Equations (8.3.4) and (8.3.7) may be solved to determine the two unknowns, viz. ϵ_{max} and x. As direct solution is difficult, an iterative procedure may be adopted. Knowing ϵ_{max} and x, the curvature at the section Φ may be determined from

$$\Phi = \frac{\epsilon_{max}}{x} \tag{8.3.8}$$

Proceeding in a similar manner, the curvatures at several sections of the beam may be computed. As curvature is the rate of change of slope or the change of slope per unit length, the slopes at several sections may be computed by integration of curvature.

$$\beta = \int \Phi \, dx \tag{8.3.9}$$

Similarly, the integration of slopes leads to deflection,

$$\Delta = \int \beta \, dx = \iint \Phi \, dx \, dx \tag{8.3.10}$$

As direct integrations of curvature and slope are difficult, numerical integrations in accordance with Newmark's procedure may be adopted. The two constants of integration may be determined from the conditions of zero deflection at ends. In case of symmetry, the condition of zero slope at mid-span may be utilized. The procedure may be used to compute the short-time slopes and deflections at any stage of loading in the post-cracking range, including the stage of incipient collapse.

8.3.1.2 Equivalent Moment of Inertia Method

The strain compatibility is cumbersome and is, therefore, justifiable in those cases in which deformations are important and are likely to control the design. In most common cases, however, only an approximate estimate of deformations is sufficient. In these cases, the approximate method based on the equivalent or effective moment of inertia I_e may be used. In this method, the deformations caused by the prestressing force and the service loads may be computed by using the appropriate equations of Sec. 8.2.1 in which the moment of inertia I is replaced by the equivalent moment of inertia I_e for a cracked beam. A well known expression for I_e is given by Eq. (8.3.11) in accordance with the American Code ACI: 318-1971. The expression is formulated in such a manner that I_e becomes equal to the moment of inertia of the gross uncracked section I_g when the maximum moment M_{max} becomes equal to the cracking moment M_{cr}. Hence I_g is the upper limit of I_e. This is appropriate because for $M_{max} = M_{cr}$, the entire beam is uncracked. On the other hand I_e approaches I_{cr} as M_{max} becomes large compared to M_{cr}. This too is appropriate because almost the entire beam is cracked as M_{max} becomes large compared to M_{cr}. Tests have shown that the short-time deformations computed using Eq. (8.3.11) for I_e give reasonably accurate results.

$$I_e = \left(\frac{M_{cr}}{M_{max}}\right)^3 I_g + \left[1 - \left(\frac{M_{cr}}{M_{max}}\right)^3\right] I_{cr} \qquad (8.3.11)$$

where $M_{cr} = $ cracking moment

$$= \frac{(f_{cr} + f_B) I_g}{y_B}$$

$$= \frac{(0.625 \sqrt{f'} + f_B) I_g}{y_B}$$

$f_B = $ compressive stress at bottom fibre due to prestress

$f_{cr} = $ modulus of rupture of concrete

$y_B = $ distance of extreme bottom fibre from the centroidal axis

The moment of inertia of a cracked rectangular section or flanged section in which neutral axis lies in the flange may be expressed as

$$I_{cr} = \frac{1}{3} bx^3 + mA_p (d_p - x)^2 + mA_s (d_s - x)^2 \qquad (8.3.12)$$

where $b = $ width of the rectangular section or width of the compression flange

$x = $ depth of neutral axis

$m = $ modular ratio

$A_p, A_s = $ areas of prestressed and non-prestressed tension steel

$d_p, d_s = $ effective depths of A_p and A_s

In the case of a flanged section with neutral axis located in the web,

$$I_{cr} = \frac{1}{12} b_f d_f^3 + b_f d_f \left(x - \frac{d_f}{2}\right)^2 + \frac{1}{3} b_w (x - d_f)^3$$
$$+ mA_p (d_p - x)^2 + mA_s (d_s - x)^2 \qquad (8.3.13)$$

where b_f = width of compression flange

$\qquad d_f$ = thickness of compression flange

$\qquad b_w$ = width (thickness) of web

The position of the neutral axis for any given value of the applied moment may be determined from the condition of compatibility of strains. In the post-cracking range, as the applied bending moment increases, the stress in prestressing steel increases and the depth of the neutral axis decreases progressively in order to cope with increasing bending moment. A decrease in the depth of neutral axis x results in a larger value of the internal lever arm jd_p. An iterative procedure based on strain compatibility described by the following steps may be used to determine the depth of neutral axis in the post-cracking range.

(1) Start by assuming a reasonable value of the internal lever arm jd_p. The internal lever arm factor j generally lies in the range of 0.80 to 0.95 in the post-cracking range.

(2) Determine the total tension in steel T from

$$T = A_p f_p = \frac{M}{jd_p} \qquad (8.3.14)$$

and the stress in steel f_p from

$$f_p = \frac{T}{A_p} \qquad (8.3.15)$$

in which A_p is the area of prestressing steel.

(3) Determine the strain ϵ_p from the stress-strain curve for prestressing steel and determine the strain ϵ_{pM} caused by applied moment M from

$$\epsilon_{pM} = \epsilon_p - \epsilon_{pe} \qquad (8.3.16)$$

in which ϵ_{pe} is the effective strain in prestressing steel at the commencement of loading.

(4) Assume a reasonable value of depth of neutral axis x. Hence determine the maximum compressive strain ϵ_{max} at top fibre from strain compatibility relationship,

$$\frac{\epsilon_{maa}}{x} = \frac{\epsilon_{pm}}{d_p - x} \qquad (8.3.17)$$

Next determine the total compression C in the compression zone using the stress-strain relationship for concrete. For the equilibrium of internal forces, C must be equal to T computed in step 2. Otherwise revise the value of depth of neutral axis x until $C = T$.

(5) Determine the position of centroid of compressive forces and hence determine the value of internal lever-arm jd_p.

(6) Compare the value of jd_p assumed in step 1. If the difference is appreciable, iterate until j and x are determined to the desired accuracy.

The iterative procedure is illustrated by Example 8.3.2.

The Indian Code IS: 456-1978 recommends the following equation for the effective or equivalent moment of inertia for a cracked beam,

$$I_e = \frac{I_{cr}}{1.2 \, \dfrac{jM_{cr}}{M_{max}} \left(1 - \dfrac{x}{d_p}\right) \dfrac{b_w}{b_f}} \tag{8.3.18}$$

in which the depth of neutral axis x may be computed as explained above and the internal lever arm factor j may be determined from Eq. (8.3.14).

8.3.2 Long-Time Deformations

The long-time deformations of a prestressed concrete cracked beam are only partly recoverable. The proportion of recoverable deformations depends upon the degree of overloading and the extent of permanent deformations in the constituent materials. The inaccuracies in the computations for long-time deformations of a cracked beam arise mainly due to (i) decreasing value of E_c, (ii) decreasing value of effective moment of inertia I_e, (iii) decreasing prestressing force, (iv) increasing concrete strength and (v) increasing creep coefficient. Consequently an accurate prediction of long-time deformations is not easy. Several computer programmes have been written which take into account the non-linearities of materials, time dependent phenomena, the extent of cracking and the consequent changes in E_c and I_e. Such sophisticated computations may be warranted in special problems. However, an approximate estimate of long-time curvature Φ_t of a cracked section may be obtained from Eq. (8.2.16) by replacing I by equivalent moment of inertia I_e.

$$\Phi_t = \frac{P_t \, e_p}{E_c \, I_e} \left[1 + 0.5 \, c_{ct} \left(1 + \frac{1}{\eta_t}\right)\right] - \frac{M_w}{E_c \, I} \tag{8.3.19}$$

in which the equivalent moment of inertia I_e may be computed from Eq. (8.3.11). The curvature may be determined at a number of equispaced sections. The numerical integration of curvature may then be carried out to detrermine the slope and deflection at any point.

EXAMPLE 8.3.1

Using the strain compatibility method, determine the midspan bending moment and the curvature in a beam having a cross-section shown in Fig. 8.3.2 (a) for a compressive strain of 0.003 at extreme top fibre. Use the

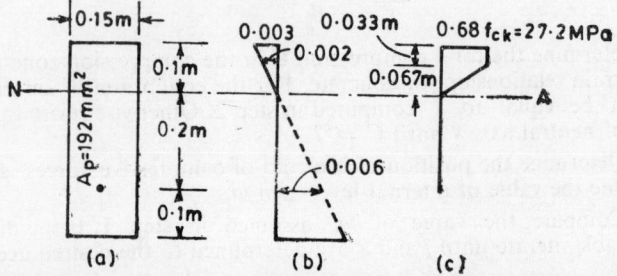

Fig. 8.3.2

idealised stress-strain relationships for concrete and steel shown in Fig. 8.3.3
(a) and (b). The initial stress in prestressing steel before the application of
moment is 1000 MPa. The grade of concrete is M-40.

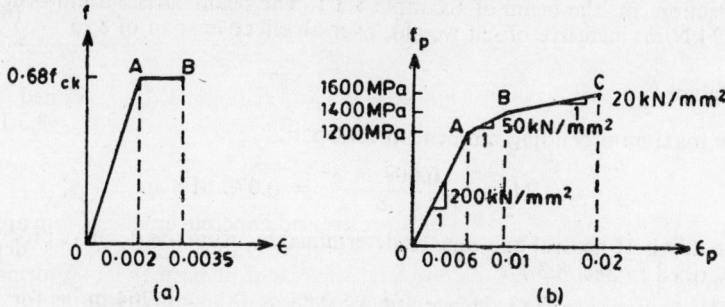

Fig. 8.3.3

Solution

The problem may be solved by using an iterative procedure which starts by
assuming a reasonable value for the depth of the neutral axis x. Taking
$x = 0.1$ m and referring to the strain and stress diagrams shown in Fig.
8.3.2 (b) and (c), the total compression in concrete,

$$C = 0.15 \times \frac{1}{30} \times (0.68 \times 40) + 0.15 \times \frac{1}{15}(0.5 \times 0.68 \times 40)$$

$$= 0.272 \text{ MN}$$

As total compression is equal to total tension, the tensile stress in prestressing steel,

$$f_p = \frac{0.272}{192 \times 10^{-6}} = 1416.6 \text{ MPa}$$

Referring to Fig. 8.3.3 (b), the corresponding strain in steel,

$$\epsilon_p = 0.01083$$

The initial strain in steel is $1000/(2 \times 10^5) = 0.005$. Hence the net strain
caused by the applied moment is $(0.01083 - 0.005) = 0.00583$. Referring to
Fig. 8.3.2 (b),

$$\frac{0.003}{x} = \frac{0.00583}{0.3 - x}$$

Hence, $x = 0.102$ m. As the new value of x is very close to the value
assumed earlier, further iteration is not required and x may be taken equal
to 0.10 m. Taking moments of the compressive forces about the centroid of
the steel, the applied bending moment,

$$M = 0.15 \times \frac{1}{30}(0.68 \times 40)\left(0.30 - \frac{1}{60}\right)$$

$$+ 0.15 \times \frac{1}{15}(0.5 \times 0.68 \times 40)\left(0.3 - \frac{1}{30} - \frac{1}{3} \times \frac{1}{15}\right)$$

$$= 0.07178 \text{ MN·m}$$

The curvature due to the applied bending moment, $\Phi = \dfrac{0.003}{0.1} = 0.03$.

EXAMPLE 8.3.2

Using the equivalent moment of inertia method, determine the midspan deflection in the beam of Example 8.3.1. The beam carries a uniform load of 9 kN/m, inclusive of self weight, over an effective span of 8 m.

Solution

The maximum bending moment at midspan,

$$M_{max} = \frac{0.009 \times 8^2}{8} = 0.072 \text{ MN} \cdot \text{m}$$

The depth of neutral axis may be determined by using the iterative procedure described in Sec. 8.3.1.2

(1) Assume $j = 0.88$. Hence, $jd_p = 0.88 \times 0.3 = 0.264$ m

(2) Total tension in steel,

$$T = \frac{M_{max}}{jd_p} = \frac{0.072}{0.264} = 0.2727 \text{ MN}$$

and the stress in prestressing steel,

$$f_p = \frac{T}{A_p} = \frac{0.2727}{192 \times 10^{-6}} = 1420 \text{ MPa}$$

(3) From the stress-strain curve of Fig. 8.3.3 (b), the strain in prestressing steel, $\epsilon_p = 0.011$ and the net strain caused by bending moment,

$$\epsilon_{pM} = \epsilon_p - \epsilon_{pe} = 0.011 - \frac{1000}{2 \times 10^5} = 0.006$$

(4) Assume the depth of neutral axis, $x = 0.1$ m. Hence from the strain compatibility relationship,

$$\frac{\epsilon_{max}}{0.1} = \frac{0.006}{0.3 - 0.1}$$

or $\epsilon_{max} = 0.003$

Using the stress-strain curve for concrete as shown in Fig. 8.3.3 (a), the total compression,

$$C = 0.15 \times \frac{1}{30}(0.68 \times 40) + 0.15 \times \frac{1}{15}\left(\frac{1}{2} \times 0.68 \times 40\right)$$

$$= 0.272 \text{ MN}$$

As this value is very close to total tension T, the assumed value of depth of neutral axis is correct and further iteration is not required.

(5) The distance of centroid of compressive forces from top,

$$\bar{x}_c = \frac{0.15 \times \frac{1}{30}(0.68 \times 40) \times \frac{1}{60} + 0.15 \times \frac{1}{15}\left(\frac{1}{2} \times 0.68 \times 40\right) \times \left(\frac{1}{30} + \frac{1}{3} \times \frac{1}{15}\right)}{0.2727}$$

$$= 0.03602 \text{ m}$$

$$jd_p = d_p - \bar{x}_c = 0.30 - 0.03602 = 0.26398 \text{ m}$$

Hence
$$j = \frac{0.26398}{0.3} = 0.8799$$

(6) Comparing the value of j computed in step (5) with the value assumed in step (1), it is noted that the two values are very close making further iteration unnecessary. Hence the values, $j = 0.88$ and $x = 0.1$ m may be adopted for computation of I_{cr} and I_e.

Using Eq. (8.3.12) and taking $m = \dfrac{2 \times 10^5}{36050} = 5.55$,

$$I_{cr} = \frac{1}{3} \times 0.15 \times 0.1^3 + 5.55 \, (192 \times 10^{-6}) \, (0.3 - 0.1)^2$$

$$= 0.000092624 \text{ m}^4$$

$$I_g = \frac{1}{12} \times 0.15 \times 0.40^3 = 0.0008 \text{ m}^4$$

Also from Example 8.3.1, $M_{cr} = 0.04971$ MN·m

(i) *ACI equation*

Using Eq. (8.3.11),

$$I_e = \left(\frac{0.04971}{0.072} \right)^3 \times 0.0008 + \left[1 - \left(\frac{0.04971}{0.072} \right)^3 \right] \times 0.000092624$$

$$= 0.00032541 \text{ m}^4$$

Hence the midspan deflection,

$$\Delta = \frac{- \, (192 \times 10^{-6} \times 1000) \times 0.1 \times 8^2}{8 \times 36050 \times 0.00032541} + \frac{5}{384}$$

$$\times \frac{0.009 \times 8^4}{36050 \times 0.00032541}$$

$$= 0.027824 \text{ m}$$

(ii) *IS code equation,*

Using Eq. (8.3.18),

$$I_e = \frac{0.000092624}{1.2 - 0.88 \times \dfrac{0.04971}{0.072} \left(1 - \dfrac{0.1}{0.3} \right)} \quad (1)$$

$$= 0.0001165 \text{ m}^4$$

Hence the midspan deflection,

$$\Delta = \frac{- \, (192 \times 10^{-6} \times 1000) \times 0.1 \times 8^2}{8 \times 36050 \times 0.0001165} + \frac{5}{384}$$

$$\times \frac{0.009 \times 8^4}{36050 \times 0.0001165}$$

$$= 0.07772 \text{ m}$$

8.4   MOMENT-CURVATURE RELATIONSHIP

The determination of the moment-curvature relationship is the first step in the computations for flexural deformations, i.e. the slope or rotation and deflection at any section. Curvature is defined as the change of slope or flexural rotation per unit length. Figure 8.4.1 (a) shows a unit length of a flexural member between two sections $A_1 A_2$ and $B_1 B_2$ subjected to bending moment M. The rotation Φ of the section $B_1 B_2$ relative to section $A_1 A_2$, being the change of slope per unit length of the member, is evidently the curvature of the element shown in the figure. As the elongation and shortening of longitudinal fibres of unit length also represent their respective strains, the angle Φ shown in the strain diagram of Fig. 8.4.1 (b) is also the curvature of the beam at the section under consideration. As Navier's hypothesis regarding linear variation of strain is assumed to be

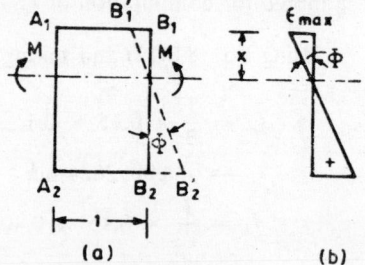

(a)   (b)

Fig. 8.4.1

valid at all stages of loading upto incipient collapse, the curvature at any stage, $\Phi = (\epsilon_{max}/x)$. At the stage of ultimate moment or collapse, $\epsilon_{max} = \epsilon_u$ and $x = \chi_u$. Hence,

$$\Phi_u = \frac{\epsilon_u}{\chi_u} \qquad (8.4.1)$$

in which ϵ_u is the crushing strain of concrete which generally lies between 0.003 and 0.004 and χ_u is the depth of neutral axis at the limit state of collapse.

In the precracking range, in which the entire section is effective, the linear elastic theory may be used. From the well known equation in the theory of simple bending,

$$\Phi = - \frac{M_r}{E_c I} \qquad (8.4.2)$$

The minus sign on the right side of Eq. (8.4.2) arises due to the sign convention adopted in Sec. 1.4. In the unloaded condition, the external bending moment M is zero. Hence resultant moment,

$$M_r = M + M_p = - Pe_p$$

Therefore, from Eq. (8.4.2), the curvature under no load condition,

$$\Phi_o = - \frac{M_p}{E_c I} = \frac{Pe_p}{E_c I} \qquad (8.4.3)$$

Point A in the M-Φ plot of Fig. 8.4.2 represents the no load condition. Let the applied moment M be increased progressively. At the stage when $M = - M_p$, $M_r = 0$ and the curvature is also zero as represented by point B. At the stage of decompression, represented by point C, $M = M_{dc}$ and hence the curvature Φ_{dc} under decompression moment M_{dc},

$$\Phi_{dc} = - \frac{M_{dc} - Pe_p}{E_c I} \qquad (8.4.4)$$

At the limit state of crcking represented by point D, $M = M_{cr}$ and hence the curvature Φ_{cr} under cracking moment M_{cr},

$$\Phi_{cr} = -\frac{M_{cr} - Pe_p}{E_c I} \qquad (8.4.5)$$

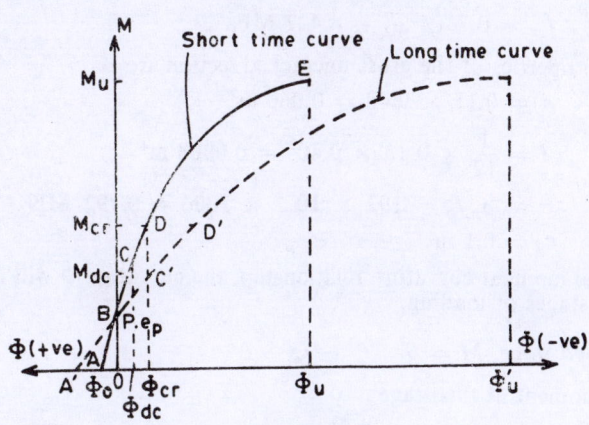

Fig. 8.4.2

The moment curvature relationship in the precracking range is linear and is represented by the straight line ABCD. The point D on the $M - \Phi$ plot marks the end of the precracking range and the beginning of the post-cracking range. The non-linear relationship in the post-cracking range is represented by the curve DE. The curvature Φ for any value of the applied bending moment M in the post-cracking range may be determined from Eq. (8.3.8) in which the maximum strain ϵ_{max} and the depth of neutral axis χ may be determined by using compatibility of strains as explained in Sec. 8.3.1.1. Finally the curvature Φ_u at the limit state of collapse may be determined from Eq. (8.4.1) in which x_u may be computed as per Sec. 5.8.3. The moment-curvature plot of Fig. 8.4.2 represents $M - \Phi$ relationship for a bonded section under short-time loading. The strains and consequently the curvatures are increased due to creep of concrete under sustained loads. In the precracking range, the curvature is increased by the factor $(1 + C_c)$ in which C_c is the creep coefficient. Hence the straight line A' B' C' D' for long time response in precracking range may be obtained by multiplying Φ in short time loading by $(1 + C_c)$. In the post-cracking range, C_c is not constant but increases progressively as ultimate moment is reached. Hence $M-\Phi$ plot corresponds to progressively increasing factro $(1 + C_c)$.

EXAMPLE 8.4.1

Plot the moment-curvature relationship for the rectangular section of Example 8.3.1. Determine the curvatures at decompression. cracking and ultimate moments. Also plot the long-time relationship taking the coefficient due to combined effect of shrinkage and creep as 1.8.

Solution

According to IS : 1343 − 1980, for M − 40 concrete,

$$E_c = 5700 \sqrt{40} = 36050 \text{ MPa}$$

$$f_{cr} = 0.7 \sqrt{40} = 4.427 \text{ MPa}$$

Also the properties of the gross uncracked section are

$$A = 0.15 \times 0.40 = 0.060 \text{ m}^2$$

$$I = \frac{1}{12} \times 0.15 \times 0.40^3 = 0.0008 \text{ m}^4$$

$$P = A_p f_p = 192 \times 10^{-6} \times 1000 = 0.192 \text{ MN}$$

$$e_p = 0.1 \text{ m}$$

To plot the moment-curvature relationship, the curvature Φ will be calculated at six stages of loading.

(i) *Unloaded stage, $M = 0$*

The net moment at this stage,

$$M_r = M_p = - P\, e_p = - 0.192 \times 0.1 = - 0.0192 \text{ MN}\text{m}$$

$$\Phi = \frac{M_r}{E_c I} = \frac{- 0.0192}{36050 \times 0.0008}$$

$$= - 0.0006657 \text{ rad/m}$$

(ii) *Zero net moment, $(M_r = 0)$, $M = - M_p$*

$$\Phi = \frac{M_r}{E_c I} = 0$$

(iii) *Decompression stage, $M = M_{dc}$*

The stress due to prestress at bottom fibre,

$$f_B = \frac{(- 0.192)}{0.06} + \frac{(-0.0192) \times 0.2}{0.0008} = - 8 \text{ MPa}$$

Hence the decompression moment,

$$M_{dc} = \frac{8\, I}{y} = \frac{8 \times 0.0008}{0.2} = 0.032 \text{ MN.m}$$

$$M_r = M + M_p = 0.032 - 0.0192 = 0.0128 \text{ MN.m}$$

$$\Phi = \frac{M_r}{E_e I} = \frac{0.0128}{36050 \times 0.0008} = 0.0004438 \text{ rad./m}$$

(iv) *Cracking stage, $M = M_{cr}$*

$$M_{cr} = (8 + f_{cr}) \frac{I}{y_B} = (8 + 4.427) \times \frac{0.0008}{0.2}$$

$$= 0.04971 \text{ MN.m}$$

$$M_r = M + M_p = 0.04971 - 0.0192 = 0.03051 \text{ MN.m}$$

$$\Phi = \frac{M_r}{E_c I} = \frac{0.03051}{36050 \times 0.0008} = 0.001058 \text{ rad./m}$$

(v) *Post-cracking stage,* $M = 0.07178$ MN.m

From Example 8.3.1, the curvature $\Phi = 0.03$ rad./m when the applied bending moment, $M = 0.07178$ MN.m

(vi) *Incipient collapse,* $M = M_u$

The extreme fibre strain in concrete $\epsilon_{max} = \epsilon_u = 0.0035$. Assuming the depth of neutral axis, $x = 0.09$ m, the strain compatibility relationship gives,

$$\frac{0.0035}{0.09} = \frac{\epsilon_p - \epsilon_{pe}}{0.3 - 0.09} \tag{a}$$

As the effective strain ϵ_{pe} before application of bending moment M is $1000/(2 \times 10^5) = 0.005$. Eq. (a) gives $\epsilon_p = 0.013168$. The corresponding stress in prestressing steel from the stress-strain relationship of Fig. 8.3.3 (b) is 1463.3 MPa. Hence total tension, $T = A_p f_p = (192 \times 10^{-6}) \times 1463.3 = 0.28095$ MN. As T is equal to C, it follows that $C = 0.28095$ MN. Hence the depth of the neutral axis may be determined from,

$$\left(0.15 \times \frac{3}{7} x \right) (0.68 \times 40) + \left(0.15 \times \frac{4}{7} x \right) \frac{(0.68 \times 40)}{2} = 0.28095$$

or
$$x = 0.0964 \text{ m}$$

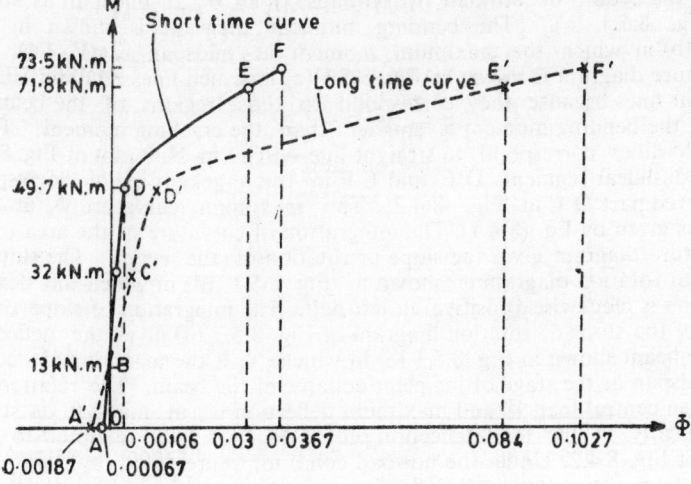

Fig 8.4.3

The second cycle of iteration may commence by assuming $x = 0.096$ m. Proceeding as before, value of x at the end of second cycle is 0.09544 m.

As this value is reasonably close to the value assumed at the beginning of the second cycle, it may be adopted for the computation of moment and curvature at incipient collapse.

$$M_u = \left(0.15 \times \frac{3}{7} \times 0.09544\right)(0.68 \times 40)\left(0.3 - \frac{1}{2} \times \frac{3}{7} \times 0.09544\right)$$

$$+ \left(0.15 \times \frac{4}{7} \times 0.09544\right)\frac{(0.68 \times 40)}{2}\left(0.3 - \frac{13}{21} \times 0.09544\right)$$

$$= 0.0735 \text{ MN.m}$$

$$\phi = \frac{0.0035}{0.09544} = 0.03667 \text{ rad./m}$$

The complete moment-curvature relationship is shown in Fig. 8.4.3 in which the points A, B, C, D, E and F correspond respectively to stages (i) to (vi). The long time moment-curvature relationship A′ B′ C′ D′ E′ F′ is obtained by multiplying the short-time curvatures by the factor $(1 + C_{cs})$ = $(1 + 1.8) = 2.8$.

8.5 LOAD-DEFLECTION RELATIONSHIP

As pointed out in Sec. 8.4, the determination of curvature at several sections of a beam is the key for the computation of its deformation, viz. slope or rotation and deflection because these deformations can be found by successive integration of curvature. Consider, for instance, a simply supported prestressed concrete beam at the limit state of incipient collapse under the action of ultimate or collapse load W_u at midspan as shown in Fig. 8.5.1 (a). The bending moment diagram is shown in Fig. 8.5.1 (b) in which the maximum moment at midspan is $(W_u \ l/4)$. The curvature diagram is shown in Fig. 8.5.1 (c) in which lines AD and BE are straight lines because they correspond to those regions of the beam in which the bending moment is smaller than the cracking moment. These straight lines correspond to straight line ABCD in M-Φ plot of Fig. 8.4.2. The non-linear segments D F and E F in the cracked region correspond to curved part D E in Fig. 8.4.2. The maximum curvature Φ_u at midspan is given by Eq. (8.4.1). The integration of curvature or the area of the curvature diagram gives the slope or rotation of the beam. The slope or flexural rotation diagram is shown in Fig. 8.5.1 (d) in which the flexural rotation is clockwise (positive) in left half. The integration of slope or the area of the slope or rotation diagram of Fig. 8.5.1 (d) gives the deflection of the beam shown in Fig. 8.5.1 (e) in which $\triangle_u$ is the maximum deflection at midspan at the stage of incipient collapse of the beam. The relationship between central load W and maximum deflection $\triangle$ at midspan is shown graphically in the load-deflection plot of Fig. 8.5.2. It is similar to M-Φ plot of Fig. 8.4.2. Under the no-load condition represented by point A, W = o, the negative (upward) deflection $\triangle_o$ represents the camber caused by prestress. The load W_o is required to offset the camber and make the net deflection equal to zero as represented by point B. At the stage of decompression marked by point C, the deflection $\triangle_{dc}$ occurs under load W_{dc} which causes decompression at midspan bottom fibre. The line A B C D is a straight line upto cracking load W_{cr}. In the post-cracking range, $W > W_{cr}$, the plot becomes non-linear with deflection increasing at a progressively higher rate as cracking progresses over larger regions of the beam. The curve becomes almost horizontal as the ultimate or collapse load W_u is

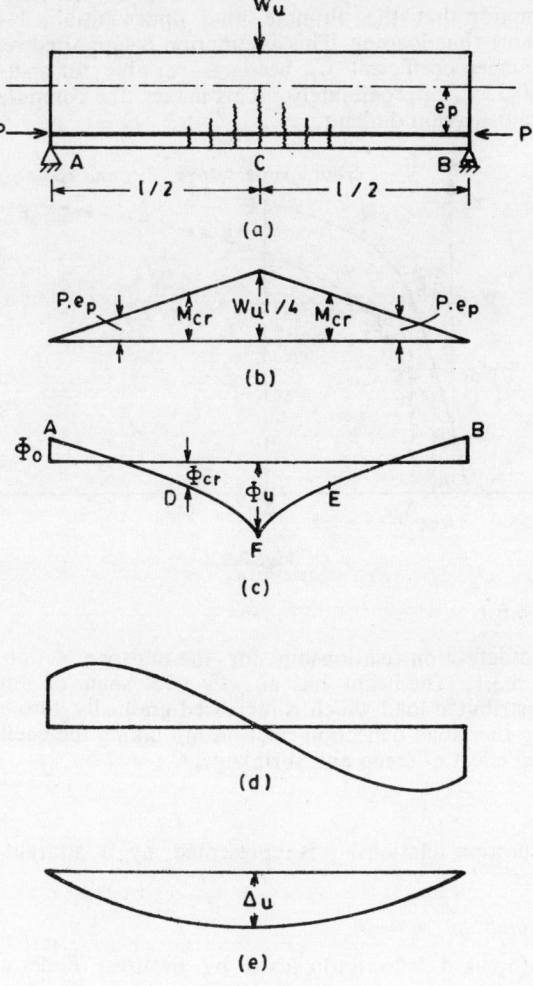

Fig. 8.5.1

approached. Curve DE in $W\text{-}\Delta$ plot of Fig. 8.5.2 represents short-time deflections. In the precracking range, the strains and the consequent deformations increase by the factor $(1 + C_c)$ due to creep of concrete under sustained loading. Hence the long-time deflection in the precracking range may be represented by the straight line A′ B′ C′ D′ obtained by multiplying the short-time deflections by $(1 + C_c)$. With increasing load in the post-cracking range, the creep coefficient C_c is not constant. The creep strains and the resulting long-time deflection increase at a progressively higher rate as the load is increased gradually from cracking load W_{cr} to collapse load W_u. The long-time response represented by the curve $D'\,E'$ is based on the assumption that the ultimate load under sustained loading is the same as in short-time loading. This assumption is supported reasonably well by tests. The creep coefficient C_c becomes variable for a sustained stress higher than $0.33\,f_{ck}$ approximately. This makes the computation of long-time ultimate deflection difficult.

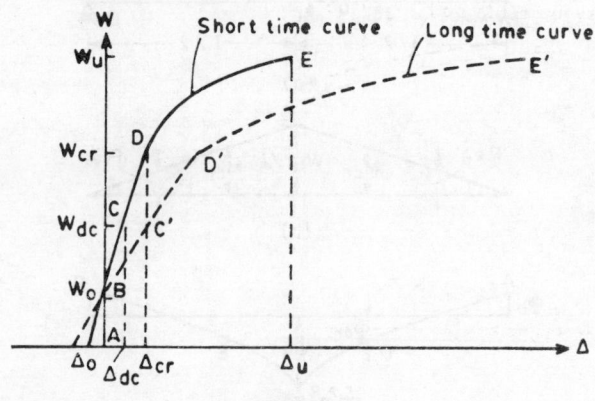

Fig. 8.5.2

EXAMPLE 8.5.1

Plot the load-deflection relationship for the midspan section of the beam of Example 8.3.1. The beam has an effective span of 8 m and carries uniformly distributed load which is increased gradually upto collapse. Also plot the long-time load-deflection relationship taking the coefficient due to the combined effect of creep and shrinkage, $C_{cs} = 1.5$.

Solution

The load-deflection relationship is represented by a straight line upto the cracking stage.

(i) *No load condition, w = 0*

The camber (upward deflection) caused by prestress under no load condition,

$$\Delta = \frac{-192 \times 10^{-6} \times 1000 \times 0.1 \times 8^2}{8 \times 36050 \times 0.0008} = -\,0.005326 \text{ m}$$

(ii) *Cracking stage, w = w_{cr}*

From Example 8.3.1, the bending moment at the stage of incipient cracking,

$$M_{cr} = 0.04971 \text{ MN. m}$$

Hence the intensity of uniform load at the stage of incipient cracking,

$$w_{cr} = \frac{0.04971 \times 8}{8^2} = 0.00621375 \text{ MN/m}$$

The net deflection at incipient cracking,

$$\Delta = -0.005326 + \frac{5}{384} \times \frac{0.00621375 \times 8^4}{36050 \times 0.0008} = 0.006165 \text{ m}$$

(iii) *Post-cracking stage, $w > w_{cr}$*

From Example 8.3.1, for a curvature equal to 0.03 or maximum strain ϵ_{max} = 0.003, the bending moment, $M = 0.07178$ MN.m. The intensity of uniform load causing this bending moment at midspan,

$$w = \frac{0.07178 \times 8}{8^2} = 0.0089725 \text{ MN/m}$$

The bending moment diagram under the action of uniform load $w = 0.0089725$ MN/m is shown in Fig. 8.5.3 (a). The corresponding curvature diagram based on the moment-curvature relationship of Fig. 8.4.3 is shown in Fig. 8.5.3 (b). The slope or flexural rotation α is obtained by integration of curvature. As the slope is zero at midspan, the slope at any point is represented by the area under the curvature diagram between the point under consideration and the midspan section. The diagram representing slope α obtained in this manner is shown in Fig. 8.5.3 (c). The deflection at any point is obtained by integration of slope. As the deflection at the ends and the slope at the midspan section are zero, the deflection at the midspan is represented by the area of the slope diagram between the end section and midspan section. The midspan deflection is found to be 0.082 m.

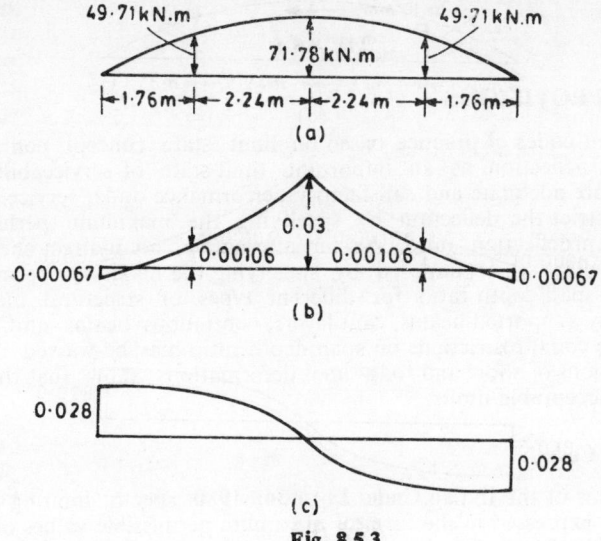

Fig. 8.5.3

(iv) *Limit state of collapse,* $w = w_u$

From Example 8.4.1, the ultimate bending moment, $M_u = 0.0735$ MN·m. Proceeding as in (iii) above, the midspan deflection at incipient collapse is found to be 0.119 m.

The load-deflection curve for midspan section is shown in Fig. 8.5.4. The stages (i) to (iv) are represented by the points A, B, C and D respectively. This plot represents short-time deflection at midspan. The long-time deflection is represented by the curve A′ B′ C′ D′ which is obtained by multiplying the short-time deflections by the factor $(1 + C_{cs}) = 2.5$.

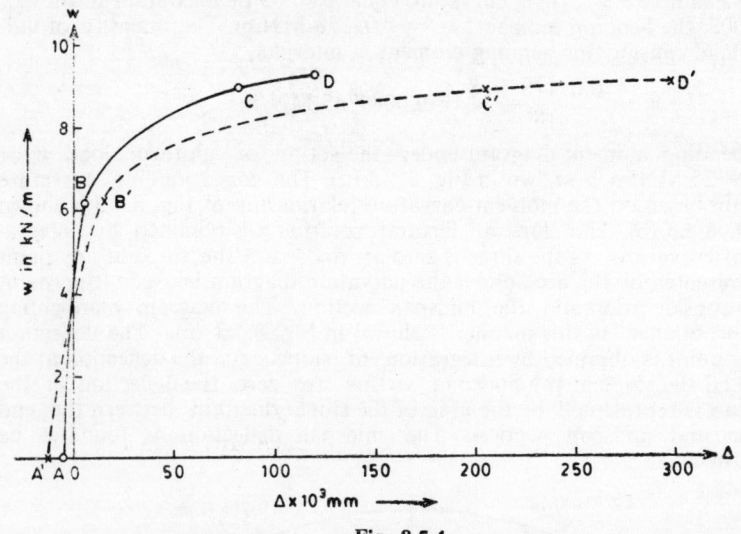

Fig. 8.5.4

8.6 CODE PROVISIONS

All prominent codes of practice based on limit state concept contain the limit state of deflection as an important limit state of serviceability. In order to ensure adequate and satisfactory performance under service loads, the codes restrict the deflection by specifying the maximum permissible values of span/deflection ratio. As an alternative, an indirect check on maximum deflection is imposed by specifying the maximum permissible values of the span/depth ratio for different types of structural members scuh as simply supported beams, cantilevers, continuous beams and slabs. However, the codal restrictions on span/depth ratio may be waived if careful computations of short and long-time deformations show that they are within the acceptable limits.

8.6.1 Indian Code

The provisions of the Indian Code IS : 1343-1980 specify limiting values of deflection expressed in the form of maximum permissible values of span/deflection ratio. The code also specifies maximum permissible values of span/depth ratio for structures in which the permanent load is greater than 25% of the design imposed load.

In order to ensure that the deflection of a structure or part thereof does not adversely affect the appearance or efficiency of the structure or finishes or partitions, the deflections are generally restricted to the following values:

(1) In prestressed concrete members to which finishes have to be applied, the total upward deflection or camber should not exceed span/300 unless uniformity of camber between adjacent units can be ensured.

(2) Taking into account the effects of creep, shrinkage and temperature changes, the maximum deflection after the erection of partitions and the application of finishes should not etceed span/350 or 20 mm, whichever is less.

(3) Taking into account the effects of creep, shrinkage and temperature, the maximum deflection due to all loads measured from the as-cast levels of supports of floors, roofs and all other horizontal members, should not normally exceed span/250.

The code provides guidelines for the computation of deflections in the three types of prestressed concrete members. For members of types 1 and 2, the short-time or instantaneous deflection due to design loads may be computed by using elastic analysis based on uncracked section and the modulus of elasticity of concrete E_c computed from

$$E_c = 5700 \sqrt{f_{ck}}$$

The long term deflections due to prestressing force, dead load and sustained imposed load, taking into account the effects of creep, shrinkage, temperature change, cracking of concrete and loss of prestress, based on elastic analysis should not exceed the limits on span/deflection ratio specified earlier.

8.6.2 American Code

The American Code ACI 318-1989 specifies that all members subjected to bending moment should be designed to have adequate stiffness to limit deflections or any deformations which may adversely affect the strength or serviceability of the structure under service loads.

(i) *Short-time or immediate deflections*

The short-time deflections under service loads may be determined by using the usual formulas in which the moment of inertia may be based on the gross uncracked section. The modulus of elasticity of normal weight concrete may be determined from

$$E_c = 4750 \sqrt{f_c'}$$

where f_c' = characteristic (cylinder) strength of concrete.

(ii) *Long-time deflections*

The additional long-time deflections should be computed taking into account the stresses in concrete and steel due to the sustained loads and prestressing force. The effects due to creep and shrinkage of concrete and the relaxation of steel should also be taken into account. In the absence of detailed computations, the additional long time deflection due to creep and shrinkage of concrete may be determined by multiplying the short-time deflection due to sustained load by a factor equal to $\frac{c}{(1 + 50 \, \rho')}$ in which ρ' is the proportion of untensioned reinforcing steel A'_s in the compression zone. The constant c depends upon the time period and may be taken as 1.0, 1.2, 1.4 and 2.0 for 3 months, 6 months, 12 months and five or more years.

(iii) *Allowable deflections*

The short-time and long-time deflections computed in accordance with the above guidelines should comply with the following provisions:

(a) In the case of flat roofs neither supporting nor attached to non-structural elements which are likely to be damaged by large deflections, the short-time or immediate deflection due to live loads should not exceed span/180. This limit on deflection does not provide a safeguard against possible ponding which should be checked by appropriate calculations of deflection taking into account the additional deflections due to ponded water, long-time effects of all sustained loads, camber, construction tolerances and reliability of provisions for drainage.

(b) In the case of floors neither supporting nor attached to non-structural elements which are likely to be damaged by large deflections, the short-time or immediate deflection due to live loads should not exceed span/360.

(c) In the case of roof or floor construction supporting or attached to non-structural elements which are not likely to be damaged by large deflections, that part of the total deflection which occurs after attachment of the non-structural elements, i.e. the sum of the long-time deflection due to all sustained loads and the short-time deflection due to any additional live load should not exceed span/240. The long-time deflection may be computed as indicated above. However, the computed deflection may be reduced by the amount of deflection which occurs before attachment of the non-structural elements. The specified limit on deflection is subject also to the maximum permissible tolerance for the non-structural elements. The specified deflection limit may be exceeded if a camber is provided so that the total deflection minus the camber does not exceed the permissible deflection.

(d) In the case of roof or floor construction supporting or attached to non-structural elements which are likely to be damaged by large deflections, that part of the total deflection which occurs after attachment of the non-structural elements, i.e. the sum of the long-time deflection due to all sustained loads and the short-time deflection due to any additional live load should not exceed span/480. The longtime deflection may be computed as indicated above. However, the computed deflection may be reduced by the amount of deflection which occurs before the attachment of the non-structural elements.

This limit may, however, be exceeded if adequate measures are taken to prevent damage to supported or attached elements.

8.6.3 European Code

The European Code CEB—FIP 1972 specifies that the designer should verify whether any special checks are needed for in-service deformations to comply with the requirements of serviceability, camber and prevention of damage. The short-time or immediate deformations should be calculated for the most unfavourable load combinations. The long-term deformations should be computed for permanent and semi-permanent combinations. The computations for camber should be based only on permanent and semi-permanent loads.

The computations for deflections are not indispensable in the following cases and where the danger of damage to surface finishes owing to the deflections or the rotations at the supports does not exist :

1. The beams and slabs with spans not exceeding 5 m

2. One or two-way slabs for which $\frac{kl}{D}$ does not exceed 30 in which k is a constant which depends upon the type of supports or end conditions

3. Beams for which $\frac{kl}{D}$ does not exceed 25

4. Floor slabs supporting partitions whose behaviour is likely to be affected by the deflections but whose slenderness ratio $\frac{kl}{D}$ does not exceed $\frac{150}{kl}$.

The values of the constant k for different types of supports or end conditions are as follows :

(i) fixed or continuous at both ends   0.6

(ii) One end fixed or continuous and the other end simply supported and discontinuous   0.8

(iii) both ends simply supported and discontinuous   1.0

(iv) One end fixed and the other unsupported, i.e. cantilever   2.4

The camber may be made equal to the mean deflection caused by the permanent effects, unless a larger camber is needed to meet the serviceability requirements.

REFERENCES

8.1 ACI Committee 435, *Deflections of prestressed concrete members,* ACI Manual of Concrete Practice, Part II, pp. 439-69.

8.2   Corley, W.G. ; Sozen, M.A. and Siess, C.P., *Time-dependent deflections of prestressed concrete beams*, Bulletin No. 307, Highway Research Board, Washington, 1961.

8.3   Fadel, A.I. ; Gamble, W.L. and Mohraz, B., *Tests of a precast post-tensioned composite bridge girder having two spans of 124 feet*, University of Illinois Engineering Experiment Station, Structural Research Series No. 439, April 1977.

8.4   Martin, L.D., *A rational method for estimating camber and deflection of precast prestressed members*, Journal of the Prestressed Concrete Institute, Vol. 22, No. 1, January-February 1977, pp. 100-8.

8.5   Sinno, R. and Furr, H.L., *Computer program for predicting prestress loss and camber*, Journal of the Prestressed Concrete Institute, Vol. 17, No. 5, September-October 1972.

PROBLEMS

8.1   A pretensioned beam having a rectangular cross-section 0.5×0.6 m carries an effective prestress of 0.9 MN at a constant eccentricity of 0.15 m. The beam has an effective span of 12 m. Determine (i) camber caused by prestress and (ii) deflection due to uniformly distributed load causing decompression at midspan. Take $E_c = 36050$ MPa.

8.2   A post-tensioned girder having a symmetrical I-section carries an effective prestressing force of 2.16 MN. The cable line is parabolic with zero eccentricities at ends and a sag of 0.45 m at midspan. The girder has an effective span of 30 m. Determine (i) camber due to prestress, (ii) the net deflection under the action of only self weight and (iii) net deflection under the action of uniform load causing incipient cracking at midspan. The cross-sectional area and moment of inertia are 0.36 m^2 and 0.0621 m^4 respectively. The modulus of rupture of concrete, $f_{cr} = 4.14$ MPa and $E_c = 33720$ MPa. Take unit weight of concrete as 24 kN/m^3.

8.3   Determine the long-time deflection at midspan in the beam of Problem 8.1 if 90 percent of the uniform load is permanent. Take the coefficient due to the combined effect of creep and shrinkage, $C_{cs} = 1.8$.

8.4   Determine the following long-time displacements at midspan in the girder of Problem 8.2 :

      (i) camber due to prestress

      (ii) net deflection due to self weight and

      (iii) net deflection under permanent load causing incipient cracking.

      Take $C_{cs} = 1.5$

8.5 Describe the procedure for the determination of position of neutral axis in a prestressed concrete section in the post-cracking range for (i) a given bending moment and (ii) given curvature.

8.6 Discuss the approaches used for the determination of the deflection of a prestressed concrete beam in the post-cracking range.

8.7 Plot the moment-curvature relationships for the beams of Problems 8.1 and 8.2. Use the idealised stress-strain curves for concrete and steel shown in Fig. 8.3.3 (a) and (b) respectively. The grade of concrete is $M - 40$.

8.8 Plot the load-deflection relationships for the beams of Problems 8.1 and 8.2. Use the idealised stress-strain curves for concrete and steel shown in Fig. 8.3.3 (a) and (b) respectively. The grade of concrete is M-35.

9

Composite Beams

9.1 INTRODUCTION

In the preceding chapters, prestressed concrete is visualised essentially as high grade concrete prestressed with the help of high tensile steel. In structural systems, prestressed concrete is often combined with other materials such as reinforced concrete for the sake of economy and efficiency. The resulting structure formed by two or more materials is called composite construction. For instance, a composite beam of T-section may comprise plain or reinforced concrete flange and prestressed concrete web. A composite bridge deck may be formed by combining reinforced concrete cast-in-situ slab and precast prestressed girders. The economy and efficiency of a composite system depend essentially on the choice of the most appropriate material for each component.

In a traditional prestressed concrete beam, the applied bending moment is resisted by pretensioned high tensile steel ; whereas in a reinforced concrete beam, it is resisted by untensioned reinforcing steel.

Although composite beams can be formed by combining a variety of materials and techniques, the composite prestressed concrete beams are usually formed by combining precast prestressed concrete in tension zone and cast-in-situ non-prestressed concrete in compression zone. As concrete in tension zone is precompressed, flexural cracking due to superimposed load occurs at a higher load level. Consequently, the working moment for no-tension or no-crack conditions is enhanced due to prestressing of concrete in tension zone. The distribution of stresses and the resulting structural response of a composite beam depend upon the technique of construction and the stage of loading. A variety of construction techniques are available for achieving greater efficiency and maximum economy depending upon the type of structural system and the conditions at site.

9.2 CONSTRUCTION TECHNIQUES

Composite constructions comprising prestressed concrete and plain or reinforced concrete may differ from each other in respect of the following :

(i) proportion, geometry and position of the prestressed concrete elements in the composite construction

(ii) proportion, geometry and position of the plain or reinforced concrete elements in the composite construction

(iii) sequence of construction and stages of loading

(iv) type of formwork or props and

(v) manner of connection between constituent materials.

The prestressed concrete elements in composite construction use high grade concrete with grades ranging from M-30 to M-60. The modulus of elasticity, therefore, varies from 31 to 44 kN/mm^2. The prestressed elements may be either pretensioned or post-tensioned. The pretensioning is usually carried out in a central plant for relatively small sized elements. The post-tensioning on the other hand is generally adopted for large elements which may be prestressed in their final positions. The in-situ concrete which is either plain or reinforced is of relatively lower grade, generally M-15 or M-20 with modulus of elasticity ranging from 22 to 26 kN/mm^2. Hence the modular ratio of in-situ concrete with reference to prestressed concrete ranges from 0.50 to 0.84.

Although plain, reinforced and prestressed concretes can be combined in a composite construction in many different ways. it is common to have prestressed elements in the flexural tension zone and plain or reinforced concrete in compression zone. As the prestressed elements constitute the costlier part of the construction, their proportion is kept to the minimum consistant with the required strength. The prestressed elements may have a variety of geometrical shapes. The rectangular and inverted, T-shapes shown in Fig. 9.2.1 (a) and (b) are commonly used for the construction of composit slabs. The prestressed elements may be placed side by side to form the soffit of the slab, thereby eliminating the need of formwork at the construction site. In order to connect the adjacent elements, they may have longitudinal grooves and corresponding projections. In order to increase the shear strength against possible failure due to slip between precast and cast-in-place concretes, castellations may be provided as shown in the Fig. 9.2.1 (a). Alternatively, the shear connectors may be in the form of stirrups projecting out from the precast prestressed elements as shown in Fig. 9.2.1 (c), (d) and (e). In Fig. 9.2.1 (c), the prestressed element of rectangular shape is connected to in-situ concrete flange by means ofstirrups projecting out from the precast element. An alternative way of ensuring adequate shear connection is indicated in Fig. 9.2.1 (d) and (e). In the composite construction of Fig. 9.2.1 (d). the top flange of the precast element is keyed into the in-situ concrete to enhance shear resistance. In Fig. 9.2.1 (e), known as the Uddal construction, the precast element having U-shape forms the lower part of the composite web. In order to reduce the dead weight of the construction, hollow blocks, known as filler blocks may be used in conjunction with the precast prestressed elements and in-situ concrete as shown in Fig. 9.2.1 (b).

The sequence of construction, the type of formwork and the corresponding stages of loading depend upon the type of construction and the conditions at site. Very often the prestressed elements are designed to carry not only their own weights but also the weight of the in-situ concrete during its placement. The main advantage of this type of construction, known as unpropped construction, is that the formwork is eliminated completely. It is particularly suitable for the construction of bridges over rivers running through deep gorges where the supporting formwork would be very expensive. At construction sites, where supporting formwork is not expensive, the props may be provided either to support the weight of the in-situ concrete or the combined weight of in-situ concrete and prestressed elements. In the latter case, the prestressed elements have to carry neither their own weights nor that of the in-situ concrete. In either case, the construction is known as propped construction. The advantage of propped construction is that the proportion of precast prestressed elements can be reduced resulting into substantial economy.

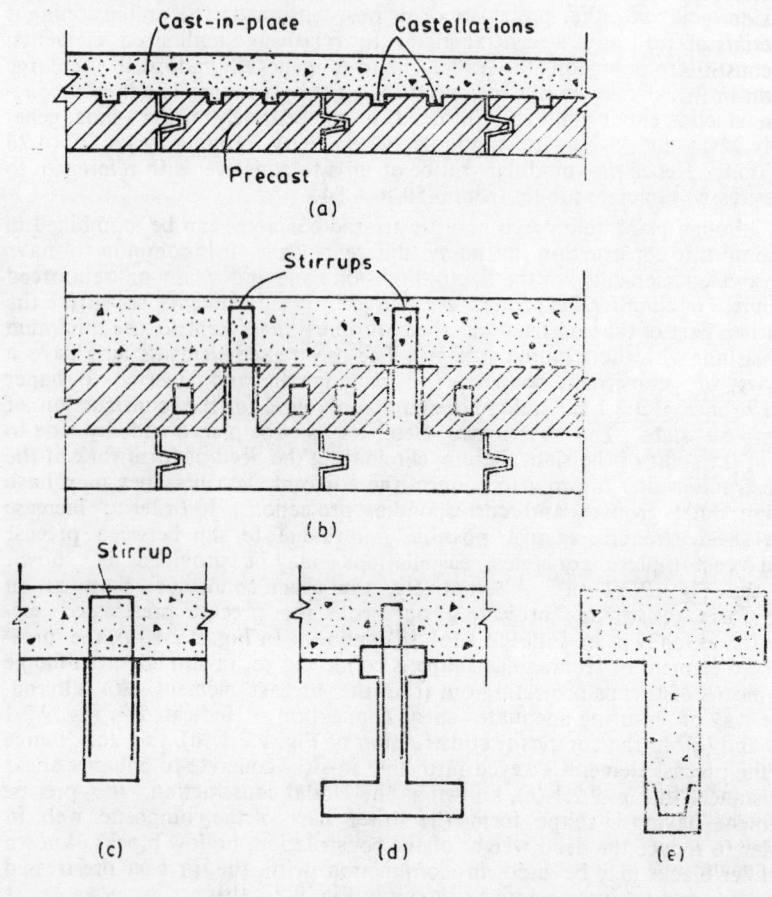

Fig. 9.2.1

In order to prevent the failure of a composite beam on account of the separation of precast and cast-in-place concretes, it is necessary to provide adequate shear connectors. These may be in the form of roughened contact surfaces, castellations and reinforcing bars projecting from the precast elements across the contact surface. The shear connectors should provide adequate shear resistance so that the conventional shear strength against the web-shear cracking or flexure-shear cracking is not lowered.

9.3 PRECRACKING RANGE

The distribution of stresses in a flexurally uncracked section of a composite beam may be determined by using linear elastic theory. The elastic analysis of composite beams in flexure is based on two main assumptions, viz. strain variation is linear and perfect bond exists between all constituent materials. Consider a composite section of any arbitrary shape comprising n different materials of areas $A_1, A_2, \ldots, A_n$ as shown in Fig. 9.3.1 (a). The strains in the constituent materials due to bending moment M are $\epsilon_1, \epsilon_2, \ldots, \epsilon_n$ as shown in Fig. 9.3.1 (b). The stresses in the constituent materials $f_1, f_2 \ldots, f_n$ are

$$f_1 = \epsilon_1 E_1 = \epsilon_1 \left(\frac{E_1}{E_r}\right) E_r = \epsilon_1 m_1 E_r$$

$$f_2 = \epsilon_2 E_2 = \epsilon_2 \left(\frac{E_2}{E_r}\right) E_r = \epsilon_2 m_2 E_r \qquad (9.3.1)$$

$$\cdots\cdots\cdots\cdots\cdots\cdots\cdots$$

$$f_n = \epsilon_n E_n = \epsilon_n \left(\frac{E_n}{E_r}\right) E_r = \epsilon_n m_n E_r$$

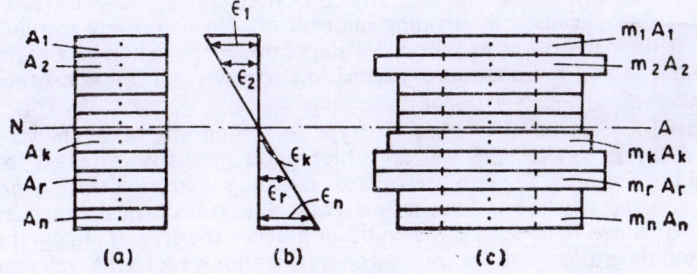

Fig. 9.3.1

in which E_r is the modulus of elasticity of the reference material r and m_1, $m_2, \ldots, m_n$ are the modular ratios of the constituent materials with respect to the reference material. Any one of the n materials may be chosen as the reference material. The longitudinal forces $F_1, F_2, \ldots, F_n$ in the constituent materials due to bending may be expressed as

$$F_1 = A_1 f_1 = (m_1 A_1) \epsilon_1 E_r \qquad (9.3.2)$$

$$F_2 = A_2 f_2 = (m_2 A_2) \epsilon_2 E_r$$

$$\cdots\cdots\cdots\cdots\cdots\cdots\cdots$$

$$F_n = A_n f_n = (m_n A_n) \epsilon_n E_r$$

Equations (9.3.2) show that the transformed equivalent section may be obtained by replacing the areas $A_1, A_2, \ldots, A_n$ by their equivalent areas

$m_1 A_1, m_2 A_2, \ldots, m_n A_n$ as shown in Fig. 9.3.1 (c). The transformed section is a homogeneous section consisting of the reference material only. The properties of the transformed section such as cross-sectional area, position of centroid and moment of inertia may be determined in the usual manner as for a homogeneous reference material.

Using Eq. (9.3.1), the stress f_k in constituent material k may be expressed as

$$f_k = m_k \, \epsilon_k \, E_r = m_k f_k'$$ (9.3.3)

in which f_k' is the stress in the transformed homogeneous section at the point at which the kth material is located. From Eq. (9.3.3) it follows that the flexural stress due to bending moment M in any constituent material k may be expressed as

$$f_k = m_k \left(\frac{My_k}{I_t} \right)$$

in which y_k is the distance of the point in material k from centroidal axis and I_t is the moment of inertia of the transformed section about its centroidal axis. Knowing the stresses and therefore the forces in all constituent materials, the first moment of these forces about centroid of the transformed section represents the internal moment of resistance which must be equal to applied bending moment M for equilibrium.

If some of the constituent materials are prestressed, the net stresses may be computed by combining algebraically the stresses due to prestressing with the stresses caused by bending moment M. The composite section is adequate if the net stresses at all crucial stages under the action of working loads do not exceed the maximum permissible stresses for all constituent materials.

Consider for instance, the common type of composite beam shown in Fig. 9.3.2 (a) in which the web is of high grade precast prestressed concrete and the flange is of non-prestressed in-situ concrete. Treating prestressed concrete as the reference material, the transformed equivalent section is obtained by reducing the width and hence the area of the flange by the modular ratio m of the in-situ concrete with respect to the reference material, i.e. precast concrete, as shown in Fig. 9.3.2 (b). The transformed

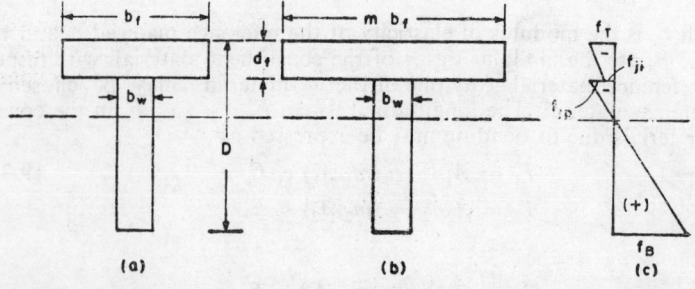

(a)   (b)   (c)

Fig. 9.3.2

section may be treated as a homogeneous section with the following cross-sectional properties. The cross-sectional area,

$$A_t = mb_f\, d_f + b_w\,(D - d_f)$$

Distance of centroid from top face,

$$\bar{x}_t = \frac{(mb_f\; d_f)\dfrac{d_f}{2} + b_w\,(D - d_f)\left(d_f + \dfrac{D - d_f}{2}\right)}{mb_f\, d_f + b_w\,(D - d_f)}$$

The moment of inertia of the transformed section, I_t about the centroidal axis may also determined by treating it as a homogeneous section.

The distribution of stresses in a composite section caused by bending moment M in the precracking range may be determined by using the flexure theory for transformed section. From Eq. (9.3.3), the stress at any point is equal to the stress in the homogeneous transformed section multiplied by the respective modular ratio. Consider, for instance, the composite section of Fig. 9.3.2 (a). Treating precast concrete as the reference material, the transformed section is shown in Fig. 9.3.2 (b). The stress diagram due to sagging bending moment M acting on the composite section is shown in Fig. 9.3.2 (c). The stress at top of the section,

$$f_T = -\,m\left(\frac{My_T}{I_t}\right)$$

Stress in the in-situ concrete flange at its junction with rib,

$$f_{jt} = -\,m\left(\frac{M\,y_j}{I_t}\right)$$

Stress in the precast rib at its junction with the flange,

$$f_{jp} = -\left(\frac{My_j}{I_t}\right)$$

Stress at bottom of the rib.

$$f_B = \frac{My_B}{I_t}$$

where $\quad m = \dfrac{E_{ci}}{E_{cp}} = $ modular ratio of in situ concrete

$\qquad E_{ci},\, E_{cp} = $ moduli of elasticity of in-situ concrete and precast concrete respectively

In the method of construction commonly adopted, the webs are precast and prestressed in a central plant, transported to site and hoisted in their final position. The webs are then used to support the formwork, the green in-situ concrete of the flange and the live load during the placement of in-situ concrete. The composite beam may be pressed into service after the in-situ concrete has gained sufficient strength. When the forgoing technique of construction is used, the stresses have to be checked at the following stages:

(1) Initial stage

At this stage, the precast prestressed web is checked for the stresses caused by prestressing force P_i before losses combined with initial dead load, usually in the form of its own weight. These stresses exist immediately after stress transfer in precast web.

(2) Final stage (A)

This load stage prevails during the placement of in-situ concrete in the top flange with the precast prestressed web serving as load bearing member. At this stage the stresses in the web due to effective prestressing force P after losses combined with dead and live load during placement of in-situ concrete are checked. This stage of loading is relevant only for unpropped construction in which the prestressed elements and in-situ concrete are not supported by formwork or props.

(3) Final stage (B)

This stage arises when the structure is pressed into service. At this stage, the composite section is subjected to its own weight and full service load. The stresses in the flange and the web constituting the composite section under this load combination are determined. The net stresses in the web are obtained by combining the stresses due to loads with the stresses due to effective prestressing force P after losses. This stage is relevant both for propped as well as unpropped constructions.

The stress diagrams which are crucial for design of propped and unpropped constructions may be obtained by superimposing the stresses due to prestressing on the stresses due to loads relevant to the stage under consideration. Consider first the transformed section of Fig. 9.3.3 (a) built by propped construction. The stress diagrams for the precast rib before and after losses are shown in Fig. 9.3.3 (b) and (c) respectively. The stress diagram for the whole section acting as a composite section caused by self weight and superimposed or service load is shown in Fig. 9.3.3. (d). The net stress diagram shown in Fig. 9.3.3 (e) is obtained by combining stress diagrams of Fig. 9.3.3 (c) and (d). In the case of unpropped construction, the stress at the initial stage (Fig. 9.3.3 b) and at the final stage (Fig. 9.3.3 e) are crucial. Hence the stress may be taken to be safe at the limit state of serviceability if the stresses in Fig. 9.3.3 (b) and (e) do not exceed the respective working stresses.

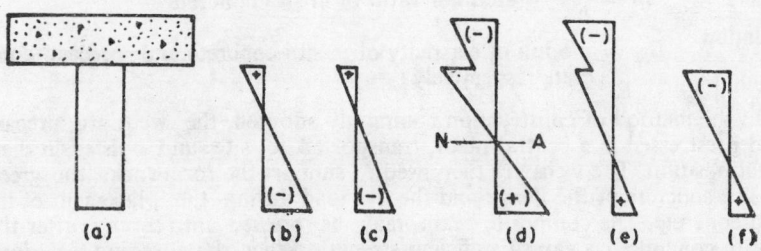

Fig. 9.3.3

In the case of unpropped construction, the stresses in the precast pre-stressed concrete rib at final stage (A) has also to be considered. At this stage, the precast element has to carry the weight of in-situ concrete in addition to its own weight. The stress diagram for the final stage (A) is shown in Fig. 9.3.3 (f). It should be checked that the stresses in Fig. 9.3.3 (f) do not exceed the appropriate working stresses.

EXAMPLE 9.3.1

A composite beam having the section shown in Fig. 9.3.4 (a) comprises top flange, web and bottom flange formed of M-15, M-25 and M-35 concretes respectively. Determine the strains, stresses and curvature caused by a bending moment of 2000 kN.m.

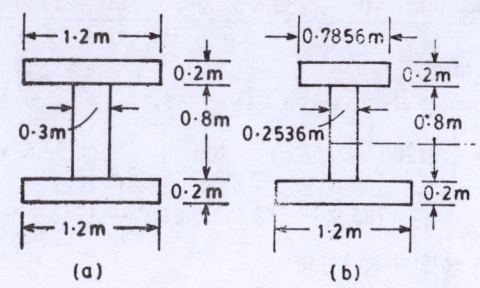

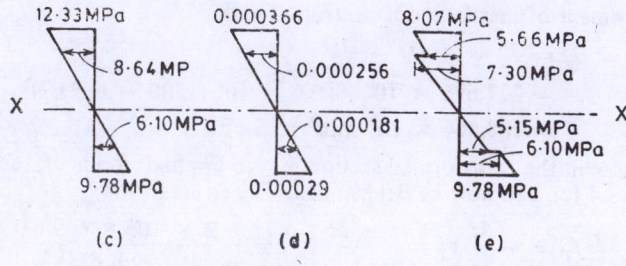

Fig. 9.3.4

Solution

Using the provisions of I.S. Code, the moduli of elasticity of M-15, M-25 and M-35 concretes are as follows:

$$E_1 = 5700 \sqrt{15} = 22075 \text{ MPa}$$

$$E_2 = 5700 \sqrt{25} = 28500 \text{ MPa}$$

$$E_3 = 5700 \sqrt{35} = 33720 \text{ MPa}$$

Using M-35 concrete as reference material, the modular ratios of M-15 and M-25 concretes are,

$$m_1 = \frac{E_1}{E_3} = \frac{22075}{33720} = 0.6547$$

$$m_2 = \frac{E_2}{E_3} = \frac{28500}{33720} = 0.8452$$

The transformed section shown in Fig. 9.3.4 (b) is obtained by multiplying the widths of the top flange and web by the respective modular ratios, viz. m_1 and m_2. The cross-sectional properties of the transformed section are as follows. The area of cross-section,

$$A_t = 785.6 \times 200 + 253.6 \times 800 + 1200 \times 200$$
$$= 0.6 \times 10^6 \text{ mm}^2$$

Distance of the centroid from top,

$$\bar{x}_t = \frac{785.6 \times 200 \times 100 + 253.6 \times 800 \times 600 + 1200 \times 200 \times 1100}{0.6 \times 10^6}$$

$$= 669.07 \text{ mm}$$

The moment of inertia about axis XX (Fig. 9.3.4 b),

$$I_X = \frac{1}{3} (1200 - 253.6) \times 200^3 + \frac{1}{3} \times 253.6 \times 1200^3$$

$$+ \frac{1}{12} (785.6 - 253.6) \times 200^3 + (785.6 - 253.6)$$

$$\times 200 \times 1100^2$$

$$= 277.696 \times 10^9 \text{ mm}^4$$

Hence moment of inertia about centroidal axis,

$$I_t = I_X - A_t (D - \bar{x}_t)^2$$
$$= 277.696 \times 10^9 - 0.6 \times 10^6 (1200 - 669.07)^2$$
$$= 108.564 \times 10^9 \text{ mm}^4$$

The stresses in the transformed section due to applied moment, as shown in Fig. 9.3.4 (c), can now be determined. The stress at top,

$$f_T = -\frac{M}{I_t} y_T = \left(\frac{-M}{I_t} \bar{x}_t \right) = -\frac{2 \times 10^9 \times 669.07}{108.564 \times 10^9}$$

$$= -12.33 \text{ MPa}$$

The stress at bottom,

$$f_B = \frac{M}{I_t} y_B = \frac{M}{I_t} (D - \bar{x}_t) = \frac{2 \times 10^9 \times 530.93}{108.564 \times 10^9}$$

$$= 9.78 \text{ MPa}$$

The corresponding strain diagram shown in Fig. 9.3.4 (d) is obtained by dividing the stresses shown in Fig. 9.3.4 (c) by the modulus of elasticity of the reference material, viz. E_3. The curvature due to the applied moment,

$$\Phi = \frac{3.6566 \times 10^{-4}}{0.66907} = 0.0005465 \text{ rad/m}$$

The stresses in the actual section shown in Fig. 9.3.4 (e) are obtained by multiplying the strains of Fig. 9.3.4 (d) by the respective moduli of elasticity. Alternatively, the stress diagram of Fig. 9.3.4 (e) may be obtained by multiplying stresses of Fig. 9.3.4 (c) by the respective modular ratios.

The same results may be obtained from first principles by using the basic equations of equilibrium, viz. (i) total compression is equal to total tension and (ii) internal moment is equal to external moment.

EXAMPLE 9.3.2

A composite beam having a T-section shown in Fig. 9.3.5 (a) formed by propped construction technique has top flange made from in-situ concrete and the web made from precast prestressed concrete. The beam has an effective span of 16 m. It carries only its self weight at the initial stage and a superimposed load of 23.56 kN/m at the final stage. Determine the stresses at the two stages. The area of prestressing steel, $A_p = 720$ mm^2. The initial and final stresses, $f_{pi} = 1200$ MPa and $f_p = 1000$ MPa. The moduli of elasticity of in-situ and precast concretes are 24 and 32 kN/mm^2 respectively.

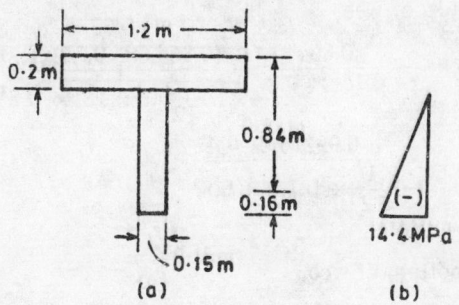

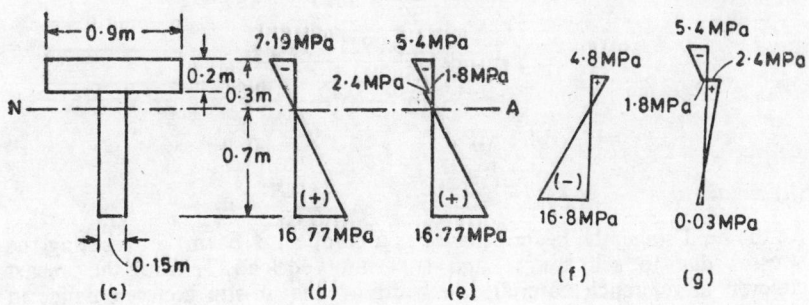

Fig. 9.3.5

Solution

(i) *Initial stage*

The stresses in the web at the initial stage before placement of in-situ concrete have to be checked. At this stage, only the self weight of the web is effective. Taking unit weight of concrete as 24 kN/m³, the self weight

$$w_b = 0.15 \times 0.8 \times 24 = 2.88 \text{ kN/m}$$

Bending moment due to self weight,

$$M_D = \frac{2.88 \times 16^2}{8} = 92.16 \text{ kN} \cdot \text{m}$$

The initial prestressing force,

$$P_i = 720 \times 1200 = 0.864 \text{ MN}$$

It acts at an eccentricity,

$$e_p = 0.40 - 0.16 = 0.24 \text{ m}$$

Initial stress at top of web,

$$f_T = -\frac{0.864}{0.15 \times 0.8} + \frac{(0.864 \times 0.24) \times 0.4}{\frac{1}{12} \times 0.15 \times 0.8^3}$$

$$-\frac{0.09216 \times 0.4}{\frac{1}{12} \times 0.15 \times 0.8^3}$$

$$= 0$$

Initial stress at bottom of web,

$$f_B = \frac{-0.864}{0.15 \times 0.8} - \frac{(0.864 \times 0.24) \times 0.4}{\frac{1}{12} \times 0.15 \times 0.8^3}$$

$$+\frac{0.09216 \times 0.4}{\frac{1}{12} \times 0.15 \times 0.8^3}$$

$$= -14.4 \text{ MPa}$$

(ii) *Final stage*

At the final stage, the beam behaves as a composite beam in resisting the stresses due to self weight and superimposed load. Treating the precast concrete as reference material, the width of the in-situ concrete flange in the transformed section,

$$b_{ft} = 1.2 \times \frac{24}{32} = 0.9 \text{ m}$$

The transformed section is shown in Fig. 9.3.5 (c). The cross-sectional properties of the transformed section are

$$A_t = 0.3 \text{ m}^2$$

$$\bar{x}_t = 0.3 \text{ m}$$

$$I_t = 0.043 \text{ m}^4$$

The self weight of the beam is $(1.2 \times 0.2 + 0.15 \times 0.8) \times 24 = 6.64$ kN/m. Hence total load is $(8.64 + 23.56) = 32.2$ kN/m. The bending moment at midspan,

$$M = \frac{32.2 \times 16^2}{8} = 1030.4 \text{ kN.m}$$

The stress at top fibre,

$$f_T = \frac{-1.0304 \times 0.3}{0.043} = -7.19 \text{ MPa}$$

The stress at bottom fibre,

$$f_B = \frac{1.0304 \times 0.7}{0.043} = 16.77 \text{ MPa}$$

The stress diagram due to applied moment for the transformed section is shown in Fig. 9.3.5 (d). The actual stresses in the in-situ concrete flange are obtained by multiplying the stresses in Fig. 9.3.5 (d) by the modular ratio, $m = 24/32 = 0.75$. The stresses in the actual section due to applied moment are shown in Fig. 9.3.5 (e). These stresses should be combined with the stresses in the web caused by prestress. The prestressing force at the final stage,

$$P = 720 \times 10^{-6} \times 1000 = 0.72 \text{ MN}$$

This effective prestressing force acting at an eccentricity of 0.24 m produces stresses in the web as shown in Fig. 9.3.5 (f). The net stresses at the final stage obtained by combining the stresses of Figs 9.3.5 (e) and (f) are shown in Fig. 9.3.5 (g).

EXAMLE 9.3.3

Determine the stresses at the critical stages in the composite beam of Example 9.3.2 if it is formed by unpropped construction technique. The precast elements have to carry not only its own weight but also the weight of the in-situ concrete flange and a uniform load of 1.5 kN/m² due to formwork and construction equipment during placement of in-situ concrete.

Solution

The stresses in the web at the initial stage are the same as in Example 9.3.2. However, the stresses have to be checked at final stages (A) and (B) in the case of unpropped construction. The stresses at final stage (B) have already been checked in Example 9.3.2. The stresses at final stage (A) are computed below.

Weight of flange per metre length of beam,

$$w_f = 1.2 \times 0.2 \times 24 = 5.76 \text{ kN/m}$$

Hence the superimposed load on the precast element,

$$w_s = 5.76 + 1.5 \times 1.2 = 7.56 \text{ kN/m}$$

Self weight of precast element

$$w_D = 0.15 \times 0.8 \times 24 = 2.88 \text{ kN/m}$$

Hence the total load on precast element,

$$w = 7.56 + 2.88 = 10.44 \text{ kN/m}$$

The bending moment due to this load,

$$M = \frac{10.44 \times 16^2}{8} = 334.08 \text{ kN·m}$$

Stress at top of precast element,

$$f_T = -\frac{0.72}{0.15 \times 0.8} + \frac{(0.72 \times 0.24)\, 0.4}{\frac{1}{12} \times 0.15 \times 0.8^3}$$

$$-\frac{0.33408 \times 0.4}{\frac{1}{12} \times 0.15 \times 0.8^3}$$

$$= -16.08 \text{ MPa}$$

Stress at bottom of precast element,

$$f_B = -\frac{0.72}{0.15 \times 0.08} = \frac{(0.72 \times 0.24) \times 0.4}{\frac{1}{12} \times 0.15 \times 0.8^3}$$

$$+\frac{0.33408 \times 0.4}{\frac{1}{12} \times 0.15 \times 0.8^3}$$

$$= 4.08 \text{ MPa}$$

These stresses may be larger than the permissible stresses. However, they may be permitted because final stage (A) is only a passing phase. The stresses may be reduced progressively as in-situ concrete develops its strength and composite action starts.

9.4 POST-CRACKING RANGE

The behaviour of a composite beam in the post-cracking range depends upon the proportions and disposition of the constituent materials. In a composite beam formed by in-situ concrete flange and precast prestressed web, the flexural cracks appear in the web as the load is increased progressively beyond the cracking load. The compressive stress in the top flange and the tensile stress in the prestressing tendons increase until the limit state of collapse is reached. In beams with relatively low proportion of steel, the failure occurs due to breaking of tendons. This type of failure is undesirable as it is abrupt, sudden and catastrophic. In a well-proportioned beam, the failure occurs by progressive crushing of concrete in the compression zone. The ultimate moment may be determined by using the principles discussed in Sec. 5.8. When the neutral axis lies in the flange at the limit state of collapse, the ultimate moment is the same as that of the corresponding prestressed concrete beam made from in-situ concrete alone. In this case, IS code method, Eq. (6.9.1) may be used. When the neutral axis is in the web, the iterative procedure based on compatibility of strains may be used. At failure, a part of the compression zone comprises precast concrete in the web portion above the neutral axis. The procedure for the computation of ultimate moment of a composite beam is illustrated by Ex. 9.4.1.

EXAMPLE 9.4.1

Determine the ultimate moment capacity of the composite section shown in Fig. 9.4.1 (a). The flange and the web are made respectively from M-20 in-situ concrete and M-40 precast concrete. Use the idealised stress-strain curves for concrete and steel shown in Fig. 9.4.1 (b) and (c) respectively. The effective prestress in steel is 1000 MPa. Also determine the ultimate moment using Indian Code equation.

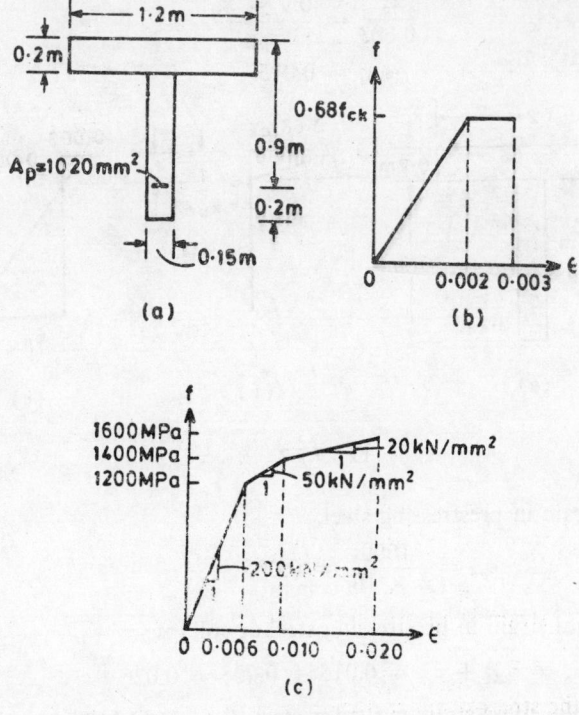

Fig. 9.4.1

Solution

(i) *Strain compatibility method*

The strain compatibility may be utilized in the following iterative procedure.

As the proportion of prestressing steel is relatively low, the stress in prestressing steel at ultimate moment may be assumed equal to the ultimate strength of the material, i.e. $f_p = f_{pu} = 1600$ MPa. Hence total tension,

$$T = A_p f_p = 1020 \times 10^{-6} \times 1600 = 1.632 \text{ MN}$$

As the total tension and compression must be equal, $C = T = 1.632$ MN, the position of neutral axis can be determined. Assuming the neutral axis to lie in the flange and referring to the stress diagram of Fig. 9.4.2(b),

$$1.2 \times \frac{x_u}{3} \times 0.68 \times 20 + 1.2 \times \frac{2x_u}{3} \times \frac{0.68}{2} \times 20 = 1.632$$

Hence $x_u = 0.15$ m

The strain in prestressing steel due to the application of ultimate moment may now be determined by using the strain diagram of Fig. 9.4.2 (c).

$$\frac{x_u}{0.003} = \frac{0.9 - x_u}{\epsilon_{pM}}$$

∴. $$\epsilon_{pM} = 0.015$$

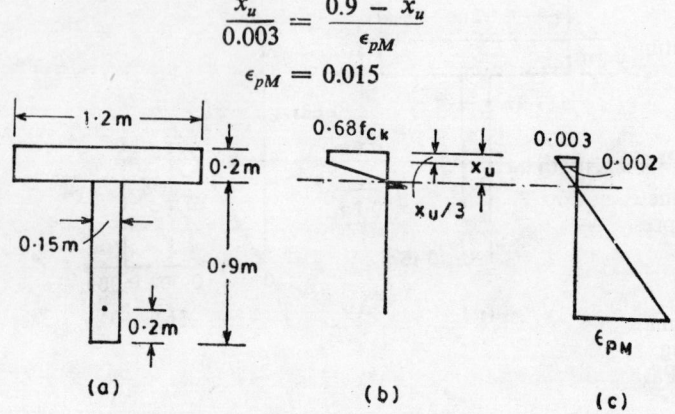

(a)   (b)   (c)

Fig. 9.4.2

The initial strain in prestressing steel,

$$\epsilon_{pi} = \frac{1000}{2 \times 10^5} = 0.005$$

Hence the total strain in prestressing steel at collapse,

$$\epsilon_p = \epsilon_{pM} + \epsilon_{pi} = 0.015 + 0.005 = 0.02$$

Referring to the stress-strain curve for steel, $f_p = 1600$ MPa. As this stress is equal to the assmed stress, further iteration is not required. The ultimate moment,

$$M_u = \left(1.2 \times \frac{0.15}{3}\right)(0.68 \times 20)\left(0.9 \times \frac{1}{2} \times \frac{0.15}{3}\right)$$

$$+ \left(1.2 \times \frac{2}{3} \times 0.15\right)\left(\frac{1}{2} \times 0.68 \times 20\right)$$

$$\times \left(0.9 - \frac{0.15}{3} - \frac{1}{3} \times \frac{2}{3} \times 0.15\right)$$

$$= 1.3804 \text{ MN.m}$$

(ii) *IS code equation*

According to IS Code 1343-1980, the ultimate moment,

$$M_u = A_p f_p (d_p - 0.42 \, x_u) \qquad \text{(a)}$$

in which f_p and x_u depend upon the effective proportion of prestressing steel.

$$\omega_p = \frac{A_p}{bd_p} \frac{f_{pu}}{f_{ck}} = \frac{1020 \times 10^{-6}}{1.2 \times 0.9} \times \frac{1600}{20}$$

$$= 0.07555$$

From Table 6.9.1,

$$f_p = 1.0 \, (0.87 f_{pu}) = 1392 \text{ MPa}$$
$$x_u = 0.1642 \, d_p = 0.1642 \times 0.9 = 0.1478 \text{ m}$$

Substituting into Eq. (a),

$$M_u = 1020 \times 10^{-6} \times 1392 \, (0.9 - 0.42 \times 0.1478)$$
$$= 1.19 \text{ MN.m}$$

EXAMPLE 9.4.2

Determine the ultimate moment of the section of Example 9.4.1 if the area of prestressing steel, $A_p = 2083$ mm^2.

Solution

The ultimate moment may be determined by using strain compatibility. Adopting an iterative approach, let the stress in steel be assumed to be 1420 MPa. Hence total tension,

$$T = A_p f_p = 2083 \times 10^{-6} \times 1420 = 2.958 \text{ MN}$$

As $C = T$, the position of the neutral axis may be determined. Assuming neutral axis lying in the web as shown in Fig. 9.4.3,

$$1.2 \times \frac{x_u}{3}(0.68 \times 20) + 1.2\left(0.2 - \frac{x_u}{3}\right)\left[\frac{\dfrac{0.68 \times 20 \, (x_u - 0.2)}{\frac{2}{3} \, x_u} + 0.68 \times 20}{2}\right]$$

$$+ \, 0.15 \, (x_u - 0.2) \times \frac{1}{2}\left[\frac{0.68 \times 40 \, (x_u - 0.2)}{\frac{2}{3} \, x_u}\right] = 2.958$$

Hence $x_u = 0.3$ m

Referring to the strain diagram of Fig. 9.4.3 (c),

$$\frac{0.3}{0.003} = \frac{0.9 - 0.3}{\epsilon_{pm}}$$

or

$$\epsilon_{pm} = 0.006$$

$$\epsilon_p = \epsilon_{pm} + \epsilon_{pi} = 0.006 + \frac{1000}{2 \times 10^5} = 0.011$$

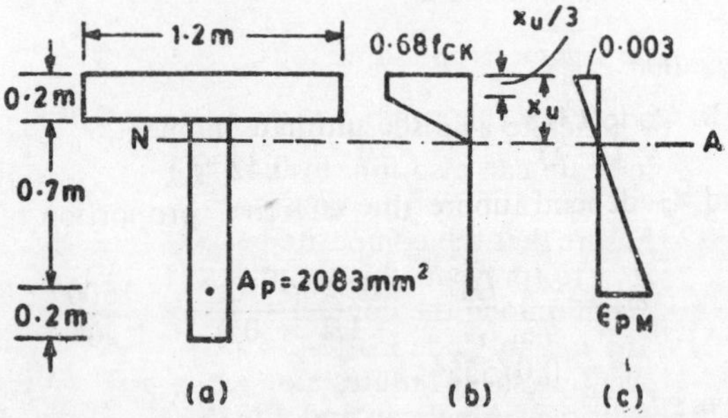

Fig. 9.4.3

Hence from the stress-strain diagram of Fig. 9.4.1 (c), $f_p = 1420$ MPa. As this stress is equal to the assumed stress, further iteration is not required. The ultimate moment,

$$M_u = 1.2 \times \frac{0.3}{3}(0.68 \times 20)\left(0.9 - \frac{1}{2} \times \frac{0.3}{3}\right)$$

$$+ 1.2\left(0.2 - \frac{0.3}{3}\right)\left[\frac{\dfrac{0.68 \times 20(0.3 - 0.2)}{\dfrac{2}{3} \times 0.3} + 0.68 \times 20}{2}\right]$$

$$\times \left[0.9 - \frac{0.3}{3} - \frac{0.1}{3}\left(\frac{0.002 + 2 \times 0.001}{0.002 + 0.001}\right)\right]$$

$$+ 0.15\left(0.3 - 0.2\right) \times \frac{1}{2}\left[\frac{0.68 \times 40(0.3 - 0.2)}{\dfrac{2}{3} \times 0.3}\right]$$

$$\left[0.9 - 0.2 - \frac{0.1}{3}\right]$$

$$= 2.40 \text{ MN.m}$$

9.5 SHEAR STRENGTH

A composite beam comprising prestressed concrete and plain or reinforced concrete may fail in one of the following two ways :

(1) It may fail in vertical shear like a conventional non-composite beam as discussed in Sec. 7.2. In order to prevent shear failure in this manner, it should be verified that the design ultimate shear does not exceed the ulti-

mate shear strength in accordance with the code provisions. The shear resisted by concrete V_{co} in the flexurally uncracked zone and V_{cr} in the flexurally cracked zone in accordance with Indian and British codes may be determined on the basis of the transformed section in which prestressed concrete may be chosen as the reference material. The transformed section may then be visualised as comprising entirely of prestressed concrete. The tensile strength f_t in the expression for V_{co} and τ_c in the expression for V_{cr} may then be taken to depend upon the characteristic compressive strength of concrete f_{ck}. As the shear resisted by web steel V_s is independent of the grade of concrete, its value is the same for composite and non-composite beams. The web steel should be designed to resist the shear force in excess of that resisted by concrete as discussed in Sec. 7.2.

(2) A composite beam can also fail in horizontal shear on account of slip and consequent separation of the constituent materials along the interface. In order to ensure that the composite beam behaves like a monolithic unit, it is necessary to prevent this kind of failure by the provision of adequate shear strength along the contact surfaces between the constituent materials. While the Indian Code IS: 134.:-1980 does not make any specific provisions for preventing shear failures along contact surfaces, certain other codes, notably the British, American and European codes, contain specific provisions.

(a) British Code

The British Code BS: 8110-1985 contains specific provisions for the permissible ultimate shear stress at the interface and the design of link reinforcement when the ultimate permissible stress is exceeded. The average horizontal shear stress at the limit state of collapse may be computed by dividing the design horizontal shear force at the interface by the product of contact width and length of the beam between point of maximum positive or negative design moment and the point of zero moment. When the interface lies in flexural tension zone, the horizontal design shear force may be taken equal to total compression (or tension) due to ultimate bending moment. When it lies in flexural compression zone, the design horizontal shear force may be taken equal to the compressive force in the area located between the interface and the compression face. The average design shear stress may then be distributed in proportion to the vertical design shear force diagram to obtain the horizontal shear stress at any point along the length of the member. The design horizontal shear stress v_h computed in this manner should not exceed the permissible value v_{ch} which depends upon the following factors as shown in Table 9.5.1.

(i) *Type of precast section*

    (a) Type A—without link reinforcement
    (b) Type B—with nominal link reinforcement projecting into in-situ concrete

(ii) *Class of contact surface*

    (a) Class 1—as-cast or as-extruded. The description as-cast covers those cases where the concrete is placed and vibrated leaving a rough finish. The surface is rougher than would be required for finishes to be applied directly without a further finishing screed but not as rough as would be obtained if tamping, brusting or other artificial roughening had taken place. The description as-extruded covers those cases in which an open-textured surface is produced direct from an extruding machine.

(b) Class 2—brushed, screeded or rough tamped. This description covers those cases where some form of deliberate surface roughening has taken place but not to the extent of exposing the aggregate.

(c) Class 3—washed to remove laitance or treated with retarder and cleaned.

(iii) *Grade of in-situ concrete*

The nominal link reinforcement, when provided, should have a minimum area of 0.15 per cent of contact area and should satisfy the following requiremets:

1. The spacing of links should not be excessive. In T-beam ribs with composite flanges, the spacing of links should not exceed four times the minimum thickness of in-situ concrete nor 600 mm.

2. The links should be anchored adequately on both sides of the interface.

When the design ultimate horizontal shear stress v_h is greater than the permissible value v_{ch} (Table 9.5.1.), the total horizontal shear should be assumed to be resisted by the link reinforcement. The area of link reinforcement A_{vf} per unit length may be determined from

$$A_{vf} = \frac{b \, v_h}{0.87 f_y} \qquad (9.5.1)$$

where b = breadth at interface

Table 9.5.1 Permissible design ultimate horizontal shear stress at interface, v_{ch} in MPa

Class of contact surface	Type of precast section					
	Type A			Type B		
	M-25	M-30	M-40 and over	M-25	M-30	M-40 and over
Class 1	0.40	0.55	0.65	1.20	1.80	2.00
Class 2	0.60	0.65	0.75	1.80	2.00	2.20
Class 3	0.70	0.75	0.80	2.10	2.20	2.50

(b) American Code

According to American Code, when the entire composite section is assumed to resist the vertical shear, the section should be designed in accordance with the provisions of Sec. 7.2.3.3 (as for a monolithically cast member of the same cross-sectional shape. The nominal ultimate horizontal shear stress along the interface may be determined from

$$v_{hu} = \frac{V_u}{\phi \, bd} \qquad (9.5.2)$$

where, V_u = design ultimate shear force

b = breadth of rectangular section or breadth (thickness) of web of flanged section

d = effective depth upto centroid of tension reinforcement

ϕ = capacity reduction factor which is 0.85 for shear

The permissible ultimate horizontal shear stresses v_{chu} for different kinds of contact surfaces are shown in Table 9.5.2.

Table 9.5.2

Type	Description of contact surface	v_{chu}
1	Clean and intentionally roughened with a full amplitude of approx. 6 mm without ties	0.56 MPa
2	Clean but not intentionally roughened with ties satisfying minimum requirements given below	0.56 MPa
3	Clean and intentionally roughened with ties satisfying minimum requirements given below	2.45 MPa

If the value of v_{hu} determined from Eq. (9.5.2) exceeds 2.45 MPa, the shear-friction reinforcement should be provided in accordance with the following equation:

$$A_{vf} = \frac{V_u}{\phi f_y \, \mu} \qquad (9.5.3)$$

in which the yield stress f_y for the shear-friction reinforcement should not be taken greater than 415 MPa. The coefficient of friction μ should be taken equal to 1.0 for concrete placed against hardened concrete and 0.7 for concrete placed against as-rolled structural steel. Equation. (9.5.3) gives the total area of shear friction reinforcement in the form of dowel reinforcement to be provided over the entire horizontal shear length which may be taken equal to half the span in simply supported beams and half the distance between the inflection points in continuous beams.

If tensile force exists in the direction normal to the interface, the shear transfer by contact may be assumed only when ties satisfying minimum requirements given below have been provided. The ties for horizontal shear may be in the form of single bars, multiple leg stirrups or the vertical legs of welded wire fabric. All ties should be anchored adequately on either side of the interface. When vertical bars or extended stirrups are used as ties, their spacing should not exceed 4 times the least dimension of the supported element nor 600 mm.

(c) European Code

The CEB—FIP Code considers composite sections in which the precast part is of prestressed concrete and in which the transmission of the tangential forces along the contact interface are in accordance with the usual techniques of reinforced concrete. The code specifies that the design should consider the state of stress existing before hardening of the second stage of concrete, the respective mechanical properties of the two concretes and the redistribution of forces resulting from shrinkage and creep. The stress conditions may be checked either at the ultimate limit states or only at the limit states of serviceability. A composite section may be designed as a monolithic section if,

$$v_{hu} \leqslant C_1 \frac{A_{sf} f_{sy}}{s} (\sin \beta + \cos \beta) + C_2 v_{chu} b \qquad (9.5.4)$$

Where v_{hu} = nominal ultimate horizontal shear stress

A_{sf} = area of shear-friction reinforcement

s = spacing of shear-friction reinforcement

f_{sy} = yield stress of shear-friction reinforcement

β = inclination of shear-friction reinforcement

b = breadth of interface

The dimensionless constants, $C_1 = 0$ and $C_2 = 1.25$ for $\frac{A_{sf}}{sb} \leqslant 0.002$ provided the contact surfaces are adequately rough. Also $C_1 = 0.9$ and $C_2 = 2.50$ for $\frac{A_{sf}}{sb} \geqslant 0.005$. For intermediate values of A_{sf}/sb, C_1 and C_2 may be computed by linear interpolation. The permissible ultimate horizontal shear v_{chu} depends upon the grade of concrete as shown in Table 9.5.3 in which the grade of concrete should be taken for the weakest of the concretes in contact.

Table 9.5.3

f_{ck}, MPa	12	16	20	25	30	35	40	45	50
v_{chu}, MPa	0.18	0.22	0.26	0.30	0.34	0.38	0.42	0.46	0.50

EXAMPLE 9.5.1

The composite section of Example 9.3.2. is subjected to a shear force of 75 kN. Determine the shear stress at the interface in the in-situ and precast concretes.

Solution

The intensity of shear stress may be determined from Equation 7.2.1 using the cross-sectional properties of the transformed section.

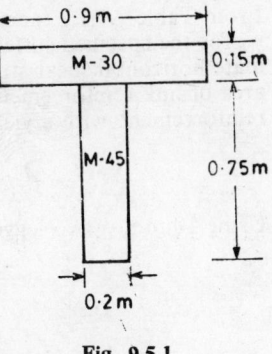

$$V_r = 0.075 \text{ MN}$$

$$I_t = 0.043 \text{ m}^4$$

$$b = 0.15 \text{ m}$$

$$(A_0 \, \bar{y}_0) = 0.9 \times 0.2 \, (0.3 - 0.1)$$

$$= 0.036 \text{ m}^3$$

$$v = \frac{0.075 \times 0.036}{0.043 \times 0.15} = 0.4186 \text{ MPa}$$

Fig. 9.5.1

EXAMPLE 9.5.2

A composite section shown in Fig. 9.5.1. comprises in-situ concrete flange of grade M-30 and precast prestressed web of grade M-45. The design ultimate horizontal shear stress at the as-cast contact surface is 1.5 MPa. Check the adequacy of the joint in accordance with the British Code.

Solution

From Table 9.5.1., the permissible design ultimate horizontal shear stress v_{ch} for the specified contact surface, (Class 1) is 0.55 MPa for Type A and 1.80 MPa for Type B. As the design ultimate horizontal shear stress v_h lies between these two values, nominal link reinforcement is necessary in order to make the joint adequate. A minimum link reinforcement of 0.15 per cent should be provided. Hence area of link reinforcement per metre length,

$$A_{vf} = \frac{0.15 \times 1000 \times 200}{100} = 300 \text{ mm}^2$$

Using 6 mm ϕ two legged stirrups, the spacing

$$s = \frac{1000}{\left(\dfrac{300}{2 \times \dfrac{\pi}{4} \times 6^2}\right)} = 189 \text{ mm}$$

EXAMPLE 9.5.3

The composite section of Example 9.5.2. has to resist a design ultimate horizontal shear stress of 2.0 MPa. Design the link reinforcement according to British Code.

Solution

From Table 9.5.1., the permissible design ultimate horizontal shear stress v_{ch} for the specified surface (Type B, class 1) is 1.8 MPa. As the design ultimate horizontal shear stress v_h is greater than the permissible value v_{ch}, the area of link reinforcement may be computed from Eq. (9.5.1). Using link reinforcement with a yield stress of 415 MPa,

$$A_{vf} = \frac{0.2 \times 2.0 \times 10^6}{0.87 \times 415} = 1108 \text{ mm}^2/\text{m}$$

Using 10 mm ϕ two legged stirrups, the spacing,

$$s = \frac{1000}{\left(\dfrac{1108}{2 \times \dfrac{\pi}{4} \times 10^2} \right)} = 142 \text{ mm}$$

EXAMPLE 9.5.4

A composite section shown in Fig. 9.5.1 carries a design ultimate shear force of 57 kN. Design the link reinforcement in accordance with ACI Code. Effective depth of the section is 0.75 m.

Solution

Using Eq. (9.5.2),

$$v_{hu} = \frac{0.057}{0.85 \times 0.20 \times 0.75} = 0.447 \text{ MPa}$$

As this shear stress is lesser than the permissible shear stress for type 2 contact surface, the surface need not be roughened intentionally. However, a minimum link reinforcement may be provided in accordance with Eq. (7.2.23). Using 6mm ϕ two-legged stirrups with a yied stress of 250 MPa,

$$s = \frac{2 \times \dfrac{\pi}{4} \times 6^2}{200} \times \frac{250}{0.34} = 208 \text{ mm}$$

EXAMPLE 9.5.5

A simply supported composite beam having a cross-section shown in Fig. 9.5.1 carries a design ultimate load of 42 kN/m over a clear span of 18 m. Design the shear friction reinforcement in accordance with ACI Code.

Solution

The critical section for shear is at a distance $\dfrac{D}{2} = 0.45$ m from the face of the support. The design ultimate shear force at the critical section,

$$V_u = 0.042 \left(\frac{18}{2} - 0.45 \right) = 0.3591 \text{ MN}$$

Hence from Eq. (9.5.2),

$$v_{hu} = \frac{0.3591}{0.85 \times 0.20 \times 0.75} = 2.816 \text{ MPa},$$

As v_{hu} exceeds 2.54 MPa, shear friction reinforcement is necessary. From Eq. (9.5.3), the total area of shear friction reinforcement using two-legged vertical stirrups with a yield stress of 250 MPa,

$$A_{vf} = \frac{0.3591 \times 10^6}{0.85 \times 250 \times 1} = 1690 \text{ mm}^2$$

This reinforcement should be provided over a length of $\frac{18}{2} = 9$ m. Adopting 6 mm diameter stirrups, the spacing

$$s = \frac{9000 \times 2 \times \frac{\pi}{4} \times 6^2}{1690} = 301.1 \text{ mm}$$

This spacing is acceptable as this is less than 4 times the least lateral dimension of in-situ concrete and is also less than 600 mm.

9.6 DEFLECTION

Similar to a monolithic beam, approach to be used for computation of deflection in a composite beam depends upon whether the stage of loading under consideration belongs to precracking range or post-cracking range.

9.6.1 Precracking Range

The deflections at different stages in the precracking range of the composite beam may be determined with reasonable accuracy using the linear elastic theory. The deflection depends upon the method of construction (propped or unpropped) and the stage of loading. In a composite beam formed by the combination of precast prestressed concrete and in-situ concrete, the moment of inertia of only the precast component I_{pc} should be taken into account in computing the initial and final camber due to prestress. The moment of inertia of the precast component should also be used in computing the deflection caused by self weight and weight of in-situ concrete in the case of unpropped construction. The weight of in-situ concrete is carried by formwork in propped construction until the beam can act as a composite section. Hence the moment of inertia of the transformed section I_t for the composite section may be used in the computation of deflection caused by weight of in-situ concrete in propped construction. As far as the service loads are concerned, they are generally imposed only after the in-situ concrete has gained sufficient strength and the beam develops composite action. Consequently, the moment of inertia I_t should be used for the computation of deflection caused by service loads both in propped as well as unpropped construction. The principle of superposition may be used to determine the deflection at any stage of loading. The loading at any stage may comprise one or more of prestressing force, weights of constituent materials and superimposed loads. The short-time deflections may be computed by using the expressions given in Sec. 8.2 or Table 8.2.1. The long-time deflections

344   PRESTRESSED CONCRETE

caused by permanent loads may be obtained by multiplying the short-time deflections by the factor $(1 + C_{cs})$ where C_{cs} is the coefficient for the combined effect of creep and shrinkage of respective concrete.

9.6.2 Post-cracking Range

The composite beams are designed such that they develop flexural cracking in the precast prestressed component located in the flexural tension zone only under overload condition, i.e. when the superimposed load exceeds the service load. The deflection caused by superimposed load in the post-cracking range may be determined by using Navier's hypothesis according to which the strains are distributed linearly across the composite section. The position of neutral axis, the distribution of strains and the resulting curvature at any cross-section for given superimposed load may be determined by adopting the iterative procedure discussed in Sec. 8.4. The deflection may then be determined by double integration of the curvature. Alternatively, an approximate value of deflection may be computed by using the equivalent moment of inertia I_e discussed in Sec. 8.3.1.2. The short-time deflection computed in this manner may be multiplied by the factor $(1 + C_{cs})$ to determine the long-time deflection caused by permanent load. The procedure is illustrated by Example 9.6.3.

EXAMPLE 9.6.1

Determine the short time deflections at the initial and final stages in the compoite beam of Example 9.3.2 formed by propped construction technique. The cable line in the precast prestressed element is parabolic with zero eccentricity at the ends and a sag of 0.24 m at midspan.

Solution

(i) *Initial stage*

Initial prestressing force, P_i = $720 \times 10^{-6} \times 1200$

= 0.864 MN

Intensity of transverse force due to prestress,

$$w_{pi} = \frac{2 P_i S}{l^2} = \frac{9 \times 0.864 \times 0.24}{16^2}$$

= 0.00648 MN/m

Moment of inertia of precast element,

$$I_p = \frac{1}{12} \times 0.15 \times 0.8^3 = 0.0064 \text{ m}^4$$

Deflection due to prestress,

$$\Delta_1 = -\frac{5}{384} \times \frac{0.00648 \times 16^4}{32000 \times 0.0064} = -0.027 \text{ m}$$

The minus sign denotes upward deflection or camber.

Self weight of precast element $= 0.15 \times 0.8 \times 24$

$$= 2.88 \text{ kN/m}$$

$$= 0.00288 \text{ MN/m}$$

Deflection due to self weight of precast element,

$$\Delta_2 = \frac{5}{384} \times \frac{0.00288 \times 16^4}{32000 \times 0.0064} = 0.012 \text{ m}$$

The net deflection of precast element,

$$\Delta_i = \Delta_1 + \Delta_2 = -0.027 + 0.012 = -0.015 \text{ m}$$

(ii) *Final stage*

Due to losses, the stress in prestressing steel decreases from 1200 to 1000 MPa. Hence, the final deflection due to prestress,

$$\Delta_3 = \Delta_1 \times \frac{1000}{1200} = -0.0225 \text{ m}$$

Moment of inertia of composite section,

$$I_t = 0.043 \text{ m}^4$$

Self weight of in-situ concrete $= 1.2 \times 0.2 \times 24$

$$= 5.76 \text{ kN/m}$$

$$= 0.00576 \text{ MN/m}$$

Deflection due to weight of in-situ concrete,

$$\Delta_4 = \frac{5}{384} \times \frac{0.00576 \times 16^4}{32000 \times 0.043} = 0.0036 \text{ m}$$

Deflection due to superimposed load,

$$\Delta_5 = \frac{5}{384} \times \frac{0.02356 \times 16^4}{32000 \times 0.043} = 0.0146 \text{ m}$$

Hence the net deflection,

$$\Delta = \Delta_2 + \Delta_3 + \Delta_4 + \Delta_5$$

$$= 0.0120 - 0.0225 + 0.0036 + 0.0146$$

$$\approx 0.0077 \text{ m}$$

EXAMPLE 9.6.2

Determine the deflections at the initial and final stages in the composite beam of Example 9.3.2 formed by unpropped construction technique. The cable line in the precast prestressed element is parabolic with zero eccentricity at the ends and a sag of 0.24 m at midspan.

Solution

(i) *Initial stage*

The deflection at this stage is the same as in the case of propped construction (Example 9.6.1),

$$\Delta_i = -\,0.015 \text{ m}$$

(ii) *Final stage (A)*

In unpropped construction, the weight of in-situ concrete is carried by precast element. The deflection due to weight of in-situ concrete,

$$\Delta_6 = \frac{5}{384} \times \frac{0.00576 \times 16^4}{32000 \times 0.0064} = 0.024 \text{ m}$$

Hence the net deflection,

$$\begin{aligned}
\Delta_A &= \Delta_2 + \Delta_3 + \Delta_6 \\
&= 0.0120 - 0.0225 + 0.0240 \\
&= 0.0135 \text{ m}
\end{aligned}$$

(iii) *Final stage (B)*

The net deflection at this stage

$$\begin{aligned}
\Delta_B &= \Delta_A + \Delta_5 \\
&= 0.0135 + 0.0146 \\
&= 0.0281 \text{ m}
\end{aligned}$$

EXAMPLE 9.6.3

Determine the midspan deflection at incipient collapse for a composite beam having the cross-section of Example 9.4.1 using ACI method of equivalent moment of inertia. The beam carries uniformly distributed load over an effective span of 22 m. Take the modular ratios of in-situ concrete and prestressing steel with respect to precast concrete as 0.675 and 6.5 respectively. Take modulus of rupture of precast concrete, $f_{cr} = 4.4$ MPa.

Solution

From Example 9.4.1, the ultimate moment,

$$M_u = 1.19 \text{ MN·m}$$

Using strain compatibility, the iterative procedure explained in the Example gives the depth of neutral axis, $x_u = 0.15$ m. Hence using Eq. (8.3.13), the moment of inertia of the cracked section at midspan,

$$\begin{aligned}
I_{cr} &= \frac{1}{3} \times 1.2 \times 0.15^3 + 6.5 \times 1020 \times 10^{-6} (0.9 - 0.15)^2 \\
&= 0.00508 \text{ m}^4
\end{aligned}$$

The transformed section may be considered in determining the properties of gross concrete section and the cracking moment M_{cr}. Treating precast concrete as reference material, the flange width of the transformed section $= 1.2 \times 0.675 = 0.81$ m. Hence the depth of centroid of gross transformed section from top,

$$\bar{x} = \frac{0.81 \times 0.2 \times 0.1 + 0.9 \times 0.15 \, (0.2 + 0.45)}{0.81 \times 0.2 + 0.9 \times 0.15}$$

$$= 0.35 \text{ m}$$

The moment of inertia about centroidal axis,

$$I_g = I_t = \frac{1}{12} \times 0.81 \times 0.2^3 + 0.81 \times 0.2 \times (0.35 - 0.1)^2$$

$$+ \frac{1}{12} \times 0.15 \times 0.9^3 + 0.15 \times 0.9 \, (0.65 - 0.35)^2$$

$$= 0.0319 \text{ m}^4$$

The effective prestressing force, $P = 1020 \times 10^{-6} \times 1000 = 1.02$ MN acts at an eccentricity, $e_p = 0.9 - 0.35 = 0.55$ m. Hence the compressive stress at bottom fibre due to prestressing,

$$f_B = \frac{1.02}{0.81 \times 0.2 + 0.9 \times 0.15} + \frac{1.02 \times 0.55 \, (1.1 - 0.35)}{0.0319}$$

$$= 16.62 \text{ MPa}$$

The cracking moment,

$$M_{cr} = \frac{(f_B + f_{cr}) \, I_g}{y_B}$$

$$= \frac{(16.62 + 4.40) \times 0.0319}{(1.1 - 0.35)}$$

$$= 0.894 \text{ MN·m}$$

Taking $M_{max} = M_u = 1.19$ MN·m, Eq. (8.3.11) gives,

$$I_e = \left(\frac{0.894}{1.19}\right)^3 \times 0.0319 + \left[1 - \left(\frac{0.894}{1.19}\right)^3\right] \times 0.00508$$

$$= 0.01645 \text{ m}^4$$

Taking $E_c = (2 \times 10^5)/6.5 = 30769$, the midspan deflection at incipient collapse,

$$\Delta_u = \frac{5}{48} \frac{M_u \, l^2}{E_c \, I_e}$$

$$= \frac{5}{48} \times \frac{1.19 \times 22^2}{30769 \times 0.01645}$$

$$= 0.1185 \text{ m}$$

9.7 DIFFERENTIAL SHRINKAGE AND CREEP

The in-situ concrete in a composite beam is of relatively lower grade (M-15 or M-20) and has correspondingly higher water-cement ratio. On the other hand, the precast prestressed concrete is of higher grade (M-30 to M-60) and most of its shrinkage has already occurred before the placement of in-situ concrete. Consequently, the in-situ concrete shrinks more than the precast concrete. It follows that there is differential shrinkage between the two types of concrete. The position in respect of creep is just the opposite. Generally the precast prestressed concrete creeps more than in-situ concrete as prestressing force is a permanent load. Consequently, the effect of differential shrinkage is partly cancelled by differential creep. Due to this reason and also due to the complexity of the problem, the effects of differential shrinkage and creep are often ignored in the computation of stresses in a composite beam. On the other hand, differential shrinkage in certain cases may have appreciable effect on the deformations, particularly the camber.

The differential shrinkage and creep induce self-straining of the composite beam in which the precast and in-situ concretes interact with each other by means of two self equilibrating equal and opposite tangential forces acting along the interface. If the differential shrinkage of in-situ concrete is larger than the differential creep of precast concrete, the in-situ concrete is subjected to a tensile force while the precast concrete is subjected to an equal compressive force along the interface of two types of concrete. The magnitude of these interactive forces is such that the continuity of strain at the interface is maintained i.e. the shortening of the two materials along the interface is the same. It follows that the magnitude of the interactive forces at the interface depends upon the relative stiffness against elongation for the in-situ and precast elements. Due to the difficulty in the determination of relative stiffnesses and also due to the fact that both shrinkage and creep are time-dependent phenomena, it appears imperative to introduce certain simplifying assumptions. It is commonly assumed for simplicity that the precast component is rigid so that the differential shrinkage of in-situ concrete is completely restrained. Consequently, the tensile force T induced in the in-situ concrete.

$$T = A_{ci} \ E_{ci} \ \epsilon_{sh} \qquad (9.7.1)$$

where A_{ci} = area of in-situ concrete

E_{ci} = modulus of elasticity of in-situ concrete

ϵ_{sh} = differential shrinkage strain

The uniform tensile stress in in-situ concrete due to restrained shrinkage is $T/A_i = E_{ci} \ \epsilon_{sh}$. On the other hand a compressive force C equal to the tensile force T acting at the centroid of the in-situ concrete is resisted by the composite section. Figure 9.7.1 (a) shows a composite section comprising in-situ concrete flange and precast prestressed rib. Taking precast flange as the reference material, the transformed section is shown in Fig. 9.7.1 (b) in which the width of the flange is m b_f where m is the modular ratio of in-situ concrete with respect to precast concrete. As the compressive force C

acts at an eccentricity e from the centroidal axis of the transformed section, the net stress at top fibre of the composite section.

$$f_T = m \left[\frac{-C}{A_t} - \frac{(Ce) y_T}{I_t} \right] + \frac{T}{A_{ci}} \qquad (9.7.2)$$

where $\quad m = \dfrac{E_{ci}}{E_{cp}}$ = modular ratio of in-situ concrete

E_{cp} = modulus of elasticity of precast concrete

A_t = area of transformed section

I_t = moment of inertia of the transformed section about its centroidal axis.

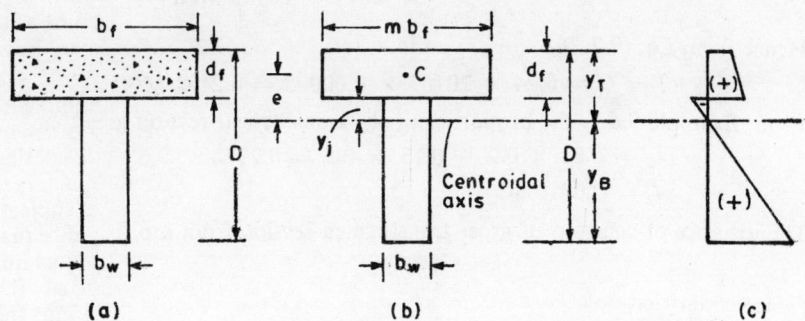

Fig. 9.7.1

The net stress in in-situ concrete at its junction with precast concrete,

$$f_{ji} = m \left[\frac{-C}{A_t} - \frac{(Ce) y_j}{I_t} \right] + \frac{T}{A_{ci}} \qquad (9.7.3)$$

The stress in precast concrete at its junction with in-situ concrete,

$$f_{jp} = \frac{-C}{A_t} - \frac{(Ce) y_j}{I_t} \qquad (9.7.4)$$

The streas at the bottom fibre of the composite section,

$$f_B = \frac{-C}{A_t} + \frac{Ce \, y_B}{I_t} \qquad (9.7.5)$$

The stresses given by Eq. (9.7.2) to (9.7.5) are caused only due to differential shrinkage. These should be added algebraically to the stresses caused by prestress and service load to obtain the net stresses. The stress diagram due to differential shrinkage in composite beams of common proportions is shown in Fig. 9.7.1 (c). While the in-situ concrete flange is in tension, the rib develops compressive stress at its top and tensile stress at bottom. The stresse due to differential shrinkage influence the elastic stresses caused by prestress and service loads. They do not affect the behaviour at the ultimate stage.

EXAMPLE 9.7.1

Determine the stresses due to differential shrinkage in the composite section of Example 9.4.1 taking $\epsilon_{sh} = 0.000015$. Modular ratios of in-situ concrete and prestressing steel with respect to precast concrete are 0.675 and 6.5 respectively.

Solution

Using the data of Examples 9.4.1 and 9.6.3,

$$A_{ci} = 1.2 \times 0.2 = 0.24 \text{ m}^2$$

$$E_{ci} = \frac{2 \times 10^5}{6.5} \times 0.675 = 20769 \text{ MPa}$$

Hence from Eq. (9.7.1),

$$T = C = 0.24 \times 20769 \times 0.00005 = 0.2492 \text{ MN}$$

From Example 9.6.3, the properties of the transformed section are

$$A_t = 0.81 \times 0.2 + 0.15 \times 0.9 = 0.297 \text{ m}^2$$
$$I_t = 0.0319 \text{ m}^4$$

The distance of centroid of gross transformed section from top,

$$\bar{x} = 0.35 \text{ m}$$

Hence eccentricity of C,

$$e = 0.35 - \frac{0.2}{2} = 0.25 \text{ m}$$

Also,

$$y_T = \bar{x} = 0.35 \text{ m}$$
$$y_B = 1.1 - 0.35 = 0.75 \text{ m}$$
$$y_j = 0.35 - 0.20 = 0.15 \text{ m}$$

From Eq. (9.7.2), the stress at top of the composite section,

$$f_T = 0.675 \left[\frac{-0.2492}{0.297} - \frac{0.2497 \times 0.25 \times 0.35}{0.0319} \right] + \frac{0.2492}{0.24}$$

$$= 0.0106 \text{ MPa}$$

From Eq. (9.7.3), the stress in in-situ concrete at its junction with precast concrete,

$$f_{ji} = 0.675 \left[\frac{-0.2492}{0.297} - \frac{0.2497 \times 0.25 \times 0.15}{0.0319} \right] + \frac{0.2492}{0.24}$$

$$= 0.274 \text{ MPa}$$

From Eq. (9.7.4), the stress in precast concrete at its junction with in-situ-concrete

$$f_{jp} = - \frac{0.2492}{0.297} - \frac{0.2497 \times 0.25 \times 0.15}{0.0319}$$

$$= - 1.132 \text{ MPa}$$

From Eq. (9.7.5), the stress at the bottom fibre of composite section,

$$f_B = -\frac{0.2497}{0.297} + \frac{0.2497 \times 0.25 \times 0.75}{0.0319}$$

$$= 0.626 \text{ MPa}$$

9.8 DESIGN PROCEDURE

As pointed out in Sec. 9.2, there are several construction techniques for composite members. Evidently there are several desing procedures appropriate to the construction tenchniques being adopted. In the following discussion, the common type of composite construction is considered in which the precast prestressed element is placed in flexural tension zone and in-situ concrete element in flexural compression zone. Similar procedure can be developed for other types of composite construction.

The precast prestressed element in an unpropped construction has to be designed for the initial stage and for the final stage (A) discussed in Sec. 9.3. For this purpose, the design procedure developed in Sec. 6.8 may be adopted. In the case of propped construction, the final stage (A) is not relevant. Hence only the initial stage is crucial for precast element in propped construction.

The size and geometrical properties of the whole section acting as a composite section are mostly controlled by final stage (B) under the action of full service load. The composite section may be designed for the limit state of collapse in flexure. For this purpose, the cross-section of the composite beam should be selected in such a way that ultimate moment of the composite section is not smaller than the design ultimate moment obtained by multiplying the working moment by the appropriate load factor. The ultimate moment of the composite section may be computed from the equation,

$$M = T\,(jd_p)$$

$$= A_p\,f_p\,(jd_p) \tag{9.8.1}$$

The stress f_p in prestressing steel of area A_p may be taken tentatively equal to the average of initial stress f_{pi} and ultimate stress f_{pu}. The lever arm factor j usually ranges from 0.8 to 0.9. For a preliminary design f_p and j may be taken tentatively equal to $0.5\,(f_{pi} + f_{pu})$ and 0.85 respectively. The area of prestressing steel A_p and effective depth of composite section d_p may then be adjusted so that ultimate moment capacity of the section is not less than the ultimate design moment. Alternatively, the ultimate design procedure of the Indian Code discussed in Sec. 6.9 may be used treating the composite section as a homogeneous one. The composite section thus chosen should then be checked for the loading stages discussed in Sec. 9.3 and also for the limit state of collapse in flexure using strain compatibility analysis outlined in Sec. 9.4. The composite construction should also be checked for shear (Sec. 9.5), deflection (Sec. 9.6) and the effect of differential shrinkage (Sec. 9.7). In particular, adequate shear connectors should be provided at the interfaces of constituent elements to ensure composite action.

EXAMPLE 9.8.1

A composite beam of T-section comprising 1.2×0.2 m flange and 0.15×0.9 m web has to be built by propped construction technique. The in-situ concrete flange and the precast post-tensioned rib are to be made of concretes of grades M-20 and M-45 respectively. The beam has to carry a service load of 7 kN/m and an ultimate load of 20 kN/m exclusive of self weight over an effective span of 18 m. Design the midspan section.

Solution

Gross cross-sectional area of concrete,

$$A_g = 1.2 \times 0.2 + 0.15 \times 0.9 = 0.375 \text{ m}^2$$

Taking the unit weight of concrete as 24 kN/m^3, the self weight,

$$w_b = 0.375 \times 24 = 9 \text{ kN/m}^3$$

Hence the design ultimate moment at midspan,

$$M_{ud} = \frac{(9 + 20) \times 18^2}{8} = 1174.5 \text{ kN·m} = 1.1745 \text{ MN·m}$$

The area of prestressing steel may be determined from Eq. (9.8.1). Taking the initial stress in prestressing steel as 1100 MPa and its characteristic ultimate strength as 1600 MPa, the stress in prestressing steel f_p at ultimate moment may be taken tentatively as $(1100 + 1600)/2 = 1350$ MPa. Similarly, the internal lever arm may be taken as $0.85\, d_p$. Allowing a concrete cover equal to 0.2 m, $d_p = 1.1 - 0.2 = 0.9$ m.

Hence the area of prestressing steel,

$$A_p = \frac{1.1745}{1350 \times (0.85 \times 0.9)} = 0.001137 \text{ m}^2 = 1137 \text{ mm}^2$$

The initial prestressing force before losses,

$$p_i = 1137. \times 10^{-6} \times 1100 = 1.25 \text{ MN}$$

Assuming 20% loss of prestress,

$$P = 0.8 \times 1.25 = 1 \text{ MN}$$

The section may now be checked for limit states of serviceability and collapse.

(i) Initial stage

At this stage the precast prestressed rib is subjected to initial prestressing force P_i and its own weight. The self weight of rib is equal to $0.15 \times 0.9 \times 1 \times 0.024 = 0.00324$ MN/m. The eccentricity of prestress, $e_p = 0.45 - 0.2 = 0.25$ m. The stress at top fibre of the rib,

$$f_T = -\frac{1.25}{0.15 \times 0.9} + \frac{1.25 \times 0.25}{\frac{1}{6} \times 0.15 \times 0.9^2} - \frac{0.00324 \times 18^2}{8 \times \frac{1}{6} \times 0.15 \times 0.9^2}$$

$$= -9.26 + 15.43 - 6.48 = -0.31 \text{ MPa}$$

The stress at bottom fibre,

$$f_B = -0.26 - 15.43 + 6.48 = -18.21 \text{ MPa}$$

Assuming the strength of concrete at transfer $f_{ci} = 40$ MPa, the permissible compressive stress according to Eq. (6.7.2),

$$\sigma_{cbci} = 0.455 \times 40 = 18.2 \text{ MPa}$$

Hence the actual stresses are acceptable.

(ii) *Final stage (B)*

At this stage the composite section is required to carry the service load. The properties of the composite section are as follows.

Modular ratio of in-situ concrete with respect to precast concrete treated as reference material,

$$m = \frac{5700 \sqrt{20}}{5700 \sqrt{45}} = 0.667$$

Hence the equivalent width of the flange in the transformed section is 1.2 × 0.667 = 0.8 m. Area of the transformed section,

$$A_t = 0.80 \times 0.20 + 0.15 \times 0.9 = 0.295 \text{ m}^2$$

The distance of centroid from the top face,

$$\bar{x} = \frac{0.80 \times 0.2 \times 0.1 + 0.15 \times 0.9 \, (0.2 + 0.45)}{0.295}$$

$$= 0.3517 \text{ m}$$

Moment of inertia about centroidal axis,

$$I_t = \frac{1}{12} \times [0.80 \times 0.2^3 + 0.80 \times 0.2 \, (0.3517 - 0.1)^2$$

$$+ \frac{1}{12} \times 0.15 \times 0.9^3 + 0.15 \times 0.9 \, (0.65 - 0.3517)^2$$

$$= 0.0318 \text{ m}^4$$

Total service load inclusive of self weight is 7 + 9 = 16 kN/m and hence service load moment at midspan,

$$M_w = \frac{0.016 \times 18^2}{8} = 0.648 \text{ MN.m}$$

Stress at top fibre,

$$f_T = -\frac{0.648}{0.0318} \times 0.3517 = -7.17 \text{ MPa}$$

Stress at bottom fibre due to prestress after losses,

$$f_{Bp} = -\frac{1.0}{0.15 \times 0.9} - \frac{1.0 \times 0.25}{\frac{1}{6} \times 0.15 \times 0.9^2} = -19.75 \text{ MPa}$$

Stress at bottom fibre due to service loads,

$$f_{Bw} = \frac{0.648\,(1.1 - 0.3517)}{0.0318} = 15.25 \text{ MPa}$$

Hence the net stress at bottom,

$$f_B = -19.75 + 15.25 = -4.50 \text{ MPa}$$

As the permissible compressive stress in flexure σ_{cbc} in M-20 concrete as per IS: 456-1978 is 7 MPa and from Eq. (6.7.1) for M-45 concrete σ_{cbc} is equal to 16.65 MPa, hence the stresses at final stage (B) are acceptable.

(iii) *Limit state of collapse*

The ultimate moment of the composite section may be determined conservatively by assuming the entire section to be formed of the weaker M-20 concrete and using the provisions of IS: 1343-1980. As the area of prestressing steel was tentatively taken equal to 1137 mm^2, the single Freyssinet cable 24 ϕ 8 having a nominal area of 1206 mm^2 may be adopted. The dia. of sheathing for this cable is 60 mm which can be conveniently accommodated in 150 mm thick web. According to Table 3.3.2, the nominal ultimate strength of the cable, $A_p f_{pu} = 1688$ kN. Hence the ratio,

$$\frac{A_p f_{pu}}{b_f\, d_p\, f_{ck}} = \frac{1.688}{1.2 \times 0.9 \times 20} = 0.078$$

From Table 6.9.1,

$$\frac{f_p}{0.87\, f_{pu}} = 1$$

or $f_p = 0.87\, f_{pu} = 0.87 \times \dfrac{1.688}{1206 \times 10^{-6}} = 1218$ MPa

Also,

$$\frac{x_u}{d_p} = 0.1695$$

or $x_u = 0.1695 \times 0.9 = 0.1526$ m

Using Eq. (6.9.1), the ultimate moment,

$$M_u = 1206 \times 10^{-6} \times 1218\,(0.9 - 0.42 \times 0.1526)$$
$$= 1.2279 \text{ MN.m}$$

As the moment capacity is greater than the design ultimate moment of 1.1745 MN.m, the design satisfies the requirement at the limit state of collapse.

Although the shear force at midspan section is zero, a minimum area equal to 0.15 percent of the contact area may be provided in accordance with the provisions of the British code. Hence area of link reinforcement per metre length of the beam,

$$A_{vf} = \frac{0.15}{100} \times 1000 \times 150 = 225 \text{ mm}^2$$

Using 6 mm ϕ two legged stirrups, the spacing of stirrups,

$$s = \frac{1000 \times 2 \times \frac{\pi}{4} \times 6^2}{225} = 251.43 \text{ mm}$$

Hence a spacing of 250 mm may be adopted. These stirrups may also serve to resist vertical shear in the composite section. According to Eq. (7.2.15), the minimum area of a single stirrup with a yield stress of 315 MPa,

$$A_v = \frac{0.4 \times 150 \times 250}{0.87 \times 315} = 54.7 \text{ mm}^2$$

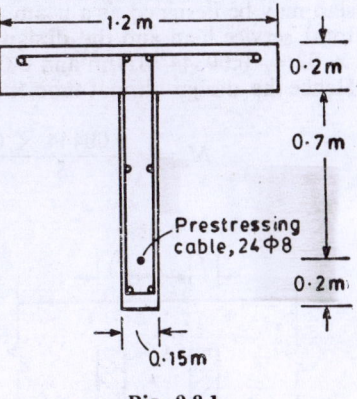

As the actual area of two vertical legs is $2 \times \frac{\pi}{4} \times 6^2 = 56.6 \text{ mm}^2$, the stirrups provided are sufficient to meet the minimum requirement. The vertical legs may extend from the precast prestressed rib into the cast-in-place flange upto a point close to the top face and bent at right angles as shown in Fig. 9.8.1. In addition to the stirrup reinforcement, 10 mm ϕ longitudinal bars may also be provided at the four corners of the stirrups. In addition, four more longitudinal bars each of 10 mm ϕ may also be provided as shown in the figure.

Fig. 9.8.1

EXAMPLE 9.8.2

The composite deck slab for a pedestrian bridge formed by unpropped construction technique comprises precast prestressed concrete elements of grade M-40 and cost-in-situ concrete of grade M-20 as shown in Fig. 9.8.2 (a). The effective span of the deck slab is 6 m. The superimposed service load and the design ultimate load are 10 and 15 kN/m² respectively. Design the deck slab.

Solution

(A) *Midspan section*

Taking the overall thickness of the composite slab equal to 1/20 th of the span,

$$D = \frac{6}{20} = 0.3 \text{ m}$$

Taking the unit weight of concrete as 0.024 MN/m³,

$$w_s = 0.3 \times 1 \times 1 \times 0.024 = 0.0072 \text{ MN/m}^2$$

Hence the total service load is $(0.010 + 0.0072) = 0.0172 \text{ MN/m}^2$

and the design ultimate load is $(0.015 + 0.0072) = 0.0222 \text{ MN/m}^2$.

The precast prestressed elements may by formed by long line system of pre-tensioning. The cross-section of the element may be in the form of inverted T comprising 0.2×0.05 m flange and 0.15×0.05 web with an overall depth of 0.2 m. The precast elements serve as formwork during the placement of in-situ concrete having a minimum thickness of 0.10 m over the top of the precast element. The precast elements are placed side by side to obtain the required overall width of the slab. For the sake of analysis a strip of slab of 0.2 m width covered by a single unit as shown in Fig. 9.8.2 (b) may be considered. Assuming one way action, the strip of composite slab may be designed as a beam independent of the adjacent strips. The total service load and the design ultimate load for the strip are 0.0172 $\times$ 02 = 0.00344 MN/m and $0.0222 \times 0.2 = 0.00444$ MN/m respectively. Hence the design ultimat moment,

$$M_{ud} = \frac{0.00444 \times 6^2}{8} = 0.01998 \text{ MN} \cdot \text{m}$$

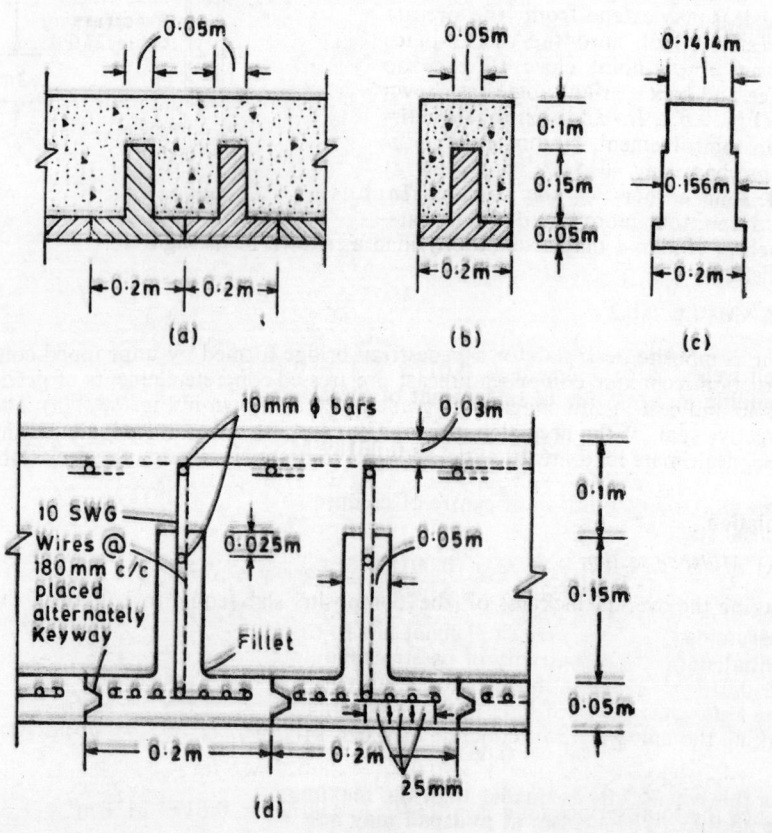

Fig. 9.8.2

For an initial stress of 1100 MPa and breaking stress of 1500 MPa, the stress in tendons at ultimate moment may be taken tentatively as $(1100 + 1500)/2 = 1300$ MPa. Taking the internal lever arm gqual to $0.65\ D$ and using a partial safety factor of 0.87 for the tendons, the area of prestressing steel,

$$A_p = \frac{M_{ud}}{(jd_p)\,f_p} = \frac{0.01998}{0\,65 \times 0.3 \times 0.87 \times 1300} \times 10^6$$

$$= 90.59\ \text{mm}^2$$

The initial prestressing force before losses,

$$P_i = A_p f_{pi} = 90.59 \times 10^{-6} \times 1100 = 0.09965\ \text{MN}$$

Assuming 20% losses, the effective prestressing force,

$$P = 0.8\ P_i = 0.8 \times 0.09965 = 0.07972\ \text{MN}$$

The properties of the precast element are as follows.

$$A = 0.2 \times 0.05 + 0.05 \times 0.15 = 0.0175\ \text{m}^2$$

Distance of centroid from top,

$$\bar{x} = y_T = \frac{0.2 \times 0.05 \times 0.175 + 0.05 \times 0.15 \times 0.075}{0.0175}$$

$$= 0.1321\ \text{m}$$

$$y_B = 0.2 - 0.1321 = 0.0679\ \text{m}$$

Moment of inertia of the element about its centroidal axis,

$$I = 5.9 \times 10^{-5}\ \text{m}^4$$

Distance of lower kern from centroidal axis,

$$e_{kl} = \frac{I/A}{yT} = \frac{5.0 \times 10^{-5}/0.1075}{0.1321} = 0.0255\ \text{m}$$

Self weight of the element is $0.0175 \times 1 \times 0.024 = 0.00042$ MN/m. The bending moment due to self weight of the element,

$$M = \frac{0.00042 \times 6^2}{8} = 0.00189\ \text{MN·m}$$

The shift of the position of centre of compression due to self weigth at the initial stage,

$$\Delta = \frac{M}{P_i} = \frac{0.00189}{0.09965} = 0.019\ \text{m}$$

Assuming that the precast element has to carry only its own weight at the initial stage, the eccentricity of prestress should not exceed $e_{kl} + \Delta = 0.0255 + 0.019 = 0.0445$ m in order to avoid tensile stress at the top fibre. Allowing a concrete cover of 0.03 m, the eccentricity of prestress,

$$e_p = 0.0679 - 0.03 = 0.0379\ \text{m}$$

As this eccentricity is smaller than the maximum permissible value, it may be adopted. The stresses at midspan may now be checked.

(i) *Initial statge*

At this stage, the prestressing force, $P_i = 0.09965$ MN and only the self weight of the precast element is effective. The stress at top fibre,

$$f_T = -\frac{0.09965}{0.0175} + \frac{0.09965 \times 0.0379 \times 0.1321}{5.9 \times 10^{-5}} - \frac{0.00189 \times 0.1321}{5.9 \times 10^{-5}}$$

$$= -1.47 \text{ MPa}$$

The stress at bottom fibre,

$$f_B = -\frac{0.09965}{0.0175} - \frac{0.09965 \times 0.0379 \times 0.0679}{5.9 \times 10^{-5}} + \frac{0.00189 \times 0.0679}{5.9 \times 10^{-5}}$$

$$= -7.87 \text{ MPa}$$

Assuming that the prestress is transferred when the strength of concrete $f_{ci} = 0.6 f_{ck} = 24$ MPa, the permissible compressive stress according to Eq. (6.7.2),

$$\sigma_{cbci} = 0.51 \times 24 = 12.24 \text{ MPa}$$

Hence the initial stresses are within permissible limits.

(ii) *Final state* (A)

At this stage, the prestressing force, $P = 0.07972$ MN and the self weight of the entire composite section is effective. The bending moment due to self weight,

$$M = \frac{(0.3 \times 0.2 \times 1 \times 0.024) \times 6^2}{8} = 0.00648 \text{ MN·m}$$

Stress at top fibre of precast element

$$f_T = -\frac{0.07972}{0.0175} + \frac{0.07972 \times 0.0379 \times 0.1321}{5.9 \times 10^{-5}} - \frac{0.00648 \times 0.1321}{5.9 \times 10^{-5}}$$

$$= -12.30 \text{ MPa}$$

Stress at bottom gbre,

$$f_B = -\frac{0.07972}{0.0175} - \frac{0.07972 \times 0.0379 \times 0.0679}{5.9 \times 10^{-5}} + \frac{0.00648 \times 0.0679}{5.9 \times 10^{-5}}$$

$$= -0.57 \text{ MPa}$$

The permissible compressive stress for M-40 concrete according to Eq. (6.7.1),

$$\sigma_{cbc} = 0.39 f_{ck} = 0.39 \times 40 = 15.6 \text{ MPa}$$

Hence the stresses in the precast element are within acceptable limits.

(iii) *Final stage* (B)

At this stage, the prestressing force $P = 0.07972$ MN and the entire service load equal to 0.00344 MN/m is resisted by the composite section. Modular ratio of in-situ concrete with respect to precast concrete,

$$m = \frac{5700 \sqrt{20}}{5700 \sqrt{40}} = 0.707$$

The properties of the transformed section shown in Fig. 9.8.2 (c) are as follows,

$$A_t = 0.1414 \times 0.1 + 0.156 \times 0.15 + 0.2 \times 0.05$$
$$= 0.04754 \text{ m}^2$$

The distance of centroid from top fibre,

$$\bar{x} = y_T = \frac{0.1414 \times 0.10 \times 0.05 + 0.156 \times 0.15 \times 0.175 + 0.2 \times 0.05 \times 0.275}{0.04754}$$

$$= 0.1589 \text{ m}$$

$$y_B = 0.3 - 0.1589 = 0.1411 \text{ m}$$
$$I_t = 3.6629 \times 10^{-4} \text{ m}^4$$

The bending moment at midspan,

$$M = \frac{0.00344 \times 6^2}{8} = 0.01548 \text{ MN·m}$$

The stress in in-situ concrete at top fibre,

$$f_T = \frac{-0.01548 \times 0.1589}{3.6629 \times 10^{-4}} \times 0.707 = -4.75 \text{ MPa}$$

The permissible stress σ_{cbc} for M-20 concrete is 7 MPa. Hence the actual stress is within permissible limit. Stress at bottom fibre due to service load,

$$f_{Bw} = \frac{0.01548 \times 0.1411}{3.6629 \times 10^{-4}} = 5.96 \text{ MPa}$$

Stress at bottom fibre due to prestressing,

$$f_{Bp} = -\frac{0.07972}{0.0175} - \frac{0.07972 \times 0.0379 \times 0.0679}{5.9 \times 10^{-5}}$$

$$= -8.03 \text{ MPa}$$

Hence the net stress at bottom fibre,

$$f_B = f_{Bw} + f_{Bp} = 5.96 - 8.03 = -2.07 \text{ MPa}$$

Hence the stress at this stage are within permissible limits.

(iv) *Limit state of collapse*

At this stage a composite section is required to resist design ultimate moment of 19.98 kN·m. The ultimate moment capacity may be computed conservatively by assuming that the entire section is formed of weaker M-20 concrete. Using IS: 1343-1980 and taking breaking stress of prestressing steel $f_{pu} = 1500$ MPa, the ratio

$$\frac{A_p f_{pu}}{bd_p f_{ck}} = \frac{0.959 \times 10^{-6} \times 1500}{0.2 \times 0.27 \times 20} = 0.1258$$

Hence from Table 6.9.1,

$$\frac{f_p}{0.87 f_{pu}} = 1$$

or $\qquad f_p = 0.87 \times 1500 = 1305$ MPa

and $\qquad \dfrac{x_u}{d_p} = 0.273$

or $\qquad x_u = 0.273 \times 0.27 = 0.0737$ m

Using Eq. (6.9.1), the ultimate moment,

$$M_u = 90.59 \times 10^{-6} \times 1305 \,(0.27 - 0.42 \times 0.0737)$$
$$= 0.0283 \text{ MN·m}$$

As the moment capacity is greater than the ultimate design moment, the section is safe.

(v) Reinforcement details

Using 4 mm diameter indented tensile wires for prestressing, the number of wires required,

$$n = \frac{90.59}{\dfrac{\pi}{4} \times 4^2} = 7.206$$

Using 7 wires, the initial stress in each wire,

$$f_{pi} = \frac{0.09965}{7 \times \dfrac{\pi}{4} \times 4^2 \times 10^{-6}} = 1132.4 \text{ MPa}$$

The effective stress after losses,

$$f_{pe} = 0.8 f_{pi} = 0.8 \times 1132.4 = 906 \text{ MPa}$$

The wires may be placed at a spacing of 25 mm in a single row at a distance of 30 mm from the soffit as shown in Fig. 9.8.2 (d). Although the shear force at midspan section is zero, a minimum area equal to 0.15 per cent of the contact area may be provided in accordance with the provisions of the British code. Hence area of link einforcement per metre length of the beam,

$$A_{vf} = \frac{0.15}{100} \times 50 \times 1000 = 75 \text{ mm}^2$$

Using 10 SWG mild steel wire having a dimeter of 3.25 mm, the spacing of link reinforcement,

$$s = \frac{1000 \times \dfrac{\pi}{4} \times 3.25^2}{75} = 110.65 \text{ mm}$$

The link reinforcement may also serve to resist vertical shear in the composite section. According to Eq. (7.2.15), the maximum spacing of 10 SWG vertical bar with a yield stress of 250 MPa,

$$s = \frac{\dfrac{\pi}{4} \times 3.25^2 \times 0.87 \times 250}{0.4 \times 50} = 90.25 \text{ mm}$$

Hence vertical bars of 10 SWG at a spacing of 90 mm may be adopted. This spacing is less than 4 times the thickness of web and 3/4th of the effective depth. The link reinforment should project from the precast element and anchored adequately in the cast-in-situ concrete. The in-situ concrete may have 0 3 per cent reinforcement to take care of shrinkage and thermal stresses. Area of in-situ concrete,

$$A_{ci} = 0.2 \times 0.3 - 0.0175 = 0.0425 \text{ m}^2$$

Hence area of steel in in-situ concrete,

$$A_{si} = \frac{0.0425 \times 0.3}{100} \times 10^6 = 127.5 \text{ mm}^2$$

Hence 10 mm reinforcing bars may be provided at 100 mm spacing in the longitudinal direction as shown in the figure. An equal amount of reinforcement may also be provided in the transverse direction. The precast elements may have suitable keyways in the flange and 45° fillets at the junction of flange and web.

(B) End Section

As the effective span is 6 m, the clear span may be taken equal to 6 − 0.27 = 5.73 m. Assuming that the transfer length for 4 mm diameter tendons is $100 \times 4 = 400$ mm, the prestressed units should extend 0.4 m beyond the edge of the support. Hence the overall length of the element should be $5.73 + 2 \times 0.4 = 6.53$ m. As the bending moment near the end section is negligible, the prestressed element and the composite section need not be checked for flexure. However, the cable line may be raised linearly to the centroid of the prestressed element at its ends in order to eliminate tensile stress at top fibre and also to gain some advantage in resisting shear. The critical section for shear is located at a distance d_p (equal to 0.27 m) from the edge of the support. The distance of critical section from midspan is $0.5 \times 5.73 - 0.27 = 2.595$ m. Hence the design ultimate shear force,

$$V_{ud} = 2.595 \times 0.00444 = 0.01152 \text{ MN}$$

The shear force has to be resisted by the composite section which may be taken conservatively as consisting entirely of weaker M-20 concrete. As the bending moment is negligible, the section may be assumed to be uncracked in flexure. Hence the shear force resisted by concrete may be taken equal to V_{c0} in accordance with Eq. (7.2.10). As only a part of the composite section is prestressed, the prestress may by ignored in computing the shear resistance of concrete. As the shear force resisted by concrete is greater than the design ultimate shear force, the shear reinforcement need not be provided.

The prestressed element should also be checked for shear at final stage (A). At this stage, the design ultimate shear force, adopting a partial safety factor $\gamma_f = 1.5$,

$$V_{ud} = 0.2 \times 0.3 \times 1 \times 0.024 \times 1.5 \times 2.595$$
$$= 0.0056 \text{ MN}$$

The shear resisted by concrete V_{co} may be computed from Eq. (7.2.10).

$$P = \frac{P}{A} = \frac{0.07972}{0.0175} = 4.555 \text{ MPa}$$

$$\sin \theta \approx \frac{e_p}{\left(\dfrac{L}{2}\right)} = \frac{0.0379}{\left(\dfrac{6.53}{2}\right)} = 0.0116$$

$$V_{co} = 0.05 \times 0.2 \times 0.67 \times 0.24 \sqrt{40} \sqrt{1 + \frac{0.8 \times 4.555}{0.24\sqrt{40}}}$$
$$+ 0.07972 \times 0.0116$$
$$= 0.01968 \text{ MN}$$

As the shear resistance of concrete is greater than V_{ud}, the minimum shear reinforcement provided in the prestressed element should suffice.

It may also be verified that the shear reinforcement provided in the prestressed element is adequate at any other section as well.

REFERENCES

9.1 Abeles, P.W.; Brown, E.I. and Hu, C.H., *Tests of composite concrete beams with prestressed planks*, Materiaux et Construction, Vol. 5, No. 5. 1972, pp. 31-40.

9.2 Dudra, J., *Design and construction of Hudson Hop Bridge*, Journal of the Prestressed Concrete Institute, Vol. 11, No. 2, April 1966.

9.3 Evans, R.H. and Parker, A.S., *Behaviour of prestressed concrete composite beams*, Journal of the American Concrete Institute, Vol. 51, 1955, pp. 861-81.

9.4 Evans, R.H. and Chung H.W., *Horizontal shear failure of prestressed composite T-beams with cast-in-situ lightweight concrete deck*, Concrete, April 1969, pp. 124-6.

9.5 Evans, R.H. and Chung, H.W., *Flexural cracks in composite prestressed lightweight concrete beams*, Civil Engineering and Public Works Review, Vol. 63, January 1968.

9.6 Gurfinkel, G., *Design of hold downs in prestressed concrete girders*, The Indian Concrete Journal, Vol. 41, No. 2, February 1967, p. 62.

9.7 Hanson, N.W., *Precast prestressed concrete bridges, horizontal shear connections*, PCA. 1960.

9.8 Lu Ln Wang, *A direct method for designing composite sections in prestressed concrete*, Journal of the Prestressed Concrete Institute, Vol. 10, No. 5, October 1965.

9.9 Mattock, A.H. *Precast prestressed concrete bridges, creep and shrinkage studies*, Journal of the PCA Research and Development Laboratories, Vol. 3. No. 2, May 1961, pp. 32-66.

9.10 Maver, J.L., *Precast and prestressed concrete bridges and structures in irrigation projects*, Journal of the Institution of Engineers (Australia), Vol. 39, No. 6, June 1967, pp. 67-74.

9.11 Mustafa, Saad, E., *Ultimate load test of a segmentally constructed concrete I-beam*, Journal of the Prestressed Concrete Institute, July-August 1974.

9.12 Ogara, M.B. and Bezouska, T.J., *Field study of curved continuous prestressed bridge*, Journal of the Prestressed Concrete Institute, Vol. 8, No. 6, December 1963.

9.13 Pretzer, C.A., *Unusual application of prestressed waffle slabs and composite beams*, Journal of the American Concrete Institute, Proceedings Vol. 69, No. 12, December 1972, pp. 765-69.

9.14 Rahman, Abdul, P.M. and Karim, Abdul, E., *Bonding precast concrete elements using epoxy resins*, Indian Concrete Journal, May 1975, pp. 142-7.

9.15 Reeves, J.S. and Morice, P.B., *Reinforced concrete joints between prestressed concrete members*, Magazine of Concrete Research, Vol. 13, No. 37, March 1961, pp. 13-20.

9.16 Samuely, F.J., *Some recent experience in composite precast in-situ concrete construction with particular reference to prestressing*, Proceedings of the Institute of Civil Engineers, Vol. 1, 1952, Part I, No. 30. pp. 222-59.

PROBLEMS

9.1 A composite beam has uniform I-section comprising top flange 1.2×0.2 m, web 0.3×0.8 m and bottom flange 1.2×0.2 m. The modular ratio of concretes in top flange and web with respect to concrete in the bottom flange are 0.6547 and 0.8452 respectively. Determine the bending moment to produce a curvature of 0.00041 radian per meter. The modulus of elasticity of concrete in bottom flange is 33720 MPa.

9.2 Determine the maximum bending moment which can be applied to the section of Prob. 9.1, if the maximum stress is not to exceed 5.87 MPa.

9.3 A composite beam formed by propped construction technique has a uniform T-section comprising in-situ concrete flange 1.2×0.2 m and precast prestressed web 0.15×0.80 m. The effective prestressing force of 720 kN acts at 0.160 m from the soffit. The moduli of elasticity of concretes in flange and web are 24 and 32 kN/mm^2 respectively. Determine the maximum superimposed load which may be carried by the beam over an effective span of 16 m if the tensile stress at soffit is not to exceed 2 MPa. Take unit weight of concrete as 24 kN/m^3.

9.4 Determine the maximum value of superimposed load which may be applied to the beam of Prob. 9.3. if the maximum permissible compressive stress in in-situ concrete is 7 MPa.

9.5　Determine the minimum value of effective prestressing force in the beam of Prob. 9.4, if it is formed by unpropped construction technique. The tensile stress at bottom fibre has not to exceed 1.5 MPa at final stage (A) when the precast prestressed element has to carry not only its own weight but also the weight of in-situ concrete and a load of 1.5 kN/m due to formwork and construction technique.

9.6　Determine the ultimate moment capacity of a composite T-section comprising in-situ concrete top flange 1.2 m × 0.2 m and precast prestressed web 0.15 × 0.9 m. The stress in prestressing tendons having an area of 1020 mm² at ultimate moment is 1600 MPa. Use the idealised stress-strain curve of concrete as shown in Fig. 9.4.1(c).

9.7　Determine the ultimate moment of section of Prob. 9.6 if the total tension in prestressing tendons at ultimate moment is 2.96 MN.

9.8　A composite T-section comprises in-situ concrete top flange 1.8 × 0.3 m of grade M-25 and precast prestressed web 0.3 × 0.6 m of grade M-45. The modular ratio of in-situ concrete with respect to precast concrete is 2/3. Determine the intensity of shear stress at the junction of flange and web when the section carries a shear force of 90 kN.

9.9　The composite section of Prob. 9.8 has a roughened contact surface at the junction of flange and web. Design the link reinforcement using British code when the section carries a shear force of (i) 45 kN and (ii) 270 kN.

9.10　Design the link reinforcement for the composite section of Prob. 9.8 in accordance with ACI code if the design ultimate shear force is (i) 100 kN and (ii) 540 kN. The effective depth of the composite section is 0.8 m.

9.11　Determine the minimum effective prestressing force in the composite beam of Example 9.3.2 if the short time deflection at final stage is restricted to 8 mm. The cable line in the precast prestressed element is parabolic with zero eccentricity at the ends and an eccentricity of 0 24 m at midspan.

9.12　Determine the maximum value of the superimposed load which may be carried by the composite section of Example 9.3.2 if the short-time deflection is not to exceed 10 mm. The cable line in the precast prestressed element is parabolic with zero eccentricity at ends and an eccentricity of 0.24 m at midspan.

9.13　Determine the minimum value of the effective prestressing force in the composite beam of Example 9.3.2 if the short-time deflection at final stage (A) has not to exceed 15 mm. The beam is formed by unpropped construction technique. The cable line in the precast prestressed element is parabolic with zero eccentricity at the ends and an eccentricity of 0.24 m at midspan.

9.14　Discuss the nature of stresses caused by differential shrinkage and creep of concrete in composite sections. How do these stresses affect the behaviour at working load level and at ultimate stage?

9.15 Design the composite slab for the roof of an assenbly hall with a clear span of 7.5 m. The soffit of the composite slab is formed by the precast prestressed elements of inverted T-sections. The top in-situ concrete is of grade M-25. The slab has to carry a live load of 1.5 kN/m². Adopt propped construction technique.

9.16 A composite beam of a bridge deck comprises precast post-tensioned girder of symmetrical I-section having an overall depth of 1 m and in-situ concrete top slab 1.8 × 0.2 m. The beam has to carry a superimposed service load which may be taken equivalent to uniform load of 25 kN/m. The service load may be assumed to occupy any part of the beam which produces the maximum effect. The clear span is 20 m. Design the composite beam adopting unpropped construction technique.

Partially Prestressed Beams

10.1 INTRODUCTION

The passive resistance of reinforcing steel and the active force delivered by prestressing steel are the basic ingredients of reinforced concrete and prestressed concrete respectively. Each type of steel has its own relative merits and demerits. The reinforcing steel can do little to prevent cracking of concrete. In fact, reinforcing steel becomes effective only after cracking of concrete. Hence cracking of concrete is essential for reinforcing steel to function. Once cracking has occurred, a well proportioned and properly detailed reinforcement is very effective in controlling the width and spacing of cracks. On the other hand, prestressing steel can prevent or delay the cracking of concrete. However once the cracking commences, the prestressing tendons, particulary the unbonded ones, exert a rather limited control on width or spacing of cracks. Due to its passive role, an excess of reinforcing steel can do little harm unless it is so excessive as to hamper proper placement and compaction of concrete. On the other hand, an excess of prestressing force can be as harmful as the lack of it. The quantum of prestressing force appropriate from considerations of stresses may not be necessarily so from the point of view of deformations. For instance, a prestressing force, which may appear to be appropriate for stresses, may produce undesirable or even unacceptable camber. In certain structures, there are sharp peaks of bending moment. For instance, such peaks occur at the interior supports in a continuous beam. If the quantum of prestressing force is chosen to counteract the peak bending moment, the stresses at other sections and the camber due to prestress may be unacceptable. In structures subjected to high live load moments as compared to dead load moments, it is generally not proper to counteract the entire dead plus live load moment by prestress because it will lead either to unacceptable stresses when the live load, is absent or to uneconomical stocky members. In all these cases, an economical design, meeting all requirements of strength and displacements may be obtained by a judicious combination of active prestressing force and passive resistance of reinforcing steel.

In respect of the manner in which the service load moment is resisted by the members of a concrete structure, they may be classified into four types. While Type 1 (full prestress) members have no flexural tension under full service load, Type 2 (limited prestress) members have only a small tensile stress which is not sufficient to cause flexural cracking. Members of Types 1 and 2 do not have any reinforcing steel for resisting flexural tension. Members of Type 3 utilize both untensioned steel and pretensioned high tensile steel to resist the service load moment. These members, called partially prestressed members, form the subject of discussion in this chapter. Type 4 members have no prestressing steel and are traditionally called reinforced concrete members. From this classification of concrete members, it is evident that Type 3 (partial prestress) members may be visualized as combinations of prestressed concrete (Type 1 or 2) and reinforced concrete (Type 4) members. The service load moment in a partially prestressed member is resisted partly by prestressing tendons and partly by untensioned reinforcement. In this way, the technique by partial prestressing attempts to exploit the advantages of both prestressed and reinforced concretes. The challenge in the design of partially prestressed structures lies in arriving at the most optimum combination of prestressing tendons and untensioned steel not only to obtain the most satisfying structure in respect of strength and serviceability but also to produce maximum possible economy.

At the commencement of his pioneering work, Freyssinet thought that optimum results, both in terms of economy and structural efficiency, could be achieved only through the use of fully prestressed concrete (Type 1) or reinforced concrete (Type 4). He changed his mind at a later stage when he realized that full prestress is unnecessary and even undesirable in the case of structures subjected to relatively high live load occuring infrequently. Following the lead given by Freyssinet, several investigators, notably Abeles, focused their attention on intermediate degrees of prestress, viz. limited prestress (Type 2) and partial prestress (Type 3). A survey of prev..iling cost of high tensile steel, including the expenses on stressing, sheathing and grouting ranges from 3 to 5 times the cost of reinforcing steel including its placement. It is, therefore, evident that substantial saving in cost could be achieved through the use of intermediate degrees of prestress. Thurlimann showed that the saving of prestressing steel through the use of partial prestressing could be as high as 30 per cent if partially prestressed concrete structures are designed according to Swiss Code SIA 162 of 1968. Partial prestressing can be achieved by adopting one of the following alternatives:

(i) Resisting part of the working moment by prestressed high tensile steel and the remaining part by untensioned high tensile steel of intermediate grade steel or mild steel.

(ii) Tensioning the entire high tensile steel to a stress smaller than the maximum permissible stress.

The first alternative is often advantageous because it leads to a reduction in the cost of stressing, sheathing and grouting. The use of high yield strength deformed bars is generally believed to offer better crack control and high ultimate strength. Early experiments by Emperger in 1939 attempted to improve reinforced concrete by introducing a few prestressing tendons. Taking Emperger's lead, Abeles carried out extensive studies on partially prestressed structures. His efforts were mainly responsible for the adoption of tensile stresses of 3.85 MPa in partially prestressed rail-road bridges by British Railways. This stress was subsequently raised to 4.55 MPa.

A partially prestressed member may also be visualized as a composite member comprising three constituent elements, viz. concrete, prestressing steel and untensioned steel. It may, therefore, be analysed in the same manner as a composite member discussed in Chapter 9. While the prestressing force may be the same all over the length of the member, the untensioned steel may vary in accordance with the requirements from section to section. For instance, untensioned steel may be provided in the regions of high bending moments. It may be reduced or eliminated completely in other regions of the member. Evidently there are immense possibilities of combining reinforced and prestressed concretes. As lower intensities of prestress are adopted in the case of partially prestressed members, it becomes more convenient to utilize electro-chemical methods of prestressing resulting in simplification and economy in the stressing operations. It is currently felt that the full potential of partial prestressing has yet to be realised.

EXAMPLE 10.1.1

Compare the expenses on steel inclusive of its placement and other relevant operations for resisting the same ultimate design moment M_u for the following three options :

  (i) M_u resisted entirely by prestressing steel
  (ii) Sixty per cent of M_u resisted by prestressing steel and the remaining by untensioned reinforcement and
  (iii) M_u resisted by entirely by untensioned reinforcement.

Assume that the internal lever arm is the same and the cost ratio, inclusive of related operations, is 4 for the two types of steel. The yield stress of untensioned reinforcement is 415 MPa and the stress in prestressing steel at ultimate moment is 1500 MPa.

Solution

The area of prestressing steel in option (i),

$$A_p = \frac{M_w}{1500 \, (jd)}$$

The areas of prestressing steel A_p and untensioned reinforcing steel A_s in option (ii),

$$A_p = \frac{0.6 \, M_w}{1500 \, (jd)}$$

$$A_s = \frac{0.4 \, M_w}{415 \, (jd)}$$

Area of untensioned reinforcing steel in option (iii),

$$A_s = \frac{M_w}{415 \, (jd)}$$

Using cost ratio of 4, the ratio of expenses in options (i) and (iii),

$$\eta_1 = \left[\frac{4M_w}{1500 \, (jd)}\right] \Big/ \left[\frac{M_w}{415 \, (jd)}\right] = 1.107$$

Similarly, the ratio of expenses in options (ii) and (iii),

$$\eta_2 = \left[\frac{4 \times 0.6 \, M_w}{1500 \, (jd)} + \frac{0.4 \, M_w}{415 \, (jd)} \right] \Big/ \left[\frac{M_w}{415 \, (jd)} \right]$$

$$= 1.064$$

It may be verified that the cost ratios η_1 and η_2 are equal to 0.667 and 0.8 respectively if the yield stress of untensioned reinforcement is 250 MPa.

10.2 OBJECTIVES AND DEFINITIONS

The designer may adopt partial prestressing to serve one or more of the following objectives.

(i) *Better stress distribution*

Partial prestressing may be adopted to achieve an overall superior stress condition at various working stages. Particularly, a member subjected to a load combination with a high ratio of temporary and permanent loads may derive significant benefit from partial prestressing. In this case only a part of the relatively large temporary load may be counteracted by prestressing leaving the remaining part to be resisted by non-prestressed steel. If the relatively large temporary load is resisted entirely by prestressing steel, i.e. if the member is designed as Type 1 or 2 member, it will have unfavourable or even unacceptable stresses when the temporary load is absent unless the cross-sectional area and the prestressing force are increased significantly.

(ii) *To reinforce regions of moment peak*

Certain members have sharp moment peaks. For instance, a continuous eam has a region of sharp moment peak at interior supports. Full prestress bnthese regions of high moment requires unduly large cross-sectional area and prestressing steel It may be more economical to let the prestress cater to average value of bending moment over the entire span and to resist the excess moment in the region of high moment by non-prestressed reinforcement. This option will lead to a superior structural performance particularly when for practical reasons, the prestressing force has to be the same over the entire length of the member.

(iii) *Control on excessive camber*

As part of the load is carried by non-prestressed steel in partially prestressed members, the area of prestressing steel and therefore the intensity of prestress are reduced. Consequently, the camber due to eccentric prestress is reduced. The problem of excessive camber is particularly important for beams having a high proportion of temporary load. These beams exhibit excessive long-time camber due to creep of concrete if a relatively large prestressing force required for full prestressing is adopted.

(iv) *Tc control crack width*

Non-prestressed high strength deformed bars can be used effectively to ensure superior post-cracking behaviour of prestressed concrete beams. Under overload condition, post-tensioned beams, particularly unbonded ones. develop relatively fewer cracks at larger spacing. The provision of bonded untensioned longitudinal bars has the effect of reducing crack spacing and, therefore the crack width. Control on crack width is important for better serviceability and higher durability. Wide cracks allow ingress of moisture and consequent corrosion of tendons.

(v) *To increase ductility and rotation capacity*

The ultimate strain of prestressing steel ranges from 3 to 5 per cent whereas it is as high as 10 to 30 per cent for different grades of reinforcing steel. Consequently, the provision of non-prestressed reinforcing steel tends to increase the ductility and toughness of partially prestressed members. The higher ductility leads to enhanced rotation capacity which is helpful in increasing the extent of favourable redistribution of moments from regions of high moment to regions of low moment in statically indeterminate structures such as continuous beams. The favourable redistribution of moments leads to higher ultimate moment capacity of the structure.

The flexural regidity decreases and the ductility increases progressively from Type 1 to Type 4 members. Fig. 10.2.1, shows the load-deflection relationship for the four types of members in which the ultimate tensile strength of the steel in the flexural tension zone is the same. Consequently, the ultimate flexural strength is also approximately the same. As the prestressing force decreases progressively in Type 1 to Type 4 members, the camber due to prestress also decreases and is zero for Type 4 members. The ductility which is represented by the area under the load-deflection curve increases progressively from Type 1 to Type 4 members.

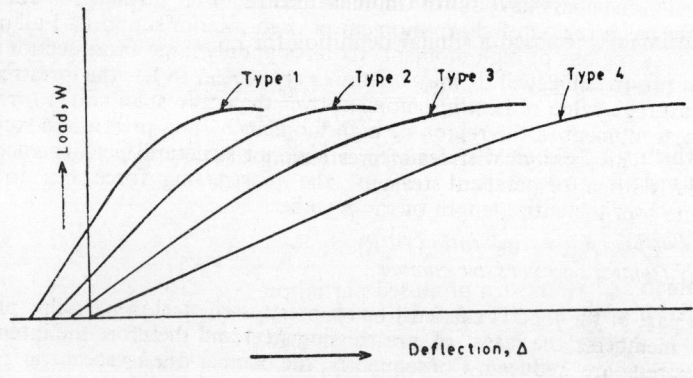

Fig. 10.2.1

(vi) *To meet code requirements at limit state of collapse*

Non-prestressed reinforcing steel may be added in regions of high moment to ensure that the ultimate moment capacity is not smaller than the design ultimate moment in accordance with the code provisions.

(vii) *To obtain overall economy*

As partially prestressed beams have relatively smaller proportion of pre-stressing steel, there is a distinct saving of costly prestressing steel. The saving of labour required for tensioning of prestressing tendons and in the cost of sheathing, grouting and end anchorages leads to additional economy in partially prestressed structures. Thus a partially prestressed structure may not only be superior in terms of overall structural action but it may also be appreciably cheaper as compared to Types 1 and 2 structures. The achievement of optimum structural performance coupled with maximum economy which is possible only through the best combination of prestressed and non-prestressed steel poses a real challenge to designer.

The extent of partial prestressing depends upon the manner in which the applied moment is resisted by reinforcing and prestressing steels. The follow-ing definitions are useful in quantifying the extent of partial prestressing.

(a) *Prestressing index (PI)*

Chaikes defined prestressing index as the ratio of ultimate strength of pre-stressing steel to that of the entire steel. It may be expressed by the equa-tion,

$$PI = \frac{A_p f_{pu}}{A_p f_{pu} + A_s f_y} \tag{10.2.1}$$

where A_t = area of prestressing steel

A_s = area of untensioned reinforcing steel

f_{pu} = ultimate strength of prestressing steel

f_y = yield strength of untensioned reinforcing steel.

Thurlimann proposed a similar definition for prestressing index.

$$PI = \frac{A_p f_{pv}}{A_p f_{pv} + A_s f_v} \tag{10.2.2}$$

in which f_{py} is the yield stress of prestressing steel corresponding to 0.2 per cent residual or permanent strain.

(b) *Partial prestressing ratio (PPR)*

Naaman and Siriaksorn proposed partial prestressing ratio as a measure of partial prestressing. It is defined by the equation,

$$PPR = \frac{M_{up}}{M_u} \tag{10.2.5}$$

where M_{up} = ultimate moment capacity corresponding to prestressing steel alone

M_u = ultimate moment capacity due to total tension reinforce-ment.

As the ultimate stress f_p is generally close to ultimate strength f_{pu} in partially prestressed beams and the internal lever arms for the two types of steel do not differ substantially, the value of PPR is close to the value of PI.

(c) Degree of Prestress (DP)

The decompression moment M_{dc} has also been utilised as a measure of partial prestress. The degree of partial prestress is defined as

$$DP = \frac{M_{dc}}{M_w} \qquad (10.2.4)$$

in which M_w is the working bending moment due to full service load.

As M_{dc} is zero for reinforced concrete and equal to M_w for full prestress, the value of degree of prestress varies from 0 to 1 for the entire range of concrete members from Type 4 to Type 1. Similarly the prestressing index and partial prestress ratio also vary from 0 to 1 for the entire range from reinforced concrete to fully prestressed concrete. The actual value of any-one of the three ratios indicates clearly the extent of partial prestressing Although beams are classified into Types 1 to 4, strictly speaking it is the sections which can really be classified in this way. As the bending moment varies along the span of a beam, the same beam may be of Type 1, 2 or 3 at different sections depending on cross-sectional properties. Fig. 10.2.2 (a) shows a partially prestressed beam under the action of working load assumed to be distributed uniformly. The bending moment diagram is parabolic with maximum midspan moment equal to the working moment M_w as shown in Fig. 10.2.2 (b). In Zones 1 located near the ends, the bending moment is smaller than the decompression moment M_{dc}. In these regions the beam is fully prestressed (Type 1). In zones 2, the bending moment M is larger than M_{dc} but is smaller than the cracking moment M_{cr}. Hence in these regions, the beam carries limited or moderate prestress (Type 2). In zone 3, the bending moment is larger than the cracking moment. Hence the

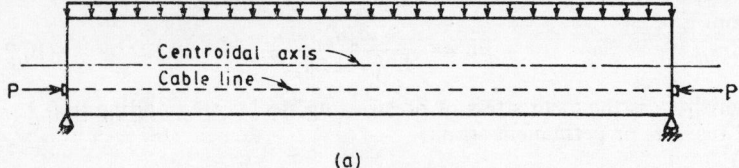

(a)

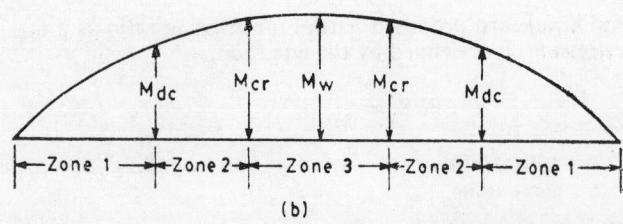

(b)

Fig. 10.2.2

beam is partially prestressed (Type 3) in zone 3. It follows that the same beam is fully, moderately and partially prestressed in zones 1, 2 and 3 respectively.

EXAMPLE 10.2.1

At a cross-section of a prestressed concrete beam shown in Fig. 10.2.3, area of prestressing steel, $A_p = 1260$ mm² and area of untensioned reinforcement, $A_s = 960$ mm². The effective stress, 0.2 percent proof stress and ultimate stress of prestressing steel are 1000 MPa, 1250 MPa and 1500 MPa respectively. The yield stress of untensioned reinforcement is 415 MPa. The internal lever arms for A_p and A_s are 0.675 m and 0.825 m respectively. The service load moment to be resisted by the section is 0.75 MN.m. The stress in prestressing steel at ultimate moment is 1420 MPa. Determine PI, PPR and DP at the section.

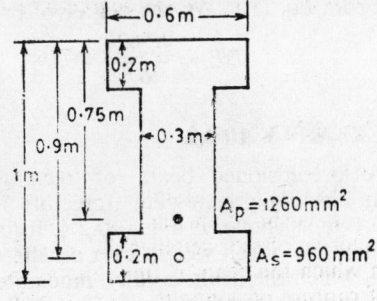

Fig. 10.2.3

Solution

(i) Prestressing index

From Eq. (10.2.1), the presstressing index,

$$PI = \frac{1260 \times 1500}{1260 \times 1500 + 960 \times 415} = 0.826$$

From Eq. (10.2.2), the prestressing index,

$$PI = \frac{1260 \times 1250}{1260 \times 1250 + 960 \times 415} = 0.798$$

(ii) Partial prestressing ratio

From Eq. (10.2.3), the partial prestressing ratio,

$$PPR = \frac{1260 \times 1420 \times 0.675}{1260 \times 1420 \times 0.675 + 960 \times 415 \times 0.825}$$

$$= 0.786$$

(iii) Degree of prestress

$$A = (0.6 \times 0.2) \times 2 + 0.3 \times 0.6 = 0.42 \text{ m}^2$$

$$I = \frac{1}{12} \times 0.6 \times 1^3 - \frac{1}{12} \times 0.3 \times 0.6^3 = 0.0446 \text{ m}^4$$

374 PRESTRTSSED CONCRETE

Intensity of prestress at bottom,

$$f_B = - \frac{1.26}{0.42} - \frac{1.26 \times 0.25 \times 0.5}{0.0446} = -6.53 \text{ MPa}$$

Hence the decompression moment,

$$M_{dc} = -6.53 \times \frac{0.0446}{0.5}$$

$$= 0.5825 \text{ MN.m}$$

From Eq. (10.2.4), the degree of prestress

$$DP = \frac{0.5825}{0.75} \, 0.7767$$

EXAMPLE 10.2.2

A pretensioned beam of rectangular cross-section 0.3×0.6 m has an effective prestressing force of 1080 kN at a constant eccentricity of 0.1 m as shown in Fig. 10.2.4. It has to carry a service load of 24 kN/m inclusive of self weight over an effective span of 12 m. Determine the zones in which the beam is fully, moderately and partially prestressed. Modulus of rupture of concrete, $f_{cr} = 6$ MPa.

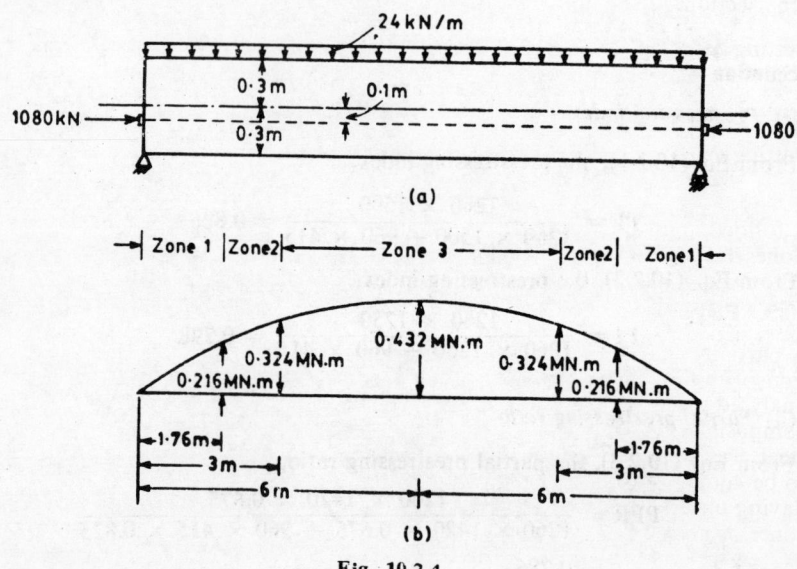

Fig. 10.2.4

Solution

Maximum service load moment at midspan,

$$M_w = \frac{0.024 \times 12^2}{8} = 0.432 \text{ MN.m}$$

The parabolic bending moment diagram due to service load is shown in Fig. 10.2.4 (b). The cross-sectional properties of the beam are

$$A = 0.3 \times 0.6 = 0.18 \text{ m}^2$$

$$Z = \frac{1}{6} \times 0.3 \times 0.6^2 = 0.018 \text{ m}^3$$

Intensity of prestress at bottom fibre,

$$f_B = -\frac{1.08}{0.18} - \frac{1.08 \times 0.1}{0.018} = -12 \text{ MPa}$$

Hence the decompression moment which decompresses the bottom fibre,

$$M_l = f_B Z = 12 \times 0.018 = 0.216 \text{ MN.m}$$

The cracking moment,

$$M_{cr} = (f_B + f_{cr}) Z = (12 + 6) \times 0.018 = 0.324 \text{ MN.m}$$

Taking the origin at the left end, the parabolic bending moment diagram due to service load may be expressed as

$$M = \frac{4M_w \, x \, (l - x)}{l^2} = \frac{4 \times 0.432 \, x \, (12 - x)}{12^2}$$

$$= 0.012 \, x \, (12 - x)$$

Putting $M = M_{dc} = 0.216$ MN.m, the decompression of bottom fibre occurs at

$$x = 6 \pm 3 \sqrt{2} = 1.76 \text{ m}, 10.24 \text{ m}$$

Putting $M = M_{cr} = 0.324$ MN.m, the cracking of concrete commences at

$$x = 3 \text{ m} \quad \text{or} \quad 9 \text{ m}$$

It follows that the beam is fully, moderately and partially prestressed in Zones 1, 2 and 3 respectively as shown in Fig. 10.2.4 (b).

10.3 PRECRACKING RANGE

A partially prestressed beam remains uncracked when it carries only a fraction of the full working load. The uncracked section in flexure may be analysed using linear elastic theory. The section may be visualised as a composite section comprising concrete, prestressing steel and reinforcing steel. Taking the modular ratio m of prestressing steel and reinforcing steel to be equal, the transformed or equivalent area of a rectangular section, having untensioned steel A_s and A_s' in flexural tension and compression zones, is given by

$$A_t = bD + (m - 1) (A_p + A_s + A_s')\qquad(10.3.1)$$

The stresses in the three constituent materials may be determined using the analysis discussed in Sec. 9.3. These stresses may be combined with the stresses caused by prestressing to evaluate the net or resultant stresses. It should, however, be noted that the stresses in reinforcing steel in an uncracked section either due to prestressing or due to external loading are generally small.

The elastic stresses in precracking range are modified on account of creep and shrinkage of concrete. The stresses in the untensioned reinforcing steel are increased by the factor $(1 + C_c)$ due to creep of concrete where C_c is the creep coefficient. The effect of shrinkage is to induce compressive stress in untensioned reinforcing steel located both in flexural tension and compression zones. Consider, for example, the partially prestressed section having untensioned reinforcing steel A_s in the tension zone and A'_s in the compression zone as shown in Fig. 10.3.1. If prestressing force P acts at eccentricity e_p, the stress (compressive) in A_s due to prestress, including the effect of creep and shrinkage, may be expressed as

$$f_s = - m \left[\frac{P}{A_t} + \frac{(P \cdot e_p)(d_s - x)}{I_t} \right] (1 + C_c) - \epsilon_{sh} \ E_s$$

(10.3.2)

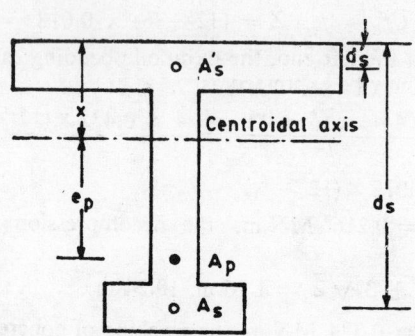

Fig. 10.3.1

Similarly, the stress (compressive) in A'_s,

$$f'_s = - m \left[\frac{P}{A_t} - \frac{(P \cdot e_p)(x - d'_s)}{I_t} \right] (1 + C_c) - \epsilon_{sh} \ E_s$$

(10.3.3)

where $m =$ modular ratio

$\qquad\quad = \dfrac{E_s}{E_c}$

$\qquad E_s =$ modulus of elasticity of untensioned steel

$\qquad I_t \ =$ moment of inertia of the transformed section.

$\qquad \epsilon_{sh} =$ shrinkage strain

while the stress f_s is definitely compressive, the stress f'_s may be compressive or tensile. Although these stresses are modified on account of working load, the stress in untensioned steel in the tension zone may still be compressive. Hence this steel does not serve its intended purpose in the precracking range. It is, therefore, not possible to design untensioned steel for the working load. In fact both types of untensioned steel, viz. A_s and A'_s

become fully effective only at the ultimate stage. Hence untensioned steel can be designed rationally only for the limit state in flexure.

EXAMPLE 10.3.1

The partially prestressed section shown in Fig. 10.3.2 (a) has areas of prestressing steel, $A_p = 1800$ mm², untensioned tension steel, $A_s = 1200$ mm² and untensioned compression steel, $A_s' = 600$ mm². The effective prestressing force is 1800 kN. Determine the stresses in steel at (i) no-load condition, (ii) limit state of decompression and (iii) limit state of cracking. The modular ratio, $m = 6$ and modulus of rupture of concrete, $f_{cr} = 6$ MPa.

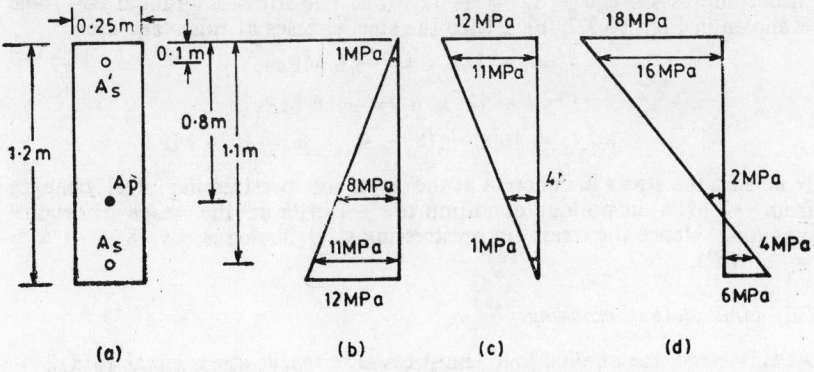

Fig. 10.3.2

Solution

The cross-sectional properties of gross concrete section are,

$$A = 0.25 \times 1.2 = 0.3 \text{ m}^2$$

$$Z = \frac{1}{6} \times 0.25 \times 1.2^2 = 0.06 \text{ m}^3$$

Intensity of prestress at top fibre,

$$f_T = -\frac{1.8}{0.3} + \frac{1.8 \times 0.2}{0.06} = 0$$

Intensity of prestress at bottom fibre,

$$f_B = -\frac{1.8}{0.3} - \frac{1.8 \times 0.2}{0.06} = -12 \text{ MPa}$$

(i) No-load condition

The stress diagram for concrete at no-load condition is shown in Fig. 10.3.2 (b). With modular ratio, $m = 6$, the stress in untensioned steel,

$$f_s = -11 \times 6 = -66 \text{ MPa}$$

Stress in untensioned compression steel,

$$f_s' = -1 \times 6 = -6 \text{ MPa}$$

Stress in prestressing steel,

$$f_p = \frac{1.8}{1800 \times 10^{-6}} = 1000 \text{ MPa}$$

(ii) *Limit state of decompression*

In order to cause decompression at the bottom fibre, the applied load must create a tensile stress of 12 MPa. Hence the stress at top fibre at the stage of decompression is $0 - 12 = -12$ MPa. The stress diagram at this stage is shown in Fig. 10.3.2 (c). Hence the steel stresses at this stage are

$$f_s = -1 \times 6 = -6 \text{ MPa}$$
$$f_s' = -11 \times 6 = -66 \text{ MPa}$$
$$f_p = 1000 + (8 - 4) \times 6 = 1024 \text{ MPa}$$

Note that the stress in concrete at the level of prestressing steel changes from -8 MPa at no-load condition to -4 MPa at the stage of decompression. Hence the stress in prestressing steel increases by $(8 - 4) \times 6 = 24$ MPa.

(iii) *Limit state of cracking*

At this stage, the applied load must cause a tensile stress equal to $(12 + f_{cr}) = 18$ MPa at the bottom fibre. Hence the stress diagram at this stage is as shown in Fig. 10.3.2 (d). The steel stresses at incipient cracking are

$$f_s = 4 \times 6 = 24 \text{ MPa}$$
$$f_s' = -16 \times 6 = -96 \text{ MPa}$$
$$f_p = 1000 + (8 - 2) \times 6 = 1036 \text{ MPa}$$

It may be checked that the steel stresses do not differ appreciably if the cross-sectional properties based on transformed section are used.

EXAMPLE 10.3.2

Determine the stresses in untensioned steel at the limit state of decompression in the section of Example 10.3.1, taking into account the effect of creep and shrinkage of concrete. The creep coefficient, $C_c = 1.5$ and shrinkage strain, $\epsilon_{sh} = 0.0002$. The applied load is permanent. Take modulus of elasticity of untensioned steel, $E_s = 2 \times 10^5$ MPa

Solution

The strains in concrete increase by the factor $(1 + C_c) = 2.5$ on account of creep of concrete. Consequently, the elastic stresses in A_s and A_s' also increase by the same factor. Assuming uniform shrinkage over the entire section, the compressive stress in A_s as well as A_s' due to shrinkage is

0.0002 × 2 × 2 × 10⁵ = 40 MPa. Hence the net stresses in A_s and A'_s taking into account creep and shrinkage of concrete,

$$f_s = -1 \times 2.5 \times 6 - 40 = -55 \text{ MPa}$$

$$f'_s = -11 \times 2.5 \times 6 - 40 = -205 \text{ MPa}$$

10.4 POST-CRACKING RANGE

The depth of neutral axis x in a partially prestressed section may be determined by treating the cracked section as eccentrically loaded column with large eccentricity. The concrete in the tension zone may be ignored and it may be assumed for simplicity that the prestressing force P remains constant due to the working moment M_w and the distribution of stresses and strains are linear. Consider, for example, the general case of a flanged section with neutral axis lying in the web as shown in Fig. 10.4.1. The centre of compression coincides with the centroid of prestressing tendons in the unloaded condition. The upward shift h of the centre of compression due to the application of sagging bending moment M_w may be expressed as

$$h = \frac{M_w}{P} = d_p + x_1 = d_p + k_p\, d_s \qquad (10.4.1)$$

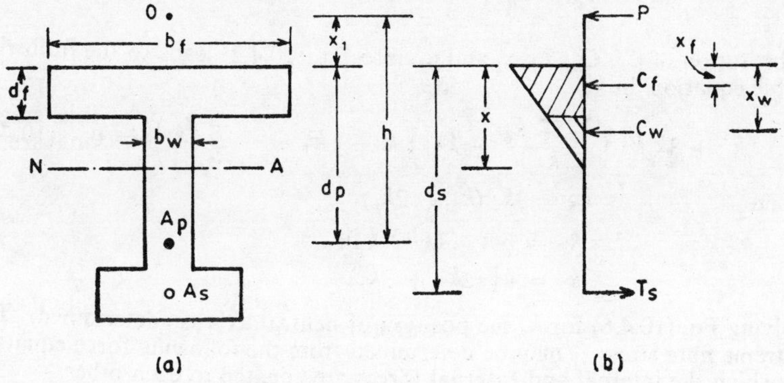

Fig. 10.4.1

in which k_p is a constant. Equating the moments of all internal forces about point 0,

$$C_f(k_p\, d_s + x_f) + C_w(k_p\, d_s + x_w) - T_s(k_p\, d_s + d_s) = 0 \qquad (10.4.2)$$

in which the compressive force in flange C_f, compressive force in web C_w and tensile force in untensioned reinforcing steel T_s may be expressed in terms of the extreme fibre stress of concrete f_c.

$$C_f = \frac{f_c b_f d_s}{2}\left(\frac{2k - k_1}{k}\right) k_1 \qquad (10.4.3)$$

$$C_w = \frac{f_c b_f d_s}{2} \frac{(k - k_1)^2}{k} k_2 \qquad (10.4.4)$$

$$T_s = f_s A_s = \frac{m(1 - k)}{k} f_c p_s b_f d_s \qquad (10.4.5)$$

where
$$k = \frac{x}{d_s}$$

$$k_1 = \frac{d_f}{d_s}$$

$$k_2 = \frac{b_w}{b_f}$$

$$p_s = \frac{A_s}{b_f d_s} = \text{proportion of untensioned steel}$$

The distances x_f and x_w of the compressive forces C_f and C_w from the top face may be expressed as

$$x_f = k_3 d_s = \left(\frac{3k - 2k_1}{2k - k_1}\right) \frac{k_1}{3} d_s$$

$$x_w = k_4 d_s = \left(\frac{2k_1 + k}{3}\right) d_s$$

substitution of C_f, C_w, T_s, x_f and x_w into Eq. (10.4.2) leads to the following cubic equation in k,

$$k^3 + 3k_p k^2 \left(\frac{k_5 + k_6}{k_2} - k_5\right) k + \left(k_7 - \frac{k_7 + k_6}{k_2}\right) = 0 \qquad (10.4.6)$$

where
$$k_5 = 3k_1 (k_1 + 2k_p)$$

$$k_6 = 6m p_s (1 + k_p)$$

$$k_7 = k_1^2 (2k_1 + 3k_p)$$

Solving Eq. (10.4.6) for k, the position of neutral axis is determined. The extreme fibre stress f_c may be determined from the following force equation in which the internal and external forces are equated to each other.

$$P = C_f + C_w - T_s \qquad (10.4.7)$$

Substituting for C_f, C_w and T_s in Eq. (10.4.7),

$$f_c = \frac{P}{b_f d_s} \left[\frac{2k}{k_1 (2k - k_1) + k_2 (k - k_1)^2 - 2m p_s (1 - k)}\right] \qquad (10.4.8)$$

Knowing f_c, the stress in untensioned steel may be determined from

$$f_s = m \left(\frac{1 - k}{k}\right) f_c \qquad (10.4.9)$$

An iterative graphical procedure for the determination of the stress f_s in untensioned steel has been given by Levi described by Guyon.

PARTIALLY PRESTRESSED BEAMS 381

EXAMPLE 10.4.1

Determine the stress at top fibre and in untensioned steel when the grouted post-tensioned section shown in Fig. 10.4.2 is subjected to a sagging bending moment of 1572 kN.m. The section carries an effective prestressing force of 1156 kN at a distance of 0.96 m from top. The modular ratio, $m = 6$.

Solution

Using Eq. (10.4.1), the shift of centre of compression due to applied moment,

$$h = \frac{1572}{1156} = 1.36 \text{ m}$$

Hence, $x_1 = k_p d_s = 1.36 - 0.96$
$$= 0.4 \text{ m}$$

or $k_p = \frac{0.4}{1.08} = 0.3704$

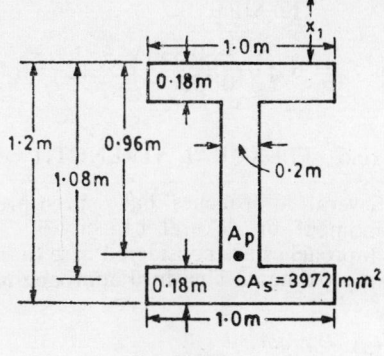

Fig. 10.4.2

The other constants in Eq. (10.4.6) may now be determined.

$$k_1 = \frac{d_f}{d_s} = \frac{0.18}{1.08} = 0.167$$

$$k_2 = \frac{b_w}{b_f} = \frac{0.2}{1} = 0.2$$

$$p_s = \frac{A_s}{b_f d_s} = \frac{3972 \times 10^{-6}}{1 \times 1.08} = 0.003678$$

$$k_5 = 3k_1(k_1 + 2k_p) = 3 \times 0.167(0.167 + 2 \times 0.3704)$$
$$= 0.4537$$

$$k_6 = 6m\,p_s(1 + k_p) = 6 \times 6 \times 0.003678(1 + 0.3704)$$
$$= 0.1815$$

$$k_7 = k_1^2(2k_1 + 3k_p) = 0.167^2(2 \times 0.167 + 3 \times 0.3704)$$
$$= 0.040125$$

Substituting the values of constants in Eq. (10.4.6),

$$k^3 + 1.1112\,k^2 + 2.7223\,k - 1.068 = 0$$

or $k = 0.333$

Depth of neutral axis,

$$x = kd_s = 0.333 \times 1.08 = 0.36 \text{ m}$$

Hence from Eq. (10.4.8) and (10.4.9), the stresses in concrete and steel are

$$f_c = \frac{1.156}{1.0 \times 1.08} \left[\frac{2 \times 0.333}{0.167 \, (2 \times 0.333 - 0.167) + 0.2 \, (0.333 - 0.167)^2 - 2 \times 6 \times 0.003678 \, (1 - 0.333)} \right]$$

$$= 12 \text{ MPa}$$

$$f_s = 6 \left(\frac{1 - 0.333}{0.333} \right) \times 12 = 144 \text{ MPa}$$

10.5 FLEXURAL STRENGTH

Several approaches have been proposed for determination of ultimate moment or flexural capacity of a partially prestressed section. Three approaches are considered here of which the first two may be regarded as approximate. The third approach is relatively more precise.

(a) *Approach 1*

In this approximate method, the ultimate moment of a partially prestressed Beam A is taken equal to the sum of the ultimate moments of the corresponding prestressed concrete Beam B having only prestressing steel and the corresponding reinforced concrete Beam C having only untensioned reinforcing steel as indicated in Fig. 10.5.1. The ultimate moments of Beams B and C may be determined using conventional ultimate strength theory and added to obtain the ultimate moment of Beam A.

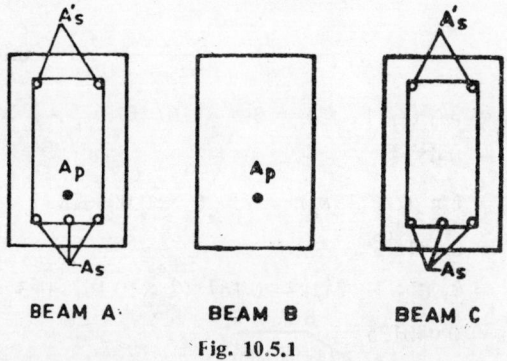

BEAM A   BEAM B   BEAM C

Fig. 10.5.1

(b) *Approach 2*

In this approximate method, the reinforcing steel in flexural tension zone is assumed to yield. The tensile force in prestressing steel is determined ignoring the presence of reinforcing steel. The total tension T at the limit state of collapse may then be expressed as

$$T = A_p f_p + A_s f_y \tag{10.5.1}$$

As total compression C is equal to total tension T, the depth of neutral axis x_u, the position of centroid of stress block and the internal lever arm may then be found. The ultimate moment may be expressed as

$$M_u = (A_p f_p + A_s f_y) \, (jd) \qquad (10.5.2)$$

This approach is less approximate as compared to the first approach.

(c) *Approach 3*

In this relatively more precise approach called the strain compatibility approach, the stresses in prestressing steel and reinforcing steel are determined from strain compatibility using the stress-strain curves for the two types of steel and for concrete. For this purpose, the iterative method discussed in Sec. 5.8.3 can be used with advantage. If the stresses in prestressing and reinforcing steels are f_p and f_s at collapse, the ultimate moment M_u may be expressed as

$$M_u = (A_p f_p + A_s f_s) \, (jd) \qquad (10.5.3)$$

EXAMPLE 10.5.1

Determine the ultimate moment of the pretensioned section shown in Fig. 10.5.2 (a). The steel areas are $A_s = A_s' = 600$ mm² and $A_p = 720$ mm². The grade of concrete is M-40. The effective initial prestress in steel, $f_{pe} =$

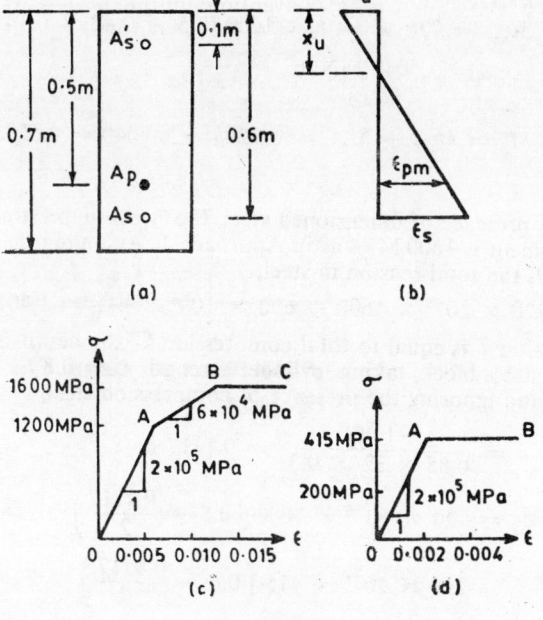

Fig. **10.5.2**

1040 MPa. The idealised stress-strain relationships for prestressing steel and untensioned steel are shown in Fig. 10.5.2 (c) and (d) respectively.

Use the equivalent rectangular stress block shown in Fig. 5.8.2. with $k_1 = k_3 = 0.85$ and $k_2 = 0.425$. Take crushing strain of concrete. $\epsilon_u = 0.003$.

Solution

Approach 1

Using steel beam theory, the ultimate moment of corresponding reinforced concrete section,
$$M_{urc} = 600 \times 10^{-6} \times 415 \times 0.5 = 0.1245 \text{ MN.m}$$

The ultimate moment of corresponding prestressed concrete section M_{upc} may be determined using IS: 1343-1980. However, the understrength factor of 0.87 may be ignored in computing the ultimate moment.
$$\frac{A_p f_{pu}}{bd f_{ck}} = \frac{720 \times 10^{-6} \times 1600}{0.3 \times 0.7 \times 40} = 0.137$$

Hence from Table 6.9.1, the stress in prestressing steel at the limit state of collapse,
$$f_p = f_{pu} = 1600 \text{ MPa}$$

and the depth of neutral axis,
$$x_u = 0.2979 \, d_p = 0.2979 \times 0.5 = 0.1489 \text{ m}$$

Using Eq. (6.9.1),
$$M_{upc} = 720 \times 10^{-6} \times 1600 (0.5 - 0.42 \times 0.1489)$$
$$= 0.5040 \text{ MN.m}$$

Hence,
$$M_u = M_{urc} + M_{upc} = 0.1245 + 0.5040 = 0.6285 \text{ MN.m}$$

Approach 2

Ignoring the presence of untensioned steel, the stress in prestressing steel at ultimate moment is 1600 MPa as in Approach 1. Assuming the untensioned steel to yield, the total tension in steel,
$$T = 720 \times 10^{-6} \times 1600 + 600 \times 10^{-6} \times 415 = 1.401 \text{ MN}$$

As total tension T is equal to total compression C, the depth of equivalent rectangular stress block, taking cylinder strength $f_c' = 0.8 f_{ck} = 0.8 \times 40 = 32$ MPa and ignoring the presence of compression steel,
$$a = \frac{1.401}{0.85 \times 32 \times 0.3} = 0.1717 \text{ m}$$

$$M_u = 720 \times 10^{-6} \times 1600 \left(0.5 - \frac{0.1717}{2}\right)$$
$$+ 600 \times 10^{-6} \times 415 \left(0.6 - \frac{0.1717}{2}\right)$$
$$= 0.6051 \text{ MN.m}$$

Approach 3

The ultimate moment may be computed more precisely by using the following iterative procedure based on linearity of strains and strain-compati-
bility.

(i) Assume the stress in prestressing steel at ultimate moment, $f_p = 1500$ MPa.

(ii) From the idealised stress-strain curve of Fig. 10.5.2 (c), the total
strain in prestressing steel, $\epsilon_p = 0.011$. Also the initial strain, $\epsilon_{pi} = 0.0052$. Hence the strain caused by applied moment, $\epsilon_{pM} = 0.011 - 0.0052$
0.0058.

(iii) *From the strain diagram of Fig. 10.5.2 (b),*

$$\frac{x_u}{0.003} = \frac{0.5 - x_u}{0.0058}$$

or $$x_u = 0.17 \text{ m}$$

(iv) *Depth of equivalent rectangular stress block,*

$$a = 2 k_2 x_u = 2 \times 0.425 \times 0.17 = 0.1445 \text{ m}$$

ordinate of equivalent rectangular stress block,

$$k' f_c' = \frac{k_1 k_3}{2 k_2} \times 0.8 f_{ck}$$

$$= \frac{0.85 \times 0.85}{2 \times 0.425} \times 0.8 \times 40$$

$$= 27.2 \text{ MPa}$$

Hence compressive force in concrete,

$$C_c = 0.3 \times 0.1445 \times 27.2 = 1.1791 \text{ MN}$$

From Fig. 10.5.2 (b), the strain at the level of compression steel,

$$\epsilon_s' = \frac{0.003 \times 0.07}{0.17} = 0.001235$$

As this strain is smaller than yield strain, the stress in compression steel,

$$f_s' = 0.001235 \times 2 \times 10^5 = 247 \text{ MPa}$$

Hence the compressive force in compression steel,

$$C_s = 600 \times 10^{-6} \times 247 = 0.1482 \text{ MN}$$

Hence total compression,

$$C = C_c + C_s = 1.1791 + 0.1482 = 1.3273 \text{ MN}$$

(v) *As total tension is equal to total compression,*

$$T = C = 1.3273 \text{ MN}$$

From the strain diagram of Fig. 10.5.2 (b),

$$\frac{\epsilon_s}{0.6 - 0.17} = \frac{0.003}{0.17}$$

Hence the strain in tension steel,

$$\epsilon_s = 0.00759$$

As this strain is greater than the yield strain, the tension steel yields. Hence stress in tension steel,

$$f_s = f_y = 415 \text{ MPa}$$

Force in tension steel,

$$T_s = 600 \times 10^{-6} \times 415 = 0.249 \text{ MN}$$
$$T = T_p + T_s$$

or
$$T_p = T - T_s = 1.3273 - 0.249 = 1.0783 \text{ MN}$$

(vi) *The stress in prestressing steel,*

$$f_p = \frac{T_p}{A_p} = \frac{1.0783}{720 \times 10^{-6}} = 1498 \text{ MPa}$$

As this stress is very close to the stress assumed in step (i), further iteration is unnecessary.

The distance of centroid of compressive force from top face,

$$\bar{x}_c = \frac{1.1791 \times \dfrac{0.1445}{2} + 0.1482 \times 0.10}{1.1791 + 0.1482} = 0.0753 \text{ m}$$

Similarly, the distance of centroid of tensile forces from top face,

$$\bar{x}_T = \frac{1.0783 \times 0.5 + 0.249 \times 0.6}{1.0783 + 0.249} = 0.5188 \text{ m}$$

Hence the internal lever arm,

$$jd = \bar{x}_T - \bar{x}_c = 0.5188 - 0.0753 = 0.4435 \text{ m}$$

Ultimate moment,

$$M_u = T (jd) = 1.3273 \times 0.4435 = 0.5887 \text{ MN.m}$$

10.6 SHEAR STRENGTH

As may be expected, the behaviour of partially prestressed beams in flexural shear is intermediate between that of fully prestressed beams and reinforced concrete beams. Due to lower intensity of prestress, cracking occurs earlier in partially prestressed beams so that both flexural cracks and diagonal tension cracks arise at a relatively lower load level as compared to fully prestressed beams. While the flexural tension cracks remain vertical, the inclination of diagonal tension cracks to horizontal increases progressively with decreasing degree of prestress until it becomes 45° at the level of neutral axis in the extreme case of reinforced concrete beams.

In accordance with the provisions of the Indian Code IS: 1343-1980, the shear resisted by concrete in the zone uncracked in flexure decreases progressively with decreasing degree of prestress. While the first term on the right side of Eq. (7.2.10) decreases with decreasing intensity of prestress p, the second term decreases due to smaller value of prestressing force P. As the effect of decreasing intensity of prestress on shear resisted by concrete in the zone cracked in flexure is not appreciable, [Eq. (7.2.11)], the contribution of concrete in partially prestressed beams decreases appreciably only if V_{co} is smaller than V_{cr}.

The formulation of first term on the right side of Eq. (7.2.11) may appear to be such as to cover intermediate degrees of prestress. However, the ratio f_{pe}/f_{pu} reflects only the stress to which the tendons are stressed. It does not necessarily indicate the intensity of prestress. If the equation has to be made applicable to all categories of beams ranging from Type 1 to Type 4, it must include the intensity of prestress or the relative contributions of prestressed and untensioned steels. The Swiss code SIA 162 : 1968 attempted to do so by including a term reflecting the relative contributions of the two types of steel. The code provides that

$$V \leqslant (V_c + V_n + V_s) \tag{10.6.1}$$

in which V is the ultimate shear force at the section. The shear resisted by concrete.

$$V_c = \left[1 + \frac{A_p f_{pe}}{A_p f_{pu} + A_s f_y} \right] v_c \, b \, d_v \tag{10.6.2}$$
$$\leqslant 1.5 \, v_c \, b \, d_v$$

where $A_p f_{pe}$ = effective prestressing force

$A_p f_{pu}$ = yield force of prestressing steel

$A_s f_y$ = yield force of untensioned steel

The term inside the parentheses ranges from 1 to 1.5 for different categories of beams being 1 for Type 4 and 1.5 for Type 1 beams.

The term V_n on the right side of Eq. (10.6.1) which has to be included only if the section is flexurally uncracked, may be expressed as

$$V_n = 0.2P \tag{10.6.3}$$

The term V_s representing the contribution of stirrups may be written as

$$V_s = \frac{0.87 \, f_{sy} \, A_v \, d_v \, (\sin \theta_s + \cos \theta_s)}{s} \tag{10.6.4}$$

The stirrups in accordance with Eq. (10.6.4) should be provided only when V is greater than V_c. The requirement of stirrup reinforcement may be expressed as

$$\frac{A_v}{s} \geqslant \frac{V - V_c}{0.87 \, f_{sy} \, d_v \, (\sin \theta_s + \cos \theta_s)} \tag{10.6.5}$$

In the case of vertical stirrups, $(\theta_s = 90°)$, Eq. (10.6.5) coincides with Eq. (7.2.14) of the Indian code. The code also provides that a minimum stirrup reinforcement to resist a shear force equal to $0.5 \, v_c \, b \, d_v$ should be provided.

EXAMPLE 10.6.1

Determine the ultimate shear strength of the section shown in Fig. 10.5.2 (a) if the section is (i) uncracked in flexure and (ii) cracked in flexure. The section is reinforced with 12 mm dia. two legged vertical stirrups at 300 mm spacing. The yield stress of stirrups, $f_{sy} = 250$ MPa.

Solution

Using the provisions of the Swiss code, the shear resisted by concrete may be determined from Eq. (10.6.2)

$$1 + \frac{A_p f_{pe}}{A_p f_u + A_s f_y} = 1 + \frac{720 \times 1040}{720 \times 1600 + 600 \times 415}$$

$$= 1.534$$

Hence the upper limit of 1.5 has to be used in computing V_c. For M-40 grade of concrete and percentage of prestressing steel $\dfrac{A_p}{bd_p} \times 100 = 0.48$, the ultimate shear stress resisted by concrete v_c in accordance with Table 7.2.1 is 0.5 MPa. Taking $d_v = 0.6$ m, Eq. (10.6.2) gives

$$V_c = 1.5 \times 0.5 \times 0.3 \times 0.6 = 0.135 \text{ MN}$$

From Eq. (10.6.3),

$$V_n = 0.2 \, P = 0.2 \, A_p f_{pe}$$
$$= 0.2 \times 720 \times 10^{-6} \times 1040$$
$$= 0.1498 \text{ MN}$$

The shear resisted by vertical stirrups ($\theta_s = 90°$) in accordance with Eq. (10.6.4),

$$v_s = 0.87 \times 250 \times \left(2 \times \frac{\pi}{4} \times 12^2 \times 10^{-6} \right) \times \frac{0.6}{0.3}$$

$$= 0.0984 \text{ MN}$$

(i) If the section is uncracked in flexure, the ultimate shear strength,

$$V = V_c + V_n + V_s$$

$$= 0.135 + 0.1498 + 0.0984$$
$$= 0.3832 \text{ MN}$$

(ii) If the section is cracked,

$$V = V_c + V_s$$

$$= 0.135 + 0.0984$$
$$= 0.2334 \text{ MN}$$

10.7 DEFLECTION

The moment-curvature relationship for a partially prestressed beam may be obtained in the same manner as for Type 1 beam in accordance with the analysis discussed in Sec. 8.4. Assuming a linear ditiribution of strains, the bending moment at any given curvature upto the limit state of collapse may be determined using the stress-strain relationships for the three constituent materials. Successive integration of curvature then gives the slope and deflection of the beam. To determine the long-time deformations, effect of creep and shrinkage of concrete should be taken into account as discussed in Sec. 8.4. As shown in Fig. 10.2.1, the initial upward deflection or camber due to prestress is smaller and the ultimate deflection is greater for a partially prestressed beam as compared to a fully prestressed one. While a smaller camber helps in satisfying the serviceability requirements, a larger ultimate deflection is useful at the limit state. because it leads to increased ductility and rotation capacity.

EXAMPLE 10.7.1

Determine the curvature at the cracked section of Example 10.4.1. Take modulus of elasticity of steel. $E_s = 2 \times 10^5$ MPa.

Solution

Strain at the level of untensioned steel in tension zone,

$$\epsilon_s = \frac{f_s}{E_s} = \frac{144}{2 \times 10^5} = 0.00072$$

Hence curvature at the section,

$$\Phi = \frac{\epsilon_s}{d - x_c} = \frac{0.00072}{1.08 - 0.36} = 0.001 \text{ rad/m}$$

EXAMPLE 10.7.2

Determine the midspan deflection of the beam of Example 10.5.1 at incipient collapse. It carries uniformly distributed load over a span of 14 m. The modulus of rupture of concrete, $f_{cr} = 4.4$ MPa and modulus of elasticity, $E_c = 0.333 \times 10^5$ MPa.

Solution

The deflection may be computed using equivalent moment of inertia I_e in accordance with Eq. (8.3.11). At the stage of incipient collapse, $M_{max} = M_u = 0.5887$ MN.m according to Example 10.5.1 (Approach 3).

$$P = A_p f_{pe} = 720 \times 10^{-6} \times 1040 = 0.7488 \text{ MN}$$

Intensity of prestress at bottom fibre,

$$f_B = - \frac{0.7488}{0.3 \times 0.7} - \frac{0.7488 (0.5 - 0.35)}{\frac{1}{6} \times 0.3 \times 0.7^2}$$

$$= - 8.15 \text{ MPa}$$

Hence cracking moment,

$$M_{cr} = (-8.15 - 4.4) \times \frac{1}{6} \times 0.3 \times 0.7^2$$

$$= 0.3075 \text{ MN.m}$$

Gross mement of inertia of the section,

$$I_g = \frac{1}{12} \times 0.3 \times 0.7^3 = 0.008575 \text{ m}^4$$

Depth of neutral axis at section of maximum moment, $x_u = 0.17$ m. Hence moment of inertia of cracked section,

$$I_{cr} = \frac{1}{3} \times 0.3 \times 0.17^3 + 6 \times 720 \times 10^{-6} (0.5 - 0.17)^2$$

$$+ 6 \times 600 \times 10^{-6} (0.6 - 0.17)^2 + (6 - 1)$$
$$\times 600 \times 10^{-6} (0.17 - 0.10)^2$$

$$= 0.002788 \text{ m}^4$$

Substituting for M_{max}, M_{cr}, I_g and I_{cr} in Eq. (8.3.11),

$$I_e = \left(\frac{0.3075}{0.5887}\right) \times 0.008575 + \left[1 - \left(\frac{0.3075}{0.5887}\right)^3\right] \times 0.002788$$

$$= 0.003605 \text{ m}^4$$

Midspan deflection at incipient collapse

$$\Delta_u = \frac{5}{48} \left[\frac{M_u \, l^2}{E_c \, I_e}\right]$$

$$\simeq \frac{5}{48} \times \frac{0.5887 \times 14^2}{0.333 \times 10^5 \times 0.003605}$$

$$= 0.10 \text{ m}$$

10.8 CRACK WIDTH

The cracking of concrete is of particular importance for partially prestressed beams because these beams crack under the normal service load. The width of crack is related closely to the serviceability requirements. A wide crack permits ingress of moisture and aggressive gases causing corrosion of steel and consequently reducing the durability of the structure. The codes, therefore, place restrictions on crack width in a partially prestressed beam under service load either directly or indirectly by limiting the hypothetical tensile stress in uncracked section. While an upper limit of 0.2 mm is commonly specified for the maximum crack width under normal environmental conditions, a more stringent limit of 0.1 mm is permitted in aggressive environments prevailing, for instance, in chemical plants and marine structures.

The width and spacing of cracks are closely inter-related. The tensile strains in the flexural tension zone are almost entirely concentrated at the crack locations. Thus the total elongation of the flexural tension face is approximately equal to the sum of widths of all flexural cracks. It follows that the larger the number of cracks, the smaller is the crack width. For better serviceability, several fine cracks at close spacing are preferable as compared to a few wide cracks. Among several factors, the width and spacing of cracks in a partially prestressed beam under service load depend mainly upon:

(i) Increased in stress in tension steel between the stage of decompression and full service load. The untensioned steel in a partially prestressed beam carries compressive stress and the prestressing steel carries tensile stress equal to effective prestress after losses under no-load condition. As the load is increased gradually, the compressive stress in untensioned steel decreses and the tensile stress in prestressing steel increases until at the stage of decompression, the stress in untensioned steel practically becomes zero. As the load is increased beyond the stage of decompression, the tensile stresses in both types of steel increase approximately by the same amount if their distances from the neutral axis are not appreciably different. The width of crack increases for larger increments of stress in tension steel beyond the stage of decompression. Consequently, the crack width can be controlled by specifying an upper limit on the increment of stress in untensioned steel δf, beyond the stage of decompression.

(ii) Type and proportion of tension steel. Tests have shown that the crack width decreases with increasing proportion of tension steel. High yield strength deformed bars are very effective in controlling the crack width. Some codes specify a minimum proportion of bonded untensioned steel in partially prestressed unbonded beams.

(iii) Bond between concrete and tension steel. Superior bond between concrete and tension steel results in closer spacing and smaller width of cracks. This may be expected because a crack can form only by a local failure of bond near the crack. It is mainly due to the superior bond of concrete with high strength deformed bars that this type of steel is highly effective in controlling the crack width.

(iv) Nature of service load. The width of crack grows progressively under repetitive loading. Consequently, a larger crack width may be expected when the load is repeated large number of times. It follows that the repetitive nature of loading should be taken into account in any prediction equation for the crack width.

Due to its importance, the crack width has been investigated extensively. As a large number of factors influence the crack width, empirical approaches have been preferred for arriving at prediction equations for the width or spacing of flexural cracks. Two main approaches have been adopted in the prediction of crack width. In the first approach, the crack width is related to the increase in stress in tension steel beyond the stage of decompression. In the second approach, the crack width is related to the hypothetical tensile stress in concrete in the fictitious uncracked section. The first approach generally gives more accurate results as compared to the second approach. However, the indirect approach of linking crack width to hypothetical tensile stress is simpler and is therefore sometimes preferred compared to the first approach which involves lengthy calculations for a reinforced concrete cracked section under combined bending and axial compression.

(i) *Bennett aud Chandrasekhar*

Bennett and Chandrasekhar reported extensive tests on partially prestressed beams using 7 mm diameter hard drawn crimped wires as prestressing steel and deformed bars or strands as untensioned steel. The mean crack width w_m was expressed by the general equation

$$w_m = (K_1 \epsilon_s + K_2 \delta \epsilon_s) c_s \qquad (10.8.1)$$

where K_1 = constant which is equal to 3.8 for deformed bars and strands and 5 for crimped wires

K_2 = constant which is equal to 20 for deformed bars and strands and 30 for crimped wires

ϵ_s = strain in untensioned steel occurring after decompression of concrete at the level of tendons

$\delta \epsilon_s$ = change of strain in untensioned steel occurring between decompression at the level of tension face and decompression at the level of tendons

c_s = minimum conerete cover over untensioned steel.

As an alternative, simpler approach to Eq. (10.8.1), Bennett and Chandrasekhar proposed the following expression:

$$w_m = K_3 f_{ft} c_s \qquad (10.8.2)$$

where K_3 = constant which is equal to 0.000435, 0.000725 and 0.001160 mm^2/N for deformed bars, strands and wire reinforcement respectively

f_{ft} = fictitious or hypothetical tensile stress in concrete at the tension face in MPa

(ii) *CEB-FIP*

The *CEB-FIP* recommended the following general equation for the maximum crack width w_{max} in mm

$$w_{max} = \left(\delta f_s - \frac{KE_s}{p_s} \right) \times 10^{-3} \qquad (10.8.3)$$

where δf_s = increase of stress in MPa in untensioned steel beyond the stage of decompression at the level of untensioned steel

K = constant which is related to the nature of bond and loading conditions (Table 10.8.1)

p_s = proportion of untensioned steel as defined in Table 10.8.1 but not less than the specified minimum value, p_s (min).

TABLE 10.8.1

Type of member	p_s	p_t (min)	KE_s in MPa
(i) Rectangular and T-section in simple bending	$\dfrac{A_s}{b_w \, d_s}$	0.010	0.37
(ii) Rectangular and T section in compression and bending	$\dfrac{A_s}{b_w \, (d_s - x)}$	0.016	0.60
(iii) Ties	$\dfrac{A_s}{\text{total area}}$	0.040	1.50
(iv) Beams with bottom flange	$\dfrac{A_s}{\text{Area of bottom flange}}$	0.040	1.50

The CEB-FIP model code recommends the following simplifying version of Eq. (10.8.3).

(a) For non repetitive loading

$$w_{max} = (\delta f_s - 40) \times 10^{-3} \text{ mm} \qquad (10.8.4)$$

(b) For repetitive loading,

$$w_{max} = f_s \times 10^{-3} \text{ mm} \qquad (10.8.5)$$

(iii) *Beeby and Taylor*

Beeby and Taylor proposed the following expression

$$w_m = \frac{1.75 f_{ft} \, (D - x)}{E_c} \qquad (10.8.6)$$

where $D =$ overall depth of the beam

$x =$ depth of neutral axis computed for a cracked section subject to compression and bending

(iv) *Nawy and Huang*

Nawy and Huang proposed the following equation for pretensioned and bonded post-tensioned beams,

$$w_{max} = 8.32 \times 10^{-6} \frac{A_{ct} \, (\delta f_p)}{\Sigma u_t} \qquad (10.8.7)$$

where $A_{ct} =$ effective area of concrete in uniform tension which may be taken as the area between soffit and a layer located at a clear distance c above the topmost layer of tension reinforcement in which c is the clear concrete cover below the lowermost layer

$\delta f_p =$ net change in stress in MPa in prestressing steel beyond the stage of decompression

$\Sigma u_t =$ sum of perimeters of all tension steel

Equation (10.8.7) gives the maximum crack width at the centroid of tension steel. The crack width at soffit may be taken as w_{max} multiplied by the ratio of distances of soffit and centroid of tension steel measured from neutral axis.

(v) Nawy and Chiang

Nawy and Chiang reported that Eq. (10.8.7) may also be used for unbonded post-tensioned beams except that the multiplier 8.32 should be replaced by 9.26.

EXAMPLE 10.8.1

Determine the crack width in the section of Example 10.4.1 at the stage of applied external moment. The prestressing steel and untensioned steel are in the form of strands and deformed bars respectively.

Solution

The crack width has to be calculated at the stage of loading at which the external bending moment, $M = 1.156$ MN.m, $f_c = 12$ MPa and $f_s = 144$ MPa.

(i) Bennett and Chandrasekhar

In Eq. (10.8.1),

$$\epsilon_s = \frac{f_s - f_{so}}{E_s}$$

in which f_{so} is the stress in untensioned steel at the stage of decompression at the level of tendons. Intensity of prestress at the level of tendons,

$$f_{pt} = -\frac{P}{A_g} - \frac{(Pe_p) e_p}{I_g}$$

$$A_g = 2 \times 1 \times 0.18 + 0.2 \times 0.84 = 0.528 \text{ m}^2$$

$$I_g = \frac{1}{12} \times 1 \times 1.2^3 - \frac{1}{12} \times 0.8 \times 0.84^3 = 0.1045 \text{ m}^4$$

Hence, $f_{pt} = -3.62$ MPa

The bending moment required to cause decompression at the level of tendons,

$$M'_{dc} = \frac{3.62 \times 0.1045}{0.36} = 1.051 \text{ MN.m}$$

The stress at bottom fibre under the combined action of prestress and M'_{dc},

$$f_B = -\frac{1.156}{0.528} - \frac{1.156 \times 0.36 \times 0.36}{0.1045} + \frac{1.051 \times 0.6}{0.1045}$$

$$= 1.46 \text{ MPa}$$

Hence the stress at the level of untensioned steel is $\dfrac{1.46 \times 0.12}{0.24} = 0.73$. It follows that

$$f_{so} = m \times 0.73 = 6 \times 0.73 = 4.38 \text{ MPa}$$
$$\epsilon_s = \frac{144 - 4.38}{2 \times 10^5} = 0.000698$$

The change of strain in untensioned steel $\delta\epsilon_s$ may be expressed as

$$\delta\epsilon_s = \frac{M'_{dc} - M_{dc}}{E_c\, I_g} \times (d_s - y_t)$$

in which M_{dc} is the moment required to cause decompression at the bottom fibre. The intensity of prestress at bottom fibre,

$$f_{pB} = \frac{1.156}{-0.528} - \frac{1.156 \times 0.36 \times 0.60}{0.1045} = -4.58 \text{ MPa}$$

Hence, $\quad M_{dc} = \dfrac{4.58 \times 0.1045}{0.6} = 0.798 \text{ MN.m}$

$$\delta\epsilon_s = \frac{(1.051 - 0.798)\, 0.48}{\left(\dfrac{2 \times 10^5}{6}\right) \times 0.1045} = 0.000035$$

Taking $K_1 = 3.8$ and $K_2 = 20$, the mean crack width in accordance with Eq. (10.8.1),

$$w_m = ((3.8 \times 0.300698 + 20 \times 0.000035) \times 100$$
$$= 0.335 \text{ mm}$$

In the simplified Eq. (10.8.2), the fictitious tensile stress at bottom fibre,

$$f_{ft} = f_{pB} + \frac{M\, y_B}{I_g} = -4.58 + \frac{1.572 \times 0.6}{0.1045}$$
$$= 4.45 \text{ MPa}$$

Taking $K_3 = 0.000435$, the mean crack width in accordance with Eq. (10.8.2),

$$w_m = 0.000435 \times 4.45 \times 100 = 0.194 \text{ mm}$$

(ii) *CEB-FIP*

Referring to Eq. (10.8.3) and Table 10.8.1,

$$p_s = \frac{3972 \times 10^{-6}}{0.18 \times 1} = 0.0221$$
$$KE_s = 1.50$$

Hence the maximum crack width in accordance with Eq. (10.8.3),

$$w_{max} = \left(144 - \frac{1.5}{0.0221}\right) \times 10^{-3} = 0.076 \text{ mm}$$

The maximum crack width for non-repetitive loading according to CEB-FIP code, Eq. (10.8.4),

$$w_{max} = (144 - 40) \times 10^{-3} = 0.104 \text{ mm}$$

and for repetitive loading, Eq. (10.8.5),

$$w_{max} = 144 \times 10^{-3} = 0.144 \text{ mm}$$

(iii) *Beeby and taylor*

The mean crack width in accordance with Eq. (10.8.6),

$$w_m = \frac{1.75 \times 4.45(1.2 - 0.36)}{\left(\dfrac{2 \times 10^5}{6}\right)}$$

$$= 0.000196 \text{ m} \quad \text{or} \quad 0.196 \text{ mm}$$

EXAMPLE 10.8.2

Determine the width of crack at the level of tension steel and at soffit in the section shown in Fig. 10.8.1 if the tendons are (i) bonded and (ii) unbonded The tension reinforcement comprises 4 tendons each having 7 wires and 3 reinforcing bars of 16 mm diameter.

The nominal diameter of the tendon is 15.2 mm. The net increase in stress in tendons beyond the stage of decompression is 150 MPa. The neutral axis is located at a distance of 200 mm from top.

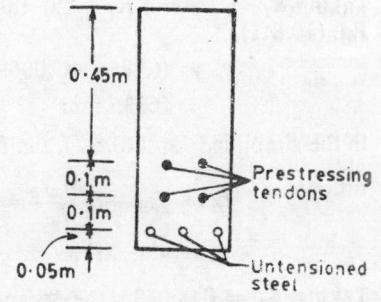

Fig. 10.8.1

Solution

(i) *Bonded tendons*

Clear concrete cover,

$$c = 50 - 8 = 42 \text{ mm}$$

Hence the effective area of concrete in uniform tension,

$$A_{ct} = 300 \left(250 + \frac{15.2}{2} + 42 \right) = 89880 \text{ mm}^2$$

$$\sum u_t = \pi (3 \times 16 + 4 \times 15.2) = 341.9 \text{ mm}$$

Centroid of tension steel is located at 131.9 mm from soffit. Hence from Eq. (10.8.7), the maximum crack width at the level of centroid of tension steel,

$$w_{max} = 8.32 \times 10^{-6} \left(\frac{89880 \times 150}{341.9} \right)$$

$$= 0.328 \text{ mm}$$

Maximum crack width at soffit,

$$w_{max} = 0.328 \times \frac{500}{500 - 131.9} = 0.446 \text{ mm}$$

(iii) *Unbonded tendons*

Maximum crack width at the level of centroid of tension steel,

$$w_{max} = 0.328 \times \frac{9.26}{8.32} = 0.365 \text{ mm}$$

Maximum crack width at soffit,

$$w_{max} = 0.446 \times \frac{9.26}{8.32} = 0.496 \text{ mm}$$

10.9 DESIGN PRINCIPLES

In order to ensure adequate serviceability, an upper limit is placed on the width of crack in a partially prestressed beam under the working load. As discussed in Sec. 10.8, the crack width may be determined either directly from the cracked section analysis or indirectly from the hypothetical tensile stress in uncracked section. Tests have established the reliability of the simpler indirect method based on uncracked section and is, therefore, commonly used in design. The other serviceability requirements such as those related to camber and deflection for Type 3 members are the same as for members of Types 1 and 2. In order to satisfy the strength requirements at the limit state of collapse, it is generally found most advantageous to supplement high tensile prestressing steel with untensioned high strength deformed bars. Whereas the prestressed steel remains the same over the entire length of the member, the quantum and position of untensioned steel differ from section to section depending upon strength requirements. This arrangement is preferable for practical reasons because prestressing tendons cannot be terminated at intermediate sections whereas the passive untensioned steel can be deployed in any region or location. At all sections in general and at the critical sections in particular, the ultimate moment of resistance developed by the joint action of prestressing tendons and untensioned reinforcement should not be less than the ultimate design moment as per code provisions.

Some additional considerations have to be made in the design of partially prestressed beams. The presence of substantial proportions of untensioned steel may lead to appreciable restraint on shrinkage and creep of concrete. Consequently, the untensioned reinforcement develops compressive stress and the loss of prestress in tendons is reduced. These stresses deserve careful consideration at the working load level. Another important consideration is the fatigue of partially prestressed beams which are relatively more vulnerable to fatigue failure under repetitive loading due to the presence of cracks and consequent upward shifting of neutral axis. In order to guard against possible fatigue failure, certain restrictions have been recommended on the increase in stress in steel beyond the limit of decompression. For instance a maximum increase of $0.12 f_{pu}$ or 180 MPa for 7 mm wires has been proposed for the increase in stress in tendons due to the service loads beyond the limit of decompression. The corresponding maximum permissible increase in stress in untensioned deformed barse is 140 MPa. Unbonded beams which develop a few wider cracks at larger spacing, are particularly prone to fatigue failure unless substantial untensioned bonded reinforcement is provided.

10.10 CODE PROVISIONS

The concept of partial prestressing is relatively new. The behaviour of partially prestressed members are still under investigation. Consequently, detailed provisions for all aspects of partial prestressing are yet to enter into prominent codes. At present, code provisions specifically for partially prestressed members relate only to a few aspects or areas of partial prestressing. Some of the notable provisions, specifically for partially prestressed members, are discussed below.

10.10.1 Indian Code

The provisions of the Indian Code IS: 1343-1980, which are similar to those of the British Code, permit cracking in partially prestressed members but it should not affect the appearance or durability of the structure. The acceptable limits of cracking vary with the type of structure and environment. The width of crack at concrete surface should not exceed 0.1 mm for members exposed to particularly aggressive environmental conditions. Structures exposed to sea water, alternate wetting and drying, freezing whilst wet and condensation or corrosive fumes belong to this category. The crack width should not exceed 0.2 mm in all other structures. In order to satisfy these serviceability requirements, the code specifies maximum values of hypothetical flexural tensile stress in concrete in MPa in an uncracked section as shown in Table 10.10.1. When additional reinforcement is distributed within the tension zone and positioned close to the tension

Tables 10.10.1   Maximum permissible value of hypothetical flexural tensile stress, σ_{ft}

Type of tendons	Limiting crack width mm	Grade of Concrete				
		M–30	M–35	M–40	M–45	M–50 and above
Pretensioned	0.1	—	—	4.1	4.4	4.8
tendons	0.2	—	—	5.0	5.4	5.8
Grouted post-tensioned	0.1	3.2	3.6	4.1	4.4	4.8
tendons	0.2	3.8	4.4	5.0	5.4	5.8
Pretensioned tendons distributed in the tensile zone and	0.1	—	—	5.3	5.8	6.3
positioned close to the tension faces of concrete	0.2	—	—	6.3	6.8	7.3

face of concrete, the stresses shown in the table may be increased in proportion to the percentage of additional reinforcement. For every additional 1 per cent reinforcement, the increase in stress should be 4 MPa in the case of members with pretensioned or grouted post-tensioned tendons and 3 MPa in the case of other members subject to the condition that total hypothetical tensile stress does not exceed $0.25 f_{ck}$. The stresses in Table 10.10.1 should be multiplied by a factor which is equal to 1.1 for members with a depth of 200 mm or less and 0.7 for members with a depth of 1000 mm or more. For members with a depth ranging from 200 mm to 1000 mm, the factor varies linearly from 1.1 to 0.7.

The code also makes specific recommendations for deflections in partially prestressed beams. When the permanent load does not exceed 25 per cent of the design imposed load, the deflection may be computed as in Sec. 8.6.1. If the permanent load exceeds 25 per cent of the design imposed load, the vertical deflection requirements for beams and slabs may be assumed to be satisfied provided the span to effective depth ratio does not exceed the following values.

(a) For span upto 10 m

    (i) cantilever   7

    (ii) simply supported   20

    (iii) continuous   26

(b) For span above 10 m, the values in (a) may be multiplied by 10/span in metres except for cantilevers in which case deflection calculations should be made.

10.10.2 British Code

The provisions of the British Code BS: 8110-1985 are similar to those of the Indian Code. The British Code also specifies that the design tensile stress at transfer for Type 3, members should not, in general, exceed that for Type 2 members (Sec. 6.7). However, the section should be assumed in design to have cracked if the tensile stress exceeds the limiting value. In respect of deflections, the code states that the approach of restricting the deflections within permissible limits by means of limiting span/depth ratios of reinforced concrete beams is not possible for prestressed concrete beams due to the major influence of the level of prestress. When computations for deflections are considered necessary, the elastic analysis based on concrete section properties may be used for determining the instantaneous and long-time deflections if the design tensile stresses do not exceed $0.45 \sqrt{f_{ck}}$ for pretensioned beams and $0.36 \sqrt{f_{ck}}$ for post-tensioned beams. When the stresses exceed these limiting values, more rigorous computations based on moment-curvature relationships for cracked sections should be used.

10.10.3 American Code

In the case of unbonded beams, the American Code requires the provision of a minimum amount of untensioned bonded reinforcement which should be distributed uniformly over the tension zone near the extreme tension fibre. The minimum area of untensioned reinforcement A_s (min) in beams and one way slab should be computed from Eq. (10.10.1).

$$A_s \text{ (min)} = 0.004 \ A_t \qquad (10.10.1)$$

in which A_t is the area of concrete between the flexural tension face and the centroid of the gross section.

10.10.3 CEB-FIP CODE

In the context of cracking and durability of concrete structures, the code classifies three categories of exposure conditions.

(i) Mild exposure

It applies to interior of buildings for normal habitation or offices and for conditions where a high level of relative humidity is reached for a short period only in any one year.

(ii) Moderate exposure

It applies to the interior of buildings where the humidity is high and where there is a risk of temporary presence of corrosive vapours and to structures exposed to running water, inclement weather and ordinary soils.

(iii) Severe exposure

It applies to structures exposed to liquids containing slight amounts of acids, saline or strongly oxygenated waters, corrosive gases or soils and corrosive industrial or maritime atmospheric conditions.

The code specifies values of computed characteristic crack width w_k for the above conditions of exposure at serviceability limit states of decompression, crack formation and cracking for reinforcements slightly and highly sensitive to corrosion. The serviceability limit state of decompression is defined as the stage at which the compression due to prestress or axial load effects are neutralized at a specified point in tension zone. The limit state of crack formation is defined as the stage at which the stress at the specified fibre in tension zone becomes equal to the tensile strength of concrete in bending. The limit state of cracking is defined as the stage at which the calculated characteristic width of the cracks at a specified level is equal to a specified value, viz. 0.1 mm, 0.2 mm and 0.4 mm. The characteristic crack width w_k may be taken equal to 1.7 times the mean crack width w_m which may be taken equal to the mean elongation δl_m occurring over the average distance between two consecutive cracks s_m. The average crack spacing s_m depends upon several factors such as concrete cover, diameter, spacing, proportion and bond properties of reinforcing bars, from of stress diagram and nature of loading.

In principle, the limit state of decompression should be checked with serviceability requirements in partially prestressed beams having unbonded tendons in the absence of supplementary untensioned reinforcement. However, the limit state of crack formation or cracking may be adopted if supplementary bonded tendons are provided to ensure that the requirements in respect of ultimate strength and serviceability are satisfied.

EXAMPLE 10.10.1

Determine the maximum value of the service load moment which may be carried by the section of Example 10.4.1 when the environmental conditions are (i) mild and (ii) aggressive. The grade of concrete is M-35.

Solution

(i) *Mild environment*

Limiting the crack width to 0.2 mm, the maximum permissible value of hypothetical tensile stress for M-35 concrete in accordance with Table 10.10.1 is 4.4 MPa. The percentage of untensioned steel,

$$\frac{A_s}{A_g} \times 100 = \frac{3972 \times 10^{-6}}{0.528} = 0.7523$$

The depth factor is 0.7 for an overall depth of 1.2 m. Hence the effective value of permissible hypothetical tensile stress,

$$\sigma_{ft} = 4.4 \times 0.7 + 4 \times 0.7523 = 6.089 \text{ MPa}$$

Consequently, maximum service load moment M_w may be determined from the equation,

$$\frac{M_w \times y_B}{I_g} - f_{pB} = \sigma_{ft}$$

From Example 10.8.1, the intensity of prestress at bottom fibre, $f_{pB} = 4.58$ MPa and $I_g = 0.1045 \text{ m}^4$. Hence the maximum permissible service load moment for mild environment,

$$M_w = 1.8582 \text{ MN.m}$$

(ii) *Aggressive environment*

In this case, the crack width should be limited to 0.1 mm. Hence

$$\sigma_{ft} = 3.6 \times 0.7 + 4 \times 0.7523 = 5.529 \text{ MPa}$$

$$\therefore \frac{M_w \times 0.6}{0.1045} - 4.58 = 5.529 \quad \text{or} \quad M_u = 1.7607 \text{ MN.m}$$

EXAMPLE 10.10.2

The section of Example 10.5.1 is required to carry a moment of 0.32 MN.m under service load and 0.48 MN.m at the limit state of collapse. Check the adequacy of the section for mild environment.

Solution

The adequacy of the section should be checked both for serviceability and ultimate strength.

(i) *Limit state of serviceability*

Referring to Table 10.10.1, the maximum permissible value of hypothetical tensile stress for M-40 concrete is 5 MPa in order to restrict the crack width to 0.2 mm. The percentage of untensioned steel in the tension zone,

$$\frac{A_s}{A_g} \times 100 = \frac{600 \times 10^{-6}}{0.3 \times 0.7} \times 100 = 0.286$$

As the depth factor for an overall depth of 0.7 m is 0.85, the effective value of maximum permissible hypothetical tensile stress,

$$\sigma_{ft} = 5 \times 0.85 + 0.286 = 5.394 \text{ MPa}$$

The intensity of prestress at bottom fibre,

$$f_{pB} = -\frac{720 \times 10^{-6} \times 1040}{0.3 \times 0.7} - \frac{720 \times 10^{-6} \times 1040 \times 0.15}{\frac{1}{6} \times 0.3 \times 0.7^2}$$

$$= -8.15 \text{ MPa}$$

Hence the hypothetical tensile stress at the bottom fibre,

$$f_{ft} = \frac{0.32}{\frac{1}{6} \times 0.3 \times 0.7^2} - 8.15 = 4.91 \text{ MPa}$$

As this value is smaller than the permissible value, the section is adequate

(ii) *Limit state of collapse*

From Example 10.5.1, the ultimate moment, M_u is equal to 0.588 MN.m based on compatibility of strains (Approach 3). Allowing a strength reduction factor, $\Phi = 0.95$ in accordance with ACI code (section 5.9.3), the ultimate moment capacity of the section may be taken equal to $0.9 \times 0.588 = 0.5292$ MN.m. As the moment capacity is greater than the ultimate design moment, the section is adequate.

EXAMPLE 10.10.3

Design the midspan of a pretensioned beam carrying a service load of 20 kN/m and an ultimate load of 36 kN/m over an effective span of 12 m. The beam has uniform rectangular section 0.3×0.6 m and is prestressed with an effective force of 1080 kN at a constant eccentricity of 0.1 m. Use concrete of grade M-40 and high yield strength deformed bars having yield strength of 415 MPa. The breaking stress of prestressing steel is 1600 MPa. Also check the serviceability requirement with a restriction of 0.2 mm as the maximum crack width.

Solution

From Example 10.1.1, the effective prestressing force, $P = 1.08$ MN. Adopting the working stress in prestressing steel as 1000 MPa, the area of prestressing steel,

$$A_p = \frac{P}{f_{pe}} = \frac{1.08 \times 10^6}{1000} = 1080 \text{ mm}^2$$

The section may be designed using Approach 1 of Section 10.5 as the basis. The ultimate moment resisted by prestressing steel alone may be computed using the Indian code.

$$\frac{A_p f_{pu}}{b d_p f_{ck}} = \frac{1080 \times 10^{-6} \times 1600}{0.3 \times 0.4 \times 40} = 0.36$$

Hence from Table 6.9.1,

$$\frac{f_p}{0.87 f_{pu}} = 0.94$$

or $f_p = 0.94 \times 0.87 \times 1600 = 1308.5$ MPa

and $\frac{x_u}{d_p} = 0.732$

or $x_u = 0.732 \times 0.4 = 0.2928$ m

Using Eq. 6.9.1,

$$M_{upc} = 1080 \times 10^{-6} \times 1308.5 \, (0.4 - 0.42 \times 0.2928)$$
$$= 0.3915 \text{ MN.m}$$

Ultimate design moment,

$$M_u = \frac{0.036 \times 12^2}{8} = 0.648 \text{ MN.m}$$

Hence, moment to be resisted by untensioned steel

$$M_{urc} = 0.6480 - 0.3915 = 0.2565 \text{ MN.m}$$

Allowing a concrete cover of 0.050 m, the internal lever arm for untensioned steel may be taken as,

$$j \, d_s = 0.6 - 0.05 - 0.42 \times 0.2928 = 0.427 \text{ m}$$

Adopting a partial safety factor for untensioned steel, $\gamma_m = 1.15$ and assuming the untensioned steel to yield at the limit state of collapse,

$$A_s = \frac{M_{ur}}{\left(\frac{f_s}{\gamma_m}\right) j d_s} = \frac{0.2565 \times 1.15}{415 \times 0.427} = 1664 \times 10^{-6} \text{ m}^2$$

Hence six high yield strength deformed bars of 20 mm diameter will suffice.

Referring to Table 10.10.1, the maximum permissible value of hypothetical tensile stress for maximum crack width of 0.2 mm is 5 MPa. For an overall depth of 600 mm, the depth factor is 0.9. The percentage of untensioned steel,

$$\frac{A_s}{bd} \times 100 = \frac{1664 \times 10^{-6} \times 100}{0.3 \times 0.6} = 0.9244$$

Hence the effective hypothetical tensile stress according to Indian code,

$$\sigma_{ft} = 5 \times 0.9 + 4 \times 0.9244 = 8.198 \text{ MPa}$$

The actual hypothetical tensile stress,

$$f_{ft} = \frac{-1.08}{0.3 \times 0.6} - \frac{1.08 \times 0.1}{\frac{1}{6} \times 0.3 \times 0.6^2} + \frac{0.020 \times 12^2}{8 \times \frac{1}{6} \times 0.3 \times 0.6^2}$$

$$= 8 \text{ MPa}$$

Hence the section satisfies the serviceability requirement in respect of crack width.

REFERENCES

10.1 Abeles, P.W., *Fully and partially prestressed concrete*, Journal of the American Concrete Institute, Vol. 16, No. 3, January 1945, pp. 181-214.

10.2 Abeles, P.W., *The use of high strength steel in ordinary reinforced and prestressed concrete beams*, Preliminary Publications, Fourth Congress, International Association for Bridge and Structural Engineers, 1952. Also supplement to above, Final Report, Fourth Congress, 1953.

10.3 Abeles, P.W., *Static and fatigue tests on partially prestressed concrete constructions*, Journal of the American Concrete Institute, Vol. 50, December 1954, pp. 361-76.

10.4 Abeles, P.W., *Design of partially prestressed members*, Journal of the American Concrete Institute, Vol. 65, 1967, pp. 669-72.

10.5 Abeles, P.W., *The practical application of prestressing*, International Association for Bridge and Structural Engineers (IABSE) Final Report 1968.

10.6 Beeby, A.W. and Taylor, H.P.J., *Cracking in partially prestressed members*, Sixth Congress FIP Prague, Preliminary Publications, Concrete Society (London), 1970.

10.7 Bennet, E.W. and Chandrasekhar, C.S., *Calculation of the width of cracks in class 3 prestressed beams*, Proceedings of the Institution of Civil Engineers (London), Vol. 49, 1971, pp. 333-46.

10.8 Bennet, E.W., *Partial Prestressing*, Chapter 4 in Developments in Prestressed Concrete Vol. I, edited by F. Sawko, Applied Science Publishers, London, 1978.

10.9 Burns. N.H., *Moment-curvature relationships for partially prestressed concrete beams*, Journal of the Prestressed Concrete Institute, Vol. 9, No. 1, February 1964, pp. 52-63.

10.10 Burns, N.H. and Pierce, D.W., *Strength and behaviour of prestressed concrete beams with unbonded tendons*, Journal of the Prestressed Concrete Institute, Vol. 12, No. 5, October 1967, pp. 15-29.

10.11 Desayi, P., *A method for determining the spacing and width of cracks in partially prestressed concrete beams*, Proceedings of the Institution of Civil Engineers (London), Vol. 59, September 1975, Part 2, pp. 411-28.

10.12 Guyon, Y., *Limit state design of prestressed concrete*, Vol. I., Applied Science Publishers (London), 1974.

10.13 Leonhardt. F., *Recommendations for the degree of prestressing in prestressed concrete structures*, FIP Notes 69, July-August 1977, pp. 9-14.

10.14 Lin, T.Y., *Partial prestressing design philosophies*, FIP Notes 69, July-August 1977, pp. 5-9.

10.15 Naaman, A.E. and Siriaksorn, A., *Serviceability based design of partially prestressed concrete beams*, Journal of the Prestressed Concrete Institute, Vol. 24, No. 2, March-April 1979, pp. 64-89.

10.16 Nawy, E.G. and Huang, P.T., *Crack and deflection control of pretensioned prestressed beams*, Journal of the Prestressed Concrete Institute, Vol. 22, No. 3, May-June 1977, pp. 30-47.

10.17 Nawy, E.G. and Chiang J.Y., *Serviceability behaviour of posttensioned beams*, Journal of the Prestressed Concrete Institute, Vol. 25, No. 1, January-February 1980, pp. 74-95.

10.18 Parmeswaran, V.S. and Annamalai, G., *Flexural behaviour of class 3 beams*, The Indian Concrete Journal, Vol. 49, No. 7, July 1975, pp. 206-12.

10.19 Thurlimann, B., *A case for partial prestressing*, Bericht No. 41, ETH, Zurich, 1971.

PROBLEMS

10.1 A pretensioned beam has a rectangular section 0.2×0.6 m. It carries a prestressing force of 0.36 MN at 100 mm from the soffit. Determine the limiting value of service load moment for (i) full prestress and (ii) moderate prestress. The modulus of rupture of concrete, $f_{cr} = 4.5$ MPa

10.2 A pretensioned rectangular section 0.2×0.6 m has prestressing tendons of area. $A_p = 360$ mm^2 located at 100 mm from the soffit and untensioned supplementary reinforcement in the form of deformed bars of Area, $A_s = 600$ mm^2 located at 50 mm from the soffit. The effective prestressing force is 360 kN. The yield stresses of deformed bars and prestressing tendons are 415 MPa and 1245 MPa respectively. The breaking stress of tendons is 1660 MPa. The ultimate moments resisted by prestressing tendons and deformed bars are 243 kN.m and 1245 kN.m respectively. The service load moment is 1600 kN.m. Determine (i) Prestressing Index, (ii) Partial Prestressing Ratio and (iii) Degree of Prestress.

10.3 Determine the stress in untensioned steel when the section of Prob. 10.2 is subjected to a moment of 144 kN.m. Take modular ratio, $m = 5.5$.

10.4 Determine the stress in untensioned steel when the section of Prob. 10.2 is subjected to a bending moment of 0.144 MN.m taking into account the effect of shrinkage and creep of concrete. The creep coefficient, $C_c = 0.8$ and shrinkage strain, $\epsilon_{sh} = 0.00025$. Take modular ratio, $m = 5.5$ and $E_s = 2 \times 10^5$ MPa.

10.5 Determine the stress in concrete at top and stress in untensioned steel when the neutral axis in the section of Prob. 10.2 is located at 300 mm from the top face. Modular ratio, $m = 5.5$. Also determine the position of centre of compression and the applied moment.

10.6 Detemine the ultimate moment of the section of Prob. 10.2 assuming that the untensioned steel yields at the ultimatestage. The stress in tendons at the limit state of collapse is 1500 MPa. Use equivalent rectangular stress block having an ordinate equal to $0.85 f_c'$. The cylinder strength of concrete, $f_c' = 32$ MPa.

10.7 Determine the ultimate moment of the rectangular section of Prob. 10.2 if the neutral axis is located at a distance of 150 mm from the top face at the limit state of collapse. The stress block is characterised by the constants $k_1 = k_3 = 0.85$ and $k_2 = 0.425$. The crushing strain of concrete, $\epsilon_u = 0.0035$. The initial prestress in prestressing tendons is 1020 MPa. Use the idealised stress-strain curves for prestressing steel and untensioned steel given in Fig. 10.5.2 (c) and (d) respectively.

10.8 Discuss the provisions of the Indian and Swiss codes in respect of shear strength of partially prestressed beams. Upto what extent do the codes take into account the lower intensity of prestress in these beams?

10.9 Determine the midspan deflection of a pretensioned beam carrying uniform load over an effective span of 12 m. The midspan section of the beam is the same as in Prob. 10.2. At the midspan section, the bending moment is 209.8 kN.m and the neutral axis is located at 0.2 m from top. Use equivalent moment of inertia, I_e as per ACI code. Take modular ratio, $m = 5.5$.

10.10 Determine the maximum crack width in accordance with CEB-FIP code if the section of Prob. 10.2 is subjected to a bending moment of 209.8 kN.m. The increase in stress in untensioned steel beyond the stage of decompression at the level of untensioned steel is 243.6 MPa. Load is non-repetitive.

10.11 Using Beeby and Taylor's equation, determine the mean crack width in the section of Prob. 10.2 when it is subjected to a bending moment of 209.8 kN.m. The neutral axis is located at 0.2 m from top. Take $E_c = 36360$ MPa.

10.12 Determine the limiting value of the service load moment which may be carried by the section of Prob. 10.2. Assume, moderate environment. The grade of concrete is M-40. Use Indian code.

10.13 Check the adequacy of the rectangular section of Prob. 10.2 if the service load moment and the design ultimate moment are 170 and 255 kN.m respectively. The environmental conditions are aggressive. Grade of concrete is M-40. Use idealised stress-strain curves for prestressing steel and untensioned steel shown in Fig. 10.5.2 (c) and (d) respectively. Use stress block for concrete as per Indian code.

10.14 A post-tensioned bonded beam has an overall depth of 1.5 m. It has symmetrical I-section comprising flanges 1.2 m wide and 0.2 m thick and web of 0.2 m thickness. The centroid of the prestressing tendons having an area of 2190 mm^2 is located at 250 mm from the soffit. The effective stress in tendons is 960 MPa. Design the section to carry a service load moment of 3.2 MN.m and an ultimate moment of 4.8 MN m. Grade of concrete is M-35. Use the provisions of the Indian code.

11

Continuous Beams

11.1 | INTRODUCTION

Prestressing concrete beams may have their load capacities increased and deflections reduced by utilising continuity as in reinforced concrete or steel beams. The hogging bending moments at interior supports reduce the sagging bending moments near midspan resulting in smaller net bending moments and deflections as compared to corresponding simply supported beams. Thus continuous beams can carry heavier loads for a given cross-section. Alternatively, for a given service load, a continuous beam requires a smaller cross-section. The economy of continuous construction is therefore evident.

In order to examine the increase in load carrying capacity at the working load level and at ultimate stage due to continuity, compare the load capacity of the continuous span AB shown in Fig. 11.1.1 (a) with that of simple span AB shown in 11.1.1 (b). For the sake of simplicity it may be assumed that the vertical axis through the midspan section C is the axis of symmetry and that the transverse load is uniform. The free bodies of left half of the

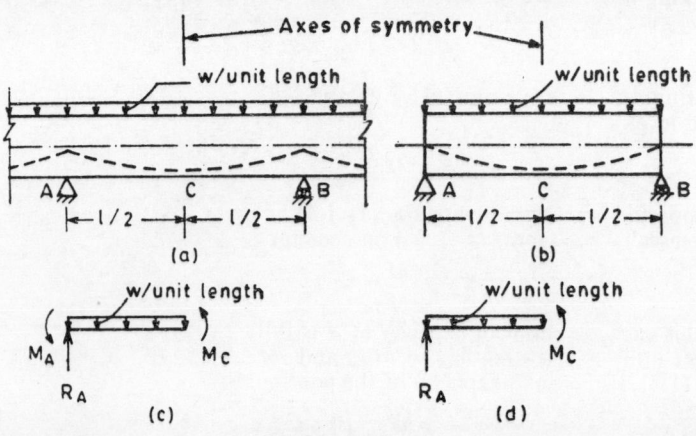

Fig 11.1.1

continuous span and simple span are shown in Fig. 11.1.1 (c) and (d) respectively. Due to symmetry, the shear force at midspan section C is zero. Considering the free body of left half of continuous span AB in Fig. 11.1.1 (c) and taking moments about A,

$$w \cdot \frac{l}{2} \cdot \frac{l}{4} - M_A - M_C = 0$$

or
$$w = \frac{8}{l^2} (M_A + M_C) \qquad (11.1.1)$$

Similarly considering the free body of left half of simple span AB in Fig. 11.1.1 (d) and taking moments about A,

$$w = \frac{8}{l^2} M_C \qquad (11.1.2)$$

(i) Working load level

As the intensity of transverse load w increases monotonically in the elastic range, the bending moments M_A at A and M_C at C increase proportionately maintaining a constant ratio, $k_e = (M_A/M_C)$. The constant ratio k_e, which depends upon the geometry of the continuous beam and its loading, can be determined by conventional elastic analysis. At the working load level, one of the following two possibilities may arise:

(a)
$$k_w = \frac{M_{Aw}}{M_{Cw}} > k_e$$

If the ratio of moment capacities at working load level M_{Aw} and M_{Cw}, which are the maximum bending moments at A and C without exceeding the permissible stresses, is greater than k_e, the moment M_C at C reaches its maximum permissible value M_{Cw} first whereas the moment M_A at A remains smaller than the maximum permissible value M_{Aw}. Inserting $M_C = M_{Cw}$ and $M_A = k_e M_C = k_e M_{Cw}$ into Eq. (11.1.1), the intensity of loading w_w at working load level in continuous span,

$$w_w = \frac{8}{l^2} M_{Cw} (1 + k_e) \qquad (11.1.3)$$

Putting $M_C = M_{Cw}$ in Eq. (11.1.2), the intensity of loading at working load level in simple span,

$$w_w = \frac{8}{l^2} M_{Cw} \qquad (11.1.4)$$

Comparing Eq. (11.1.3) with Eq. (11.1.4), it is seen that the load capacity increases by the factor $(1 + k_e)$ on account of continuity.

(b)
$$k_w = \frac{M_{Aw}}{M_{Cw}} < k_e$$

In this case, the moment capacity at A is fully utilized whereas that at C is under utilized. Inserting $M_A = M_{Aw}$ and $M_C = (M_A/k_e) = (M_{Aw}/k_e)$ into Eq. (11.1.1), the load capacity of the continuous span,

$$w_w = \frac{8}{l^2} M_{Aw} \left(1 + \frac{1}{k_e}\right) \qquad (11.1.5)$$

Comparing Eq. (11.1.5) with Eq. (11.1.4), it is seen that the load capacity is increased by the factor $k_w \left(1 + \dfrac{1}{k_e}\right)$ due to continuity.

(ii) *Ultimate stage*

As the load intensity is increased beyond the elastic range, inelastic action starts and eventually a plastic hinge forms either at A or at C. However, the continuous span collapses only when plastic hinges form at A, B and C as a result of redistribution of moments. Hence inserting into Eq. (11.1.1), $M_A = M_{Au}$ and $M_C = M_{Cu}$ where M_{Au} and M_{Cu} are the respective ultimate moment capacities at A and C, the moment capacity of the continuous span at the limit state of collapse,

$$w_u = \frac{8}{l^2}(M_{Au} + M_{Cu}) \qquad (11.1.6)$$

As the simple span collapses as soon as the plastic hinge forms at C, putting $M_C = M_{Cu}$ in Eq. (11.1.2), the collapse load intensity of the simple span,

$$w_u = \frac{8}{l^2} M_{Cu} \qquad (11.1.7)$$

Comparing Eq. (11.1.6) with Eq. (11.1.7), it is seen that the load capacity at the ultimate stage is increased by the factor $(M_{Au} + M_{Cu})/M_{Cu}$ on account of continuity.

It must be kept in mind that continuous beams are very sensitive to unequal settlements of supports particularly in the working range. The distribution of bending moments under working loads may change drastically due to unequal settlement of supports. On the other hand, unequal settlements may not affect ultimate strength appreciably if the beam has sufficient ductility and the collapse mechanism remains unchanged. As the behaviour under working load is important, prestressed concrete continuous beams are generally used only when the possibility of appreciable unequal settlement of supports does not exist.

Continuous beams represent the simplest type of statically indeterminate structures. A continuous beam is statically indeterminate externally because it has more than three support reaction components which are essential for static determinacy and stability. The problem of prestressing of statically indeterminate structures has been introduced in Sec. 1.12. Secondary forces due to prestressing arise because the deformation caused by prestress is restrained by the additional reaction components at the supports. These secondary forces complicate the analysis and design of continuous beams.

It may be concluded that continuity in prestressed concrete beams offers both advantages as well as disadvantages. Among the advantages are (i) better distribution of moments giving smaller net moments, (ii) larger moment capacity at working and ultimate stages. (iii) greater stiffness leading to smaller camber and deflection, (iv) smaller cross-sectional area with consequent saving of concrete, (v) smaller dead load moment, (vi) saving of end anchorages, (vii) saving of labour cost in stressing operations and (viii) larger shear resistance of tendons due to their greater curvature. Among the disadvantages are (i) reversal of bending moment which makes use of highly eccentric cable line difficult and thereby offsets a part of the major advantage of prestressing, (ii) undulating tendons make stressing more difficult, (iii) greater curvature of tendons increases frictional losses, (iv) greater complexity in analysis, design and construction, (v) sensitivity to unequal settlements of supports, particularly at the working load level, (vi) concurrence of maximum shear force and large bending moments near interior supports, (vii) axial shortening of long continuous beams leading to lateral forces and bending moments in supporting columns built monolithically with the beam and, (viii) moment peaks requiring more tendons or untensioned steel near interior supports.

EXAMPLE 11.1.1

The central span AB of a prestressed concrete continuous beam symmetrical about midspan section C carries uniformly distributed load over an effective span of 10 m. The moment capacities of sections A and C are 300 kN.m and 200 kN.m at working load level and 450 kN.m and 300 kN.m at ultimate stage. The ratio of the elastic moments $M_A/M_C = 2$. Compare the moment capacity of the continuous span with that of the corresponding simple span.

Solution

(i) *Working load level*

$$k_e = \frac{M_A}{M_C} = 2$$

$$k_w = \frac{M_{Aw}}{M_{Cw}} = \frac{300}{200} = 1.5$$

As $k_w < k_e$, the moment capacity at A is fully utilised while that at C is only partially utilised. Hence, $M_A = M_{Aw} = 300$ kN.m and $M_C = M_A/k_e = M_{Aw}/k_e = 300/2 = 150$ kN.m.

Substituting into Eq. (11.1.1).

$$w_w = \frac{8}{10^2} (300 + 150) = 36 \text{ kN.m}$$

From Eq. (11.1.2), moment capacity of simple span,

$$w_w = \frac{8}{10^2} (200) = 16 \text{ kN.m}$$

Hence the load capacity of continuous span is $36/16 = 2.25$ times that of simple span.

(ii) *Ultimate stage*

Putting $M_A = M_{Au} = 450$ kN.m and $M_C = M_{Cu} = 300$ kN.m in Eq. (11.1.6), ultimate load capacity of continuous span,

$$w_u = \frac{8}{10^2} (450 + 300) = 60 \text{ kN.m}$$

From Eq. (11.1.7), the ultimate load capacity of simple span,

$$w_u = \frac{8}{10^2}(300) = 24 \text{ kN.m}$$

Hence the ultimate load capacity of continuous span is $60/24 = 2.5$ times that of the simple span.

11.2 ELASTIC ANALYSIS

The elastic analysis based on the assumptions of homogeneity, isotropy and linear elasticity of the material can be used to determine the internal forces and displacements in the precracking range with reasonable accuracy. It is, therefore, used for the working stress design of prestressed concrete continuous beams. To simplify the elastic analysis, the following assumptions are commonly made:

1. The cable line is very flat so that the slope of cable line θ at any point is very small. Hence as an approximation, $\dfrac{de_p}{dx} = \tan \theta = \sin \theta = \theta$ and $\cos \theta = 1$.

2. The prestressing force P is the same along the entire length of the continuous beam.

Unless stated otherwise, these simplifying assumptions will be applicable throughout this section. Where these assumptions are not valid, the elastic analysis and the theorems based thereon can be amended suitably.

11.2.1 Secondary Forces and Concordance

The nature of secondary forces in continuous beams arising due to prestressing may be illustrated by means of a simple example. Consider a pretensioned beam of length L resting on unyielding supports A, B and C at the same level as shown in Fig. 11.2.1 (a). The two spans are equal and the cable line has constant eccentricity e_o throughout the length. The beam carries no external load and the self weight of the beam is ignored. Thus the effect of a constant prestressing force P on otherwise unloaded beam is under consideration. The free body of the concrete beam is shown in Fig. 11.2.1 (b). Treating support reaction R_B as the redundant force and the simply supported beam AC as the released structure, it is evident that the hogging couples of magnitude pe_o at A and C produce an upward (negative) deflection at B. Hence R_B must be downward (negative) in order to make the net deflection at B equal to zero,

$$-\frac{Pe_o L^2}{8EI} - \frac{R_B L^3}{48EI} = 0 \qquad (11.2.1a)$$

or

$$R_B = -\frac{6Pe_o}{L} \qquad (11.2.1b)$$

The minus sign shows that R_B is downward. By symmetry $R_A = R_C$ and for equilibrium of vertical forces, $R_A + R_B + R_C = 0$. Hence it follows that

$$R_A = R_C = \frac{3Pe_o}{L} \qquad (11.2.2)$$

The support reactions R_A, R_B and R_C in the unloaded condition represent the secondary forces due to prestressing. As their resultant is zero, they represent a self equilibrating system of forces. The secondary forces in the form of support reactions give rise to internal forces. The secondary bending moments in this case are sagging (positive). The secondary bending moment diagram due to prestressing is shown in Fig. 11 2.1 (c). Combining it with the primary bending moment diagram shown in Fig. 11.2.1 (d), the resultant bending moment diagram shown in Fig. 11.2 1 (e) is obtained. At any section of the continuous beam, the net or resultant bending moment due to prestressing M_p is the sum of primary bending moment M_{pp} and secondary bending moment M_{ps}.

$$M_p = M_{pp} + M_{ps} \tag{11.2.3}$$

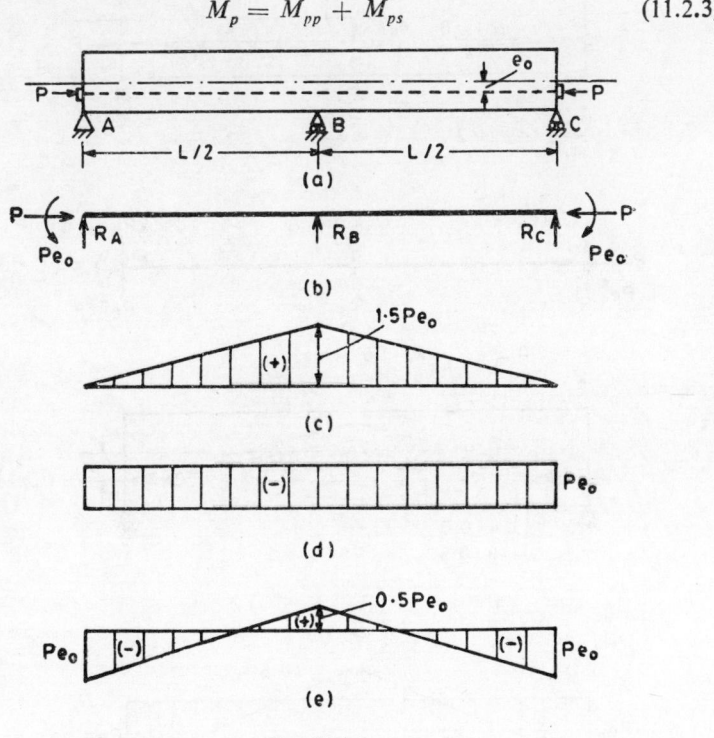

Fig. 11.2.1

The distribution of net bending moment due to prestressing shown in Fig. 11.2.1(e) is derived from the compatibility condition of zero deflection at B represented by Eq. (11.2.1a).

Consider next the preceding continuous beam in which the cable line is depressed through vertical distance ke_o at B as shown in Fig. 11.2.2 (a). The change of direction of cable line at B, $2\theta = \dfrac{4ke_o}{L}$. Hence using Eq. (5.4.17), the upward force imposed on concrete by the cable line is $\dfrac{4Pke_o}{L}$. The free body of the concrete beam is shown in Fig. 11.2.2 (b). As the net deflection at B must be zero,

$$-\frac{Pe_o L^2}{8EI} - \left(R_B + \frac{4Pke_o}{L}\right)\frac{L^3}{48EI} = o \qquad (11.2.4a)$$

or

$$R_B = -P(6 + 4k)\frac{e_o}{L} \qquad (11.2.4b)$$

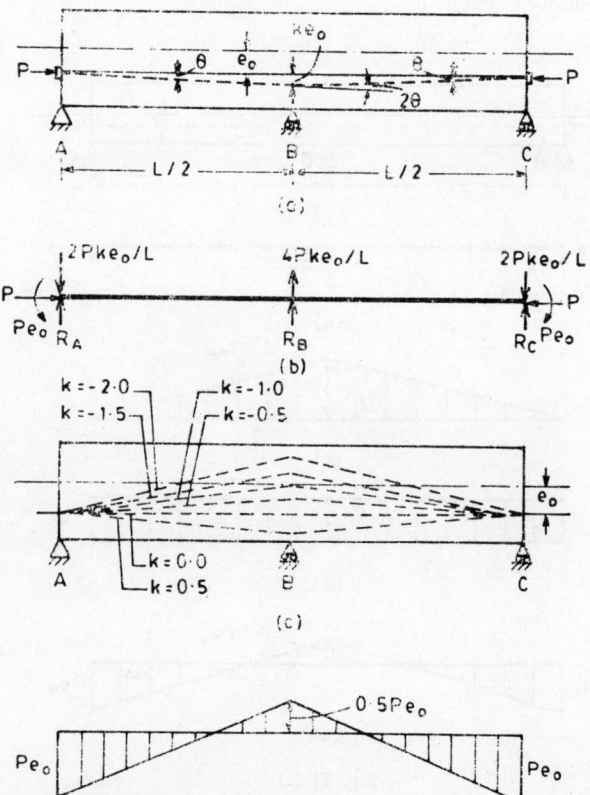

Fig. 11.2.2

If $k = o$, the cable line becomes straight and $R_B = -\dfrac{6Pe_o}{L}$ as in the preceding example. On the other hand, if $k = -1.5$, the reaction $R_B = 0$. As $R_A = R_C$ by symmetry and $R_A + R_B + R_C = o$ for equilibrium, it follows that $R_A = R_B = R_C = o$. As the secondary forces in the form of support reactions are zero, the secondary shear forces and bending moments are also zero. Consequently, the cable profile with $k = -1.5$ makes all secondary forces zero. Any cable profile which does not give rise to secondary forces is called concordant cable profile and the corresponding state is called the state of concordance. As in the state of concordance, the secondary moments M_{ps} vanish, net or resultant bending moment M_p at any section is equal to the primary bending moment M_{pp} according to Eq. (11.2.3). Hence for concordant cable profile, the net or resultant bending moment diagram due to prtstress is the same as the primary bending moment diagram. Figure 11.2.2 (c) shows six cable profiles for $k = 0.5, 0, -0.5, -1.0, -1.5$ and -2. Among the six cable profiles, only one with $k = -1.5$ is concordant. All others are non-concordant. It may be easily verified that although the six cable profiles differ in respect of their primary and secondary bending moment diagrams, the net or resultant bending moment diagram is the same for all of them and is shown in Fig. 11.2.2 (d). It is the same as the primary bending moment diagram for the concordant cable profile with $k = -1.5$. For any arbitrary value of k, the primary bending moment at B,

$$M_{pp} = -P(1 + k)e_o \qquad (11.2.5)$$

Using Eq. (11.2.4 b) for the value of R_B, the secondary bending moment at B,

$$M_{ps} = P(1.5 + k)e_0 \qquad (11.2.6)$$

combining primary and secondary bending moments according to Eq. (11.2.3), the net bending moment,

$$M_p = M_{pp} + M_{ps} = 0.5\, Pe_0 \qquad (11.2.7)$$

Equation (11.2.7) shows that the net bending moment M_p is independent of k. It follows that the net bending moment diagram is the same for the six profiles and is shown in Fig. 11.2.2 (d). In particular the straight cable line of preceding example is obtained by putting $k = 0$ and it is noted that the net bending moment diagram for $k = o$ is the same as for any value of k. Although the net bending moment M_p can always be determined by combining the secondary moment M_{ps} with the primary moment M_{pp}, it must be emphasised that it is possible and in fact more convenient to determine the net moment M_p due to prestress and the resultant moment M_r at any stage in the precracking range directly without the computation of secondary moment M_{ps}. This can be done by considering the free body of the concrete beam acted upon by the forces imposed on it by prestressing tendons and external loads, if any. The continuous beam may then be analysed as ordinary non-prestressed beam by any one of the conventional methods such as moment distribution, slope-deflection equations or unit load method. For hand computation, the method of moment distribution is particularly appropriate for analysing continuous beams. The free body approach for the determination of net moment M_p due to prestress or the resultant moment M_r at any stage of loading is applicable to other skeletal structures such as arches, interconnected girders and rigid frames and also to surface structures such as slabs and shells.

Consider again the pretensioned beam shown in Fig. 11.2.1 (a) in its unloaded condition. The free body of the concrete beam acted upon by forces imposed on it by prestressing tendons is shown in Fig. 11.2.1 (b). This beam may now be analysed by the method of moment distribution as shown in Table 11.2.1. As the two spans have the same relative stiffness, the distribution factors at joint B are 0.5 and 0.5 as shown in line 1. As the beam has no transverse loads, the fixed-end moments are zero as shown in line 2. Joints A and C may next be released permanently by applying hogging couples Pe_0 which are the moments caused by prestressing at A and C and carrying over half the moments to the other end of each span as shown in line 3. In moment distribution method, it is convenient to adopt the frame sign convention according to which clockwise couples are taken positive and anticlockwise negative. The net initial moments shown in line 4 are obtained by summing up the moments in lines 2 and 3. As the unbalanced moment at joint B is zero, the moments in the balancing operations are zero as shown in line 5. Consequently, the moment distribution closes automatically. The final moments shown in line 6 indicate that the bending moment at B is $0.5 Pe_0$ (sagging) which is the same as that given by Eq. (11.2.7). Hence the net or resultant bending moment diagram due to prestressing shown in Fig. 11.2.1 (c) is obtained directly without computation of secondary moments.

Table 11.2.1

	A	B		C
1. D.F	—	0.5	0.5	—
2. FEM	0	0	0	0
3. Release A and C permanently and C.O.	$-Pe_0$ →	$-\frac{1}{2} Pe_0$	$\frac{1}{2} Pe_0$ ←	Pe_0.
4. Initial moments	$-Pe_0$	$-\frac{1}{2} Pe_0$	$\frac{1}{2} Pe_0$	Pe_0,
5. Balance	0	0	0	0
6. Final moments	$-Pe_0$	$-\frac{1}{2} Pe_0$	$\frac{1}{2} Pe_0$	Pe_0.

Table 11.2.2

	A	B		C
1. DF	—	0.5	0.5	—
2. FEM	$-\dfrac{wL^2}{48}$	$\dfrac{wL^2}{48}$	$-\dfrac{wL^2}{48}$	$\dfrac{wL^2}{48}$
3. Release A and C permanently and C.O.	$\left(\dfrac{wL^2}{48}-Pe_0\right)$	$\frac{1}{2}\left(\dfrac{wL^2}{41}-Pe_0\right)$	$\frac{1}{2}\left(-\dfrac{wL^2}{48}+Pe_0\right)$	$\left(-\dfrac{wL^2}{48}+Pe_0\right)$.
4. Initial moments	$-Pe_0$	$\left(\dfrac{wL^2}{32}-\dfrac{1}{2}Pe_0\right)$	$-\left(\dfrac{wL^2}{32}-\dfrac{1}{2}Pe_0\right)$	Pe_0.
5. Balance	0	0	0	0
6. Final moments	$-Pe_0$	$\left(\dfrac{wL^2}{32}-\dfrac{1}{2}Pe_0\right)$	$-\left(\dfrac{wL^2}{32}-\dfrac{1}{2}Pe_0\right)$	Pe_0.

Table 11.2.3

	A	B	C	
1. DF	–	$\dfrac{K_1}{K_1 + K_2}$	$\dfrac{K_2}{K_1 + K_2}$	–
2. EFM	$\dfrac{2}{3} PS_1$	$-\dfrac{2}{3} PS_1$	$\dfrac{2}{3} PS_2$	$-\dfrac{2}{3} PS_2$
3. Release A and C permanently and C.O.	$-\dfrac{2}{3} PS_1$	$-\dfrac{1}{3} PS_1$	$\dfrac{1}{3} PS_2$	$\dfrac{2}{3} rS_2$
4. Initial moments	0	$- PS_1$	PS_2	0
5. Balance	–	$\dfrac{P(S_1 - S_2) K_1}{K_1 + K_2}$	$\dfrac{P(S_1 - S_2) K_2}{K_1 + K_2}$	–
6. Final moments	0	$\dfrac{-P(S_1 K_2 + S_2 K_1)}{K_1 + K_2}$	$\dfrac{P(S_1 K_2 + S_2 K_1)}{K_1 + K_2}$	0

In a similar way, the resultant moments M_r may be determined when the beam carries any given transverse loading. Consider, for instance, the prestensioned beam carrying uniform load of intensity w as shown in Fig. 11.2.3 (a). The free body of the concrete beam is shown in Fig. 11.2.3 (b). The moment distribution is carried out in Table 11.2.2. The magnitudes of the fixed-end moments (FEM) for uniform loading is $\dfrac{w\,(L/2)^2}{12} = \dfrac{wL^2}{48}$.

The fixed end moments are shown in line 2. The correction moments at A and C required to release these ends permanently and the carry-over moments are shown in line 3. The initial moments shown in line 4 do not require balancing as shown in line 5. The final moments are shown in line 6 which indicate that the sagging bending moment at support B is $(0.5\,Pe_0 - wL^2/32)$. The final bending moment diagram is shown in Fig. 11.2.3 (c). Fig. 11.2.3 (d) shows the bending moment diagram due to transverse loading acting on an ordinary non-prestressed beam. It may be verified that the resultant moment diagram of Fig. 11.2.3 (c) may be obtained by adding the bending moment diagram due to prestress shown in Fig. 11.2.1 (e) and the bending moment diagram due to w shown in Fig. 11.2.3 (d) in accordance with the principle of superposition. It is needless to say that computation of prestressing moment M_p is unnecessary if only the resultant moment M_r is of interest.

11.2.2 Linear Transformation

The cable line in a continuous beam is said to be transformed linearly when it is raised or lowered only at interior supports without altering the intrinsic shape characterised by its curvature and sag. As a simple example of linear

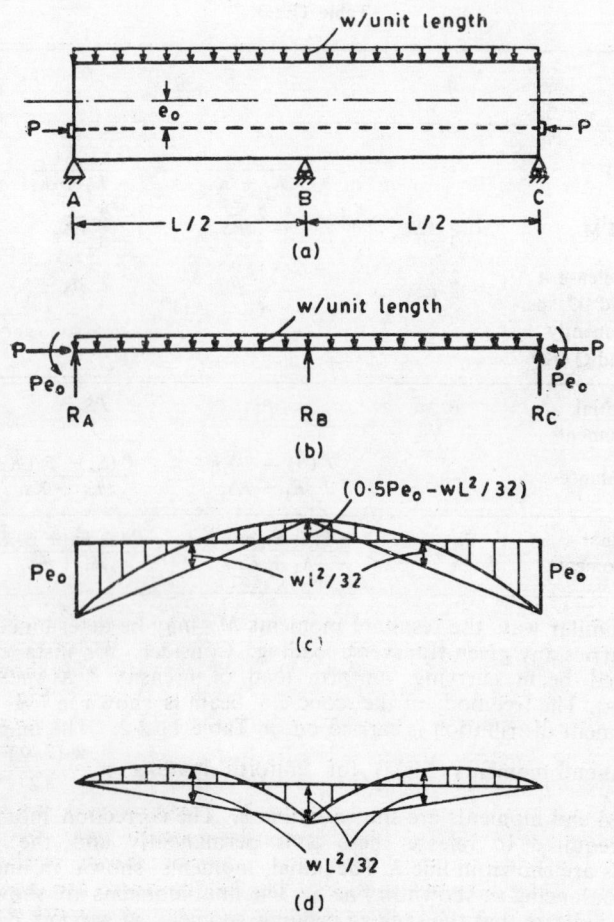

Fig. 11.2.3

transformation, consider the cable line in a three span continuous beam shown in Figure 11.2.4 (a). Figure 11.2.4 (b) shows another cable line for the same beam. It has zero eccentricity at end supports A and D and any arbitrary eccentricities $\triangle_5$ and $\triangle_6$ at interior supports B and C. The cable line in each of the three spans is straight. The eccentricities of the cable lines in Fig. 11.2.4 (a) and (b) are added algebraically to obtain the eccentricities of the cable line shown in Fig. 11.2.4 (c). In this case the cable line of Fig. 11.2.4 (c) is called the linear transformation of the cable line of Fig. 11.2.4 (a) and vice-versa. A careful comparison of the linearly transformed cable line of Fig. 11.2.4 (c) with that of Fig. 11.2.4 (a) shows that the following rules must be obeyed in carrying out linear transformation of cable line in continuous beam:

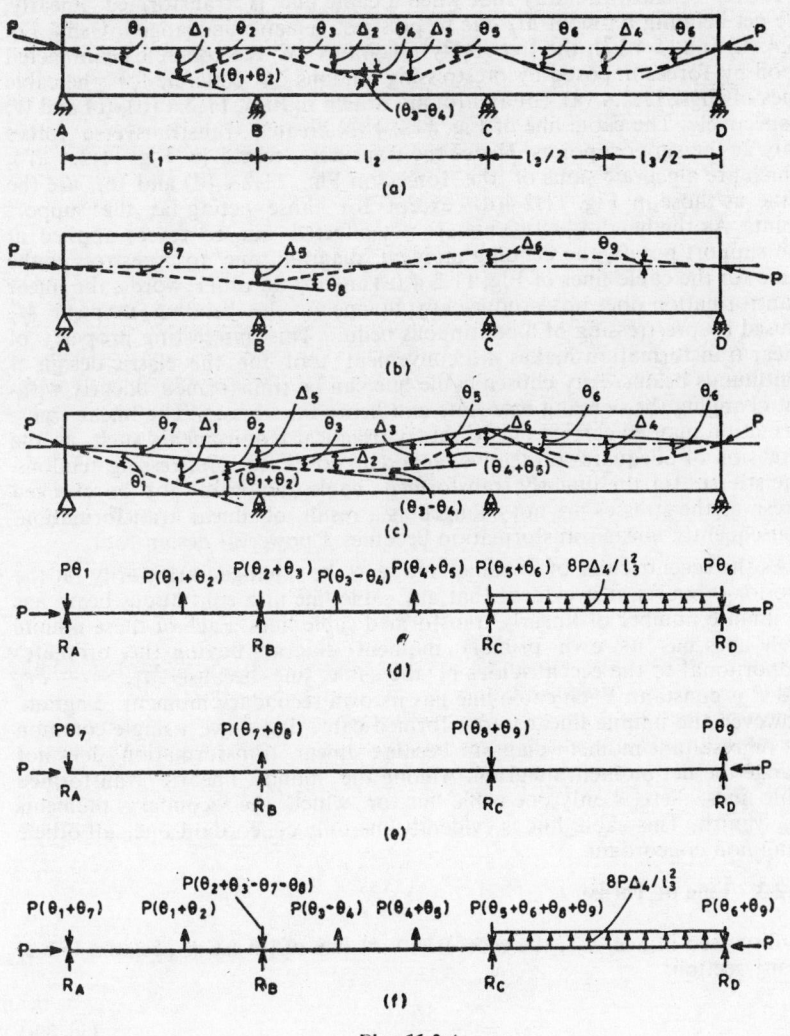

Fig. 11.2.4

(i) The eccentricities of cable line at the two outermost supports must not be changed.

(ii) The intrinsic shape of cable line characterised by its curvature and sag in each span must not be changed.

(iii) The eccentricities of cable line at some or all interior supports may be changed arbitrarily. The operation is equivalent to giving a rigid body rotation to the cable line in two or all spans.

It may be shown readily that when a cable line is transformed linearly, the net bending moment M_p due to prestress remains unchanged. Using Eq. (5.4.10) and (5.4.17), the free body diagrams of the concrete beam acted upon by forces imposed by prestressing tendons on concrete for the cable lines of Figs. 11.2.4 (a), (b) and (c) are drawn in Figs. 11.2.4 (d), (e) and (f) respectively. The cable line of Fig. 11.2.4 (b) creates transtransverse forces only at the support points. Hence the transverse forces in Fig. 11.2.4 (f), which are algebraic sums of the forces in Fig. 11.2.4 (d) and (e), are the same as those in Fig. 11.2.4(d) except for those acting at the support points. As the bending moments are not affected due to forces applied at the support points, the bending moment diagram due to prestress is the same for the cable lines of Fig. 11.2.4 (a) and (c). In other words, the linear transformation does not produce any change in the bending moment M_p caused by prestressing of a continuous beam. This interesting property of linear transformation makes it a convenient tool for the elastic design of continuous beams. Any chosen cable line can be transformed linearly without changing the bending moments and hence the stresses. The linear transformation may be carried out to satisfy practical requirements such as the provision of adequate protective concrete cover to prestressing tendons. The stresses for the linearly transformed cable line need not be checked afresh as the stresses are not changed as a result of linear transformation. Consequently linear transformation becomes a powerful design tool.

As the eccentricities of a cable line can be changed arbitrarily at the interior supports, it is evident that any cable line in a continuous beam has an infinite number of linearly transformed cable lines. Each of these infinite cable lines has its own primary moment diagram having the ordinates proportional to the eccentricities of the cable line because $M_{pp} = -Pe_p$ and P is constant. Each cable line has its own secondary moment diagram. However, the infinite linearly transformed cable lines have a single common net or resultant moment diagram because linear transformation does not change the net moment diagram. Among the infinite linearly transformed cable lines, there is only one cable line for which the secondary moments M_{ps} vanish. This cable line is evidently the only concordant one, all others being non-concordant.

11.2.3 Line of Thrust

As discussed in Sec. 1.7, the eccentricity of line of thrust or pressure line e_c at any section,

$$e_c = -\frac{M_r}{P} \qquad (11.2.8)$$

in which M_r is the resultant bending moment at the stage of loading under consideration.

(i) *Unloaded beam*

In an unloaded beam having no external load, $(M = o)$,

$$M_r = M_p = M_{pp} + M_{ps} = -Pe_p + M_{ps} \qquad (11.2.9)$$

Hence the eccentricity of pressure line at the unloaded stage,

$$e_c = - \frac{M_r}{P} = e_p - \frac{M_{ps}}{P} \qquad (11.2.10)$$

If the cable line is concordant, $M_{ps} = o$ and hence $e_c = e_p$. It follows that in the state of concordance. the pressure line in a continuous beam at the unloaded stage coincides with the cable line similar to that in a simply supported beam. If the cable line is non-concordant, the intercept between the cable line and the pressure line in an unloaded beam is equal to M_{ps}/P in which M_{ps} is the secondary moment at the section under consideration.

(ii) *Loaded beam*

If the beam carries external loads causing bending moment M,

$$M_r = M_p + M = M_{pp} + M_{ps} + M$$
$$= - Pe_p + M_{ps} + M \qquad (11.2.11)$$

Hence the eccentricity of pressure line at any section subjected to bending moment M due to external loads,

$$e_c = - \frac{M_r}{P} = e_p - \frac{M_{ps}}{P} - \frac{M}{P} \qquad (11.2.12)$$

It follows that for any non-concordant cable line, the vertical intercept between the pressure line and cable line is equal to $(M_{ps} + M)/P$. In the particular case of a concordant cable line for which the secondary moment $M_{ps} = o$, the vertical intercept between the pressure line and cable line is equal to M/P similar to that in a simply supported beam. Knowing the position of pressure line, the stresses in continuous beam can be computed readily. The net stresses at any section for any stage of loading may be determined by combining the direct compressive stress (P/A) with the bending stress $(M_r y/I)$ in which the resultant or net bending moment $M_r = - Pe_c$. As all linearly transformed cable lines have the same net bending moment diagram and also because the ordinates of net bending moment diagram are proportional to eccentricities of the pressure line, all linearly transformed cable lines have the same pressure line. In other words, the pressure line in a continuous beam remains unaffected by linear transformation at any stage of loading.

11.2.4 Elastic Theorems

Some interesting properties of prestressed concrete continuous beams in the precracking or elastic range which are related to concordance. linear transformation and pressure line may be expressed in the form of theorems. Basically there are only two elastic theorems. Several others can, however, be stated as corollaries of the basic theorems.

Theorem 1

Any cable line in a continuous beam is necessarily concordant if its ordinates or eccentricities are proportional to the ordinates of the bending moment diagram for the corresponding non-prestressed beam resting on unyielding supports and carrying any arbitrary transverse loading.

Theorem 2

This theorem first enunciated by Guyon in 1945 may be stated as follows :

The linear transformation of the cable line in a continuous beam does not produce any change in the net or resultant bending moment diagram, the pressure line or the stresses at any stage of loading.

To readers familiar with the analysis of ordinary non-prestressed continuous beams, the derivation of Theorem 1 will appear to be very simple. The distribution of bending moments in a transversely loaded non-prestressed continuous beam resting on unyielding supports is such that the deflections at supports are zero. Consequently, if the ordinates or eccentricities of the cable line are proportional to the ordinates of the bending moment diagram for the corresponding non-prestressed beam subjected to any arbitrary transverse loading, the primary moments due to prestressing will produce no deflections at the supports. As a result, the secondary forces will not arise and the condition of concordance is attained.

The reasons why linear transformation does not change the net bending moment diagram or the pressure line or the stresses as stated in Theorem 2 are discussed in Secs. 11.2.2 and 11.2.3. Several interesting properties related to concordance, linear transformation and pressure line in continuous beams can be derived from the two basic theorems stated above. Some of them are as follows:

1. If all ordinates of a concordant cable line are multiplied by a constant, the resulting cable line is also concordant. The non-concordant cable lines too have the same property.

2. If the ordinates of any two concordant cable lines are added. the resulting cable line is concordant.

3. If the ordinates of a concordant cable line are added to the ordinates of a non-concordant cable line, the resulting cable line is non-concordant.

4. Any cable line, whether concordant or non-concordant, imposes a set of transverse forces which produce a bending moment diagram which is the same as the net or resultant moment (M_p) diagram for the cable line.

5. If a concordant cable line is transformed linerly, the resulting cable line is non-concordant.

6. Any continuous beam has infinite number of cable lines which are linear transformations of each other. Among them only one cable line is concordant whereas all others are non-concordant.

7. The infinite cable lines, which are linear transformations of each other, have the same net or resultant bending moment diagram, the same pressure line and produce the same stresses in a continuous beam.

8. The pressure line in an unloaded continuous beam represents a linear transformation of the non-concordant cable line in the beam. The

linearly transformed cable line coinciding with the pressure line is concordant.

EXAMPLE 11.2.1

A beam of uniform section is restrained against rotation at both ends. It has a draped cable line as shown in Fig. 11.2.5 (a). The prestressing force is 500 kN. Draw the primary, secondary and resultant bending moment diagrams due to prestressing.

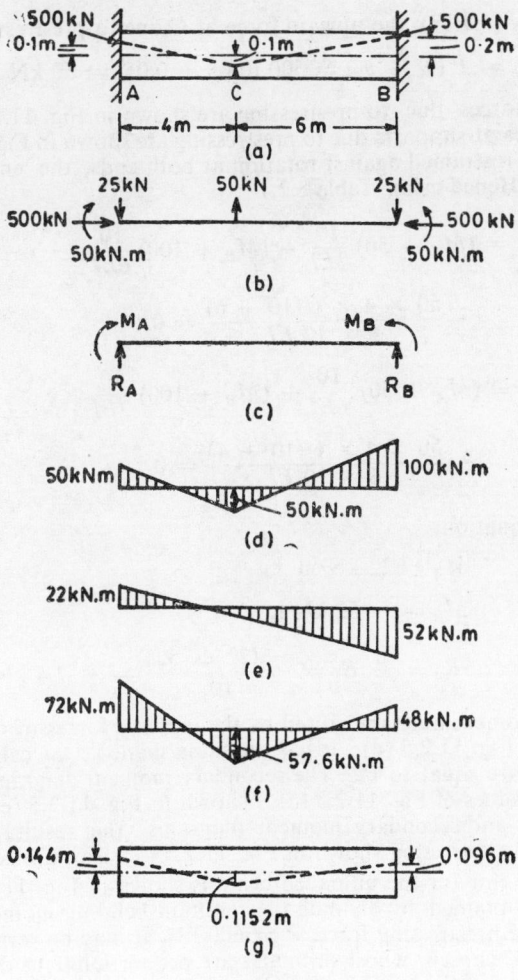

Fig. 11.2.5

Solution

Slope of cable line in portion AC,

$$\theta_1 = \frac{0.1 + 0.1}{4} = 0.05$$

Slope of cable line in portion CB,

$$\theta_2 = \frac{0.1 + 0.2}{6} = 0.05$$

Hence form Eq. (5.4.17), the upward force at C due to prestressing,

$$W_p = P\,(\theta_1 + \theta_2) = 500\,(0.05 + 0.05) = 50 \text{ kN}$$

The primary forces due to prestressing are shown in Fig. 11.2.5 (b). The secondary forces at supports due to prestressing are shown in Fig. 11.2.5 (c). As the beam is restrained against rotation at both ends, the end rotations $\beta_A = \beta_B = 0$. Hence using Table 8.2.1

$$\beta_A = (M_A + 50)\,\frac{10}{3EI} + (M_B + 100)\,\frac{10}{6EI}$$

$$-\,\frac{50 \times 4 \times 6\,(10 + 6)}{6 \times 10\,EI} = 0$$

$$\beta_B = (M_A + 50)\,\frac{10}{6EI} + (M_B + 100)\,\frac{10}{3EI}$$

$$-\,\frac{50 \times 4 \times 6\,(10 + 4)}{6 \times 10\,EI} = 0$$

Solving these equations,

$$M_A = 22 \text{ kN·m}$$

$$M_B = -52 \text{ kN·m}$$

$$R_A = -\,R_B = \frac{-\,(22 + 52)}{10} = -7.4 \text{ kN}$$

The primary moment diagram caused by the primary forces of Fig. 11.2.5 (b) is shown in Fig. 11.2.5 (d). It has the same shape as the cable line and the ordinates are equal to Pe_p. The secondary moment diagram caused by the secondary forces of Fig. 11.2.5 (c) is shown in Fig. 11.2.5 (e). Combining the primary and secondary moment diagrams, the resultant moment diagram due to prestressing shown in Fig. 11.2.5 (f) is obtained.

The pressure line for the unloaded beam is shown in Fig. 11.2.5 (g). Its ordinates are obtained by dividing the resultant bending moments in Fig. 11.2.5 (f) by the prestressing force, $P = 500$ kN. It may be verified that if a cable line is chosen whose ordinates are proportional to the resultant bending moments or the ordinates of the pressure line, it is necessarily concordant.

EXAMPLE 11.2.2

A contionus beam ABC of uniform section has a parabolic cable line with zero eccentricity at the extreme ends A and C as shown in Fig. 11.2.6. The prestressing force P is constant over the entire length of the beam. Determine the condition for concordance.

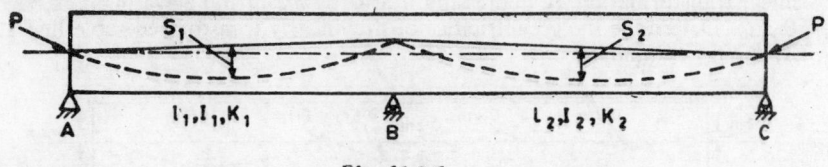

Fig. 11.2.6

Solution

Using Eq. (5.4.10), the uniform upward forces caused by prestressing in spans AB and BC are,

$$w_{p1} = \frac{8PS_1}{l_1^2}$$

$$w_{p2} = \frac{8PS_2}{l_2^2}$$

The moments caused by these forces may be determined by using moment-distribution method as shown in Table 11.2.3. The distribution factors (D.F.) are shown in line 1. The fixed-end moments in spans AB and BC are $-\dfrac{w_{p1} \, l_1^2}{12} = -\dfrac{2}{3} \, PS_1$ and $-\dfrac{w_{p2} \, l_2^2}{12} = -\dfrac{2}{3} \, PS_2$ respectively as shown in line 2. The ends A and C may be released permanently by applying appropriate moments at these ends and carrying-over half the moments to the far ends as shown in line 3. The initial moments shown in line 4 are obtained by summing up the moments of lines 2 and 3. Joint B may next be balanced by applying the correction moments proportional to flexural stiffnesses K_1 and K_2 as shown in line 5. The moment distribution closes approximately at this step as there are no carry-over moments and consequently no correction or balancing moments. The final moments caused by prestressing are shown in line 6. While the moments at A and C are zero, the sagging moment at support B,

$$M_B = \frac{P \, (S_1 \, K_2 + S_2 \, K_1)}{(K_1 + K_2)}$$

The cable line produces concordance if secondary moments vanish or if the final moments are equal to the primary moments. The primary moment at B is Pe_B (sagging). Equating the primary and final moments at B, the condition for concordance may be expressed as,

$$e_B = \frac{S_1 \, K_2 + S_2 \, K_1}{K_1 + K_2}$$

In the case of symmetry about midsupport B, i.e., $l_1 = l_2$, $S_1 = S_2 = S$ and $K_1 = K_2$, the eccentricity at B, $e_B = S$. It follows that for concordance, the eccentricity at central support should be twice the eccentricity at midspan sections.

EXAMPLE 11.2.3

The cable line in a continuous beam shown in Fig. 11.2.7 (a) is it given a linear transformation by depressing it at C by 0.2 m and elevating it at F by 0.3 m. Determine the eccentricities of the linearly transformed cable line at A, B, D, E, G and H.

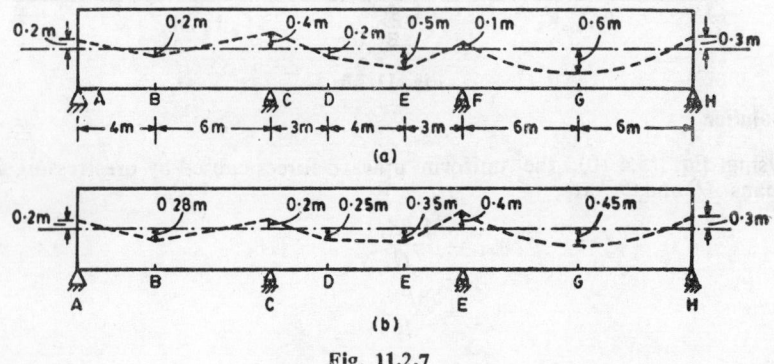

Fig. 11.2.7

Solution

The sags S_B, S_D, S_E and S_G at sections B, D, E and G in the cable line of Fig. 11.2.7 (a) are 0.48, 0.51, 0.69 and 0.80 m respectively. The sags must not change during linear transformation. Also the eccentricities at the end supports A and H must remain unchanged. Consequently, the eccentricities at A, B, D, F, G and H are $-0.2, 0.28, 0.25, 0.35, 0.45$ and -0.3 m respectively as shown in Fig. 11.2.7 (b).

EXAMPLE 11.2.4

Give a linear transformation to the cable line shown in Fig. 11.2.8 (a) so that the cable line has a minimum concrete cover of 0.1 m.

Solution

In order to meet minimum cover requirement at E, the cable line at B must be raised through a minimum distance equal to $2(0.65 - 0.50) = 0.30$ m. The linearly transformed cable line is shown in Fig. 11.2.8 (b).

EXAMPLE 11.2.5

If the parabolic cable lines shown in Fig. 11.2.9 (a), (b) and (c) are linear transformations of each other, determine the eccentricities of prestress at supports and midspan sections.

Solution

As the end eccentricities e_A and e_D and the sags S_{AB}, S_{BC} and S_{CD} do not change during linear transformation, they are invariants in all the three cable lines. Hence,

$$e_A = -0.20 \text{ m} \qquad e_D = -0.10 \text{ m}$$

$$S_{AB} = 0.35 \text{ m} \qquad S_{BC} = 0.40 \text{ m}$$

$$S_{CD} = 0.40 \text{ m}$$

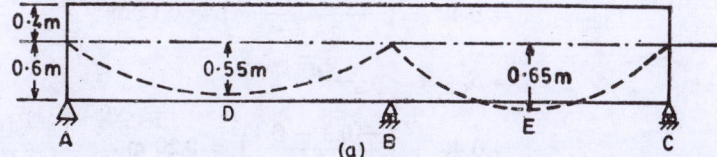

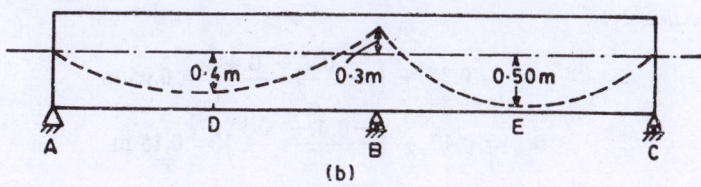

Fig. 11.2.8

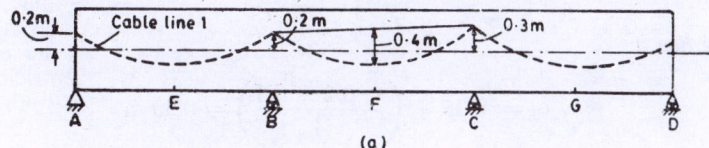

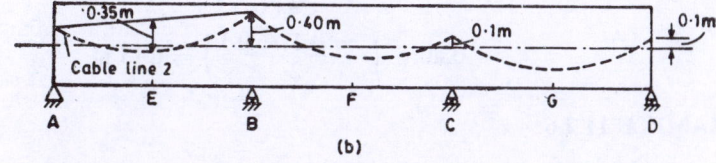

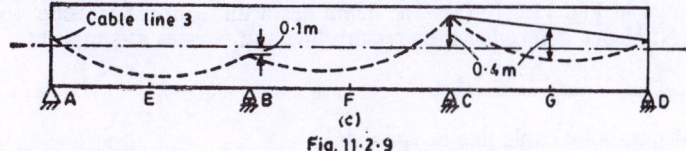

Fig. 11.2.9

Cable line 1

$$e_E = S_{AB} + \frac{e_A + e_B}{2}$$

$$= 0.35 + \left(\frac{-0.2 - 0.2}{2}\right) = 0.15 \text{ m}$$

$$e_F = S_{BC} + \frac{e_B + e_C}{2}$$

$$= 0.40 + \left(\frac{-0.2 - 0.3}{2}\right) = 0.15 \text{ m}$$

$$e_G = S_{CD} + \frac{e_C + e_D}{2}$$

$$= 0.40 + \left(\frac{-0.3 - 0.1}{2}\right) = 0.20 \text{ m}$$

Cable line 2

$$e_E = 0.35 + \left(\frac{-0.2 - 0.4}{2}\right) = 0.05 \text{ m}$$

$$e_F = 0.40 + \left(\frac{-0.4 - 0.1}{2}\right) = 0.15 \text{ m}$$

$$e_G = 0.40 + \left(\frac{-0.1 - 0.1}{2}\right) = 0.30 \text{ m}$$

Cable line 3

$$e_E = 0.35 + \left(\frac{-0.2 + 0.1}{2}\right) = 0.30 \text{ m}$$

$$e_F = 0.40 + \left(\frac{+0.1 - 0.4}{2}\right) = 0.25 \text{ m}$$

$$e_G = 0.40 + \left(\frac{-0.4 \div 0.1}{2}\right) = 0.15 \text{ m}$$

EXAMPLE 11.2.6

Analyse the continuous beam of uniform rectangular section 0.2×0.6 m as shown in Fig. 11.2:10 (a). The beam has a uniform prestressing force of 600 kN. Hence determine the pressure line and stresses at support C.

Solution

Sag of parabolic cable line in span AC,

$$S_{AC} = 0.1 + \frac{0 + 0.1}{2} = 0.15 \text{ m}$$

Hence from Eq. (5.4.10), upward pressure due to prestressing,

$$w_p = \frac{8 \times 600 \times 0.15}{10^2} = 7.2 \text{ kN/m}$$

The net downward load in span AC is therefore $14.4 - 7.2 = 7.2$ kN/m. The change of direction of cable line at D,

$$\delta\theta = \frac{0.1 + 0.1}{6} + \frac{0 + 0.1}{6} = 0.05 \text{ rad.}$$

Hence from Eq. (5.4.17), the upward force at D due to prestressing,

$$W_p = 600 \times 0.05 = 30 \text{ kN}$$

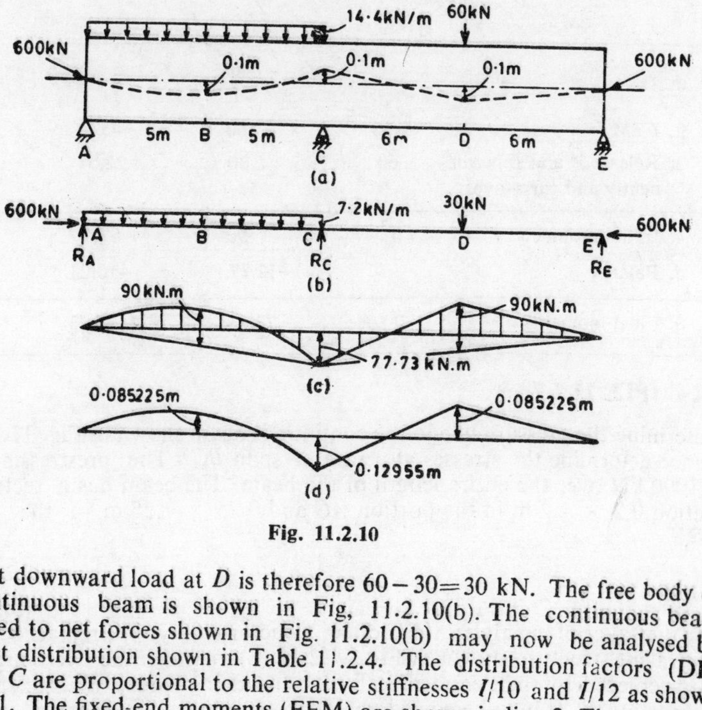

Fig. 11.2.10

The net downward load at D is therefore $60-30=30$ kN. The free body of the continuous beam is shown in Fig, 11.2.10(b). The continuous beam subjected to net forces shown in Fig. 11.2.10(b) may now be analysed by moment distribution shown in Table 11.2.4. The distribution factors (DF) at joint C are proportional to the relative stiffnesses $I/10$ and $I/12$ as shown in line 1. The fixed-end moments (FEM) are shown in line 2. The moments required to release ends A and E permanently and the consequent carry-over moments are shown in line 3. The initial moments in line 4 are obtained by summing up the moments in lines 2 and 3. Joint C is balanced in line 5. As there are no further carry-over moments, the moment distribution terminates automatically at this stage. From line 6 it is seen that the hogging moment at C is 77.73 kN.m. The net or resultant bending moment diagram is shown in Fig. 11.2.10 (c). The line of thrust or pressure line of Fig. 11.2.10 (d) is obtained by dividing the moments in Fig. 11.2.10 (c) by the prestressing force, $P = 600$ kN.

The stresses at section C are,

$$f_T = \frac{-0.6}{0.2 \times 0.6} + \frac{0.07773}{\frac{1}{6} \times 0.2 \times 0.6^2} = 1.48 \text{ MPa}$$

$$f_B = \frac{-0.6}{0.2 \times 0.6} - \frac{0.07773}{\frac{1}{6} \times 0.2 \times 0.6^2} = -11.48 \text{ MPa}$$

Table 11.2.4

	A		C	E
1. DF	–		$\frac{6}{11}$	$\frac{5}{11}$	–
2. FEM	−60		60	−45	45
3. Release A and E permanently and carry over	60	→	30	−22.5 ←	−45
4. Initial moments	0		90	−67.5	0
5. Balance	–		−12.27	−10.23	–
6. Final moments	0		77.73	−77.73	0

EXAMPLE 11.2.7

Determine the pressure line in the continuous beam shown in Fig. 11.2.11(a). Hence determine the stresses at centre of span BC. The prestressing force is 1000 kN over the entire length of the beam. The beam has a rectangular section 0.2×1.2 m in the portion AC and 0.25×1.25 m in the portion CE.

Solution

For the sake of analysis, the actual structure may be replaced by the equivalent structure shown in Fig. 11.2.11 (b) in which the overhangs have been replaced by their equivalent forces. In other words, the free body of the portion BCD may be considered. Taking moments at B and D,

$$M_B = 1000 \times 0.1 - 35 \times 2 \times 1 = 30 \text{ kN. m (sagging)}$$
$$M_D = 1000 \times 0.4 - 28 \times 3 \times 1.5 = 274 \text{ kN.m (sagging)}$$

Using Eq. (5.4.10), the upward force due to prestressing in span BC,

$$w_p = \frac{8 \times 1000 \times 0.64}{16^2} = 20 \text{ kN/m}$$

Similarly, the upward force due to prestressing in span CD,

$$w_p = \frac{8 \times 1000 \times 0.8}{20^2} = 16 \text{ kN/m}$$

Hence the net uniform forces in spans BC and CD are $35 - 20 = 15$ kN/m and $28 - 16 = 12$ kN/m respectively. The fixed-end moments in spans BC and CD are $15 \times 16^2/12 = \pm 320$ kN.m and $12 \times 20^2/12 = \pm 400$ kN.m respectively. The relative stiffnesses of spans BC and CD are $I/16$ and $1.25 \, I/20 = I/16$. Hence the distribution factors at joint C are 0.5 and 0.5. The moment distribution is carried out in Table 11.2.5. The fixed-end moments are shown in line 2. The moments to be applied at B and D to release them permanently and the resulting carry-over moments are shown in line 3. As the net couple at B is 30 kN.m (clockwise), the correcting moment required is $320 + 30 = 350$ kN.m with 175 kN.m as carry-over moment. Similarly, the correcting moment required at D is -674 kN.m with -337 kN.m as the carry-over moment. The net initial moments after the permanent release of joints B and D are shown in line 4. Next, joint C may be balanced as shown in line 5. As there are no carry-over moments, the moment distribution terminates automatically at this stage. The final moments are shown in line 6. It follows that the moments at B, C and D are 30 kN.m (sagging), 616 kN.m (hogging) and 274 kN.m (sagging) respectively.

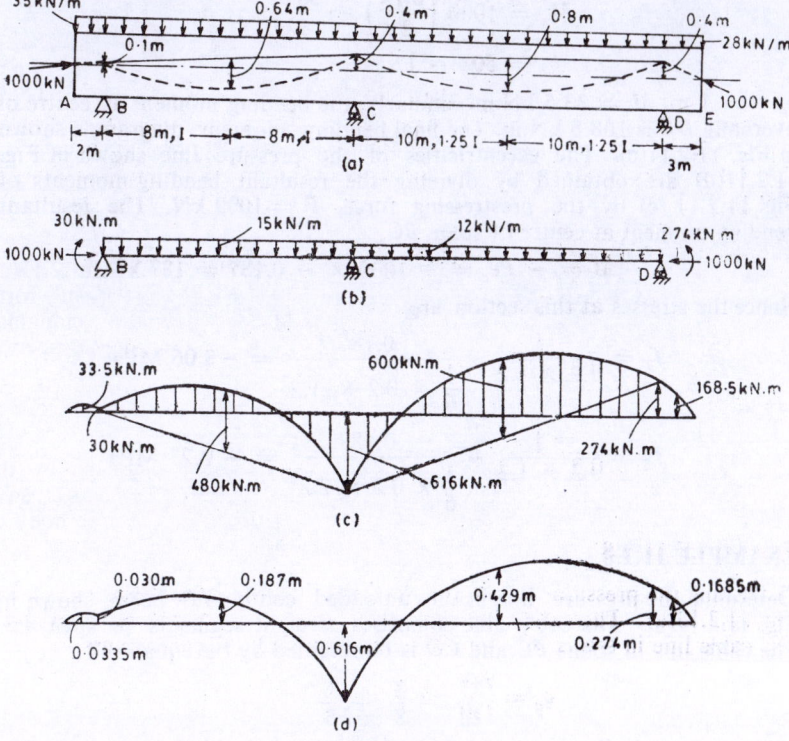

Fig. 11.2.11

Table 11.2.5

	B	C	C	D
1. DF	—	0.5	0.5	—
2. FEM	—320	320	—400	400
3. Release B and D permanently and carry-over	350 →	175	—737 ←	—674
4. Initial moments	30	495	—737	—274
5. Balance	—	121	121	—
6. Final moments	30	616	—616	—274

The eccentricity of cable line at a distance x from A in the portion AB is $0.1x/2$. Hence the bending moment at x,

$$M_x = 1000 \left(\frac{0.1x}{2} \right) - \frac{35x^2}{2}$$
$$= 50x - 17.5x^2$$

At $x = 1$ m, $M = 33.5$ kN·m. Similarly, the bending moment at centre of overhang DE is 168.5 kN.m. The final bending moment diagram is shown in Fig. 11.2.11(c). The eccentricities of the pressure line shown in Fig. 11.2.11(d) are obtained by dividing the resultant bending moments of Fig. 11.2.11 (c) by the prestressing force, $P = 1000$ kN. The resultant bending moment at centre of span BC,

$$M = - Pe_c = - 1000 \times - 0.187 = 187 \text{ kN.m}$$

Hence the stresses at this section are,

$$f_T = \frac{-1}{0.2 \times 1.2} - \frac{0.187}{\frac{1}{6} \times 0.2 \times 1.2^2} = -8.06 \text{ MPa}$$

$$f_B = \frac{-1}{0.2 \times 1.2} + \frac{0.187}{\frac{1}{6} \times 0.2 \times 2.2^2} = - 0.27 \text{ MPa}$$

EXAMPLE 11.2.8

Determine the pressure line in the unloaded continuous beam shown in Fig. 11.2.12(a). The cable line comprises straight segments in span AB. The cable line in spans BC and CD is represented by the equations,

$$e_p = \frac{7x^2}{720} - \frac{x}{8} + \frac{1}{5}$$

$$e_p = \frac{x^3}{3840} + \frac{x^2}{320} - \frac{x}{15} + \frac{1}{10}$$

The beam has uniform prestressing force of 1200 kN.

Solution

The vertical forces at E, F and G due to prestressing are determined by using Eq. (5.4.17).

$$W_E = 1200 \left[\frac{0.1 - (-0.1)}{2} - \frac{0.2 - 0.1}{3} \right] = 80 \text{ kN}$$

$$W_F = 1200 \left[\frac{0.2 - 0.1}{3} - \frac{0.1 - 0.2}{2} \right] = 100 \text{ kN}$$

$$W_G = 1200 \left[\frac{0.1 - 0.2}{2} - \frac{-0.2 - 0.1}{3} \right] = 60 \text{ kN}$$

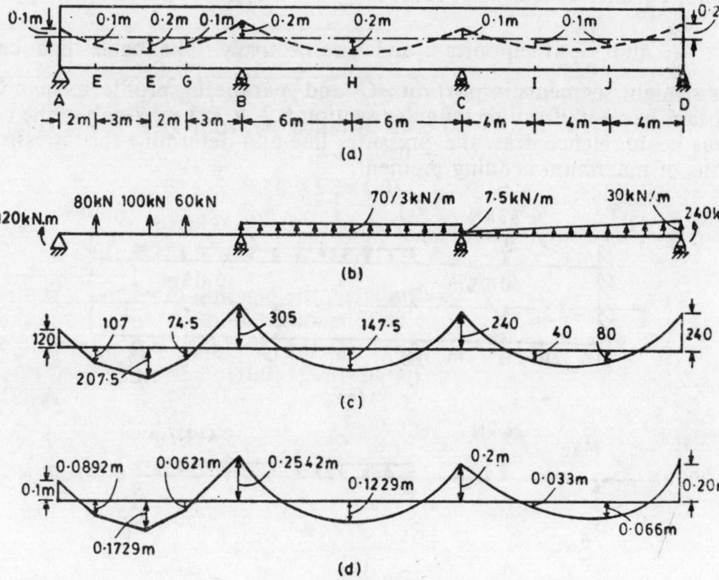

Fig. 11.2.12

Using Eq. (5.4.4), the force in span BC due to prestressing,

$$w_p = - 1200 \left(\frac{14}{720} \right) = - \frac{70}{3} \text{ kN/m}$$

The minus sign shows that the force is upward. It is uniform over the entire span because it is independent of x. Similarly, using Eq. (5.4.4), the force in span CD due to prestressing

$$w_p = - 1200 \left(\frac{x}{640} + \frac{1}{160} \right)$$

This equation shows that the load is trapezoidal with intensities equal to 7.5 kN/m and 30 kN/m at C and D respectively. The forces caused by prestressing are shown in Fig. 11.2.12 (b). Analysing the beam for the forces shown in Fig. 11.2.12 (b) by moment distribution or otherwise, the sagging bending moments at the supports A, B, C and D are 120, 305, 240 and 240 kN.m respectively. Hence the resultant bending moment diagram shown in Fig. 11.2.12 (c) is obtained. Dividing the resultant moments due to prestressing by the prestressing force, $P = 1200$ kN, the pressure line in the unloaded beam shown in Fig. 11.2.12 (d) is obtained.

EXAMPLE 11.2.9

The continuous beam restrained against rotation at A as shown in Fig. 11.2.13 (a) carries uniform prestressing force of 200 kN. The yielding of supports permits a clockwise rotation. $\beta_A = \dfrac{4}{EI}$ at A and vertical settlements $\dfrac{24}{EI}$ and $\dfrac{12}{EI}$ at supports C and E respectively. The cable line comprises straight segments in portion AC and parabolic profile in span CE. The beam has a uniform rectangular section 0.2×0.3 m. Analyse the continuous beam. Hence draw the pressure line and determine the stresses at sections of maximum bending moment.

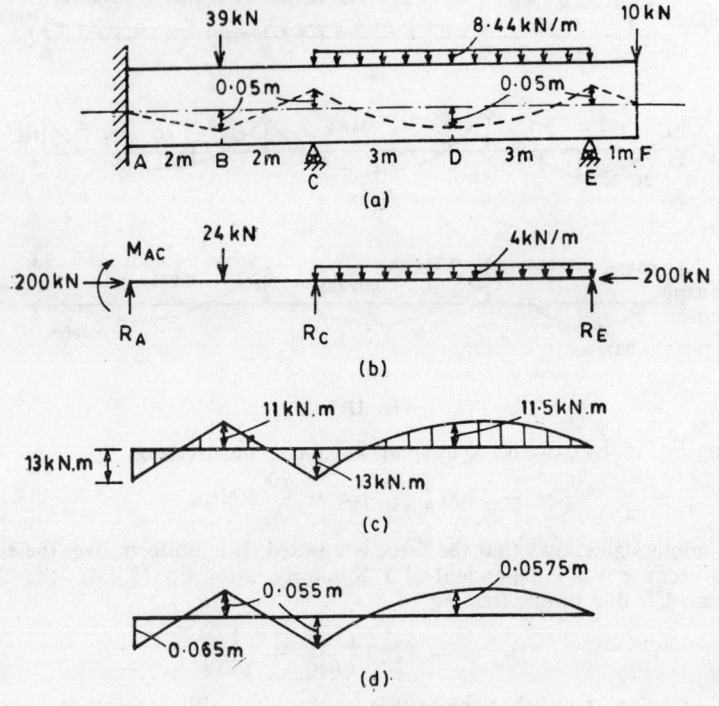

Fig. 11.2.13

Solution

Using Eq. (5.4.17), the upward force at B due to prestressing,

$$W_p = 200 \left(\frac{0.05 - 0}{2} - \frac{-0.05 - 0.05}{2} \right) = 15 \text{ kN}$$

Hence the downward force at B is $39 - 15 = 24$ kN. Using Eq. (5.4.10), the uniform upward force due to prestressing in span CE

$$w_p = \frac{8 \times 200 \times 0.1}{6^2} = 4.44 \text{ kN/m}$$

Hence the net downward uniform loading in span CE is $8.44 - 4.44 = 4$ kN/m. The bending moment at E,

$$M_E = 200 \times 0.05 - 10 \times 1 = 0$$

The free body of portion AE shown in Fig. 11.2.13 (b) may be analysed by using the method of slope-deflection equations.

$$M_{AB} = M_{fAB} + \frac{2EI}{L} (2\beta_A + \beta_B - 3\psi)$$

$$M_{BA} = M_{fBA} + \frac{2EI}{L} (\beta_A + 2\beta_B - 3\psi)$$

where, $M_{fAB}, M_{fBA} =$ fixed-end moments at A and B

$\beta_A, \beta_B =$ rotations at ends A and B

$\psi =$ member rotation

$\quad =$ difference of level between A and B divided by L

In these equations, all clockwise moments and rotations are taken positive. The fixed-end moments in spans AC and CE are ± 12 and ± 12 kN.m respectively. The slope-deflection equations for span AC may be written as

$$M_{AC} = -12 + \frac{2EI}{4} \left[2 \times \frac{4}{EI} + \beta_C - 3 \left(\frac{\frac{24}{EI} - 0}{4} \right) \right] \qquad \text{(a)}$$

$$M_{CA} = 12 + \frac{2EI}{4} \left[\frac{4}{EI} + 2\beta_C - 3 \left(\frac{\frac{24}{EI} - 0}{4} \right) \right] \qquad \text{(b)}$$

Similarly, the slope-deflection equations for span CE may be written as

$$M_{CE} = -12 + \frac{2EI}{6} \left[2\beta_C + \beta_E - 3 \left(\frac{\frac{12}{EI} - \frac{24}{EI}}{6} \right) \right] \qquad \text{(c)}$$

$$M_{EC} = 12 + \frac{2EI}{6} \left[\beta_C + 2\beta_E - 3 \left(\frac{\frac{12}{EI} - \frac{24}{EI}}{6} \right) \right] \qquad \text{(d)}$$

The unknown rotations β_C and β_E in the slope-deflection equations may be determined by using the following equations of equilibrium for joints C and E

$$M_{CA} + M_{CE} = 0 \qquad (e)$$

$$M_{EC} = 0 \qquad (f)$$

Substituting from Eq. (b), (c) and (d) into Eq. (e) and (f) and solving the resulting equations, $\beta_C = \dfrac{8}{EI}$ and $\beta_E = -\dfrac{25}{EI}$. Substituting for β_C and β_E into Eq. (a), (b), (c) and (d),

$$M_{AC} = -13 \text{ kN.m}$$

$$M_{CA} = 13 \text{ kN.m}$$

$$M_{CE} = -13 \text{ kN.m}$$

$$M_{EC} = 0$$

The net or resultant bending moment diagram is shown in Fig. 11.2.13(c). The pressure line shown in Fig. 11.2.13 (d) is obtained by dividing the resultant bending moments by the prestressing force, $P = 200$ kN. The maximum bending moment occurs at supports A and C. The extreme fibre stresses at these sections are

$$f_T = \frac{-0.2}{0.2 \times 0.3} + \frac{0.013}{\dfrac{1}{6} \times 0.2 \times 0.3^2} = 1 \text{ MPa}$$

$$f_B = \frac{-0.2}{0.2 \times 0.3} - \frac{0.013}{\dfrac{1}{6} \times 0.2 \times 0.3^2} = -7.67 \text{ MPa}$$

11.3 CRACKING AND DEFORMATION

Upto the limit state of cracking, the flexural rotations and deflections under short-time loading may be computed with reasonable accuracy using the conventional elastic theory in which the cross-sectional properties are based on the gross uncracked section. The long-time deformations including the effect of creep of concrete may be evaluated by multiplying the short-time deformations by the factor $(1 + C_c)$ in which C_c is the creep coefficient. As soon as the load exceeds the limit state of cracking, the beam develops cracks at the most critical points. In most beams, the flexural cracks appear first at interior supports at which the bending moments are largest in magnitude. At these interior support sections, the flexural cracks originate from the top face which is in tension under the action of hogging moments. These cracks propagate downwards with increase in load. In the mean time, flexural cracks originate from the bottom face near midspan sections at which the sagging moments are largest. These cracks propagate upwards with increasing load. The position of inflexion points depends upon the span lengths, the magnitude and position of loads and the net moments induced by prestressing. The distance of inflexion points often lies between $l/6$ to $l/4$ from an interior support where l is the span in which the inflexion point occurs. Sections near the inflexion points remain uncracked in flexure upto the limit state of collapse. These sections develop inclined web shear or diagonal tension cracks due to excessive diagonal tension. On the other hand, sections located between midspan points and inflexion points, which are subjected to a combination of bending and shear, develop inclined flexure-shear cracks. Fig. 11.3.1 shows the orientation of cracks in a two span continuous beam carrying uniform transverse loading. Cracks marked as 1, 2 and 3 represent respectively the flexure cracks, web shear cracks and flexure-shear cracks.

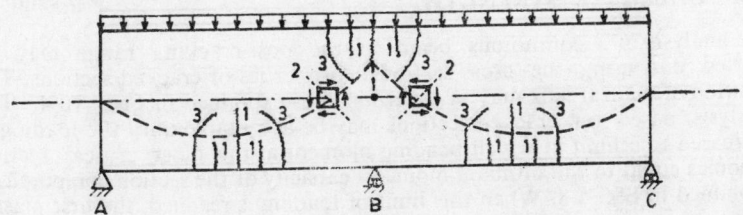

Fig. 11.3.1

In the post-cracking range, the effective cross-sectional properties depend upon the extent of cracking at the section under consideration. The effective cross-sectional area A and the moment of intertia I may be based on gross section in the uncracked portion and the transformed section in the cracked portion of the beam. Both effective area and moment of inertia change continuously not only with the increase in load but also from section to section along the beam. The extent of cracking at any section depends upon the magnitude of bending moment at the section which in turn depends upon the extent of cracking and effective values of A and I because the continuous beam is statically indeterminate. As direct solution of the problem is not feasible, an iterative procedure with the following steps may be adopted:

1. Start with an assumed distribution of moments. For this purpose, the moments at interior supports may be given reasonable values.

2. Divide the beam into several parts and determine the effective values of A and I at all nodal points consistent with the distribution of moments assumed in step 1.

3. Determine the moments at supports and nodal points using the effective sectional properties computed in step 2. Numerical approach, such as the Newmarks' method, may be used for the analysis of the continuous beam.

4. The moments determined in step 3 may be compared with those assumed in step 1 and iteration carried out to desired accuracy.

5. For the moments at the nodal points determined in step 4, the curvatures at all nodal points may be computed as explained in Sec. 8.2.

6. The flexural rotations and deflections may then be determined by numerical integration of the curvatures computed in step 5.

It is evident that the computations for the determination of deformations in a continuous beam in the post-cracking range are time consuming. The determination of long-time deformations are still more difficult. The long-time displacements computed by multiplying the short-time displacements by the factor $(1 + C_c)$ may under-estimate the actual displacements significantly because the creep coefficient C_c is not a constant but increases progressively with increasing stress under high stress condition. Hence the theoretical methods can only provide a rough estimate of the short and long-time deformations of continuous beams in the post-cracking range.

11.4 ULTIMATE STRENGTH

The analysis of a continuous beam in the post-cracking range may be carried out using the cross-sectional properties of cracked sections. The position of neutral axis may be determined as explained in Sec. 10.4. The analysis based on cracked sections may be acceptable until the loading is increased to a limit at which bending moment at the most critical section becomes equal to the ultimate moment capacity of the section computed as explained in Sec. 5.8. When this limit of loading is reached, the first plastic hinge forms at the most critical section. With further increase in loading, the moment at the most critical section remains constant at its moment capacity whereas the moments at other sections keep on increasing provided the beam has adequate ductility to permit plastic rotations necessary for redistribution of moments. The successive formation of plastic hinges due to redistribution of moments continues until the beam is converted into a mechanism. This stage gives the ultimate strength or collapse load of the beam. If the beam has little ductility or if the limited ductility is ignored, the ultimate strength of the beam is equal to that at the formation of the first plastic hinge. On the other hand, the ultimate strength corresponds to the formation of a collapse mechanism if the beam possesses sufficient ductility to allow successive plastic hinge formation. It follows that the limits of loading at the formation of first plastic hinge and the collapse mechanism represent respectively the lower bound and the upper bound of the ultimate strength of the continuous beam. The analysis for the most critical collapse mechanism is the same as for ordinary non-prestressed continuous beams.

In order to examine the ultimate strength of a continuous beam, consider a typical continuous span AB carrying transverse loading as shown in Fig. 11.4.1. The span collapses when plastic hinges form at the interior supports A and B and at a section near midspan C where the sagging moment is the largest. The section of maximum sagging moment may be assumed to coincide with C without appreciable error unless the loading is extremely eccentric or unsymmetrical with respect to the central section C. The collapse mechanism formed due to the formation of plastic hinges at A, B and C may be given a small virtual displacement shown by the broken line. Equating the internal virtual

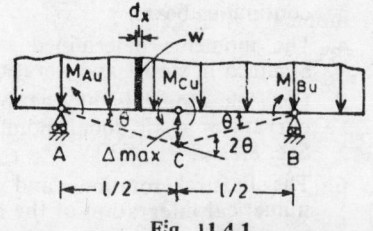

Fig. 11.4.1

work at the plastic hinges to the external virtual work due to the displacement of loading,

$$M_{Au}\,\theta + M_{Cu}\,(2\theta) + M_{Bu}\,\theta = \int (w\,dx)\,\Delta \qquad (11.4.1)$$

in which M_{Au}, M_{Bu}, and M_{Cu} are the ultimate moment capacities at sections A, B and C respectively. Two specific cases of loading may now be considered.

(i) Central concentrated load

If the continuous span carries a central concentrated load W_u at the limit state of collapse, Eq. (11.4.1) may be written as

$$(M_{Au} + 2M_{Cu} + M_{Bu})\,\theta = W_u\,\Delta_{\max} \qquad (11.4.2)$$

Putting $\qquad \Delta_{\max} = \theta\,\dfrac{l}{2}$,

$$W_u = \frac{2}{l}\,(M_{Au} + 2M_{Cu} + M_{Bu}) \qquad (11.4.3)$$

In the case of a continuous beam having uniform section or constant moment capacity,

$$M_{Au} = M_{Bu} = M_{Cu} = M_u,$$

$$W_u = \frac{8M_u}{l} \qquad (11.4.4)$$

(ii) Uniform load

If the collapse load w_u is uniformly distributed, Eq. (11.4.1) may be written as

$$(M_{Au} + 2M_{Cu} + M_{Bu})\,\theta = (w_u\,l)\,\frac{\Delta_{\max}}{2} \qquad (11.4.5)$$

Putting $\Delta_{\max} = \theta\,\dfrac{l}{2}$,

$$w_u = \frac{4}{l^2}\,(M_{Au} + 2M_{Cu} + M_{Bu}) \qquad (11.4.6)$$

In the case of a continuous beam having uniform section or constant moment capacity,

$$M_{Au} = M_{Bu} = M_{Cu} = M_u,$$

$$w_u = \frac{16M_u}{l^2} \qquad (11.4.7)$$

In order to determine the weakest or the most critical span which collapses first, all loads acting on a continuous beam may be expressed in terms of the characteristic load W_k. Thus the loads of the first span may be expressed as $c_1'\,W_k$, $c_1''\,W_k$, ..., those of the second span as $c_2'\,W_k$, $c_2''\,W_k$, ..., and so on. Here c_1', c_1'', ..., c_2', c_2'', are absolute constants which remain unchanged during monotonic increase in all loads. Thus all loads acting on the continuous beam increase proportionately as the characteristic load W_k increases monotonically, until at collapse or ultimate stage, the characteristic load W_k becomes the characteristic ultimate load W_{ku}. For a continuous beam with given moment capacities at all critical points, the characteristic ultimate load W_{ku} for each span can be calculated by using Eq. (11.4.1). The weakest or the most critical span which collapses first is the one for which the characteristic ultimate load W_{ku} is the smallest. As the weakest span controls the ultimate strength of a continuous beam, the collapse load corresponds to the minimum characteristic ultimate load W_{ku} (min). The ultimate strength of the continuous beam is reached when the loads are $c_1' W_{ku}$ (min), $c_1'' W_{ku}$ (min), . . . on first span $c_2' W_{ku}$ (min), $c_2'' W_{ku}$ (min), . . . on second span and so on. When some of the spans fail while others remain intact, the failure is called partial failure. The complete failure, in which all spans fail simultaneously, is possible only when all spans have the same value of the characteristic ultimate load W_{ku}. In a complete failure, the moment capacities at all critical sections are fully utilised at the collapse stage. On the other hand, the moment capacities at only some of the critical points are fully utilised in partial failure while those at other critical points remain under-utilised.

The influence of secondary bending moments due to prestressing on the ultimate flexural strength of a continuous beam deserves due consideration. If the beam possesses sufficient ductility so that failure occurs only after the formation of sufficient plastic hinges required to degenerate the structure into a mechanism, the ultimate load is controlled by the moment capacities in the plastic hinging regions. In this case, therefore, the secondary moments have no influence on the load carrying capacity of the beam. Consequently, even a non-concordant cable may be assumed to behave like a concordant one as long as the beam possesses sufficient ductility for the formation of a collapse mechanism. On the other hand, the neglect of secondary moments may be unsafe for beams with limited ductility. If the secondary moments are negative (hogging) at support sections and positive at midspan sections, the load carrying capacity of the continuous beam is reduced. The reverse effect may be expected if the secondary moments are of opposite signs. Consider for instance the cable line of Fig. 11.2.1 which produces a positive (sagging) secondary moment at the support section B. Suppose that the beam possesses limited ductility so that it collapses as soon as a plastic hinge forms at B. In this case the external load will have to produce a negative bending moment at B equal to the sum of the moment capacity M_{Bu} and the secondary bending moment M_{ps} at B. Consequently, the load carrying capacity of the beam is increased by the ratio $(M_{Bu} + M_{ps})/M_{Bu}$. If the secondary moment M_{ps} at B is negative for a non-concordant cable profile, the effect of secondary moments is to reduce the load capacity by the ratio $(M_{Bu} - M_{ps})/M_{Bu}$. This is subject to the condition that moment capacities at midspan sections are large enough to eliminate the formation of plastic hinges at these sections.

EXAMPLE 11.4.1

A continuous beam of constant moment capacity M_u having two spans, each of length l, carries uniformly distributed load as shown in Fig. 11.4.2

(a). Determine the lower and upper bounds of collapse load assuming that failure occurs at (i) formation of first plastic hinge and (ii) formation of a collapse mechanism.

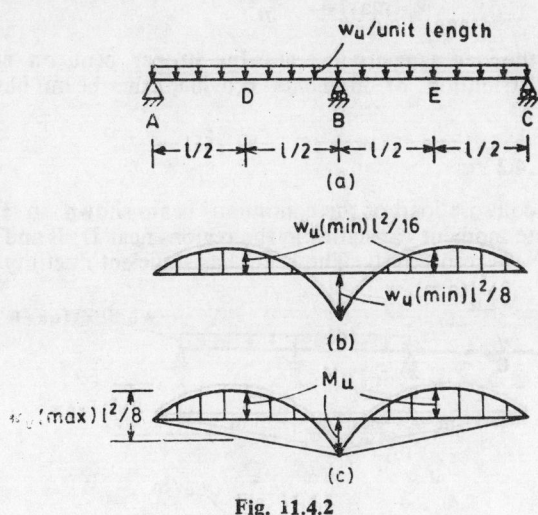

Fig. 11.4.2

Solution

The bending moment diagram based on elastic theory is shown in Fig. 11.4.2 (b). As the maximum bending moment occurs at the central support, the first plastic hinge forms at B. If the beam has little ductility, it fails at the formation of the first plastic hinge at B. On the other hand, plastic hinges will appear at midspan sections D and E at a higher load level if the beam has sufficient ductility.

(i) Lower bonud.

The lower bound of collapse load corresponds to the formation of first plastic hinge at B which is given by

$$M_u = \frac{w_u \, l^2}{8}$$

or

$$w_u \, (\text{min}) = \frac{8M_u}{l^2}$$

(ii) Upper bound

The bending moment diagram shown in Fig. 11.4.2 (c) corresponds to the stage at which plastic hinges form at D, B and E thereby converting the structure into a mehanism. This stage of loading corresponds to the upper bound of collapse load given by

$$M_u + \frac{1}{2} M_u = w_u \,(\text{max}) \, \frac{l^2}{8}$$

or
$$w_u \,(\text{max}) = \frac{12 M_u}{l^2}$$

It follows that the load capacity increases by 50 per cent on account of favourable redistribution of moments provided the beam has sufficient ductility.

EXAMPLE 11.4 2

Determine the collapse load of the continuous beam shown in Fig. 11.4.3 (a). The ultimate moment capacities in the regions near D, B and E are 400, 600 and 600 kN·m respectively. The beam has sufficient ductility.

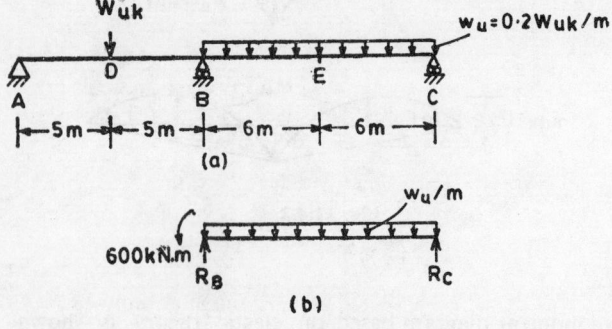

Fig. 11.4.3

Solution

Using Eq. (11.4.3), the characteristic ultimate load of span AB,

$$W_{ku} = \frac{2}{10} \, (0 + 2 \times 400 + 600) = 280 \text{ kN}$$

Similarly using Eq. (11.4.6), for span BC,

$$W_u = 0.2 \, W_{ku} = \frac{4}{12^2} \, (600 + 2 \times 600 + 0) = 50 \text{ kN/m}$$

or
$$W_{ku} = 250 \text{ kN}$$

Hence the characteristic collapse load, $W_{ku} = 250$ kN. The span BC collapses first when the uniform load on span BC becomes 50 kN/m. At this stage the load at D is 250 kN. As the ultimate load capacity of span AB is 280 kN, it does not collapse.

The collapse load of 50 kN/m for span BC obtained from Eq. (11.4.6) is based on the approximation that the plastic hinge forms at the centre of span BC. Actually the plastic hinge forms slightly to the right of E at the section where the sagging bending moment is largest. To locate the position of maximum bending moment, consider the free body of span BC shown in Fig. 11.4.3 (b). Taking moments about B.

$$w_u \times 12 \times 6 - 13\,R_C - 600 = 0$$

or

$$R_C = 6\,w_u - 50$$

The distance x from C at which shear force is zero and therefore the bending moment is maximum is determined from

$$R_B - w_u\,x = 0$$

or

$$x = \frac{R_C}{w_u} = 6 - \frac{50}{w_u} \qquad (a)$$

Hence the maximum bending moment,

$$M_{\max} = R_C\,x - \frac{w_u\,x^2}{2}$$

$$= \frac{(6w_u - 50)^2}{2w_u} \qquad (b)$$

Putting $M_{\max} = M_{Eu} = 600$ kN·m in Eq. (b) and solving for w_u, the collapse load, $w_u = 48.57$ kN/m and from Eq. (a), $x = 4.97$ m. The plastic hinge therefore forms at a distance of 1.03 m to the right of midspan section E. Comparing the collapse load of 48.57 kN/m with the approximate value of 50 kN/m computed earlier, it is seen that the error on account of the assumption that plastic hinge forms at midspan section is less than 3 per cent.

EXAMPLE 11.4.3

Calculate the collapse load of uniform intensity w_u for the continuous beam shown in Fig. 11.4.4. The beam has a uniform moment capacity of 400 kN·m and has sufficient ductility.

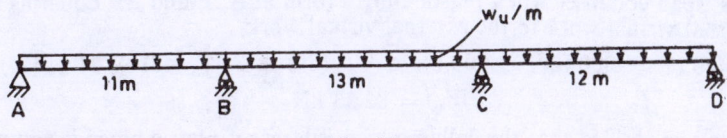

Fig. 11.4.4

Solution

Using Eq. (11.4.6), the collapse load of span AB,

$$w_u = \frac{4}{11^2}\,(0 + 2 \times 400 + 400) = 39.67 \text{ kN/m}$$

The collapse load of span BC,

$$w_u = \frac{4}{13^2}\,(400 + 2 \times 400 + 400) = 37.87 \text{ kN/m}$$

Similarly the collapse load of span CD,

$$w_u = \frac{4}{12^2} (400 + 2 \times 400 + 0) = 33.33 \text{ kN/m}.$$

The weakest span CD collapses first because its load capacity is the smallest, $w_u = 33.33$ kN/m.

EXAMPLE 11.4.4

Determine the characteristic collapse load W_{ku} for the continuous beam having sufficient ductility shown in Fig. 11.4.5 (a). The ultimate moment capacities in the regions near A, E, B and G are 400, 300, 500 and 360 kN·m respectively.

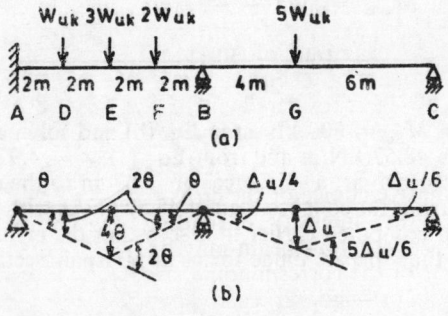

Fig. 11.4.5

Solution

(i) Span AB

This span collapses when plastic hinges form at A, E and B. Equating the internal virtual work to the external virtual work,

$$400 \,(\theta) + 300 \,(2\theta) + 500 \,(\theta) = W_{ku} \,(2\theta) + 3W_{ku} \,(4\theta) + 2W_{ku} \,(2\theta)$$

or

$$W_{ku} = 83.33 \text{ kN}$$

It may be verified that the collapse load is larger if plastic hinge is assumed to form either at D or F instead of that at E. Consequently, the plastic hinge forms at the most critical section E.

(ii) Span BC

This span collapses when plastic hinges form at B, G and C. Equating the internal virtual work to the external virtual work,

$$500 \left(\frac{\Delta_u}{4}\right) + 360 \left(\frac{5}{6} \, \Delta_u\right) = 5 W_{ku} \,(\Delta_u)$$

or

$$W_{ku} = 85 \text{ kN}$$

The failure is controlled by the weaker span AB. The characteristic collapse load, $W_{k_u} = 83.33$ kN.

11.5 DESIGN APPROACHES

Similar to the approach commonly adopted for the design of other statically indeterminate structures, a prestressed concrete continuous beam may be designed using the working stress method based on elastic theory and checked to verify that the design provides adequate safety against collapse as per provisions of the code. Alternatively, it is possible to design the continuous beam using the ultimate strength theory and to verify that the stresses and deformations at the working load level are within limits specified by the code.

11.5.1 Working Stress Method

Although continuous beams are often non-prismatic and may carry irregular loading, the simpler case of a continuous beam of uniform section carrying uniform load may be considered first. The working stress design for a prismatic beam with uniformly distributed load may be described by the following steps:

(1) Make a tentative choice of the cross-sectional dimensions appropriate for the span lengths, intensity of loading and the type of structure. The overall depth of a continuous beam may be taken to be 20 to 30 per cent smaller than that of the corresponding simply supported beam. While the beam is taken to be prismatic in each span, the cross-section may vary from span to span keeping in view the span lengths. The cross-section changes only at supports. While open flaged sections may be used for buildings, box sections may be adopted for bridges, particularly in the case of curved and skew bridges in which torsion is significant.

(2) Determine the cross-sectional properties sucn as area A and moment of inertia I in each span. Also determine the self weight and other dead loads acting at the initial stage. Also determine the dead load moment M_D at the midspan sections of all spans treating them as simply supported spans.

(3) Determine the live load per unit length including the effect of impact, if any. If the loading comprises a series of concentrated loads or somewhat irregular loads, it may be converted into its statically equivalent uniformly distributed load.

(4) Take a design decision in respect of the portion cf transverse load to be counteracted by prestress. Generally an attempt is made to counteract the entire dead load w_D and a fraction of the live load w_L. The fraction of live load to be counteracted by prestress depends upon the nature, duration and frequency of live load. If the ratio of live and dead load (w_L/w_D) is high and the live load is relatively infrequent, a smaller fraction of live load should be balanced by prestress. On the other hand, if live load is relatively small and is of more or less permanent nature, a larger fraction of live load may be counteracted by prestress. The choice of the magnitude and eccentricity of prestress for balancing a desired fraction of transverse load may also be guided by the consideration of permissible camber at initial stage and permissible deflection at the final stage.

A criterion, which is generally found to be satisfactory is, to select pre-stress in such a manner that the net unbalanced upward load $(w_{pi} - w_D)$ at the initial stage is equal to the net downward unbalanced load $(w_D + w_L - w_p)$ at the final stage.

$$w_{pi} - w_D = w_D + w_L - w_p \tag{11.5.1}$$

in which w_{pi} and w_p are the initial and final transverse loads (upward) imposed by prestressing tendons on the concrete beam at the initial and final stages respectively. If this criterion is adopted, the bending moments at the initial and final stages will be equal in magnitude and opposite in sign. Consequently, the design moments at all critical sections of the continuous beam are minimised. Inserting $w_p = \eta \, w_{pi}$ into Eq. (11.5.1),

$$w_{pi} = \frac{2w_D + w_L}{1 + \eta} \tag{11.5.2}$$

in which η is equal to reduction factor for effective prestress or loss factor. If the loss of prestress is ignored as an approximation, $\eta = 1$, Eq. (11.5.2) becomes

$$w_{pi} = w_p = w_D + 0.5 \, w_L \tag{11.5.3}$$

Equation (11.5.2) shows that the prestress should approximately counteract the entire dead load w_D and half of the live load w_L in order to minimise design moments in the working stress design. It is interesting to note that if Eq. (11.5.3) is satisfied, the unbalanced load is $0.5 \, w_L$ (upward) at the initial stage and $0.5 \, w_L$ (downward) at the final stage. Consequently, the moments are roughly equal and opposite at the two crucial stages in the working stress design. If w_D and w_L are not the same for all spans, determine w_{pi} and w_p for each span.

(5) Determine the maximum practical values of sags of the cable line in all spans allowing for adequate minimum protective cover of concrete, c. Adopting parabolic cable profiles, the maximum practical sag in the end spans for zero end eccentricities as shown in Fig. 11.5.1 is

$$S_1 \text{ (max)} = (y_B - c) + \tfrac{1}{2}(y_T - c)$$
$$= y_B + 0.5 \, y_T - 1.5c \tag{11.5.4}$$

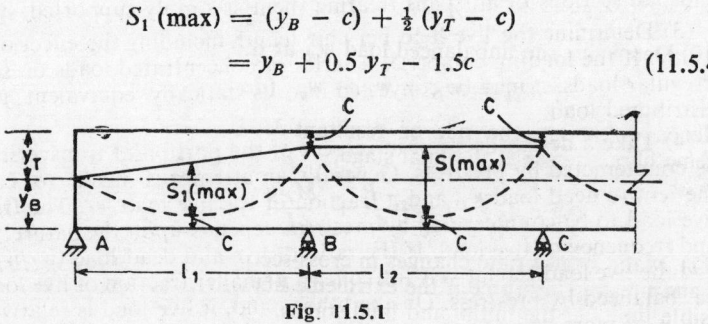

Fig. 11.5.1

Similarly, the maximum practical sag in an interior span shown in the figure is

$$S \text{ (max)} = D - 2c \tag{11.5.5}$$

(6) Using the values of w_p and S (max) computed in steps 4 and 5, determine the effective prestressing force P in each span from the equation

$$P = \frac{w_p \, l^2}{8S \, (\text{max})} \tag{11.5.6}$$

Equation (11.5.6) may give different values of P in different spans. If P has to be the same in all spans, the largest value of P determined from Eq. (11.5.6) may be adopted because there is no scope to increase the sag S without infringing on the requirement of minimum concrete cover.

(7) Using the values of w_p and P computed in steps 4 and 6, determine the sag S in each span from

$$S = \frac{w_p \, l^2}{8P} \tag{11.5.7}$$

Hence choose the cable line giving the sags computed from Eq. (11.5.7).

(8) Determine w_{pi} and w_p for each span using values of P and S adopted in steps 6 and 7 from the equations,

$$w_p = \frac{8PS}{l^2} \tag{11.5.8a}$$

and

$$w_{pi} = \frac{w_p}{\eta} \tag{11.5.8b}$$

(9) Dtermine the unbalanced load w_{ubi} at the initial stage in each span

$$w_{ubi} = w_{pi} - w_D \quad \text{(upward)}$$

Analyse the continuous beam for the unbalanced load at the initial stage using moment distribution or any other method. Hence determine the net or resultant bending moment M_{ri} at initial stage at all critical points (support and midspan points) and compute the extreme fibre stresses,

$$f = \frac{-P_i}{A} + \frac{M_{ri} \, y}{I} \tag{11.5.9}$$

(10) Determine the unbalanced load w_{ub} at the final stage

$$w_{ub} = w_D + w_L - w_p \quad \text{(downward)} \tag{11.5.10}$$

Hence determine the net or resultant M_r at all critical points and the extreme fibre stresses at the final stage.

$$f = \frac{-P}{A} + \frac{M_r \, y}{I} \tag{11.5.11}$$

(11) Make appropriate changes in cross-sections at critical points, cable line and prestressing force if the extrmeme fibre stresses are not within permissible limits at the initial and final stages.

(12) Draw shear force diagrams for initial and final stages and provide adequate shear reinforcement.

(13) Check that the cambers at initial stage and deflections at final stage are within permissible limits.

(14) The chosen cable line may be given linear transformation to satisfy all practical requirements. The positions of individual tendons may be selected so as to obtain the desired cable line.

(15) Finally check that the elastic design based on working loads at initial and final stages also satisfies the design requirements at all other relevant stages of loading. In particular, it should be verified that the design offers adequate safety against collapse.

11.5.2 Ultimate Strength Method

As an alternative approach, a continuous beam may be designed for the limit state of collapse and then checked for its adequacy at all other relevant limit states, particularly at stress transfer and at the working load level. In the ultimate strength design, the moment capacities at the critical sections (midspan and support sections) may be chosen in such a manner that under the action of design ultimate load, all spans reach the point of incipient collapse simultaneously. In this manner, the moment capacities at all critical sections are fully utilized at the limit state of collapse. The virtual work method or mechanism method of plastic analysis and Eq. (11.4.1) may be used to determine the moment capacities at critical sections to satisfy the criterion of simultaneous collapse of all spans in a continuous beam. Care should be taken to see that all cross-sections of the continuous beam are sufficiently under-reinforced to ensure ductile behaviour. The redistribution of moments leading to simultaneous collapse of all spans is possible only when the beam possesses adequate ductility.

In the ultimate strength design of a continuous beam, it is desirable to provide ultimate moment capacities at all critical points in such a manner that the characteristic collapse load is the same for all spans. If this condition is satisfied, partial failure is avoided and moment capacities at all critical sections are fully utilized. Consider a continuous beam with n spans and p critical points. Using Eq. (11.4.1), a set of n simultaneous equations involving moment capacities at critical points may be written. Among the moment capacities at p critical points, moment capacities at $(p - n)$ critical points may be assigned suitable arbitrary values while the remaining n moment capacities may be determined by solving the set of n simultaneous equations. Proceeding in this manner, it may be ensured that all spans reach the stage of incipient collapse simultaneously and the moment capacities at all critical points are fully utilized under the action of design ultimate load. If the continuous beam designed in this manner has sufficient ductility, it may be expected to have the maximum possible extent of favourable redistribution of moments before collapse.

EXAMPLE 11.5.1

A post-tensioned two span continuous beam ABC has spans AB and BC equal to 15 m and 20 m. It has a uniform rectangular cross-section 0.2 × 0.9 m. It carries a uniform dead load of 6 25 kN/m inclusive of self weight at the initial stage and an additional live load of 10 kN/m at the final stage. Sixty per cent of the live load is permanent. The loss of prestress is 20 per cent. Determine the minimum uniform prestressing force required to equalise the unbalanced loads at the initial and final stages. Also determine the appropriate parabolic cable line having a minimum concrete cover of 0.1 m and zero eccentricities at A and C. Determine the extreme fibre stresses at intermediate support B.

Solution

Using Eq. (11.5.2), the upward force due to prestressing at initial stage,

$$W_{pi} = \frac{2 \times 6.25 + 10}{1 + 0.8} = 12.5 \text{ kN/m}$$

and $$w_p = \eta \, w_{pi} = 0.8 \times 12.5 = 10 \text{ kN/m}$$

The minimum prestressing force is controlled by the sag of the cable line in the larger span. Maximum practical sag in larger span BC as shown in Fig. 11.5.2 (a),

$$S_{BC} = (0.45 - 0.1) + \frac{1}{2} [(0.45 - 0.1) + 0] = 0.525 \text{ m}$$

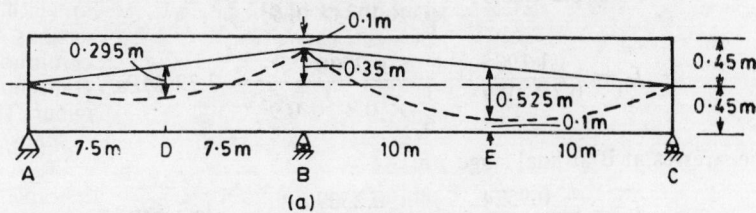

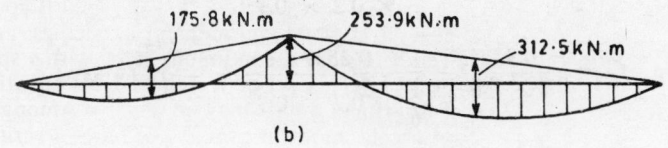

(b)

Fig. 11.5.2

Hence from Eq. (11.5.6), the minimum prestressing force at initial stage,

$$P_i = \frac{12.5 \times 20^2}{8 \times 0.525} = 1190.5 \text{ kN}$$

and $$P = \eta \, P_i = 0.8 \times 1190.5 = 952.4 \text{ kN}$$

In order to obtain the same upward force due to prestressing in both spans, the sag in the smaller span,

$$S_{AB} = S_{BC} \times \left(\frac{L_{AB}}{L_{BC}}\right)^2$$

$$= 0.525 \times \frac{15^2}{20^2} = 0.295 \text{ m}$$

Hence eccentricity of cable line at midspan point D of span AB,

$$e_D = 0.295 - \frac{1}{2} (0.35 + 0) = 0.12 \text{ m}$$

The parabolic cable line having the desirable eccentricities and sags is shown in Fig. 11.5.2 (a). The unbalanced loads at initial and final stage are

$$w_{ubi} = w_{pi} - w_D = 12.5 - 6.25 = 6.25 \text{ kN/m (upward)}$$
$$w_{ub} = w_D + w_L - w_p = 6.25 + 10 - 10$$
$$= 6.25 \text{ kN/m (downward)}$$

Using moment distribution or otherwise, the bending moment diagram at the initial stage shown in Fig. 11.5.2 (b) is obtained. The bending moments at the final stage are equal in magnitude but opposite in sign. The maximum bending moment at both stages occurs at intermediate support section B. The stresses at B at initial stage are

$$f_T = \frac{-1.1905}{0.2 \times 0.9} - \frac{0.2539}{\frac{1}{6} \times 0.2 \times 0.9^2} = -16.02 \text{ MPa}$$

$$f_B = \frac{-1.1905}{0.2 \times 0.9} + \frac{0.2539}{\frac{1}{6} \times 0.2 \times 0.9^2} = 2.79 \text{ MPa}$$

The stresses at B at final stage are

$$f_T = \frac{-0.9524}{0.2 \times 0.9} + \frac{0.2539}{\frac{1}{6} \times 0.2 \times 0.9^2} = 4.11 \text{ MPa}$$

$$f_B = \frac{-0.9524}{0.2 \times 0.9} - \frac{0.2539}{\frac{1}{6} \times 0.2 \times 0.9^2} = -14.69 \text{ MPa}$$

Utilising concrete of grade M-40 and designing the beam as Type-2 (limited prestress) member, the permissible compressive stresses at initial and final stages, taking f_{ci} equal to 35 MPa, are 16.92 and 15.6 MPa respectively. Hence the actual compressive stresses are within permissible limits. The Indian code allows a tensile stress of 4.5 MPa provided under the permanent component of the service load, the stress remains compressive. Unbalanced load at final stage when only the permanent component of the service load is present,

$$w_{ub} = 6.25 + 0.6 \times 10 - 10 = 2.25 \text{ kN/m (downward)}$$

It may be verified that the entire section at B is under compression on account of unbalanced load of 2.25 kN/m.

EXAMPLE 11.5.2

Select a suitable symmetrical I-section and a parabolic cable profie with zero eccentricity at the extreme ends A and C for a two span post-tensioned continuous beam ABC having left span equal to 20 m and right span equal to 25 m as shown in Fig. 11.5.3 (a). The beam carries a uniform dead load of 10 kN/m inclusive of self weight at the initial stage and an additional superimposed live load of 13.75 kN/m at the final stage. Use concrete of grade M-35.

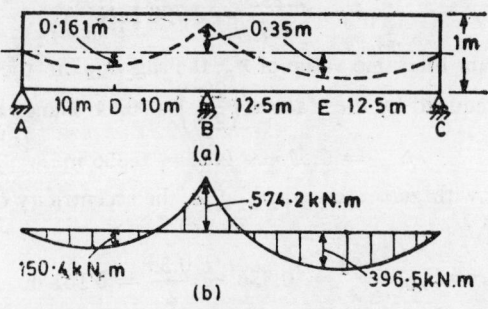

(a)

(b)

Fig. 11.5.3

Solution

Considering the intensity of loading and the spans, the overall depth of the symmetrical I-section may be taken equal to 1 m which is $\frac{1}{25}$ th of the larger span. From the data, dead load $w_D = 10$ kN/m and live load $w_L = 13.75$ kN/m. Adopting a minimum protective cover of 0.15 m on the centroid of prestressing tendons, the maximum permissible eccentricity is $0.50 - 0.15 = 0.35$ m. In order to maximise the sag in the larger span BC, adopt the eccentricity at B, $e_B = -0.35$ m, midspan eccentricity $e_E = 0.35$ m and end eccentricity $e_C = 0$. Consequently, the largest permissible sag of cable line in span BC, $S_{BC} = 0.35 + 0.35/2 = 0.525$ m. For an economic design, the bending moments at initial and final stages may be equalised. Assuming the prestress losses to be 20 per cent,

$$w_p = \frac{2w_D + w_L}{1 + \frac{1}{\eta}} = \frac{2 \times 10 + 13.75}{1 + \frac{1}{0.8}} = 15 \text{ kN/m}$$

$$w_{pi} = \frac{15}{0.8} = 18.75 \text{ kN/m}$$

Unbalanced load at initial stage,

$$w_{ubi} = 18.75 - 10 = 8.75 \text{ kN/m (upward)}$$

Unbalanced load at final stage,

$$w_{ub} = 10 + 13.75 - 15 = 8.75 \text{ kN/m (downward)}$$

Adopting uniform prestressing force over the entire length of the continuous beam for practical reasons, the prestressing force is controlled by the larger span BC. The required uniform prestressing force,

$$P = \frac{w_p \, l_{BC}^2}{8S_{BC}} = \frac{15 \times 25^2}{8 \times 0.525} = 2232 \text{ kN}$$
$$= 2.232 \text{ MN}$$

$$P_i = \frac{2.232}{0.8} = 2.790 \text{ MN}$$

In order to obtain the same value of w_p, the sag S_{AB} in the smaller span AB should be reduced by the factor $\left(\frac{20^2}{25^2}\right) = 0.64$. Hence sag in span AB,

$$S_{AB} = 0.525 \times 0.64 = 0.336 \text{ m}$$

It follows that with zero eccentricity at A, the eccentricity of cable line at midspan section D,

$$e_D = S_{AB} - \frac{e_B}{2} = 0.336 - \frac{0.35}{2} = 0.161 \text{ m}$$

The cable line thus adopted is shown in 11.5.3 (a). With a uniform initial prestressing force of 2.79 MN and final prestressing force of 2.232 MN, the unbalanced load in both spans at initial as well as final stages is 8.75 kN/m. The bending moments due to unbalanced load at initial stage may be computed by moment distribution. The bending moments at the final stage are equal in magnitude but opposite in sign. The bending moment diagram for the initial stage is shown in Fig. 11.5.3 (b). The largest bending moment of 574.2 kN·m (sagging) occurring at the intermediate support B controls the size of uniform section. Taking $f_{pi} = f_{ck} = 35$ MPa, the permissible or working stresses for M-35 concrete in accordance with the Indian code IS: 1343-1980 are as follows:

$$\sigma_{cbci} = 17.91 \text{ MPa}$$
$$\sigma_{cbc} = 13.75 \text{ MPa}$$

Adopting fully prestressed (Type 1) design,

$$\sigma_{cbti} = \sigma_{tbt} = 0$$

The direct stresses due to prestressing are $\left(-\dfrac{P_i}{A}\right) = -\dfrac{2.79}{A}$ at initial stage and $\left(-\dfrac{P}{A}\right) = -\dfrac{2.232}{A}$ at final stage. For optimum conditions, Equations (a) and (b) pertaining to initial stage and Equations (c) and (d) pertaining to final stage should be satisfied.

$$f_T = -\frac{2.79}{A} - \frac{0.5742}{Z} = -17.91 \tag{a}$$

$$f_B = -\frac{2.79}{A} + \frac{0.5742}{Z} = 0 \tag{b}$$

$$f_T = -\frac{2.232}{A} + \frac{0.5742}{Z} = 0 \tag{c}$$

$$f_B = -\frac{2.232}{A} - \frac{0.5742}{Z} = 13.75 \tag{d}$$

Solving Eqs (a) and (b),

$$A = 0.3116 \text{ m}^2$$
$$Z = 0.0641 \text{ m}^3$$

Similarly, solving Eqs. (c) and (d),

$$A = 0.3247 \text{ m}^2$$
$$Z = 0.0835 \text{ m}^3$$

The requirements on A and Z are more stringent for the final stage which may, therefore, be taken to control the size of the section. Adopting a symmetrical I-section having flange width, flange thickness, web thickness and overall depth equal to 0.62 m, 0.20 m, 0.14 m and 1 m respectively, the cross-sectional properties are, $A = 0.332 \text{ m}^2$ and $Z = 0.08605 \text{ m}^3$. For this section, the stresses at the most critical section B are, $f_T = -15.08$ MPa and $f_B = -1.73$ MPa at initial stage and $f_T = -0.05$ MPa and $f_B = -13.4$ MPa at final stage. As these stresses are within permissible limits, the section adopted is safe under service loads.

EXAMPLE 11.5.3

Design the reinforcement for the section of Example 11.5.2 if the effective prestressing force after losses is chosen in such a way that the girder has zero deflection under the dead load of 10 kN/m. Check for the durability of the structure in aggressive environment in accordance with the Indian code.

Solution

In order to have zero deflection under the action of dead load of 10 kN/m, the upward load w_p due to prestressing should be of equal magnitude.

$$w_p = \frac{8P (0.525)}{25^2} = 10 \text{ kN/m}$$

or

$$P = 1488 \text{ kN}$$

and

$$P_i = \frac{P}{0.8} = \frac{1488}{0.8} = 1860 \text{ kN}$$

Upward load due to prestressing at initial stage,

$$w_{pi} = \frac{w_p}{0.8} = \frac{10}{0.8} = 12.5 \text{ kN/m}$$

Unbalanced load at initial stage,

$$w_{ubi} = 12.5 - 10 = 2.5 \text{ kN/m (upward)}$$

Unbalanced load at final stage,

$$w_{ub} = 10 + 13.75 - 10 = 13.75 \text{ kN/m (downward)}$$

The unbalanced load at initial stage is small and will produce a slight cambre at stress transfer. This camber will reduce progressively with time. The unbalanced load at final stage is larger and therefore controls the amount of reinforcement. The moments in Fig. 11.5.3 (b) caused by an unbalanced load of 8.75 kN/m may be multiplied by the factor $\left(\dfrac{-13.75}{8.75} \right)$ in order to obtain the following moments of the final stage:

$$M_B = -0.9023 \text{ MN·m}$$
$$M_D = 0.2363 \text{ MN·m}$$
$$M_E = 0.6231 \text{ MN·m}$$

(i) *Section B*

The hypothetical tensile stress at the top fibre at section B at final stage,

$$f_T = -\frac{1.488}{0.332} + \frac{0.9023}{0.08605} = 6 \text{ MPa}$$

The modulus of rupture of concrete,

$$f_{cr} = 0.7 \sqrt{f_{ck}} = 0.7 \sqrt{35} = 4.14 \text{ MPa}$$

As the hypothetical tensile streess is greater than modulus of rupture concrete will crack. Using high yield strength deformed bars of grade F_e 415 as supplementary untensioned reinforcement and assuming the internal lever arm equal to 0.81 m and working stress of 230 MPa, the area of supplementary reinforcement,

$$A_s = \frac{0.9023}{0.8 \times 2301} = 0.004843 \text{ m}^2 = 4843 \text{ mm}^2$$

The percentage of steel is therefore equal to $(0.004843 \times 100)/0.332$ = 1.458. Referring to Table 10.10.1, the maximum permissible hypothetical tensile stress in aggressive environment (maximum crack width = 0.1 mm),

$$\sigma_{ft} = 3.6 \times 0.7 + 1.458 \times 4$$
$$= 8.452 \text{ MPa}$$

As the fictitious tensile stress is smaller than the maximum permissible value, the durability of the structure is verified. The area of supplementary reinforcement can be curtailed suitably on either side of support section B as the hogging bending moment decreases. The supplementary reinforcement computed above is on the conservative side because it is based on the requirement for flexure without taking into account the presence of direct compressive force. A precise reinforcement requirement may be computed by carrying out the cracked section analysis as explained in Section 10.4.

(ii) *Section D*

The extreme fibre stresses at section D are

$$f_T = -\frac{1.488}{0.332} - \frac{0.2363}{0.08605} = -7.23 \text{ MPa}$$
$$f_B = -\frac{1.488}{0.332} + \frac{0.2363}{0.08605} = -1.74 \text{ MPa}$$

These stresses are within permissible limits. Also no supplementary reinforcement is required because the entire section is in compression.

(iii) *Section E*

The extreme fibre stresses at section E are

$$f_T = -\frac{1.488}{0.332} - \frac{0.06231}{0.08605} = -11.72 \text{ MPa}$$

$$f_B = -\frac{1.488}{0.332} + \frac{0.06231}{0.08605} = 2.76 \text{ MPa}$$

The tensile stress at bottom fibre is less than modulus of rupture of concrete. Hence the section will not crack. Also the tensile stress is less than the permissible stress of 3 MPa for Type 2 members. Consequently, there is no need to provide supplementary reinforcement near centre of span BC.

In conclusion, it may be emphasised that the beam can be designed for the given service load with a reduced prestressing force resulting in several advantages associated with partial prestressing. It is only necessary to provide supplementary reinforcement over a short length in the vicinity of section B.

EXAMPLE 11.5.4

A post-tensioned prestressed concrete continuous girder shown in Fig. 11.5.4 (a) has to carry a superimposed load of 16 kN/m in addition to its self weight. The girder has unsymmetrical I-section as shown in Fig. 11.5.4 (b). While the width and thickness of flanges remain constant, the

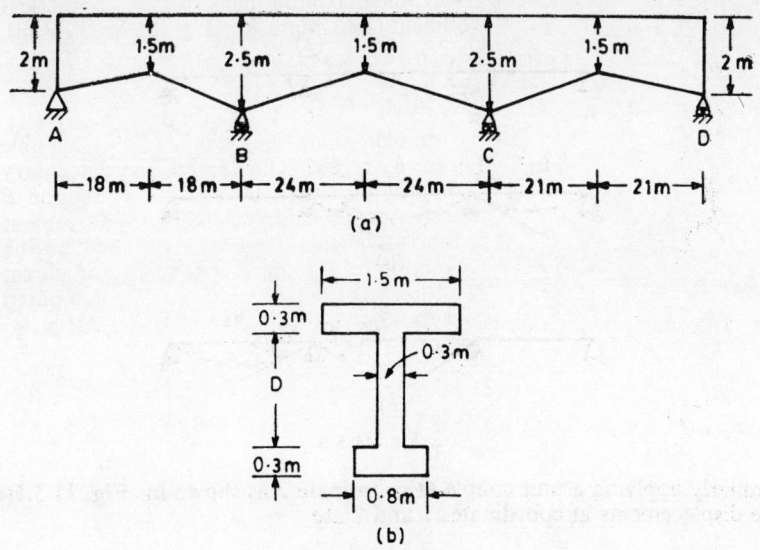

(a)

(b)

Fig. 11.5.4

overall depth of the girder decreases linearly in each span being maximum at supports and minimum at midspan. The relevant cross-sectional properties are shown in Table 11.5.1. The initial and effective prestressing forces are 7.72 and 6.72 MN respectively. The grade of concrete is M-35. Check the stresses and deflections. Take modulus of elasticity of concrete, $E_c =$ 33675 MPa and strength of concrete at stress transfer, $f_{ci} = 30.5$ MPa.

Solution

As the girder is non-prismatic, the analysis may be carried out using numerical integration and Newmark's procedure. For this purpose each girder is divided into panels of 6 m length. Hence there are 6, 8 and 7 panels in spans AB, BC and CD. The cross-sectional properties at nodal points, viz. overall depth D, cross-sectional area A, moment of inertia I and section moduli Z_T and Z_B are shown in Table 11.5.1. The eccentricity of prestress e_p at each nodal point is also listed in the table.

The continuous girder may be analysed using the flexibility approach of matrix analysis. If the bending moments at interior supports B and C are chosen as redundant forces, the released structure is a series of three simply supported beams AB, BC and CD as shown in Fig. 11.5.5 (a). Assigning coordinates 1 and 2 to the redundant forces M_B and M_C, the elements of the flexibility matrix may be obtained by applying a unit couple successively at coordinates 1 and 2 and computing the displacements at the coordinates. Applying a unit couple at coordinate 1 as shown in Fig. 11.5.5 (b), the displacements at coordinates 1 and 2 are

$$\delta_{11} = \theta_1 + \theta_2$$
$$\delta_{21} = \theta_3$$

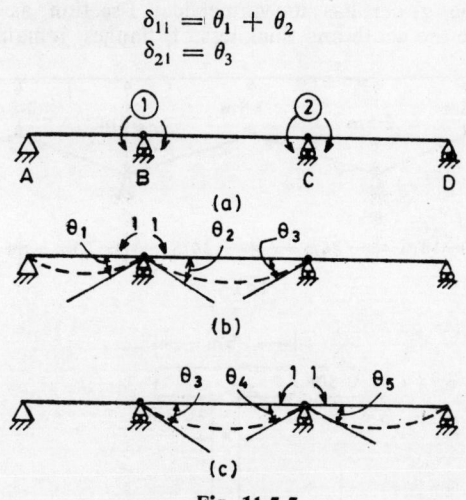

Fig. 11.5.5

Similarly applying a unit couple at coordinate 2 as shown in Fig. 11.5.5(c), the displacements at coordinates 1 and 2 are

$$\delta_{12} = \theta_3$$
$$\delta_{22} = \theta_4 + \theta_5$$

Table 11.5.1

Span	Property	Panel Points								
		1	2	3	4	5	6	7	8	9
AB	D, m	2.0000	1.8333	1.6666	1.5000	1.8333	2.1666	2.5000		
	A, m^2	1.1100	1.0600	1.0100	0.9600	1.0600	1.1600	1.2600		
	I, m^4	0.5436	0.4330	0.3375	0.2553	0.4330	0.6695	0.9691		
	Z_T, m^3	0.6478	0.5663	0.4881	0.4126	0.5663	0.7321	0.9085		
	Z_B, m^3	0.4683	0.4053	0.3459	0.2897	0.4053	0.5345	0.6761		
	e_r, m	0.4000	—0.0666	—0.4000	—0.6000	—0.2666	0.2000	0.8000		
BC	D, m	2.5000	2.2500	2.0000	1.7500	1.5000	1.7500	2.0000	2.2500	2.5000
	A, m^2	1.2600	1.1850	1.1100	1.0350	0.9600	1.0350	1.1100	1.1850	1.2600
	I, m^4	0.9691	0.7380	0.5436	0.3835	0.2553	0.3835	0.5436	0.7380	0.9691
	Z_T, m^3	0.9085	0.7750	0.6478	0.5269	0.4126	0.5269	0.6478	0.7750	0.9085
	Z_B, m^3	0.6761	0.5687	0.4683	0.3752	0.2897	0.3752	0.4683	0.5687	0.6761
	e_r, m	0.8000	0.3190	—0.0750	—0.3810	—0.6000	—0.3810	—0.0750	0.3190	0.8000
CD	D, m	2.5000	2.2140	1.9280	1.6420	1.5710	1.7140	1.8570	2.0000	
	A, m^2	1.2600	1.1742	1.0884	1.0026	0.9813	1.0242	1.0671	1.1100	
	I, m^4	0.9691	0.7071	0.4941	0.3243	0.2886	0.3631	0.4480	0.5436	
	Z_T, m^3	0.9085	0.7563	0.6123	0.4766	0.4444	0.5099	0.5778	0.6478	
	Z_B, m^3	0.6761	0.5538	0.4407	0.3373	0.3132	0.3624	0.4142	0.4683	
	e_r, m	0.8000	0.2780	—0.1470	—0.4730	—0.5310	—0.3180	—0.0080	0.4000	

As the released structure comprises simple spans, the end rotations θ_1 to θ_5 may be calculated readily by using conjugate beam and Newmark's procedure. The flexibility matrix thus computed may be written as

$$[\delta] = \frac{1}{E} \begin{bmatrix} 47.727 & 19.106 \\ 19.106 & 51.040 \end{bmatrix}$$

In accordance with the flexibility method, the redundant forces M_B and M_C may be determined from the equation,

$$\begin{bmatrix} M_B \\ M_C \end{bmatrix} = \begin{bmatrix} \delta_{11} & \delta_{12} \\ \delta_{21} & \delta_{22} \end{bmatrix}^{-1} \left\{ \begin{bmatrix} \Delta_1 \\ \Delta_2 \end{bmatrix} - \begin{bmatrix} \Delta_{1L} \\ \Delta_{2L} \end{bmatrix} \right\} \qquad \text{(a)}$$

As the net displacements at coordinates 1 and 2 must be zero for continuity,

$$\Delta_1 = \Delta_2 = 0$$

The displacements Δ_{1L} and Δ_{2L} are the displacements at coordinates 1 and 2 caused by the loading acting on the released structure. Equation (a) may be written as

$$\begin{bmatrix} M_B \\ M_C \end{bmatrix} = - E_c \begin{bmatrix} 47.727 & 19.106 \\ 19.106 & 51.040 \end{bmatrix}^{-1} \begin{bmatrix} \Delta_{1L} \\ \Delta_{2L} \end{bmatrix} \qquad \text{(b)}$$

To determine the stresses and deflections at initial and final stages, the following loadings have to be considered.

(i) Self weight

Taking the unit weight of concrete equal to 24 kN/m^3, the intensity of load at any panel point is equal to 24x cross-sectional area at the panel point in kN/m. The equivalent concentrated forces at panel points may be computed next using parabolic law. Determining Δ_{1L} and Δ_{2L} by Newmark's procedure and using Eq. (b), the redundant forces due to self weight are

$$M_B = - 5.599 \text{ MN.m}$$
$$M_C = - 7.106 \text{ MN.m}$$

(ii) Superimposed Load

The redundant forces due to the uniform superimposed load of 16 kN/m may be determined using Eq. (b) and numerical integration for computing Δ_{1L} and Δ_{2L}.

$$M_B = - 3.517 \text{ MN.m}$$
$$M_C = - 4.472 \text{ MN.m}$$

(iii) Prestressing force

(a) *Initial stage.* At this stage, the prestressing force of 7.72 MN acts at eccentricities given in Table 11.5.1. The primary moment, which is equal to the product of prestressing force and eccentricity of prestress, divided by E_c I may be treated as the elastic load on the conjugate beam. Proceeding in thismanner and using numerical integration, the displacements Δ_{1L} and Δ_{2L} at coordinates 1 and 2 caused by prestressing may be determined. Using Eq. (a), the redundant forces at coordinates 1 and 2 may be determined.

$$M_B = 8.717 \text{ MN.m}$$
$$M_C = 8.452 \text{ MN.m}$$

(b) *Final stage.* The initial prestressing force of 7.72 MN reduces to 6.72 MN at final stage. Consequently, the redundant forces due to prestressing also decrease in the same proportion.

$$M_B = 7.588 \text{ MN.m}$$
$$M_C = 7.357 \text{ MN.m}$$

Knowing the redundant forces, the bending moments at all nodal points may be determined by statics. The bending moments due to self weight, superimposed load, initial prestressing force and final prestressing force are shown in lines 1 to 4 of Table 11.5.2. The bending moments at initial stage are obtained by adding the moments in lines 1 and 3. Similarly, the moments at the final stage are obtained by adding the moments in lines 1, 2 and 4. The extreme fibre stresses at initial stage shown in lines 5 and 6 are obtained by combining the flexural stresses at initial stage with the direct stresses due to prestressing. Similarly, the extreme fibre stresses at final stage shown in lines 7 and 8 may be obtained. Deflections at all nodal points at initial and final stages obtained by numerical integration are shown in lines 9 and 10 respectively. According to Indian Code, the maximum permissible stresses in compression are $\sigma_{cbci} = 15.61$ MPa and $\sigma_{cbc}=13.75$ MPa. Hence the compressive stresses at initial and final stages are within permissible limits. Tensile stresses occur only at two nodal points at the final stage. These stresses are small and may therefore be permitted. The deflections at inital and final stages are also within permissible limits.

EXAMPLE 11.5.5

The continuous beam shown in Fig. 11.5.6(a) has to carry the design ultimate load of 30 kN/m. The ultimate moment capacities in the regions near midspan sections F and G are 500 kN.m and 600 kN.m respectively. Determine the ultimate moment capacities at sections E, B and C so that the ultimate moment capacities at all critical sections are fully utilized.

Solution

In order that the ultimate moment capacities of all critical sections are fully utilised, the three spans must reach the collapse stage simultaneously. It follows that plastic hinges should form at E, B, F, C and G simultaneously to form the collapse mechanism shown in Fig. 11.5.6(b). Considering span CD and equating the internal and external works,

$$M_{Cu}(\theta_3) + M_{Gu}(2\theta_3) = 30 \times 18 \times \frac{\delta_3}{2}$$

Table 11.5.2

Line	Span			1	2	3	4	5	6	7	8	9
			Panel Points									
1		M_D,	MN·m	0.000	1.310	1.704	1.225	−0.106	2.551	−5.599	−3.615	−7.106
2		M_L,	MN·m	0.000	0.854	1.131	0.834	−0.040	−1.491	−3.517	−2.336	−4.472
3		M_{p_i},	MN·m	3.088	−0.093	−2.241	−3.361	−0.366	3.661	8.717	4.771	8.452
4		M_p,	MN·m	2.688	−0.081	−1.951	−2.920	−0.319	3.187	7.588	4.153	7.357
5		f_{T_i},	MPa	−11.722	−9.433	−6.543	−2.865	−6.451	−8.444	−9.559	−8.007	−7.609
6	AB	f_{B_i},	MPa	−0.356	−4.281	−9.195	−15.415	−8.449	−4.204	−1.515	−4.482	−4.136
7		f_{T_i},	MPa	−10.203	−10.012	−8.464	−4.899	−5.519	−4.898	−3.651	−3.351	−0.069
8		f_{B_i},	MPa	−0.314	−1.201	−4.097	−9.993	−7.487	−7.018	−7.593	−8.833	−11.576
9		Δ_i,	mm	0.000	−2.900	−8.700	−12.600	−8.800	−3.400	0.000	0.800	0.000
10		Δ,	mm	0.000	3.700	2.400	−1.300	−2.100	−1.600	0.000	−2.300	0.000
1		M_D,	MN·m	−5.599	−2.484	−0.395	−0.737	0.974	0.360	−1.148	−3.615	−7.106
2		M_L,	MN·m	−3.517	−1.621	−0.299	0.445	0.614	0.206	−0.778	−2.336	−4.472
3		M_{p_i},	MN·m	8.717	4.971	1.896	−0.500	−2.244	−0.566	1.762	4.771	8.452
4		M_p,	MN·m	7.588	4.327	1.650	−0.435	−1.953	−0.493	1.534	4.153	7.357
5		f_{T_i},	MPa	−9.559	−9.724	−9.272	−7.909	−5.012	−7.068	−7.903	−8.007	−7.609
6	AC	f_{B_i},	MPa	−1.515	−2.142	−3.750	−6.827	−12.357	−8.008	−5.644	−4.482	−4.136
7		f_{T_i},	MPa	−3.651	−5.937	−7.530	−7.911	−6.157	−6.632	−5.449	−3.351	−0.069
8		f_{B_i},	MPa	−7.593	−5.281	−4.013	−4.502	−8.201	−6.298	−6.891	−8.833	−11.576
9		Δ_i,	mm	0.000	3.400	3.300	0.400	−2.900	−1.800	0.000	0.800	0.000
10		Δ,	mm	0.000	1.600	3.000	2.500	0.300	−0.800	−2.000	−2.300	0.000

Table 11.5.2 (Contd. .)

Line	Span			Panel Points								
				1	2	3	4	5	6	7	8	9
1		M_D,	MN·m	−7.106	−3.298	−0.504	1.350	2.329	2.450	1.686	0.000	
2		M_L,	MN·m	−4.472	−2.106	−0.314	0.901	1.539	1.602	1.080	0.000	
3		M_{Pi},	MN·m	8.452	4.097	0.491	−2.352	−3.124	−1.805	0.263	3.088	
4		M_i,	MN·m	7.357	3.566	0.427	−2.047	−2.719	−1.571	0.229	2.688	
5	CD	f_{T1},	MPa	−7.609	−7.631	−7.072	−5.598	−6.078	−8.803	−0.608	−11.722	
6		f_{B1},	MPa	−4.136	−5.132	−7.122	−10.671	−10.405	−5.758	−2.530	−0.361	
7		f_{Tr},	MPa	−0.687	−3.293	−5.535	−7.131	−9.434	−11.427	−11.498	−10.203	
8		f_{Br},	MPa	−11.576	−9.042	−7.061	−6.098	−3.179	0.285	0.958	−0.314	
9		Δ_i,	mm	0.000	−0.900	−2.800	−4.600	−3.300	0.400	2.500	0.000	
10		Δ,	mm	0.000	2.3	7.400	13.300	18.400	19.300	13.200	0.000	

Putting $M_{Gu} = 600$ kN.m and $\delta_3 = 9\theta_3$,

$$M_{Cu} = 1230 \text{ kN.m}$$

Considering next the central span BC,

$$M_{Bu}(\theta_2) + M_{Fu}(2\theta_2) + M_{Cu}(\theta_2) = 30 \times 20 \times \frac{\delta_2}{2}$$

Putting $M_{Fu} = 500$ kN.m, $M_{Cu} = 1230$ kN.m and $\delta_2 = 10\theta_2$,

$$M_{Bu} = 770 \text{ kN·m}$$

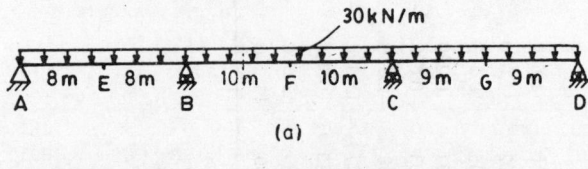

(a)

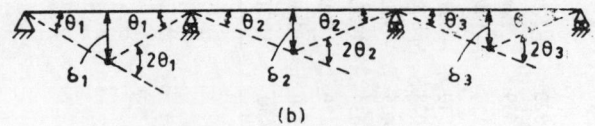

(b)

Fig. 11.5.6

Finally considering span AB,

$$M_{Eu}(2\theta_1) + M_{Bu}(\theta_1) = 30 \times 16 \times \frac{\delta_1}{2}$$

Putting $M_{Bu} = 770$ kN.m and $\delta_1 = 8\theta_1$,

$$M_{Eu} = 575 \text{ kN.m}$$

EXAMPLE 11.5.6

A continuous beam has to carry the design collapse load shown in Fig. 11.5.7(a). The prestressing tendons provide uniform positive and negative

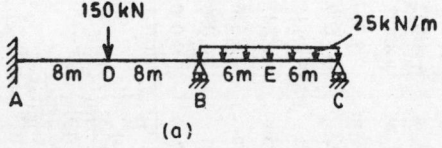

(a)

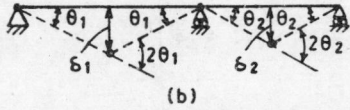

(b)

Fig. 11.5.7

moment capacities equal to 250 kN.m over the entire length of the beam. Determine the additional negative moment capacities to be developed by supplementary untensioned reinforcement in the regions near A and B in order to avoid partial failure.

Solution

In order to avoid partial failure, plastic hinges should form at A, D, B and E simultaneously to form the collapse mechanism shown in Fig. 11.5.7(b). Considering span BC and equating the internal and external works,

$$M_{Bu}\,(\theta_2) + M_{Eu}\,(2\theta_2) = 25 \times 12 \times \frac{\delta_2}{2}$$

Putting $M_{Eu} = 250$ kN.m and $\delta_2 = 6\theta_2$,

$$M_{Bu} = 400 \text{ kN.m}$$

Similarly, considering span AB,

$$M_{Au}\,(\theta_1) + M_{Du}\,(2\theta_1) + M_{Bu}\,(\theta_1) = 150 \times \delta_1$$

Putting $M_{Du} = 250$ kN.m, $M_{Bu} = 400$ kN.m and $\delta_1 = 8\theta_1$,

$$M_{Au} = 300 \text{ kN.m}$$

As the prestressing tendons provide a moment capacity of 250 kN.m, the additional negative moment capacities to be developed by supplementary untensioned reinforcement is $300 - 250 = 50$ kN.m at A and $400 - 250$ $=150$ kN.m at B

11.6. PRACTICAL ASPECTS

Practical considerations dominate at almost every stage of planning, design and construction of prestressed concrete continuous beams. They often outweigh theoretical considerations. The major practical aspects may be discussed in respect of the following items:

(i) *Type of construction*

As the total length and the self weight of prestressed concrete continuous beams are generally large, it is necessary to make a careful choice in respect of the type of construction. In this regard the beams may be classi- fied into (a) fully continuous beams and (b) partially continuous beams. The fully continuous beams are those in which all tendons generally extend from one end to the other end of the beam although some of them could be terminated at intermediate points if so desired. Fully continuous beams are generally cast-in-place in their final position so as to eliminate lifting and launching operations. Partially or semi-continuous beams are those in which each span is cast separately as a simple span and then connected to each other at the supports to form a continuous beam. In this type of construction, the individual spans are generally precast and prestressed as simply supported beams with due consideration given to the stresses during lifting and launching. In the case of long spans, it is preferable to adopt segmental approach in which each span is cast into several segments. These segments are lifted to their final positions and connected to each other by means of prestressing cables to form the individual spans. The continuity at the supports may be achieved in any one of the following ways.

(a) By the provision of short prestressing tendons over the supports as shown in Fig. 11.6.1(a). In this method, short tendons are inserted through the two adjacent ends of the precast beams and the in-situ concrete after the erection of the precast beams. The short tendons are then prestressed to obtain the desired continuity.

(b) By the provision of cap cables as shown in Fig. 11.6.1(b). The cap cables are in the form of short curved cables in the intermediate support regions. They are accommodated in the end regions of two adjacent spans and are generally anchored at the soffits of these beams.

(c) By the provision of untensioned steel or precast prestressed elements in the cast-in-place top slab to form a composite construction. The ends of the precast beams should have adequate shear connectors projecting from them into in-situ concrete for obtaining the desired continuity as shown in Fig. 11.6.1(c).

(d) By the provision of reinforced or prestressed concrete planks as shown in the plan view of Fig. 11.6.1(d). These planks are placed on both sides of the precast beams and the entire assembly is prestressed in transverse (horizontal) direction.

(e) By the provision of couplers as shown in Fig. 11.6.1 (e). In this method, the prestressing tendons, particularly the high tensile rods, are connected to each other by means of couplers and the stressing operation is simplified by carrying it out from span to span. For example, the tendon in the extreme left span may be stressed first. The right end of the stressed tendon may be connected to the left end of the unstressed tendon in the second span by means of a coupler and the second tendon stressed from its right end. The sequence may be repeated for the tendon in the third span and so on. In this manner, the precast beams may be erected and stressed span by span from one end of the continuous beam to the other end.

(ii) *Girder profile and cable layout*

As the regions near the midspan and interior supports are subjected to sagging and hogging bending moments respectively, the cable line has to be placed near the soffit at midspan and near the top at supports in order to take full advantage of eccentric prestress. This requirement leads to undulating cable line if the beam is prismatic. Undulating tendons create practical problems in stressing and also lead to increased frictional losses. Prismatic continuous beams are undesirable also because the bending moments at supports are generally much greater than those at midspan. Consequently, the girder profile may be chosen to offer a larger cross-section at supports as compared to that at midspan. Several types of non-prismatic girder profiles are possible. For example, the top of the girder may be horizontal while the soffit may be either curved as shown in Fig. 11.6.2 (a) or segmental as shown in Fig. 11.6.2 (b). The cross-section near supports can also be increased by the provision of adequate haunches as shown in Fig. 11.6.2 (c). In either case, the cable line may be straight (Fig.

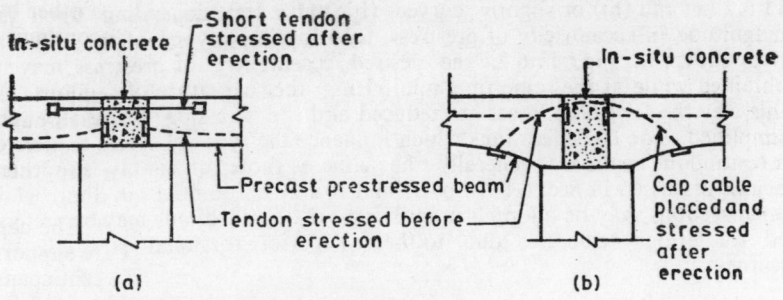

(a)   (b)

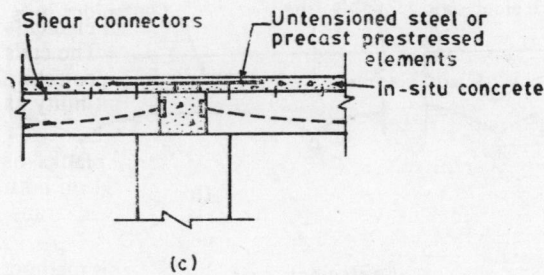

(c)

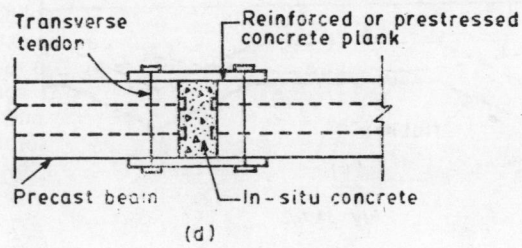

(d)

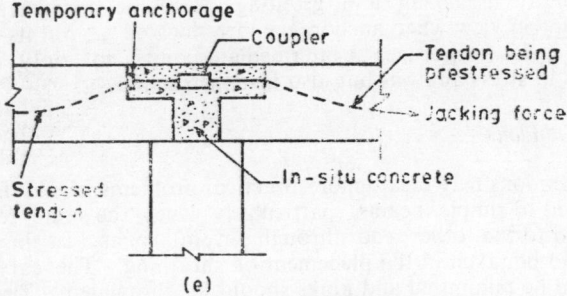

(e)

Fig. 11.6.1

11.6.2 (a) and (b)) or slightly curved (Fig. 11.6.2 c) depending upon the magnitude of eccentricity of prestress. By adopting anyone of the alternatives shown in Fig. 11.6.2, the desired eccentricity of prestress may be obtained while at the same time minimising the curvature of tendons. In this way the frictional losses are reduced and the stressing operations are simplified. The considerations which influence the type of cross-section of a continuous beam are generally the same as those for simply supported beams discussed in Sec. 6.8 In particular it may be pointed out that while open sections may be adopted in buildings, the box sections may be preferred for bridge structures due to their superior torsional strength and stiffness.

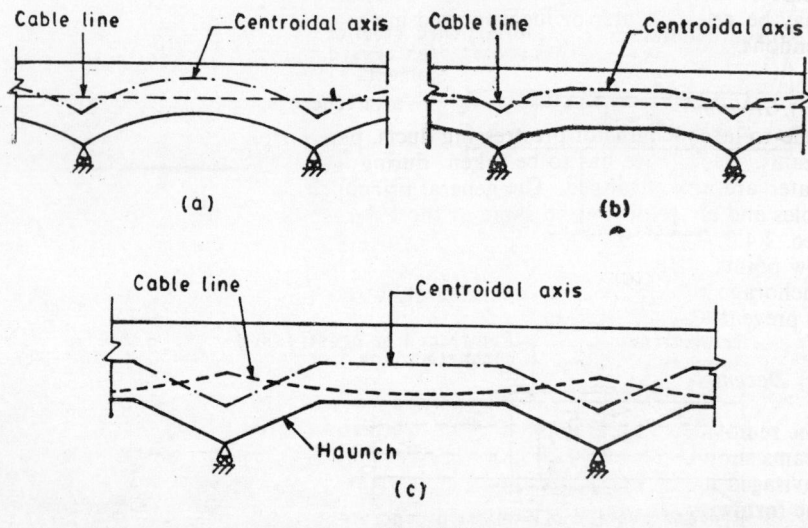

(a)

(b)

(c)

Fig. 11.6.2

When some of the tendons are terminated at interior supports or other intermediate points, care should be taken to locate the anchorages so that there is no difficulty in jacking and grouting. Aesthetic considerations should also be kept in view when anchorages are located at points other than the extreme ends. Anchorages at intermediate points not only introduce complexities in stress analysis but also in construction.

(iii) *Stressing of tendons*

The stressing of tendons may create more practical problems in continuous beams as compared to simple beams, particularly when the same tendon runs from one end to the other end through several spans. In this case special care should be taken in the placement of sheathing The curvature of tendons should be minimised and kinks should be eliminated. The frictional losses are reduced at stressing operations simplified to some extent by jacking at both ends. In the case of large spans, it is preferable to adopt partially continuous beams in which stressing operations can be carried out span by span as discussed in (i) above. The sequence of stressing depends upon the type of construction. If each span is precast separately, the stressing of tendons should enable the girders to carry their own weight and other dead load as simply supported beam as well as to resist the stresses during lifting and launching. For this purpose only a few selected tendons located at appropriate positions may be stressed. The remaining cables and those required for continuity may be stressed according to a well planned sequence. Usually several stages of loading have to be taken into account for arriving at the appropriate sequence in which the tendons have to be stressed. Depending upon the construction technique, some of the tendons may have to be stressed temporarily. These tendons may be released later or incorporated in the overall scheme of stressing of tendons.

(iv) *Grouting*

Due to large lengths of prestressing ducts, particularly in fully continuous beams, special care has to be taken during grouting to ensure that air or water are not entrapped. The general principles for the location of grout holes and air vents are the same as those for other structures discussed in Sec. 2.4.3. The grouting of ducts for cap cables, whose ends are located at low points, should be carried out from both ends simultaneously. The anchorage pockets of cap cables should be properly sealed before grouting to prevent leakage of grout.

(v) *Decentring*

The removal of formwork during decentring of cast-in situ continuous beams should be carefully planned so as to eliminate stresses which are not envisaged in design. For instance, an adverse stress condition may arise if the formwork of only one of the intermediate spans is completely removed.

Instead it would be better to take up decentring of all spans according to a well planned sequence. For instance, it may be desirable to start decentring from midspan sections of all spans simultaneously and to proceed progressively towards the supports in each span. In any case it would be desirable to decide about the sequence of decentring in advance and to ensure that the stresses caused during decentring are fully taken care of in the design procedure. This requirement introduces additional complexity in the analysis and design of statically indeterminate prestressed concrete continuous structures.

11.7 CODE PROVISIONS

The code provisions for prestressed concrete continuous beams are in general the same as those for simple beams. In respect of design for shear, it should be noted that the critical sections for shear located at a distance equal to effective depth from the face of intermediate supports are generally subjected concurrently to large negative bending moments. Consequently, these critical sections are usually cracked in flexure. The shear resisted by concrete is therefore given by V_{cr} according to Eq. (7.2.11). In respect of bond and anchorage zone stresses, continuous beams require special consideration with regard to cap cables, short cables and similar other cables terminating at intermediate sections. Due to greater stiffness of continuous beams, the code requirements in respect of deflection are more liberal for continuous beams as compared to simple beams. For instance IS: 1343-1980 specifies (Sec. 8.6.1) that in the case of Type 3 members, the vertical deflection limits may generally be assumed to be satisfied if span to effective depth ratio does not exceed 26 in the case of continuous beams and 20 in the case of simple beams. Similar allowance for continuity is made in European code as discussed in Sec. 8.6.3.

While ductility is desirable in all types of concrete members, it is particularly important in the case of continuous beams and other statically indeterminate structures. The favourable redistribution of moments leading to larger load capacity is possible only if the continuous beam possesses adequate ductility. The sections in the hinging regions must possess adequate rotation capacity to permit successive formation of plastic hinges leading to a collapse mechanism. In the absence of adequate ductility or rotation capacity, the beam fails prematurely before the formation of sufficient plastic hinges required for a collapse mechanism.

Consider for example a two span continuous beam ABC shown in Fig. 11.7.1 (a). The first plastic hinge may form at intermediate support B where the elastic moment is generally the largest. The plastic hinge at B can form only if the rotation capacity θ_{Bc} of section B is greater than θ_{Bu} which is the flexural rotation at the formation of plastic hinge at B as shown in the idealised moment-rotation relationship for section B in Fig. 11.7.1 (b). In the idealised moment-rotation relationship of Fig. 11.7.1 (b), it is assumed that the entire rotation in the plastic hinging region in the neighbourhood of section B is concentrated at B. At point P which marks

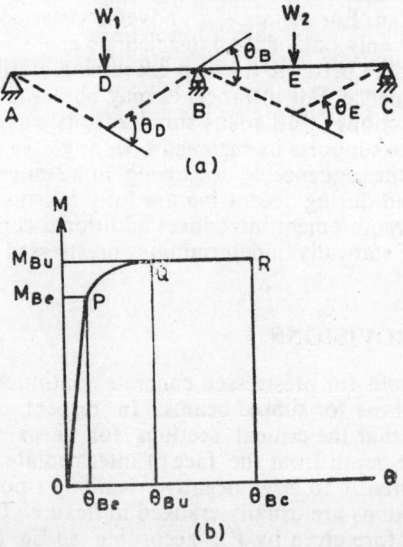

Fig. 11.7.1

the end of elastic action, the moment M_{Bc} produces elastic rotation θ_{Be} which is small compared to plastic rotations. The portion PQ of moment-rotation relationship represents the region of inelastic action, significant cracking and eventual formation of a plastic hinge at ultimate moment M_{Bu} and rotation θ_{Bu}. The plateau QR which corresponds to plastic rotation $(\theta_{Bc} - \theta_{Bu})$ represents the ductility of section B. While a prolonged plateau represents high ductility, very limited ductility is indicated if the section fails in the neighbourhood of point Q. In the absence of adequate ductility, the beam may fail even before the formation of plastic hinge at B. On the other hand, the plastic hinge at B may undergo additional rotation at constant moment M_{Bu} to permit the formation of a plastic hinge either at D or at E provided the beam has sufficient ductility. If the continuous beam is designed so that the moment capacities at critical sections B, D and E are fully utilised, the plastic hinges will appear at these sections provided that the rotations θ_B, θ_D and θ_E shown in Fig. 11.7.1 (a) are not larger than the respective rotation capacities θ_{Bc}, θ_{Dc} and θ_{Ec}. It is evident that in the absence of adequate ductility or rotation capacity, the maximum possible load carrying capacity of the continuous beam cannot be utilised.

To ensure ductile behaviour and prevent possible premature failure, code provisions usually specify restrictions on the proportion of tension steel. A lower limit on proportion of tension steel or reinforcing index is specified to ensure that the member does not fail suddenly at cracking when concrete transfers its tensile stress to steel. On the other hand, an upper limit is specified to prevent compression failure by crushing of concrete before appreciable inelastic action in tension steel. For instance, the Indian Code IS: 1343-1980 specifies that a minimum longitudinal reinforcement of 0.2 per cent of the total concrete area should be provided except in the case of pretensioned units of small sections. This reinforcement may be reduced to 0.15 per cent if high yield strength deformed bars are used.

The American code specifies a lower limit on the proportion of tension steel to ensure that the design ultimate moment is not less than 1.2 times the cracking moment. The code specifies an upper limit on effective reinforcing index for rectangular sections in accordance with equations

$$\omega_p \leqslant 0.48 \, k_2 \tag{11.7.1}$$

and

$$\omega_p + (\omega - \omega') \, \frac{d}{d_p} \leqslant 0.48 \, k_2 \tag{11.7.2}$$

where $\omega_p = \dfrac{A_p}{b \, d_p} \dfrac{f_p}{f_c'}$

$\omega = \dfrac{A_s}{b \, d} \dfrac{f_y}{f_c'}$

$\omega' = \dfrac{A_s'}{bd} \dfrac{f_y}{f_c'}$

k_2 = constant related to equivalent rectangular stress block (Eq. 6.9.7).

$d = $ effective depth of untensioned steel A_s measured from compression face.

$f_p = $ stress in prestressing steel under design ultimate moment.

In the case of flanged sections, the breadth b should be replaced by the thickness of web b_w. The code recommends that prestressed concrete continuous beams and other statically indeterminate structures should be analysed by elastic theory for reactions, moments, shears and axial forces produced by prestressing, temperature changes, creep and shrinkage of concrete, axial deformation, restraint from attached structural elements and foundation settlements. The negative moments at intermediate supports due to design loads in prestressed concrete continuous beams with sufficient bonded reinforcement to ensure control of cracking may be increased or decreased by not more than

$$20\left[1 - \frac{\omega_p + (\omega - \omega')\dfrac{d}{d_p}}{0.72\,k_2}\right]$$

per cent subject to the following conditions:

(i) The modified negative moments at supports are also used for final calculations of the moments at other sections in the span corresponding to the same loading condition.

(ii) The sections at which the moments are modified are designed so that Eq. (11.7.1) & (11.7.2) are satisfied.

EXAMPLE 11.7.1

Check the ductility of the bonded post-tensioned section shown in Fig. 11.7.2 in accordance with ACI code. The effective stress in prestressing tendons after losses, $f_{pe} = 1000$ MPa.

The particulars of prestressing steel and supplementary untensioned reinforcement comprising high strength deformed bars are as follows:

$A_p = 0.00090$ m² $A_s = 0.00045$ m²

$A_s' = 0.003$ m²

$f_p = 1600$ MPa $f_y = 415$ MPa.

$f_c' = 28$ MPa $f_{cr} = 3.5$ MPa

$M_u = 1.15$ MN.m

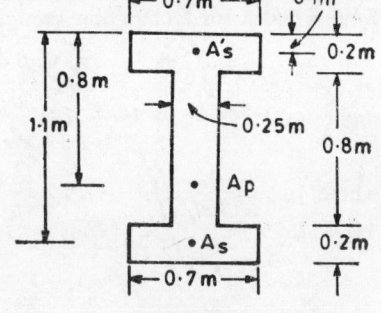

Fig. 11.7.2

Solution

The code specifies that the ultimate moment should be at least 20 per cent greater than cracking moment. The effective prestressing force,

$$P = 0.00090 \times 1000 = 0.90 \text{ MN}$$

The cross-sectional properties of the section are

$$A = 0.48 \text{ m}^2$$

$$I = 0.0816 \text{ m}^4$$

The stress at bottom fibre due to prestressing,

$$f_B = - \frac{0.90}{0.48} - \frac{0.90 \times 0.2 \times 0.6}{0.0816} = -3.2 \text{ MPa}$$

Hence the cracking moment,

$$M_{cr} = (3.2 + 3.5) \times \frac{0.0816}{0.\ } = 0.9112 \text{ MN.m}$$

$$\frac{M_u}{M_{cr}} = \frac{1.15}{0.9112} = 1.262$$

Hence the lower limit on reinforcement for ductility is satisfied.
According to ACI code, the reinforcement indices

$$\omega_p = \frac{0.0009}{0.25 \times 0.8} \times \frac{1600}{28} = 0.2571$$

$$\omega = \frac{0.00045}{0.25 \times 1.1} \times \frac{415}{28} = 0.0242$$

$$\omega' = \frac{0.0003}{0.25 \times 1.1} \times \frac{415}{28} = 0.0162$$

$$\omega_p + (\omega - \omega') \frac{d}{d_p} = 0.2571 + (0.0242 - 0.0162) \times \frac{1.1}{0.8}$$

$$= 0.2681$$

$$0.48 \, k_2 = 0.48 \times 0.425 = 0.204$$

It follows that the proportions of reinforcement exceed the limits specified
by the code to ensure ductility.

REFERENCES

11.1 Baker, A.L.L., *A plastic theory of design for ordinary reinforced and prestressed concrete including moment redistribution in continuous members*, Magazine of Concrete Research, Vol. 1, No. 2, June 1949

11.2 Bennett, E.W.; Cooke, N. and Naughton, L.P., *Deformation of continuous prestressed concrete beams and its effect on the ultimate load*, Proceedings of the Institution of Civil Engineers (London), Vol. 37, May 1967, pp. 57-74 and December 1967, pp. 769-74.

11.3 Burns, N., *Development of continuity between precast prestressed concrete beams*, Journal of the Prestressed Concrete Institute, Vol. 11, No. 3, June 1966, pp. 23-36.

11.4 Corley, W.G., *Rotational capacity of reinforced concrete beams*, Journal of the Structural Division, ASCE, Vol. 92, 1966, pp. 121-46.

11.5   Fiesenheiser, E.I., *Rapid design of continuous prestressed members*, Journal of the American Concrete Institute, Proceedings Vol. 50, April 1954, pp. 669-76.

11.6   Guyon, Y., *A study of continuous beams and statically redundant systems in prestressed concrete*, CACC Translation No. 33, London, 1951.

11.7   Harris, A.J., *Continuity of prestressed concrete structures: the practical aspect*, Proceedings of the Symposium on Prestressed concrete Statically Indeterminate Structures, 1951, p. 1, Cement and Concrete Association (London), 1953.

11.8   Hawkins, N.M.; Sozen, M.A. and Siess, C.P., *Strength and behaviour of two span continuous prestressed concrete beams*, Structural Research Series No. 225, University of Illinois, September 1961.

11.9   Lin, T.Y., *Strengths of continuous prestressed concrete beams under static and repeated loads*, Journal of the American Concrete Institute, Proceedings Vol. 51, June 1955.

11.10 Lin, T.Y., *A new concept for prestressed concrete*, Journal of the American Concrete Institute, December 1961.

11.11 Lin, T.Y., *Load balancing method for design and analysis of prestressed concrete structures*, Journal of the American Concrete Institute, Proceedings Vol. 60, June 1963, pp. 719-41.

11.12 Lin, T.Y. and Thornton, K., *Secondary moments and moment redistribution in prestressed concrete continuous beams*, Journal of the Prestressed concrete Institute, January-February 1972, pp. 8-20.

11.13 Macchi, G., *Experimental study of continuous prestressed concrete beams in the plastic range and at failure*, Fed. Int. Precontr. Second Congress, Amsterdam, 1955.

11.14 Morice, P.B. and Lewis, H.E., *The ultimate strength of two span continuous prestressed concrete beams as affected by tendon transformation and untensioned steel*, Fed. Int. Precontr. Second Congress, Amsterdam, 1955.

11.15 Nasser, G.D., *A practical approach to the design of continuous prestressed concrete structures*, The Indian Concrete Journal, Vol. 42, May 1968, pp. 199-209.

11.16 Nawy, E.G. and Salek, F., *Moment-rotation relationships of non-bonded prestressed flanged sections confined with rectangular spirals*, Journal of the Prestressed Concrete Institute, August 1968, pp. 40-55.

11.17 Newmark, N.M., *Numerical procedure for computing deflections, moments and buckling loads*, Transactions of ASCE, Vol. 108, 1943, pp. 1161-1234.

11.18 Pandit, G.S. and Gupta, S.P., *Free-body approach for analysis of prestressed concrete beams*, The Indian Concrete Journal, Vol. 54, No. 6, June 1980, pp. 154-61.

11.19 Pandit, G.S. and Gupta, S.P., *Numerical analysis and design of prestressed concrete continuous beams*, The Indian Concrete Journal, Vol. 55, No. 4, April 1981, pp. 99-105.

11.20 Parme, A.L. and Paris, G.H., *Designing for continuity in prestressed concrete structures*, Journal of the American Concrete Institute, Proceedings Vol. 47, September 1951, pp. 45-64.

11.21 *Symposium on prestressed concrete continuous and framed structures*, Cement and Concrete Association (London), 1951.

11.22 Visvesvaraya, H.C. and Raghavendra, N., *Continuity in prestressed concrete construction*, Seminar on Problems of Prestressing, Madras, 1970, pp. I-91-4.

PROBLEMS

11.1 A beam restrained at both ends has a cable line comprising straight segments as shown in Fig. P 11.1. Derive expressions for the secondary moments at A and B due to prestressing.

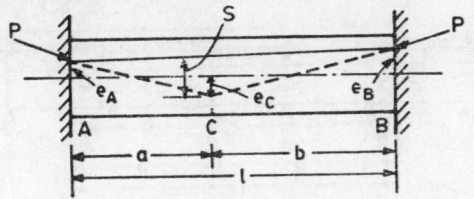

Fig. P 11.1

11.2 For the beam of Fig. P 11.1, show that the cable line is concordant if $e_A = 0.144$ m, $e_B = 0.096$ m, $e_C = 0.1152$ m, $a = 4$ m and $b = 6$ m.

11.3 A cantilever beam AB is fixed at A and propped at B to the same level as the fixed end A. The cable line is parabolic as shown in Fig. P 11.3. Determine the secondary moment at A due to prestressing.

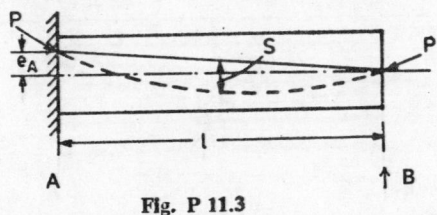

Fig. P 11.3

11.4 Show that if $S = e_A$ in Fig. P 11.3, the cable line is concordant.

11.5 Which of the cable lines shown in Fig. P 11.4 (a), (b), (c) and (d) is a linear transformation of the cable line shown in Fig. (e).

11.6 Determine the eccentricities of cable line at midspan sections D and E if the cable line shown in Fig. P 11.6 is transformed linearly by (a) raising it at B by 0.1 m, (b) lowering it at B by 0.2 m and (c) raising it at E by 0.1 m.

11.7 Define linear transformation of cable line and discuss its application in design.

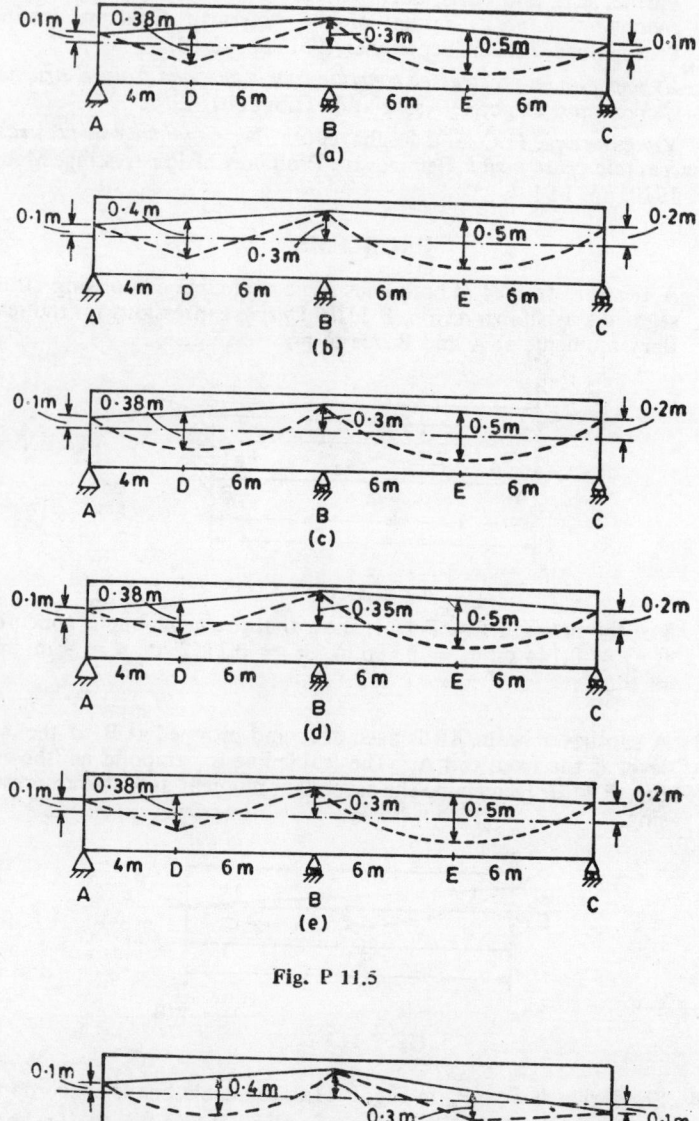

Fig. P 11.5

Fig. P 11.6

11.8 Analyse the continuous beam of Fig. P 11.8. Hence determine the bending moment at B. EI_c is constant.

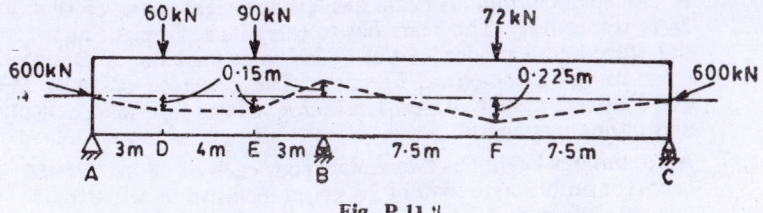

Fig. P 11.8

11.9 Analyse the continuous beam of Fig. P 11.9. Hence determine the eccentricity of pressure line at B, E_c I is constant.

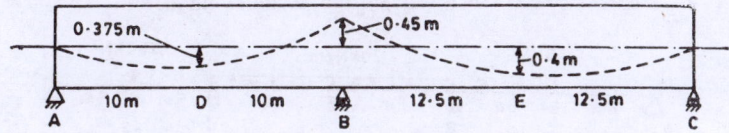

Fig. P 11.9

11.10 Determine the extreme fibre stresses at section B of the beam of Prob. 11.9 if the beam has a symmetrical I-section with cross-sectional area, $A = 0.5$ m² and section modulus, $Z = 0.216$ m³.

11.11 Determine the bending moment at B in the beam of Prob. 11.8 if the beam has an overhang of 2 m to the left of A. In addition to the loads shown in Fig. P 11.8, a concentrated load of 30 kN acts at the free end.

11.12 Determine the eccentricity of line of pressure at B in the beam of Prob. 11.9 if the beam has an overhang of 4 m to the right of C carrying a uniformly distributed load of 50 kN/m.

11.13 Determine the value of the collapse load W_{ku} for the beam of Fig. P 11.13. The ultimate moment capacities of the sections D, B and E are 400, 800 and 600 kN.m respectively.

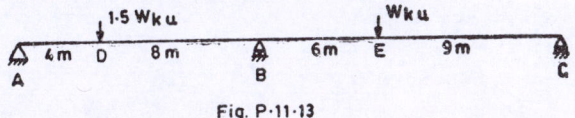

Fig. P·11·13

11.14 Determine the intensity of uniformly distributed collapse load for a three span continuous beam having left, central and right spans equal to 12 m, 18 m and 15 m respectively. The beam has a constant positive and negative moment capacity of 600 kN.m.

11.15 A continuous beam has two equal spans each of 12 m. It has to carry its own weight at the initial stage and a service load of 18 kN/m at the final stage. The beam has a uniform rectangular section 0.3 × 0.6 m. Design the beam using Magnel Blaton system and concrete of grade M-35.

11.16   A two span continuous beam has left and right spans of 18 m and 24 m respectively. The beam has to carry its self weight at the initial stage and a service load of 21 kN/m at final stage. Design the beam using a symmetrical I-section. Use concrete of grade M-40 and Fressinet system of post-tensioning. Check for shear, deflection and ultimate strength.

11.17   A continuous beam has two equal spans each of 18 m. Design the beam for an ultimate load of 30 kN/m inclusive of self weight. Use concrete of grade M-45.

11.18   Determine the moment capacities at sections B, F and C of the continuous beam shown in Fig. P 11.18. The ultimate moment capacities at Section E and G are 300 and 500 kN.m respectively. A partial failure of the beam has to be avoided.

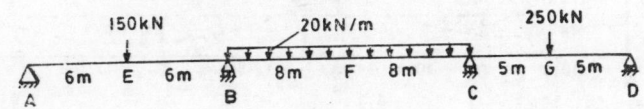

Fig. P 11.18

11.19   A continuous beam has two equal spans each of 15 m. The beam has a constant positive and negative moment capacity of 500 kN.m. Determine the moment to be resisted by untensioned supplementary reinforcement at the central support. The beam has to carry the design ultimate load of 30 kN/m.

11.20   Using ACI code, check the ductility requirement of a bonded post-tensioned rectangular section 0.2×0.6 m having the following particulars:

$A_p = 240 \text{ mm}^2$ $f_{pe} = 1000 \text{ MPa}$ $f_p = 1500 \text{ MPa}$

$A_s = 120 \text{ mm}^2$ $A'_s = 60 \text{ mm}^2$ $f_y = 250 \text{ MPa}$

$M_u = 0.144 \text{ MN.m}$ $f_{cr} = 3.3 \text{ MPa}$ $f'_c = 20 \text{ MPa}$

$d = 0.55 \text{ mm}$ $d_p = 0.4 \text{ m}$

12

Tension Members

12.1 INRODUCTION

Prestressed concrete has found extensive applications in the domain of tension members due to distinct advantages over reinforced concrete and structural steel. A reinforced concrete tension member cracks over the entire cross-section even before reaching the working load level. Consequently, the entire working load is carried by reinforcing steel whereas concrete serves merely as protective cover against corrosion of reinforcement. However, due to the presence of cracks, the role of concrete as protective cover is severely impaired, particularly in aggressive enviroment. Structural steel suffers from serious maintenance problems particularly when it is used in marine structures, containment structures and in hostile environment. In contrast to the two traditional materials, prestrssed concrete has maintenance free long life because it can be designed to remain crack-free in its entire service range. Due to the use of high tensile steel, a prestressed concrete tension member may have superior strength characteristics. In addition, the technique of prestressing can be used effectively to achieve better deformation characteristics which may be more important in some cases. The superior deformation characteristics of a prestressed concrete tension member may be illustrated through an example. For a working stress of 150 MPa in reinforcement or structural steel, a 10 m long tension member of reinforced concrete or structural steel has an elongation of 7.5 mm under working load. When prestressed concrete having modular ratio, $m = 5$ is used, the initial contraction is only 2.5 mm if the intensity of initial prestress in concrete is 10 MPa. If the residual compressive stress under full working load is 1 MPa, the short-time contraction at working load level is only 0.25 mm. Even when the entire working load is permanent, the long-time contraction under working load is 0.75 mm for a creep coefficient, $C_c = 2.0$. This deformation is only one-tenth of the deformation of the corresponding tension member made of reinforced concrete or structural steel. The prestressed concrete tension member may undergo an additional contraction of 2 mm due to shrinkage of concrete having shrinkage coefficient, $\epsilon_{sh} = 0.0002$. Even then, the total deformation of 2.75 mm compares very favourably with 7.5 mm in a reinforced concrete or structural steel member. In fact, the technique of prestressing gives considerable flexibility to the designer in choosing the desired deformations of the member.

Tension members belong to plane and spatial skeletal structures as well as to surface structures such as domes and shells. Ties of roof trusses, lattice girders, bow string girders and cable stayed bridges; suspenders of bow-string girders and suspension bridges; edge members including ring beams of domes and shells; walls of stroage tanks, bunkers and silos; pressure pipes and containment or reactor pressure vessels of nuclear plants are some of the prominent examples of tension members. Direct tension is the predominant internal force caused by service loads of these members. When direct tension due to service loads is practically axial, concentric prestress is evidently the logical option. The advantage of eccentric prestress can be taken whenever direct tension acts at a definite eccentricity. In this case, the centroid of prestressing force may be made coincident with that of the eccentric tension. The concentric prestress may also have to be adopted when the eccentricity of direct tension is indefinite or reversible. Tension members with large eccentricity which essentially are beam-ties, may be designed for flexure and additional prestressing tendons may be provided to take care of co-existing direct tension.

12.2   TIES

A tie is a structural member in which the predominant internal force is direct tension. Ties are generally straight and prismatic. A varying cross-section may be required if the tensile force varies along the length of the member. The ties are generally prestressed concentrically. However, some eccentricity of prestress may be desirable when direct tension is accompanied by non-reversible flexural moment. The following stages have to be considered in the analysis for stresses and deformations of tension members.

(i) Initial stage

At the initial stage or at stress transfer, the prestressing force is P_i before the occurrence of time dependent losses and the compressive strength of concrete is f_{ci}. In some cases, a part of the service load T_D may become effective at the initial stage. The initial stress in concrete f_i under the combined action of compressive force P_i and tensile force T_D may be expressed as

$$f_i = -\frac{P_i}{A_c} + \frac{T_D}{A_t} \qquad (12.2.1)$$

where A_c = net area of concrete

$\qquad = A_g - A_p$ for pretensioned members

$\qquad = A_g - A_d$ for post-tensioned members

A_g = gross area of concrete

A_d = total area of ducts

The transformed area A_t may be expressed as

$$A_t = A_c + mA_p$$

in which $m = (E_s/E_c)$ is the modular ratio at the appropriate level of stress. The initial stress f_i should not exceed the permissible compressive stress in concrete σ_{cc1} at initial stage. A value ranging from $0.15 f_{ck}$ to $0.25 f_{ck}$ is often used.

The instantaneous change in length (shortening) of a tension member of length L at stress transfer may be expressed as

$$\delta_i = \frac{f_i L}{E_c} = \frac{L}{E_c} \left(-\frac{P_i}{A_c} + \frac{T_D}{A_t} \right) \qquad (12.2.2)$$

If the member is restrained from its free shortening on account of adjoining structure which imposes a restraining tensile force T_{ri} at the initial stage, the compressive stress and longitudinal shortening at stress transfer are reduced by (T_{ri}/A_c) and $(T_{ri}/A_c E_c)$ respectively. As tension members are rarely isolated, the effect of adjoining structure on the tension member and vice versa should be given due consideration.

(ii) *Final stage*

At this stage, the initial prestressing force P_i reduces to the effective prestressing force P and the remaining part T_L is superimposed on the initial part T_D of the total service load $T_w = (T_D + T_L)$. The stress in concrete at the final stage,

$$f = -\frac{P}{A_c} + \frac{T_w}{A_t} \qquad (12.2.3)$$

and the corresponding stress in steel,

$$f_F = f_{pe} + m \left(\frac{T_w}{A_t} \right) \qquad (12.2.4)$$

where f_{pe} = effective stress in prestressing steel after losses

The prestressing force P should be selected in such a manner that concrete is not fully decompressed under the action of service load T_w. The desirable value of the residual compressive stress f_{rc} depends upon the type and importance of the structure. A higher value is prescribed for important structures particularly for containment structures which have to be free from cracks. Similarly, a higher residual compressive stress is prescribed for structures which are prone to overloading. For instance, a low residual compressive stress may be prescribed for building ties whereas a higher value may be justified for a bridge tie. The values commonly recommended range from 0 to 0.1 f_{ck}. A value of 0.70 MPa (100 *psi*) is often used.

The net instantaneous change of length (shortening) may be expressed as

$$\delta = \frac{fL}{E_c} = \frac{L}{E_c} \left(-\frac{P}{A_c} + \frac{T_w}{A_t} \right) \qquad (12.2.5)$$

In the computation of eventual long-time change of length, the effect of creep and shrinkage of concrete should be considered. The creep of concrete occurs due to average prestressing force $0.5 (P_i + P)$ and the sustained or permanent part of service load T_{ws}. The deferred change of length (shortening) of the member due to creep and shrinkage of concrete,

$$\delta_d = \frac{L}{E_c} \left(-\frac{P_i + P}{2A_c} + \frac{T_{ws}}{A_t} \right) C_c - \epsilon_{sh} L \qquad (12.2.6)$$

where $C_c = \dfrac{\text{Creep strain}}{\text{elastic strain}} = $ creep coefficient

$\epsilon_{sh} = $ shrinkage strain

The total eventual long-time change of length (shortening) is the sum of instantaneous and deferred changes of length.

$$\delta_{LT} = \delta + \delta_d \qquad (12.2.7)$$

The initial restraining tensile force T_{rl} changes to T_r at the final stage. The long-time elongation of the member due to the restraining tensile force imposed by the adjoining structure may be expressed as

$$\delta_r = \frac{L}{E_c} \left(\frac{T_{rl} + T_r}{2A_t} \right) (1 + C_c) \qquad (12.2.8)$$

The net long-time change of length of a restrained tension member is $(\delta_{LT} + \delta_r)$. It should be checked that the net long-time deformation is within prescribed limit. The long-time effect of restraining structure on the tension member and vice versa should be duly examined.

(iii) Stage of decompression

At this stage, the tensile force applied to the tension member is just large enough to overcome the compressive stress induced by prestressing. The decompression tensile force T_{dc} may be determined by replacing T_w by T_{dc} and equating the stress in concrete f in Eq. (12.2.3) to zero.

$$T_{dc} = P \left(\frac{A_t}{A_c} \right) = P \left(1 + \frac{mA_p}{A_c} \right) \qquad (12.2.9)$$

As the ratio (A_p/A_c) is small (generally not exceeding 0.01), Eq. (12.2.9) shows that the tensile force required to cause decompression is only slightly greater than the prestressing force P. The decompression force T_{dc} should be larger than the service load T_w so that decompression does not occur under normal service condition but only in the event of overloading. This design requirement is satisfied by adopting a suitable value of residual compressive stress f_{rc} when the member carries its full service load T_w.

(iv) Cracking stage

Cracking occurs when the tensile stress in concrete becomes equal to the tensile strength of concrete in direct tension f_t. Replacing T_w by T_{cr} and equating f in Eq. (12.2.3) to f_t, the tensile force required to cause cracking,

$$T_{cr} = \left(f_t + \frac{P}{A_c} \right) A_t \qquad (12.2.10)$$

The cracking strength T_{cr} given by Eq. (12.2.10) applies only to those members which are not precracked by accidental overloading or by shrinkage. Once cracked, the tensile strength of concrete f_t is lost for ever. For precracked members, therefore, the cracking strength is equal to the tensile force at decompression T_{dc}. In order to obtain a crack-free structure, it is necessary to adopt a suitable margin of safety against cracking and also to ensure that the member is not precracked due to any reason.

(v) *Ultimate stage*

At failure, the concrete is cracked extensively and, therefore, contributes nothing to ultimate strength. The total load being carried by steel or the nominal ultimate strength,

$$T_u = A_p f_{pu} \qquad (12.2.11)$$

If the supplementary untensioned reinforcement A_s having yield stress f_y is also provided.

$$T_u = A_p f_{pu} + A_s f_y \qquad (12.2.12)$$

Applying a capacity reduction factor Φ, the design ultimate strength may be expressed as

$$T_u = \Phi \left(A_p f_{pu} + A_s f_y \right) \qquad (12.2.13)$$

The design ultimate strength should not be less than the design ultimate load. In practice, the ultimate strength given by Eq. (12.2.11) or (12.2.12) can seldom be utilized because the elongation of the member at its ultimate load is so large that it can rarely occur without causing failure of the adjoining structure. Hence a part of the available ultimate strength remains as non-usable reserve. The ultimate strength should provide a reasonable margin of safety on service load T_w, decompression load T_{dc} and cracking load T_{cr}.

In addition to the foregoing stages of loading, a check may by required for other relevant stages such as lifting, transportation, hoisting and launching. These stages seldom control the design except in the case of very long members.

EXAMPLE 12.2.1

A tension member is 21 m long and has a square cross-section of 0.5 m side. It is post-tensioned by means of four prestressing tendons placed concentrically with an initial stress of 1200 MPa and an effective stress of 960 MPa. Each tendon has 12 wires of 7 mm diameter with ultimate stress $f_{pu} = 1720$ MPa and is housed in a duct of 40 mm diameter. The strength properties of concrete are, $f_{ck} = 35$ MPa, $f_c' = 28$ MPa, $f_{ci} = 21$ MPa, $f_t = 2$. MPa and $E_c = 36000$ MPa. The member carries a dead load of 270 kN at initial stage and an additional load of 1430 kN at final stage. Sixty per cent of the total service load is permanent. Determine the stresses and deformations at initial and final stages. Also determine the margins of safety on decompression, cracking and ultimate strength. Take $E_s = 2 \times 10^5$ MPa, $C_c = 1.5$ and $\epsilon_{sh} = 0.0002$. Assume that the restraint from adjoining structure is negligible.

Solution

$$A_p = 4 \times 12 \times \frac{\pi}{4} \times 7^2 = 1848 \text{ mm}^2 = 0.001848 \text{ m}^2$$

$$A_c = 0.5 \times 0.5 - 0.001848 = 0.248152 \text{ m}^2$$

$$m = \frac{2 \times 10^5}{36000} = 5.55$$

$$A_t = 0.248152 + 5.55 \times 0.001848 = 0.258408 \text{ m}^2$$

(i) *Initial stage*

$$P_i = 0.001848 \times 1200 = 2.2176 \text{ MN}$$

From Eq. (12.2.1), the stress in concrete

$$f_i = -\frac{2.2176}{0.248152} + \frac{0.270}{0.258408} = -7.89 \text{ MPa}$$

This value lies between $0.2 f_c'$ and $0.3 f_c'$. Also the permissible compressive stress σ_{cci} according to Indian code

$$\sigma_{cci} = c_2 f_{ci} = 0.511 \times 21 = 10.73 \text{ MPa}$$

From Eq. (12.2.2), the change in length,

$$\delta_l = \frac{-7.89 \times 21}{36000} = -0.0046 \text{ m}$$

The stress in prestressing steel,

$$f_p = f_{pi} + m \frac{T_D}{A_t}$$

$$= 1200 + 5.55 \times \frac{0.27}{0.258408}$$

$$= 1205.8 \text{ MPa}$$

(ii) *Final stage*

$$P = 0.001848 \times 960 = 1.7741 \text{ MN}$$
$$T_w = T_D + T_L = 0.27 + 1.43 = 1.70 \text{ MN}$$

From Eq. (12.2.3), the stress in concrete,

$$f = -\frac{1.7741}{0.248152} + \frac{1.70}{0.258408}$$

$$= -0.57 \text{ MPa}$$

From Eq. (12.2.4), the stress in prestressing steel,

$$f_p = 960 + 5.55 \times \frac{1.7}{0.258408} = 996.5 \text{ MPa}$$

From Eq. (12.2.5), the instantaneous change in length,

$$\delta = -\frac{0.57 \times 21}{36000} = -0.00033 \text{ m}$$

From Eq. (12.2.6), the deferred change in length,

$$\delta_d = \frac{21}{36000} \left[-\frac{2.2176 + 1.7741}{2 \times 0.248152} + \frac{0.6 \times 1.7}{0.258408} \right]$$
$$\times 1.5 - 0.0002 \times 21$$
$$= -0.00778 \text{ m}$$

Hence the total long-time deformation,

$$\delta_{LT} = \delta + \delta_d$$
$$= -0.00033 - 0.00778 = -0.00811 \text{ m}$$

(iii) *Stage of decompression*

From Eq. (12.2.9), the decompression tensile force,

$$T_{dc} = 1.7741 \times \frac{0.258408}{0.248152} = 1.8474 \text{ MN}$$

Hence the overload factor for decompression,

$$\frac{T_{dc}}{T_w} = \frac{1.8474}{1.7} = 1.09$$

(iv) *Cracking stage*

From Eq. (12.2.10), the cracking tensile force,

$$T_{cr} = \left(2.8 + \frac{1.7741}{0.248152}\right) \times 0.258408 = 2.5710 \text{ MN}$$

Hence the overload factor for cracking,

$$\frac{T_{cr}}{T_w} = \frac{2.5710}{1.7} = 1.51$$

(v) *Ultimate stage*

From Eq. (12.2.11), the nominal ultimate strength,

$$T_u = 0.001848 \times 1720 = 3.1786 \text{ MN}$$

Hence the overload factor for ultimate strength,

$$\frac{T_u}{T_w} = \frac{3.1786}{1.7} = 1.87$$

The margin of safety on decompression,

$$\frac{T_u}{T_{dc}} = \frac{3.1786}{1.8474} = 1.72$$

The margin of safety on cracking,

$$\frac{T_u}{T_{cr}} = \frac{3.1786}{2.5710} = 1.24$$

EXAMPLE 12.2.2

Determine the stresses and deformations at initial and final stages in the tension member of Example 12.2.1 assuming that the restraining tensile force imposed by the adjoining structure is (a) $T_{ri} = 400$ kN and $T_r = 600$ kN and (b) 160 kN per millimetre of shortening.

Solution

Case (a) $T_{ri} = 400$ kN and $T_r = 600$ kN

(i) *Initial stage*

Stress in concrete in restrained member,

$$f_{ir} = f_i + \frac{T_{ri}}{A_t}$$

$$= -7.89 + \frac{0.400}{0.258408} = -6.34 \text{ MPa}$$

Hence the instantaneous change of length,

$$\delta_{ir} = \frac{f_{ir}L}{E_c} = \frac{-6.34 \times 21}{36000} = -0.0037 \text{ m}$$

Stress in prestressing steel,

$$f_{pir} = f_{pi} + \frac{m(T_D + T_{ri})}{A_t}$$

$$= 1200 + 5.55 \left(\frac{0.27 + 0.40}{0.258408} \right) = 1214.4 \text{ MPa}$$

(ii) *Final stage*

Stress in concrete in restrained member,

$$f_r = f + \frac{T_t}{A_t}$$

$$= -0.57 + \frac{0.6000}{0.258408} = 1.75 \text{ MPa}$$

Hence the long-time change of length of restrained member,

$$\delta_{LTr} = \delta_{LT} + \frac{L}{E_c} \left(\frac{T_{ri} + T_r}{2A_t} \right) (1 + C_c)$$

$$= -0.00811 + \frac{21}{36000} \left(\frac{0.40 + 0.60}{2 \times 0.258408} \right) (1 + 1.5)$$

$$= -0.0053 \text{ m}$$

Stress in prestressing steel,

$$f_{pr} = f_{pe} + \frac{m(T_w + T_r)}{A_t}$$

$$= 960 + 5.55 \left(\frac{1.70 + 0.60}{0.258408} \right) = 1009.4 \text{ MPa}$$

Case (b) Restraining force = 160 kN/mm

(i) Initial stage

$$- 160\left[\delta_t + \frac{L}{E_c}\left(\frac{T_{ri}}{A_t}\right)\right] = T_{ri}$$

or

$$- 160\left[- 0.0046 + \frac{T_{ri} \times 21}{0.258408 \times 36000}\right] = T_{ri}$$

or $\quad T_{ri} = 0.5412$ MN

Using this value of $T_{,i}$.

$$f_{ir} = - 7.89 + \frac{0.5412}{0.258408} = - 5.80 \text{ MPa}$$

$$\delta_{ir} = - \frac{5.80 \times 21}{36000} = - 0.0034 \text{ m}$$

$$f_{pir} = 1200 + 5.55 \left(\frac{0.270 + 0.5412}{0.258408}\right) = 1217.4 \text{ MPa}$$

(ii) Final stage

$$- 160\left[\delta_{LT} + \frac{L}{E_c}\left(\frac{T_{ri} + T_r}{2A_t}\right)(1 + C_c)\right] = T_r$$

or

$$- 160\left[- 0.00811 + \frac{21}{36000}\left(\frac{0.5412 + T_r}{2 \times 0.258408}\right)(1 + 1.5)\right] = T_r$$

or $\quad T_r = 0.7264$ MN

Using this value of T_r,

$$f_r = - 0.57 + \frac{0.7264}{0.258408} = 2.24 \text{ MPa}$$

$$f_{LTr} = - 0.00811 + \frac{21}{36000}\left(\frac{0.5412 + 0.7264}{2 \times 0.258408}\right)(1 + 1.5)$$

$$= - 0.0045 \text{ m}$$

$$f_{pr} = 960 + 5.55 \left(\frac{1.70 + 0.7264}{0.258408}\right) = 1012.1 \text{ MPa}$$

EXAMPLE 12.2.3

A 15 m long tension member has to carry an initial service load of 120 kN at stress transfer and an additional service load of 720 kN at final stage Seventy-five percent of the total service load may be assumed to be permanent. Design a suitable cross-section for the member using Freyssine. system of post-tensioning. The initial shortening and the final shortening inclusive of the effect of creep and shrinkage of concrete should not except 5 mm and 10 mm respectively. Assume $C_c = 1.2$ and $\epsilon_{sh} = 0.0003$. The residual compressive stress under full service load should not be less than 0.5 MPa. The overload capacity in respect of decompression and cracking should not be less than 1.05 and 1.25 times the service load. The margins of safety on decompression and cracking should not be less than 1.50 and 1.25 respectively.

Solution

Using M-40 concrete, the physical properties may be taken as

$$f_{ck} = 40 \text{ MPa}$$
$$f_c' = 0.8 f_{ck} = 32 \text{ MPa}$$
$$f_t = 0.24 \sqrt{f_{ck}} = 1.52 \text{ MPa}$$
$$E_c = 5700 \sqrt{f_{ck}} = 36050 \text{ MPa}$$
$$m = \frac{E_s}{E_c} = \frac{2 \times 10^5}{36050} = 5.55$$

As the minimum specified value of overload factor for decompression is 1.10, the tensile force required to cause decompression,

$$T_{dc} = 1.1 \, T_w = 1.1 \, (0.120 + 0.720) = 0.924 \text{ MN}$$

Using Eq. (12.2.9) and taking $A_p/A_c = 0.01$ tentatively,

$$0.924 = P \, (1 + 5.55 \times 0.01)$$

or
$$P = 0.8754 \text{ MN}$$

Using 5 mm diameter high tensile steel wires with characteristic ultimate strength, $f_{pu} = 1700$ MPa and an effective stress, $f_{pe} = 930$ MPa, the number of wires,

$$n = \frac{0.8754}{\dfrac{\pi}{4} \times 5^2 \times 10^{-6} \times 930} = 47.92$$

Hence 4 Fressi cables each having 12 wires of 5 mm diameter may be adopted giving an effective prestressing force,

$$P = 48 \times \frac{\pi}{4} \times 5^2 \times 10^{-6} \times 930 = 0.8769 \text{ MN}$$

Assuming the losses to be 20 per cent,

$$P_i = \frac{0.8769}{0.8} = 1.0961 \text{ MN}$$

Taking the initial stress in concrete, $f_i = 0.25 f_c' = 8$ MPa and $A_c = A_t$ as an approximation in Eq. (12.2.1),

$$-8 = \frac{-1.0961}{A_c} + \frac{0.12}{A_c}$$

or
$$A_c = 0.1220 \text{ m}^2$$

Hence a square cross-section of 360 mm side may be adopted. Assuming the diameter of the cable duct to be 40 mm,

$$A_c = 0.36^2 - 4 \times 0.040^2 = 0.1232 \text{ m}^2$$

$$A_c = A_c + m A_p$$

$$= 0.1232 + 5.55 \times \frac{\pi}{4} \times 5^2 \times 10^{-6}$$

$$= 0.1284 \text{ m}^2$$

The section adopted may now be checked at different stages of loading.

(i) *Initial stage*

From Eq. (12.2.1), the stress in concrete,

$$f_i = - \frac{1.0961}{0.1232} + \frac{0.12}{0.1284} = - 7.96 \text{ MPa}$$

This stress lies within $0.2 f_c'$ and $0.3 f_c'$. From Eq. (12.2.2), the inst antaneous change of length,

$$\delta_i = \frac{- 7.96 \times 15}{36050} = - 0.0033 \, m = - 3.3 \text{ mm}$$

which is less than the specified limit of $- 5$ mm.

(ii) *Final stage*

From Eq. (12.2.3), the stress in concrete,

$$f = - \frac{0.8769}{0.1232} + \frac{0.84}{0.1284} = - 0.58 \text{ MPa}$$

Hence the residual compressive stress is greater than the minimum specified limit of 0.5 MPa. From Eq. (12.2.5), the instantaneous change of length,

$$\delta = - \frac{0.58 \times 15}{36050} = 0.0002 \text{ m} = - 0.2 \text{ m}$$

From Eq. (12.2.6), the deferred change of length,

$$\delta_d = \frac{15}{36050} \left(- \frac{1.0961 + 0.8769}{2 \times 0.1232} + \frac{0.84 \times 0.75}{0.1284} \right) \times 1.2 - 0.0003 \times 15$$

$$= - 0.006 \, m = - 6 \text{ mm}$$

Hence the total long-time deformation,

$$\delta_{LT} = \delta + \delta_d = - 0.2 - 6 = - 6.2 \text{ mm}$$

This deformation is less than the maximum specified limit of $- 10$ mm.

(iii) *State of decompression*

From Eq. (12.2.9), the tensile force required to cause decompression,

$$T_{dc} = 0.8769 \left(\frac{0.1284}{0.1232} \right) = 0.9139 \text{ MN}$$

Hence the overload factor for decompression.

$$\frac{T_{dc}}{T_w} = \frac{0.9139}{0.84} = 1.088$$

It is greater than the minimum specified value of 1.05.

(iv) Cracking stage

From Eq. (12.2.10), the cracking tensile force,

$$T_{cr} = \left(1.52 + \frac{0.8769}{0.1232}\right) \times 0.1284 = 1.1091$$

Hence the overload factor for cracking,

$$\frac{T_{cr}}{T_w} = \frac{1.1091}{0.84} = 1.320$$

which is greater than the minimum specified value of 1.25.

(v) Ultimate stage

According to Indian code, the design ultimate strength of tension member,

$$T_u = 0.87 \, A_p \, f_{pu}$$

$$= 0.87 \times 48 \times \frac{\pi}{4} \times 5^2 \times 1700 \times 10^{-6}$$

$$= 1.3945 \text{ MN}$$

The design ultimate load is equal to $1.5 \, T_w = 1.5 \times 0.84 = 1.26$ MN which is less than the design ultimate strength. The margin of safety on decompression,

$$\frac{T_u}{T_{dc}} = \frac{1.3945}{0.9139} = 1.526$$

and the margin of safety on cracking,

$$\frac{T_u}{T_{cr}} = \frac{1.3945}{1.1091} = 1.257$$

These values are greater than the minimum specified values.

12.3 PRESSURE PIPES

Large diameter concrete pipes, which have to carry high internal pressure, are generally prestressed using the technique of circular prestressing. The common range of internal fluid pressure is generally from 0.5 MPa (5 atmospheres or 50 m head of water) to 2 MPa (20 atmospheres or 200 m head of water). According to the Indian Code IS : 784 and the British Code BS : 4625, prestressed pressure pipes may be mainly classified into the following two types:

(i) Cylinder pipes

In these pipes a welded steel cylinder is prestressed circumferentially by winding high tensile steel wires around it. The steel cylinder has steel sockets and spigot rings welded to its ends. For the sake of protection, the steel cylinder is lined with suitably compacted concrete on the inner side and coated with cement mortar. These pipes are widely used in countries such as United States where steel is cheap and labour is costly. Due to their superior impermeability, these pipes are also preferred for the transportation of poisonous and obnoxious gases.

(ii) Non-Cylinder Pipes

In these pipes, a concrete core of high grade concrete is pretensioned longitudinally by means of high tensile steel wires embedded in concrete and circumferentially post-tensioned using the technique of circular prestressing. The pipes are subsequently coated with cement mortar to serve as protective cover for circumferential prestressing tendon.

The discussion in this section refers mainly to non-cylinder pipes.

12.3.1 Fabrication Processes

Prestressed concrete pressure pipes are manufactured in plants having a high degree of precision control and automation. The manufacturing processes are highly sophisticated and have been patented by leading manufacturers. From the point of view of manufacturing process, prestressed concrete non-cylinder pipes may be divided mainly into the following two types:

(a) Monolyte Construction

In monolyte construction, the prestressing and manufacturing of the pipes are carried out simultaneously in a single step. It is based on the principle that if steel is placed in a mass of fresh concrete which is triaxially compressed, and the concrete is deformed while maintaining the triaxial pressure, the embedded steel also experiences the deformation.

(i) Freyssinet method

The technique developed by Freyssinet in early thirties is still regarded as one of the most sophisticated process of monolyte construction. In this process, high grade concrete is placed in cylindrical mould with its axis vertical and compacted by means of high frequency vibration. The mould comprises an external filtering shell and an internal longitudinally vibrating mandrel. The outer shell comprises longitudinal segments held together by means of springs. The inner shell is provided with expansible rubber membrane. Both shells are therefore expansible. High compaction of concrete is achieved under pressure and vibration. The compacted concrete is then expanded. The stiffness of the compacted concrete is such that the longitudinal reinforcement under tension already placed in the mould is also expanded and tensioned circumferentially. With the aid of steam curing it is possible to reduce the period of one cycle of manufacture to one hour.

(ii) *Sentab (Swedish) method*

The method uses a simpler version of Freyssinet mould by replacing the hydraulic system for the expansion of outer shell by a mechanical system. The end pieces are cast monolithically with the pipe. The longitudinal wires are deflected through the end pieces. This procedure enables the pipe to be connected to spigot and socket joints instead of Johnson type couplings used in the Freyssinet pipe.

(iii) *Russian method*

In Russia, the technique of chemical prestressing was developed for the self stressing of small diameter pipes. A preformed spiral of high strength steel is embedded in high grade concrete using expanding cement. The curing of concrete is carried out under highly controlled conditions in order to achieve the desired degree of prestressing.

(b) Two-stage Construction

In the first stage the concrete pipe is cast and prestressed longitudinally. In the second stage, the circumferential prestress is applied by winding high tensile steel wires under tension around the pipe and coated with a protective cover. The technique was developed by Lewistown Pipe Corporation around 1930. Several other techniques of two stage construction have been developed and patented all over the world. These methods differ mainly in respect of the devices used for achieving longitudinal prestress. In French (Socoman) method, the longitudinal bars are anchored to lugs in cast iron spigot and socket ends. The bars are prestressed by jacking the lugs apart before casting the pipe. Pretensioned high tensile steel wires are used in French (Bonna) and Dutch (N.V. Betondak) methods. The longitudinal bars are pretensioned by means of nuts in Belgian (Socol) method. The joints in this method comprise prestressed rings sealed by a special rubber gasket.

The pipe line is formed by joining together precast prestressed pipe segments of appropriate lengths by means of well designed joints. The joints should be able to develop the necessary internal fluid pressure without showing any leakage. A typical joint of a prestressed concrete pressure pipe is shown in Fig. 12.3.1. A rubber gasket is inserted in the annular space between the small end called the spigot and the enlarged end known as the socket. In order to obtain a taper and the consequent wedge action, the outer diameter of the spigot end is reduced gradually towards the pipe. As an additional device to prevent leakage, the joint is sealed externally by a sealing compound. As the circumferential winding for circular prestressing cannot be carried out upto the ends of the pipe and must terminate slightly short of the ends, both the spigot as well as the socket ends of the pipe are usually reinforced by means of high tensile steel helix with a close spacing.

12.3.2   Analysis

In order to check the adequacy of prestressed concrete pipe, it is necessary to analyse the stresses at different stages of loading.

(i) *Initial stage*

In monolyte construction, the longitudinal and circumferential stresses are transferred simultaneously. In two-stage construction, it is the longitudinal stress which is transferred first usually by the technique of pretensioning. The circumferential prestress is applied subsequently by means of wire winding using the technique of circular prestressing when concrete has gained sufficient strength. The longitudinal as well as circumferential prestress should not exceed the permissible compressive stresses consistent with the strength of concrete at the respective stress transfer.

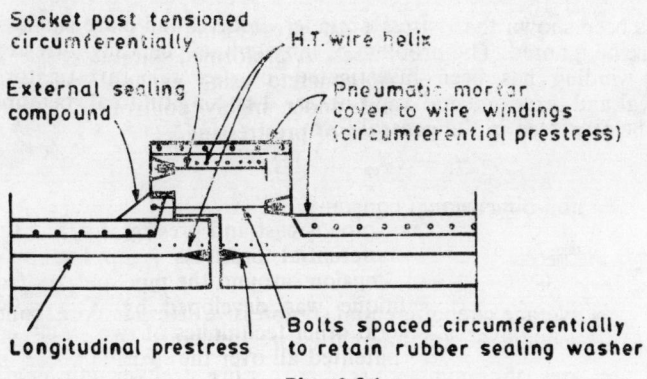

Socket post tensioned circumferentially

H.T. wire helix

External sealing compound

Pneumatic mortar cover to wire windings (circumferential prestress)

Longitudinal prestress

Bolts spaced circumferentially to retain rubber sealing washer

Fig. 12.3.1

(a) Longitudinal prestress

The object of longitudinal prestressing is to counteract the tensile stresses due to (1) partial winding during circumferential prestressing, (2) effect of water hammer caused by closure of valves and (3) beam action during lifting and transportation of pipe under its own weight. The pipe has to act like a beam in carrying its self weight, weight of water and any incidental load when the pipeline is supported by pillars or trestles above ground level. Although the longitudinal stresses due to water hammer and beam action are generally small, they should be taken into account in final design. Significant longitudinal stresses arise during circumferential prestressing due to partial winding. These stresses may be large enough to cause circumferential cracks during circumferential wire winding operation unless the pipe is prestressed adequately in the longitudinal direction. Three types of stresses arise in the concrete core during the wire winding operation.

1. Longitudinal bending stress caused by longitudinal bending moment which arises because the diameter of the wound portion of the pipe decreases whereas that of the unwound portion remains undiminished as shown in Fig. 12.3.2. The walls of the pipes are bent longitudinally and a longitudinal stress reaches its maximum value near the point at which the wire is being wound on the cylinder.

2. Bending stress due to tangertial load of the tensioned wire acting on the pipe which behaves like a beam simply supported between the bearings on which it is rotated during the wire winding operation.

3. Shear stresses due to the torque induced in the pipe during circumferential winding.

Pipe before prestressing

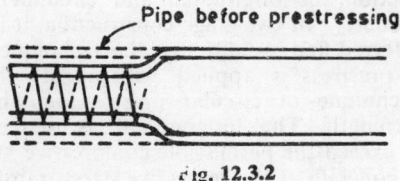

Fig. 12.3.2

It has been shown that stresses under items 2 and 3 are small and may therefore be ignored. The problem of longitudinal bending stresses during partial winding has been investigated by several investigators using analytical and experimental approaches. The longitutdinal bending stress f_l may be expressed in the general form,

$$f_l = c f_{cp} \tag{12.3.1}$$

where c = non-dimensional constant

$$f_{cp} = \frac{A_{pc} f_{pic}}{t_c} \tag{12.3.2}$$

= average cireumferential compressive stress in core concrete on account of circumferential prestressing

A_{pc} = area of circumferential prestressing steel per unit length

f_{pic} = initial stress in circumferential prestressing steel at the time of wire winding

t_c = thickness of concrete core.

Morice and Cooley reported that the longitudinal bending stress due to winding of wire from one end may be computed from Eq. (12.3.1) in which constant c may be taken as 0.284. Curtis and Cowan showed that the problem of longitudinal stresses due to partial winding is similar to that of a partially filled tube. They gave a general solution of the differential equation and showed that the longitudinal bending stress may be determined from Eq. (12.3.1) in which the constant $c = 0.275$. Stephenson studied the problem experimentally and found that the longitudinal bending stress due to wire winding may be determined from Eq. (12.3.1) with constant $c = 0.3$. Doanides reported that the longitudinal bending stresses are of transient nature when winding is carried out from one extreme end to the other. The maximum value of the transient stress near the end of the pipe may be determined from Eq. 12.3.1 in which constant c may be taken as

$$c = c_1 \frac{R_i}{(R_i + t_c)} \tag{12.3.3}$$

in which $c_1 = 0.355$. The transient stress reduces towards the middle region of the pipe and may be determined from Eq. (12.3.1) and (12.3.3) in which $c_1 = 0.283$. For practical reasons, however, the wire winding generally starts at some distance from the spigot end and proceeds towards the other end. In this case maximum transient and permanent stresses occurring at the spigot end may be determined by taking $c_1 = 0.6$ and 0.355 respectively.

The main object of longitudinal prestress is to counteract the longitudinal bending stress discussed above. The intensity of longitudinal compressive stress due to prestressing may be expressed as

$$f_{lip} = \frac{A_{pl} f_{pil}}{2\pi R_c t_c} \tag{12.3.4}$$

where A_{pl} = total area of longitudinal prestressing steel

f_{pil} = initial stress in longitudinal prestressing steel at the time of transfer of longitudinal prestress

R_c = mean radious of concrete core

The initial longitudinal stress f_{lp} should not exceed the permissible stress in concrete at transfer and should be large enough to reduce the longitudinal tensile stress during circumferential wire winding to a permissible value.

(b) Circumferential prestress

The object of circumferential prestress is to counteract the hoop or ring tension caused by (1) internal fluid pressure and (2) statically indeterminate transverse bending moment due to self weight, lateral earth pressure, and weight of earthfill and contained fluid. The surcharge due to live loads on earthfill over the pipeline also causes additional transverse bending moment and circumferential tensile stresses. The initial circumferential prestress in concrete core at stress transfer may be determined by considering a unit length of pipe and considering the equilibrium of vertical forces acting on the upper or lower half of the pipe. It is given by Eq. (12.3.2).

(ii) *Final stage*

At this stage, the internal fluid pressure is resisted by the composite pipe section comprising prestressed concrete core and non-prestressed protective cover or shell. At this stage the pipe carries its full service loading. Considering unit length of pipe and equilibrium of vertical forces acting on the upper or lower half of the pipe, the circumferential stress in core concrete due to combined action of prestressing and internal pressure may be expressed as

$$f_c = -\frac{A_{pc} f_{pec}}{t_c} + \frac{p \, R_i}{t_c + m \, A_{pc} + m_s \, t_s} \tag{12.3.5}$$

where p = internal fluid pressure

f_{pec} = effective stress in circumferential prestressing steel

m = modular ratio of prestressing steel

m_s = modular ratio of protective shell material

t_s = thickness of protective shell.

The transformed area represented by the term $(t_c + mA_{pc} + m_s t_s)$ includes the contribution made by protective cover. This contribution may be ignored conservatively by taking $m_s = o$. This approach may appear justifiable because unlike core concrete, the protective cover is not prestressed. In order to ensure crack free and therefore leak proof pipe, it is desirable that the net circumferential stress in core concrete under full service load is compressive. The magnitude of residual compressive stress may range from 0 to 3 MPa depending upon the type of fluid conveyed and the importance of the pipeline.

(iii) State of decompression

As internal fluid pressure is increased progressively, a stage is reached at which the circumferential prestress is overcome completely. In order to obtain an expression for the internal pressure p_{dc} causing decompression, the circumferential stress f_c given by Eq. (12.3.5) may be equated to zero. Hence,

$$p_{dc} = \left(\frac{A_{pc} f_{pec}}{R_i} \right) \left(\frac{t_c + mA_{pc} + m_s t_s}{t_c} \right) \tag{12.3.6}$$

In order to ensure a residual compressive stress under service condition, the decompression pressure p_{dc} should be larger than the full service pressure inclusive of the effect of water hammer.

(iv) Cracking stage

As the overload is stepped up beyond the stage of decompression, the pipe develops hoop tension until it reaches the stage of cracking. At this stage, the hoop tension is equal to the tensile strength of core concrete f_t. Hence equating the circumferential stress f_e given by Eq. (12.3.5) to f_t, the internal pressure causing cracking may be expressed as

$$p_{cr} = \left(\frac{f_t t_c + A_{pc} f_{pec}}{R_i} \right) \left(\frac{t_c + mA_{pc} + m_s t_s}{t_c} \right) \tag{12.3.7}$$

An appropriate margin of safety against cracking may be obtained by specifying the minimum permissible value of the ratio of cracking pressure p_{cr} and the working pressure p_w.

(v) Bursting stage

The core concrete cracks extensively as the internal pressure is increased to the ultimate or bursting stage. As the internal pressure is resisted entirely by prestressing tendons at the stage of bursting, the expression for bursting pressure may be written as

$$p_{bst} = \frac{A_{pc} f_{puc}}{R_i} \tag{12.3.8}$$

where f_{puc} = ultimate breaking stress of circumferencial prestressing tendons.

The margin of safety against bursting represented by ratio (p_{bst}/p_w) is chosen depending upon the extent and frequency of overloading, the type of pipe and the consequences of bursting.

(vi) *Other stages and loadings*

The longitudinal stresses during lifting and transportation due to self weight should be computed by treating the pipe as a beam with support conditions consistent with the manner in which the pipe is handled during these operations.

The pipe, which behaves like a closed ring in resisting lateral loads such as live load, weight of earth fill, weight of fluid and self weight, is subjected to axial force F_p, radial shear force Q_p and transverse bending moment M_p at any point P as shown in Fig. 12.3.3. These statically indeterminate internal forces may be determined from the following equations of compatibility of deformation.

$$\int M \frac{\partial M}{\partial F_p} \, ds = \int M \frac{\partial M}{\partial Q_p} \, ds = \int M \frac{\partial M}{\partial M_p} \, ds = 0 \qquad (12.3.9)$$

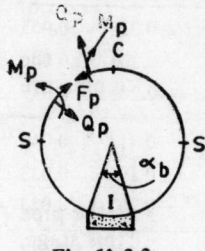

Fig. 12.3.3

in which the integration should cover the entire curve length of the pipe. Once the internal forces at any point P are known, those at any other point are determined readily by statics. The internal forces depend upon the type of loading and the bending angle α_b. The stresses due to radial shear force Q are small and may, therefore, be ignored. The direct force F per unit length and the bending moment M per unit length at any point may be expressed in the general form

$$F = C_1 W \qquad (12.3.10a)$$
$$M = C_2 WD \qquad (12.3.10b)$$

in which D is the mean diameter and W represents the load per unit length, viz. weight of earth fill W_E, live load W_L weight of fluid M_F or self weight W_p. The values of constants C_1 and C_2 for computing the internal forces at the invert (Point I), side (Point S) and crown (Point C) for various values of bedding angle α_b are given in Table 12.3.1 based on Olander's equations. The bedding angle α_b depends upon the preparation of the bottom of the trench. It may be taken equal to 90° under moderately good conditions of placing. The load W_E due to earth fill depends upon the height H of fill above crown, the width of trench B and manner of backfilling and type of soil. To compute the load transferred to the pipe, the following cases may be considered.

(a) $(B/D) < 2$

In this case the load transmitted to the pipe increases with (H/B) upto a certain limit. It is found that due to arch action, the load transferred to the pipe remains practically constant for $H > 6B$ approximately. Hence for $H < 6B$, W_E may be computed from

$$W_E = \frac{2}{3} w_E HB \qquad (12.3.11a)$$

where w_E = unit weight of earth or soil in the backfill

Table 12.3.1 Values of Coefficients C_1 and C_2

α_b	Type of loading	At crown (C) C_1	C_2	At Side (S) C_1	C_2	At Invert (I) C_1	C_2
30°	W_L or W_E	−0.487	0.031	−0.633	−0.043	−0.376	0.087
	W_F	0.237	0.040	0.058	0.050	0.352	0.097
	W_P	0.078	0.040	−0.302	−0.050	−0.125	0.097
60°	W_L or W_E	−0.425	0.033	−0.576	−0.043	−0.345	0.073
	W_F	0.230	0.038	0.059	−0.048	0.310	0.078
	W_P	0.071	0.038	−0.302	−0.048	−0.168	0.078
90°	W_L or W_E	−0.382	0.035	−0.539	−0.045	−0.324	0.063
	W_F	0.220	0.035	0.062	−0.044	0.272	0.061
	W_P	0.061	0.035	−0.295	−0.044	−0.207	0.061
120°	W_L or W_E	−0.351	0.037	−0.516	−0.046	−0.316	0.055
	W_F	0.205	0.032	0.065	−0.039	0.233	0.046
	W_P	0.046	−0.032	−0.282	−0.077	−0.245	0.046
150°	W_L or W_E	−0.331	0.040	−0.504	−0.047	−0.313	0.049
	W_F	0.185	0.027	0.067	−0.032	0.196	0.033
	W_P	0.026	0.027	−0.267	−0.032	−0.282	0.033
180°	W_L or W_E	−0.318	0.043	−0.500	−0.048	−0.318	0.044
	W_F	0.159	0.022	0.068	−0.024	0.159	0.022
	W_P	0.0	0.022	−0.250	−0.024	−0.318	0.022

For $H \geqslant 6B$, W_E may be computed from

$$W_E = 4w_E B^2 \tag{12.3.11b}$$

(b) $(B/D) > 2$

When $B > 2D$ or when the pipe is partially buried. the entire load of the soil over the pipe may be assumed to be transferred to the pipe. For more detailed treatment of the load W_E, reference may be made to IS: 783. The load W_L transmitted to the pipe on account of live loads or surcharge on the top of the backfill may be computed using Boussinesqu's equation. Alternatively, the actual live loads may be converted into their equivalent uniformly distributed loads w_e per unit area. In this case the effect of surcharge due to live loads may be taken equal to additional backfill of height, $H_e = w_e/w_E$. The horizontal pressure exerted by the soil may be determined by using the theory of soil mechanics. However, because the horizontal pressure may or may not be present, it should not be taken into account whenever it tends to relieve the stresses caused by vertical loads.

The minus sign for coefficient C_1 indicates that the direct force F is compressive. The bending moment M is taken positive if it creates tension on the inner face of the pipe.

For analysing the stresses at the forgoing stages of loading, it is necessary to make appropriate allowance for the losses of prestress. The losses incurred during manufacture depend upon the methods used. The following losses may be assumed when the technique of pretensioning is used for longitudinal prestress and wire winding for circumferential prestress:

(a) Loss of circumferential prestress equal to 5 per cent of initial tension immediately due to anchorage slip, 5 per cent under test conditions and 20 per cent under working conditions.

(b) Loss of longitudinal prestress equal to 5 per cent at transfer, 10 per cent under test conditions and 30 per cent under working conditions.

12.3.3 Design and Specifications

As pressure pipes are manufactured under highly controlled conditions, they are expected to satisfy stringent requirements. The Indian Code IS: 784 and the British Code, BS: 4625 lay down detailed specifications for the manufacture and testing of cylinder and non-cylinder prestressed pressure pipes. Detailed specifications for prestressed concrete cylinder pipes are also provided by the American Water Works Association (AWWA). Unless mentioned otherwise, the specifications given below are as per IS: 784.

(i) *Core concrete*

The minimum grade of concrete should be M-45. The minimum concrete cover on all reinforcement in any finished pipe should be 12 mm.

(ii) *Cover coating*

The corrosion of circumferential prestressing steel due to ingress of water is prevented by the application of outer cover coating made of cement mortar, concrete or any other approved material with a minimum thickness of 18 mm. The protective coating should be applied within 16 hours of circumferential wire winding. If cement mortar is used, the ingredients should be thoroughly mixed and applied by means of rotary brushes or belts or by other approved methods. Penumatic process in which mixing of materials occurs in the nozzle or gun is not permitted. The tensile stress in cover coating should not exceed its modulus of rupture. The thickness of cover coating may be reduced to 15 mm in the case of processes in which the cover coating is also subjected to induced compression.

(iii) *Protective coatings*

The protective coatings in the form of bitumen, epoxy or other approved materials may be applied on the inner side of core concrete and the outside of the cover coating. When the pipe is used for conveying potable water, the inner protective layer should not contain any constituent soluble in water or any ingredient which could impart any taste or odour.

(iv) *Longitudinal prestre ;sing*

The core concrete should be prestressed throughout its length including the socket by means of high tensile steel wires anchored in permanent anchorages within the joint portion at each end. The spacing of longitudinal wires along the circumference of the pipe should not be greater than twice the core thickness or 150 mm whichever is the greater. The transfer of longitudinal presstress should not occur before the strength of core concrete becomes equal to $0.55 f_{ck}$ or 1.8 times the induced longitudinal prestress. The wires should be stressed sufficiently taking into account the yield and slip in temporary anchorages on the pipe moulds and held by positive means during curing of core concrete.

In order to guard against longitudinal tensile stresses during circumferential wire winding, Curtis and Cowan have recommended the following equation for total longitudinal prestressing steel,

$$A_{pl} = \frac{2\pi t_c \, (R_l + 0.5t_c) \, (0.275 f_{cp} - \sigma_{ct})}{f_{pl}} \qquad (12.3.12)$$

where f_{pl} = stress in longitudinal steel at the time of wire winding.

f_{cp} = average circumferential stress in core wall on account of circumferential prestressing

σ_{ct} = permissible tensile stress in core concrete at the time of wire winding.

Doanides has suggested a design criterion for longitudinal prestress according to which the net longitudinal tensile stress at the time of wire winding, taking into account the longitudinal prestress, should not exceed $0.8 \sqrt{f_{ci}}$ for transient stress and $0.5 \sqrt{f_{ci}}$ for permanent stress in which f_{ci} is the strength of core concrete at the time of wire winding. Hence

$$(0.6 f_{cp} - f_{lp}) \leqslant 0.8 \sqrt{f_{ci}} \qquad (12.3.13 \text{ a})$$

and

$$(0.355 f_{cp} - f_{lp}) \leqslant 0.5 \sqrt{f_{ci}} \qquad (12.3.13 \text{ b})$$

in which f_{cp} is the average circumferential compressive stress in core concrete on account of circumferential prestressing (Eq. 12.3.2) and f_{lp} is the longitudinal compressive stress due to prestressing in core concrete at the time of circumferential wire winding.

(v) *Circumferential prestressing*

Concrete should develop sufficient compressive strength but not less than 32 MPa at transfer of circumferential prestress. The intensity of circumferential prestress should not exceed 55 per cent of the strength of core concrete at transfer. The clear spacing between the successive turns of the circumferential prestressing wire should not be less than (a) diameter of wire, (b) 1.5 times the maximum nominal size of the aggregate used for outer coat concrete and (c) 5 mm, whichever is the greatest.

(vi) *Design criteria*

The design criteria for circumferential and longitudinal prestressing are shown in Tables 12.3.2 and 12.3.3 in which f_{c1}, f_{c2} and f_{c3} are the strengths of concrete at transfer of longitudinal prestress, circumferential prestress and at operating conditions respectively. As the tensile strength of concrete in bending f_{cr} is significantly greater than the tensile strength of concrete in direct tension f_t, Neville has suggested that the bending stress may be multiplied by a factor 0.6 while checking the tensile stress under load combinations which produce direct as well as bending tensile stresses.

Table 12.3.2 Permissible circumferential stresses in core concrete

S. No.	Load combination	Permissible stress (MPa)	
		Tension	*Compression*
1.	Operating conditions (self weight + weight of water + earth fill load + working pressure + final prestress)	0	$0.33 f_{ck}$
2.	Operating conditions plus surge or water hammer pressure.	$0.56 \sqrt{f_{ck}}$	$0.4 f_{ck}$
3.	Operating conditions plus live load (with impact)	$0.56 \sqrt{f_{ck}}$	$0.4 f_{ck}$
4.*	Hydrostatic proof test pressure + self weight + weight of water + final prestress	$0.56 \sqrt{f_{ck}}$	$0.4 f_{ck}$
5.	Self weight + initial prestress	0	$0.5 f_{c2}$

*The hydrostatic proof test pressure should be taken equal to 1.5 times the working pressure or working pressure plus allowance for surge, whichever is greater.

EXAMPLE 12.3.1

A prestressed concrete pressure pipe has internal radius, $R_i = 450$ mm and concrete core thickness, $t_c = 50$ mm. It is prestressed in the longitudinal direction by means of 22 wires each of 4 mm diameter and in circumferential direction by wire of 5 mm diameter wound with a pitch of 40 mm. The initial and effective prestress in circumferential wire are 1200 MPa and 960 MPa respectively. Determine the service pressure under which the concrete core has a residual compressive stress of 1 MPa. Also determine the internal pressure at decompression, cracking and bursting. The thickness of cement mortar cover coating is 25 mm. Take modular ratios, $m = 6.5$ and $m_s = 0.5$, tensile strength of concrete, $f_t = 1.8$ MPa and ultimate stress of prestressing steel, $f_{pu} = 1800$ MPa. Also determine the longitudinal tensile stress during wire winding if the prestress in longitudinal steel at the time of wire winding is 1050 MPa.

Table 12.3.3 Permissible longitudinal stresses in core concrete

S. No.	Load combination	Permissible stress (MPa)	
		Tension	Compression
1.	Longitudinal prestress after losses	—	Minimum residual compression, $f_{rc} = 1$ MPa
2.	Initial prestress less the loss of longitudinal prestress at the time of wire winding.	$0.56 \sqrt{f_{c2}}$	$0.5 f_{c2}$
3.*	Longitudinal prestress after losses plus beam action.	$0.34 \sqrt{f_{c3}}$	$0.5 f_{c3}$
4.	Handling before winding.	0.67 MPa	$0.5 f_{c1}$

*To compute the longitudinal stresses due to beam action caused by uneven bearing, the pipe should be assumed to span its full length and support a total load equal to the self weight of the pipe, the weight of water in the pipe and an earth load equal to 2.2 times the outer diameter of the pipe (in metres) in kN per linear metre.

Solution

$$A_{pc} = \frac{\pi}{4} \times 5^2 \times 10^{-6} \left(\frac{1000}{40} \right) = 0.000491 \ \text{m}^2/\text{m}$$

Hence from Eq. (12.3.5),

$$-1 = - \frac{0.000491 \times 960}{0.05} + \frac{p_w \times 0.45}{0.05 + 6.5 \times 0.000491 + 0.5 \times 0.025}$$

or service pressure, $p_w = 1.23$ MPa.

From Eq. (12.3.6), the decompression pressure,

$$p_{dc} = \left(\frac{0.000491 \times 960}{0.45} \right) \left(\frac{0.05 + 6.5 \times 0.000491 + 0.5 \times 0.025}{0.05} \right)$$

$$= 1.38 \ \text{MPa}$$

From Eq. (12.3.7), the cracking pressure,

$$p_{cr} = \left(\frac{1.8 \times 0.05 + 0.000491 \times 960}{0.45} \right) \left(\frac{0.05 + 6.5 \times 0.000491 + 0.5 \times 0.025}{0.05} \right)$$

$$= 1.64 \ \text{MPa}$$

From Eq. (12.3.8), the bursting pressure,

$$p_{bst} = \frac{0.000491 \times 1800}{0.45} = 1.96 \ \text{MPa}$$

Longitudinal stresses during wire winding

From Eq. (12.3.2).

$$f_{cp} = \frac{0.000491 \times 1200}{0.05} = 11.78 \text{ MPa}$$

(i) *Morice and Cooley*

From Eq. (12.3.1), the longitudinal bending stress,

$$f_l = 0.284 \times 11.78 = 3.35 \text{ MPa}$$

From Eq. (12.3.4), the intensity of longitudinal compressive stress due to prestressing at the time of wire winding,

$$f_{llp} = \frac{22 \times \frac{\pi}{4} \times 4^2 \times 10^{-6} \times 1050}{2 \times \pi \times 0.475 \times 0.05}$$

$$= 1.94 \text{ MPa}$$

Hence the net tensile stress at the time of wire winding is $3.35 - 1.94 = 1.41$ MPa

(ii) *Curtis and Cowan*

$$f_l = 0.275 \times 11.78 = 3.24 \text{ MPa}$$

Hence the net tensile stress is $3.24 - 1.94 = 1.30$ MPa

(iii) *Stephenson*

$$f_l = 0.3 \times 11.78 = 3.53 \text{ MPa}$$

Hence the net tensile stress is $3.53 - 1.94 = 1.59$ MPa

(iv) *Doanides*

(a) Transient stress a spigot end,

$$f_l = 0.6 \left(\frac{0.45}{0.45 + 0.025} \right) \times 11.78 = 6.70 \text{ MPa}$$

(b) Permanent stress at spigot end,

$$f_l = 0.355 \left(\frac{0.45}{0.45 + 0.025} \right) \times 11.78 = 3.96 \text{ MPa}$$

Hence the net transient and permanent tensile stresses at the spigot end are $6.70 - 1.94 = 4.66$ MPa and $3.96 - 1.94 = 2.02$ MPa respectively.

EXAMPLE 12.3.2

Determine the longitudinal stresses in the pipe of Example 12.3.1 when it it lifted by holding it at (i) ends and (ii) mid point. The length of pipe is 9 m. Take the unit weight of the pipe material as 24 kN/m^3. Take impact factor equal to 1.0. Take effective stress in longitudinal steel after losses as 980 MPa.

Solution

Unit weight of pipe,

$$w = \frac{\pi}{4} (1.05^2 - 0.9^2) \times 24 = 5.52 \text{ kN/m}$$

Hence the effective weight inclusive of impact factor is $5.52 \times 2 = 11.04$ kN/m.

Polar moment of intertia of transformed section of the pipe,

$$I_z = 2\pi \times 0.475 \times 0.05 \times 0.475^2$$

$$+ 6.5 \times 22 \times \frac{\pi}{4} \times 4^2 \times 10^{-6} \times 0.502^2$$

$$+ 0.5 \times 2\pi \times 0.5125 \times 0.025 \times 0.5125^2$$

$$= 0.04471 \text{ m}^4$$

Hence the moment of inertia,

$$I_x = I_y = \frac{I_z}{2} = \frac{0.04471}{2} = 0.02236 \text{ m}^4$$

Effective compressive stress in core concrete due to prestressing,

$$f_{lp} = \frac{22 \times \frac{\pi}{4} \times 10^{-6} \times 980}{2\pi \times 0.475 \times 0.050} = 1.82 \text{ MPa}$$

(i) *Pipe held at ends*

Maximum bending moment at midlength,

$$M = \frac{11.04 \times 9^2}{8} = 111.78 \text{ kN.m}$$

Hence stress in core concrete at top,

$$f_T = -1.82 - \frac{0.11178 \times 0.5}{0.02236} = -4.32 \text{ MPa}$$

The stress in core concrete at bottom,

$$f_B = -1.82 + \frac{0.11178 \times 0.5}{0.02236} = 0.68 \text{ MPa}$$

The stress in longitudinal steel at top,

$$f_{pT} = 980 - 6.5 \times \frac{0.11178 \times 0.502}{0.02236} = 963.7 \text{ MPa}$$

The stress in longitudinal steel at bottom,

$$f_{pB} = 980 + 6.5 \times \frac{0.11178 \times 0.502}{0.02236} = 996.3 \text{ MPa}$$

The stress in cover coating at top,

$$f_{sT} = -0.5. \times \frac{0.11178 \times 0.525}{0.02236} = -1.31 \text{ MPa}$$

The stress in cover coating at bottom,

$$f_{sB} = 1.31 \text{ MPa}$$

(ii) *Pipe held at midpoint*

In this case, the maximum bending moment which occurs at midpoint is $\frac{wl^2}{8}$ which is numerically the same as in (i) above but is opposite is sign. Hence the stresses are the same as in (i) except that top and bottom are interchanged.

EXAMPLE 12.3.3

Determine the transverse stresses in the pipe of Example 12.3.1 on account of self weight, weight of water and earthfill including surcharge which produces a load $W_E + W_L = 15$ kN/m on the pipe. The bedding angle α_b may be taken equal to $90°$.

Solution

$$W_E + W_L = 15 \text{ kN/m}$$

$$W_F = \frac{\pi}{4} \times 0.9^2 \times 10 = 6.36 \text{ kN/m}$$

$$W_p = \frac{\pi}{4} (1.05^2 - 0.9^2) \times 24 = 5.52 \text{ kN/m}$$

(i) *Stresses at invert (I)*

The direct force at invert,

$$F = -0.324 \times 15 + 0.272 \times 6.36 - 0.207 \times 5.52$$
$$= -4.27 \text{ kN}$$

The bending moment at invert,

$$M = (0.063 \times 15 + 0.061 \times 6.36 + 0.061 \times 5.52) \times 0.975$$
$$= 1.628 \text{ kN.m}$$

Stress at the inner surface of core concrete,

$$f_1 = \frac{-427}{1 \times 0.05} + \frac{1.628}{\frac{1}{6} + 1 \times 0.05^2} = 3822 \text{ kN/m}^2$$

$$= 3.822 \text{ MPa}$$

Stress at the external surface of core concrete,

$$f_2 = \frac{-4.27}{1 \times 0.05} - \frac{1.628}{\frac{1}{6} \times 1 \times 0.05^2} = -3993 \text{ kN/m}^2$$

$$= -3.993 \text{ MPa}$$

(ii) *Stresses at side (S)*

The direct force at side,

$$F = -0.539 \times 15 + 0.062 \times 6.36 - 0.297 \times 5.52$$
$$= -9.33 \text{ kN}$$

The bending moment at side,

$$M = (-0.045 \times 15 - 0.044 \times 6.36 - 0.044 \times 5.52) \times 0.975$$
$$= -1.168 \text{ kN.m}$$

Stress at the inner surface of core concrete,

$$f_1 = \frac{-9.33}{1 \times 0.05} - \frac{1.168}{\frac{1}{6} \times 1 \times 0.05^2} = -2990 \text{ kN/m}^2$$

$$= -2.99 \text{ MPa}$$

Stress at the external surface of core concrete,

$$f_2 = \frac{-9.33}{1 \times 0.05} + \frac{1.168}{\frac{1}{6} \times 1 \times 0.05^2} = 2617 \text{ kN/m}^2$$

$$= 2.617 \text{ MPa}$$

EXAMPLE 12.3.4

Design a suitable cross-section for the prestressed concrete non-cylinder pipe having an internal diameter of 1.5 m for a water supply scheme. The maximum service pressure inclusive of the effect of water hammer is equivalent to a head of 120 m. The minimum factors of safety on cracking and bursting are 1.25 and 1.50 respectively. Take $f_t = 2.1$ MPa and $f_{pu} = 1750$ MPa.

Solution

Internal service pressure including surcharge,

$$p_w = 10 \times 120 = 1200 \text{ kN/m}^2 = 1.2 \text{ MPa}$$

Using concrete of grade M-45 and assuming that its strength at transfer of circumferential prestress is 32 MPa, the maximum permissible value of the initial compressive stress in core concrete,

$$\sigma_{cci} = 0.55 \times 32 = 17.6 \text{ MPa}$$

Hence it is safe to adopt an initial compressive stress of 15 MPa during wire winding operation. In order to determine the thickness of core concrete, Eq. (12.3.5) may be written in the form,

$$f_c = - \frac{A_{pc}(\eta f_{pic})}{t_c} + \frac{p R_i}{t_c + m A_{pc} + m_s t_s} \tag{a}$$

Taking $\eta = 0.80$, $\dfrac{A_{pc} f_{pic}}{t_c} = 15$ MPa, $f_c = -1$ MPa to ensure a residual compressive stress of 1 MPa, $p = p_w = 1.2$ MPa and ignoring the contribution of steel and coating conservatively,

$$-1 = -0.8 \times 15 + \frac{1.2 \times 0.75}{t_c}$$

or
$$t_c = 0.082 \text{ m}$$

Adopting a core thickness, $t_c = 0.08$ m, the area of prestressing steel A_{pc} per metre length of the pipe may be determined from

$$\frac{A_{pc} f_{pic}}{t_c} = 15$$

Taking initial stress in prestressing steel $f_{pic} = 1200$ MPa,

$$A_{pc} = \frac{15 \times 0.08}{1200} = 0.001 \text{ m}^2/\text{m}$$

Using 5 mm diameter wire, the number of turns of circumferential steel per metre,

$$n = \frac{1000}{\frac{\pi}{4} \times 5^2} = 50.9$$

Hence a pitch of 20 mm may be adopted giving $A_{pc} = 982 \text{ mm}^2$. A cement mortar cover coating of 25 mm thickness may be used. Equation (a) may now be used to check the residual compressive stress under full service pressure. The modulus of elasticity of core concrete,

$$E_c = 5700 \sqrt{45} = 38237 \text{ MPa}$$

Hence the modular ratio,

$$m = \frac{2 \times 10^5}{38237} = 5.23$$

Ignoring the contribution of cover coating,

$$f_c = - \frac{0.000982 \times 0.8 \times 1200}{0.08} + \frac{1.2 \times 0.75}{0.08 + 5.23 \times 0.000982}$$

$$= -1.21 \text{ MPa}$$

which is satisfactory.

The transfer of longitudinal prestress may be assumed to occur when the strength of concrete, $f_{ci} = 0.55 f_{ck} = 0.55 \times 45 = 24.75$ MPa. Hence using Eq. (12.3.13 b), the longitudinal compressive stress during wire winding,

$$f_{lp} = 0.355 f_{cp} - 0.5\sqrt{f_{ci}}$$

$$f_{cp} = \frac{A_{pc} f_{ptc}}{t_c} = \frac{0.000982 \times 1200}{0.08} = 14.73 \text{ MPa}$$

Hence $f_{lp} = 0.355 \times 14.73 - 0.5 \sqrt{24.75}$

$$= 2.74 \text{ MPa}$$

Total longitudinal prestressing force at wire winding,

$$P_l = 2.74 \times \frac{\pi}{4} (1.66^2 - 1.50^2)$$

$$= 1.0885 \text{ MN}$$

Using 5 mm diameter wires with an effective stress of 1080 MPa at the time of wire winding, the total number of wires in longitudinal direction,

$$n_l = \frac{1.0885}{\frac{\pi}{4} \times 5^2 \times 10^{-6} \times 1080} = 51.3$$

Hence 52 wires may be provided giving a spacing along the circumference equal to 95.5 mm which is satisfactory.

From Eq. (12.3.7), ignoring the contribution of cover coating, the cracking pressure,

$$p_{cr} = \frac{2.1 \times 0.08 + 0.000982 \times 0.8 \times 1200}{0.75} = 1.576 \text{ MPa}$$

Hence factor of safety against crackin, $p_{cr}/p_w = 1.31$ which is greater than the minimum permissible value of 1.25.

Using Eq. (12.3.8) and inserting the capacity reduction factor $\Phi = 0.87$, the bursting pressure,

$$p_{bst} = \frac{0.87 \times 000982 \times 1750}{0.75} = 1.993 \text{ MPa}$$

Hence the factor of safety against bursting, $p_{bst}/p_w = 1.66$ which is greater than the minimum permissible value of 1.5.

The chosen cross-section may now be checked for the load combinations given in Tables 12.3.2 and 12.3.3 and changes incorporated, if necessary.

12.4 CYLINDRICAL CONTAINERS

If there is a field in construction industry in which the technique of pre-stressing ushered in a revolution, it is undoubtedly in the domain of super-size cylindrical containers. A prestressed cylindrical container represents one of the best combinations of structural form and material for the storage of fluids and solids. Since the development of the technique of prestressing in 1940s, several thousand containment structures have been built. The technique of prestressing has made it a practical reality to build cylindrical water tanks with capacities of the order of 50 million litres. Prestressed containment structures with diameter as large as 100 m and filling depth as large as 36 m have been built. In the domain of water tanks, prestressed concrete combines the relative merits both of structural steel and concrete. It possesses the strength and impermeability of the former and the maintenance free long life of the latter. The relative advantages and economy of prestressed concrete are so great that they hold a virtual monopoly in the domain of large containment structures.

Prestressed concrete cylindrical containers have been used extensively as storage tanks for water, petroleum products and other fluids, as bunkers and silos for storing grains, coal, chemicals, fertilizers and other solids and also as pressure vessels for nuclear power stations. Although the cross-section of the cylindrical container may be square, pentagonal, hexagonal or any other regular polygon, the circular shape is by far the most popular. A circular cylindrical container placed with its axis vertical represents an axisymmetric structure whose loading due to contained material is also symmetric about the vertical axis. The main advantage of this structural form is that the vertical walls of the containers are not subjected to bending moments and the associated flexural shear forces in the horizontal planes. The predominant structural action is that of horizontal slices of the vertical walls acting like closed rings carrying hoop tension due to internal pressure and hoop compression due to circumferential prestressing. The container wall is, therefore, essentially a tension member and the primary object of prestressing is to counteract the hoop tension caused by containment. Bending moments and the associated shear forces arise in the radial vertical planes when the bottom and top edges of the cylindrical walls are restrained from free radial expansion. The object of vertical or longitudinal prestressing is to counteract the tensile stresses caused by these bending moments.

12.4.1 Construction Techniques

Several ingeneous methods have been used for the construction of cylindrical walls and for achieving the desired circumferential prestress. The methods of construction of cylindrical walls can be classified into two main types.

(i) Monolithic construction

In this technique, high strength concrete is generally placed in vertical slices keyed together. Pneumatic mortar has also been used, particularly when the wall thickness is less than 150 mm. The concrete or pneumatic mortar is allowed to develop sufficient strength before the application of circumferential and vertical prestress. This technique is suitable when circumferential prestress is achieved by means of circular prestressing.

(ii) Precast construction

In this technique, the cylindrical walls are constructed by assembling together several precast segments or units. Each segment is generally stiffened by means of longitudinal and circumferential ribs, particularly along the edges of the segments. The prestressing tendons are housed in ducts formed by placing sheathing at appropriate locations. To minimise the joints, the segments may be as large as possible, keeping in view the problems associated with handling the segments. The technique of construction is suitable when circumferential prestress is achieved by means of linear prestressing.

Although cylindrical tanks are also prestressed vertically, the main object of prestressing is to counteract hoop tension by means of circumferential prestressing. There are two main methods used for circumferential prestressing.

(a) *Circular prestressing*

Due to the pioneering work of Preload Engineering Company of United States, this method is also known as the preload method. In this method a high tensile steel wire is wound under requisite tension continuously in the form of spiral around the cylindrical wall. The circumferential wire winding is achieved by means of a traction machine commonly referred to as merry-go-round. The traction machine is suspended from a trolley which moves in a circular path along the top of the cylindrical wall. The requisite tension in the prestressing wire is achieved by passing it through a die. In order to guard against the possibility of the loops wound earlier losing their tension suddenly in the event of wire fracture, it is secured to the cylindrical wall by means of clips at regular intervals. In one complete revolution of the traction machine, it is lifted by the trolley through a vertical distance equal to the desired pitch of the spiral. The wire winding commences from the bottom of the cylindrical wall and progresses upwards. The splicing of the wire is achieved by means of spring loaded torpedo splices capable of developing full strength of the wire. The machine which can wind more than 10 km of wire per hour is capable of completing the circumferential prestressing of a cylindrical tank of 5 million litres capacity in a couple of days.

(b) *Linear prestressing*

In this technique, the prestressing tendons are not continuous but are in the form of equal segments along the arc of a circle. The prestressing tendons are housed in preformed ducts and are stressed from both ends by means of hydraulic jacks resting against the vertical pylons cast at uniform spacing along the cylindrical wall. The jacking points for adjacent tendons may be staggered in order to equalise the frictional loss of prestress around the circumference of the container. The technique is gaining popularity progressively and is now being used extensively all over the world. The container walls may be cast-in-situ or prefabricated in suitable segments. The main advantage is that the entire wall thickness is prestressed, whereas in circular prestressing the outer coating or gunite remains unprestressed and consequently does not offer as much protection. In order to reduce the frictional loss, the curved length of the tendon should be limited by placing the jacking points at distances not greater than one-third of the perimeter of the cylindrical wall. It follows that the number of anchorages and the labour entailed in stressing operation increase which represents the main drawback of the technique.

12.4.2 Analysis

In order to understand the nature of internal forces in a prestressed concrete circular cylindrical wall, it may be visualised as series of closed circular rings lying in horizontal planes. If the net or resultant internal pressure inclusive of the effect of prestress is constant and all rings, particularly those at top and bottom, are free to expand in radial direction, all rings will expand equally. Consequently, the deformed shape of the wall remains a circular cylinder. The vertical strips of the wall remain straight. Hence they do not develop any bending moment or shear force. If the distribution of net internal pressure is linear and the edges of the wall are unrestrained, the deformed wall takes a conical shape. The vertical strips in this case too, have no bending moment or shear force because they remain straight after deformation. However, if the net internal pressure is non-linearly distributed in the vertical direction and/or the edges (top or bottom) are restrained in any manner, the vertical strips become curved. Hence the vertical strips behave like transversely loaded beams carrying bending moments and the associated flexural shear forces in radial vertical planes. It follows that the net internal pressure p is resisted partly by hoop action of a closed ring p_h and partly by beam action of the vertical strips p_b.

$$p = p_h + p_b \qquad (12.4.1)$$

The net internal pressure p may be expressed as

$$p = p_i - p_p \qquad (12.4.2)$$

in which p_i is the internal pressure due to contained material and p_p is the external pressure due to circumferential prestress. The internal pressure p_i is constant if the contained material is in gaseous form. In the case of liquids, it varies linearly from zero value at top to a maximum value equal to wH at bottom in which w is the density or unit weight of liquid and H is the height of cylindrical wall or maximum depth of filling. At any depth x measured from top,

$$p_i = wx \qquad (12.4.3)$$

To derive an expression for the external pressure p_p due to circumferential prestress, consider an element of cylindrical wall of unit height and unit curved length as shown in Fig. 12.4.1. Considering free body of the element and resolving all forces in vertical direction,

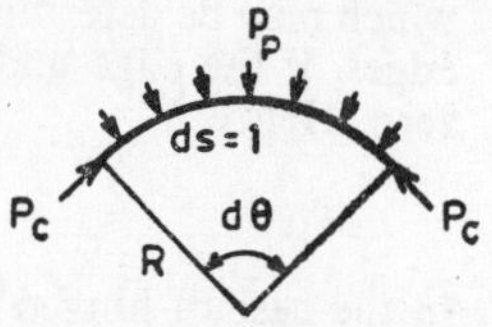

Fig. 12.4.1

$$p_p = 2P_c \sin\left(\frac{d\theta}{2}\right)$$

$$\approx P_c \, d\theta$$

$$= \frac{P_c}{R} \qquad (12.4.4)$$

in which P_c is the effective circumferential prestressing force per unit height and R is the radius of the cylindrical wall. The pitch of cirumferential prestressing tendons generally increases linearly from bottom towards top. Hence p_p varies linearly and at depth x from top,

$$p_p = \frac{1}{R}\left[P_{Tc} + (P_{cB} - P_{cT})\frac{x}{H} \right] \qquad (12.4.5)$$

in which P_{cT} and P_{cB} are the values of P_c at top and bottom. The appropriate values of p_i and p_p may be inserted into Eq. (12.4.2) to obtain the value of net internal pressure p inclusive of the effect of prestress.

General expressions for the three non-zero internal forces, viz. horizontal hoop tension per unit height of wall T, bending moment M and its associated flexural shear force Q in radial vertical planes per unit circumference of the wall, may be obtained from the general solution of the fourth order differential equation derived from Eq. (12.4.1). Let y be the outward radial displacement of the wall at depth x from top. As circumferential and radial strains are equal,

$$\frac{\delta R}{R} = \frac{y}{R} = \frac{p_h R}{t_c E_c}$$

or

$$p_h = \frac{t_c E_c y}{R^2} \qquad (12.4.6)$$

in which t_c is the concrete core wall thickness, E_c is the modulus of elasticity of concrete and p_h is the internal pressure resisted by hoop action. The internal pressure p_b resisted by beam action is related to displacement y by the equation,

$$p_b = - E_c I \frac{d^4 y}{dx^4} \qquad (12.4.7)$$

where $I = \dfrac{t_c^3}{12}$

= moment of inertia per unit circumference of wall

Substituting from Eq. (12.4.6) and (12.4.7) into Eq. (12.4.1),

$$- E_c I \frac{d^4 y}{dx^4} + \frac{t_c E_c y}{R^2} - p = 0 \qquad (12.4.8)$$

The general solution of Eq. (12.4.8) involves 4 constants of integration which may be determined from the support conditions at the top and bottom edges. If the edge under consideration is fixed, the deflection and slope are zero. Hence

$$y = \frac{dy}{dx} = 0 \qquad (12.4.9)$$

In the case of hinged edge, the deflection and bending moment are zero. Hence,

$$y = \frac{d^2 y}{dx^2} = 0 \qquad (12.4.10)$$

If the edge is free, the bending moment and shear force are zero. Hence,

$$\frac{d^2 y}{dx^2} = \frac{d^3 y}{dx^3} = 0 \qquad (12.4.11)$$

After obtaining the general expression for the lateral displacement y, the hoop tension T may be determined from

$$T = p_h R = \frac{t_c E_c y}{R} \qquad (12.4.12)$$

and the bending moment M from,

$$M - E_c I \frac{d^2 y}{dx^2} \tag{12.4.13}$$

The radial shear force is small and therefore does not influence the design in any way.

The form of Eq. (12.4.8) shows that hoop tension T and bending moment M depend upon the following factors:

(i) Support condition at top—free, hinged or partially restrained.

(ii) Support condition at bottom—free (frictionless sliding joint), hinged, partially restrained or fixed.

(iii) Distribution of internal pressure—rectangular, triangular or trapezoidal.

(iv) Thickness of wall—uniform or varying.

(v) Form factor $- \dfrac{H^2}{2 R t_c}$

(vi) Position of point under consideration represented by depth x.

The hoop tension T and bending moment M which are important for the design of cylindrical walls may be expressed as

$$T = C_t (p_{max} R) \tag{12.4.14}$$

and
$$M = C_m (p_{max} H^2) \tag{12.4.15}$$

in which C_t and C_m arc the influence cofficients for T and M respectively and p_{max} is the maximum internal pressure at base. To serve as useful design aids, standard tables are available for the influence coefficients C_t and C_m. These tables are available in Indian Code IS: 3370 Part IV and also in Concrete Information Series ST-57 of Portland Cement Association of United States. Values of influence coefficients C_t and C_m are given in Tables 12.4.1 to 12.4.4. These tables are applicable for walls of uniform thickness. Positive values of C_t correspond to hoop tension. Similarly positive values of C_m correspond to bending moments which produce tension on the outer face. The tables may be used directly for the compution of T and M at any point due to rectangular or triangular distribution of internal pressure. The edge conditions in these tables correspond to fixed base and free top. The principle of superposition may be used to determine T and M for the case in which the base is hinged and the top is free. For this purpose Tables 12.4.5 and 12.4.6 have been prepared. In these tables, values of C_t and C_m are given when moment M_o is applied at the base which is hinged, the top being free. Using these tables, T and M are given by

$$T = C_t \frac{M_o R}{H^2} \tag{12.4.16}$$

and
$$M = C_m M_o \tag{12.4.17}$$

where M_o is the bending moment in the radial vertical plane applied at the base per unit length along the circumference. By combining Tables 12.4.6 with the relevant tables (Tables 12.4.1 or 12.4.4), the moment at the base can be released to obtain the hinged base condition as illustrated in Example

Table 12.4.1 Hoop Tension Coefficients, C_t

(Fixed base, free top, rectangular internal pressure)

$x \quad \frac{H^2}{2Rt_c} \rightarrow$ $\downarrow H$	0.4	1.2	2.0	4.0	6.0	10.0	14.0	20.0	32.0	48.0
0.0 H	0.582	1.218	1.253	1.085	1.010	0.989	0.997	—	—	—
0.1 H	0.505	1.078	1.144	1.073	1.024	0.998	0.998	—	—	—
0.2 H	0.431	0.946	1.041	1.057	1.038	1.010	1.000	—	—	—
0.3 H	0.353	0.808	0.929	1.029	1.045	1.023	1.007	—	—	—
0.4 H	0.277	0.665	0.806	0.997	1.034	1.039	1.022	—	—	—
0.5 H	0.206	0.519	0.667	0.887	0.986	1.040	1.040	—	—	—
0.6 H	0.145	0.378	0.514	0.746	0.879	0.996	1.035	—	—	—
0.7 H	0.092	0.246	0.345	0.553	0.694	0.859	0.949	—	—	—
0.75 H	—	—	—	—	—	—	—	0.949	1.026	1.043
0.80 H	0.046	0.127	0,186	0.322	0.430	0.591	0.705	0.825	0.953	1.022
0.85 H	—	—	—	—	—	—	—	0.629	0.788	0.911
0.90 H	0.013	0.034	0.055	0.105	0.149	0.226	0.294	0.379	0.519	0.652
0.95 H	—	—	—	—	—	—	—	0.128	0.189	0.262

Table 12.4.2 Moment Coefficients, C_m

(Fixed base, free top, rectangular internal pressure)

$x \downarrow$ $\frac{H^2}{2Rt_c} \rightarrow$	0.4	1.2	2.0	4.0	6.0	10.0	14.0	20.0	32.0	48.0
0.1 H	−0.0023	0.0008	0.0010	0.0004	0.0001	0.0	0.0	—	—	—
0.2 H	−0.0093	0.0026	0.0036	0.0015	0.0004	−0.0001	0.0	—	—	—
0.3 H	−0.0227	0.0037	0.0066	0.0033	0.0011	0	0.0	—	—	—
0.4 H	−0.0439	0.0029	0.0088	0.0052	0.0022	0.0002	0.0	—	—	—
0.5 H	−0.0710	−0.0009	0.0089	0.0068	0.0036	0.0009	0.0002	—	—	—
0.6 H	−0.1018	−0.0089	0.0059	0.0075	0.0049	0.0021	0.0010	—	—	—
0.7 H	−0.1455	−0.0227	−0.0019	0.0053	0.0048	0.0030	0.0018	—	—	—
0.8 H	−0.2000	−0.0468	−0.0167	−0.0013	0.0017	0.0026	0.0021	0.0015	0.0008	0.0004
0.85 H	—	—	—	—	—	—	—	0.0013	0.0009	0.0006
0.9 H	−0.2593	−0.0815	−0.0389	−0.0145	−0.0073	−0.0022	−0.0007	0.0002	0.0006	0.0006
0.95 H	—	—	—	—	—	—	—	−0.0024	−0.0010	−0.0003
H	−0.3310	−0.1178	−0.0719	−0.0365	−0.0242	−0.0147	−0.0105	−0.0073	−0.0046	−0.0031

Table 12.4.3 Hoop Tension Coefficients, C_t
(Fixed base, free top, triangular internal pressure)

$x \downarrow \quad \frac{H^2}{2Rt_c} \rightarrow$	0.4	1.2	2.0	4.0	6.0	10.0	14.0	20.0	32.0	48.0
0.0 H	0.149	0.283	0.234	0.067	0.018	−0.011	−0.002	—	—	—
0.1 H	0.134	0.271	0.251	0.164	0.119	0.098	0.098	—	—	—
0.2 H	0.120	0.254	0.273	0.256	0.234	0.208	0.200	—	—	—
0.3 H	0.101	0.234	0.285	0.339	0.344	0.323	0.306	—	—	—
0.4 H	0.082	0.209	0.285	0.403	0.441	0.437	0.420	—	—	—
0.5 H	0.066	0.180	0.274	0.429	0.504	0.542	0.539	—	—	—
0.6 H	0.049	0.142	0.232	0.409	0.514	0.608	0.639	—	—	—
0.7 H	0.029	0.099	0.172	0.334	0.447	0.589	0.666	—	—	—
0.75 H	—	—	—	—	—	—	—	0.716	0.782	0.791
0.80 H	0.014	0.045	0.104	0.210	0.301	0.440	0.541	0.654	0.768	0.828
0.85 H	—	—	—	—	—	—	—	0.520	0.663	0.785
0.90 H	0.004	0.016	0.031	0.073	0.112	0.179	0.241	0.325	0.459	0.593
0.95 H	—	—	—	—	—	—	—	0.115	0.182	0.254

Table 12.4.4 Moment Coefficients C_m

(Fixed base, free top, triangular internal pressure)

$x \downarrow$ $\quad \dfrac{H^2}{2Rt_c} \rightarrow$	0.4	1.2	2.0	4.0	6.0	10.0	14.0	20.0	32.0	48.0
0.1 H	0.0005	0.0012	0.0010	0.0003	0.0001	0.0	0.0	—	—	—
0.2 H	0.0014	0.0042	0.0035	0.0015	0.0003	0.0	0.0	—	—	—
0.3 H	0.0021	0.0077	0.0068	0.0028	0.0008	0.0001	0.0	—	—	—
0.4 H	0.0007	0.0103	0.0099	0.0047	0.0019	0.0004	0.0	—	—	—
0.5 H	—0.0042	0.0112	0.0120	0.0066	0.0032	0.0007	0.0001	—	—	—
0.6 H	—0.0150	0.0090	0.0115	0.0077	0.0046	0.0019	0.0008	—	—	—
0.7 H	—0.0302	0.0022	0.0075	0.0069	0.0051	0.0029	0.0019	—	—	—
0.8 H	—0.0529	—0.0108	—0.0021	0.0023	0.0029	0.0028	0.0023	0.0015	0.0007	0.0
0.85 H	—	—	—	—	—	—	—	0.0014	0.0009	0.0001
0.90 H	—0.0816	—0.0311	—0.0185	—0.0080	—0.0041	—0.0012	—0.0001	0.0005	0.0007	0.0006
0.95 H	—	—	—	—	—	—	—	—0.0018	—0.0008	—0.0003
H	—0.1205	—0.0602	—0.0436	—0.0268	—0.0187	—0.0122	—0.0090	—0.0063	—0.0040	—0.0026

Table 12.4.5 Hoop Tension Coefficients, C_t
(Hinged base, free top, moment M_o applied at base)

x ↓ $\dfrac{H^2}{2Rt_c}$ →	0.4	1.2	2.0	4.0	6.0	10.0	14.0	20.0	32.0	48.0
0.0 H	2.70	1.06	−0.68	−1.87	−1.04	0.21	0.26	—	—	—
0.1 H	2.50	1.42	0.22	−.00	−0.86	−0.23	0.04	—	—	—
0.2 H	2.30	1.79	1.10	−0.08	−0.59	−0.64	−0.28	—	—	—
0.3 H	2.12	2.03	2.02	1.04	−0.05	−0.94	−0.76	—	—	—
0.4 H	1.91	2.46	2.90	2.47	1.21	−0.73	−1.29	—	—	—
0.5 H	1.69	2.65	3.69	4.31	3.34	0.82	−0.87	—	—	—
0.6 H	1.41	2.80	4.30	6.34	6.54	4.79	2.29	—	—	—
0.7 H	1.13	2.60	4.54	8.19	10.28	11.63	10.55	—	—	—
0.75 H	—	—	—	—	—	—	—	15.30	8.1	−0.7
0.80 H	0.80	2.22	4.08	8.82	13.08	19.48	23.50	25.9	23.2	14.1
0.85 H	—	—	—	—	—	—	—	36.9	45.9	45.1
0.90 H	0.44	1.37	2.75	6.81	11.41	20.87	30.34	43.3	65.4	87.2
0.95 H	—	—	—	—	—	—	—	35.3	63.6	103.0

Table 12.4.6 Moment Coefficients, C_m
(Hinged base, free top, moment M, applied at base)

x ↓ / $\frac{H^2}{2Rt_c}$ →	0.4	1.2	2.0	4.0	6.0	10.0	14.0	20.0	32.0	48.0
0.1 H	0.013	0.006	-0.002	-0.008	-0.005	0.0	0.0	—	—	—
0.2 H	0.051	0.027	-0.002	-0.026	-0.018	-0.002	0.0	—	—	—
0.3 H	0.109	0.063	0.012	-0.044	-0.040	-0.009	0.0	—	—	—
0.4 H	0.196	0.125	0.034	-0.051	-0.058	-0.028	-0.008	—	—	—
0.5 H	0.296	0.206	0.096	-0.034	-0.065	-0.053	-0.029	—	—	—
0.6 H	0.414	0.316	0.193	0.023	-0.037	-0.067	-0.059	—	—	—
0.7 H	0.547	0.454	0.340	0.150	0.057	-0.031	-0.060	—	—	—
0.8 H	0.692	0.616	0.519	0.354	0.252	0.123	0.048	-0.015	-0.062	-0.064
0.85 H	—	—	—	—	—	—	—	0.095	0.002	-0.049
0.90 H	0.843	0.802	0.748	0.645	0.572	0.467	0.387	0.296	0.178	0.081
0.95 H	—	—	—	—	—	—	—	0.606	0.515	0.424
H	1.000	1.000	1.000	1.000	1.000	1.000	1.000	1.000	1.000	1.000

Table 12.4.7 Hoop Tension Coefficients, C,
(Fixed base, free top, shear Q_0 applied at top)

$x \downarrow$ $\frac{H^2}{2Rt_c} \rightarrow$	0.4	1.2	2.0	4.0	6.0	10.0	14.0	20.0	32.0	48.0
0.0 H	−1.57	−3.95	−5.12	−7.34	−9.02	−11.67	−13.77	16.44	−20.84	−25.52
0.05 H	—	—	—	—	—	—	—	−9.98	−10.72	−10.82
0.10 H	−1.32	−3.17	−3.83	−4.73	−5.17	−5.43	−5.34	−4.90	−3.70	−0.26
0.15 H	—	—	—	—	—	—	—	−1.59	−0.04	−2.06
0.20 H	−1.08	−2.44	−2.68	−2.60	−2.27	−1.43	−0.68	0.22	1.26	1.66
0.30 H	−0.86	−1.79	−1.74	−1.10	−0.50	0.36	0.80	—	—	—
0.40 H	−0.65	−1.25	−1.02	−0.19	0.34	0.78	0.81	—	—	—
0.50 H	−0.47	−0.81	−0.52	0.26	0.59	0.62	0.42	—	—	—
0.60 H	−0.31	−0.48	−0.21	0.38	0.53	0.33	0.13	—	—	—
0.70 H	−0.18	−0.25	−0.05	0.33	0.35	0.12	0.00	—	—	—
0.80 H	−0.08	−0.10	0.01	0.19	0.17	0.02	−0.03	—	—	—
0.90 H	−0.02	−0.02	0.01	0.06	0.01	0.0	−0.01	—	—	—

Table 12.4.8 Moment coefficients, C_m
(Fixed base, free top, shear Q_0 applied at top)

x ↓ $\dfrac{H^2}{2Rt_c}$ →	0.4	1.2	2.0	4.0	60.0	10.0	14.0	20.0	32.0	4.80
0.05 H	—	—	—	—	—	—	—	0.032	0.028	0.024
0.1 H	0.093	0.082	0.077	0.068	0.062	0.053	0.046	0.039	0.029	0.024
0.15 H	—	—	—	—	—	—	—	0.033	0.020	0.011
0.20 H	0.172	0.132	0.115	0.088	0.070	0.049	0.036	0.323	0.011	0.003
0.25 H	—	—	—	—	—	—	—	0.014	0.004	0.000
0.3 H	0.240	0.157	0.126	0.081	0.056	0.029	0.017	—	—	—
0.4 H	0.300	0.164	0.119	0.063	0.036	0.012	0.004	—	—	—
0.5 H	0.354	0.159	0.103	0.043	0.018	0.002	—0.001	—	—	—
0.6 H	0.402	0.145	0.080	0.025	0.006	—0.002	—0.002	—	—	—
0.7 H	0.448	0.127	0.056	0.010	0.0	—0.002	—0.001	—	—	—
0.8 H	0.492	0.106	0.031	—0.001	—0.003	—0.002	—0.001	—	—	—
0.9 H	0.535	0.084	0.006	—0.010	—0.005	—0.001	0.0	—	—	—
H	0.578	0.062	—0.091	—0.019	—0.006	0.0	0.0	—	—	—

12.4.2. Tables 12.4.7 and 12.4.8 have been prepared to deal with the case in which the top of the wall is subjected to radial horizontal shear Q_o on account of the top dome. In this case T and M can be determined from

$$T = C_t \left(\frac{Q_o R}{H} \right) \tag{12.4.18}$$

and

$$M = C_m (Q_o H) \tag{12.4.19}$$

in which, Q_o is taken positive when it acts inwards. It should be noted that Tables 12.4.5 to 12.4.8 can still be used when the conditions at top and bottom are interchanged provided x is measured from bottom instead of from top.

12.4.3 Design and Specifications

Except for the simplest case of frictionless sliding base, the problem of cylindrical wall is statically indeterminate which leads to a complex distribution of internal forces even when several simplifying assumptions are made. The design tables given in Sec. 12.4.2 facilitate to a great extent the computations of internal forces for various edge conditions and distributions of internal pressures. These internal forces should be resisted adequately by the container wall by means of suitable prestress in circumferential and vertical directions. In addition, the design should satisfy the relevant specifications and codes of practice.

12.4.3.1 Shape and Proportions

The shape and proprtions are guided by structural, functional, aesthetic and economic considerations. The polygonal shape has the merit of having simpler and therefore cheaper formwork. However, it has the demerit of having bending moments and shear forces in vertical as well as horizontal planes. The circular shape corresponds to the best structural form due to the absence of bending moments in horizontal planes. However, it requires curved and therefore costlier formwork. Based on the experience of a large number of containers, the Preload Engineering Company of United States has given the economic proportions for circular cylindrical water tanks of different sizes. The economic ratio of radius to height (R/H) is approximately equal to 2. The hoop action is predominant in deep tanks except near the base whereas beam action dominates in shallow squatty tanks. The minimum wall thickness at top is approximately 120 mm for water tightness and other practical considerations. The wall thickness often increases towards the base, particularly in large tanks. The wall thickness may be kept constant for capacities upto 2000 m^3. The wall may have a thickness of approximately 0.2 m at top and 0.75 m at bottom for a capacity of the order of 36000 m^3.

12.4.3.2 Edge Conditions

In actual practice, the cylindrical wall is partially restrained at the top edge due to roof dome and its ring beam as well as at the bottom edge due to the base slab However, for simplicity in design, certain idealisation have tobe made in view of the complexity of the problem. The top edge of the container is generally treated as free edge except that the horizontal radial shear transferred by the roof dome is duly allowed for in the design. The edge condition at the base is assumed to depend upon the type of connection between the wall and the base slab. Three types of base connections are often adopted in practice.

(i) *Sliding base*

In a free, movable or sliding joint shown in Fig. 12.4.2 (a), the cylindrical wall is supported by a rubber or neoprene pad which allows horizontal movement relative to base slab by its shear deformation upto a critical value of 30°. The horizontal force against the movement developed by the pads depends upon its thickness and the shear modulus of clasticity of its material. To ensure water tightness, a water stop comprising a copper or PVC plate is provided as shown in Fig. 12.4.2 (a). In the sliding joint, the reinforcements in the wall and the base slab are completely delinked as shown in Fig. 12.4.2 (b). The main advantage of a sliding joint is the elimination of vertical bending moments and shear forces resulting in great simplification in analysis and design.

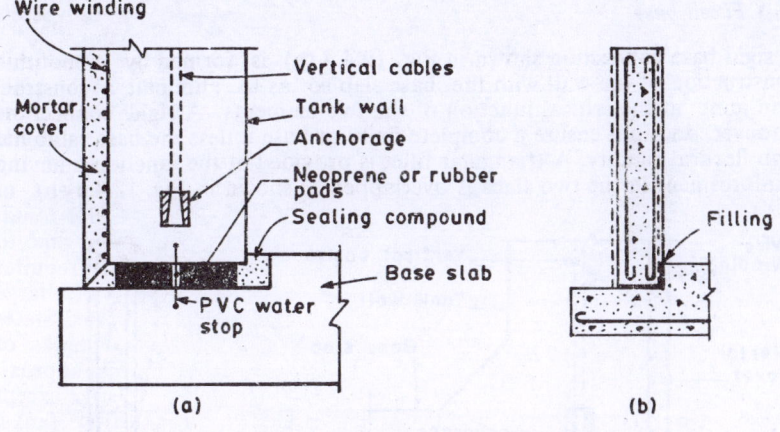

Fig. 12.4.2

(ii) *Hinged base*

In a hinged base connection, which is not very common in practice, the cylindrical wall rests in a groove in the base slab as shown in Fig. 12.4.3(a). The groove is packed with cement mortar after the completion of circumferential wire winding. Alternatively, the wall is supported on an annular bearing resting on the footing from which the base slab is isolated by a joint filled with a compressible material. In order to permit free rotation of the wall at its base, the reinforcement is provided so as to make the flexural capacity of the joint negligible as shown in Fig. 12.4.3 (b).

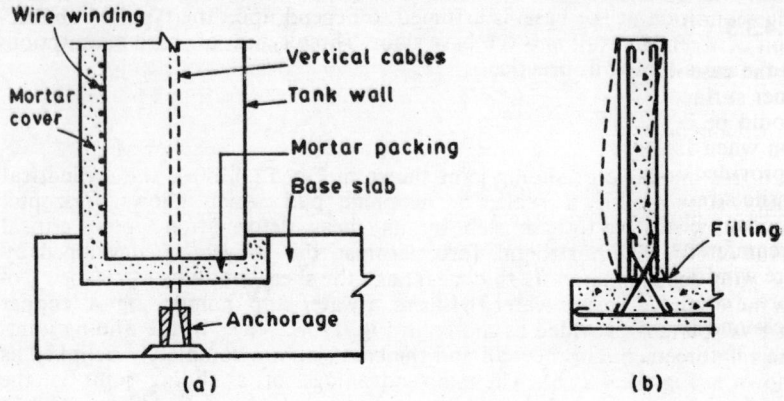

Fig. 12.4.3

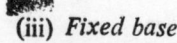

(iii) *Fixed base*

A rigid base connection shown in Fig. 12.4.4 (a) is formed by monolithic construction of the wall with the base slab so as to eliminate a construction joint at the critical junction of the two elements. A rigid connection, however, does not ensure a complete fixity at base unless the base slab has high flexural rigidity. A triangular fillet is provided at the junction and the reinforcement in the two slabs is overlapped as shown in Fig. 12.4.4 (b).

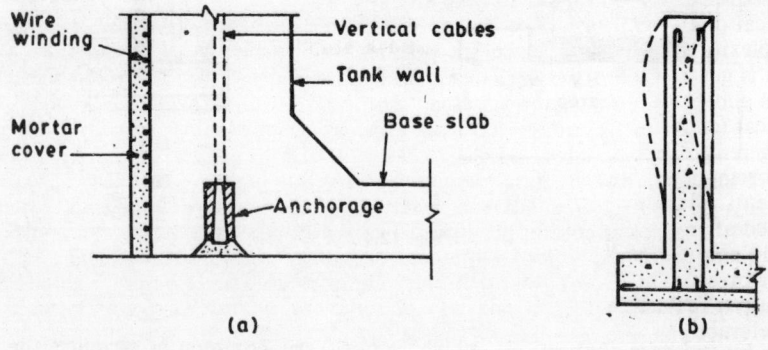

Fig. 12.4.4

The Indian Code IS: 3370 recommends that whenever the sliding or hinged conditions prevail at the base of the tank, any advantage due to restraining effect should be ignored. The floor slab may be designed as a thin membrane provided with a foundation ring under the wall. In the case of a sliding base connection, the vertical bending moment in the tank wall may be computed by taking the restraint at the base equal to one-half of that provided by a hinged base.

12.4.3.3 Circumferential Prestressing

In the case of liquid retaining structures, it is necessary to ensure that the inner surface is free from cracks. Hence the circumferential prestressing should be designed to ensure a certain amount of residual hoop compression when the tank is full. The free board may be ignored unless the tank is provided with adequate and dependable openings for overflow. The tensile stress in vertical direction due to partial winding of circumferential wire may be taken equal to 0.3 times the average hoop compression due to circumferential prestressing. In the circumferential prestress by means of wire winding, the prestressing wire should be protected by a pneumatic mortar coating of not less than 40 mm thickness. For aggressive atmosphere and in marine structures, the prestressing tendons should be placed inside the concrete and protected by means of effective grouting.

12.4.3.4 Vertical Prestressing

The object of vertical or longitudinal prestress is to counteract the tensile stresses caused by (i) partial winding of circumferential prestressing tendons, (ii) vertical bending moments due to hugging external pressure exerted by circumferential prestressing and (iii) vertical bending moments due to bursting pressure exerted by the contained material. The distribution of vertical moments depends upon the edge conditions and proportions of the cylindrical wall. As the edge conditions cannot be determined precisely, it is difficult to determine the vertical moments accurately, particularly for walls of varying thickness. Also the vertical moments caused by hugging external pressure and bursting internal pressure are opposite in sign. Consequently, it is difficult to take any advantage of eccentric prestress in vertical direction. Besides, eccentric vertical prestress gives rise to further complexity in analysis. Hence, in practice, vertical prestress of container walls is generally concentric. The vertical tendons are placed in preformed ducts and post-tensioned to produce the required vertical prestress. The vertical tendons may be placed near the inner and outer surfaces and staggered to reduce congestion. The container wall may require vertical reinforcing steel for connection at base and also to counteract the vertical moments due to circumferential prestressing if vertical prestressing is applied after circumferential prestress. The grouting of the vertical ducts should commence from the bottom and continued until the grout emerges under pressure at the top of the duct. The vertical prestressing is costly because it requires numerous anchorages. Hence a sliding base connection is preferable because it tends to reduce the vertical moments.

12.4.3.5 Permissible Stresses and Load Factors

The Indian Code IS: 3370 recommends a minimum residual hoop compression of 0.7 MPa when the tank is full. The code also provides that the tensile stress at any point, when the tank is empty, should not exceed 1 MPa. The worst condition of stresses resulting from the pressure of contained liquid, surrounding pressure if any, temperature, shrinkage and restraint from roof should be duly considered. The pressure exerted by the surrounding soil should be ignored whenever it produces a less severe stress condition unless there is no risk of slip in the embankment and the assumed soil pressure is the minimum. The shrinkage stresses become important when the tank is to be emptied and filled at frequent intervals or may be left empty for a long period. Similarly, the thermal stresses may be significant in structures exposed to direct sun. A special analysis for thermal stresses may be necessary for tanks containing hot fluids. The loss of prestress in tendons should be computed in accordance with the provisions of IS: 1343. However, the loss due to friction may be reduced to 80 per cent when linear prestressing is used to produce circumferential prestress.

Relatively smaller values of load factors against cracking and failure may appear to be justifiable for containers which have a lesser probability of overloading. The Indian Code IS: 3370 recommends a minimum load factor of 1.2 against cracking as compared to 1.25 recommended by the British Code. In computing the cracking load, the strength of concrete in direct tension may be taken equal to $0.267 \sqrt{f_{cu}}$ in which f_{cu} is the cube strength of concrete. The tensile strength in bending f_{cr} may be taken equal to twice the strength in direct tension. The Indian Code prescribes a minimum load factor equal to 2.0 against collapse as compared to 2.5 recommended by the British Code.

EXAMPLE 12.4.1

A cylindrical container having radius $R = 10$ m, heigh $H = 6$ m and wall thickness $t_c = 0.3$ m is fixed at base and free at top. It contains a gaseous chemical under a pressure of 0.2 MPa. It is prestressed circumferentially by means of 5 mm wire at a constant pitch of 10 mm under an effective stress of 950 MPa. Determine the maximum values of hoop tension T and bending moment M and the points at which they occur. Also determine the maximum circumferential and vertical tensile stresses and the minimum vertical concentric prestressing force required to eliminate tensile stress due to M. Also determine the vertical prestressing force required to overcome the tensile stress caused by partial circumferential winding.

Solution

Circumferential prestressing force per metre height,

$$P_c = \frac{\pi}{4} \times 5^2 \times \frac{1000}{10} \times 950 \times 10^{-6} = 1.866 \text{ MN}$$

Hence from Eq. (12.4.4), the external pressure due to prestress,

$$p_p = \frac{1.866}{10} = 0.1866 \text{ MPa}$$

From Eq. (12.4.2), the net internal pressure,

$$p = 0.2 - 0.1866 = 0.0134 \quad \text{MPa}$$

The form factor,

$$\frac{H^2}{2R\,t_c} = \frac{6^2}{2 \times 10 \times 0.3} = 6$$

The pressure distribution is rectangular. Hence using Tables 12.4.1 and 12.4.2, the maximum values of hoop tension and bending moment are

$$T\,(\text{max}) = 1.045 \times 0.0134 \times 10 = 0.14 \text{ MN/m}$$

$$M\,(\text{max}) = 0.0049 \times 0.0134 \times 6^2 = 0.00236 \text{ MN. m/m}$$

Maximum hoop tension and maximum bending moment occur respectively at 1.8 m and 3.6 m from the top of the container wall. Maximum tensile stress in longitudinal or vertical direction due to M (max),

$$f_l = \frac{0.00236}{\dfrac{1}{6} \times 1 \times 0.3^2} = 0.1573 \text{ MPa}$$

The longitudinal or vertical prestressing force per metre length required to eliminate tensile stress due to M (max),

$$P_l = 0.1573\,(1 \times 0.3) = 0.0472 \text{ MN/m}$$

Tensile stress in vertical direction due to partial winding,

$$f_l = 0.3 \left(\frac{P_c}{t_c \times 1} \right) = 0.3 \times \frac{1.866}{0.3 \times 1} = 1.866 \text{ MPa}$$

The longitudinal prestressing force per metre length required to eliminate tensile stress due to partial winding,

$$P_l = 1.866\,(1 \times 0.3) = 0.56 \text{ MN/m}$$

EXAMPLE 12.4.2

A prestressed concrete water tank has radius $R = 25$ m, height $H = 10$ m and wall thickness $t_c = 0.2$ m. It is prestressed circumferentially by means of 7 mm diameter wire under effective stress of 1020 MPa at a pitch which varies linearly from 14 mm at bottom to a large value at top. Determine the hoop tension T, bending moment M and corresponding stresses at 2 m from bottom assuming that the wall is free at top and is

(i) free at bottom
(ii) fixed at bottom and
(iii) hinged at bottom.

Solution

Maximum water pressure at base,

$$p_i = wH = 0.01 \times 10 = 0.1 \text{ MPa}$$

Circumferential prestressing force at base per metre height of wall,

$$P_c = \frac{\pi}{4} \times 7^2 \times \frac{1000}{14} \times 1020 \times 10^{-6} = 2.805 \text{ MN}$$

Using Eq. (12.4.4), external pressure due to prestressing,

$$p_p = \frac{2.805}{25} = 0.1122 \text{ MPa}$$

Hence the net internal pressure at base,

$$p_{max} = p_i - p_p = 0.1 - 0.1122 = -0.0122 \text{ MPa}$$

The minus sign shows that the net pressure is external. The distribution of p_i as well as p_p is triangular. Hence the distribution of net pressure p is also triangular. The form factor,

$$\frac{H^2}{2R\, t_c} = \frac{10^2}{2 \times 25 \times 0.2} = 10$$

(i) *Free at base*

In this case bending moment $M = 0$ and hoop tension at 2 m from base $(x = 0.8\, H)$,

$$T = (-0.0122 \times 0.8) \times 25 = -0.244 \text{ MN/m}$$

Hence the circumferential stress,

$$f_c = \frac{-0.244}{0.2 \times 1} = -1.22 \text{ MPa}$$

(ii) *Fixed at base*

Using Tables 12.4.3 and 12.4.4, the hoop tension T and bending moment M at 2 m from base $(x = 0.8\, H)$,

$$T = 0.440 \,(-0.0122) \times 25 = -0.1342 \text{ MN/m}$$
$$M = 0.0028 \,(-0.0122) \times 10^2 = -0.00342 \text{ MN.m/m}$$

Hoop stress due to T,

$$f_c = \frac{-0.1342}{0.2 \times 1} = -0.671 \text{ MPa}$$

Flexural stress due to M at outer face,

$$f_l = \frac{-0.00342}{\frac{1}{6} \times 1 \times 0.2^2} = -0.513 \text{ MPa}$$

The minus sign shows that the outer face is in compression. The inner face has a tensile stress of 0.513 MPa.

(iii) *Hinged at base*

The internal forces T and M for hinged base may be determined by using the principle of superposition. Using Table 12.4.4, the bending moment at fixed base,

$$M = (-0.0122)(-0.0122) \times 10^2 = 0.01488 \text{ MN.m}$$

Hence the obtain the hinged base condition, an equal and opposite moment should be applied at the base. Using Tables 12.4.5 and 12.4.6, the consequent internal forces,

$$T = 19.48 \, (- \, 0.01488) \times \frac{25}{10^2} = - \, 0.0725 \text{ MN/m}$$

$$M = 0.123 \, (- \, 0.01488) = - \, 0.00183 \text{ MN.m/m}$$

Combining these internal forces with the internal forces computed in (ii) above, the net internal forces,

$$T = - \, 0.1342 - 0.0725 = - \, 0.2067 \text{ MN/m}^{\cdot}$$

$$M = - \, 0.00342 - 0.00183 = - \, 0.00525 \text{ MN.m/m}$$

Hoop stress due to T,

$$f_c = \frac{- \, 0.2067}{0.2 \times 1} = - \, 1.0335 \text{ MPa}$$

Flexural stress due to M at outer face,

$$f_l = \frac{- \, 0.00525}{\frac{1}{6} \times 1 \times 0.2^2} = - \, 0.7875 \text{ MPa}$$

EXAMPLE 12.4.3

A prestressed concrete water tank fixed at base and free at top has radius $R = 20$ m, height $H = 10$ m and wall thickness $t_c = 0.25$ m. It carries a radial outward shear force of 20 kN per meter at the top of the cylindrical wall on account of the roof dome. Determine T and M at 6 m from top. Also determine the maximum values of these internal forces and their locations.

Solution

The form factor,

$$\frac{H^2}{2R \, t_c} = \frac{10^2}{2 \times 20 \times 0.25} = 10$$

As the radial shear force is directed outward, $Q_o = - \, 20$ kN/m. Using Tables 12.4.7 and 12.4.8, the internal forces at 6 m from top ($x = 0.6 \, H$) on account of Q_o,

$$T = 0.33 \, (- \, 0.02) \times \frac{20}{10} = - \, 0.0132 \text{ MN/m}$$

$$M = (- \, 0.002) \, (- \, 0.02) \times 10 = 0.0004 \text{ MN.m/m}$$

The maximum values of internal forces are,

$$T_{\max} \text{ (positive)} = - \, 11.67 \, (- \, 0.02) \times \frac{20}{10}$$

$$= 0.4668 \text{ MN at top}$$

$$T_{max} \text{ (negative)} = 0.78 \; (-\; 0.02) \times \frac{20}{10}$$

$$= 0.0312 \text{ MN/m at 4 m from top}$$

$$M_{max} \text{ (positive)} = -\; 0.002 \; (-\; 0.02) \times 10$$

$$= 0.0004 \text{ MN.m/m approximately at 7 m from top}$$

$$M_{max} \text{ (negative)} = 0.053 \; (-\; 0.02) \times 10$$

$$= 0.0106 \text{ MN.m/m at 1 from top}$$

EXAMPLE 12.4.4

The vertical wall of a prestressed concrete water tank having radius $R = 36$ m, height $H = 12$ m and thickness $t_c = 0.5$ m is fixed at base and free at top. The effective circumferential prestressing force increases linearly from 0.9 MN/m at top to 4.5 MN/m at bottom. The effective vertical prestressing force applied concentrically is 0.8 MN/m. Determine the net circumferential and vertical stresses when the tank is

(i) empty and
(ii) full.

Solution

The form factor,

$$\frac{H^2}{2R\,t_c} = \frac{12^2}{2 \times 36 \times 0.5} = 4$$

(i) *Empty tank*

Minimum external pressure due to prestressing at top,

$$p_p \text{ (min)} = \frac{P_c}{R} = \frac{0.9}{36} = 0.025 \text{ MPa}$$

Maximum external pressure due to prestressing at bottom,

$$p_p \text{ (max)} = \frac{4.5}{36} = 0.125 \text{ MPa}$$

Distribution of external pressure due to prestressing is trapezoidal which can be broken into rectangular and triangular distributions.

(a) *Rectangular distribution*

As the pressure is external, $p_{max} = -\; 0.025$ MPa (constant). Using Tables 12.4.1 and 12.4.2, the internal forces at 9.6 m $(x = 0.8 \; H)$ from top,

$$T = 0.322 \; (-\; 0.025) \times 36 = -\; 0.2898 \text{ MN/m}$$

$$M = -\; 0.0013 \; (-\; 0.025) \times 12^2 = 0.00468 \text{ MN.m/m}$$

(b) Triangular distribution

As the pressure is external, $p_{max} = -0.1$ MPa. Using Tables 12.4.3. and 12.4.4, the internal forces at 9.6 m ($x = 0.8\,H$) from top,

$$T = 0.210\,(-0.1) \times 36 = 0.756\ \text{MN/m}$$
$$M = 0.0023\,(-0.1) \times 12^2 = -0.03312\ \text{MN.m/m}$$

Combining (a) and (b), the net internal forces are

$$T = -0.2898 - 0.756 = -1.0458\ \text{MN/m}$$
$$M = 0.00468 - 0.03312 = -0.02844\ \text{MN.m/m}$$

Hoop stress at 9.6 m from top,

$$f_c = \frac{-1.0458}{0.5 \times 1} = -2.092\ \text{MPa}$$

As the vertical prestressing force is 0.8 MN/m, the longitudinal stress at the outer face,

$$f_l = \frac{-0.02844}{\frac{1}{6} \times 1 \times 0.5^2} - \frac{0.8}{0.5 \times 1} = -2.283\ \text{MPa}$$

Similarly longitudinal stress at the inner face,

$$f_l = \frac{0.02844}{\frac{1}{6} \times 1 \times 0.5^2} - \frac{0.8}{0.5 \times 1} = -0.917\ \text{MPa}$$

(ii) Tank full

Maximum water pressure at base,

$$p_{max} = 0.01 \times 12 = 0.12\ \text{MPa}$$

As the pressure distribution is triangular, the internal forces using Tables 12.4.3 and 12.4.4,

$$T = 0.210\,(0.12) \times 36 = 0.9072\ \text{MN/m}$$
$$M = 0.0023\,(0.12) \times 12^2 = 0.03974\ \text{MN.m/m}$$

Hoop stress due to water pressure,

$$f_c = \frac{0.9072}{0.5 \times 1} = 1.814\ \text{MPa}$$

Longitudinal stress at outer face due to water pressure,

$$f_l = \frac{0.03974}{\frac{1}{6} \times 1 \times 0.5^2} = 0.954\ \text{MPa}$$

Longitudinal compressive stress at inner face is 0.954 MPa. Combining these stresses due to water pressure with the stresses due to prestressing computed in (i), the net stresses for tank full condition,

9788123901534

$$f_c = -2.092 + 1.814 = -0.278 \text{ MPa}$$
$$f_l \text{ (outer face)} = -2.283 + 0.954 = -1.329 \text{ MPa}$$
$$f_l \text{ (inner face)} = -0.917 - 0.954 = -1.871 \text{ MPa}$$

EXAMPLE 12.4.5

Design a prestressed concrete water tank of circular cylindrical shape for a capacity of 20.8 million litres. The minimum permissible residual compressive stress in the tank full condition is 0.7 MPa. The minimum permissible values of load factors against cracking and collapse are 1.2 and 2.0. The tank wall may be assumed to be fixed at base and free at top and to receive a radial shear force 75 kN/m due to the roof dome. Use M-40 concrete. The maximum permissible compressive stress at stress transfer is 15 MPa.

Solution

Adopting the economical ratio, $(R/H) = 2$ and allowing a free board of 0.5 m,

$$\pi (2H)^2 (H - 0.5) = 20800$$

Hence, $H = 12$ m approximately and therefore $R = 24$ m. Adopting an average wall thickness, $t_c = 0.3$ m tentatively, the form factor,

$$\frac{H^2}{2R \, t_c} = \frac{12^2}{2 \times 24 \times 0.3} = 10$$

(i) Circumferential prestressing

Ignoring the free board, maximum water pressure at bottom,

$$p_{max} = 0.01 \times 12 = 0.12 \text{ MPa}$$

As the pressure distribution is triangular, the maximum hoop tension using Table 12.4.3,

$$T = 0.608 (0.12) \times 24 = 1.751 \text{ MN/m}$$

This maximum hoop tension occurs at $0.6 H = 7.2$ m approximately from top. Using Table 12.4.7, the hoop tension at 7.2 m from top on account of shear force applied at top,

$$T = 0.33 (-0.075) \times \frac{24}{12} = -0.0495 \text{ MN/m}$$

Hence net hoop tension at 7.2 m from top,

$$T = 1.751 - 0.0495 = 1.7015 \text{ MN/m}$$

Hence hoop stress at 7.2 m from top,

$$f_c = \frac{1.7015}{1 \times 0.3} = 5.672 \text{ MPa}$$

In order to obtain a residual compressive stress of 0.7 MPa, the prestressing force should produce a hoop compression equal to $5.672 + 0.7 = 6.372$ MPa. Assuming that the circumferential prestress p_p increases linearly from

0 at top to maximum value at bottom and using Table 12.4.3, the hoop tension at 7.2 m from top,

$$T = -\ 0.608 \times p_p\ (\text{max}) \times 24 = -\ 14.592\ p_p\ (\text{max})$$

Hence,

$$-\ 14.592\ p_p\ (\text{max}) = -\ 6.372 \times 0.3 \times 1$$

or

$$p_p\ (\text{max}) = 0.131\ \text{MPa}$$

The circumferential prestressing force per metre height at base,

$$P_c = 0.131 \times 24 = 3.144\ \text{MN/m}$$

This circumferential prestressing force may be applied using either of the following two techniques:

(a) Circular prestressing (Preload method)

Adopting 8 mm diameter high tensile steel wire with an effective stress of 1020 MPa with a pitch of 16 mm at base,

$$P_c = \frac{\pi}{4} \times 8^2 \times 1020 \times \frac{1000}{16} \times 10^{-6} = 3.206\ \text{MN/m}$$

As it is slightly greater than the minimum required value, it may be adopted. The pitch of circumferential prestressing wire may be increased to a large value, say 160 mm, at the top.

(b) Linear prestressing

Adopting Magnel Blaton system of post-tensioning and using cables having 16 wires of 5 mm diameter at an effective stress of 1020 MPa, the prestressing force per cable is 0.3206 MN. Hence the number of cables per metre height,

$$n_c = \frac{3.144}{0.3206} = 9.81$$

Hence the spacing of the cables at bottom may be 100 mm giving the prestressing force per metre at bottom,

$$P_c = 16 \times \frac{\pi}{4} \times 5^2 \times 1020 \times \frac{1000}{100} \times 10^{-6} = 3.206\ \text{MN/m}$$

The spacing may be increased linearly to a large value, say 1000 mm, near the top. The cables may be stretched from both ends against vertical pylons or buttresses placed at 120° along the circumference of the tank. The cables may be housed inside concrete in preformed square ducts of 55 mm side.

(ii) Longitudinal or vertical prestressing

When the tank is full, the external pressure due to prestressing and the internal water pressure approximately balance each other. Consequently, the maximum bending moment in vertical direction occurs in the empty tank condition. Maximum external pressure at base due to circumferential pre-stressing,

$$p_p \text{ (max)} = \frac{3.206}{24} = 0.1336 \text{ MPa}$$

Using Table 12.4.4, the maximum bending moment,

$$M \text{ (max)} = -0.0122\,(-0.1336) \times 12 = 0.0196 \text{ MN.m/m}$$

This maximum bending moment occurs at base. Maximum tensile stress in longitudinal or vertical direction,

$$f_l = \frac{0.0196}{\dfrac{1}{6} \times 1 \times 0.3^2} = 1.307 \text{ MPa}$$

In order to obtain a residual compressive stress of 0.7 MPa, the longitudinal or vertical prestressing is required to produce a compressive stress of 1.307 + 0.7 = 2.007 MPa. According to Indian Code, the longitudinal bending stress due to partial winding.

$$f_l = 0.3\,f_c = 0.3 \times \frac{P_c}{\eta t_c}$$

Taking loss factor, $\eta = 0.75$,

$$f_l = 0.3 \times \frac{3.206}{0.75 \times 0.3} = 4.275 \text{ MPa}$$

As the stress due to partial winding is greater than the stress due to vertical bending moment, the longitudinal prestress may be designed for the larger stress of 4.275 MPa. Hence vertical prestressing force per metre,

$$P_l = 4.275 \times 0.3 \times 1 = 1.2825 \text{ MN/m}$$

Adopting Magnel Blaton system of post-tensioning and using cables having 16 wires of 5 mm diameter at an effective stress of 1020 MPa, the number of cables per meter of circumference,

$$n_l = \frac{1.2825}{0.3206} = 4$$

Hence total number of longitudinal or vertical cables,

$$N_l = 2\pi \times 24 \times 4 = 603.4$$

The longitudinal prestress may be achieved by placing 604 cables at uniform spacing concentrically along the circumference.

(iii) *Untensioned reinforcement*

Untensioned reinforcement is required to resist stresses due to shrinkage of concrete and thermal changes. Providing 0.2 per cent reinforcement, the area per metre length,

$$A_s = \frac{0.2}{100} \times 1000 \times 300 = 600 \text{ mm}^2$$

Hence 8 mm diameter bars at 160 mm spacing may be provided near both faces in the longitudinal and circumferential directions with a clear cover of 20 mm.

(iv) *Stresses and load factors*
Initial circumferential stress at base in empty tank condition,

$$f_c = \frac{3.206}{0.75 \times 0.3 \times 1} = 14.25 \text{ MPa}$$

Initial longitudinal stress at base in empty tank condition,

$$f_l = 4.275 \text{ MPa} \qquad \text{from (ii) above}$$

These stresses are acceptable as they are smaller than the permissible value at stress transfer. The strength of concrete in direct tension,

$$f_t = 0.267 \sqrt{40} = 1.7 \text{ MPa}$$

As maximum hoop tension due to water pressure for fixed base condition occurs at 7.2 m from top at which the hoop compressive stress due to pre-stressing is 6.372 MPa from (i) above, the hoop tension required to cause cracking,

$$T_{cr} = (6.372 + 1.7) \times 1 \times 0.3 = 2.422 \text{ MN/m}$$

Maximum hoop tension due to water pressure at 7.2 m from top is 1.751 MN/m from (i) above. Hence load factor against cracking,

$$\frac{T_{cr}}{T_{max}} = \frac{2.422}{1.751} = 1.383$$

At ultimate stage, concrete cracks extensively and entire hoop tension may be assumed to be carried by circumferential steel. The foregoing analysis based on elastic theory is no longer valid. Assuming that the entire water pressure is resisted by hoop action, the maximum hoop tension at base,

$$T_{max} = 0.01 \times 12 \times 24 = 2.88 \text{ MN/m}$$

The ultimate strength of circumferential steel per metre height of wall near the base,

$$T_u = \frac{\pi}{4} \times 8^2 \times 1850 \times 10^{-6} \times \frac{1000}{16} = 5.814 \text{ MN/m}$$

Hence load factor against collapse,

$$\frac{T_u}{T_{max}} = \frac{5.814}{2.880} = 2.042$$

(v) *General*

The average wall thickness, $t_c = 0.3$ m assumed tentatively is found to be satisfactory. However, for theoretical and practical reasons, it is better to adopt varying thickness. The wall may have a thickness of 0.2 m at top and 0.4 m at bottom. For the sake of simplicity, however, the computations may be based on average thickness, $t_c = 0.3$ m. The foregoing computations are based on gross-section of the wall ignoring the presence of ducts and cover coating. For greater precision, the net area of concrete should be considered in the computation of stresses caused by prestressing. On the other hand, the transformed area should be used in computing the stresses due to water pressure. A triangular fillet of 0.3 m side may be provided at the junction of wall and base slab. To obtain a fixed base condition, the two elements should be cast monolithically and sufficient reinforcement connecting the two elements should be provided.

12.5 NUCLEAR PLANT PRESSURE VESSELS

One of the latest major applications of prestressed concrete is in the domain of reactor vessels and containment vessels for nuclear power plants. Prestressed concrete has clearly demonstrated its efficiency and economy in the construction of both types of pressure vessels. The spectacular pace of development in the field of prestressed concrete nuclear plant pressure vessels was the result of innovation, technical skill and joint effort of engineers working in a variety of disciplines. The pace of development may be gauged from the fact that three generations of containment vessels were witnessed in a relatively brief span of fifteen years commencing with early sixties.

The prestressed concrete reactor vessel (PCRV) is the main pressure vessel which performs the dual function of being a pressure container for the reactor coolant and a biological shield against high energy X-rays, γ-rays and neutrons. The integrated PCRV accommodates the gas cooled reactor core containing stainless steel clad fuel elements with carbon dioxide as the coolant at working pressures of 3.5 to 4.2 MPa and entire gas circuit including boilers, stand pipes and air blowers within a single pressure vessel. The shape such as spherical or cylindrical for a PCRV is determined by the manner in which the boiler units are located with respect to the reactor core. A steel liner plate, a thermal shield and a cooling system are provided at the inner surface of PCRV to reduce thermal gradient across prestressed concrete wall. PCRVs have been used in France and United Kingdom mainly for gas cooled reactors.

The prestressed concrete containment vessel (PCCV), which was first developed in France and subsequently used in United States, India and other countries, is a secondary containment and the final barrier between nuclear fission products and the environment. PCCVs have been widely used for light water cooled reactors in which the working pressures are of a lower order than gas cooled reactors. A PCCV comprises a cylindrical concrete vessel capped with a dome and supported on a flat foundation slab. A steel liner plate or non-metallic coating like vinyl/epoxy paint is provided at the inner surface.

In addition to the general advantages of prestressed concrete enumerated in Sec. 1.14, the following advantages relate particularly to nuclear plant pressure vessels:

(i) The high degree of static redundancy of prestressing tendons in pressure vessels ensures a ductile behaviour which eliminates the possibility of sudden catastrophic failure.

(ii) The unbonded tendon system allows continuous surveillance and monitoring the entire useful life.

(iii) Prestressing induces triaxial compression in concrete of pressure vessels and thereby increases the cracking and ultimate strengths.

(iv) Prestressing facilitates anchoring of high pressure piping directly to the concrete through penetration sleeves.

(v) Prestressed concrete offers excellent radiation absorption characteristics.

(vi) The structural response of prestressed concrete pressure vessels during proof loading is more predictable because the concrete remains uncracked and tendon stresses remain practically unchanged. Consequently, the testing risk is reduced considerably.

(vii) Prestressing permits greater optimisation in the selection of vessel diameter, volume and pressure. Consequently, larger vessels with greater internal pressure are feasible with the aid of prestressing.

12.5.1 Structural Form

The current structural form of containment vessels for nuclear power plants may be seen in the light of three distinct generations of these vessels witnessed in United States.

(i) *First generation*

Prior to the development of first generation prestressed concrete containment vessels, steel vessels of various configurations were used for nuclear power plants having capacities ranging from 50 to 400 MW. The shell thickness without stress relieving ranged up to 40 mm. The need to double the installed capacity brought the steel shell thickness almost to its practical limit. Some of the early installations used a steel containment vessel and a separate nuclear shield building comprising concrete cylinder and precast concrete hemi-spherical dome. For increased installed capacity and more effective nuclear shielding, it was considered prudent to combine the containment vessel and shield building into a single composite steel lined concrete vessel.

In the process of evaluation of concrete vessels in mid-sixties, reinforced concrete vessels with cylindrical wall and hemi-spherical dome having thicknesses of the order of 1.35 m and 0.75 m were studied. These vessels were found to present some unique design and construction problems. As a result, the designers in United States turned their attention towards prestressed concrete which was already being used for reactor vessels in some European countries. After a thorough feasibility study, prestressed concrete was adopted as the main construction material for ten containment vessels in the first generation. These vessels comprised a cylindrical wall together with six equispaced vertical buttresses, an ellipsoidal dome and ring beam at the junction of the two shells. The vessel was also provided with stiffened steel liner plate of 6 mm to 10 mm thickness to serve as a leak-barrier and also as interior form.

(ii) *Second generation*

With the experience and confidence gained with the design and construction of first generation vessels, it became possible to make some major structural changes for the second generation vessels for plants having installed capacity upto 900 MW. The number of vertical buttresses was reduced from six to three. Larger prestressing tendons having twice as many wires as in the 90 − 6 mm wire tendons used in the first generation were adopted. In order to reduce the number of anchorages, 240° tendons were placed and tensioned so that a set of three consecutive tendons formed a complete ring. Each prestressing tendon had an ultimate strength as high as 9 MN and could therefore deliver an initial force of 7.2 MN which reduced to 5.4 MN due to losses occurring during 40 year—useful life of the structure.

(iii) *Third generation*

Major structural changes incorporated in the third generation vessels for plants having installed capacities above 1000 MW included replacement of low profile ellipsoidal dome by a hemi-spherical dome and elimination of ring girder at the top of cylindrical wall. The number of anchorages were reduced and tensioning operations were simplified by combining the tendons in the dome with the vertical tendons in the cylindrical wall to obtain inverted U-shaped tendons. The Trojan Nuclear Plant was the first in the third generation containment vessels having hemi-spherical dome and 70 inverted U-shaped vertical tendons and 150 hoop tendons. The inverted U-shaped tendons were divided into two parallel sets of tendons oriented at 90° to each other. The central tendon in each set was placed in the diametric plane of the dome as well as the cylindrical wall. The 240° hoop tendons were anchored against three equispaced vertical buttresses. The saving in respect of number of tendons was 625 as compared to first generation and 115 as compared to second generation.

12.5.2 Analysis and Design

The complexity of ʃanalysis is enhanced due to the presence of several penetrations or openings for functional reasons. Hence both theoretical and experimental techniques of analysis were utilized. The finite element technique was found to be most suitable theoretical approach. The theoretical results have been verified by extensive tests on scale models. In addition, the actual prototype structures have been kept under constant surveillance and monitoring devices to study their actual performance.

The analysis and design of PCRV which is an integral part of a complex system is greatly influenced by the overall design of the plant. The main stages to be considered in a PCRV design are as follows:

(i) During construction before application of prestress
(ii) Non-pressurised vessel after prestressing
(iii) Non-pressurised vessel after initial heating
(iv) Pressurised vessel after initial heating
(v) Pressurised vessel at start of operations
(vi) Pressurised vessel during useful life
(vii) Non-pressurised vessel during useful life

The design philosophy for a PCRV has the following main features:

(i)   The vessel should have an elastic response for all possible combinations of design loads. The stresses developed in the materials under test pressure should be restricted to the values specified in relevant codes.

(ii) The vessel should exhibit ductile behaviour, i.e. gradual failure under overloads accompanied by large deformations before ultimate failure.

(iii) The ultimate pressure should have a minimum load factor of 2.5 on the design pressure.

The following are the main stages for stress analysis of containment vessels:

(i)   Prior to application of prestress
(ii) During application of prestress in a predetermined sequence
(iii) Under sustained prestress and appropriate losses
(iv) Under service loads
(v)   Under overload conditions, particularly at cracking and failure

The second stage is generally the most critical among the first four stages. During this stage, the structure is subjected to most severe stresses which are never attained later under service conditions. Due to the risk of nuclear pollution in the event of failure, the structure should also be designed for environmental hazards such as tornado loading and missile or aircraft impact.

12.5.3   Construction Features

The trend is towards the use of larger prestressing tendons with capacities as high as 12 MN. The prestressing systems which have been used extensively are Freyssinet, BBRV and Gifford-Uddal CCL. The prestressing tendons are deflected either side of the opening in order to maintain their continuity. The number or spacing of tendons in regions pierced by an opening is kept the same as in the neighbouring uninterrupted region. In order to keep the tendons under constant surveillance and to monitor their stress continuously, the cable ducts are filled with grease or petroleum based jelly instead of cement grout. This also facilitates restressing of tendons at any subsequent time. It also enables the removal of a tendon for inspection and its replacement if the corrosion has gone beyond an acceptable limit. The containment vessel has several small and large openings or penetrations. The largest opening for the equipment hatch is approximately 6 m in diameter. An opening of approximately 9 m × 12 m is required to provide access for construction equipment and material. This opening is closed subsequently by suitable placement of tendon sheathing, reinforcing bars and high grade concrete.

The overall dimensions, material quantities and construction features of nuclear pressure vessels vary over a wide range. However, some particulars of a typical PCRV and a PCCV may be mentioned here. The Hartlepool

PCRV in Britain for a nuclear plant with an installed capacity of 622 MW has a design internal pressure of 4.53 MPa. The cylindrical wall of 13.12 m diameter is provided with a cap of 5.60 m thickness. The jacking force of each Gifford Uddal CCL tendon comprising 28 strands of 18 mm diameter was 7.92 MN. The Kalpakkam nuclear power plant in Tamilnadu (India) with a pressurised heavy water moderated and cooled reactor of 200 MW installed capacity has a PCCV designed for a working pressure of 0.115 MPa. It has a double containment system comprising a prestressed concrete inner cylinder of 600 mm thickness and an outer masonry wall of 711 mm thickness separated by an annular space of 1 m width. The vessel having inner diameter of 39.7 m and height 52.8 m utilised 3.5 MN of prestressing steel.

12.5.4   Current and Future Trends

The feasibility of having only two buttresses located on diametrically opposite sides is being examined. The two buttress system may have $360°$ hoop tendons, the ends of which are anchored in the same buttress. The consecutive tendons may have their anchorages staggered between the two buttresses. In view of the severity of environmental pollution resulting from structural failure, future installations would require permanent devices for surveillance and monitoring of tendon stresses, possible corrosion and malfunctioning of any constituent. A complete inspection and testing may be prescribed at 1, 3 and 5 years of construction and every 5 years thereafter. Accordingly ten inspections may be prescribed towards the useful life of forty years.

12.6   RING BEAMS

A ring beam is often provided at the springing of a dome to stiffen the edge and to resist the tensile force produced by the horizontal outward component of meridional force in the dome at its springing. When the ring beam is supported uniformly over its entire length on a vertical wall, all other internal forces in the ring beam except hoop tension may be assumed to be absent. However, when the ring beam is supported only at discrete points by means of columns, it is subjected to flexure, shear and torsion in addition to hoop tension. In either case, hoop tension is an essential component of the internal forces in ring beam supporting a dome. The hoop tension in the ring beam depends upon the geometry of the dome and its loading. Treating the dome as a thin shell, the internal forces may be determined by membrane theory. Considering the dome as a surface of revolution, generated by revolving a curved meridian called generator about the vertical axis passing through the crown, any point on the dome surface may be identified by the slope of the dome surface ψ to norizontal and its meridian θ measured from a reference meridian. The slope ψ varies from zero at crown to ψ_s at the springing. The meridian θ varies from 0 to $360°$ similar to the meridians on the globe. Consider a small element DEFG cut out by two consecutive latitudes and meridians in the neighbourhood of point P as shown in Fig. 12.6.1 (a). The membrane forces acting on the element shown in Fig. 12.6.1 (b) may be determined from equations of equilibrium of the element. In the case of axi-symmetry in which the loads and support reactions are functions of ψ only and are independent of θ, the membrane shears $N_{\psi\theta}$ and $N_{\psi\theta}$ are absent from consideration of symmetry. The magnitude and nature (tensile or compressive) of the membrane direct forces N_ψ and N_θ depend upon the shape of dome, its loading and the position of the point considered defined by its coordinates (ψ, θ). In a spherical dome of radius R_d, the meridonal force N_ψ and the circumferential force N_θ per unit width on account of self weight w_D per unit surface area and uniform live load w_L per unit projected (horizontal) area may be expressed as

$$N_\psi = - R_d \left(\frac{w_D}{1 + \cos \psi} + \frac{w_L}{2} \right) \qquad (12.6.1)$$

$$N_\theta = - R_d \cos \psi \, (w_D + w_L \cos \psi) - N_\psi' \qquad (12.6.2)$$

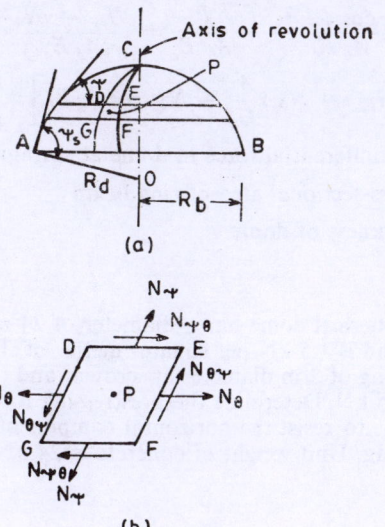

(a)

(b)

Fig. 12.6.1

While the meridional force N_ψ is always compressive $(-ve)$, the circumferential force N_θ is compressive in the upper part of the dome and may become tensile in the lower part. If the dome has an opening at the crown extending upto latitude $\psi = \psi_o$ and a vertical load W_o is distributed uniformly along the edge of the opening, the membrane forces are

$$N_\psi = - N_\theta = - \frac{W_o}{2 R_d \sin^2 \psi} \qquad (12.6.3)$$

Equation (12.6.3) shows that while N_ψ is always compressive, N_θ is always tensile. The meridonial force at springing N_{ψ_s} obtained by putting $\psi = \psi_s$ in Eq (12.6.1) or (12.6.3) exerts a horizontal outward force equal to $(-N_{\psi_s} \cos \psi_s)$ on the supporting structure. Equations (12.6.1) to (12.6.3) are valid only when the dome is simply supported and the supporting structure provides the horizontal reaction required to resist the horizontal component of N_{ψ_s}. If an effective prestressing force, $P = (-N_{\psi_s} \cos \psi_s) R_b$ is applied along the rim of the dome, the supporting structure has to carry only the vertical component of N_{ψ_s}. The dome prestressed in this manner may be placed on a masonry wall which is incapable of resisting any lateral (horizontal) force. In actual practice, however, the prestressing force P is generally too large to be accommodated in the rim of the dome. Consequenty, it becomes necessary to provide a ring beam which makes the problem statically indeterminate. The hoop tension in the ring beam depends upon the manner in which the dome is connected to it. Consider the simple case in which the meridonal force passes through the centroid of the concentrically prestressed ring beam. For compatibility of circumferential deformation, the circumferential strain of the ring beam should be equal to that of the edge of the dome. Hence,

$$ -\frac{N_{\psi_s} \cos \psi_s \, R_b}{A_b E_c} - \frac{P}{A_b E_c} = \frac{N_{\theta_s} - \nu N_{\psi_s}}{t_d E_c} $$

or $$ P = - \left[(N_{\theta_s} \eta - N_{\psi_s}) \frac{A_b}{t_d} + N_{\psi_s} \cos \psi_s R_b \right] \tag{12.6.4} $$

where N_{θ_s} = circumferential force in dome at springing

A_b = cross-sectional area of ring beam

t_d = thickness of dome

EXAMPLE 12.6.1

A 150 mm thick spherical dome has a diameter of 24 m at base and a rise of 6 m. The live load is 1.5 kN per square metre of horizontal area. The dome has an opening of 2 m diameter at crown and is provided with a lantern weighing 25 kN. Determine the prestressing force to be provided at the rim of the dome to resist the horizontal component of meridional compression at springing. Unit weight of concrete is 24 kN/m³.

Solution

Self weight of dome,

$$ w_D = 0.15 \times 24 = 3.6 \text{ kN/m}^3 $$

Radius of dome R_d may be determined from

$$ 6 (2R_d - 6) = 12 \times 12 $$

or $$ R_d = 15 \text{ m} $$

Rise of dome h_o above the level of opening,

$$ h_o (2 \times 15 - h_o) = 1 \times 1 $$

or $$ h_o = 0.0334 \text{ m} $$

Hence surface area of dome lost due to opening,

$$ \pi \times 2 \times 15 \times 0.0334 = 3.149 \text{ m}^2 $$

Effective weight of lantern may be taken as

$$W_o = 25 - 3.149 \times 3.6 = 13.66 \text{ kN}$$

$$\sin \psi_s = \frac{12}{15} = 0.8 \quad \text{and} \quad \cos \psi_s = 0.6$$

Using Eq. (12.6.1) and (12.6.3), the meridional force at springing due to combined effect of self weight, live load and weight of lantern,

$$N_{\psi_s} = -15 \left(\frac{3.6}{1 + 0.6} + \frac{1.5}{2} \right) - \frac{13.66}{2\pi \times 15 \times 0.64}$$

$$= -45.23 \text{ kN/m}$$

The effective prestressing force P at rim of the dome should be equal to the hoop tension due to horizontal component of N_{ψ_s}.

$$P = 45.23 \times 0.6 \times 12 = 325.7 \text{ kN}$$

EXAMPLE 12.6.2

A ring beam of rectangular section 0.3×0.6 m is provided at the springing of dome of Example 12.6.1 in such a way that N_{ψ_s} passes through its centroid. Determine the concentric prestressing force to be provided in the ring beam to resist hoop tension due to meridional thrust. Poisson's ratio for concrete, $\nu_c = 0.15$.

Solution

Using Eq. (12.6.2) and (12.6.3), the circumferential force at springing due to combined effect of self weight, live load and weight of lantern,

$$N_{\theta_s} = -1.5 \times 0.6 (3.6 + 15 \times 0.6) + 45 + \frac{13.66}{2\pi \times 15 \times 0.64}$$

$$= -40.5 + 45 + 0.23 = 4.73 \text{ kN/m}$$

Using Eq. (12.6.4), the prestressing force in ring beam,

$$P = -\left[\frac{4.73 - 0.15(-45.23) \times 0.3 \times 0.6}{0.15} + (-45.23) \times 0.6 \times 12 \right]$$

$$= 311.8 \text{ kN}$$

REFERENCES

12.1 ACI Committee 344, *Design and construction of circular prestressed concrete structures*, ACI 344 R, American concrete Institute, Detroit, 1970.

12.2 ACI Committee 344, *Design and construction of circular wire and strand wrapped prestressed concrete structures*, ACI 344-1R, American Concrete Institute, Detroit. 1989.

12.3 ACI Committee 344, *Design and construction of circular prestressed concrete structures with circumferential tendons*, ACI 344-2R, American Concrete Institute, Detroit, 1989.

12.4 Brondum Nielsen, T., *Prestressed tanks*, Journal of the American Concrete Institute, July-August 1985, pp. 500-9.

12.5 Crom, J.M., *Design of prestressed tanks*, Transactions, ASCE, Vol. 117, 1952, pp. 89-118.

12.6 Curtis, A.R. and Cowan, H.J., *Design of longitudinal cables in circumferentially wound prestressed tanks*, Magazine of Concrete Research, Vol. 5, March 1954, pp. i23-6.

12.7 Doanides, P.J., *Some notes on the prestressed concrete pipes of the Veeranam aqueduct for the water supply of Madras*, International Symposium on Prestressed Concrete Pipes, Poles, Pressure Vessels and Sleepers, Proceedings Vol. 2, Madras 1972 pp. 3/1-40.

12.8 Evans, R H., *Applications of prestressed concrete to water supply and drainage*, Proceedings of the Institution of Civil Engineers (London), Part 3, Vol. 4, 1955, pp. 725-75.

12.9 FIP, *Reaommendations for the design of prestressed concrete oil storage tanks*, FIP CACA, Wexham Springs, Slough SL3 6 PL, England, 1977.

12.10 FIP, *The design and construction of prestressed concrete reactor vessels*, FIP, England, 1977.

12.11 Grom, J.M., *Design of prestressed tanks*, Proceedings of ASCE, Separate No. 37, October 1950.

12.12 Halligan, D.W., *Prestressed concrete nuclear plant containment structurers*, Journal of the Prestressed Concrete Institute, September-October 1976, pp. 158-75.

12.13 Hendrickson, J.G., *Prestressed concrete pipe*, Proceedings of First U.S. Conference on Prestressed Concrete, Massachusetts, 1951.

12.14 Joshi, N.G., *Prestressed concrete pipes-state of art*, International Symposium on Prestressed Concrete Pipes, Poles, Pressure Vessels and Sleepers, Proceedings Vol. 1, Madras, 1972, pp. 0/1-29.

12.15 Ooykaas, G.A.P. *Discontinuity in prestressed concrete pipes*, Magazine of Concrete Research. Vol. 3, March 1952, pp. 131-8.

12.16 Ooykaas, G.A.P., *Prestressed concrete pipes*, Journal of the Institution of Water Engineers, Vol. 6, 1952, pp. 85-109.

12.17 Preload Engineers Company Inc., *Design of preload tanks*, Bulletin T-19. New York.

12.18 PCA, *Circular concrete tanks without prestressing*, Concrete Information Series ST-57, Portland Cement Association, Skokie, III, 1957, 32 p.

12.19 PCI Committee on Precast Prestressed Concrete Storage Tanks, *Recommended Practice for precast prestressed concrete circular storage tanks*, Prestressed Concrete Institute, Chicago, 1987.

12.20 Robert, M.E. and Lebelle, M.P., *Prestressed concrete reservoir at Orleans*, Cement Concrete Association Library Translation No. 10, London, 1949.

12.21 Vessey, J.V. and Preston, R.L., *A critical review of code requirements for circular prestressed concrete reservoirs*, Paris: FIP, 1978.

PROBLEMS

12.1 A 20 m long beam of rectangular section 150 × 300 mm is prestressed concentrically by means of 25 wires of 4 mm diameter carrying an initial stress of 1200 MPa and an effective stress of 960 MPa. Determine the minimum tensile force which may act at stress transfer and at final stage in order to restrict the initial instantaneous contraction to 4 mm and final long-time contraction to 8.09 mm. Take $C_c = 1.2$, $\epsilon_{sh} = 0.00025$, $E_c = 36000$ MPa and m = 6. Assume that 50 per cent of the total load is permanent.

12.2 A post-tensioned beam of rectangular section 300 × 600 mm is prestressed concentrically by means of six cables each having 12 wires of 5 mm diameter. The initial and final stresses in the wires are 1150 and 980 MPa. Determine the tensile force at decompression, cracking and collapse. Take $f_{pu} = 1650$ MPa, $f_t = 1.8$ MPa and m = 6.5.

12.3 If the tension member of Prob. 12.2 has to carry a service load of 1.30 MN, determine the residual compressive stress under service load. Also determine the overload factors for decompression, cracking and collapse.

12.4 Fifteen metre long tension member has to carry a service load of 200 kN at stress transfer and an additional service load of 1300 kN at the final stage. Design a suitable post-tensioned prestressed concrete section for the member. The instantaneous contraction at stress transfer and eventual long-time contraction are not to exceed 5 mm and 10 mm respectively. The restraining force imposed by the adjoining structure may be taken as 150 kN per mm of contraction. The creep coefficient $C_c = 1.5$ and shrinkage strain, $\epsilon_{sh} = 0.00015$. The residual compressive stress under full service load should not be less than 0.7 MPa. The overload factors at decompression and at cracking should not be less than 1.1 and 1.25 respectively. Also check the adequacy of the section in respect of collapse.

12.5 A prestressed concrete pressure pipe has internal radius $R_i = 375$ mm, core thickness $t_c = 40$ mm and protective shell thickness $t_s = 20$ mm. It is prestressed by 17 wires of 4 mm diameter in longitudinal direction and 5 mm diameter spiral at 60 mm pitch in circumferential direction. The effective prestress in spiral wire is 980 MPa. Determine the residual compressive stress in core concrete under internal service pressure of 0.8 MPa. Also determine the internal pressure which will cause decompression, cracking and bursting of the pipe. Take modular ratios m = 6 and $m_s = 0.5$. The tensile strength of concrete $f_t = 2.1$ MPa and ultimate stress of circumferential steel $f_{puc} = 1800$ MPa.

12.6 Determine the longitudinal tensile stress in the pipe of Prob. 12.5 taking the hoop tensile stress due to partial winding equal to 0.284 times the average hoop compression in core concrete due to circumferential prestressing. The stresses in longitudinal and circumferential steel during wire winding are 1020 and 1200 MPa respectively.

12.7 Determine the net longitudinal stresses at the bottom of the inner surface of core concrete of the pipe of Prob. 12.5 due to weight of water, self weight and a load of 1.83 kN/m due to backfill. The pipe line has simple supports at 7.5 m spacing. Take unit weight of pipe material as 24 kN/m³. The effective stress in longitudinal steel is 950 MPa.

12.8 Design a prestressed concrete pressure pipe of 0.9 m internal diameter to resist a service pressure of 1.2 MPa. The pipe line has an earth fill of 0.6 m and live load of 10 kN/m². The bedding angle $\alpha_b = 90°$. The minimum margins of safety against cracking and bursting are 1.5 and 2.0. The minimum permissible compression under service pressure is 0.7 MPa.

12.9 A prestressed concrete gas container with a sliding base connection and free top has radius, $R = 20$ m and wall thickness, $t_c = 0.15$ m. The effective circumferential prestressing force is 1.575 MN/m. Determine the gas pressure for (i) residual compressive stress of 1 MPa, (ii) cracking and (iii) bursting. The tensile strength of concrete, $f_t = 1.8$ MPa. The breaking strength of circumferential steel is 2.835 MN per metre height of the container wall.

12.10 A prestressed concrete water tank with fixed base and free top has radius $R = 16$ m, height $H = 8$ m and wall thickness $t_c = 0.2$ m. The effective circumferential prestressing force increases linearly from zero at top to 1.6 MN/m at bottom. The effective concentric vertical prestressing force is 0.6 MN/m. Determine the net circumferential and vertical stresses at 5.6 m from top when the tank is (i) empty and (ii) full.

12.11 A container wall hinged at base and free at top has radius $R = 18$ m, height $H = 9$ m and thickness $t_c = 0.375$ m. Determine the hoop tension and vertical bending moment at 5.4 m from top due to bending moment $M_o = 1.2$ MN.m per metre applied at base of the wall.

12.12 Solve Prob. 12.11 if the wall is hinged at top, free at bottom and moment M_o is applied at top.

12.13 A container wall fixed at base and free at top has radius $R = 7$ m, height $H = 7$ m and thickness $t_c = 0.25$ m. Determine the ring tension and vertical bending moment at 4.9 m from top of wall due to outward radial shear force of 50 kN/m applied at top.

12.14 Solve Prob. 12.13 if the wall is fixed at top, free at bottom and the shear force Q_o is applied at bottom.

12.15 Design the cylindrical wall of a prestressed concrete water tank of 15 million litre capacity adopting provisions of Indian code. Use M-40 concrete. Wall may be assumed to be fixed at base and free at top.

13

Compression Members

13.1 INTRODUCTION

The prestressing of a compression member appears like carrying coal to New Castle. If there were a compression member whose service load is a compressive force exactly coincident with the member axis, it is sure enough to preclude the possibility of buckling and it carries no other action such as flexure, shear or torsion, then the application of prestress may be inconsequential if not harmful. As prestress induces a compressive force in the member, it takes away a part of the load carrying capacity of the member. The larger the prestressing force, the greater is the reduction in the load carrying capacity. For instance, if the intensity of prestress is 4, 6 or 8 MPa, the reduction in the working load capacity of an axially loaded short compression member for a working stress in concrete equal to 12.5 MPa is approximately 32, 48 and 64 per cent as compared to the corresponding non-prestressed member. In a flexural or tension member, the application of service load elongates the prestressing tendons and consequently increases the prestressing force. In contrast to these members, the service load in a compression member causes contraction of the tendons resulting in a reduction or loss of prestressing force. With increasing axial load, the prestressing force decreases progressively until at the crushing stage, the prestressing force may become a small part of its initial value. Consequently, the reduction in load capacity on account of prestress is relatively small at the limit state of crushing as compared to that at the working stage as illustrated by Example 13.1.1.

In actual practice a compression member rarely carries pure compression. Columns of building frames and those of rigid frames for bridges carry significant bending moments on account of their rigid connections to beams. Even those columns, which are not monolithic with the beams, carry eccentric load which causes bending moment in addition to axial compression. All codes of practice specify a certain minimum eccentricity for the design of columns. The object of prestressing a compression member is actually to resist the tensile stresses due to bending moment which invariably accompanies axial compression. It is, therefore, evident that the benefit of prestressing a compression member increases directly with increase in bending moment or the eccentricity of loading. Although a pile, which is another prominent compression member, is generally subjected to axali compression due to service load, it may develop substantial tensile stresses and may even crack during handling and driving unless it is prestressed. Piers and abutments of bridges are subjected to significant bending moments due to water currents or pressure exerted by retained soil. Prestress may be used with advantage to counteract these bending moments. It may also be used to resist the uplift caused by subsoil water pressure. Compression members may also be prestressed to resist the tensile stresses caused by shrinkage of concrete and thermal changes.

Can prestress cause buckling of a slender member ? To answer this crucial question, it is necessary to distinguish between

(i) internally prestressed member with bonded tendons,

(ii) internally prestressed member with unbonded tendons and

(iii) externally prestressed member.

Figure 13.1.1 (a) shows a straight slender member subjected to axial compression W. If the member is given a small lateral displacement as indicated by the broken line, the member springs back to its original straight configuration as soon as the disturbing force causing lateral displacement is removed. However, the member does not do so when the axial load is increased to W_E called Euler's crippling or buckling load. In this case, the bending moment $W_E \, e_0$ caused on account of the eccentricity e_0 of the buckling load W_E is responsible for preventing the return of the member to its original form. If the applied load were to remain coincident with the axis of the member, there would be no eccentricity of loading and consequently no buckling would occur. Figure 13.1.1 (b) shows an internally prestressed slender member with bonded tendon coincident with the axis of the member. When the member is given a small lateral displacement as indicated by the broken line, the bonded tendon is also displaced equally. Consequently, it remains coincident with the member axis and no eccentricity is produced. It follows that prestress will not induce buckling in this case. On the contrary, prestress will prevent buckling because the bonded tendon, which elongates further due to buckling tends to snap back to its original straight configuration. Figure 13.1.1 (c) shows an internally prestressed slender member with unbonded tendon. Here the member can buckle without displacing the tendon until it comes in contact with the inner surface of the duct in which the unbonded tendon is housed. Beyond this stage, the tendon is displaced along with the member thereby preventing further increase in tendon eccentricity and at the same time elongating the tendon further. Consequently prestress will prevent buckling as soon as the tendon comes in contact with the duct. The extent of possible buckling due to prestress depends upon the gap or clearance between the tendon and duct surfaces. Figure 13.1.1 (d) shows an externally prestressed member. As the tendon is placed outside the body of the member, it remains straight and undisplaced during buckling of the member. The situation is identical to that of buckling due to external load as in Fig. 13.1.1 (a). Hence external prestress with tendon laterally unattached to the body of the member causes buckling in the same manner as the external load. If the relatively infrequent external prestress and unbonded tendons are excluded from consideration, it may be concluded that prestress does not induce buckling but, on the contrary, discourages buckling. It is likely that prestress prevents lateral buckling of beams but conclusive tests in this regard are lacking.

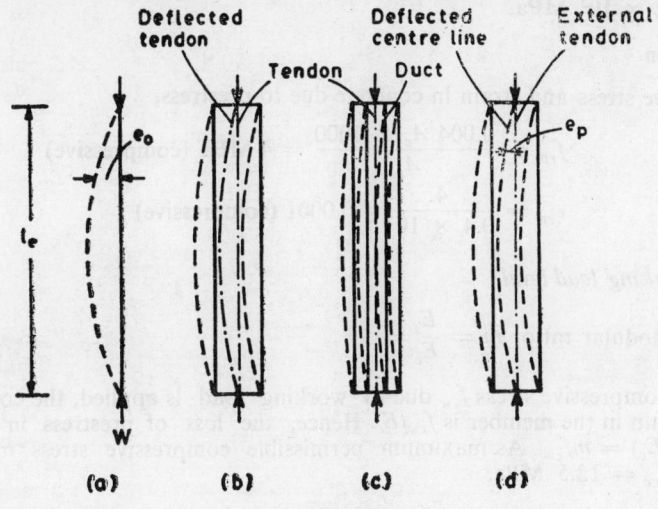

Fig. 13.1.1

Should a compression member be prestressed concentrically or eccentrically? The answer depends upon the type of member and nature of its buckling. Full advantage of eccentric prestress can be taken only when the loadings at all critical stages of loading produce tension on the same face of the member. This is possible only when the bending moments at all critical loading stages are non-reversible and are of the same sign. As these conditions are not commonly satisfied, the compression members are frequently prestressed concentrically. Columns in buildings and bridges and piles are generally subjected to reversible bending moments and are, therefore, generally prestressed concentrically. Sometimes it may be advantageous to adopt a combination of concentric and eccentric prestress to counteract reversible and irreversible bending moments. For instance, a sheet pile may be prestressed concentrically to resist reversible moments during handling and prestressed eccentrically to resist irreversible moments under service loading. Sometimes it may be prudent to apply concentric or eccentric prestress temporarily and to withdraw it when it is no long needed. For instance, a long and heavy precast column may be temporarily prestressed during handling and the prestress may be withdrawn when the column is placed in its final position.

EXAMPLE 13.1.1

An axially loaded short compression member contains 0.4 per cent prestressing steel carrying an effective stress of 1000 MPa. Determine the reduction in load capacity of the member on account of prestress at (i) working load level and (ii) crushing load level. The permissible direct compressive stress is 12.5 MPa. At crushing the average stress in concrete is 30 MPa and strain is 0.003. The moduli of elasticity, $E_c = 0.4 \times 10^5$ MPa and $E_s = 2 \times 10^5$ MPa.

Solution

Effective stress and strain In concrete due to prestress,

$$f_{cp} = \frac{0.004 \, A_c \times 1000}{A_c} = 4 \text{ MPa (compressive)}$$

$$\epsilon_{cp} = \frac{4}{0.4 \times 10^5} = 0.0001 \text{ (compressive)}$$

(i) *Working load level*

Modular ratio, $m = \dfrac{E_s}{E_c} = 5$

When compressive stress f_{cw} due to working load Is applied, the compressive strain in the member is f_{cw}/E_c. Hence, the loss of prestress in steel is $(E_s f_{cw}/E_c) = mf_{cw}$. As maximum permissible compressive stress in concrete, $\sigma_{cc} = 12.5$ MPa,

$$\left(4 - \frac{mf_{cw} A_p}{A_c}\right) + f_{cw} = 12.5$$

or $f_{cw} = 8.67$ MPa

Net stress in prestressing steel at working load level,

$$f_p = 1000 - mf_{cw} = 1000 - 5 \times 8.67 = 956.65 \text{ MPa}$$

Reduction in load capacity due to prestress,

$$\frac{12.5 - 8.67}{12.5} \times 100 = 30.64 \text{ per cent}$$

(ii) *Crushing load level*

Crushing strain of concrete, $\epsilon_u = 0.003$. As effective prestrain, $\epsilon_{cw} = 0.0001$, the crushing load causes strain equal to $0.003 - 0.0001 = 0.0029$. Hence net stress in prestressing steel at the crushing load level,

$$f_p = 1000 - 0.0029 \times 2 \times 10^5 = 420 \text{ MPa}$$

Reduction in load capacity due to prestress,

$$\frac{420 \, A_p}{30 \, A_c} \times 100 = 5.6 \text{ per cent}$$

13.2 COLUMNS

Prestressed concrete columns like reinfoced ones are called either short or long depending upon the ratio of effective length to least lateral dimension or the minimum radius of gyration. They are loaded uniaxially or biaxially depending upon whether the load is eccentric with respect to one or both principal axes. The design of prestressed concrete columns which constitute the most prominent category of compression members is based on extensive test results which are reflected in the current code provisions.

13.2.1 Short Columns

As short columns do not buckle, all sections may be assumed to have the same internal forces and identical distribution of stresses and strains.

13.2.1.1 Precracking Range

The elastic theory based on uncracked section and linear distribution of stresses and strains may be used with reasonable accuracy to determine the stresses and curvatures in the precracking range at stress transfer and under working loads. Consider a rectangular section of a reinforced and prestressed concrete column having untensioned steel A_s and A_s' and prestressing steel A_{p1} and A_{p2} as shown in Fig. 13.2.1. The stresses in concrete at top and bottom faces at the stage of stress transfer,

$$f_{Ti} = \frac{-P_i}{A_{ti}} + \frac{(A_{p1}\,e_{p1i} - A_{p2}\,e_{p2i})\,f_{pi}\,y_{Ti}}{I_{ti}} \qquad (13.2.1)$$

$$f_{Bi} = \frac{-P_i}{A_{ti}} - \frac{(A_{p1}\,e_{p1i} - A_{p2}\,e_{p2i})\,f_{pi}\,y_{Bi}}{I_{ti}} \qquad (13.2.2)$$

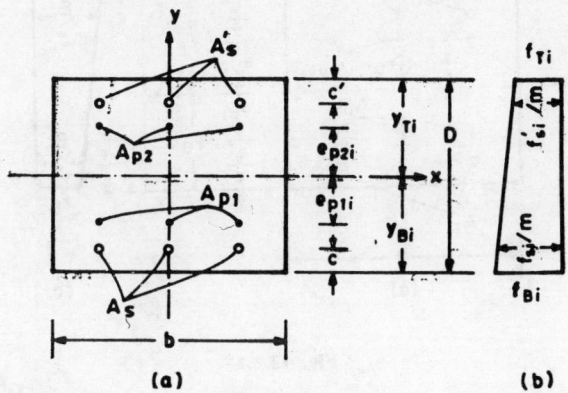

Fig. 13.2.1

The stress diagram at stress transfer is shown in Fig. 13.1.1 (b). From the stress diagram, the stresses in untensioned steel A_s and A_s' may be determined by proportion.

$$f_{si} = m\left[f_{Ti} + (f_{Bi} - f_{Ti})\,\frac{(D-c)}{D}\right] \qquad (13.2.3)$$

$$f_{si}' = m\left[f_{Ti} + (f_{Bi} - f_{Ti})\,\frac{c'}{D}\right] \qquad (13.2.4)$$

where $P_i = (A_{p1} + A_{p2})\,f_{pi}$

f_{pi} = initial stress in prestressing steel at stress transfer.

The area A_{ti} of the transformed section at initial stage may be expressed as

$$A_{ti} = bD + (m - 1)(A_s + A'_s) \qquad (13.2.5)$$

The symbol I_{ti} represents the moment of inertia of the section A_{ti} about its centroidal axis xx.

The stresses in the column section on account of load W acting at an eccentricity e with respect to centroidal axis xx shown in Fig. 13.2.2 may be determined by using transformed section A_t in which the area of pre-stressing steel is also included. The initial stresses due to prestressing are reduced by the loss factor η due to loss of prestress. Hence the net stresses in concrete at top and bottom faces may be expressed as

$$f_T = \eta f_{Ti} - \frac{W}{A_t} - \frac{We y_T}{I_t} \qquad (13.2.6)$$

$$f_B = \eta f_{Bi} - \frac{W}{A_t} + \frac{We y_B}{I_t} \qquad (13.2.7)$$

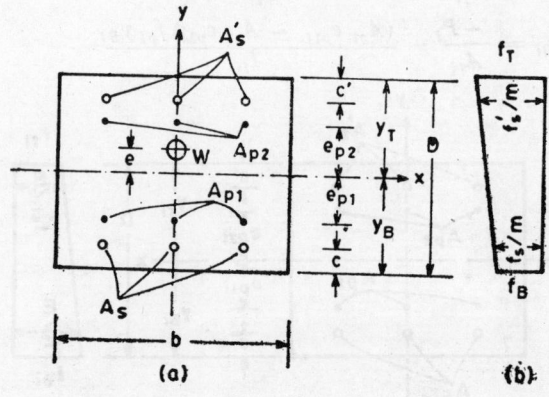

Fig. 13.2.2

Using the stress diagram of Fig. 13.2.2 (b). the net stresses in untensioned steel A_s and A'_s may be expressed as

$$f_s = m \left[f_B + (f_T - f_B) \frac{c}{D} \right] \qquad (13.2.8)$$

$$f'_s = m \left[f_B + (f_T - f_B) \frac{(D - c')}{D} \right] \qquad (13.2.9)$$

As the curvature is equal to the difference of strains at top and bottom faces divided by D, the initial curvature at the stage of stress transfer,

$$\Phi_i = \frac{f_{Bi} - f_{Ti}}{DE_c} \qquad (13.2.10)$$

Similarly, the net curvature under load W,

$$\Phi = \frac{f_B - f_T}{DE_c} \qquad (13.2.11)$$

The foregoing equations are simplified considerably if the cross-section of the column is symmetrical and the prestressing force is concentric. In this case, $A_s = A_s'$, $c = c'$, $A_{p1} = A_{p2}$. $e_{p1} = e_{p2}$ and $y_T = y_B = \dfrac{D}{2}$. A symmetrical section with concentric prestress is preferred when the eccentricity of loading e is reversible.

The effect of shrinkage of concrete is included at least partially in the loss factor η which accounts for the reduction of prestressing force from its initial value P_i to its effective value P. When the cross-section of the column has symmetrical reinforcing and prestressing steel, the shrinkage of concrete may be assumed to be uniform. The restrained uniform shrinkage induces a compressive stress $\epsilon_{sh}' E_s$ in both types of steel where the restrained shrinkage strain ϵ_{sh}' may be taken approximately equal to the free shrinkage strain ϵ_{sh}. The restrained shrinkage induces tensile stress in concrete which may be determined by equating total tension in concrete to total compression in steel. When steel is placed unsymmetrically, the more heavily reinforced side offers greater restraint to shrinkage. Consequently, it suffers smaller shrinkage strain as compared to the other side causing a strain gradient. The strains and the corresponding stresses may be determined by equating the force and bending moment in concrete to those in steel because shrinkage can induce only a self-equilibrating system of forces.

The application of service load W causes a further loss of prestress in a compression member unlike flexure or tension member in which the service load induces an increase in prestressing force. In addition to instantaneous loss of prestress, the compression member suffers loss due to creep of concrete if service load W is permanent. The loss of prestress increases with increase in W. It may become a major part of the initial prestress as the limit state of collapse is approached.

The effect of creep of concrete is included again only partially in the loss factor η as it reflects only the reduction in prestressing force. The increase in compressive stress in untensioned steel due to creep of concrete may be allowed for by visualising a softened concrete with a smaller modulus of elasticity E_c. Hence the larger stress in untensioned steel may be determined by taking a larger value of modular ratio, $m = E_s/E_c$. The appropriate increase in m to reflect the creep effect evidently depends upon creep coefficient C_c which in turn depends upon several factors as discussed in Sec. 4.3. Under normal conditions C_c ranges from 1 to 2. Hence m may be increased by a factor ranging from 2 to 3 to allow for the effect of creep of concrete.

13.2.1.2 Post-cracking Range

A prestressed concrete column section cracks under overload condition. The stresses and curvatures at any stage of loading in the post-cracking range may be determined using Navier's hypothesis, i.e. plane sections remain plane after loading. The equations of equilibrium between internal and external forces as well as internal and external moments are sufficient for the determination of stresses and resulting curvatures. Appropriate stress-strain curves for the constituent materials may be used in the computation of internal forces.

Consider the rectangular section of a prestressed concrete short column subjected to an eccentric load W acting at a distance x_1 from the top face as shown in Fig. 13.2.3(a). The corresponding strain diagram on the basis of linear distribution of strains in shown in Fig. 13.2 3.(b). Concrete below the neutral axis located at a distance x from top face may be assumed to have cracked and therefore rendered ineffective. The magnitude and position of the resultant compressive force in concrete in the compression zone C_c depends upon the shape of stress-strain curve of concrete. The tensile forces T_{p1} and T_{p2} in prestressing steels A_{p1} and A_{p2} depend upon their effective prestresses and the shape of the stress-strain curve of prestressing steel. Similarly, the tensile force T_s in untensioned steel A_s and compressive force C_s' in untensioned steel A_s' depend upon their respective stress-strain curves. The condition of equilibrium of vertical forces may be expressed by equating the external and internal vertical forces.

$$C_s' - T_{p2} + C_c - T_{p1} - T_s = W \qquad (13.2.12)$$

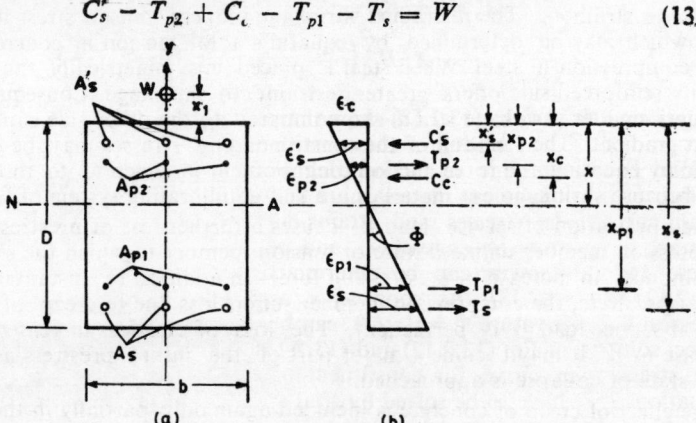

Fi_g. 13.2.3

For the equilibrium of moments, the resultant of all internal forces must coincide with the line of action of W, i.e. the resultant of C_s', T_{p2}, C_c, T_{p1} and T_s should pass through the point 0 at which the eccentric load W acts. Hence taking moments about 0, the condition of equilibrium of moments may be expressed as

$$C_s' (x_1 + x_s') - T_{p2} (x_1 + x_{p2}) + C_c (x_1 + x_c)$$
$$- T_{p1} (x_1 + x_{p1}) - T_s (x_1 + x_s) = 0 \qquad (13.2.13)$$

The internal forces acting on the cross-section of the column may be expressed as

$$C_s' = A_s' E_{ss}' \epsilon_s' = A_s' E_{ss}' \frac{x - x_s'}{x} \epsilon_c \qquad (13.2.14)$$

$$T_{p2} = A_{p2} E_{ps2} (\epsilon_{pe2} - \epsilon_{p2})$$

$$= A_{p2} E_{ps2} \left[\epsilon_{pe2} - \frac{x - x_{p2}}{x} \epsilon_c \right] \tag{13.2.15}$$

in which E_{ps2} is the secant modulus of prestressing steel A_{p2} at the level of its net strain $(\epsilon_{pe2} - \epsilon_{p2})$ and ϵ_{pe2} is the effective prestrain in A_{p2}.

$$C_c = \int_0^{\epsilon_c} \epsilon E_{cs} \cdot dA \tag{13.2.16}$$

in which E_{cs} is the secant modulus of concrete at the level of strain ϵ which varies from 0 to ϵ_c.

$$T_{p1} = A_{p1} E_{ps1} (\epsilon_{pe1} + \epsilon_{p1})$$

$$= A_{p1} E_{ps1} \left[\epsilon_{pe1} + \frac{x_{p1} - x}{x} \epsilon_c \right] \tag{13.2.17}$$

in which E_{ps1} is the secant modulus of prestressing steel A_{p1} at the level of its net strain $(\epsilon_{pe1} + \epsilon_{p1})$ and ϵ_{pe1} is the effective prestrain in A_{p1}.

$$T_s = A_s E_{ss} \epsilon_s$$

$$= A_s E_{ss} \left(\frac{x_s - x}{x} \epsilon_c \right) \tag{13.2.18}$$

Equations (13.2.14) to (13.2.18) show that the internal forces depend upon ϵ_c, x and the secant moduli of the constituent materials. If the magnitude of load W and its line of action defined by x_1 are given and the secant moduli of the constituent materials are known from their respective stress-strain curves, the stresses and strains depend on two unknowns, viz. the extreme fibre strain in concrete ϵ_c and the depth of the neutral axis x. These two unknowns can be determined from equilibrium equations (13.2.12) and (13.2.13). As soon as ϵ_c and x are determined, all strains, stresses and curvature $\Phi = (\epsilon_c/x)$ can be determined readily. However, a direct solution of Eq. (13.2.12) and (13.2.13) is difficult, particularly when the stress-strain curve for constituent materials are non-linear. Hence the equations may have to be solved by trial and error or by iteration.

The foregoing analysis shows that the problem of eccentrically loaded prestressed concrete short column in the post-cracking range involves only four variables, viz. eccentric load W, its eccentricity $e = (0.5 D + x_1)$, extreme fibre strain in concrete ϵ_c and depth of neutral axis x provided that the cross-sectional properties and stress-strain curves of constituent materials are given. If any two of the four variables are known, the remaining two may be determined from Eq. (13.2.12) and (13.2.13). As pointed out earlier, the determination of ϵ_c and x for given values of W and x_1 is difficult and may require trial and error. On the other hand, W and x_1 can be determined directly from Eq. (13.2.12) and (13.2.13) if ϵ_c and x are given.

13.2.1.3 Ultimate Strength

As the eccentric load W on the column section increases in the post-cracking range, the extent of cracking increases progressively marked by upward shift of neutral axis and decreasing area of compression zone. The consequent increase in compressive strain in concrete at extreme fibre in compression zone leads eventually to crushing of concrete at its ultimate strain ϵ_u resulting in collapse of the column. The magnitude of collapse load W_u acting at a given eccentricity can be determined by using the analysis of Sec. 13.2.1.2. Out of the four variables, viz. W, x_1, ϵ_c, and x, two variables viz. x_1 and ϵ_c are specified. The remaining two viz. W and x can be determined from Eq. (13.2.12) and (13.2.13). At the limit state of collapse $\epsilon_c = \epsilon_u$, $x = x_u$ and $W = W_u$. An interative procedure described by the following steps may be found to be convenient in the computation of collapse load W_u

1. Assume a reasonable value of depth of neutral axis x_u.
2. Taking $\epsilon_c = \epsilon_u$ and $x = x_u$ assumed in step 1, determine all internal forces, viz. C'_s, T_{p2}, C_c, T_{p1} and T_s from prescribed stress-strain curves of constituent materials, strain diagram and geometrical properties of the section. The effective prestrain should be added while computing the forces T_{p1} and T_{p2} in prestressing steel.
3. Determine collapse load $W = W_u$ from Eq. (13.2.12).
4. Determine x_1 and hence eccentricity e_u of W_u from Eq. (13.2.13).
5. Compare eccentricity e_u computed in step 4 with the given eccentricity. If the computed eccentricity is smaller than the given eccentricity, the next cycle of iteration should be carried out with a value of x_u smaller than the value assumed in step 1 of first cycle. The cycles should be repeated until the computed eccentricity agrees with the given value to the desired degree of accuracy.

13.2.1.4. Interaction Diagrams

An eccentric load with eccentricity ranging from zero to infinity represents the whole spectrum of combination of axial load and bending moment. The interaction between axial compression and bending moment is best illustrated by means of interaction diagrams. The interaction between the two internal forces is characterized by a dual criterion. Whereas axial compression produces a compressive stress, the bending moment induces both compressive and tensile stresses. If the failure of column is controlled by a limiting compressive stress, the two types of internal forces assist each other in causing failure. Consequently, the presence of any one of the two internal forces reduces the magnitude of other internal force required to cause failure. On the other hand, if the failure is controlled by a limiting tensile stress, the two types of internal forces oppose each other in causing failure. Consequently, the presence of any one of the two internal forces increases the magnitude of the other internal force required to cause failure. The interaction relationship may be studied both at the working load level and at the limit state of collapse.

(i) Working Load Level

Let σ_{cc} and σ_{ct} be the limiting or permissible compressive and tensile stresses in concrete in an eccentrically loaded column having an effective concentric prestressing force P. When the load capacity of the column is controlled by σ_{cc},

$$\frac{P}{A} + \frac{W}{A} + \frac{My_T}{1} = \sigma_{cc}$$

or

$$W = A\left[\sigma_{cc} - \frac{P}{A} - \frac{My_T}{1}\right]$$ (13.2.19)

When the load capacity of the column is controlled by σ_{ct},

$$\frac{My_B}{1} - \frac{P}{A} - \frac{W}{A} = \sigma_{ct}$$

or

$$W = A\left[\frac{My_B}{1} - \frac{P}{A} - \sigma_{ct}\right]$$ (13.2.20)

Equations (13.2.19) and (13.2.20) show that in a compression failure, W decreases linearly with the increase in M whereas in tension failure, it increases linearly with M.

Equations (13.2.19) and (13.2.20) may be expressed in non-dimensional form by noting that if W_o and M_o represent the strengths in simple compression and simple flexure respectively,

$$W_o = \sigma_{cc} A$$ (13.2.21)

and

$$M_o = \sigma_{ct} \frac{I}{y_B}$$ (13.2.22)

Combining Eq. (13.2.19) with Eq. (13.2.21) and (13.2.22), the non-dimensional interaction relationship for compression failure may be expressed as

$$\frac{W}{W_o} = 1 - \frac{P}{W_o} - \frac{M}{M_o} \cdot \frac{\sigma_{ct}}{\sigma_{cc}} \cdot \frac{y_T}{y_B}$$ (13.2.23)

Similarly, combining Eq. (13.2.20) with Eq. (13.2.21) and (13.2.22), the non-dimensional interaction relationship for tension failure may be expressed as

$$\frac{M}{M_o} = 1 + \frac{P}{W_o} \cdot \frac{\sigma_{cc}}{\sigma_{ct}} + \frac{W}{W_o} \cdot \frac{\sigma_{cc}}{\sigma_{ct}}$$ (13.2.24)

Using non-dimensionalised prestressing force (P/W_o) as a parameter, the interaction relationship is represented by a family of straight lines. For example, taking $(\sigma_{cc}/\sigma_{ct}) = 5$ and $(y_T/y_B) = 1$, a set of 5 linear interaction relationships are obtained for 5 values of parameter (P/W_o), viz. $(P/W_o) = 0, 0.1, 0.2, 0.3$ and 0.4 as shown in Fig. 13.2.4. In order to distinguish between compression and tension failures, the latter are represented by broken lines. The plot shows that in the case of compression failure, the effect of increase in prestrss is to reduce the load capacity progressively. On the other hand, in the case of tension failure, the effect is just the reverse, i.e. the load capacity increases with increase in prestressing force. The plot also shows that for the assumed values of permissible stresses, the tension failure is ruled out when non-dimensionalised prestressing force (P/W_o) is equal to or greater than 0.4.

(ii) Limit State of Collapse

As the limit state of collapse is approached, the constituent materials exhibit considerable non-linearity. In addition, a part of the section cracks in the case of medium and large eccentricities. Consequently, the cross-sectional properties, viz. y_B, y_T, A and I change progressively with increase in load. Due to these factors, the interaction relationship between axial compression and bending moment is no longer linear. However, the general trend remains the same as in the elastic precracking range. For a given cross-section and prescribed stress-strain curves for constituent materials, the interaction relationship at the limit state of collapse may be obtained by using the analysis of Sec. 13.2.1.3. The iterative procedure enunciated in Sec. 13.2.1.3 may be used to determine the collapse load and its associated bending moment giving one point on the interaction diagram. The computations may be repeated for several values of eccentricities or bending moments to obtain the interaction curve.

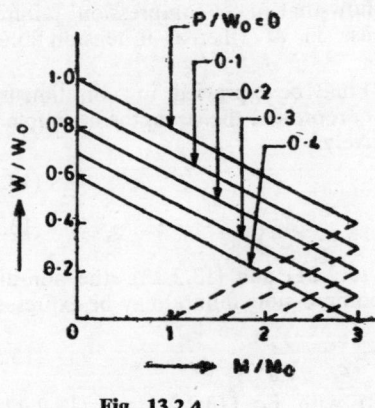

Fig. 13.2.4   Fig. 13.2.5

13.2 2. Long Columns

As internal prestressing does not contribute to buckling, the behaviour of internally prestressed long columns is similar to that of a reinforced concrete column. The effect of buckling of a long column is to increase the eccentricity of applied load. Consequently, the bending moment caused by eccentricity of loading is also increased and the load carrying capacity is decreased correspondingly. The decreased load capacity of a long column due to its tendency to buckle may be determined by adopting one of the following approaches.

13.2.2.1. Secant Formula Method

Figure 13.2.5 shows a long column of effective length l_e subjected to load W with an eccentricity e_o at the ends of the column. The eccentricity e_o may be taken to include the initial crookedness of the column. Due to buckling of the column represented by broken line, the eccentricity at midheight is increased to e_{max} which according to well known secant formula may be expressed as

$$e_{max} = e_0 \sec \left(\frac{\pi}{2} \sqrt{\frac{\overline{W}}{\overline{W}_E}} \right) \tag{13.2.25}$$

in which W_E is the Eulere's crippling load given by

$$W_E = \frac{\pi^2 E_c I_t}{l_e^2}$$ (13.2.26)

where E_c = modulus of elasticity of concrete

I_t = moment of inertia of transformed section

The maximum compressive and tensile stresses occur at midheight of the column. When the column is prestressed concentrically, the maximum compressive and tensile stresses may be expressed as

$$f_c \text{ (max)} = \frac{P}{A_c} + \frac{W}{A_t} + \frac{We_{max} y_c}{I_t}$$ (13.2.27)

$$f_t \text{ (max)} = \frac{We_{max} y_t}{I_t} - \frac{P}{A_c} - \frac{W}{A_t}$$ (13.2.28)

where P = effective prestressing force

A_c = area of concrete

A_t = transformed area

y_c, y_t = extreme fibre distances on compression and tension faces.

The column may be prestressed eccentrically, if the end eccentricity e_0 is definite and is not reversible. In this case, the eccentricity of prestress e_p should logically be oppasite to e_0 in sign. The maximum compressive and tensile stresses in eccentrically prestressed long column may then be expressed as

$$f_c \text{ (max)} = \frac{P}{A_c} + \frac{W}{A_t} + \frac{We_{max} y_c}{I_t} - \frac{Pe_p y_c}{I_t}$$ (13.2.29)

$$f_t \text{ (max)} = \frac{We_{max} y_t}{I_t} - \frac{P}{A_c} - \frac{W}{A_t} - \frac{Pe_p y_t}{I_t}$$ (13.2.30)

The column cracks when the tensile stress given by Eq. (13.2.28) or (13.2.30) reaches the tensile strength of concrete in bending f_{cr}. The foregoing anaylsis may be used with reasonable accuracy in the precracking range. As load W is increased in the post-cracking range, part of the section cracks and therefore becomes ineffective. The error involved in using the above equations increases progressively and may become significant as faiiure is approached. For a precise computation oi collapse load of a long column, the cracked section analysis based on strain compatibility and plasticity of materials may be used.

13.2.2.2 Reduction Factor Method

The reduction factor method appeared in the earliest versions of most codes and is a simplest approach for design of long columns. However, it suffers from several limitations. The approach is conceptually erroneous and often leads to unduly conservative design.

As the effect of buckling is to reduce the load capacity of a long column, it may be determined by applying a suitable reduction factor on the load capacity of the corresponding short column. The Indian code IS: 456-1978 gives reduction factors for working load capacity of long columns. These reduction factors may be applied to the working load capacity of corresponding short column computed in accordance with Sec. 13.2.1.

A column is defined as a long or slender column when the ratio of effective length and least lateral dimension (l_e/b) exceeds 12 or when the ratio of effective length and least radius of gyration (l_e/k_{min}) exceeds 40. The reduction coefficient or reduction factor may be taken as

$$C_r = 1.25 - \frac{l_e}{48 \, b} \qquad (13.2.31)$$

For more precise computations, the reduction factor may be taken as

$$C_r = 1.25 - \frac{l_e}{160 \, k_{min}} \qquad (13.2.32)$$

13.2.2.3 Additional Moment Method

In this method, the effect of buckling of a long column is taken into account by an additional moment in the plane of buckling. The approach is based on CEB—FIP recommendations of 1963. The additional deflection due to buckling of a long column depends upon the distribution of curvature which may be assumed to be rectangular, triangular or half sine wave. For a rectangular distribution, the deflection due to buckling at midheight,

$$\Delta_u = \frac{l_e^2}{8} \, \Phi_{max} \qquad (13.2.33)$$

whereas for a triangular distribution,

$$\Delta_u = \frac{l_e^2}{12} \, \Phi_{max} \qquad (13.2.34)$$

where l_e = effecitve length of column

The symbol Φ_{max} represents maximum curvature which may be expressed as

$$\Phi_{n\,ax} = \frac{\epsilon_c + \epsilon_t}{D} \qquad (13.2.35)$$

In the case of a balanced failure, the extreme fibre strains ϵ_c and ϵ_t on the compression and tension sides reach their limiting values simultaneously. Taking $\epsilon_c = 0.0035$. $\epsilon_t = 0.002$ and $\Delta_u = (l_e^2 \, \Phi_{max})/10$ in lieu of Eq. (13.2.33) and (13.2.34), the deflection due to buckling,

$$\Delta_u = \frac{l_e^2}{1820 \, D} \qquad (13.2.36)$$

where D = overall depth of the column cross-section in the plane of buckling

The additional moment due to buckling,

$$M_a = W_u \, \Delta_u = \frac{W_u \, l}{1820 \, D} \qquad (13.2.37)$$

For the limit state design of long columns, the Indian Code IS: 456-1978 gives the following expressions for additional moments M_{ax} and M_{ay} about the strong and weak axes due to buckling based on Eq. (13.2.37),

$$M_{ax} = \frac{W_u D}{2000} \left(\frac{l_{ex}}{D}\right)^2 \tag{13.2.38}$$

$$M_{ay} = \frac{W_u b}{2000} \left(\frac{l_{ey}}{b}\right)^2 \tag{13.2.39}$$

The additional moment given by Eq. (13.2.38) or (13.2.39) should be added to the initial moment which depends upon whether the column is braced or unbraced and whether it carries any lateral load over its height.

For a braced column without any transverse loads occurring in its height, the additional moment given by Eq. (13.2.38) or (13.2.39) should be added to the initial moment M_{ui} given by

$$M_{ui} = (0.4\, M_{u1} + 0.6\, M_{u2}) \geqslant 0.4\, M_{u2} \tag{13.2.40}$$

subject to the condition that $(M_{ui} + M_a)$ is not less than M_{u2}. The moments M_{u1} and M_{u2} are the end moments in which larger end moment M_{v2} has to be taken positive and the smaller end moment M_{u1} is positive for single curvature and negative for double curvature. A column bends into double curvature when M_{u1} and M_{u2} are both clockwise or anticlockwise and into single curvature when one of them is clockwise and the other is anticlockwise.

For unbraced columns, the additional moment should be added to the end moments. When the column carries lateral loads at any given level or storey, all columns at that level or storey sway equally. Consequently, the slenderness ratio of each column in the plane of buckling may be taken equal to the average value of all columns at that level or storey.

13.2.2.4 Moment Magnification Method

A bukling failure of a prestressed concrete long column, as distinct from material failure, occurs at a concrete strain less than i s ultimate or crushing strain ϵ_u due to member instability. According to ACI code, the consideration of buckling effect of slender columns or stability analysis is necessary only when the effective slenderness ratio (l_e/k) of a compression member belonging to a frame exceeds the following limits.

$$\frac{l_e}{K} = 34 - 12\left(\frac{M_{u1}}{M_{u2}}\right) \qquad \text{for braced frames}$$

$$= 22 \qquad\qquad \text{for unbraced frames} \tag{13.2.41}$$

in which M_{u1} and M_{u2} are numerically smaller and larger factored bending moments at the ends of the column computed from elastic frame analysis. The ratio (M_{ui}/M_{u2}) is to be taken positive for single curvature and negative for double curvature. The radius of gyration, $k = \sqrt{\dfrac{I_g}{A_g}}$ may be taken as 0.3 times the overall dimension in the direction in which stability is being considered for rectangular compression members and 0.25 times the diameter for circular compression members. The effective length, $l_e = c l_u$ is the distance between points of contraflexure in the buckled shape of the column. It is obtained by multiplying the unsupported length l_u by the coefficient c called the effective length factor which depends upon the end conditions of the column. For an isolated column,

(i) both ends hinged $c = 1$

(ii) both ends fixed $c = 0.5$

(iii) fixed at base and free at top $c = 2$

(iv) both ends fixed and lateral motion exsists $c = 1$ (13.2.42)

When the column belongs to a frame, the following values of c may be adopted.

(i) *braced frames*

$$c = (0.7 + 0.1\ \psi_{ave}) \leqslant 1$$

or $\qquad\qquad c = (0.85 + 0.05\ \psi_{min}) \leqslant 1$ (13.2.43)

whichever is smaller. The symbols ψ_{ave} and ψ_{min} represent the average value and smaller value of ψ at the two ends of the column and is the ratio of the sum of stiffnesses of all compression members and all flexural members.

$$\psi = \frac{\Sigma \left(\dfrac{EI}{l_u}\right)_{columns}}{\Sigma \left(\dfrac{EI}{l_b}\right)_{beams}}$$ (13.2.44)

The summation in numerator should include stiffnesses of all columns and that in the denominator should include stiffnesses of all beams lying in the plane of buckling at the column end under consideration.

(ii) *Unbraced Frames*

(a) **Both ends fixed**

$$c = (1 - 0.05\ \psi_{ave})\ \sqrt{1 + \psi_{ave}} \qquad \text{for} \quad \psi < 2$$

$$= 0.9\ \sqrt{1 + \psi_{ave}} \qquad\qquad \text{for} \quad \psi \geqslant 2 \quad (13.2.45)$$

(b) **One end hinged and other fixed**

$$c = 2 + 0.3\ \psi$$ (13.2.46)

where ψ is the value at fixed end.

The moment magnification method may be used when the effective slenderness ratio exceeds the limits specified by Eq. (13.2.41), provided it does not exceed 100 in which case a second order analysis, based on computations of deflections due to buckling, is mandatory. The moment magnification factor δ is broken up into two parts, viz. δ_b caused by vertical or gravity loads producing little or no sidesway and δ_s caused by lateral loads mainly responsible for sidesway. The factors δ_b and δ_s are related to the slenderness of the member, the stiffness of the entire frame, the restraint or applied moment at its ends and the cross-section of the column.

$$\delta_b = \frac{C_m}{1 - \dfrac{W_u}{\Phi \, W_E}} \geqslant 1 \tag{13.2.47}$$

and

$$\delta_s = \frac{1}{1 - \dfrac{\sum W_u}{\Phi \sum W_E}} \geqslant 1 \tag{13.2.48}$$

where

$$C_m = \left[0.6 + 0.4 \left(\frac{M_{u1}}{M_{u2}}\right)\right] \geqslant 0.4$$

for single curvature $\tag{13.2.42}$

$= 1$ for double curvature

$\Phi = $ capacity reduction factor

$\sum W_u, \sum W_E = $ summations of factored axial loads W_u and Euler crippling loads W_E for all columns in a storey.

All columns have to be designed for a minimum eccentricity of $(15 + 0.03 \, h)$ millimetres in which h is overall dimension of the column cross-section in millimetres in the plane of buckling. If the end eccentricities computed from frame analysis are smaller than the minimum permissible value, the ratio (M_{u1}/M_{u2}) in the computation of C_m may be based on the computed values of M_{u1} and M_{u2}. In the extreme case when M_{u1} and M_{u2} are practically zero, the ratio (M_{u1}/M_{u2}) may be taken equal to 1. The flexural rigidity $E_c I_t$ in the computation of Euler buckling load W_E may be determined from

$$E_c I_t = \frac{E_s I_s + 0.2 \, E_c I_g}{1 + \beta_d} \tag{13.2.50}$$

Alternatively, it may be determined conservatively by the simplified equation,

$$E_c I_t = \frac{0.4 \, E_c I_g}{1 + \beta_d} \tag{13.2.51}$$

where $E_s, E_c = $ moduli of elasticity of steel and concrete

$I_s, I_g = $ moment of intertia of reinforcement and gross concrete section about the centroidal axis

$\beta_d = $ ratio of factored dead load moment and factored total moment. It is taken to be alawys positive with an upper limit of 1.

The magnified moment M_m may be determined by applying the magnification factor δ to the larger end moment M_{u2}.

$$M_m = \delta \, M_{u2} = \delta_b \, M_{u2b} + \delta_s \, M_{u2s} \tag{13.2.52}$$

where $M_{u2b} = $ larger end moment due to vertical or gravity loading which produces little or no sidesway

$M_{u2s} = $ larger end moment due to lateral loading mainly responsible for sidesway.

The magnification of moments on account of lateral loads causing appreciable sideway is more serious than that caused by gravity loads. For frames braced by shear walls against sidesway, the entire moment acting on the column is considered to be M_{u2b} and δ_s is therefore assumed to be zero. If the lateral deflection or sidesway of the frame is less than (span/1500), the frame may be taken to be braced.

The computations for the design of a long column based on moment magnification method may proceed along the following steps:

(i) Determine whether or not the frame has appreciable sidesway. If it has appreciable sidesway, determine δ_b and δ_s. If sidesway is negligible, take $\delta_s = 0$.

(ii) Assume a cross-section and determine the eccentricity based on larger end moment M_{u2} computed from frame analysis. If it is less than the minimum permissible value of $(15 + 0.03\ D)$ millimetre, adopt the minimum permissible value.

(iii) Determine the effective length factor c and hence the effective slenderness ratio, $(c\ l_u/k)$ in the plane of buckling.

(iv) Determine the magnified moment M_m.

(v) Design the cross-section for a short column to carry the magnified moment M_m together with the applied axial load.

(vi) Adopt for the long coumn the same cross-section as that designed in step (v) for short column.

13.2.3. Biaxially Loaded Columns

Interior columns of building frames having substantially unequal bays and corner columns of any building frame are generally subjected to significant bending moments in two orthogonal directions in addition to axial compression. In actual practice most columns are, therefore, loaded biaxially. Biaxial moments have also to be considered on account of the code provisions in respect of minimum eccentricity. As the minimum eccentricity clause applies to both orthogonal directions, the column necessarily becomes biaxially loaded.

13.2.3.1. Short Column

In the precracking range, a biaxially loaded short column may be analysed with reasonable accuracy using elastic theory. The analysis is similar to that presented in Sec. 13.2.1.1 except that the effect of bending moment in the other orthogonal direction has also to be taken into account. Consider, for example, a prestressed concrete short column of rectangular cross-section subjected to eccentric load W as shown in Fig. 13.2.6. The stress at any point X having coordinates (x, y) may be determined from

$$f_x = -p_x - \frac{W}{A_t} - \frac{(W \cdot e_y)y}{I_{tx}} - \frac{(W \cdot e_x)x}{I_{ty}} \qquad (13.2.53)$$

where p_x = effective intensity of compressive stress at x due to prestresing

A_t = transformed area of cross-section

I_{tx}, I_{ty} = moments of inertia of transformed section with respect to x and y axes passing through the centroid of the transformed section.

The maximum compressive stress occurs at corner A,

$$f_A = -p_A - \frac{W}{A_t} - \frac{(W \cdot e_y)y_t}{It_x} - \frac{(W e_x)x_c}{It_y} \qquad (13.2.54)$$

The maximum tensile stress occurs at corner C.

$$f_C = - p_C - \frac{W}{A_t} + \frac{(W \cdot e_y) y_t}{I_{tx}} + \frac{(W \cdot e_x) x_t}{It_y} \qquad (`3.2.55)$$

The maximum permissible load or working load of the column may be determined by equating the compressive stress at A to the maximum permissible compressive stress in concrete. Alternatively, it may be determined by equating the tensile stress at C to the maximum permissible tensile stress in concrete. The smaller of the two values evidently corresponds to the working load capacity of the column. The effective prestress in concrete p_X at any point X depends upon the effective prestressing force P and its eccentricities with respect to x and y axes. The prestressing force is generally concentric when the eccentricities of load W are not definite or are reversible. In this case, the intensity of prestress is uniform at all points and is equal to (P/A_c) in which A_c is the net area of concrete.

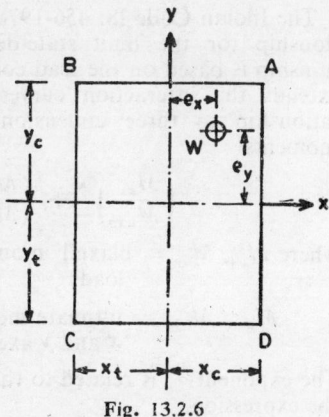

Fig. 13.2.6

The column enters the post-cracking range as soon as the fictitious tensile stress at C determined from Eq. (13.2.55) exceeds the tensile strength of concrete. The cracking load of the column may, therefore, be determined by equating the tensile stress at C to tensile strength of concrete. As soon as the section cracks, a part of the section lying in the tension zone may be assumed to be rendered ineffective. The analysis of a cracked section of a biaxially loaded column may be based on Navier's hypothesis, i.e. linearity of strain distribution and equations of equilibrium of internal and external forces. It is similar to the analysis presented in Sec. 13.2.1.2 except that (i) the neutral axis is no longer horizontal. The inclination of neutral axis to x-axis or y-axis depends upon the geometry of the cross-section, the eccentricities of W and physical properties of materials. and (ii) instead of two equations of equilibrium for uniaxially loaded columns, there are three equations of equilibrium in the case of biaxially loaded columns. The three equations represent the equilibrium of internal and external vertical forces and bending moments about any two orthogonal axes.

In most practical cases of prestressed concrete columns, a direct solution of the three equations is extremely difficult. The position of inclined neutral axis is often determined by trial and error until the equations of equilibrium are satisfied. Alternatively, design charts may be used. The internal forces may be computed from idealised stress-strain curves for constituent materials. The limit state of collapse is reached when the extreme fibre strain in concrete reaches the crushing strain ϵ_u which may be assumed to be equal to 0.003 or 0.0035.

13.2.3.2 Long Columns

Due to the effect of buckling and consequent magnification of the end eccentricities at midheight of the column, the strength of long column is smaller than that of the corresponding short column. The general approach for the design of biaxially loaded long columns is to apply a suitable factor to the load capacity of the corresponding short column in order to allow for buckling effect.

The Indian Code IS: 456-1978 proposes the following interaction relationship for the limit state design of biaxially loaded columns. The relationship is based on the load contour method given by Bresler. The method extends the interaction curves representing load-moment interaction relationship to three dimensional interaction surface for load and biaxial moments.

$$\left(\frac{M_{ux}}{M_{uxo}}\right)^{\alpha_n} + \left(\frac{M_{uy}}{M_{uyo}}\right)^{\alpha_n} \leqslant 1 \qquad (13.2.56)$$

where M_{ux}, M_{uy} = biaxial moments about x and y axes due to design loads

M_{uxo}, M_{uyo} = ultimate moment capacities for uniaxial bending along x and y axes respectively with axial load W_u.

The exponent α_n is related to the ratio (W_u/W_{uo}) in which W_{uo} is given by the expression,

$$W_{uo} = 0.45 f_{ck} A_c + 0.75 f_y A_{sc} \qquad (13.2.57)$$

where A_c = net area of concrete

f_{ck} = characteristic compressive (cube) strength of concrete

A_{sc} = area of longitudinal steel in column

f_y = yield stress of longitudinal steel.

For values of $(W_u/W_{uo}) = 0.2$ to 0.8, the values of α_n vary linearly from 1 to 2. For values of (W_u/W_{uo}) less than 0.2, $\alpha_n = 1$ and for values of (W_u/W_{uo}) greater than 0.8, $\alpha_n = 2$.

Equation (13.2.57) for a reinforced concrete column may be written as follows for a prestressed concrete column adopting the general approach in the tentative recommendations of *PCI* Committee.

$$W_{uo} = 0.45 (A_c f_{ck} - P) + 0.75 A_{sc} f_y + A_p E_s (\epsilon_u - \epsilon_{ce}) \qquad (13.2.58)$$

where P = effective prestressing force

ϵ_u = crushing strain of concrete which may be taken equal to 0.002 for axial compression

ϵ_{ce} = effective strain in concrete due to prestressing.

13.2.4. Design Specifications

The Indian Code IS: 456-1978 defines columns and struts as compression members whose effective length exceeds three times the least lateral dimension. A column may be called short when the slenderness ratio with respect to major axis, (l_{ex}/D) and that with respect to minor axis, (l_{ey}/b) do not exceed 12. The effective length of a column which depends upon its support conditions at the two ends, is equal to unsupported length multiplied by a factor called effective length factor. The code gives values of effective length factor for 7 sets of support conditions and also for columns belonging to frames without and with lateral sway. The unsupported length is defined as the clear distance between end restraints except that (i) the unsupported length in the case of flat slab is the clear distance between the floor and the bottom of column head or drop panel, (ii) the unsupported length in the case of beam-slab construction is the clear distance between floor and the soffit of shallower beam framing into the column, (iii) in the case of columns restrained laterally by horizontal struts of adequate size and strength in vertical planes not differing from 90° by more than 30°, the unsupported length is the clear distance between struts in each vertical plane and (iv) the effective length is the clear distance between the floor and the bottom of the bracket when a beam or strut frames into the column through a bracket whose width is not less than that of the strut or beam and half the width of the column.

The use of extremely slender columns is prohibited by restricting the unsupported length to 60 times the least lateral dimension. When one end of the column is unrestrained in any plane, the unsupported length should not exceed $(100\ b^2/D)$ in which D is the dimension of the column in the plane under consideration. The code stipulates a minimum eccentricity equal to (unsupported length /500) plus (lateral dimension/30), subject to a minimum of 20 mm.

In the limit state design, the maximum strain in concrete should not exceed 0.002 in the case of axial compression. In combined flexure and axial compression, the strain in concrete at most highly compressed extreme fibre should not exceed 0.0035 minus 0.75 times the strain at the least compressed extreme fibre, when entire cross-section is in compression. The code stipulates detailed specifications for proportioning of tied and helically reinforced columns and specifies that the strength of helically reinforced columns may be taken 5 per cent greater than that of the tied column.

In order to qualify as the prestressed concrete column, the British Code on prestressed concrete CP 115 prescribes a minimum prestress of 2.7 MPa. The provision is to guard against the tendency to provide a negligibly small prestress in order to circumvent the provisions of reinforced concrete code CP 114 and take advantage of higher permissible stresses prescribed in prestressed concrete code CP 115. The maximum permissible stress in direct compression according to CP 114 is 10.6 MPa. In the absence of provision in respect of minimum prestress, it would be possible to claim a higher strength of a column made of high grade concrete by using the provisions of CP 115, even when the prestress is so small that it is practically a reinforced concrete column. The comparison of the two codes shows that appreciable reduction in the size of the column can be obtained only when the bending stress is equal to or greater than the direct stress. The prestressed concrete code CP 115 specifies that the maximum compressive stress in concrete may be permitted to reach the maximum permissible stress in bending compression σ_{cbc} only when the bending stress is at least equal to four times the direct stress, otherwise it should be restricted to permissible direct stress σ_{cc}. In the absence of sufficient test results, the code permits the use of reduction factors for long columns as prescribed in CP 114. Whenever a prestressed concrete column contains untensioned longitudinal reinforcement, it is necessary to provide lateral ties or spirals in accordance with the provisions of reinforced concrete code CP 114. The effect of creep of concrete on the loss of prestress should be carefully considered in order to avoid appreciable error in the computation of cracking load.

The ACI Code specifies the following values of capacity reduction factor for the design of compression members:

(i) For columns with lateral ties, $\phi = 0.7$

(ii) For columns with spiral reinforcement, $\Phi = 0.75$

(iii) The capacity reduction factor Φ may be increased linearly from 0.7 (for tied column) or 0.75 (for helically reinforced column) to 0.9 as the axial load decreases from $0.1 f_c A_g$ to zero, provided that (a) yield stress of steel $f_y \not> 415$ MPa, (b) the column is symmetrically reinforced and (c) the effective depth, i.e. the distance between compression and tension reinforcement is not less than 0.7 times the overall depth of the section.

(iv) The capacity reduction factor Φ may be increased linearly from 0.7 or 0.75 (as the case may be) to 0.9 as the axial load decreases from $0.1 f_c' A_g$ or W_b whichever is smaller, to zero for sections with small axial compression not satisfying (iii) above. The column load W_b represents the balanced load causing simultaneous yielding of tension steel and crushing of concrete.

The ACI Code specifies that at least minimum untensioned reinforcement as prescribed for a reinforced concrete column should be used whenever the effective prestress, defined as effective prestressing force divided by gross-cross-sectional area, is less than 1.6 MPa. The requirement of minimum reinforcement may be waived when effective prestress exceeds 1.6 MPa, provided the structural analysis shows adequate strength and stability. All longitudinal steel, including prestressing steel, should be enclosed by lateral ties or spiral reinforcement in accordance with the provisions for reinforced concrete columns.

The Prestressed concrete Institute (PCI) of United States has in its tentative recommendations divided prestressed concrete columns into short, intermediate and long columns depending upon whether effective slenderness ratio (l_e/k) is less than 30, lies between 30 and 100 and is greater than 100.

(i) Short Column

The crushing load of an axially loaded short column is reduced due to the presence of effective prestressing force P. The change of stress in prestressing tendons is equal to the difference between the crushing strain of concrete ϵ_u and the effective strain in concrete due to prestressing ϵ_{ce}. Consequently, the ultimate or crushing load of an axially loaded prestressed concrete short column W_{uo} may be expressed as

$$W_{uo} = 0.85\, f_{cy}\, A_c + A_{sc} f_y - P + A_p\, E_s\, (\epsilon_u - \epsilon_{ce}) \qquad (13.2.59)$$

Putting $P = A_p\, E_s\, \epsilon_{pe}$

$$W_{uo} = 0.85\, f_{cy}\, A_c + A_{sc} f_y - A_p\, E_s\, [\epsilon_{pe} - \epsilon_u + \epsilon_{ce}] \qquad (13.2.60)$$

where f_{cy} = compressive (cylinder) strength of concrete

A_p = area of prestressing steel

ϵ_{pe} = effective strain in prestressing steel after losses

The crushing strain of concrete ϵ_u may be equal to 0.003 and $\epsilon_{ce} = P/(A_c/E_c)$. Equation (13.2.59) may be used for helically reinforced column. The strength may be reduced by 15 per cent in the case of tied columns.

In the case of an eccentrically loaded short column, the following two conditions are considered.

(a) Eccentricity e less than e'

where $e' = 0.23\, D \left(1 + \dfrac{200\, P}{A_c\, f_{pu}}\right) \leqslant 0.5D$ \qquad\qquad (13.2.61)

The ultimate load is given by

$$W_u = \frac{W_{uo}}{1 + \left(\dfrac{W_{uo} - W'_{uo}}{W'_{uo}}\right)\dfrac{e}{e'}} \qquad (13.2.62)$$

where $W'_{uo} = 0.425\, f_{cy}\, A_c \left(1 - \dfrac{95\, P}{A_c\, f_{pu}}\right) \geqslant 0.2\, f_{cy}\, A_c$ \qquad (13.2.63)

(b) Eccentricity e greater than e'

The ultimate load W_u varies linearly from 0 to W_{uo} as the applied moment $(W_u \cdot e)$ varies from $0.4\, M_{uo}$ to M'_u in which M_{uo} and M'_u are defined as follows. The moment M_{uo} is the ultimate strength in pure flexure which may be determined in accordance with Sec. 6.9.3 ignoring the capacity reduction factor Φ. The moment M'_u is the product of W'_{uo} and e'.

(ii) *Intermediate Column*

The ultimate load of an intermediate column may be determined by applying the following reduction factor to the ultimate load of a short column.

$$C_r = 1.22 - 0.007\, (1_e/k) \leqslant 1 \qquad (13.2.64)$$

(iii) *Long Column*

In the case of long columns, the bending moment due to initial crookedness and eccentricity of applied load is magnified due to buckling. The magnified moment may be determined by using the secant formula,

$$M_{\max} = W \left(e + \frac{l_e}{400}\right) \sec\left(\frac{\pi}{2} \sqrt{\frac{W}{W_E}}\right) \qquad (13.2.65)$$

The total eccentricity including initial 'crookedness $\left(e + \dfrac{l_e}{400} \right)$ should not be taken less than 0.1 D. The elastic stresses due to the combined action of axial load W and magnified moment $M_{\max}$ should not exceed the permissible stresses for the materials.

EXAMPLE 13.2.1

A prestressed concrete short column having square cross-section of 0.5 m side has to carry a load of 1.2 MN with a reversible eccentricity of 0.15 m. It is symmetrically reinforced with 1 per cent untensioned steel at an effective cover of 0.05 m. The percentage of prestressing steel. also placed symmetrically, is 0.4 per cent at an effective cover of 0.1 m. The initial and final stresses in prestressing steel are 1200 and 960 MPa respectively. The modular ratio, $m = 6$. Determine the initial and final stresses in concrete and steel.

Solution

$$A_s = A_s' = 0.5 \left(\frac{0.5 \times 0.5}{100} \right) = 0.00125 \text{ m}^2$$

$$A_{p1} = A_{p2} = 0.2 \left(\frac{0.5 \times 0.5}{100} \right) = 0.0005 \text{ m}^2$$

$$c = c' = 0.05 \text{ m}$$

$$e_{p1} = e_{p2} = 0.15 \text{ m}$$

(i) *Initial Stage*

$$P_i = 1200 \, (0.0005 + 0.0005) = 1.2 \text{ MN}$$

$$A_{ti} = 0.5 \times 0.5 + (6-1) \, (0.00125 + 0.00125) = 0.2625 \text{ m}^2$$

$$I_{ti} = \frac{1}{12} \times 0.5 \times 0.5^3 + 2 \, (6-1) \, (0.00125) \, (0.25 - 0.05)^2$$

$$= 0.005708 \text{ m}^4$$

Initial stresses at top and bottom fibres,

$$f_{Ti} = f_{Bi} = - \frac{1.2}{0.2625} = - 4.571 \text{ MPa}$$

Initial stress in untensioned steel,

$$f_{si} = f_s' = 6 \, (- 4.571) = - 27.426 \text{ MPa}$$

(ii) *Final Stage*

$$A_t = 0.2625 + (6-1) \, (0.0005 + 0.0005) = 0.2675 \text{ m}^2$$

$$I_t = 0.005708 + 2 \, (6-1) \, (0.0005) \, (0.15)^2 = 0.00582 \text{ m}^4$$

Loss factor, $\eta = \dfrac{960}{1200} = 0.8$

Stress at top (compression) fibre,

$$f_T = 0.8\,(-4.571) - \frac{1.2}{0.2675} - \frac{1.2' \times 0.15 \times 0.25}{0.00582}$$

$$= -15.875 \text{ MPa}$$

$$f_B = 0.8\,(-4.571) - \frac{1.2}{0.2675} + \frac{1.2 \times 0.15 \times 0.25}{0.00582}$$

$$= -0.411 \text{ MPa}$$

Stress in non-prestressed compression steel,

$$f_s = 6\left[-0.411 + (-15.875 + 0.411) \times \frac{0.05}{0.5}\right]$$

$$= -11.744 \text{ MPa}$$

$$f'_s = 6\left[-0.411 + (-15.875 + 0.411) \times \frac{0.45}{0.5}\right]$$

$$= -85.972 \text{ MPa}$$

Stress in prestressing steel at top fibre,

$$f_{p1} = 960 + 6\left[-0.411 + (-15.875 + 0.411) \times \frac{0.1}{0.5}\right]$$

$$= 939 \text{ MPa}$$

Stress in prestressing steel at bottom fibre,

$$f_{p2} = 960 + 6\left[-0.411 + (-15.875 + 0.411) \times \frac{0.4}{0.5}\right]$$

$$= 883.3 \text{ MPa}$$

EXAMPLE 13.2.2

Determine the eventual long-time stresses in untensioned and prestressing steel in the column of Exa. 13.2,1 if the load is permanent. Take the coefficient due to the combined effect of creep and shrinkage of concrete, $C_{cs} = 1.5$.

Solution

The effect of shrinkage and creep may be allowed for by taking the reduced value of E_c or correspondingly higher value of modular ratio,

$$m = 6\,(1 + C_{cs}) = 6\,(1 + 1.5) = 15$$

Proceeding as in Example 13.2.1,

$$f_s = -29.35 \text{ MPa}$$

$$f'_s = -214.9 \text{ MPa}$$

$$f_{p1} = 907.5 \text{ MPa}$$

$$f_{p2} = 768.3 \text{ MPa}$$

EXAMPLE 13.2.3

Determine the magnitude and line of action of the eccentric load on a prestressed concrete short column of rectangular cross-section shown in Fig.
13.2.7 (a). The area of untensioned steel $A_s = A_s' = 0.0009$ m² and that
of prestressing steel, $A_{p1} = A_{p2} = 000036$ m². The effective stress in prestressing steel is 960 MPa. The depth of neutral axis is 0.3 m and the
strain in concrete at top (compression) fibre is 0.0018. The idealised stress-
strain curves for concrete, untensioned reinforcement and prestressing steel
are shown in Fig. 13.2.7 (c), (d) and (e) respectively.

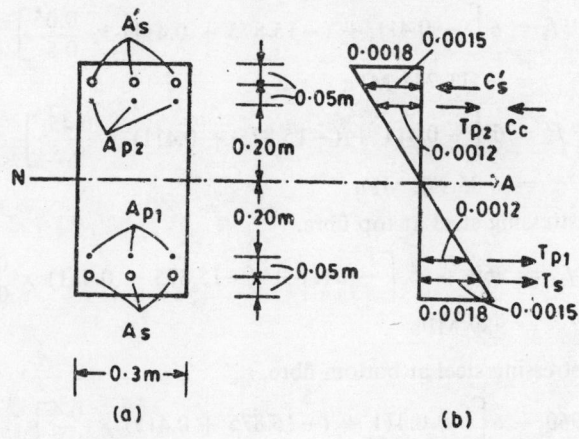

(a)   (b)

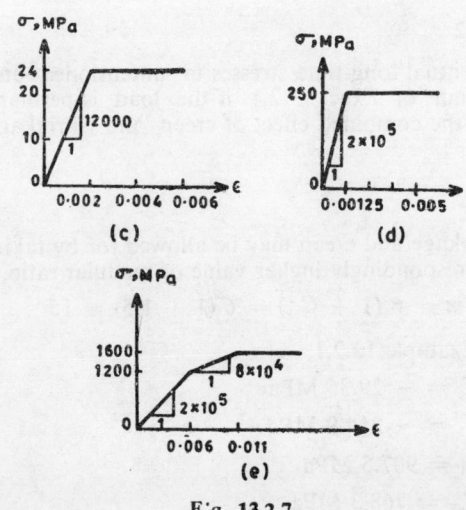

F'g. 13.2.7

Solution

For the given position of neutral axis and extreme fibre strain of 0.0018, the strain diagram based on the assumption of linear distribution of strains is shown in Fig. 13.2.7 (b). The internal forces can be determined from the relevant strains and corresponding stress-strain relationships.

$$\epsilon'_s = \frac{0.3 - 0.05}{0.3} \times 0.0018 = 0.0015$$

$$E'_{ss} = \frac{250}{0.0015} = 1.667 \times 10^5 \text{ MPa}$$

Hence from Eq. (13.2.14), the force in non-prestressed compression steel,

$$C'_s = 0.0009 \times 1.667 \times 10^5 \times 0.0015 = 0.225 \text{ MN}$$

$$\epsilon_{pe2} = \frac{960}{2 \times 10^5} = 0.0048$$

$$\epsilon_{p2} = \frac{0.3 - 0.1}{0.3} \times 0.0018 = 0.0012$$

Net strain, $\epsilon_{pe2} - \epsilon_{p2} = 0.0036$

$$E_{ps2} = 2 \times 10^5 \text{ MPa}$$

Hence from Eq. (13.2.15), the force in prestressing steel on compression side,

$$T_{p2} = 0.00036 \times 2 \times 10^5 \times 0.0036 = 0.2592 \text{ MN}$$

From Eq. (13.2.16), the force in concrete in compression zone,

$$C_c = 0.3 \times 0.3 \left(\tfrac{1}{2} \times 12000 \times 0.0018\right) = 0.972 \text{ MN}$$

$$\epsilon_{pe1} = \frac{960}{2 \times 10^5} = 0.0048$$

$$\epsilon_{p1} = \frac{0.3 - 0.1}{0.3} \times 0.0018 = 0.0012$$

Net strain, $\epsilon_{pe1} + \epsilon_{p1} = 0.006$

$$E_{ps1} = 2 \times 10^5 \text{ MPa}$$

Hence from Eq. (13.2.17), the force in prestressing steel on tension side,

$$T_{p1} = 0.00036 \times 2 \times 10^5 \times 0.006 = 0.432 \text{ MN}$$

$$\epsilon_s = \frac{0.3 - 0.05}{0.3} \times 0.0018 = 0.0015$$

$$E_{ss} = \frac{250}{0.0015} = 1.667 \times 10^5 \text{ MPa}$$

Hence from Eq. (13.2.18), force in non-prestressed steel on tension side,

$$T_s = 0.0009 \times 1.667 \times 10^5 \times 0.0015 = 0.225 \text{ MN}$$

From Eq. (13.2.12), the magnitude of eccentric load,
$$W = 0.225 - 0.2592 + 0.972 - 0.432 - 0.225$$
$$= 0.2808 \text{ MN}$$

Using Eq. (13.2.13),
$$0.225 \,(x_1 + 0.05) - 0.2592 \,(x_1 + 0.1) + 0.972 \,(x_1 + 0.1)$$
$$- 0.432 \,(x_1 + 0.5) - 0.225 \,(x_1 + 0.55) = 0$$

or $x_1 = 0.916$ m

Hence the eccentricity of applied load,
$$e = \frac{D}{2} + x_1 = \frac{0.6}{2} + 0.916 = 1.216 \text{ m}$$

EXAMPLE 13.2.4

Determine the collapse load for the column of Example 13.2.3 if the strain in concrete at extreme compression fibre is 0.003 and depth of neutral axis is 0.2 m at the limit state of collapse.

Solution

Proceeding as in Example 13.2.3, the strain in non-prestressed compression steel is more than its yield strain. Hence.
$$C_s' = 0.0009 \times 250 = 0.225 \text{ MN}$$

The compressive strain in prestressing steel A_{p2} due to loading is $\epsilon_{p2} = 0.0015$. Hence the net strain is $0.0048 - 0.0015 = 0.0033$. The corresponding stress, $f_{p2} = 660$ MPa. Hence,
$$T_{p2} = 0.00036 \times 660 = 0.2376 \text{ MN}$$

From the stress-strain curve for concrete, it is evident that the stress diagram for concrete in compression zone is rectangular over a depth of $(0.2/3)$ m from top and triangular in the remaining depth of $(0.4/3)$ m. Compressive force corresponding to rectangular stress diagram,
$$C_{c1} = \left(0.3 \times \frac{0.2}{3}\right) \times 24 = 0.48 \text{ MN}$$

C_{c1} acts at $1/30$ m from top. Compressive force corresponding to triangular stress diagram,
$$C_{c2} = \left(0.3 \times \frac{0.4}{3}\right)\left(\frac{1}{2} \times 24\right) = 0.48 \text{ MN}$$

C_{c2} acts at $\left(\frac{0.2}{3} + \frac{1}{3} \times \frac{0.4}{3}\right) = \frac{1}{9}$ m from top. Strain in prestressing steel A_{p1} due to load, $\epsilon_{p1} = 0.0045$. Hence the net strain is $(0.0048 + 0.0045) = 0.0093$. From the stress-strain curve for prestressing steel, the corresponding $f_{p1} = 1464$ MPa.

Hence, $$T_{p1} = 0.00036 \times 1464 = 0.52704 \text{ MN}$$

The strain in non-prestressed tension steel is more than its yield strain. Hence,

$$T_s = 0.009 \times 250 = 0.225 \text{ MN}$$

From Eq. (13.2.12), the magnitude of eccentric load,

$$W = 0.225 - 0.2376 + 0.96 - 0.52704 - 0.225$$
$$= 0.1954 \text{ MN}$$

Using Eq. (13.2.13),

$$0.225\,(x_1 + 0.05) - 0.2376\,(x_1 + 0.1) + 0.48\left(x_1 + \frac{1}{30}\right)$$
$$+ 0.48\left(x_1 + \frac{1}{9}\right) - 0.52704\,(x_1 + 0.5) - 0.225\,(x_1 + 0.55) = 0$$

or $\qquad x_1 = 1.6915 \text{ m}$

Hence the eccentricity of applied load,

$$e = \frac{D}{2} + x_1 = \frac{0.6}{2} + 1.6915 = 1.9915 \text{ m}$$

The collapse load of the column is 195.4 kN with an eccentricity of 1.9915 m.

EXAMPLE 13.2.5

A prestressed concrete short column of rectangular cross-section $b \times D$ has an effective prestress P at constant eccentricity e_{po} below the centroidal axis. It carries an eccentric load W at eccentricity e above the centroidal axis. Derive expression for maximum working load for compression and tension failures if the maximum permissible stresses in compression and tension are σ_{cc} and σ_{ct}. Hence determine the load for a balanced failure. Also derive corresponding expression for the case of concentric prestress.

Solution

The column reaches the point of failure in compression when the compressive stress at top (compression) fibre becomes equal to σ_{cc}. Hence, if W_c is the load causing compression failure,

$$\frac{P}{bD} - \frac{P \cdot e_{po}}{\frac{1}{6} b D^2} + \frac{W_c}{bD} + \frac{W_c \cdot e}{\frac{1}{6} b D^2} = \sigma_{cc} \qquad (13.2.66)$$

or $\qquad W_c = \dfrac{\sigma_{cc}\, b\, D - P\left(1 - \dfrac{6e_{po}}{D}\right)}{\left(1 + \dfrac{6e}{D}\right)} \qquad (13.2.67)$

Similarly, the load W_t corresponding to tension failure may be determined from,

$$- \frac{P}{bD} - \frac{P \cdot e_{po}}{\frac{1}{6} bD^2} - \frac{W_t}{bD} + \frac{W_t \cdot e}{\frac{1}{6} bD^2} = \sigma_{ct} \qquad (13.2.68)$$

or

$$W_t = \frac{\sigma_{ct} bD + P \left(1 + \frac{6e_{po}}{D} \right)}{\left(\frac{6}{D} \frac{e}{D} - 1 \right)} \qquad (13.2.69)$$

The condition for a balanced failure, in which the column attains the maximum permissible stresses in compression and tension simultaneously, may be obtained by equating W_c and W_t. Hence the eccentricity of load e_b for a balanced failure may be expressed as

$$e_b = \frac{(\sigma_{cc} + \sigma_{ct}) \frac{bD^2}{6} + 2P \cdot e_{po}}{(\sigma_{cc} - \sigma_{ct}) bD - 2P} \qquad (13.2.70)$$

An expression for load W_b for a balanced failure may be obtained by substituting from Eq. (13.2.70) into Eq. (13.2.67) or (13.2.69). The associated bending moment M_b at balanced failure is evidently equal to $W_b e_b$. In the case of concentric prestress, $e_{po} = 0$.

$$e_b = \frac{(\sigma_{cc} + \sigma_{ct}) \frac{bD^2}{6}}{(\sigma_{cc} - \sigma_{ct}) bD - 2P} \qquad (13.2.71)$$

If no tension is permitted, $(\sigma_{ct} = 0)$ and the section is concentrically prestressed, $(e_{po} = 0)$,

$$e_b = \frac{\sigma_{cc} \frac{bD^2}{6}}{\sigma_{cc} bD - 2P} \qquad (13.2.72)$$

From Eq. (13.2.72), it follows that in the absence of prestressing force P, the eccentricity of load e_b reduces to $D/6$ in agreement with well known middle third rule.

EXAMPLE 13.2.6

A prestressed concrete short column of rectangular cross-section 0.2×0.6 m is prestressed concentrically with an effective force of 0.6 MN. Determine the working load corresponding to balanced failure, given that $\sigma_{cc} = 15$ MPa and $\sigma_{ct} = 3$ MPa. Hence determine the working loads corresponding to compression and tension failures if the bending moment due to eccentric load is 0.102 MN.

Solution

Using Eq. (13.2.71) the eccentricity of load for a balanced failure,

$$e_b = \frac{(15 + 3) \times \frac{0.2 \times 0.6^2}{6}}{(15 - 3) \times 0.2 \times 0.6 - 2 \times 0.6} = 0.9 \text{ m}$$

Substituting this eccentricity into Eq. (13.2.67) or (13.2.69), the load corresponding to a balanced failure, $W_b = 0.12$ MN. The associated bending moment at balanced failure is $W_b e_b = 0.12 \times 0.9 = 0.108$ MN.m. For bending moment, $M = W_c \cdot e = 0.102$ MN.m, the load for a compression failure, according to Eq. (13.2.66), $W_c = 0.18$ MN. Similarly for $M = W_t \cdot e = 0.102$ MN.m. the load for a tension failure according to Eq. (13.2.68), $W_t = 0.06$ MN. It may be verified that the same results are obtained from non-dimensionalised interaction diagram of Fig. 13.2.4.

EXAMPLE 13.2.7

A long column hinged at both ends has an unsupported length of 6 m. It has a rectangular cross-section 0.2×0.6 m. It is prestressed concentrically with an effective force of 0.36 MN. It carries a load of 0.48 MN at an eccentricity of 0.045 m with respect to the minor axis. Determine the extreme fibre stresses in concrete at midheight of the column. Assume an initial crookedness of 10 mm. Take $E_c = 30000$ MPa.

Solution

Euler buckling load,

$$W_E = \frac{\pi^2 E_c I}{1_e^2} = \frac{\left(\frac{22}{7}\right)^2 \times 30000 \times \frac{1}{12} \times 0.6 \times 0.2^3}{6^2}$$

$$= 3.293 \text{ MN}$$

Using Eq. (13.2.25),

$$e_{max} = (0.045 + 0.010) \sec\left(\frac{\pi}{2}\sqrt{\frac{0.48}{3.293}}\right)$$

$$= 0.0666 \text{m}$$

Extreme fibre stress on compression side,

$$f_T = -\frac{0.36}{0.2 \times 0.6} - \frac{0.48}{0.2 \times 0.6} - \frac{0.48 \times 0.0666}{\frac{1}{6} \times 0.6 \times 0.2^2}$$

$$= -14.99 \text{ MPa}$$

Extreme fibre stress on tension side,

$$f_B = -\frac{0.36}{0.2 \times 0.6} - \frac{0.48}{0.2 \times 0.6} + \frac{0.48 \times 0.0666}{\frac{1}{6} \times 0.6 \times 0.2^2}$$

$$= 0.99 \text{ MPa}$$

EXAMPLE 13.2.8

Determine the eccentricity of cracking load for the column of Example 13.2.7. The tensile strength of concrete is 2.7 MPa. Also determine the minimum prestressing force P_{min} required to prevent cracking and maximum prestressing force P_{max} to restrict compressive stress to 15.6 MPa when eccentricity of loading is 45 mm.

Solution

taking f_t (max) = 2.7 MPa, $I_t = \dfrac{1}{12} \times 0.6 \times 0.2^3 = 0.0004 \text{ m}^4$ and

$A_c \approx A_t = 0.2 \times 0.6 = 0.12 \text{ m}^2$, Eq. (13.2.28) may be written as

$$2.7 = \left[\frac{0.48 \, (e + 0.010) \text{ secant} \left(\dfrac{\pi}{2} \sqrt{\dfrac{0.48}{3.293}} \right) \times 0.1}{0.0004} - \frac{0.36}{0.12} - \frac{0.48}{0.12} \right]$$

Hence, $e = 0.0567$ m

The minimum prestressing force P_{min} required to prevent cracking may be determined from Eq. (13.2.28) in which f_t (max) = 2.7 MPa and e_{max} = 0.0666 m from Example 13.2.7.

$$2.7 = \frac{0.48 \times 0.0666 \times 0.1}{0.0004} - \frac{P_{min}}{0.12} - \frac{0.48}{0.12}$$

or $\qquad P_{min} = 0.1548 \text{ MN}$

From Eq. (13.2.27), taking f_c (max) = 15.6 MPa,

$$15.6 = \frac{P_{max}}{0.12} + \frac{0.48}{0.12} + \frac{0.48 \times 0.0666 \times 0.1}{0.0004}$$

or $\qquad P_{max} = 0.433 \text{ MN}$

EXAMPLE 13.2.9

A column of rectangular cross-section 0.3×0.5 m is concentrically presressed with an effective force of 0.6 MN. The working stress in compression is 16.5 MPa. The column which has an unsupported length of 6.6 m may be assumed to be fixed at base and hinged at top. Determine the working load of the column using provision of the Indian Code.

Solution

According to IS: 456-1978, the effective length factor for the given end conditions, $c = 0.7$. Hence effective length, $l_e = 0.7 \times 6.6 = 4.62$ m and the slenderness ratio $(l_e/b) = 4.62/0.3 = 15.4$ which is more than 12. Also $l_e/k_{min} = 4.62/0.0866 = 53.3$ which is more than 40. Hence the column is long. According to Eq. (13.2.31), the reduction factor,

$$C_r = 1.25 - \frac{15.4}{48} = 0.929$$

According to Eq. (13.2.32). the reduction factor ,

$$C_r = 1.25 - \frac{53.3}{160} = 0.917$$

The working load of the corresponding short column W_w is controlled by the working stress in compression (16.5 MPa) and minimum eccentricity according to Sec. 13.2.4,

$$e_{min} = \frac{l_u}{500} + \frac{b}{30} = \frac{6.6}{500} + \frac{0.3}{30} = 0.0232 \text{ m}$$

Hence,

$$16.5 = \frac{0.6}{0.3 \times 0.5} + \frac{W_w}{0.3 \times 0.5} + \frac{W_w \times 0.0232}{\frac{1}{6} \times 0.5 \times 0.3^2}$$

or $W_w = 1.281$ MN

Applying the reduction factor, the working load of the long column is $1.281 \times 0.929 \times 1.19$ MN or $1.281 \times 0.917 = 1.175$ MN.

EXAMPLE 13.2.10

A prestressed concrete column of a braced frame has a rectangular cross-section 0.2×0.3 m. It carries a design ultimate load of 0.36 MN together with end moments of 0.018 MM.m and 0.036 MN.m about major axis and 0.0096 MN.m and 0.024 MN.m about minor axis. The column bends into double curvature about both axes. The effective lengths for buckling about major and minor axes are 7.2 m and 5.0 m respectively. Determine the design ultimate moments including the effect of buckling.

Solution

(i) *Buckling about major axis*

Using Eq. (13.2.38), additional moment about major axis,

$$M_{ax} = \frac{0.36 \times 0.3}{2000}\left(\frac{7.2}{0.3}\right)^2 = 0.0311 \text{ MN.m}$$

As the column bends into double curvature, M_{u1} is negative. Hence $M_{u1} = -0.018$ MN.m and $M_{u2} = 0.036$ MN.m. Using Eq. (13.2.40), the initial moment,

$$M_{ui} = 0.4 \ (-0.018) + 0.6 \ (0.036) = 0.0144 \text{ MN.m}$$

which is not less than $0.4 M_{u2}$. The net design ultimate moment,

$$M_{ux} = M_{ui} + M_{ax} = 0.0144 + 0.0311 = 0.0455 \text{ MN.m}$$

which is not less than M_{u2}.

(ii) *Buckling about minor axis*

Using Eq. (13.2.39), additional moment about minor axis,

$$M_{ay} = \frac{0.36 \times 0.2}{2000}\left(\frac{5.0}{0.2}\right)^2 = 0.02225 \text{ MN.m}$$

Using Eq. (13.2.40), the initial moment,

$$M_{u1} = 0.4 \ (-0.0096) + 0.6 \ (0.024) = 0.01056 \text{ MN.m}$$

which is not less than $0.4\ M_{u2}$. The net design ultimate moment,

$$M_{uy} = M_{ui} + M_{ay} = 0.01056 + 0.02225 = 0.03281 \text{ MN.m.}$$

which is not less than M_{u2}.

EXAMPLE 13.2.11

Determine the design uitimate moments for the coumn of Example 13.2.10 if it belongs to an unbraced frame and bends into single curvature about both axes.

Solution

(i) *Buckling about major axis*

Additional moment for buckling about major axis, $M_{ox} = 0.0311$ MN·m as in Example 13.2.10. As the column bends into single curvature, $M_{u1} = 0.018$ MN.m and $M_{u2} = 0.036$ MN.m. Hence the net design ultimate moments are $(0.018 + 0.0311) = 0.0491$ MN.m and $(0.036 + 0.0311) = 0.0671$ MN.m.

(ii) *Buckling about minor axis*

Additional moment for buckling about minor axis, $M_{ay} = 0.02225$ MN.m as in Example 13.2.10. As the column bends into single curvature, $M_{u1} = 0.0096$ MN.m and $M_{u2} = 0.024$ MN.m. Hence the net design ultimate moments are $(0.0096 + 0.02225) = 0.03185$ MN.m and $(0.024 + 0.02225) = 0.04625$ MN.m.

EXAMPLE 13.2.12

A tied prestressed concrete column of square cross-section of side 0.3 m has an effective length of 7.2 m. The design ultimate load and end moments are 0.36 MN, 0.018 MN.m and 0.036 MN.m. Determine the magnified moments including the effects of cracking and creep when it belongs to

(i) a braced frame and bends into double curvature.

(ii) an unbraced frame and bends into single curvature.

Assume that 70 per cent and 30 per cent of the end moments are caused by gravity loads and sway loads respectively and the dead load moment is 40 per cent of the total moment. Take $E_c = 30000$ MPa and $f_c' = 28$ MPa. All columns of the storey are identical.

Solution

(i) *Braced frame*

As the column bends into double curvature, $c_m = 1$ according to Eq. (13.2.49).

$$0.1\ bD f_c' = 0.1 \times 0.3 \times 0.3 \times 28 = 0.252 \text{ MN}$$

As $W_u = 0.36$ MN is greater than 0.252 MN, capacity reduction factor $\Phi = 0.7$. From Eq. (13.2.51), flexural rigidity,

$$E_c I_t = \frac{0.4 \times 30000 \times \frac{1}{12} \times 0.3 \times 0.3^3}{1 + 0.4} = 5.786 \text{ MN.m}^2$$

Hence Euler crippling load,

$$W_E = \frac{\left(\frac{22}{7}\right)^2 \times 5.786}{7.2^2} = 1.102 \text{ MN}$$

From Eq. (13.2.47),

$$= \frac{1}{1 - \dfrac{0.36}{0.7 \times 1.102}} = 1.875$$

From Eq. (13.2.48),

$$\delta_s = \frac{1}{1 - \dfrac{0.36}{0.7 \times 1.102}} = 1.875$$

From Eq. (13.2.52), the magnified moment,

$$M_m = 1.875 (0.7 \times 0.036) + 1.875 (0.3 \times 0.036) = 0.0675 \text{ MN.m.}$$

(ii) *Unbraced frame*

As the column bends into single curvature, $(M_{u1}/M_{u2}) = (0.018/0.036) = 0.5$. Hence according to Eq. (13.2.49),

$$C_m = 0.6 + 0.4 \times 0.5 = 0.8$$

As in (i) $\Phi = 0.7$ and $W_E = 1.102$ MN. From Eq. (13.2.47) and (13.2.48),

$$\delta_b = \frac{0.8}{1 - \dfrac{0.36}{0.7 \times 1.102}} = 1.5$$

$$\delta_s = \frac{1}{1 - \dfrac{0.36}{0.7 \times 1.102}} = 1.875$$

From Eq. (13.2.52), the magnified moment,

$$M_m = 1.5 (0.7 \times 0.036) + 1.875 (0.3 \times 0.36)$$
$$= 0.05805 \text{ MN.m.}$$

EXAMPLE 13.2.13

A prestressed concrete column of rectangular cross-section 0.45×0.60 m is fixed at base and hinged at top. The unsupported length is 4 m. It carries a load of 1.35 MN at an eccentricity of 84 mm and 69 mm with respect to major and minor axes. The column is prestressed concentrically with an effective force of 0.729 MN. Determine the maximum compressive and tensile stresses and the position of netural axis. Also determine (i) maximum prestressing force if compressive stress is restricted to 17.2 MPa, (ii) minimum prestressing force to eliminate tension and (iii) minimum prestressing force to prevent cracking if the tensile strength of concrete is 2.4 MPa.

Solution

Referring to Fig. 13.2.8, the coordinates of the point P at which the eccentric load acts are (0.069 m, 0.084 m). With cross-sectional properties based on gross area of concrete, the stress at any point X having coordinates (x, y) in accordance with Eq. (13.2.53),

$$f_X = -\frac{0.729}{0.45 \times 0.6} - \frac{1.35}{0.45 \times 0.6} - \frac{1.35 \times 0.084\,y}{\frac{1}{12} \times 0.45 \times 0.6^3}$$

$$- \frac{1.35 \times 0.069\,x}{\frac{1}{12} \times 0.6 \times 0.45^3}$$

$$= -7.7 - 14\,y - 20.444\,x$$

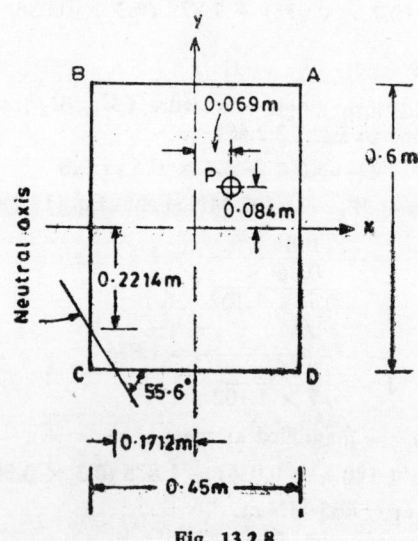

Fig. 13.2.8

At A ($x = 0.225$, $y = 0.3$), maximum compressive stress,

$$f_A = 16.5 \text{ MPa}$$

At C ($x = -0.225$, $y = -0.3$), maximum tensile stress,

$$f_C = 1.1 \text{ MPa}$$

As the stress at any point lying on neutral axis is zero, the equation for neutral axis may be written as

$$f_X = -7.7 - 14y - 20.444x = 0$$

or

$$2.655x + 1.818\dot{y} = -1$$

The position of neutral axis is shown in the figure. It is inclined at 55.6° to x-axis.

(i) If P_{max} is the maximum prestressing force without exceeding the compressive stress of 17.2 MPa at A,

$$-17.2 = -\frac{P_{max}}{0.45 \times 0.6} - \frac{1.35}{0.45 \times 0.6} - \frac{1.35 \times 0.084 \times 0.3}{\frac{1}{12} \times 0.45 \times 0.6^3}$$

$$-\frac{1.35 \times 0.069 \times 0.225}{\frac{1}{12} \times 0.6 \times 0.45^3}$$

or $P_{max} = 0.918$ MN

(ii) If P_{min} is the minimum prestressing force for no tensile stress at C,

$$0 = -\frac{P_{min}}{0.45 \times 0.6} - \frac{1.35}{0.45 \times 0.6} + \frac{1.35 \times 0.084 \times (-0.3)}{\frac{1}{12} \times 0.45 \times 0.6^3}$$

$$+\frac{1.35 \times 0.069 \times (-0.225)}{\frac{1}{12} \times 0.6 \times 0.45^3}$$

or $P_{min} = 1.026$ MN

(iii) If P_{min} is the minimum prestressing force to prevent cracking,

$$2.4 = -\frac{P_{min}}{0.45 \times 0.6} - \frac{1.35}{0.45 \times 0.6} + \frac{1.35 \times 0.084 (-0.3)}{\frac{1}{22} \times 0.45 \times 0.6^3}$$

$$+\frac{1.35 \times 0.069 (-0.225)}{\frac{1}{12} \times 0.6 \times 0.45^3}$$

or $P_{min} = 0.378$ MN

EXAMPLE 13.2.14

A prestressed concrete column of rectangular cross-section 0.36×0.54 m carries a design ultimate load of 1112 kN together with biaxial moments of 0.2 MN.m and 0.16 MN.m with respect to major and minor axes. It is prestressed concentrically by means of prestressing tendons having an area of 800 mm^2 at an effective stress of 960 MPa. The uniaxial ultimate moment capacities about major and minor axes are 0.4 MN.m and 0.25 MN.m respectively. Check the adequacy of the section, given that $f_{ck} = 35$ MPa, $E_c = 33000$ MPa and $\epsilon_u = 0.003$.

Solution

$$A_c \approx A_g = 0.36 \times 0.45 = 0.1944 \text{ m}^2$$

$$P = 800 \times 10^{-6} \times 960 = 0.768 \text{ MN}$$

$$\epsilon_{ce} = \frac{0.768}{0.1944 \times 33000} = 0.00012$$

Using Eq. (13.2.58), the ultimate strength of corresponding axially loaded short column,

$$
\begin{aligned}
W_{uo} &= 0.45 \,(0.1944 \times 35 - 0.768) \\
&\quad + 0.0008 \times 2 \times 10^5 \,(0.003 - 0.00012) \\
&= 3.177 \text{ MN}
\end{aligned}
$$

$$\frac{W_u}{W_{uo}} = \frac{1.112}{3.177} = 0.35$$

Hence the exponent, α_n in Eq. (13.2.56) is 1.25

$$
\begin{aligned}
\left(\frac{M_{ux}}{M_{uxo}}\right)^{\alpha_n} + \left(\frac{M_{uy}}{M_{uyo}}\right)^{\alpha_n} &= \left(\frac{0.2}{0.4}\right)^{1.25} + \left(\frac{0.16}{0.25}\right)^{1.25} \\
&= 0.993
\end{aligned}
$$

Hence the column section is adequate in accordance with Eq. (13.2.56).

EXAMPLE 13.2.15

Design a spirally reinforced prestressed concrete circular column of 0.3 m diameter having an unsupported length of 4.5 m to carry the design ultimate load of 520 kN at an eccentricity of 75 mm using PCI tentative recommendations. The column may be assumed to be hinged at both ends.

Solution

Using M-40 concrete, the cylinder strength,

$$f_{cy} = 0.8 \times 40 = 32 \text{ MPa}$$

Using 0.8 per cent untensioned steel with $f_y = 250$ MPa,

$$A_{sc} = 0.008 \times \frac{\pi}{4} \times 0.3^2 = 0.00057 \text{ m}^2$$

Adopting 6 bars of 12 mm diameter,

$$A_{sc} = 6 \times \frac{\pi}{4} \times 0.012^2 = 0.00068 \text{ m}^2$$

These bars may be arranged along a concentric circle with a clear concrete cover of 25 mm. Adopting an effective prestress of 3 MPa, the effective prestressing force,

$$P = 3 \times \frac{\pi}{4} \times 0.3^2 = 0.212 \text{ MN}$$

Assuming an effective stress of 960 MPa, the area of prestressing steel,

$$A_p = \frac{0.212}{960} = 0.000221 \text{ m}^2$$

Using 6 wires of 7 mm diameter,

$$A_p = 6 \times \frac{\pi}{4} \times 0.007^2 = 0.000231 \text{ m}^2$$

$$P = 0.000231 \times 960 = 0.222 \text{ MN}$$

$$A_c = \frac{\pi}{4} \times 0.3^2 - 0.00068 - 0.000231 = 0.0698 \text{ m}^2$$

Taking $\epsilon_u = 0.003$, $E_c = 36000$ MPa and $E_s = 2 \times 10^5$ MPa, the ultimate strength of the corresponding axially loaded short column in accordance with Eq. (13.2.59),

$$W_{uo} = 0.85 \times 32 \times 0.0698 + 0.00068 \times 250 - 0.222$$

$$+ 0.000231 \times 2 \times 10^5 \left(0.003 - \frac{0.222}{0.0698 \times 36000} \right)$$

$$= 1.981 \text{ MN}$$

From Eq. (13.2.61), taking $f_{pu} = 1800$ MPa,

$$e' = 0.23 \times 0.3 \left(1 + \frac{200 \times 0.222}{0.0698 \times 1800} \right) = 0.0934 \text{ m}$$

which is less than 0.5 D. As e is less than e', the ultimate strength of eccentrically loaded short column W_u is given by Eq. (13.2.62) in which,

$$W'_{uo} = 0.425 \times 32 \times 0.0698 \left(1 - \frac{95 \times 0.222}{0.0698 \times 1800} \right)$$

$$= 0.79 \text{ MN}$$

Hence from Eq. (13.2.61),

$$W_u = \frac{1.981}{1 + \left(\dfrac{1.981 - 0.79}{0.79} \right) \times \dfrac{0.075}{0.0934}} = 0.896 \text{ MN}$$

For a column hinged at both ends, $1_e = 1_u = 4.5$ m. For a circular column, the radius of gyration, $k = \dfrac{D}{4} = 0.075$ m..

$$\frac{1_e}{k} = \frac{4.5}{0.075} = 60$$

As the slenderness ratio is more than 30 but less than 100, the column is intermediate. From Eq. (13.2.64), the reduction factor,

$$C_r = 1.22 - 0.007 \times 60 = 0.8$$

Applying the reduction factor and capacity reduction factor of 0.75, the design ultimate strength of spirally reinforced intermediate column,

$$W_u = 0.896 \times 0.8 \times 0.75 = 0.5376 \text{ MN}$$

As the design strength is greater than the design ultimate load, the section is adequate. The untensioned longitudinal bars and the prestressing wires may be enclosed by a spiral reinforcement made from 6 mm diameter wire of mild steel. For a clear concrete cover of 20 mm on the spiral, the mean diameter of the spiral should be 0.254 m. A pitch of 40 mm for the spiral may be adopted.

13.3 PILES

Properly designed prestressed concrete piles have proved to be more economical as compared to steel, timber and reinforced concrete piles. They have found wide applications in almost all types of foundations, piers, abutments, bulkheads and marine structures such as wharves, mooring and fender dolphins Piles may be classified according to the purpose served by them. A load bearing pile serves the most common purpose of transferring the load of the superstructure and substructure and foundation to the soil. While a bearing pile transfers its load to the supporting rock by point bearing, a friction pile develops its load capacity by skin friction with the surrounding soil. Sheet piles are used to retain earth and hydrostatic pressure in embankments and excavations. They are widely used for the construction of retaining walls, cut off walls, wave baffles and water front bulkheads. Fender piles are used in marine structures such as wharves and jetties for absorbing the energy due to the impact of ships during berthing. Anchor piles serve the object of anchoring down a structure to the bedrock. The prestressing tendons in an anchor pile are anchored to the bedrock at one end and secured to the structure through the pile to the other end. Piles are usually categorised as compression members because axial compression is generally the dominant internal force due to service loads. However, there are instances where bending moment is more dominant. For example, a short pile is primarily a flexural member. Eccentric prestress may be used with advantage in these piles if bending moment is not reversible. On the other hand, a larger concentric prestress may be required in the case of reversible bending moment. Soldier piles used for bulkheads, sea walls and retaining walls are yet another example in which bending moment is predominant.

Although there are many cross-sectional shapes which may be adopted for a pile such as triangle, square, regular polygon, circle, rectangle with semi-circular ends. I and star, the most common shapes are square and circles. The relative merits and demerits of the cross-sectional shapes may be evaluated in terms of (i) ratio of surface area for skin friction to area of cross-section, (ii) resistance to bending about the two principal axes, (iii) resistance to wave and current forces, (iv) ease of placement and compaction of concrete, (v) simplicity of formwork and (vi) overall economy in construction. Sheet piles may also have different cross-sectional shapes but the rectangular shapes with or without longitudinal core openings is most common.

Prestressed concrete piles are concentrically prestressed with a relatively low prestress. The advantages of prestressing discussed in Sec. 1.14 apply also to prestressed concrete piles. However, the following advantages apply particularly to prestressed concrete piles.

(i) Superior characteristic during handling and driving
(ii) Resistance to tensile stresses due to uplift
(iii) Possibility of using thin walled hollow sections
(iv) Ease of connection with pile cap and
(v) Reduced tendency of buckling due to prestress

13.3.1 Manufacture

Prestressed concrete piles are generally precast and prestressed either by pretensioning or post-tensioning.

(a) Pretensioned piles

Pretensioned piles are manufactured by using either long line system or individual mould system taking advantge of the respective merits discussed in Sec. 3.2. Piles with overall cross-sectional dimension as 750 mm have been pretensioned. Low water-cement ratio and an appropriate water reducing admixture are generally utilised to obtain high strength concrete. It is preferable to use low alkali cement, particularly when piles are used in brackish waters. Special care is taken for achieving good compaction in the portions near the head and the toe which usually have closely spaced spirals or ties and may have dowels and tubes for preformed ducts. Stiff paper tubes or inflatable rubber mandrels used in the case of hollow piles should be secured firmly to the mould in order to prevent excessive floatation during vibration. An appropriate system of accelerated curing (Sec. 2.2) is commonly used for quicker release of moulds and consequent increase in production capacity. The main disadvantage of a pretensioned pile is that it cannot be extended easily and consequently leads to the adoption of unduly long piles. However, this disadvantge is offset by the technical advantages of long line system.

(b) Post-tensioned piles

The main advantage of a post-tensioned pile is that it can be more easily extended by releasing the prestress, forming the additional length and retensioning the prestressing tendons. Alternatively a pile may be cast in segments of appropriate length and then assembled and post-tensioned to obtain the necessary length. This technique is particularly suitable for piles of variable cross-section. For instance, the segments near the head and toe may be solid while the remaining segments may be of hollow cross-section. In order to prevent the failure of a hollow segment during driving, the diameter of the hollow core should generally not exceed half the side of a square pile. The hollow segment is vulnerable to torsional failure mainly due to the rigidity of the driving rig and the restraint imposed by the helmet. Hence it is advisable to ensure free torsional rotation of the pile by making it circular over a short length near the top. The change of cross-section from solid to hollow should be gradual to avoid stress concentration. The portions of the pile near the head and the toe should be reinforced by untensioned steel and little or no prestress should be applied in these regions. The joints between precast segments should be wide enough to facilitate satisfactory compaction of concrete.

13.3.2 Driving Stresses

During driving piles are subjected to most severe stresses which are seldom exceeded in their entire useful life. They may be damaged during the driving process unless adequate care is taken in their design, manufacture and driving operation itself. The capacity of the pile to resist driving stresses is enhanced by prestress of relatively low intensity generally believed to be in the range of 4 to 5 MPa. The dynamic stresses during driving depend upon the impact energy released by the hammer, rigidity of the driving rig, restraint imposed by the helmet and the physical characteristics of the cushion, the soil through which the pile is driven and those of the pile itself. Although a number of semi-empirical equations have been suggested, none of them can predict the stresses with reasonable accuracy due to the influence of a large number of parameters. The impact of the hammer causes a shock which travels from the head to the toe and then rebounds towards the head. If the main paramenters are know, precisely, the effects of the forward and reflected shock waves and the stresses caused by them can be determined with reasonable accuracy. The propagation of shock waves is accompanied by dynamic compressive and tensile stresses in the longitudinal direction. While the compressive stresses are seldom critical, the tensile stresses due to inertial rebound of the pile may be large enough to cause cracking unless the pile is suitably prestressed. The tensile stresses become critical with increasing length of the pile. Consequently, the importance of prestress increases for longer piles. In long and heavy piles, the compressive stress at the toe may be twice as great as that at the head due to the reflection of the shock wave. In order to prevent possible damage during driving, the following precautions may be taken.

(i) As the pile is required to store a large amount of energy during driving, the resilience or ductility of the pile is as important as its strength. While high strength can be achieved by means of accelerated curing, it is not advisable to commence driving unless sufficient resilience is achieved. A period of 3-4 weeks after casting is considered necessary for this purpose. The intensity of prestress should be relatively low bacause highly prestressed pile have very limited ductility.

(ii) The top of the pile should be plane and normal to the axis. Irregular and inclined top may eccentuate the driving stresses leading to cracking or spalling near the pile head. The prestressing strands should be cut flush with the top surface because even slightly projecting tendons are known to cause trouble in spite of the cushion. Chamfering of edges is found to be useful in preventing spalling near the pile head.

(iii) If the jetting is used to facilitate driving, it is preferable to use internal jetting as compared to external jetting. The latter tends to enhance the possibility of wandering of the pile tip resulting in increascd eccentricity during driving.

(iv) As the driving stresses are greatly influenced by the type of cushion, it is important to use appropriate material and thickness for the cushion. Experience has shown that laminated softwood 100 mm to 200 mm in thickness generally gives the best results. The cushion should be replaced well before it is compressed and consequently loses its elasticity.

(v) To prevent torsional distress, the pile head should be free to twist inside the driving helmet.

13.3.3 Design Principles

The precast prestressed pile may pass through several stages of loading depending upon the technique and location of its manufacture vis-a-vis the construction site and the manner of transportation and handling.

(i) Casting

The concrete mix for the fabrication of piles is usually designed to give the 28-day strength in the range from 30 to 50 MPa. The strength of concrete at stress transfer is chosen so that the pile can be lifted and transported to curing yard immediately after stress transfer. A strength of 24 MPa is usually considered to be minimum permissible at the release of presrress. The intensity of effective prestress in long and heavy piles depends upon the stresses during lifting, transportation and hoisting and to a lesser extent on the magnitude of tensile stress during driving. It is both undesirable and uneconomical to adopt a prestress larger than the minimum because it may reduce its load capacity as well as ductility. The option of temporary prestress, which may be removed partially or fully after it is driven, is worth considering particularly for point bearing piles. If the axial load capacity of the pile section is greater than the resistance developed by skin friction in a friction pile and the handling stresses are not critical, the need for prestress may be questionable although it may still be justified to prevent cracking due to shrinkage.

(ii) Handling

The stresses during lifting, transportation and hoisting are generally critical only for long and heavy piles. When a pile of length l and weight w per unit length is lifted by holding it at two points, the bending moment due to self weight is the minimum when these points are located at a distance of $0.207\ l$ from the ends. The maximum bending moment in this case is $0.0214\ w\ l^2$. During hoisting the tip of the pile rests on the ground while it is lifted by holding it at a single point some distance away from the head. The bending moment during hoisting is minimum when the pile is held at a distance of $0.293\ l$ from the top of the pile. The maximum bending moment in this case is $0.0429\ w\ l^2$. Due to impact during handling, the bending moments are increased by the factor $(1 + i)$ in which i is the impact factor. The impact factor during handling of piles usually varies from 0.33 to 1.0. Under normal conditions, a value equal to 0.5 may be used. While the compressive stresses during handling are rarely critical, the tensile stresses should be kept within permissible limits by adopting a suitable intensity of uniform prestress. In aggressive environment in which the cracking has to be avoided, the tensile stress should be reasonably below the tensile strength of concrete in bending.

(iii) Driving

The nature of driving stresses is discussed in Sec. 13.2.2. Due to the complexity of the problem and the difficulty in determining the values of large number of relevant parameters, an empirical approach based on past experience is adopted in choosing a suitable intensity of prestress. As a check against possible damage during driving, the rated energy of the hammer released per blow is related to the size of the pile. Hammer energy ratings ranging from 10 to 70 kN.m have been recommended for piles with overall cross-sectional dimension varying from 250 mm to 750 mm.

(iv) Design Loading

A pile is categorised as a compression member because it is usually subjected to axial compression due to service loads. The axial load capacity of a pile is equal to its strength as a column or the resistance offered by the supporting soil whichever is smaller. In the case of a point bearing pile resting on firm rock, it may be assumed that the pile capacity is controlled by the section of the pile behaving like a column. In the case of a friction pile, the soil resistance due to skin friction may be determined from the established principles of soil mechanics. In the case of a combined bearing-cum-friction pile, the load capacity is contributed both by bearing and friction. Depending upon its length and cross-sectional dimensions, a pile may behave like a short or long column. The capacity of the pile, when it is controlled by its strength as a column, may be determined by using the analysis presented in Sec. 13.2, provided the effective length of the columns known. The effective length depends upon the conditions of support at pile ends and the restraint imposed by the surrounding soil against buckling. As the top of the pile is usually embedded in a relatively thick cap, it is reasonable to assume a fixed end condition at top. The tip of the pile may be assumed to be hinged in the case of a bearing pile. In a friction pile driven through relatively loose soil, the tip may be assumed to have no restraint against rotation or translation. The restraint imposed by the surrounding soil over the entire length of pile depends upon the nature of soil, particularly its resistance against lateral motion. It is evident that the determination of effective length of a pile behaving like a column is by no means easy. However, it may be stated that under normal conditions, the effective length varies between $0.5\ l$ and l where l is the length of the pile.

The permissible stresses in concrete, prestressing steel and untensioned reinforcement in a pile may be the same as in other prestressed structures. However, it may be stated that the permissible tensile stresses should be chosen with care because piles are usually exposed to subsoil water. In marine structures and other aggressive environments, the tensile stresses should be small enough to prevent cracking in order to ensure adequate durability. The Prestressed Concrete Institute (PCI) of United States has recommended permissible stresses shown in Table 13.3.1. Using the permissible compressive stress in concrete, the safe load on a pile may be expressed as

$$W = A_c \left(0.33 f_{cy} - 0.27 \frac{P}{A_c} \right) \tag{13.3.1}$$

where A_c = net area of concrete

f_{cy} = compressive (cylinder) strength of concrete

P = effective prestressing force

Table 13.3.1 Permissible Stresses in Prestressed Concrete Piles

Nature of Stress	Recommended value
(i) Stresses in Concrete	
(a) Uniform axial compression in foundation piles fully embedded in soils providing lateral support	$0.33 f_{cy} - 0.27 \dfrac{P}{A_c}$
(b) Uniform axial tension	
1. Permanent and repetitive	0
2. Transient	$0.5\sqrt{f_{cy}}$
(c) Compression due to bending	
1. Bridge and marine structures	$0.4 f_{cy}$
2. Other structures	$0.45 f_{cy}$
(d) Tension due to bending	
1. Permanent and repetitive	$0.34\sqrt{f_{cy}}$ for normal environment 0 for corrosive environment
2. Transient	$0.5\sqrt{f_{cy}}$
(e) Effective prestress	
1. For piles shorter than 12 m	2.8 to 4.9 MPa
2. For piles 12 m to 52 m	4.9 to 8.4 MPa
(ii) Stresses in Steel	
(a) Temporary stresses	
1. Due to temporary jacking force	$0.8 f_{pu}$ but not greater than the maximum value recommended by manufactturer
2. Pretensioning tendons immediately after transfer or post-tensioning tendons immediately after anchoring	$0.7 f_{pu}$
(b) Effective prestress	$0.6 f_{pu}$ or $0.8 f_{py}$ whichever is smaller.
(c) Unstressed prestressing steel	$0.5 f_{py}$ but not greater than 210 MPa.

in the case of a pile subjected to eccentric load or combined axial compression and bending, it should be checked that the following interaction relationship is satisfied:

$$\left(\frac{f_{cc}}{\sigma_{cc}} + \frac{f_{cbc}}{\sigma_{cbc}}\right) \leqslant 1 \qquad (13.3.2)$$

where f_{cc} = direct compressive stress in concrete

f_{cbc} = compressive stress in concrete due to bending.

The following points are of general interest in the selection of suitable cross-section, appropriate technique for manufacture and design of piles.

(i) A solid section, generally square, is preferable for short pile. A square, hexagonal or octagonal or section with hollow core may be advantageous for medium and long piles, particularly in soft and medium driving conditions. Hollow sections are liable to damage under hard driving conditions unless water jetting is used. The wall thickness of hollow sections should not be less than 50 mm.

(ii) When piles are standardised for the sake of mass production, the technique of pretensioning using long line system is most economical. Factory production has the advantage of high quality control. Transportation of pile upto a length of 30 m usually presents no special problems.

(iii) Piles may be pretensioned in small segments and assembled together by post-tensioning. This technique facilitates addition or removal of a segment and consequently avoids the situation when a pile is found either too long or too short.

(iv) In the case of sheet piles which are subjected to bending moment, it may be preferable to use post-tensioning wherein only a part of the prestress is applied during handling and full prestress is applied only when the pile is in position. Alternatively, sheet piles may be pretensioned to resist handling stresses and subsequently post-tensioned eccentrically to resist bending moment under service conditions.

13.3.4 Detailing

As the development of a prestressed concrete pile is largely empirical, its detailing assumes special importance.

(i) *Prestressing tendons*

Prestressing tendons in the form of high tensile steel wires or multiwire strands have usually an area not less than 0.5 per cent. The tendons are arranged in a circular configuration along a concentric circle to facilitate a use of simpler spiral reinforcement.

(ii) *Untensioned reinforcement*

The untensioned reinforcement is in the form of longitudinal bars and spiral.

(a) Longitudinal bars

Untensioned longituoinal reinforcing bars are usually provided in the end regions of the pile which are subjected to most severe stresses and are therefore more liable to failure during driving. The provision of untensioned reinforcement increases the capacity of the pile to absorb a greater amount of energy during shock loading. According to American Association of State Highway Officials (AASHO) and PCI, the percentage of reinforcement should be between 1.0 and 1.5 of the gross concrete section.

(b) Spiral reinforcement

While the spiral is circular in circular, hexagonal and octagonal piles, it may be circular or square in a square pile. However, due to the ease of fabrication, greater efficiency and economy, the circular spiral is usually preferred. A minimum of 0.6 per cent spiral reinforcement in the end regions (not less than three times the least lateral dimension of the pile) has been recommended by British Concrete Society. Spiral reinforcement not less than 1.0 per cent over a length of 300 mm from the ends, 0.6 per cent in the next 600 mm and 0.3 per cent for solid piles or 0.4 per cent for hollow piles in the body of the pile has been recommended by Gerwick. The pitch of the spiral should be reasonably close. Four or five tight turns of the spiral at a pitch of 25 mm near the head and the toe followed by a pitch of 75 mm in the end regions and 150 mm in the body of the pile are usually considered to be satisfactory.

(iii) Pile shoes

A steel shoe is provided at the tip to facilitate driving in relatively stiff soil. It is unnecessary in soft driving conditions through sand, silt or clay. The shoe is in the form of a thick steel plate or stub attached firmly to the tip of the pile. Field experience shows that a square shoe with sloping sides like that of a hopper bottom is preferable to a pointed or wedge shaped shoe. The latter tends to deflect the tip sideways leading to high bending stresses and possible failure of the shoe and tip. The untensioned longitudinal bars in the bottom end region of the pile are extended upto the shoe and secured to it by means of fillet or plug welds.

(iv) Connections

An adequate connection between the pile head and the pile cap may be achieved in a number of ways. When a pile is ordered to a length longer than the exact length which is usually the case, the prestressing tendons and reinforcing bars can be exposed by chipping off concrete in the extra length and embedding them in the pile cap. If a tensioned connection is desired, the prestressing tendons may pass through preformed or field holes in the pile cap and suitably anchored at the top of the cap. In untensioned connection, the reinforcing bars and prestressing tendons should have appropriate bond lengths. A bond length equal to 50 times the diameter of the tendon is usually sufficient. When the piles are ordered to the correct length and are consequently driven to grade, the prestressing tendons and reinforcing bars should be extended by the appropriate bond length beyond the head. In this case special driving head of follower is required so that the projecting reinforcement does not interfere in the driving operations. A very simple cap connection is possible in the case of a pile with hollow core. A reinforcing cage may be fabricated and partly inserted into the hollow core and the remaining portion in the pile cap and later embedded in cast-in-place concrete.

(v) *Splicing*

Due to the difficulty of transportation and for the sake of ease in driving, it is nesirable to cast long piles into suitable number of segments and connect them together by means of splices. The desirable properties of a splice are as follows:

(a) The compressive, flexural and shear strengths of the splice should be at least equal to those of the pile segments.
(b) The splice should quickly develop adequate strength so that driving can be resumed within reasonable time.
(c) The splice should ensure proper alignment of the pile axis as misalignment of the axis leads to additional bending moment.
(d) The splice should be durable and economical.

Various types of splices have been developed to connect precast pretensioned pile segments.

(a) Pile segments may be connected by means of dowel bars and plasticised cement. Holes are driven at appropriare locations in the upper region of the already driven lower segment to receive dowel bars precast in the lower region of the upper segment. Plasticised cement in molten state is used to fill the space between the two segments and the annular space between the holes and the dowel bars. Due to high strength and fast set of plasticised cement, the driving operation can be resumed after a relatively short curing time.
(b) Relatively slow setting epoxy compounds have been used in place of plasticised cement. Extensive test results have established their superiority.
(c) A splice may be formed by welding together steel pipe sleeves or anchored plates embedded at the ends of the two segments to be connected to each other.
(d) A simpler splice connection may be used when full moment capacity is not required. For instance, a splice conection may be formed by driving the upper segment into n snugly fitting steel pipe sleeve which has previously been driven onto the upper portion of the lower segment. A central dowel bar may be used for proper alignment of the axis.

EXAMPLE 13.3.1

A pile with a uniform square section of side a weighs w per unit length. Determine the minimum prestressing force to prevent (a) tensile stress and (b) cracking during (i) two point lifting and (ii) single point hoisting.

Solution

(i) *Two point lifting*

Maximum bending moment in this case, $M = 0.0214\ w\ l^2$.

(a) No Tensile Stress

The minimum effective prestressing force P to prevent tensile stress may be determined from

$$\frac{0.0214\ (1 + i)\ w\ l^2}{\frac{1}{6}\ a^3} - \frac{P}{a^2} = 0$$

or

$$P = \frac{0.1284\ (1 + i)\ w\ l^2}{a}$$

(b) No cracking

The minimum effective prestressing force P to prevent cracking may be determined from

$$\frac{0.0214\ (1 + i)\ w\ l^2}{\frac{1}{6}\ a^3} - \frac{P}{a^2} = f_{cr}$$

or

$$P = \frac{0.1284\ (1 + i)\ w\ l^2}{a} - f_{cr}\ a^2$$

where, f_{cr} = tensile strength of concrete in bending

(ii) *Single point hoisting*

Maximum bending moment in this case, $M = 0.0429\ w\ l^2$

(b) No Tensile Stress

The minimum effective prestressing force P to prevent tensile stress may be determined from

$$\frac{0.0429\ (1 + i)\ w\ l^2}{\frac{1}{6}\ a^3} - \frac{P}{a^2} = 0$$

or

$$P = \frac{0.2574\ (1 + i)\ w\ l^2}{a}$$

(b) No cracking

The minimum effective prestressing force P to prevent cracking may be determined from

$$\frac{0.0429 \ (1 + i) \ w \ l^2}{\frac{1}{6} \ u^3} - \frac{P}{a^2} = f_{cr}$$

or
$$P = \frac{0.2574 \ (1 + i) \ w \ l^2}{a} - f_{cr} \ a^2$$

EXAMPLE 13.3.2

A point bearing pile has a uniform circular section of 450 mm diameter. It is prestressed concentrically with an effective force of 716 kN. The length of the pile is 11.25 m. The pile may be assumed to be attached rigidly to a thick pile cap at top and fixed in position but not in direction at the tip. The lateral support from the surrounding loose soil may be ignored. Determine *the safe* load for the pile. The cylinder strength of concrete is 32 MPa.

Solution

Gross cross-sectional area A_g and radius of gyration k are,

$$A_g = \frac{\pi}{4} \times 0.45^2 = 0.1591 \ \text{m}^2$$

$$k = \frac{D}{4} = \frac{0.45}{4} = 0.1125 \ \text{m}$$

Using Eq. (13.3.1) and taking $A_c = A_g$ as an approximation,

$$W = 0.1591 \left(0.33 \times 32 - 0.27 \times \frac{0.716}{0.1591} \right)$$

$$= 1.487 \ \text{MN}$$

For the given end conditions, the effective length, $l_e = 0.7 \times 11.25$ =7.875 m. Hence the slenderness ratio,

$$\frac{l_e}{k} = \frac{7.875}{0.1125} = 70$$

Hence the reduction factor according to PCI recommendations, Eq. (13.2.64),

$$C_r = 1.22 - 0.007 \times 70 = 0.73$$

Hence the safe load for the pile is $0.73 \times 1.487 = 1.086$ MN.

EXAMPLE 13.3.3

A friction pile has a uniform section of regular hexagonal shape and circular hollow core. The size of the hexagon, i.e. the distance between its parallel sides is 500 mm. The safe permissible skin friction over its length of 5.8 m is 1.38 MN. Design a suitable section assuming an effective length factor of 0.7.

Solution

The cross-section of the pile may be chosen in such a way that its safe load as a column is equal to the safe permissible skin friction of 1.38 MN. Adopting prestress of 4.9 MPa and concrete with cylinder strength of 35 MPa, Eq. (13.3.1) gives,

$$1.38 = A_c (0.33 \times 35 - 0.27 \times 4.9)$$

or

$$A_c = 0.1349 \ m^2$$

Area of solid hexagon of 500 mm size is 0.2165 m^2. Hence the hollow core diameter, D_c may be determined from

$$0.2165 - \frac{\pi}{4} D_c^2 = 0.1349$$

or

$$D_c = 0.322 \ m$$

Minimum wall thickness is equal to $250 - 161 = 89$ mm. As it is greater than 50 mm, it is satisfactory. Effective length, $l_e = 0.7 \times 5.8 = 4.06$ m. The least radius of gyration of the hollow hexagonal section is 0.1548 m giving $(i_e/k) = (4.06/0.1548) = 26.22$. As the ratio is less than 30, the pile may be assumed to behave like a short column. The pile may be pretensioned using long line system. Total effective prestressing force,

$$P = 4.9 \times 0.1349 = 0.661 \ MN$$

Using 5 mm diameter wires with an effective stress of 970 MPa, the number of wires,

$$n = \frac{0.661}{\frac{\pi}{4} \times 5^2 \times 10^{-6} \times 970} = 34.7$$

Hence provide 35 wires arranged along a concentric circle of 400 mm diameter. The spacing between the wires is $(\pi \times 400/35) = 35.9$ mm which is satisfactory.

EXAMPLE 13.3.4

A prestressed concrete bearing pile has a square cross-section of 600 mm side with hollow circular core. The length of pile is 7.2 m. Design the pile for a safe load of 2.2 MN. The pile may be assumed to be fixed at top, hinged at bottom and laterally unsupported over its entire length. Use concrete of grade M-40 and long line system of prestressing. The pile has to be driven in aggressive environment and no tension is therefore permitted.

Solution

Using an effective prestress of 4 MPa and taking the cylinder strength of concrete $f_{cy} = 0.8 \times 40 = 32$ MPa, Eq. (13.3.1) gives,

$$2.2 = A_c (0.33 \times 32 - 0.27 \times 4)$$

or

$$A_c = 0.2321 \ m^2$$

If D_c is the diameter of the hollow core,

$$0.2321 = 0.6^2 - \frac{\pi}{4} D_c^2$$

or
$$D_c = 0.403 \text{ m}$$

Hence hollow core of 400 mm diameter may be adopted. The minimum wall thickness is 100 mm which is satisfactory. The least radius of gyration $k = 202$ mm. The slenderness ratio, $l_e/k = (0.7 \times 7.2)/0.202 = 24.95$. Hence the pile may be assumed to behave like a short column. The effective prestressing force,

$$P = 4 \left(0.6^2 - \frac{\pi}{4} \times 0.4^2 \right) = 0.9371 \text{ MN}$$

Using 7 mm diameter high tensile steel wires with effective stress of 960 MPa, the number of wires,

$$n = \frac{0.9371}{\frac{\pi}{4} \times 7^2 \times 10^{-6} \times 960} = 25.35$$

Hence use 25 wires arranged along a concentric circle of 500 mm diameter. These wires should run over the entire length of the pile. Additional untensioned reinforcing bars should be provided in the two end regions each of 1 m length. The end regions may have a solid square cross-section of 600 mm side. For 1.0 per cent reinforcement,

$$A_s = 0.6 \times 0.6 \times 0.01 = 0.0036 \text{ m}^2$$

Adopting 26 bars of 14 mm diameter, the area actually provided is 0.00i4 m² giving a percentage of 1.11. As this percentage lies between 1 and 1.5 t is satisfactory. These bars may be placed along a concentric circle of 493 mm diameter. Placing a reinforcing bar between every two consecutive tendons, the centre to centre spacing between the bar and the tendon is $\pi \times 500/52 = 30.2$ mm which is satisfactory. The tendons and bars have been placed in such a way that the same helix encloses them. Using mild steel wire of 10 mm diameter for the helix, the mean diameter of the helix should be 517 mm. Adopting pitch of the helix equal to 130 mm in the middle region or body of the pile, the volume percentage of helix,

$$\frac{\frac{\pi}{4} \times 0.01^2 \times \pi \times 0.517}{0.13 \left(0.6^2 - \frac{\pi}{4} \times 0.4^2 \right)} \times 100 = 0.419$$

which, being more than 0.4, is satisfactory. In the solid end regions, the pitch of the helix may be reduced to 80 mm over a length of 600 mm, 50 mm in the next 300 mm and 25 mm in the remaining 100 mm near the ends.

A hole of 25 mm diameter may be provided at a distance of 1.49 m from each end for two point lifting of the pile. A similar hole may be provided at 2.11 m from the top for hoisting.

EXAMPLE 13.3.5

A precast pretensioned pile has a hollow circular section with internal and external diameters equal to 320 mm and 480 mm respectively. It is pre-stressed concentrically by means of 22 wires of 5 mm diameter with an effective stress of 950 MPa. The wires are placed at equal spacing along a concentric circle of 400 mm diameter. The critical section of the pile is subjected to an axial compression of 200 kN and bending moment of 63 kN. m. The permissible compressive stresses in direct compression and bending are 12 and 15 MPa respectively. The cracking of concrete has to be prevented. The tensile strength of concrete in bending is 3.6 MPa. Take modular ratio, $m = 6$. Check the adequacy of the section.

Solution

Gross area of section,

$$A_g = \frac{\pi}{4} (0.48^2 - 0.32^2) = 0.1006 \text{ m}^2$$

Area of prestressing steel,

$$A_p = 22 \times \frac{\pi}{4} \times 0.005^2 = 0.00043 \text{ m}^2$$

Hence the net area of concrete,

$$A_c = 0.1006 - 0.00043 = 0.10017 \text{ m}^2$$

Area of transformed section,

$$A_t = 0.10017 + (6 - 1) \times 0.00043 = 0.10232 \text{ m}^2$$

Moment of inertia of transformed section which is half of polar moment of inertia is 0.002262 m⁴. Effective prestressing force,

$$P = 0.00043 \times 950 = 0.4085 \text{ MN}$$

Direct stress in concrete due to axial load and prestress,

$$f_{cc} = \frac{-0.4085}{0.10017} - \frac{0.2}{0.10232} = -6.03 \text{ MPa}$$

Extreme fibre stresses in concrete due to applied bending moment,

$$f_{cbc} = \pm \frac{0.063 \times 0.24}{0.002262} = \pm 6.68 \text{ MPa}$$

Referring to Eq. (13.3.2),

$$\frac{f_{cc}}{\sigma_{cc}} + \frac{f_{cbc}}{\sigma_{cbc}} = \frac{6.03}{12} + \frac{6.68}{15} = 0.948$$

Hence the pile is safe in compression. Net tensile stress is equal to $6.68 - 6.03 = 0.65$ MPa. As this stress is less than the tensile strength of concrete, cracking is prevented. The section is therefore adequate.

EXAMPLE 13.3.6

A prestressed concrete sheet pile has a rectangular section 1.2×0.2 m. It is subjected to maximum tensile stress of 6.7 MPa during handling and to a non-reversible bending moment of 109 kN. m due to service load. The cracking of concrete has to be prevented. Design suitable prestress.

Solution

Using concrete of grade M-40, the tensile strength in bending,

$$f_{cr} = 0.7\sqrt{40} = 4.4 \text{ MPa}$$

For a factor of safety equal to 2, the tensile stress should be restricted to 2.2 MPa in order to prevent cracking. The tensile stress of 6.7 MPa during handling may be resisted by means of concentric prestress using pretensioning method. If P_1 is the effective concentric prestressing force,

$$6.7 - \frac{P_1}{1.2 \times 0.2} = 2.2$$

or $\qquad P_1 = 1.08 \text{ MN}$

Using 7 mm diameter wires with effective stress of 960 MPa, total number of wires,

$$n = \frac{1.08}{\frac{\pi}{4} \times 7^2 \times 10^{-6} \times 960} = 29.2$$

Hence provide 15 wires near each face with a clear concrete cover of 20 mm. The non-reversible bending moment of 109 kN. m may be resisted by the combined action of concentric pretensioning and eccentric post-tensioning using Lee McCall system. Adopting an eccentricity of 50 mm for the post-tensioning force P_2,

$$\frac{0.109}{\frac{1}{6} \times 1.2 \times 0.2^2} - \frac{1.08}{1.2 \times 0.2} - \frac{P_2}{1.2 \times 0.2}$$

$$- \frac{P_2 \times 0.050}{\frac{1}{6} \times 1.2 \times 0.2^2} = 2.2$$

or $\qquad P_2 = 0.6648 \text{ MN}$

Using 12 mm diameter high tensile steel bars at an effective stress of 840 MPa, the number of bars,

$$in = \frac{0.6648}{\frac{\pi}{4} \times 12^2 \times 10^{-6} \times 840} = 7$$

These bars should be placed at a distance of 50 mm from tension face in order to obtain the desired eccentricity of 50 mm.

REFERENCES

13.1. ACI Committee 543, *Recommendations for design, manufacture and installation of concrete piles*, Journal of the American Concrete Institute, Vol. 70, No. 8, August 1973, pp. 509-644.

13.2. Aroni, S., *The strength of slender prestressed concrete columns*, Journal of the Prestressed Concrete Institute, Vol. 13; No. 2, April 1968, pp. 19-33.

13.3. Bertero, V.V. et. al., *A seismic design of prestressed concrete piling* FIP congress, New York, May 1974, pp. 9.

13.4. Bruce, R.N. and Hebert, D.C., *Splicing concrete piles: Part 1— Review aud performance of splices*, Journal of the Prestressed Concrete Institute, Vol. 19, No. 5, September-October 1974, pp. 70-97.

13.5. Bruce, R.N. and Hebert, D.C., *Splicing of precast prestressed concrete piles: Part 2—Test and analysis of cement-dowel splice*, Journal of the Prestressed Concrete Institute, Vol. 19, No. 6, November-December 1974, pp. 40-66.

13.6. Gerwick, B.C., *Torslon in concrete piles during driving*, Journal of the Prestressed Concrete Institute, Vol. 4, June 1959, pp. 58-63.

13.7. Glanville, W.H. et. al., *An investigation of the stresses in reinforced concrete piles during driving*, HMSO (London), 1938, 111 p.

13.8. Hall, A.S., *Buckling of prestressed columns*, Cement and Concrete Association, Sydney, Australia, 1961.

13.9 Kabaila, A.P. and Hall, A.S., *Analysis of instability of unrestrained prestressed concrete columns with end eccentricities*, Symposium on Reinforced Concrete Columns, ACI special publication SP-13, 1965, pp. 179-192.

13.10. Lin, T.Y. and Itaya, R., *A prestressed concrete column under eccentric loading*, Journal of the Prestressed Concrete Institute, December 1957.

13.11. Lin, T.Y. and Lakhwara, T.R., *Ulttmate strength of eccentrically loaded partially prestressed columns*, Journal of the Prestressed Concrete Institute, Vol. 31, 1986, pp. 37-49.

13.12. Nathan, N.D., *Slenderness of prestressed concrete beam-columns*, Journal of the Prestressed Concrete Institute, Vol. 17. No. 6, November-December 1972, pp. 45-57.

13.13. PCI Committee Report, *Tentative recommendations for the design of prestressed concrete columns*, Journal of the Prestressed Concrete Institute, Vol. 13, No. 5. October 1968, pp. 12-21.

13.14. PCI Committee Report, *Recommended practice for design, manufacture and installation of prestressed concrete piling*, Journal of the Prestressed Concrete Institute, Vol. 22, March-April 1977. pp. 20-49.

13.15. PCI Publication, *Design of prestressed concrete piles*, Prestressed Concrete Institute, Chicago, Illinois, 1977.

13.16. Shu-Tien Li and Tony Chen-Yeh Liu, *Prestressed concrete piling-contemporary design practice and recommendations*, Journal of the

American Concrete Institute, Proceedings Vol. 67, No. 3. March 1970, pp. 201-20.

13.17 Shu-Tien Li, *Functional optimum prestress for different classes of prestressed concrete piling*, Seminar on Problems of Prestressing, Preliminary Publication, Indian National Group of IABSE, Madras. 1970, pp. I-1-22.

13.18 Smith, E.A., *Pile calculations by the wave equation*, concrete and Constructional Engineering, Technical paper No. 20, June 1958.

13.19 Strobel, G.C. and Heald, J., *Theoretical and practical discussion of the design, testing and use of pretensioned prestressed concrete piling*, Journal of the Prestressed concrete Institute, Vol. 5, September 1961, pp. 22-33.

13.20 Weller, N.H.E., *Prestressed concrete piles*, Journal of the Prestressed Concrete Institute, Vol. 7, No. 5, October 1962, pp. 46-55.

13.21 Zia, P. and Moreadith, F.L., *Ultimate load capacity of prestressed concrete columns*, Journal of the American Concrete Institute, Vol, 63, No. 7, July 1966, pp. 767-88.

13.22 Zia, P. and Guillermo, E.C., *Combined bending and axial load in prestressed concrete column*, Journal of the Prestressed Concrete Institute, Vol. 12, No. 3, June 1967, pp. 52-9.

PROBLEMS

13.1 In the column of Example 13.2.1, determine the initial and final stresses with cross-sectional properties based on gross concrete section, i.e. $A_{ti}=A_t=A_c=A_g=bD$ and $I_{ti}=I_t=I_g=\frac{1}{12}bD^3$.

13.2 Determine the initial and final curvatures in the column of Example 13.2.1, taking $E_c=33000$ MPa.

13.3 Determine the stresses in concrete and steel due to restrained creep and shrinkage in the column of Example 13.2.1, given that the unrestrained creep coefficient $C_c=1.5$ and unrestrained shrinkage strain, $\epsilon_{sh}=0.0002$. The column carries a permanent axial load of 1.8 MN. Take $E_c=33000$ MPa and $m=6$. Also determine the approximnte stresses in concrete and steel, taking restrained creep and shrinkage equal to unrestrained values.

13.4 Determine the stresses in concrete and steel when the column of Example 13.2.3 carries a load of 280.8 kN at an eccentricity of 1.216 m.

13.5 Determine the stresses in concrete and steel when the column of Example 13.2.3 is loaded to its collapse load of 195.4 kN at an eccentricity of 1.992 m. The crushing strain of concrete, $\epsilon_u=0.003$.

13.6 A prestressed concrete short column of rectangular section 0.3 × 0.9 m carries an effective concentric prestressing force of 1.08 MN. Determine the load together with a bending moment of 0.26 MN. m corresponding to compression and tension failures, given that $\sigma_{cc}=12$ MPa and $\sigma_{ct}=1.8$ MPa. Also determine the oad and eccentricity corresponding to a balanced failure.

13.7 A prestressed concrete [?]long column of rectangular section 0.2 × 0.6 m hinged at both ends has an unsupported length of 6 m. The effective concentric prestressing force is 0.036 MN. The extreme fibre stresses in concrete at midheight are −15 MPa and 1 MPa. Using the secant formula, determine the load and its eccentricity with respect to the minor axis assuming that the initial crookedness is 10 mm. Take $E_c = 30000$ MPa.

13.8 Using the secant formula, determine the cracking load acting at an eccentricity of 56.7 mm with respect to minor axis for the column of Prob. 13.7.

13.9 The effective concentric prestressing force in the column of Prob. 13.7 is 155 kN. Using the secant formula, determine the axial load acting at an eccentricity of 45 mm with respect to minor axis when the column is at the stage of incipient cracking.

13.10 A square column of side 0.4 m is concentrically prestressed with an effective force of 0.48 MN. The working stress in compression is 15 MPa. The column, which has an unsupported length of 6 m, may be assumed to be hinged at both ends. Determine the working load of the column using slenderness ratio based on (i) least lateral dimension and (ii) least radius of gyration.

13.11 A prestressed concrete column of a braced frame has a rectangular section. It carries a design ultimate load of 0.45 MN together with end moments of 0.0225 MN. m and 0.045 MN. m about major axis and 0.012 MN. m and 0.03 MN. m about minor axis. The column bends into double curvature about both axes. The effective lengths for buckling about major and minor axes are 7.5 m and 5.4 m respectively. Design a suitable cross-section for the column using additional moments method.

13.12 Design a suitable cross-section for the column of Prob. 13.11 if it belongs to an unbraced frame and bends into single curvature about both axes.

13.13 A prestressed concrete column with lateral ties has a square cross-section. The unsupported length is 7.5 m and both ends are hinged. The design ultimate load and end moments are 0.40 MN., 0.02 MN. m and 0.04 MN. m. Using moment magnification method, design a suitable cross-section for the column if it belongs to a braced frame and bends into double curvature. Make due allowance for cracking and creep of concrete. Assume that the gravity load moment is 60 per cent and sway load moment is 40 per cent. The dead load moment is 50 per cent of the total moment. All columns of the storey are identical.

13.14 Design a suitable cross-section for the column of Prob. 13.13 if it belongs to an unbraced frame and bends with single curvature.

13.15 A prestressed concrete column of square cross-section is fixed at base and hinged at top. The unsupported length is 4.5 m. It has to carry a working load of 1.50 MN at eccentricities of 90 mm and 72 mm with respect to the two principal axes. Design a suitable cross-section of the column if the maximum permissible compressive stress is 15.9 MPa and no tension is permitted.

13.16 A prestressed concrete column of square cross-section has an effective length of 6.4 m. It has to carry a design ultimate load of 1.5 MN together with biaxial moments of 0.75 MN. m and 0.6 MN. m about principal axes. Design a suitable cross-section for the column using provisions of the Indian Code.

13.17 An eighteen metre long pile has a uniform square section of 600 mm side. Determine the minimum prestressing force to prevent tensile stress during (i) two-point lifting and (ii) single point hoisting. Unit weight of concrete is 24 kN/m^3. Take impact factor, $i = 0.5$.

13.18 Determine the limiting length of a square pile of 500 mm side to prevent cracking during (i) lifting and (ii) hoisting. The effective prestressing force is 800 kN and the modulus of rupture, $f_{cr} = 4.5$ MPa. Unit weight of concrete is 24 kN/m^3. Take impact factor, $i = 0.5$,

13.19 The cross-section of a pile is a regular hexagon of 240 mm side. It is prestressed by means of 36 wires of 5 mm diameter with an effective prestress of 950 MPa. The cylinder strength of concrete is 36 MPa. Determine the safe load of the pile.

13.20 A bearing pile hinged at tip and fixed at top is 21 m in length. Design the pile for a capacity of 1500 kN. The pile has a square section with hollow circular core. Use concrete of grade M-35 and Lee McCall system of post-tensioning.

13.21 The critical section of a precast post-tensioned pile is subjected to an axial load of 180 kN and a reversible bending moment of 50 kN. m. Design a suitable cross-section using concrete of grade M-40. The cracking of concrete has to be prevented.

13.22 A sheet pile of rectangular section has to resist a maximum tensile stress of 6.5 MPa during handling and a non-reversible bending moment of 100 kN. m due to service load. Design a suitable cross-section for the pile using concrete of grade M-45.

14
Slabs

14.1 INTRODUCTION

The slab is an important member of the class of structures known as surface structures. These structures can be idealised to a surface. When the
surface is plane, the structure is called a slab or a plate. Although slabs of
other shapes are also sometimes used, the circular and rectangular slabs
are by far the most common. In circular slabs, the support reactions and
loads are usually axi-symmetric and loads in the plane of the slab are
absent. Under these circumstances, the internal membrane forces lying in
the plane of the slab are absent. With reference to polar coordinates (r, θ),
suitable for the analysis of circular slabs, the transverse internal forces
acting on the slab are radial moment M_r and the corresponding flexural
shear Q_r in radial planes normal to the slab and circumferential moment
M_θ in the circumferential plane normal to the slab. The other transverse
internal forces, viz. circumferential shear Q_θ and twisting moments are
zero due to symmetry.

Rectangular slabs are commonly located in horizontal or xy-plane and
carry transverse loads in vertical or z-direction. In the absence of in-plane
loading, the membrane forces are zero. The slab is called one-way when
the deflected surface is cylindrical with zero curvature in one of the two
orthogonal directions. A common example of one-way slab is the long
corridor slab supported on its long edges and carrying uniform loading.
These slabs develop bending moment and shear force and consequently
carry their loading in only one direction. Their structural action is therefore similar to that of a beam. In slabs supported on all four edges or
carrying non-uniform loading, the structural response is invariably two way.
However, when the aspect ratio, i.e. the ratio of long span to short span is
greater than 3, the two way action is confined to relatively small area near
the short edges whereas one-way action prevails in the major portion of the
slab. When the aspect ratio is less than 2, or the slab carries non-uniform
loading, the two way action dominates. Such slabs bend in the shape of a
saucer with curvatures in both orthogonal directions. The internal forces in
transversely loaded two way slabs are bending moment M_x and shear force
Q_y in yz-plane. Q_x in x-z plane, bending moment M_y and shear force and
twisting moments in both the orthogonal planes. The shear forces Q_x and
Q_y in slabs, particularly prestressed ones, are seldom critical. Hence the

design of prestressed concrete slab is generally controlled by the biaxial moments M_x and M_y. For convenience in design, the codes prescribe appropriate coefficients for the determination of these moments for different aspect ratios and edge conditions.

Due to their aesthetic value, flat slabs supported directly on columns are extensively used in buildings. The nature of internal forces in flat slabs is the same as in two-way slabs. However, flat slabs have to resist column reactions which are localised over a relatively small area of the slab. Due to the complexity of distribution of internal forces in flat slabs, they are usually designed using empirical coefficients based on long experience and good engineering practice.

The thickness of slab is generally small compared to its length and width. Consequently, it may be idealised to a thin plate. In the precracking range, the slab may be analysed on the basis of classical thin plate theory for homogeneous elastic material. The stresses and deformations in prestressed concrete slabs with full (type 1) and limited (type 2) prestress on account of service loads may be determined with reasonable accuracy by means of thin plate theory. The cracking and non-linearity of materials may lead to significant errors if elastic theory is used in overload conditions. However, as the state of collapse is reached, a definite yield line pattern develops justifying the use of Johansen's yield line theory for the determination of collapse load of the slab. Tests have shown that the overload capacities of slabs can be computed with reasonable accuracy through the use of yield line theory.

14.2. CIRCULAR SLABS

Circular slabs find wide applications in containment structures, pressure vessels, towers, pylons and raft foundations. They may be analysed in the precracking range by elastic theory treating them as thin plates. The internal forces due to transverse loads on prestressed concrete circular slabs in the precracking range may be determined with reasonable accuracy using the classical theory of thin plates. The forces imposed on the concrete slab by the prestressing tendons may be determined by using the free body approach discussed in Sec. 5.4 The forces caused by prestressing may be combined with the external loads to obtain the net forces acting on the concrete slab which may then be analysed like a non-prestressed thin plate.

Figure 14 2.1 (a) shows a circular slab of radius R and thickness t lying in the horizontal plane and carrying transverse loading. Consider a small element ABCD cut out by two radial vertical planes AB and CD at angular distance $d\theta$ apart and circumferential vertical planes BC and AD at distance dr apart. The small element ABCD is shown isometrically in Fig. 14.2.1 (b). The internal forces acting on the element are biaxial bending moments M_r and M_θ, the associated transverse shear forces Q_r and Q_θ and twisting moments $T_{r\theta}$ and $T_{\theta r}$. When the slab carries only transverse loads, the membrane shear forces $N_{r\theta}$ and $N_{\theta r}$, acting in the plane of the element, i.e. in the horizontal plane are absent. The non-zero internal forces, viz. M_r, M_θ, Q_r, Q_θ, $T_{r\theta}$ and $T_{\theta r}$, may be determined by solving the differential equation based on equilibrium and strain compatibility. In most practical applications, however, the transverse loading is axi-symmetric, i.e. symmetrical about the vertical axis passing through the centre of the circular slab. Such a loading, therefore, varies with r but not with θ. In the case of axi-symmetric transverse loading, the transverse shear force Q_θ and the twisting moments $T_{r\theta}$ and $T_{\theta r}$ are absent. Hence there are only three non-zero internal forces, viz. radial bending moment M_r, circumferential bending moment M_θ and radial shear force Q_r, all of which are functions only of r and independent of θ due to symmetry. As the external loads and support reactions are axi-symmetric, the deformation of the slab is also likewise. Thus the vertical deflections of all points lying on a concentric circle are equal. It follows that the contours of deflected slab are concentric circles. The deflected surface has the shape of a saucer.

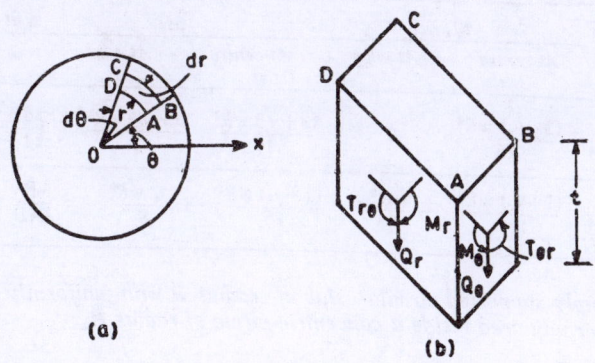

(a) (b)

Fig. 14.2.1

The simplest case of axi-symmetry is that of a circular slab carrying uniformly distributed load of intensity w per unit area. The vertical deflection Δ at a distance r from the centre of the slab when it is simply supported along the edge,

$$\Delta = \frac{w\,(R^2 - r^2)}{64\,D}\left(\frac{5 + \nu_c}{1 + \nu_c}\,R^2 - r^2\right) \tag{14.2.1}$$

When the slab is fully restrained along the edge,

$$\Delta = \frac{w\,(R^2 - r^2)^2}{64\,D} \tag{14.2.2}$$

where $D = $ plate rigidity

$$= \frac{E_c\,t^3}{12\,(1 - \nu_c^2)} \tag{14.2.3}$$

E_c, ν_c = modulus of elasticity and Poisson's ratio for concrete.

In the case of axi-symmetry, the moments M_r and M_θ may be determined from

$$M_r = -D\left(\frac{\partial^2 \Delta}{\partial r^2} + \frac{\nu_c}{r}\,\frac{\partial \Delta}{\partial r}\right) \qquad \ldots (14.2.4)$$

$$M_\theta = -D\left(\nu_c\frac{\partial^2 \Delta}{\partial r} + \frac{1}{r}\,\frac{\partial \Delta}{\partial r}\right) \qquad \ldots (14.2.5)$$

Substituting from Eq. (14.2.1) and (14.2.2), general expressions for M_r and M_θ may be derived. The deflection at centre of slab and the moments at centre and edge are shown in Table 14.2.1. The following are some general cases of axi-symmetry from which several other cases of practical interest may be derived as corollaries. The expressions are simplified considerably by taking Poisson's ratio, $v_c = 0$. This assumption, which does not generally produce appreciable error in design, is used in writing the following expressions:

Table 14.2.1 Circular Slab with Uniform Load

Support condition	M_r		M_θ		Δ at centre
	At centre	At edge	At centre	At edge	
Simply supported	$\dfrac{(3+v_c)\, wR^2}{16}$	0	$\dfrac{(3+v_c)\, wR^2}{16}$	$\dfrac{(1-v_c)\, wR^2}{8}$	$\left(\dfrac{5+v_c}{1+v_c}\right)\dfrac{wR^4}{64D}$
Fully restrained	$\dfrac{(1+v_c)\, wR^2}{16} - \dfrac{RR^2}{8}$		$\dfrac{(1+v_c)\, wR^2}{16} - \dfrac{v_c\, wR^2}{8}$		$\dfrac{wR^4}{64D}$

(i) *Simply supported circular slab of radius R with uniformly distributed load w per unit area inside a concentric circle of radius R_w*

(a) $r > R_w$

$$M_r = -\frac{3wr^2}{16} + \frac{w R_w^2}{4}\left[1 - \log_e\left(\frac{R_w}{R}\right) - \frac{R_w^2}{4r^2}\right] \qquad (14.2.6)$$

$$M_\theta = -\frac{wr^2}{16} + \frac{w R_w^2}{4}\left[1 - \log_e\left(\frac{R_w}{R}\right) - \frac{R_w^2}{4R^2}\right] \qquad (14.2.7)$$

(b) For $r > R_w$

$$M_r = -\frac{\pi R_w^2}{4}\left[\log_e\left(\frac{r}{R}\right) + \frac{R_w^2}{4}\left(\frac{1}{R^2} - \frac{1}{r^2}\right)\right] \qquad (14.2.8)$$

$$M_\theta = -\frac{w R_w^2}{4}\left[\log_e\left(\frac{r}{R}\right) - 1 + \frac{R_w^2}{4}\left(\frac{1}{R^2} + \frac{1}{r^2}\right)\right] \qquad (14.2.9)$$

(ii) *Simply supported circular slab of radius R with a concentric opening of radius R_o carrying uniform load w per unit area*

$$M_r = -\frac{3wr^2}{16} + \frac{wR_o^2}{4}\left[\log_e\left(\frac{r}{R}\right) + 0.75\left(1 + \frac{R^2}{R_o^2} - \frac{R^2}{r^2}\right)\right.$$
$$\left. + \frac{(R^2 - r^2)\, R_o^2}{(R^2 - R_o^2)\, r^2} \log_e\left(\frac{R}{R_o}\right)\right] \qquad (14.2.10)$$

$$M_\theta = -\frac{w r^2}{16} + \frac{w R_o^2}{4}\left[\log_e\left(\frac{r}{R}\right) + 0.75\left(-\frac{1}{3} + \frac{R^2}{R_o^2} + \frac{R^2}{r^2}\right)\right.$$
$$\left. - \frac{(R^2 + r^2)\, R_o^2}{(R^2 - R_o^2)\, r^2} \log_e\left(\frac{R}{R_0}\right)\right] \qquad (14.2.11)$$

(iii) *Simply supported circular slab of radius R with a concentric opening of radius R_o carrying total line load W along a concentric circle of radius R_w*

(a) For $r < R_w$

$$M_r = \frac{W R^2}{4 \pi (R^2 - R_o^2)} \left(1 - \frac{R_o^2}{r^2}\right) \lambda \qquad (14.2.12)$$

$$M_\theta = \frac{W R^2}{4 \pi (R^2 - R_o^2)} \left(1 + \frac{R_o^2}{r^2}\right) \lambda \qquad (14.2.13)$$

(b) For $r > R_w$

$$M_r = \frac{W}{4\pi} \left[\log_e \left(\frac{R_w}{r}\right) - 0.5 + \frac{(r^2 - R_o^2) R^2 \lambda}{(R^2 - R_o^2) r^2} + \frac{R_w^2}{2r^2} \right] \qquad (14.2.14)$$

$$M_\theta = \frac{W}{4\pi} \left[\log_e \left(\frac{R_w}{r}\right) + 0.5 + \frac{(r^2 + R_o^2) R^2 \lambda}{(R^2 - R_o^2) r^2} - \frac{R_o^2}{2r^2} \right] \qquad (14.2.15)$$

where $\lambda = \log_e \left(\frac{R}{R_w}\right) + 0.5 - \frac{R_w^2}{2R^2}$ $\qquad (14.2.16)$

(iv) *Simply supported circular slab of radius R with concentric opening of radius R_o carrying radial moment M_e per unit length along outer edge and radial moment M_i per unit length along inner edge*

$$M_r = \frac{R^2 M_e - R_o^2 M_i}{R^2 - R_o^2} - \frac{R^2 R_o^2}{r^2} \left(\frac{M_e - M_i}{R^2 - R_o^2}\right) \qquad (14.2.17)$$

$$M_\theta = \frac{R^2 M_e - R_o^2 M_i}{R^2 - R_o^2} + \frac{R^2 R_o^2}{r^2} \left(\frac{M_e - M_i}{R^2 - R_o^2}\right) \qquad (14.2.18)$$

Several particular cases of axi-symmetry can be obtained by assigning appropriate values to radius of the openinng R_o and radius R_w characterising the loading. For instance, the expressions for a circular slab without opening can be obtained by putting $R_o = o$ in cases (ii), (iii) and (iv). The case of uniformly loaded circular slab is obtained by putting $R_w = R$ in case (i). The moments in a circular slab carrying central concentrated load W may be obtained from case (i) by putting $w = w/\pi R_w^2$ and $R_w = o$. The moments in a circular slab carrying uniform load on annular area bounded by two concentric circles may be obtained by applying the principle of superposition to case (i). Similarly, the moments in a circular slab carrying several concentric line ioads may be obtained from case (iii). The principle of superposition may also be used to determine moments in a circular slab with free edge resting on a concentric ring beam or on a central post.

Circular slabs are generally prestressed biaxially by means of two orthogonal sets of prestressing tendons. The spacing of tendons in the two orthogonal directions x and y may be equal so that the prestressing forces per unit width, $P_x = P_y = P$. The prestressing tendons may be straight as shown in Fig. 14.2.2(a) or parabolic as shown in Fig. 14.2.2(b). In the case of straight tendons, the circular slab is in a state of isotropic compression P and isotropic hogging bending moment Pe_0 due to prestressing in which e_0 is the constant eccentricity of tendons. In the case of parabolic tendons, the slab is subjected not only to isotropic compression P and hogging bending moment Pe_0 but also to uniform upward force $(16\ PS/l^2)$ due to prestressing provided that the ratio of sag S and square of span length l for all cables is the same. Cable layouts shown in Fig. 14.2.2 (a) and (b) are suitable for slabs carrying uniform external load. The tendon profile may be changed suitably for other distributions of transverse loading. The orthogonal layout of tendons may also be used in circular slab with small or medium opening as shown in Fig. 14.2.2(c). In the case of large openings, it may be convenient to place the tendons in radial and circumferential directions as shown in Fig. 14.2.2(d).

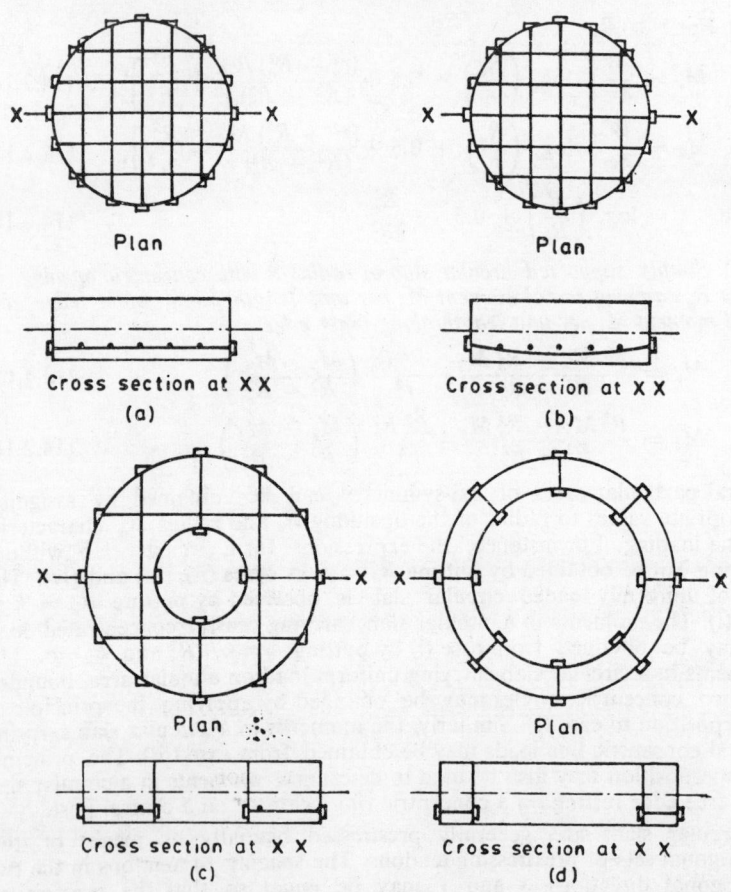

Plan

Cross section at X X
(a)

Plan

Cross section at X X
(b)

Plan

Cross section at X X
(c)

Plan

Cross section at X X
(d)

Fig. 14.2.2

EXAMPLE 14.2.1

A circular slab having a diameter of 12 m and overall thickness of 200mm is simply supported along the edge. It is prestressed biaxially by means of two sets of straight tendons placed orthogonal to each other. The prestressing force in each tendon is 240 kN. The first set of tendons is placed in two layers located at 30 mm and 90 mm from the bottom face. The second set of tendons is also placed in two layers at 50 mm and 70 mm from the bottom face. The spacing of tendons in each of the four layers is 400 mm. The slab carries uniformly distributed load of 12 kN/m² inclusive of self weight. Determine the stresses at centre of slab taking Poisson's ratio for concrete, $v_c = 0.15$.

Solution

(i) *Stresses due to prestressing*

The number of tendons per metre width of slab in each of the two orthogonal directions is 5. Hence, $P_x = P_y = 5 \times 0.24 = 1.2$ MN. The slab is, therefore, subjected to isotropic compression of $1.2/(1 \times 0.2) = 6$ MPa. The eccentricity of tendons in each direction is 40 mm. Hence the slab is subjected to an isotropic hogging bending moment of $1.2 \times 0.040 = 0.048$ MN.m. The stresses at top and bottom of slab due to prestressing,

$$f_T = -6 + \frac{0.048}{(1/6) \times 1 \times 0.2^2} = 1.2 \text{ MPa}$$

$$f_B = -6 - \frac{0.048}{(1/6) \times 1 \times 0.2^2} = -13.2 \text{ MPa}$$

(ii) *Stresses due to uniform load*

Referring to Table 14.2.1, the bending moment at centre of slab,

$$M_r = M_\theta = (3 + 0.15) \frac{0.012 \times 6^2}{16} = 0.085 \text{ MN.m/m}$$

Hence the stresses at centre of slab due to load,

$$f_T = -f_B = -\frac{0.085}{(1/6) \times 1 \times 0.2^2} = -12.75 \text{ MPa}$$

Combining the stresses in (i) and (ii), the net stresses are

$$f_T = 1.2 - 12.75 = -11.55 \text{ MPa}$$

$$f_B = -13.2 + 12.75 = -0.45 \text{ MPa}$$

EXAMPLE 14.2.2

A circular slab having a diameter of 18 m and overall thickness of 240 mm is fully restrained along the edge. It is prestressed biaxially by means of two orthogonal sets of parabolic tendons. The end eccentricity is zero and (S/l^2) is equal to 0.00028 for all tendons. The spacing of tendons in each set is 500 mm and effective prestressing force in each tendon is 720 kN. The slab carries only its self weight at initial stage and a superimposed load of 6.24 kN/m² at final stage. Determine the stresses at the two stages taking Poisson's ratio, $v_c = 0.15$ and loss factor, $\eta = 0.8$. Also determine the short time deflection at centre at final stage, taking $E_c = 33000$ MPa

Solution

(i) *Initial stage*

Prestressing force per unit width in each of the two orthogonal directions,

$$P_{xi} = P_{yi} = P_i = \frac{2 \times 0.72}{0.8} = 1.8 \text{ MN}$$

Hence initial isotropic compression is $(1.8/0.24) = 75$ MPa. Upward pressure exerted by the tendons on concrete,

$$w_{pi} = \frac{16 \, P_i \, S}{l^2}$$

$$= 16 \times 1.8 \times 0.00028 = 0.00806 \text{ MN/m}^2$$

Self weight of slab is $0.24 \times 0.024 = 0.00576 \text{ MN/m}^2$. Hence net upward load is $0.00806 - 0.00576 = 0.0023 \text{ MN/m}^2$. Referring to Table 14.2.1. bending moment at centre of slab,

$$M_r = M_\theta = \frac{(1+0.15)\,(-0.0023) \times 9^2}{16}$$

$$= -0.0134 \text{ MN.m/m}$$

Hence extreme fibre stresses at centre,

$$f_T = -7.5 + \frac{0.0134}{(1/6) \times 1 \times 0.24^2} = -6.1 \text{ MPa}$$

$$f_B = -7.5 - \frac{0.0134}{(1/6) \times 1 \times 0.24^2} = -8.9 \text{ MPa}$$

Radial bending moment at edge,

$$M_r = -\frac{(-0.0023) \times 9^2}{8} = 0.0233 \text{ MN.m/m}$$

Hence extreme fibre stresses in radial direction at edge,

$$f_T = -7.5 - \frac{0.0233}{(1/6) \times 1 \times 0.24^2} = -9.93 \text{ MPa}$$

$$f_B = -7.5 - \frac{0.0233}{(1/6) \times 1 \times 0.24^2} = -5.07 \text{ MPa}$$

Circumferential bending moment at edge,

$$M_\theta = -\frac{0.15\,(-0.0023) \times 9^2}{8} = 0.0035 \text{ MN.m/m}$$

Hence extreme fibre stresses in circumferential direction at edge,

$$f_T = -7.5 - \frac{0.0035}{(1/6) \times 1 \times 0.24^2} = -7.86 \text{ MPa}$$

$$f_B = -7.5 + \frac{0.0035}{(1/6) \times 1 \times 0.24^2} = -7.14 \text{ MPa}$$

(ii) *Final stage*

Prestressing force per unit width in each of the two orthogonal directions,

$$P_x = P_y = P = 2 \times 0.72 = 1.44 \text{ MN}$$

Hence final isotropic compression is $(1.44/0.24) = 6$ MPa. Upward pressure exerted by the tendons on concrete,

$$w_p = 16 \times 1.44 \times 0.000\,28 = 0.00645 \text{ MN/m}^2$$

Total load on slab including self weight is $0.00624 + 0.00576) = 0.012$ MN/m^2. Hence net downward load,

$$w = 0.012 - 0.00645 = 0.00555 \text{ MN/m}^2$$

Referring to Table 14.2.1, bending moment at centre of slab,

$$M_r = M_\theta = \frac{(1 + 0.15) \times 0.00555 \times 9^2}{16} = 0.0323 \text{ MN.m/m}$$

Hence extreme fibre stresses at centre,

$$f_T = -6 - \frac{0.0323}{(1/6) \times 1 \times 0.24^2} = -9.36 \text{ MPa}$$

$$f_B = -6 + \frac{0.0323}{(1/6) \times 1 \times 0.24^2} = -2.64 \text{ MPa}$$

Radial bending moment at edge,

$$M_r = -\frac{0.00555 \times 9^2}{8} = -0.0562 \text{ MN.m/m}$$

Hence extreme fibre stresses in radial direction at edge,

$$f_T = -6 - \frac{(-0.0562)}{(1/6) \times 1 \times 0.24^2} = -0.15 \text{ MPa}$$

$$f_B = -6 + \frac{(-0.0562)}{(1/6) \times 1 \times 0.24^2} = -11.85 \text{ MPa}$$

Circumferential bending moment at edge,

$$M_\theta = -\frac{0.15 \times 0.00555 \times 9^2}{8} = -0.00843 \text{ MN.m/m}$$

Hence extreme fibre stresses in circumferential direction at edge,

$$f_T = -6 - \frac{(-0.00843)}{(1/6) \times 1 \times 0.24^2} = -5.12 \text{ MPa}$$

$$f_B = -6 + \frac{(-0.00843)}{(1/6) \times 1 \times 0.24^2} = -6.88 \text{ MPa}$$

From Eq. (14.2.3), the plate rigidity,

$$D = \frac{33000 \times 0.24^3}{12 (1 - 0.15^2)} = 38.89 \text{ MN.m}$$

Referring to Table 14.2.1, the central deflection,

$$\Delta = \frac{0.00555 \times 9^4}{64 \times 38.89} = 0.0146 \text{ m}$$

EXAMPLE 14.2.3

A circular slab having a radius of 8.64 m and thickness of 210 mm is simply supported on a ring beam of 5.76 m radius. It has a concentric opening of 1.44 m radius. The slab has to carry a live load of 4 kN/m². Design the prestressing tendons using Lee-Mc Call system of post-tensioning to eliminate tensile stresses.

Solution

Taking unit weight of concrete as 24 kN/m³, the self weight of slab is $0.21 \times 24 = 5.04$ kN/m². Hence total uniform load is $5.04 + 4 = 9.04$ kN/m². The bending moments in the given slab due to this load may be determined by combining cases (ii) and (iii).

(a) *Simply supported circular slab of radius R = 8.64 m with a concentric opening of radius $R_0 = 1.44$ m carrying uniform downward load, w=9.04 kN/m²*

The radial and circumferential moments may be determined from Eq. (14.2.10) and (14.2.11). At $r = 1.44$, 3.60, 5.76 and 8.64 m, the radial moment $M_r = 0$, 84-87, 64.30 and 0 kN.m/m and the circumferential moment $M_\theta = 233.45$, 132.56, 111.84 and 86.22 kN.m/m.

(b) *Simply supported circular slab of radius R = 8.64 m with a concentric opening of radius $R_0 = 1.44$ m carrying total upward line load W = π $(8.64^2 - 1.44^2) \times 9.04 = 2062$ kN.*

It may be noted that the total upward load W is the same as the total downward load due to weight of slab and line load. The radial and circumferential moments may be determined from Eq. (14.2.12) to (14.2.16). At $r = 1.44$, 3.60, 5.76 and 8.64 m, the radial moment $M_r = 0$, -96.83, -108.06 and 0 kN.m/m and $M_\theta = -230.54$, -133.71, -122.47 and -97.53 kN.m/m.

Combining (a) and (b), the moments at $r = 1.44$, 3.60. 5.76 and 8.64 m are

$$M_r = 0, \ -11.96, \ -43.76 \text{ and } 0 \text{ kN.m/m}$$

and

$$M_\theta = 2.91, \ -1.15, \ -10.63 \text{ and } -11.31 \text{ kN.m/m}$$

The largest bending moment of 43.76 kN.m/m (hogging) occurs in the radial direction at the ring beam. The corresponding extreme fibre stress,

$$f_T = -f_B = \frac{0.04376}{(1/6) \times 1 \times 0.21^2} = 5.95 \text{ MPa}$$

Adopting orthogonal layout of tendons and concentric prestressing, the effective prestressing force required in each of the two orthogonal directions to eliminate tensile stress,

$$P_x = P_y = 5.95 \times 0.21 = 1.25 \text{ MN}$$

Using 20 mm diameter high tensile steel bars with an effective stress of 800 MPa, the spacing of bars in each direction

$$s = \frac{\frac{\pi}{4} \times 20^2 \times 10^{-6} \times 800}{1.25} = 0.201 \text{ m}$$

Two layers of bars may be provided in each direction. The bars in x-direction may be placed at 40 mm from top and bottom faces at a spacing of 400 mm. The bars in y-direction may be placed at 80 mm from top and bottom faces at a spacing of 400 mm.

14.3 RECTANGULAR SLABS

Rectangular slabs are most extensively used in buildings and bridge decks to support transverse loads. As the predominant structural action in slabs is biaxial bending, they may be prestressed biaxially to increase their load capacity and to reduce their deflection. Consider a rectangular slab panel having spans l_x and l_y in x and y directions carrying transverse load w per unit area as shown in Fig. 14.3.1 (a). A small element of the slab cut out by vertical planes AB and DC parallel to x-axis and vertical planes BC and AD parallel to y-axis is shown isometrically in Fig. 14.3.1 (b). The element

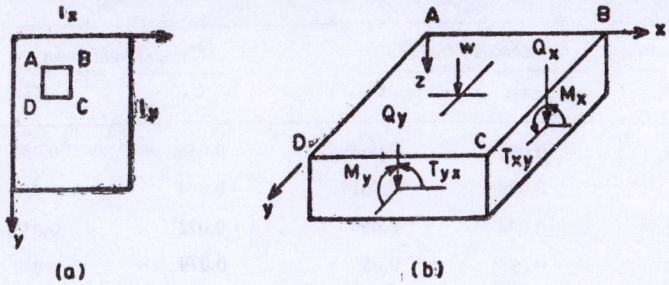

(a)   (b)

Fig. 14.3.1

is acted upon by biaxial moments M_x in xz-plane and M_y in yz-plane, the associated transverse shear forces Q_x and Q_y together with twisting moments T_{xy} and T_{yx}. However, in slabs whose thickness is small as compared to spans l_x and l_y, the stresses due to shear forces and twisting moments are generally too small to influence the design which, therefore, is controlled primarily by the biaxial bending moments M_x and M_y. The internal forces can be determined by solving a fourth order partial differential equation based on equilibrium and strain compatibility of the small element for appropriate boundary conditions in accordance with the classical theory of thin plates. The internal forces at any point depend mainly upon (i) position of the point considered, (ii) aspect ratio (l_y/l_x), (iii) magnitude and distribution of transverse loads and (iv) support conditions, i.e. whether the edges are free, simply supported, fixed or continuous. The computations of internal forces and displacements based on standard plate equation are complex and time consuming. However, solutions are available for the common cases of loading and support conditions. The Indian Code IS: 456-1978 gives maximum positive and negative values of bending moments M_x and M_y for the most common case of uniform loading. The corners of an isolated rectangular slab panel have the tendency to lift on account of twisting moments T_{xy} and T_{yx} unless they are prevented from doing so by external constraints. When a slab panel is continuous on one or more edges, the corresponding corners of the panel are prevented from lifting due to continuity. The maximum positive bending moment along short span is reduced whenever the corners are prevented from lifting. However, additional moments are induced near the corners which may require the provision of torsion reinforcement. Table 14.3.1 gives coefficients C_x and C_y for the determination of maximum positive values of elastic biaxial bending moments M_x and M_y in accordance with Eq. (14.3.1) and (14.3.2) for an isolated slab panel.

$$M_x = C_x \, wl_x^2 \qquad (14.3.1)$$

$$M_y = C_y \, wl_x^2 \qquad (14.3.2)$$

where w = uniform load per unit area

Table 14.3.1 Values of C_x and C_y for an isolated rectangular slab panel discontinuous on all edges

Aspect ratio, l_y/l_x	Corners free to lift		Corners held down	
	C_x	C_y	C_x	C_y
1.0	0.062	0.062	0.056	0.056
1.1	0.074	0.061	0.064	0.056
1.2	0.084	0.059	0.072	0.056
1.3	0.093	0.055	0.079	0.056
1.4	0.099	0.051	0 085	0.056
1.5	0.104	0.046	0.089	0.056
1.75	0.113	0.037	0.100	0.056
2.0	0.118	0.029	0.107	0.056
2.5	0.122	0.020	—	—
3.0	0.124	0.014	—	—

The more common cases of slab panels with one or more edges continuous are considered in Table 14.3.2. The two way action by which a slab carries the applied transverse load by developing bending moments in two orthogonal directions is prominent when the aspect ratio l_y/l_x is less than 2. It behaves practically like a one way slab with significant moment only along the short span when the aspect ratio exceeds 3.

Rectangular slabs are generally post-tensioned biaxially by means of two orthogonal sets of prestressing tendons parallel to x and y axes as shown in the plan view of Fig. 14.3.2 (a). The tendons may be straight as shown in the midspan sections of Fig. 14.3.2 (b) and (c) or curved as shown in Fig. 14.3.2 (d) and (e). The forces induced in the slab on account of prestressing may be determined conveniently by using the free body or load balancing approach discussed in Sec. 5.4. In the case of straight tendons, the slab is subjected to axial compression P_x and hogging edge moment $P_x e_x$ per unit width on account of prestressing tendons in x-direction in which P_x is the prestressing force per unit width acting at constant eccentricity e_x. Similarly the prestressing tendons in y-direction impose an axial compression P_y and hogging edge moment $P_y e_y$ per unit width in which P_y is the prestressing force per unit width acting at constant eccentricity e_y. In the case of curved parabolic tendons, they not only impose biaxial

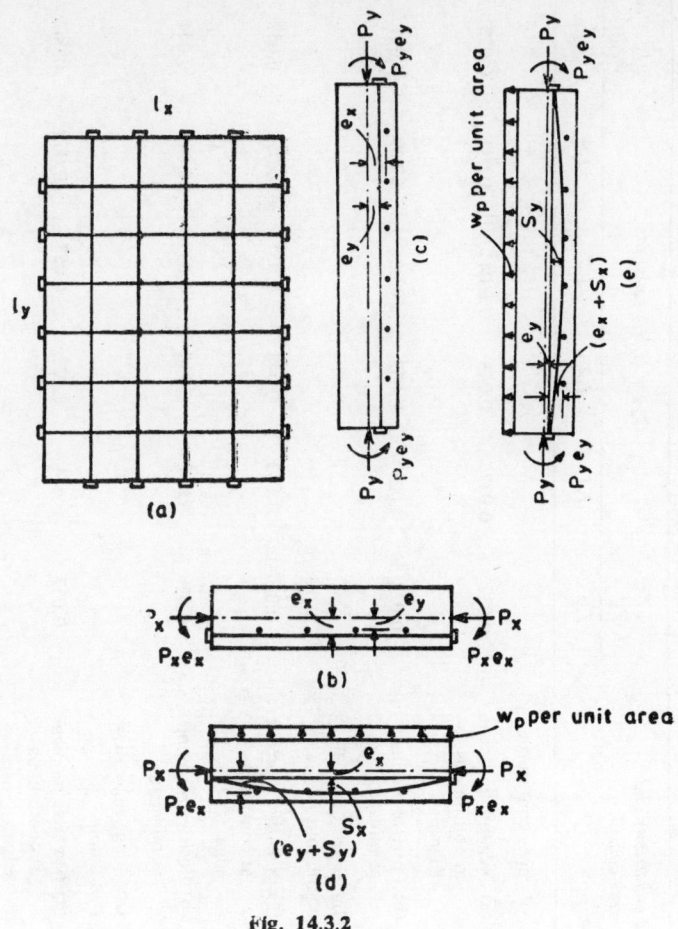

Fig. 14.3.2

Table 14.3.2 Values of C_x and C_y for continuous rectangular slab panels with corners held down

Edge conditions and moments considered	C_x for (l_y/l_x) equal to									C_y for all values of (l_y/l_x)	
	1.0	1.1	1.2	1.3	1.4	1.5	1.75	2.0		1.75	2.0
(i) All edges continuou											
(a) Negative moment at continuous edge	0.032	0.037	0.043	0.047	0.051	0.053	0.060	0.065			0.032
(b) Positive moment at midspan	0.024	0.028	0.032	0.036	0.039	0.041	0.045	0.049			0.024
(ii) One short edge discontinuous											
(a) Negative moment at continuous edge	0.037	0.043	0.048	0.051	0.055	0.057	0.064	0.068			0.037
(b) Positive moment at midspan	0.028	0.032	0.036	0.039	0.041	0.044	0.048	0.052			0.028
(iii) One long edge discontinuous											
(a) Negative moment at continuous edge	0.037	0.044	0.052	0.057	0.063	0.067	0.077	0.085			0.037
(b) Positive moment at midspan	0.028	0.033	0.039	0.044	0.047	0.051	0.059	0.065			0.028

(Contd.)

Table 14.3.2 (Contd.)

Edge conditions and moments considered	C_x for (l_y/l_x) equal to									C_y for all values of (l_y/l_x)
	1.0	1.1	1.2	1.3	1.4	1.5	1.75	2.0		
(iv) Two adjacent edges discontinuous										
(a) Negative moment at continuous edge	0.047	0.053	0.060	0.065	0.071	0.075	0.084	0.091		0.047
(b) Positive moment at midspan	0.035	0.040	0.045	0.049	0.053	0.056	0.063	0.069		0.035
(v) Two short edges discontinuous										
(a) Negative moment at continuous edge	0.045	0.049	0.052	0.056	0.059	0.060	0.065	0.069		—
(b) Positive moment at midspan	0.035	0.037	0.040	0.043	0.044	0.045	0.049	0.052		0.035
(vi) Two long edges discontinuous										
(a) Negative moment at continuous edge	—	—	—	—	—	—	—	—		0.045
(b) Positive moment at midspan	0.035	0.043	0.051	0.057	0.063	0.068	0.080	0.088		0.035

Contd.

Table 14.3.2 (Contd.)

Edge conditions and moment considered	C_x for (l_y/l_x) equal to								C_y for all values of (l_x/l_y)
	1.0	1.1	1.2	1.3	1.4	1.5	1.75	2.0	
(vii) Three edges discontinuous (one short edge discontinuous)									
(a) Negative moment at continuous edge	0.057	0.064	0.071	0.076	0.080	0.084	0.091	0.097	—
(b) Positive moment at midspan	0.043	0.048	0.053	0.057	0.060	0.064	0.069	0.073	0.043
(viii) Three edges discontinuous (one long edge discontinuous)									
(a) Negative moment at continuous edge	—	—	—	—	—	—	=	—	0.057
(b) Positive moment at midspan	0.043	0.051	0.059	0.065	0.071	0.076	0.087	0.096	0.043

compressions P_x, P_y, hogging edge moments $P_x\,e_x$ and $P_y\,e_y$ but also upward force w_p per unit area given by the following equation, which is an extension of Eq. (5.4.10) for beams.

$$w_p = \frac{8P_x\,S_x}{l_x^2} + \frac{8P_y\,S_y}{l_y^2} \tag{14.3.3}$$

where, $S_x, S_y = $ Sags of prestressing tendons in x and y directions respectively.

Tendon layouts shown in Fig. 14.3.2 may be found appropriate for rectangular slab with simply supported edges carrying uniform load. The layout of tendons may be modified to suit other types of edge conditions or loading. For instance, the tendons may have negative (upward) eccentricity at a fixed or continuous edge. When a slab carries uniform load on a portion of its area, the tendons may be parabolic in loaded portion and straight in unloaded portion of the slab. When the prestressing tendons have zero eccentricity at a discontinuous edge, they do not impose any moment at this edge. As the bending moment normal to a discontinuous edge is zero, the option of zero eccentricity appears justifiable on theoretical grounds. However, if an end eccentricity is provided for practical reasonsn as for instance in the case of straight tendons, the bending moments oer account of edge moments have to be evaluated. Timoshenko and Kriegly have given values of midspan bending moments M_x and M_y for a simpng supported thin plate subjected to edge moments. The midspan bendith moments on account of uniform sagging edge moment M_0 per unit widrs acting on two opposite edges based cn Poisson's ratio equal to 0.3 at shown in Table 14.3.3. The table shows that the midspan bending momene are always numerically smaller than the applied edge moments. Thϑ Poisson's ratio for concrete varies from 0 to 0.2. However, the values given in the table may be used for guidance or as an approximation.

Table 14.3.3 Bending moments in simply supported slabs due to edge moments M_0

Aspect ratio l_y/l_x	Short edges loaded		Long edges loaded	
	M_x	M_y	M_x	M_y
1.0	$0.394\,M_0$	$0.256\,M_0$	$0.256\,M_0$	$0.394\,M_0$
1.33	—	—	$0.476\,M_0$	$0.424\,M_0$
1.5	$0.264\,M_0$	$0.046\,M_0$	—	—
2.0	$0.153\,M_0$	$-0.010\,M_0$	$0.77\,M_0$	$0.387\,M_0$
∞	—	—	$1.0\,M_0$	$0.3\,M_0$

EXAMPLE 14.3.1

A simply supported isolated rectangular slab 6 m × 8 m with unrestrained corners has an overall thickness of 120 mm. It is post-tensioned by means of 12 mm high tensile steel bars with an effective stress of 800 MPa. The tendons along short span have a constant eccentricity of 20 mm and a spacing of 150 mm. The tendons along long span are concentric and are spaced at 225 mm. Determine the stresses at midspan when the slab carries a superimposed load of 4 kN/m². Unit weight of concrete is 24 kN/m³.

Solution

Self weight $= 0.12 \times 0.024 = 0.00288$ MN/m²

Total uniform load $= 0.004 + 0.00288 = 0.00688$ MN/m²

Aspect ratio, $\dfrac{l_y}{l_x} = \dfrac{8}{6} = 1.33$

Hence from Table 14.3.1 and Eqs (14.3.1) and (14.3.2),

$$M_x = 0.095 \times 0.00688 \times 6^2 = 0.02353 \text{ MN} \cdot \text{m/m}$$

$$M_y = 0.054 \times 0.00688 \times 6^2 = 0.01337 \text{ MN} \cdot \text{m/m}$$

Effective prestressing forces in two orthogonal directions,

$$P_x = \frac{\pi}{4} \times 12^2 \times 10^{-6} \times 800 \times \frac{1000}{150} = 0.603 \text{ MN/m}$$

$$P_v = \frac{\pi}{4} \times 12^2 \times 10^{-6} \times 800 \times \frac{1000}{225} = 0.402 \text{ MN/m}$$

$$e_x = 0.020 \text{ m}$$

$$e_y = 0$$

The prestress induces hogging edge moment $M_o = P_x e_x = 0.01206$ MN·m/m on long edges and no moment on short edges. Using Table 14.3.3 as an approximation, the midspan moments on account of edge moments on long edges are

$$M_x = -0.476 \times 0.01206 = -0.00574 \text{ MN} \cdot \text{m/m}$$

$$M_y = -0.424 \times 0.01206 = -0.00511 \text{ MN} \cdot \text{m/m}$$

Hence the net moments at midspan,

$$M_x = 0.02353 - 0.00574 = 0.01779 \text{ MN} \cdot \text{m/m}$$

$$M_y = 0.01337 - 0.00511 = 0.00826 \text{ MN} \cdot \text{m/m}$$

The extreme fibre stresses in x-direction,

$$f_T = -\frac{0.603}{1 \times 0.12} - \frac{0.01779}{(1/6) \times 1 \times 0.12^2} = -12.44 \text{ MPa}$$

$$f_B = -\frac{0.603}{1 \times 0.12} + \frac{0.01779}{(1/6) \times 1 \times 0.12^2} = -2.39 \text{ MPa}$$

The extreme fibre stresses in y-direction,

$$f_T = - \frac{0.402}{1 \times 0.12} - \frac{0.00826}{(1/6) \times 1 \times 0.12^2} = - 6.79 \text{ MPa}$$

$$f_B = - \frac{0.402}{1 \times 0.12} + \frac{0.00826}{(1/6) \times 1 \times 0.12^2} = 0.09 \text{ MPa}$$

EXAMPLE 14.3.2

A simply supported isolated rectangular slab 8 m × 12 m with restrained corners has an overall thickness of 150 mm. It is prestressed biaxially by means of a class I—12.5 mm—7 ply strands at a spacing of 150 mm along the short span and at a spacing of 180 mm along the long span. All tendons have parabolic shape with zero eccentricity at ends The sags of the tendons along short and long spans are 45 mm and 25 mm respectively. All tendons have an initial stress of 1200 MPa and the loss factor, $n = 0.8$. The slab carries only self weight at initial stage and a live load of 4 kN/m² at final stage. Determine extreme fibre stresses at midspan at the two stages.

Solution

Taking unit weight of concrete as 24 kN/m³,

Self weight $= 0.15 \times 0.024 = 0.0036$ MN/m²,

(i) Initial stage

Referring to Table 2.3.2, the nominal area of each strand is 92.9 mm². Hence the initial prestressing forces in the two orthogonal directions,

$$P_{xi} = 92.9 \times 10^{-6} \times 1200 \times \frac{1000}{150} = 0.7432 \text{ MN/m}$$

$$P_{yi} = 92.9 \times 10^{-6} \times 1200 \times \frac{1000}{180} = 0.6193 \text{ MN/m}$$

Using Eq. (14.3.3), the upward force due to prestressing,

$$w_{pi} = \frac{8 \times 0.7432 \times 0.045}{8^2} + \frac{8 \times 0.6193 \times 0.025}{12^2}$$

$$= 0.00504 \text{ MN/m}^2$$

As slab carries only self weight at initial stage, the unbalanced load,

$$w_{ubi} = 0.00504 - 0.0036 = 0.00144 \text{ MN/m}^2 \text{ (upward)}$$

From Table 12.3.1, the midspan biaxial moments,

$$M_{xi} = 0.089 \times (- 0.00144) \times 8^2 = - 0.0082 \text{ MN} \cdot \text{m/m}$$

$$M_{yi} = 0.056 \times (- 0.00144) \times 8^2 = - 0.0052 \text{ MN} \cdot \text{m/m}$$

The extreme fibre stresses along short span,

$$f_T = -\frac{0.7432}{1 \times 0.15} + \frac{0.0082}{(1/6) \times 1 \times 0.15^2} = -2.77 \text{ MPa}$$

$$f_B = -\frac{0.7432}{1 \times 0.15} - \frac{0.0082}{(1/6) \times 1 \times 0.15^2} = -7.14 \text{ MPa}$$

The extreme fibre stress along span,

$$f_T = -\frac{0.6193}{1 \times 0.15} + \frac{0.0052}{(1/6) \times 1 \times 0.15^2} = -2.74 \text{ MPa}$$

$$f_B = -\frac{0.6193}{1 \times 0.15} - \frac{0.0052}{(1/6) \times 1 \times 0.15^2} = -5.52 \text{ MPa}$$

(ii) *Final stage*

Effective prestressing forces in the two orthogonal directions,

$$P_x = 0.8 \times 0.7432 = 0.5946 \text{ MN/m}$$

$$P_y = 0.8 \times 0.6193 = 0.4954 \text{ MN/m}$$

Using Eq. (14.3.3), the upward force due to prestressing,

$$w_p = 0.8 \, w_{pi} = 0.00403 \text{ MN/m}^2$$

Total uniform load $= 0.0036 + 0.004 = 0.0076$ MN/m^2

Hence the unbalanced load,

$$w_{ub} = 0.0076 - 0.00403 = 0.00357 \text{ MN/m}^2 \text{ (downard)}$$

From Table 14.3.1, the midspan biaxial moments,

$$M_x = 0.089 \times 0.00357 \times 8^2 = 0.02033 \text{ MN·m/m}$$

$$M_y = 0.056 \times 0.00357 \times 8^2 = 0.01280 \text{ MN·m/m}$$

The extreme fibre stresses along short span,

$$f_T = -\frac{0.5946}{1 \times 0.15} - \frac{0.02033}{(1/6) \times 1 \times 0.15^2} = -9.39 \text{ MPa}$$

$$f_B = -\frac{0.5946}{1 \times 0.15} + \frac{0.02033}{(1/6) \times 1 \times 0.15^2} = 1.46 \text{ MPa}$$

The extreme fibre stresses along long span,

$$f_T = -\frac{0.4954}{1 \times 0.15} - \frac{0.0128}{(1/6) \times 1 \times 0.15^2} = -6.72 \text{ MPa}$$

$$f_B = -\frac{0.4954}{1 \times 0.15} + \frac{0.0128}{(1/6) \times 1 \times 0.15^2} = 0.11 \text{ MPa}$$

EXAMPLE 14.3.3

A rectangular slab panel 10 m × 12 m having an overal thickess of 180 mm is continuous on two adjacent edges and discontinuous on the other two edges. It has to carry only self weight at initial stage and a live load of 4.5 kN/m² at final stage. Design the prestressing tendons adopting Freyssinet system and concrete of grade M-35.

Solution

Taking unit weight of concrete as 24 kN/m³, self weight

$$w_D = 0.18 \times 0.024 = 0.00432 \ \text{MN/m}^2$$

Live load, $w_L = 0.0045 \ \text{MN/m}^2$

For equal unbalanced loads at initial and final stages in accordance with load balancing method,

$$w_{ub} = w_{pi} - w_D = w_D + w_L - w_p$$

Taking loss factor, $\eta = 0.8$,

$$w_{ub} = w_{pi} - 0.00432 = 0.00432 + 0.0045 - 0.8 \, w_{pi}$$

or $\qquad w_{pi} = 0.0073 \ \text{MN/m}^2$

$\qquad\qquad w_p = 0.00584 \ \text{MN/m}^2$

and $\qquad w_{ub} = 0.00298 \ \text{MN/m}^2$

From Table 14.3.2, the moments due to unbalanced load,

M_x at midspan $\qquad = 0.045 \times 0.00298 \times 10^2 = 0.01341 \text{MN·m/m}$

M_x at cantinuous edge $= - 0.06 \times 0.00298 \times 10^2$

$\qquad\qquad\qquad\qquad = - 0.01788 \ \text{MN·m/m}$

M_y at midspan $\qquad\quad = 0.035 \times 0.00298 \times 10^2$

$\qquad\qquad\qquad\qquad = 0.01043 \ \text{MN·m/m}$

M_y at continuous edge $= - 0.047 \times 0.00298 \times 10^2$

$\qquad\qquad\qquad\qquad = - 0.014 \ \text{MN·m/m}$

Considering the type of loading and the edge conditions, it is proposed to adopt parabolic shape for the tendons in both directions with zero eccentricity at discontinuous edges, maximum permissible positive (downward) eccentricity at midspan and maximum permissible negative (upward) eccentricity at continuous edges. Using 12 ϕ 5 Freyssi cables and 33 mm diameter sheathing with a clear cover of 20 mm over the sheathing, the minimum permissible distance of centre of cable from the surface is $20 + 16.5 = 36.5$ mm. Hence maximum permissible eccentricity is $90 - 36.5 = 53.5$ mm. Hence the cables along short span (x-direction) may have a downward eccentricity of 53 mm at midspan, upward eccentricity of 53 mm at continuous edge and zero eccentricity at discontinuous edge giving a sag,

$$S_x = 53 + \frac{53}{2} = 79.5 \ \text{mm}$$

In order to prevent the cables along long span (y-direction) from interfering with the placement of cables in the other direction, the sag S_y may be taken as

$$S_y = 79.5 - 33 = 46.5 \text{ mm}$$

From Eq. (14.3.3),

$$0.00584 = \frac{8 \times 0.0795 \, P_x}{10^2} + \frac{8 \times 0.0465 \, P_y}{12^2} \tag{a}$$

In order to obtain the same ratio of direct and bending stresses in the two orthogonal directions, P_x/P_y may be taken equal to M_x/M_y. As the negative moments are numerically greater and therefore critical, it is reasonable to take

$$\frac{P_x}{P_y} = \frac{-0.01788}{-0.0140}$$

or

$$P_x = 1.2766 \, P_y$$

Substituting into Eq. (a)

$$P_y = 0.5457 \text{ MN/m}$$
$$P_x = 0.6966 \text{ MN/m}$$

Referring to Table 3.3.2, the effective force in $12 \, \phi \, 5$ cable may be taken as 0.2256 MN corresponding to an effective stress of 960 MPa. Hence the spacing of cables along short span,

$$s_x = \frac{0.2256}{0.6966} \times 1000 = 324 \text{ mm}$$

Similarly, the spacing of cables along long span,

$$s_y = \frac{0.2256}{0.5457} \times 1000 = 413 \text{ mm}$$

The extreme fibre stresses for the two stages at the critical points, i.e. the continuous edges may now be checked.

(i) *Initial stage*

$$P_{xi} = \frac{P_x}{0.8} = 0.87075 \text{ MN/m}$$

$$P_{yi} = \frac{P_y}{0.8} = 0.6821 \text{ MN/m}$$

$$w_{ub} = 0.00298 \text{ MN/m}^2 \text{ (upward)}$$

Extreme fibre stresses in x-direction,

$$f_T = -\frac{0.87075}{1 \times 0.18} - \frac{0.01788}{(1/6) \times 1 \times 0.18^2} = -8.15 \text{ MPa}$$

$$f_B = -\frac{0.87075}{1 \times 0.18} + \frac{0.01788}{(1/6) \times 1 \times 0.18^2} = -1.53 \text{ MPa}$$

Extreme fibre stresses in y-direction,

$$f_T = -\frac{0.6821}{1 \times 0.18} - \frac{0.014}{(1/6) \times 1 \times 0.18^2} = -6.38 \text{ MPa}$$

$$f_B = -\frac{0.6821}{1 \times 0.18} + \frac{0.014}{(1/6) \times 1 \times 0.18^2} = -1.20 \text{ MPa}$$

(ii) *Final stage*

$$P_x = 0.6966 \text{ MN/m}$$
$$P_y = 0.5457 \text{ MN/m}$$
$$w_{ub} = 0.00298 \text{ MN/m}^2 \text{ (downward)}$$

Extreme fibre stresses in x-direction,

$$f_T = -\frac{0.6966}{1 \times 0.18} + \frac{0.01788}{(1/6) \times 1 \times 0.18^2} = -0.56 \text{ MPa}$$

$$f_B = -\frac{0.6966}{1 \times 0.18} - \frac{0.01788}{(1/6) \times 1 \times 0.18^2} = -7.18 \text{ MPa}$$

Extreme fibre stresses in y-direction,

$$f_T = -\frac{0.5457}{1 \times 0.18} + \frac{0.014}{(1/6) \times 1 \times 0.18^2} = -0.44 \text{ MPa}$$

$$f_B = -\frac{0.5457}{1 \times 0.18} - \frac{0.014}{(1/6) \times 1 \times 0.18^2} = -5.63 \text{ MPa}$$

As there is no tensile stress and the compressive stresses are within permissible limits, the design is satisfactory.

14.4 FLAT SLABS

Due to its aesthetic value, flat slab construction has been used extensively all over the world. When a slab is supported directly on columns instead of beams or walls, it is called a flat slab. Consequently, a flat slab behaves like a thin plate with support reactions concentrated at discrete points. The slab carries the applied transverse load by developing bending moments in two orthogonal directions and shear transfer directly to columns. The slab also transfers bending moment to the columns when the slab is cast monolithically with the supporting columns. In a flat slab construction, the diameter or lateral dimensions of the column are often increased near the top in the form of column heads or capitals. At the same time the thickness of the slab is increased slightly in the region surrounding the column head. The thickened part of the slab is known as the drop. The stresses and deformations in a flat slab may be evaluated by using the classical plate theory. However, due to the complexity of the problem, reinforced concrete flat slabs are designed by means of semi-empirical methods. The design of prestressed concrete flat slabs also follows the same approaches except that the forces due to prestressing tendons have also to be taken into account.

14.4.1 Indian Code

The Indian Code IS: 456-1978 specifies that the drops, if provided, should be rectangular in plan and should have a length in each direction not less than one-third of the panel width in that direction. The width of drop in the exterior panel, normal to the discontinuous edge, measured from the centre line of the column should be equal to one-half the width of drop in the interior panels. Whenever column heads are provided, only that portion of the column head should be considered in the computations which is included within the largest possible right circular cone or pyramid with a vertex angle of 90° contained entirely within the boundaries of the column and the column head. Like other slabs, the thickness of flat slab is controlled by considerations of deflections discussed in Sec. 14.6, except that in the case of flat slabs without drops, the maximum permissible span/depth ratios should be multiplied by 0.9. The span/depth ratio should be based on the longer span.

For the computations of bending moments in a flat slab, one of the following two methods may be used:

(A) Direct design method

As the method is empirical, it is subject to limitations imposed by the following provisions:

(a) The slab should have a minimum of three continuous spans in each of the two orthogonal directions.

(b) The panels should be rectangular. The aspect ratio for each panel, i.e. ratio of longer span to shorter span should not exceed 2.

(c) A column may be offset from the centre of the column, provided that the offset does not exceed 10 per cent of the span in the direction of the offset.

(d) The successive spans in each direction should not differ by more than one-third of the longer span. The end span should not be greater than the interior span. Also the ratio of design live load to dead load should not exceed three. These provisions guard against the possibility of negative moments near midspan.

Referring to Fig. 14.4.1 which shows a typical interior panel of a flat slab with span l_x smaller than span l_y, the total static moment M_{ox} along the shorter span may be expressed as

$$M_{ox} = \frac{Wl_{nx}}{8} \qquad (14.4.1)$$

where M_{ox} = total static moment which is equal to the absolute sum of positive moment M_1 across the section EF and average negative moment M_2 across the section AD

$$= M_1 + M_2$$

W = total design load on area $l_y \, l_{nx}$

$$= w l_y \, l_{nx}$$

$w =$ uniformly distributed design load

$l_{nx} =$ clear span extending from face to face of column capitals, brackets or walls but not less than $0.65 \, l_x$.

When the support is circular, it should be replaced, in the computation of l_n, by a square support having the same area. In computing the total panel moment M_{ox}, the panel width l_y may be taken as the distance between the centres of adjacent columns as shown in Fig. 14.4.1 or as the distance between the centre lines of two adjacent panels. In the former case the panel comprises a middle strip and two half column strips. In the latter case, the panel comprises a column strip and two half middle strips. When the span l_x being considered is adjacent and parallel to an edge, l_y should be taken as the distance between the edge and the centre line of the panel. The total static moment M_{oy} along the longer span may be considered in a similar manner. The expression for total static moment M_{ox}, which is simi-to the static moment in a uniformly loaded beam, can be derived by introducing small approximations in well known Nichols analysis based on

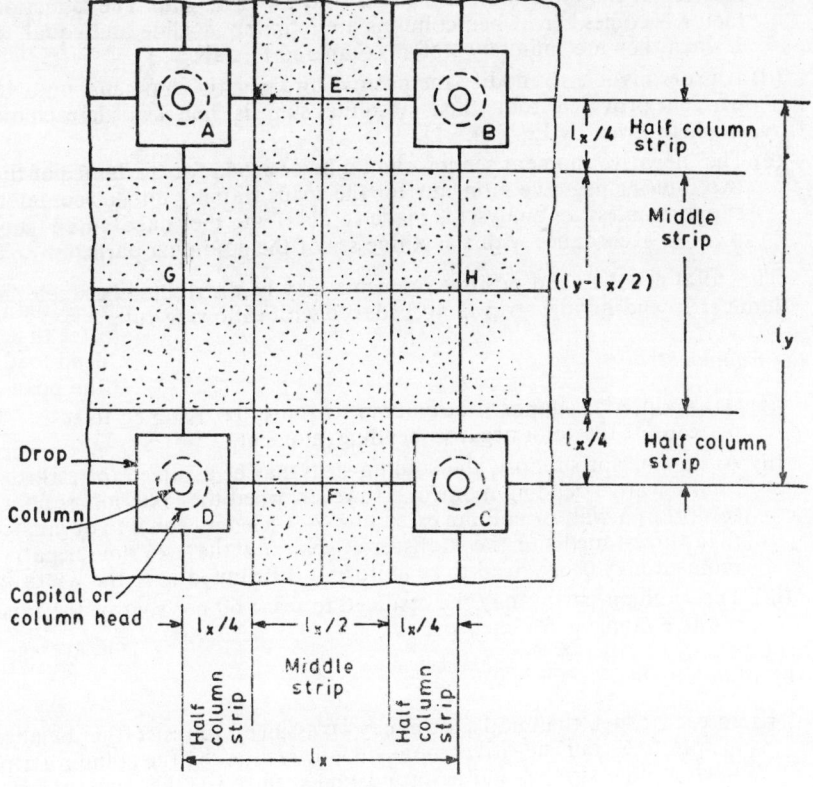

Fig. 14.4.1

equilibrium of free body of left half of the slab panel shown in Fig. 14.4.1. The total static moment M_{ox} may be assumed to be divided into total positive moment across section EF and total negative moment across section AD according to the following rules:

(a) The negative moment may be assumed to act along the face of the rectangular support.

(b) In the case of an interior span, the total negative and positive moments may be taken equal to 65 and 35 per cent of the total static moment in the direction being considered.

(c) In the case of an end span, the interior negative moment, the positive moment and exterior negative moment, may be taken respectively equal to $(0.75 - 0.10 \, R_f)$, $(0.65 - 0.28 \, R_f)$ and $0.65 \, R_f$ times the total static moment in the direction being considered. The reduction factor,

$$R_f = \frac{\alpha_c}{1 + \alpha_c} \qquad (14.4.2)$$

where α_c is the ratio of the sum of flexural stiffnesses of the columns meeting at the joint and the flexural stiffness of the slab. The reduction factor becomes zero when columns are infinitely flexible and equal to 1 when they are infinitely stiff as compared to slab.

(d) It is permissible to modify the positive and negative moments upto 10 per cent provided that their absolute sum is not less than static moment given by Eq. (14.4.1).

(e) The negative moment section should be designed for the larger of the two interior negative moments for the spans framing into a common support unless an analysis is made to distribute the unbalanced moment in accordance with the stiffnesses of the adjoining parts.

The total negative and positive moments may be distributed between the column strip and middle strip in accordance with following rules:

(a) *Column strip*

(i) At an interior support, the column strip may be assumed to resist 75 per cent of the total negative bending moment.

(ii) At an exterior support, the column strip may be assumed to resist the total negative bending moment. However, when the exterior support comprises a wall or column extending over a length equal to or greater than three-fourths of the transverse span l_y, the exterior negative moment may be assumed to be uniformly distributed over the width l_y.

(iii) The column strip may be assumed to resist 60 per cent of the total positive moment for each panel.

(b) *Middle strip*

(i) In each panel, the middle strip may be assumed to resist the balance of positive and negative moments not resisted by the column strip. Each middle strip should be proportioned to resist the sum of moments assigned to its two half middle strips.

(ii) The middle strip adjacent and parallel to an edge supported on a wall should be proportioned to resist twice the moment assigned to the half middle strip corresponding to the first row of interior columns.

The distribution of moments in flat slab panels is summarised in Table 14.4.1 in which the moment M_o should be taken equal to M_{ox} or M_{oy} depending upon the direction in which moment is being considered.

Table 14.4.1 Distribution of moments in slab panels

Location	Column Strip	Middle strip	Total
End support	0.65 $R_f M_o$	0	0.65 $R_f M_o$
Exterior midspan	$(0.378-0.168R_f) M_o$	$(0.292-0.112R_f) M_o$	$(0.63-0.28R_f) M_o$
First interior support			
(a) Outer	$(0.5625-0.075R_f) M_o$	$(0.1875-0.025R_f) M_o$	$(0.75-0.10R_f) M_o$
(b) Interior	0.4875 M_o	0.1625 M_o	0.65 M_o
Interior midspan	0.21 M_o	0.14 M_o	0.35 M_o
Second interior support	0.4875 M_o	0.1625 M_o	0.65 M_o

(B) Equivalent frame method

A more realistic distribution of moments may be determined by dividing the structure into series of equivalent frames in two orthogonal directions on the basis of following assumptions:

(a) In each of the two orthogonal directions, an interior frame comprises a row of equivalent columns or supports bounded laterally by the centre line of the panel on each side of the centre line of the columns or supports. The exterior frame adjacent and paraller to an edge is bounded by the edge and the centre line of the adjacent panel.

(b) The equivalent frames thus obtained may be analysed in their entirety to evaluate the bending moments in slab panels. Alternatively, the slab moments due to vertical loading may be determined by analysing the substitute frame comprising the slab under consideration and the columns immediately above and below the slab assumed to be fixed at their remote ends. In determining the bending moment in the slab at a given support, the slab may be assumed to be fixed at any support two panels distant therefrom, provided the slab continues beyond that point.

(c) The variation in moment of inertia due to drops should be considered. However, the stiffening effect of coiumn heads may be ignored.

(d) The moments of inertia of slabs and columns may be based on gross concrete section.

When the live load is variable but does not exceed three-fourths of the design dead load or when the nature of loading is such that all panels are loaded simultaneously, the maximum moments at all sections may be assumed to occur when full design live load acts on the entire slab system. When these conditions are not satisfied, the maximum positive moment near midspan of a panel may be assumed to occur when the span under consideration and alternate spans carry three-fourths of the design live load. Similarly the maximum negative moment in the slab at a support may be assumed to occur when the panels either side of the support carry three-fourths of the design live load. In no case, however, the design moments at all sections should be taken less than the moments when all panels are loaded with full design load.

The critical section for negative moment in an interior panel may be taken at the face of the rectangular support but not at a distance greater than 0.175 times the distance between the centres of supports. The critical section for negative moment in an exterior panel near the edge may be taken at a distance equal to half the projection of the bracket or column head beyond the face of the support.

When the slab satisfies the limitations in direct design method, the moments determined by equivalent frame method may be reduced in such proportion that the absolute sum of positive and average negative moments in a panel is not less than the static moment given by Eq. (14.4.1). The distribution of positive and negative moments across the panel width may be assumed to be the same as in direct design method.

14.4.2 Other Codes

The British Code BS: 8110 Part 1-1985 gives prominence to the equivalent frame method for the determination of bending moments in flat slabs. The guidelines for dividing the building structure into equivalent frames and those for simpler substitue frames are similar to those in the Indian code. The code also gives simplified method based on moment and shear coefficients subject to the conditions that there are at least three approximately equal spans in the direction in which the moments are being computed and all spans are loaded with maximum design ultimate load. The panel is divided into column strips and middle strips for the sake of proportioning the moments. The width of the column strip is taken equal to that of the drop However, the drop is ignored when its width is less than one-third of the width of panel. The additional thickness of the drop may still be taken into account in computing the punching shear along the periphery of the column head.

The American Code ACI: 318-1989 gives a unified treatment of two way slabs with or without beams. The moments may be determined either by direct design method or the equivalent frame method, although the American code like the Indian code gives precedence to former. The rules for dividing the structure into equivalent frames or the simpler substitute frames are also similar. When the slab is supported on an orthogonal system of beams resting on columns, the beams become a part of the respective column strips. The beams are assumed to resist 85 per cent of the moments in the respective column strips subject to the condition that the ratio of the flexural stiffness of the beam to that of the slab bounded by the centre lines of adjacent panels is not less than 1. In case, the stiffness ratio is less than 1, the percentage should be reduced linearly from 85 per cent to zero. In addition to uniform load, the beams should be proportioned to resist concentrated loads applied directly to beams and also the self weight of the stem projecting above or below the slab.

14.4.3 Distribution of Prestressing Tendons

The following points regarding the nature and distribution of bending moment in flat slabs are relevant in connection with the profile and distribution of prestressing tendons:

(i) Referring to Fig. 14.4.2, the portion of the slab marked A, which is common to two intersecting middle strips, is subjected to biaxial moments both of which are positive (sagging). The portions marked B, which are common to two intersecting column strips, are subjected to biaxial moments both of which are negative (hogging). The remaining portions of the slab marked C, which are common to a middle strip and a column strip, are subjected to biaxial moments of opposite sign. The flexural tensile stresses in the three aforementioned zones tend to create flexural cracks as shown in the figure. The cracks indicated by full and broken lines appear on the bottom face and top face respectively.

(ii) The total positive and negative bending moments in a panel are distributed between the column and middle strips with the column strip carrying approximately double the moment carried by the middle strip.

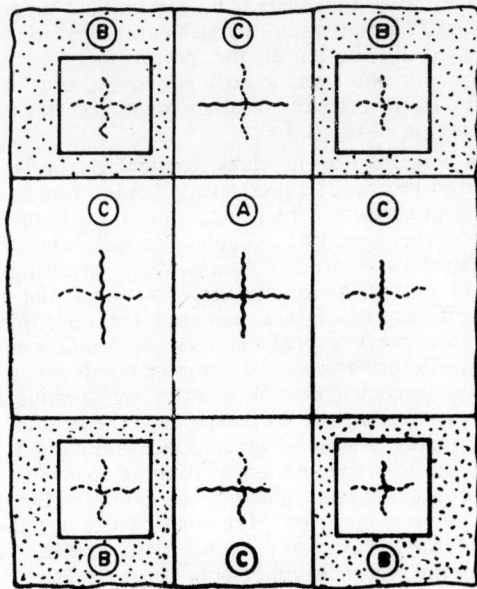

Fig. 14.4.2

(iii) The distribution of moments between the column and middle strips changes as the loading progresses in the post-cracking range and approaches the limit state of collapse. Due to a favourable redistribution of moments, the middle strips tend to carry greater share of moments as compared to the distribution indicated in (ii) above.

In view of the distribution of moments and the nature of stresses caused by them, it would appear logical to prestress a flat slab biaxially with the tendons having their high points at lines connecting the column centres. As the slabs are most commonly subjected to uniform loading, the shape of the tendons should be parabolic because it gives rise to uniform upward pressure due to prestressing. Adoption of curved tendons in both orthogonal directions may create practical problems in their placement unless the slab is considerably thick and the tendon spacing is relatively large. In order to avoid the placement problems and ensure adequate clearances between orthogonal tendons, it may become necessary to adopt concentric prestress with straight tendons in one of the two orthogonal directions. The total number of tendons or the quantum of prestressing force in a panel may be selected to counteract approximately the dead load and one half of the live load acting on the panel in accordance with Eq. (11.5.2). The total number of tendons in a panel may then be distributed between the column and middle strips in proportion to the moments carried by them. It may, therefore, appear logical to place 65 to 75 per cent tendons in the column strips and the remaining in the middle strips. It may, however, be stated that the distribution of tendons between the column and the middle strips is not critical. Tests on slabs with banded tendons and uniformly spaced tendons have shown that there is no substantial change in behaviour as the total number of tendons and other slab parameters remain unchanged. However, the banded distribution of the prestressing tendons with 65 to 75 per cent tendons in the column strip is perhaps most effective for flat slabs, particularly in respect of shear-moment transfer capacity at the column support section of the slab.

The problem of concordance in slabs, particularly in flat slabs, is more complex than that in beams. The prestressing tendon forces acting on an unloaded slab should produce zero deflection at all column support points in order to create the condition of concordance. When the slab is cast monolithically with the supporting columns, the prestressing forces should not tend to create any slope or flexural rotation at the support points. When these conditions, which are essential for concordance are not satisfied, the supports exert vertical reactions and bending moments on the slab to ensure zero deformation at all support points giving rise to secondary forces. If the eccentricities of prestressing tendons or the primary moments due to prestressing are proportional to the bending moments due to any arbitrary loading, a state of concordance is obtained. However, due to the complexity in the determination of bending moment distribution in flat slabs even for the simplest loading, the determination of concordant tendon profile is by no means easy. The concordance in slabs, as in beams, is no virtue. It is futile to search for a concordant profile. Nor is it necessary to determine the secondary moments due to prestress when a non-concordant tendon profile is adopted. The resultant moments at any stage of loading can be determined directly by using the free body approach of

Sec. 5.4. As the determination of only the resultant forces is sufficient for purpose of design, the break-up into primary and secondary forces is only of academic interest. A convenient practical approach in design of a flat slab is to select tendon profile and spacing in accordance with the afore-mentioned guidelines and to determine the resultant forces at critical stages of loading. The selected cable layout should satisfy all practical require-ments of clearances and concrete cover.

EXAMPLE 14.4.1

A flat slab of 150 mm thickness is supported by columns spaced at 6 m and 7.2 m centre to centre in two orthogonal directions. It is prestressed biaxially by effective prestressing forces equal to 1.35 MN/m and 0.9 MN/m applied concentrically in the directions of long and short spans respec-tively. The slab carries a uniform load of 6.88 kN/m^2 inclusive of self weight. Using the direct design method of Indian code, determine the critical extreme fibre stresses in column and middle strips in an interior panel. The diameter of the column head is 0.9 m.

Solution

Area of column head is $\frac{\pi}{4} \times 0.9^2 = 0.6364$ m^2. Hence the side of equiva-lent square column head is 0.8 m.

(i) *Stresses along short span (x-direction)*

Width of panel is 7.2 m which may be divided into column and middle strips each of 3.6 m width. The clear span,

$$l_{nx} = 6 - 0.8 = 5.2 \text{ m}$$

From Eq. (14.4.1), total static moment for the panel,

$$M_{ox} = \frac{0.00688 \times 7.2 \times 5.2 \times 5.2}{8} = 0.1674 \text{ MN.m}$$

For an interior panel, total negative moment is $0.65\,M_{ox} = 0.65 \times 0.1674 = 0.1088$ MN.m. Average negative moment in column strip $= 0.75 \times 0.1088 = 0.0816$ MN.m. Hence the extreme fibre stresses in column strip,

$$f_T = -\frac{0.9}{1 \times 0.15} + \frac{0.0816}{(1/6) \times 3.6 \times 0.15^2} = 0.04 \text{ MPa}$$

$$f_B = -\frac{0.9}{1 \times 0.15} + \frac{0.0816}{(1/6) \times 3.6 \times 0.15^2} = -12.04 \text{ MPa}$$

Average negative moment in middle strip $= 0.25 \times 0.1088 = 0.0272$ MN.m. Hence the extreme fibre stresses in middle strip,

$$f_T = -\frac{0.9}{1 \times 0.15} + \frac{0.0272}{(1/6) \times 3.6 \times 0.15^2} = -4.0 \text{ MPa}$$

$$f_B = -\frac{0.9}{1 \times 0.15} - \frac{0.0272}{(1/6) \times 3.6 \times 0.15^2} = -8.0 \text{ MPa}$$

The positive moment for the panel is $0.35\,M_{cx} = 0.0586$ MN.m. Positive moment in column strip $= 0.6 \times 0.0586 = 0.0352$ MN.m. Hence the extreme fibre stresses in column strip,

$$f_T = - \frac{0.9}{1 \times 0.15} - \frac{0.0352}{(1/6) \times 3.6 \times 0.15^2} = - 8.61 \text{ MPa}$$

$$f_B = - \frac{0.9}{1 \times 0.15} + \frac{0.0352}{(1/6) \times 3.6 \times 0.15^2} = - 3.39 \text{ MPa}$$

Positive moment in middle strip $= 0.4 \times 0.0586 = 0.0234$ MN.m. Hence the extreme fibre stresses in middle strip,

$$f_T = - \frac{0.9}{1 \times 0.15} + \frac{0.0234}{(1/6) \times 3.6 \times 0.15^2} = - 7.73 \text{ MPa}$$

$$f_B = - \frac{0.9}{1 \times 0.15} + \frac{0.0234}{(1/6) \times 3.6 \times 0.15^2} = - 4.27 \text{ MPa}$$

(ii) *Stresses along long span (y-direction)*

Width of panel is 6 m. Hence width of column and middle strip is 3 m.

$$l_{ny} = 7.2 - 0.8 = 6.4 \text{ m}$$

$$M_{oy} = \frac{0.00688 \times 6 \times 6.4 \times 6.4}{8} = 0.2114 \text{ MN.m}$$

Proceeding as in (i) above, the extreme fibre stresses due to negative moments are, $f_T = 0.16$ MPa and $f_B = - 18.16$ MPa in column strip and $f_T = - 5.95$ MPa and $f_B = - 12.05$ MPa in middle strip. The extreme fibre stresses due to positive moment are $f_T = - 12.95$ MPa and $f_B = - 5.05$ MPa in column strip and $f_T = - 11.63$ MPa and $f_B = - 6.37$ MPa in middle strip.

EXAMPLE 14.4.2

The edge panel of a 150 mm thick flat slab is 6.4 m $\times$ 7.2 m measured on the centre lines of columns. The slab cantilevers out by 2.4 m in the direction of long span. The prestressing tendons in the direction of long span comprise high tensile steel bars of 20 mm diameter spaced at 335 mm carrying an effective stress of 800 MPa. The tendons, which are parabolic in the end span and straight in the cantilever portion, have upward eccentricity of 35 mm at column supports, 35 mm downward eccentricity at midpoint of end span and zero eccentricity at free end of the cantilever. The diameters of column and column head are 0.5 m and 0.75 m respectively. The height of columns above and below the slab is 3.6 m. The moduli of elasticity of concretes in slab and column are 25500 and 33700 MPa respectively. The slab carries uniform load of 10.4 kN/m² inclusive of self weight. Determine the extreme fibre stresses in the column strip at the exterior support in accordance with the Indian code.

Solution

The circular column head of 0.75 m ¡diameter may be replaced by square column head of 0.665 m side. Hence clear longer span,

$$l_{ny} = 7.2 - 0.665 = 6.535 \text{ m}$$

Prestressing force per metre width in the direction of long span,

$$P_y = \frac{\pi}{4} \times 20^2 \times 10^{-6} \times \frac{1000}{335} \times 800 = 0.75 \text{ MN}$$

Sag of parabolic tendons,

$$S = 0.035 + 0.035 = 0.070 \text{ m}$$

Hence uniform upward force due to parabolic tendons,

$$w_p = \frac{8 \times 0.75 \times 0.070}{7.2^2} = 0.0081 \text{ MN/m}^2$$

Net downward load on slab,

$$w = 0.0104 - 0.0081 = 0.0023 \text{ MN/m}^2$$

Total design load on area $l_x \, l_{ny}$,

$$W = 0.0023 \times 6.4 \times 6.535 = 0.0962 \text{ MN}$$

Total static moment,

$$M_{oy} = \frac{0.0962 \times 6.535}{8} = 0.0786 \text{ MN.m}$$

The exterior negative moment is $0.65 \, R_f \, M_{oy}$.
Relative stiffness of column,

$$K_c = \frac{E_{cc} I_c}{L_c} = \frac{25500 \times \frac{\pi}{64} \times 0.5^4}{3.6} = 21.74 \text{ MN.m}$$

Relative stiffness of slab,

$$K_s = \frac{E_{cs} I_s}{L_s} = \frac{33700 \times \frac{1}{12} \times 0.15^3}{7.2} = 8.425 \text{ MN.m}$$

Hence the ratio of stiffnesses,

$$\alpha_c = \frac{\sum K_c}{K_s} = \frac{2 \times 21.74}{8.425} = 5.16$$

From Eq. (14.4.2), the reduction factor,

$$R_f = \frac{5.16}{1 + 5.16} = 0.838$$

Exterior negative moment,

$$M_1 = -0.65 \times 0.838 \times 0.0786 = -0.0428 \text{ MN.m}$$

In the absence of supporting wall, the entire negative moment may be assumed to be carried by the column strip. In addition, the column strip has to carry a part of the hogging moment M_2 due to cantilever portion.

$$M_2 = \left(-\frac{1}{2} \times 0.0104 \times 2.4^2 + 0.75 \times 0.035 \right) \times \frac{6.4}{2}$$

$$= -0.0118 \text{ MN.m}$$

The part of M_2 carried by slab,

$$M_2' = M_2 \left(\frac{K_s}{K_s + \sum K_c} \right)$$

$$= - 0.0118 \times \frac{8.425}{8.425 + 2 \times 21.74}$$

$$= - 0.0019 \text{ MN.m}$$

Hence the total moment to be resisted by the column strip,

$$M = M_1 + M_2' = - 0.0428 - 0.0019$$

$$= - 0.0447 \text{ MN.m}$$

Extreme fibre stresses in column strip at exterior supports,

$$f_T = - \frac{0.75}{1 \times 0.15} + \frac{0.0447}{(1/6) \times 3.2 \times 0.15^2}$$

$$= - 1.275 \text{ MPa}$$

$$f_B = - \frac{0.75}{1 \times 0.15} - \frac{0.0447}{(1/6) \times 3.2 \times 0.15^2}$$

$$= - 8.725 \text{ MPa}$$

EXAMPLE 14.4.3

The supporting columns of a flat slab are located at spacing of 6 m and 7.5 m in the two orthogonal directions. The slab has to carry its own weight at initial stage and a service load of 8.4 kN/m² at final stage. The diameter of columns is 0.5 m. Design a typical interior panel with suitable column heads. Adopt Lee-Mc Call system and provisions of the Indian Code.

Solution

Taking the shorter span to depth ratio equal to 40, the thickness of the slab is $6/40 = 0.15$ m. Assuming the unit weight of concrete as 0.024 MN/m³, the self weight of slab, $w_D = 0.024 \times 0.15 = 0.0036$ MN/m². In order to avoid difficulties is placement of tendons, eccentric prestress with parabolic tendons in the direction of long span and concentric prestress with straight tendons in the direction of short span may be adopted.

(i) Long span (y-direction)

The parabolic tendons in the direction of long span may be designed so that the unbalanced loads at initial and final stages are equal. Hence using Eq. (11.5.2) and taking loss factor $\eta = 0.8$,

$$w_{pi} = \frac{2 \times 0.0036 + 0.0084}{1 + 0.8} = 0.00867 \text{ MN/m}^2$$

$$w_p = 0.8 \times 0.000867 = 0.00693 \text{ MN/m}^2$$

Unbalanced load at initial stage,

$$w_{ubi} = w_{pi} - w_D$$

$$= 0.00867 - 0.0036 = 0.00507 \text{ MN/m}^2 \text{ (upward)}$$

Unbalanced load at final stage,

$$w_{ub} = w_D + w_L - w_p$$

$$= 0.0036 + 0.0084 - 0.00693$$

$$= 0.00507 \text{ MN/m}^2 \text{ (downward)}$$

Adopting a column head of 0.9 m diameter, the side of equivalent square column head is 0.798 m. Hence,

$$l_{ny} = 7.5 - 0.798 = 6.702 \text{ m}$$

$$W = 0.00507 \times 6 \times 6.702 = 0.204 \text{ MN}$$

$$M_{oy} = \frac{0.204 \times 6.702}{8} = 0.171 \text{ MN.m}$$

The moment in the column strip at the interior support which is the largest one,

$$M_{max} = 0.65 \times 0.75 \times 0.171 = 0.0834 \text{ MN.m}$$

Hence the maximum bending stress in extreme fibres,

$$f_b = \frac{0.0834}{(1/6) \times (0.5 \times 6) \times 0.15^2} = 7.4 \text{ MPa}$$

In order to avoid tensile stress, the effective direct stress due to prestressing may be taken at least equal to 7.4 MPa. Hence minimum required value of P_y is $7.4 \times 0.15 = 1.11$ MN. Adopting 20 mm diameter bars with an effective stress of 800 MPa, the spacing of bars may be 225 mm.

$$P_y = \frac{\pi}{4} \times 20^2 \times 800 \times 10^{-6} \times \frac{1000}{225} = 1.117 \text{ MN}$$

$$P_{yi} = \frac{1.117}{0.8} = 1.396 \text{ MN}$$

The sag of the parabolic cables for an upward force of

$$w_p = 0.00693 \text{ MN/m}^2,$$

$$S = \frac{0.00693 \times 7.5^2}{8 \times 1.117} = 0.0436 \text{ m}$$

Hence the parabolic cables may have an upward eccentricity of 21.8 mm at ends and an equal downward eccentricity at midspan, The concrete cover is therefore $75 - 21.8 = 53.2$ mm which is sufficient. The extreme fibre stresses at the end of column strip at initial stage,

$$f_{Ti} = - \frac{1.396}{1 \times 0.15} - 7.4 = - 16.71 \text{ MPa}$$

$$f_{Bi} = - \frac{1.396}{1 \times 0.15} + 7.4 = - 1.91 \text{ MPa}$$

The extreme fibre stresses at the end of column strip at final stage,

$$f_T = - \frac{1.117}{1 \times 0.15} + 7.4 = - 0.05 \text{ MPa}$$

$$f_B = - \frac{1.117}{1 \times 0.15} - 7.4 = - 14.85 \text{ MPa}$$

Adopting concrete of grade M-40 with a strength of 35 MPa at stress transfer, the permissible compressive stresses at initial and final stages are 16.9 and 15.7 MPa respectively. Hence the actual stresses are within permissible limits. The bending moment in the middle strip and at the centre of the column strip are smaller than M_{max}. Hence the stresses are safe at all locations.

(ii) *Short span (x-direction)*

$$l_{nx} = 6 - 0.798 = 5.202 \text{ m}$$

$$W = 0.00507 \times 7.5 \times 5.202 = 0.1978 \text{ MN}$$

$$M_{0x} = \frac{0.1978 \times 5.202}{8} = 0.1286 \text{ MN.m}$$

The moment in the column strip at interior support, which is the largest one,

$$M_{max} = 0.65 \times 0.75 \times 0.1286 = 0.0627 \text{ MN.m}$$

Hence the maximum bending stress in extreme fibres,

$$f_b = \frac{0.0627}{(1/6) \times (0.5 \times 7.5) \times 0.15^2} = 4.46 \text{ MPa}$$

In order to avoid tensile stress, the effective direct stress due to prestressing may be taken at least equal to 4.46 MPa. Hence minimum required value of P_x is $4.46 \times 0.15 = 0.669$ MN. Adopting 20 mm diameter bars with an effective stress of 800 MPa, the spacing of bars may be taken as 375 mm. The straight bars in the direction of short span may be provided in two layers, one near the top and the other near the bottom face. Each layer may comprise 20 mm diameter bars at 750 mm spacing with an effective cover of 30 mm.

Instead of adopting uniform spacing of the prestressing tendons, the spacing may be increased gradually from the centre line of columns towards the centre line of the panel in such a manner that approximately 65 per cent of the total tendons in a panel are located in the column strip. The bars may be greased and wrapped with oil paper to avoid bond with concrete.

14.5 CRACKING AND ULTIMATE STRENGTHS

With a proper layout of prestressing tendons, the bending moments due to prestressing are in general opposite in sign to those caused by external service loads. Consequently, the net bending moments for a given load are smaller in a prestressed concrete slab as compared to its corresponding reinforced concrete slab. The load required to cause cracking is, therefore, larger for a prestressed concrete slab. The first flexural crack appears at the location of largest bending moment in the direction normal to the plane of the largest moment. The cracking load may be determined by equating the stress at the location of largest moment to the tensile strength of concrete in bending.

In a uniformly loaded circular slab simply supported along its edge, the maximum radial and circumferential moments occur at the centre. Hence cracking commences by the appearance of radial cracks at centre on the bottom face. When the edge is restrained, the cracking generally commences by the appearance of circumferential cracks adjacent to the restrained edge on the top face. In the case of a rectangular slab simply supported along its edges, the maximum bending moment occurs at the centre of slab in the direction of short span. Hence in simply supported rectangular slabs, the first crack appears at the centre of the slab parallel to longer span at the bottom surface. In a rectangular slab with restrained edges, the first crack appears at the top surface adjacent and parallel to one of the two longer edges. In the case of a flat slab which is supported directly on columns, the largest moment is generally the hogging moment at the edge of the column head. Hence cracking usually commences by the appearance of cracks on the top face radiating from the edge of the column head.

As overloading continues beyond the cracking load, the first crack propagates on either side and new cracks appear at other locations wherever the net stress reaches the tensile strength of concrete. The progressive cracking eventually leads to a distinct yield line pattern in which the so called yield lines are actually relatively narrow bands which are cracked extensively on the tension side and fully plasticised in compression on the other side. These bands are idealised to straight lines called yield lines for the sake of simplicity. As almost the entire curvature is concentrated within the narrow bands, it is assumed that the segments of the slab into which the slab is divided by the yield lines are plane. It is assumed that the slab attains its ultimate strength as soon as the yield line pattern develops fully to form the collapse mechanism. Although the slab may have some additional strength on account of the membrane action, the slab is so badly deformed before the commencement of membrane action that the surplus strength may be treated as an unutilisable reserve. The collapse load of the slab may be determined from the yield line pattern by equating the internal virtual work at the yield lines to the external virtual work due to the vertical deflection of applied external loading. The most critical yield line pattern is the one which minimises the collapse load. In accordance with the upper bound or kinematic theorem, the smallest collapse load corresponding to the most critical yield line pattern is the true collapse load. The most critical yield line pattern may be determined either by trial or by equating partial derivatives of collapse load with respect to the pattern parameters to zero. For a small virtual characteristic deflection Δ, the internal virtual work U_i at yield lines may be expressed as

$$U_i = (ml) \, k_1 \left(\frac{\Delta}{l} \right) f_1 \, (\alpha, \beta, \ldots) \qquad (14.5.1)$$

where m = characteristic ultimate moment capacity of yield lines per unit length

l = characteristic length

k_1 = dimensionless constant which depends upon the geometry of the slab

$f_1 \, (\alpha, \beta, \ldots)$ = function of yield line pattern parameters $\alpha, \beta, \ldots$

The external virtual work done by the characteristic ultimate load W_u in undergoing small virtual displacement,

$$U_e = W_u k_2 \, \Delta \, f_2 \, (\alpha, \beta, \ldots) \qquad (14.5.2)$$

where k_2 = dimensionless constant which depends upon the distribution of external load

$f_2 \, (\alpha, \beta, \ldots)$ = function of yield line pattern parameters $\alpha, \beta, \ldots$

Equating the internal and external virtual works, the collapse load,

$$W_u = mkf \, (\alpha, \beta) \qquad (14.5.3)$$

where

$$k = \frac{k_1}{k_2} \qquad (14.5.3)$$

$$f (\alpha, \beta, \ldots) = \frac{f_1 \, (\alpha, \beta \ldots)}{f_2 \, (\alpha, \beta \ldots)}$$

The true collapse load is the smallest value obtained by minimisation of the pattern parameter function $f \, (\alpha, \beta, \ldots)$. Hence the most critical yield line pattern and the corresponding true collapse load w_u may be determined from the equations

$$\frac{\partial f}{\partial \alpha} = \frac{\partial f}{\partial \beta} = \ldots = 0 \qquad (14.5.4)$$

The following points are useful in the simplified yield line analysis for slabs:

(i) The positive yield lines due to sagging moment and negative yield lines due to hogging moments are idealised as straight lines. Hence the entire yield line pattern comprises straight lines only. The yield lines divide the slab into number of segments. All segments are assumed to be plane so that the entire curvature due to the deflection of slab is concentrated at the yield lines. The plane segments rotate about their respective axes of rotation to form a collapse mechanism. The idealised deflected shape of the slab at collapse comprises one or more inverted or upright pyramids with positive yield lines forming the valleys and negative yield lines forming the ridges.

(ii) The axes of rotation coincide with the supported edges of the slab. As the column supports act as pivots, the axes of rotation may pass through them in any direction.

(iii) The positive yield lines pass through the points of intersection of axes of rotation of plane segments. The negative yield lines generally run adjacent and parallel to the restrained edges.

(iv) The ultimate moment capacity of the slab $m_{\theta u}$ in a direction inclined at an angle θ to x-axis may be expressed as

$$m_{\theta u} = m_{xu} \cos^2 \theta + m_{yu} \sin^2 \theta \qquad (14.5.5)$$

where m_{xu}, m_{yu} = ultimate moment capacities per unit width along x and y-axes respectively.

Hence the ultimate moment capacity across or normal to the yield line inclined at an angle θ to y-axis is given by Eq. (14.5.5).

(v) In a slab having constant moment capacity m_u in all directions, the internal virtual work U_i at positive line OA shown in Fig. 14.5.1 may be expressed as

$$U_i = m_u \, (\overline{OB} \, \psi_1 + \overline{OC} \, \psi_2) \qquad (14.5.6)$$

where $\overline{OB} = l \cos \theta_1$

$\qquad$ = projection of yield line on axis 1

$\overline{OC} = l \cos \theta_2$

$\qquad$ = projection of yield line on axis 2

l = length of yield line OA

ψ_1, ψ_2 = rotations of segments (1) and (2) about axes 1 and 2

(vi) Figure 14.5.2 shows the positive yield line OA passing through the intersection of two adjacent supported edges along x and y-axes. which act as axes of rotation. The internal virtual work at the yield line may be expressed as

$$U_i + m_{xu} \, l_y \, \psi_y + m_{yu} \, l_x. \psi_x \qquad (14.5.7)$$

where l_x, l_y = projections of yield line along x and y-axes respectively

ψ_x, ψ_y = rotations of segments (1) and (2) about x and y-axes respectively.

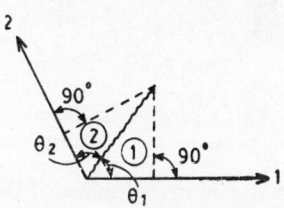

Fig. 14.5.1

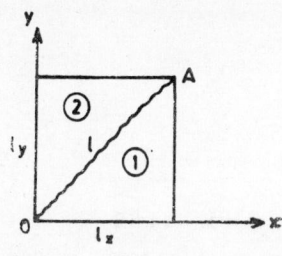

Fig. 14.5.2

(vii) The external virtual work done due to a load W_{un} distributed uniformly on nth segment, in undergoing a small virtual displacement,

$$U_e = W_{un} \, \Delta_n \qquad (14.5.8)$$

where Δ_n = virtual vertical deflection of the segment at its centroid.

Some typical yield line patterns for slabs carrying uniformly distributed load are shown in Fig. 14.5.3.

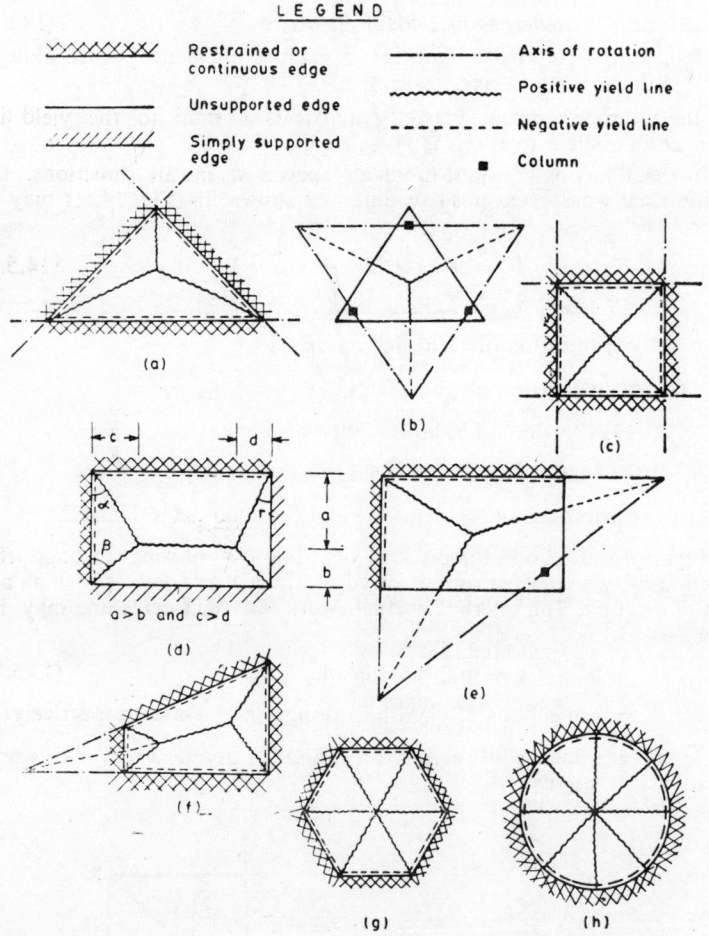

Fig. 14.5.3

EXAMPLE 14.5.1

Determine the cracking load for the prestressed concrete slab of Example 14.2.1. The modulus of rupture of concrete is 4.2 MPa.

Solution

From Table 14.2.1, the cracking moment M_{cr}, which produces cracking at centre of slab, may be expressed as

$$M_{cr} = \frac{(3 + 0.15)\, w_{cr} \times 6^2}{16} = 7.0875 \ w_{cr}$$

From Example 14.2.1, the compressive stress at bottom fibre at the centre of slab due to prestressing is 13.2 MPa. Hence the cracking load w_{cr} or the cracking moment M_{cr} should produce a tensile stress equal to $13.2 + 4.2 = 17.4$ MPa. Hence,

$$\frac{7.0875 \ w_{cr}}{(1/6) \times 1 \times 0.2^2} = 17.4$$

or $w_{cr} = 0.0164$ MN/m² or 16.4 kN/m²

EXAMPLE 14.5.2

Derive an expression for the collapse load of a 'slab having the shape of regular polygon of n sides with restrained edges and constant positive and negative ultimate moment capacities in all directions. The slab carries uniform load on a concentric polygon. Hence derive the corresponding expression for collapse load of a circular slab carrying uniform load on a concentric circle.

Solution

The yield line pattern comprises positive yield lines radiating from centre towards all corners of the polygon and negative yield lines parallel and adjacent to the restrained edges, there by dividing the slab into n triangular segments. The idealised deflected shape at collapse is an inverted pyramid having n faces with apex at centre. The edges of the slab are the axes of rotation. The total length of projection of all positive yield lines on the axes of rotation is equal to the perimeter of the polygon or n times the side of the polygon.

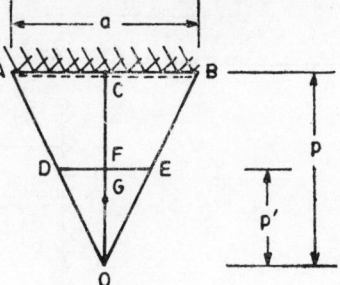

Fig. 14.5.4

Figure 14.5.4 shows one of the segments enclosed by two consecutive positive yield lines OA and OB. If Δ is the virtual deflection at centre O, then the deflection at the centroid G of the loaded area ODE is given by

$$\Delta_G = \left(\frac{p - \frac{2}{3}p'}{p}\right) = \Delta\left(1 - \frac{2}{3}\frac{p'}{p}\right) \tag{a}$$

Using Eqs. (14.5.6) and (14.5.8) and equating the internal and external virtual works,

$$na\,(m_u + m_u')\,\frac{\Delta}{p} = W_u\,\Delta_G$$

Substituting for Δ_G from Eq. (a), the collapse load,

$$W_u = \frac{na\,(m_u + m_u')}{p\left(1 - \dfrac{2}{3}\dfrac{p'}{p}\right)} \tag{14.5.9}$$

In the case of a circular slab of radius r carrying uniform load on a concentric circle of radius ρ, replacing p and p' by r and ρ respectively and putting $na = 2\pi r$,

$$W_u = \frac{2\pi\,(m_u + m_u')}{\left(1 - \dfrac{2}{3}\dfrac{\rho}{r}\right)} \tag{'4.5.10}$$

The ratio $\dfrac{\rho}{r}$ may vary from 0 to 1. For $\dfrac{\rho}{r} = 0$, i.e. for a concentrated central load,

$$W_u = 2\pi\,(m_u + m_u') \tag{14.5.11}$$

For $\dfrac{\rho}{r} = 1$, i.e. for uniformly distributed load over the entire slab,

$$W_u = 6\pi\,(m_u + m_u') \tag{14.5.12}$$

EXAMPLE 14.5.3

The uniformly loaded prestressed concrete rectangular slab with restrained edges shown in Fig. 14.5.5 has positive ultimate moment capacities m_{xu}

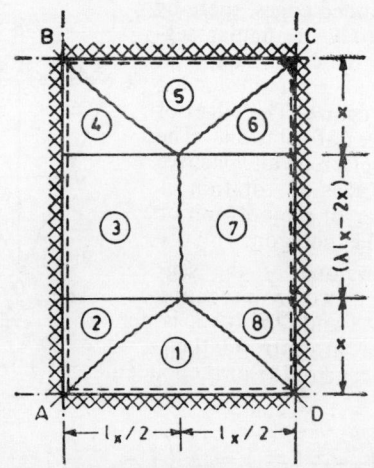

Fig. 14.5.5

and m_{v_u} and negative ultimate moment capacities m'_{x_u} and m'_{y_u} along x and y-directions respectively. Determine the geometry of the most critical yield line pattern. Hence derive expression for the intensity of collapse load.

Solution

As the slab and the support conditions are symmetrical with respect to both x and y-axes, the yield line pattern is likewise doubly symmetric. Consequently it is fully defined by a single parameter α or x which are related to each other by the equation

$$x = \frac{l_x}{2} \tan \alpha$$

The yield line pattern divides the slab into four segments. However, for convenience in writing expressions for external and internal virtual works, the slab may be divided into eight segments as shown in the figure. Assuming that Δ is the maximum virtual deflection at all points on the yield line EF and noting that the axes of rotation coincide with the restrained edges, the internal virtual work at positive yield line AE in accordance with Eq. (14.5.7) may be written as

$$u_{i1} = \frac{1}{2} l_x m_{yu} \frac{\Delta}{x} + x m_{xu} \frac{\Delta}{(1/2) l_x}$$ (b)

The internal virtual work at positive yield line EF,

$$u_{i2} = 2 (\lambda l_x - 2x) m_{xu} \frac{\Delta}{(1/2) l_x}$$ (c)

The internal virtual work at all negative yield lines,

$$u_{i3} = 2 l_x m'_{yu} \frac{\Delta}{x} + 2\lambda l_x m'_{xu} \frac{\Delta}{(1/2) l_x}$$ (d)

Hence total internal virtual work,

$$U_i = 4u_{i1} + u_{i2} + u_{i3}$$ (e)

Using Eq. (14.5.8), the external virtual work due to displacement of load acting on the segment (1),

$$u_{e1} = \frac{1}{2} l_x w_u \frac{\Delta}{3}$$ (f)

The external virtual work due to displacement of load acting on segment (2),

$$u_{e2} = \frac{1}{2} x \frac{l_x}{2} w_u \frac{\Delta}{3}$$ (g)

The external virtual work due to displacement of load acting on segment (3),

$$u_{e3} = (\lambda l_x - 2x) \frac{l_x}{2} w_u \frac{\Delta}{3}$$ (h)

Hence the total external virtual work,

$$U_e = 2u_{e1} + 4u_{e2} + 2u_{e3} \tag{i}$$

Equating the internal and external virtual works given by Eq. (e) and (i), the expression for the intensity of collapse load w_u may be obtained. The true collapse load corresponding to the most critical yield line pattern is obtained by equating the partial derivative of w_u with respect to x to zero. Solving the resulting equation,

$$x = \frac{l_x}{2} \left[\sqrt{3\mu + \frac{\mu^2}{\lambda^2}} - \frac{\mu}{\lambda} \right] \tag{14.5.13}$$

where

$$\mu = \frac{m_{yu}}{m_{xu}} = \frac{m'_{yu}}{m'_{xu}}$$

$$= \text{orthotropy ratio}$$

The intensity of collapse load is given by

$$w_u = \frac{6\mu}{x^2} (m_{xu} + m'_{xu}) \tag{14.5.14}$$

EXAMPLE 14.5.4

A prestressed concrete rectangular slab 6 m × 9 m having an overall depth of 150 mm is simply supported along its edges. It is prestressed biaxially by means of straight high tensile steel bars of 20 mm diameter at 300 mm spacing placed at constant eccentricity of 45 mm along short span and straight bars of 16 mm diameter at 300 mm spacing placed at constant eccentricity of 25 mm along long span. Determine the uniformly distributed collapse load assuming that all tendons attain their yield stress of 1500 MPa and the internal lever arm is equal to 90 per cent of the effective depth at the stage of collapse.

Solution

The effective depths of prestressing tendons along short and long spans are $\frac{150}{2} + 45 = 120$ mm and $\frac{150}{2} + 25 = 100$ mm respectively. Hence the ultimate moment capacities along short span (x-direction) and long span (y-direction) are

$$m_{xu} = \frac{1000}{300} \times \frac{\pi}{4} \times 20^2 \times 10^{-6} \times 1500 \times 0.9 \times 0.120$$

$$= 0.1697 \text{ MN} \cdot \text{m/m}$$

$$m_{yu} = \frac{1000}{300} \times \frac{\pi}{4} \times 16^2 \times 10^{-6} \times 1500 \times 0.9 \times 0.100$$

$$= 0.0905 \text{ MN.m/m}$$

Hence the orthotropy ratio,

$$\mu = \frac{m_{yu}}{m_{xu}} = \frac{0.0905}{0.1697} = 0.533$$

Hence from Eq. (14.5.13) of Example 14.5.3,

$$x = \frac{6}{2} \left[\sqrt{3 \times 0.533 + \frac{0.533^2}{1.5^2}} - \frac{0.533}{1.5} \right]$$

$$= 2.876 \text{ m}$$

$$\tan \alpha = \frac{2.876}{3} = 0.959 \quad \text{or} \quad \alpha = 43.8°$$

As the slab is simply supported, $m'_{xu} = m'_{yu} = 0$. Hence from Eq. (14.5.14), the intensity of collapse load,

$$w_u = \frac{6 \times 0.533}{2.876^2} \times 0.1697 = 0.0656 \text{ MN/m}^2$$

EXAMPLE 14.5.5

A prestressed concrete flat slab has retangular panels 6 m $\times$ 7.5 m. It has constant positive and negative ultimate moment capacities equal to 200 kN.m/m The slab is supported on circular columns with column heads having a diameter of 0.75 m. Determine the collapse load of the slab.

Solution

Assuming that the column reaction in a flat slab is an inverted concentrated load in the continuous plate field, the collapse load may be determined from Eq. (14.5.10) of Example 14.5.2. The radius of the loaded area due to column reaction is equal to the radius of the column head. The radius r may be taken approximately equal to shorter diamension of the panel. Hence putting $\rho = 0.5 \times 0.75 = 0.375$ m and $r = 6$ m in Eq. (14.5.10),

$$W_u = \frac{2\pi (0.2 + 0.2)}{1 - \frac{2}{3} \times \frac{0.375}{6}} = 2.624 \text{ MN}$$

Hence the collapse load of the panel

$$w_u = \frac{2.624}{6 \times 7.5} = 0.0583 \text{ MN/m}^2$$

14.6 CODE PROVISIONS

The general provisions in respect of strength and serviceability of reinforced concrete slabs also apply to prestressed concrete slabs. However, the codes also contain certain special provisions applicable to prestressed concrete slabs.

The provisions of Indian Code IS: 134?-1980 and British Code BS: 8110: Part 1-1985 for prestressed concrete slabs in respect of limit state of serviceability (deflection) are the same as for prestressed concrete beams. These provisions are discussed in Sec. 8.6 for Type 1 and Type 2 members and in Sec. 10.10 for Type 3 members. The American Code ACI : 318-1989 recommends the use of elastic theory for checking all serviceability limits. The span/depth ratio should not exceed 42 for floors and 48 for roofs in the case of prestressed concrete flat slabs continuous over two or more spans in each direction. These limits may be increased to 48 and 52 respectively when computations show that both short and long term deflections, camber and vibration frequency and amplitude are within acceptable limits. The distance between construction joints in prestressed concrete slabs should generally not exceed 30 to 45 m in order to minimise the effects of slab shortening and to avoid excessive loss of prestress due to friction. When the live loads are normal and the loads are distributed uniformly, the spacing of the prestressing tendons or groups of tendons in any direction should not exceed eight times the slab thickness nor 1.5 m The spacing of tendons should also ensure a minimum average prestress of 0.88 MPa on the slab section tributary to the tendon or tendon group. In slabs with unbonded tendons, untensioned bonded reinforcement should be provided in accordance with Sec. 10.10. In two way flat slabs of uniform thickness, containing unbonded tendons, the minimum area of bonded untensioned reinforcement in the positive moment areas may be determined from

$$A_s \text{ (min)} = \frac{2T_c}{f_y} \qquad (14.6.1)$$

where T_c = total tension in concrete in uncracked section caused by unfactored dead and live loods.

f_y = yield stress of bonded untensioned reinforcement which should not be taken more than 415 MPa.

The bonded untensioned reinforcement specified by Eq. (14.6.1) is not necessary when computed tensile stress in concrete under service loads after all losses is not greater than $0.17\sqrt{f_c'}$. The bonded reinforcement should be uniformly distributed over the precompressed tensile zone as close as practicable to tension face. The bonded reinforcement should be centered within the positive moment area and should have a length not less than one-third of the clear span. In the negative moment areas near the column supports, the minimum area of bonded reinforcement may be determined from

$$A_s \text{ (min)} = 0.00075 \ lD \qquad (14.6.2)$$

where l = length of span in the direction parallel to that of the reinforcement being determined

D = overall thickness of flat slab.

The reinforcement computed by Eq. (14.6.2) should be distributed within the slab width enclosed by lines located at a distance of 1.5 D from the opposite faces of the column. At least four reinforcing bars should be provided in each direction at a spacing not exeeeding 300 mm. The reinforcing bars should extend over a distance not less than one-sixth of clear span on either side of the column.

RREFRENCES

14.1 ACI-ASCE Commentary 423, *Tentative recommendations for prestressed concrete flat plates*, Journal of the American Concrete Institute, Vol. 71, No. 2, February 1974, pp. 61-71.

14.2 Burn, N.H. and Hemabom, R., *Test of scale model post-tensioned flat plate*, Journal of the Structural Division, ASCE, Vol. 103, 1977, pp. 1237-55.

14.3 Burn. N.H.; Charney, F.A. and Vines, W.R., *Tests of one-way post-tensioned slabs with unbonded tendons*, Journal of the Prestressed Concrete Institute, Vol. 23, No. 5, September-October 1978.

14.4 Dowrick, D.J. and Narasimhan, N., *Prestressed versus reinforced concrete flat slabs*, Concrete, Vol. 12, No. 9, September 1978, pp. 16-9.

14.5 Hognestad, E., *Yield line theory for the flexural strength of reinforced concrete slabs*, Journal of the American Concrete Institute, Proceedings Vol. 49, March 1953, pp. 637-58.

14.6 Lin, T.Y., *Load-balancing method for design and analysis of prestressed concrete stuctures*, Journal of the American Concrete Institute, Vol. 60, 1963, pp. 719-42.

14.7 Louis, L.G. and Burns, N.H., *Ultimate strength tests of post-tensioned flat plates*, Journal of the Prestressed Concrete Institute, Vol. 16, No. 6, November-December 1971, pp. 40-58.

14.8 Nawy, E.G. and Chakrabarti, P., *Deflection of prestressed concrcte flat plates*, Journal of the Prestressed Concrete Institute, Vol. 21, 1976, pp. 86-102.

14.9 Nawy, E.G. and Chakrabarty, P., *Serviceability deflection behaviour of two way action prestressed concrete plates*, International Conference on Prestressed Concrete, Sydney, Australia, Concrete Institute of Australia, 1976, pp. 1-10.

14.10 Nislon, A.H. and Walters, D.B., *Deflection of two way floor systems by the equivalent frame method*, Journal of the American Concrete Institute, Vol. 72, 1975, pp. 210-8.

14.11 Scordelis, A.C.; Pister, K.S. and Lin, T.Y., *Strength of a concrete slab prestressed in two directions*, Journal of the American Concrete Institute, Proceedings Vol, 53, September 1956, pp. 241-56.

14.12. Scordelis, A.C.; Lin, T.Y. and Itaya, R., *Behaviour of a continuous slab prestressed in two directions*, Journal of the American Concrete Institute, Vol. 56, 1959, pp. 441-59.

PROBLEMS

14.1 Derive expressions for radial and circumferential moments in a circular slab due to a central concentrated load W when the slab is (i) simply supported and (ii) restrained at the edge. Hence design the prestressing tendons for a circular slab of 12 m diameter and 180 mm thickness carrying a central concentrated load of 200 kN. The maximum permissible compressive stress is 15 MPa and no tension is allowed.

14.2 Using the principle of superposition or otherwise, derive expressions for radial and circumferential moments in a uniformly loaded circular slab supported by a central column whose upward reaction is assumed to be uniform. Hence design the prestressing tendons for a circular slab of 6 m diameter and 150 mm thickenss carrying a superimposed load of 1.5 kN/m^2 in addition to self weight. The slab is supported on a central circular column of 1 m diameter. While no tension is permitted, the compressive stress is restricted to to 14 MPa.

14.3 Derive expressions for radial and circumferential moments in a simply supported circular slab carrying uniformly distributed load on a concentric band of finite width. Hence design the prestressing tendons for a simply supported circular slab having diameter of 9 m and overall thickness of 160 mm to carry a superimposed load of 4 kN/m^2 over a central concentric area having a rodius of 2.25 m in addition to self weight. The maximum permissible tensile and compressive stresses are 1.5 MPa and 15 MPa respectively.

14.4 A 120 mm thick rectangular slab 4.8 × 9.6 m is simply supported along edges without corner restraint. It is prestressed biaxially by means of 16 mm bars at 230 mm spacing and constant eccentricity of 30 mm along short span and 12 mm bars at 190 mm spacing and constant eccentricity of 10 mm along long span. The effective prestress in all tendons is 820 MPa. The slab carries a load of 12.88 kN/m^2 inclusive of self weight. Determine the extreme fibre stresses at the centre of short span.

14.5 A 160 mm thick rectangular slab 8.0 × 9.6 m is simply supported along edges with restrained corners. It is prestressed along the short span by means of parabolic 12 ϕ 5 mm cables with zero eccentricity at ends and 45 mm at midspan. The cables have a spacing of 200 mm and an effective stress of 950 MPa. The slab is prestressed concentrically with an effective stress of 5.6 MPa along the long span. Determine the maximum superimposed load which can be applied to the slab without causing tensile stress. Concrete weighs 24 kN/m^3.

14.6 A rectangular slab 6 × 9 m has two adjacent edges continuous over the supports. It is prestressed concentrically along the long span by an effective prestressing force of 0.75 MN/m. The parabolic tendons with zero eccentricity at ends and 40 mm at midspan produce an effective prestressing force 0.9 MN/m along the short span. Determine the minimum thickness of slab to enable it to carry a superimposed load of 10 kN/m^2 without exceeding a compresive stress of 10 MPa. The unit weight of concrete is 24 kN/m^3.

14.7 A rectangular slab panel 7.5 × 9 m with long edges continuous has to carry its self weight at stress transfer and a live load of 4 kN/m^2 at final stage. Design the slab panel using concrete of grade M-35.

14.8 Discuss the distinguishing features of Indian, British and American codes in respect of the provisions for the design of flat slab.

14.9 Describe the main steps in the design of a prestressed concrete flat slab using (i) direct design method and (ii) equivalent frame method.

14.10 A prestressed concrete flat slab of 100 mm thickness without drops and column heads is supported on square columns of 0.6 m side at a spacing of 6.6 m in both orthogonal directions. The effective concentric prestressing force in both directions is 0.75 MN/m. The slab carries a uniform load of 6.4 kN/m² inclusive of self weight. Calculate the extreme sfibre stresses at (i) end of column strip and (ii) centre of middle strip in an intrior panel of the slab.

14.11 A prestressed concrete flat slab with appropriate drops and column heads has to be supported on circular columns of 0.4 m diameter at spacing of 7.2 and 9.6 m in the two orthogonal directions. The slab projects out by 2.7 m beyond the centre line of exterior columns on all four sides. The slab has to carry its own weight at stress transfer and a service load of 6 kN/m² at final stage. Design (i) an interior panel, (ii) an edge panel and (iii) a corner panel using the provisions of Indian code.

14.12 A prestressed concrete circular slab restrained along its edge has a diameter of 12 m and an overall thickness of 180 mm. It is prestressed isotropically with a constant compressive stress of 6 MPa in all directions. Modulds of rupture of concrete is 4 MPa. The slab carries uniformly distributed load. Determine the cracking load.

14.13 Derive Equations (14.5.5) to (14.5.8).

14.14 A uniformly loaded prestressed concrete slab simply supported along its edges has the shape of a regular hexagon. Each side of the hexagon is 4 m. The slab has constant positive ultimate moment capacity of 150 kNm/m in all directions. Determine the collapse load.

14.15 A prestressed concrete slab of regular octagonal shape with restrained edges, carries a central concentrated load. The slab has positive and negative ultimate moment capacities equal to 120 and 180 kN.m/m in all directions. Determine the collapse load if the side of the octagon is 3 m.

14.16 A prestressed concrete slab of 9.6 m side with restrained edges carries uniform load over its entire area. The slab has constant positive and negative moment capacities each equal to 160 kN·m/m in all directions. Sketch the yield line pattern and determine the collapse load.

14.17 A prestressed concrete rectangular slab 4.8 m × 7.2 m carrying uniformly distributed load is simply supported along its edges. The positive ultimate moment capacities along short and long spans are 180 and 90 kN·m/m respectively. Determine the collapse load.

14.18 Determine the collapse load of a prestressed concrete flat slab having square panels of 7.5 m side carrying uniform loading. The diameter of the column head is 0.96 m. The positive and negative ultimate moment capacities of the slab are 150 and 175 kN·m/m respectively.

15

Precast Elements

15.1 INTRODUCTION

Precasting of structural concrete elements or units connotes an element of change in time or place or both in some form or the other. While an interval of time between manufacture of a precast unit and its deployment in the structural system is evident from the prefix 'Pre', a change of place or location is implicit in the fact that precast concrete is traditionally taken as the opposite of cast-in-place concrete. Precasting of structural concrete elements may be carried out either in a central plant or in a casting yard at work site. The use of precast elements in construction entails the operations of lifting, transportation, stacking, hoisting and launching to widely varying extent or degree. Smaller units such as railway sleepers and transmission line poles may be more advantageously cast in a central plant and then transported over long distances to their final position. On the other hand, site precasting may be preferred for large and heavy members such as shells, folded plates and bridge girders. In some cases, for instance lift slabs, site precasting may be the only option. The precasting of concrete units is closely associated with prestressing and standardisation. As precasting of concrete, particularly in a central plant, produces a high quality concrete, the fullest benefit can be taken only through prestressing. The standardisation of precast units facilitates mass production leading to numerous advantages and significant economy. The discussion in this chapter is largely devoted to precasting of standardised units in a central plant. The following are the main advantages of mass production of standardised units in a central plant:

(i) high quality product with close dimensional and physical tolerances, good finish and long useful life

(ii) repetitive use of moulds, equipment and other facilities. The reuse makes it feasible to go in for steel or fibre glass moulds

(iii) feasibility of adopting complex cross-sectional shapes resulting in reduced self weight and greater structural efficiency

(iv) significant saving of materials and on-site labour

(v) independence from adverse weather conditions such as rain, snow, direct sun and extreme cold

(vi) faster construction which may be of particular advantage in busy or congested areas.

Precast units form the essential ingredients of prefabricated structures. In these structures, the precast units and the connections are usually standardised to ensure satisfactory performance. In cases, where standardisation is not possible, careful attention is required in designing and detailing the joints. A close liaison between the design engineer and builder is fruitful in this regard. The continuity and monolithicity of prefabricated structures is often achieved by a judicious combination of precast and cast-in-place concretes. Both precast concrete and cast-in-place concrete have their relative merits and demerits. An optimum composite design, using both types of concretes, may not only prove to be structurally more efficient but also more economical.

15.2. POLES, MASTS AND PYLONS

Precast prestressed concrete poles, masts and pylons are being used extensively all over the world for telephone lines, street lighting and high voltage power transmission, railway power and signalling lines, substation towers and transmission antennas.

15.2.1. Early Applications and Scope

Realising the suitability of precast prestressed concrete, the pioneers used it for the manufacture of poles, masts and pylons for extensive development of overhead transmission lines and railway electrification. In 1933 Freyssinet designed prestressed concrete poles for railway signalling equipment in France suitable for mass production. These poles gave excellent service and are still in use. The poles were also used in Algeria where they were found suitable in the desert environment. The German National Railways are among the early railways which utilised prestressed concrete masts for power transmission. Prestressed concrete spun poles are being used extensively in Japan, Soviet Union, Czechoslovakia, France and Germany. In Soviet Union prestressed concrete poles are being used extensively for high voltage transmission upto 330 kV. At present precast prestressed concrete units are being used in India as lighting poles and 11 kV transmission lines. With extensive rural electrification envisaged in five year plans, the demand for prestressed concrete poles is bound to increase greatly. With the ambitious electrification plans of Indian Railways, the prestressed concrete pylons are likely to have considerable scope in the coming years.

15.2.2. Advantages

Prestressed concrete poles have to compete with poles made from Pres-alternative materials such as timber, reinforced concrete and steel. The general advantages of prestressed concrete discussed in Sec. 1.14 are also applicable to prestressed concrete poles. In particular, the following advantages offered by prestressed concrete poles have established their superiority over other types.

(i) Due to their slender section, prestressed concrete poles are aesthetically more pleasing.

(ii) Due to the use of high concrete under permanent compression, prestressed concrete poles require minimum maintenance.

(iii) Prestressed concrete poles are lighter than reinforced concrete and steel poles. Consequently the cost of handling and transportation is reduced.

(iv) Due to the absence of cracks, the entire section of prestressed concrete pole is effective. Hence it has higher stiffness and superior response to dynamic loads.

(v) Due to the higher resistance of prestressed concrete against deterioration under alternate freezing and thawing, prestressed concrete poles are more durable in cold countries.

(vi) As prestressed concrete poles are made from high grade concrete and also because they are crack-free under service loads, they have a longer useful life. They offer high resistance against corrosion in hot, humid and aggressive environment and against erosion by blowing sand under desert conditions.

(vii) Due to greater fire resistance of prestressed concrete, these poles are liable to lesser damage on account of grass, bush or forest fires.

(viii) For the same height and design load, the prestressed concrete pole may provide the cheapest design.

15.2.3. Manufacturing Techniques

In view of the demand for large number of identical units, prestressed concrete poles are most eminently suited for mass production. They are almost invariably manufactured by using the technique of pretensioning. Due to their considerable length, the stress transfer can easily be achieved through bond thereby eliminating the need for end anchorages. The critical section for bending moment, which is the dominant internal force, is generally located at some distance away from the ends. Hence the development length for transfer of stress through bond poses no problem. The choice of long line system or individual mould system for the manufacture of poles depends upon several factors such as availability of physical facilities at the plant, the desirability of steam curing and economic considerations. The general discussion on the relative merits of the two techniques of pretensioning given in Sec. 3.2 applies also to prestressed concrete poles. Due to the possibility of repetitive use, steel moulds may prove to be more economical in the long run as compared to timber moulds. Strong steel moulds are also helpful in achieving better compaction and closer dimensional tolerances. Concrete of grade M-45 or above is usually recommended.

15.2.4. Shapes and Cross-sectional Properties

Poles, masts and pylons may have a variety of shapes and cross-sectional properties. They are tapered from the base towards the top because the bending moment, which is the dominant internal force, decreases towards the top. Short poles upto 10 m length may have a solid square or rectangular cross-section. Their main advantage is simpler formwork. They occupy less space and are therefore more conveniently stored and transported. The main disadvantage is the smaller stiffness for a given cross-sectional area. The stiffness may be increased by adopting an I-section. However, an I-section is weak in bending about the minor axis and also in torsion. Poles having hollow tubular cross-section have the advantage of high strength and stiffness in all directions both in bending as well as in torsion. The tubular sections which are in common use are hollow square, rectangle, regular polygon, ellipse and circle. The main disadvantages of tubular sections are difficult formwork and fabrication. Due to their higher strength and stiffness, the hollow sections are generally preferred for longer poles. B.B.R.V. spun prestressed concrete poles are being manufactured in 9,12, 15 and 18 m length and in diameter up to 750 mm. The self weight of poles of rectangular cross-section may be reduced by providing web openings. These vierendeel poles consume lesser material but are liable to breakage and corrosion due to their larger surface area and thinner elements. In South Africa, masts having V-section have been used for supporting overhead power conductors for railways. The two legs of V-section which are 50 mm thick are thickened at the ends and also at their junctions. The section provides good stiffness in both directions. As a modification of the V-seection obtained by omitting its webs, the tripod pylon was developed in France. The three legs of the pylon are precast and connected to precast head, foot and spacing diaphragms by means of post-tensioned cables running along the axes of the legs.

15.2.5. Design Loads

Poles, masts and pylons which support overhead transmission lines are generally required to resist the following design loads.

(i) Wind load

The wind force acting on conductors, other fixtures and the pole itself is the most dominant force which primarily controls the geometry and cross-sectional properties of the pole. The design intensity of wind pressure depends upon the maximum wind velocity in the region. For maximum effect, the wind is assumed to blow in a direction normal to the transmission line. As the poles bend in the direction of wind, they are placed such that they are strongest in bending in windward direction. Hence, for example a pole of rectangular section is placed with its larger cross-sectional dimension normal to the transmission line. As the pole are secured to the ground and are free at the top, they behave like vertical cantilevers. The bending moment decreases from ground level towards top. Hence the cross-section is generally reduced from the bottom to the top of the pole.

(ii) Load due to conductor breakage

Although poles normally act as vertical props, they are subjected to bending in the direction of transmission line and torsion when one of the conductors breaks. The force due to snapping of conductor may be minimised or even eliminated by the provision of rotating supports for the conductors. Alternatively, the force due to conductor breakage may be resisted by anchor towers located at suitable distances along the transmission line and intermediate poles serving only as vertical props.

(iii) *Self weight*

The stresses due to bending moment caused by self weight during lifting and transportation should also be checked although they are seldom critical. When the poles have a tapering section, the load due to self weight is trapezoidal. Appropriate impact factor should also be included as in Section 12.3.

15.2.6. Analysis

A pole supporting a transmission line near its top stands upright with its lower end planted in ground over a length known as the planting length l_p. For the sake of analysis, the actual wind load may be replaced by its resultant force F acting at a height h above ground level as shown in Fig. 15.2.1(a). The planting length of the pole is either embedded in a block of concrete or directly in soil. The distribution of the passive resistance of the soil depends upon the nature of soil and that of the embedment. If the distribution of soil reaction is assumed to be linear as shown in Fig. 15.2.1(b), it may be broken into a rectangular distribution having ordinate p_1 as showing in Fig. 15.2.1(c) and a triangular distribution having ordinate p_2 as shown in Fig. 15.2.1(d). For the equilibrium of the pole,

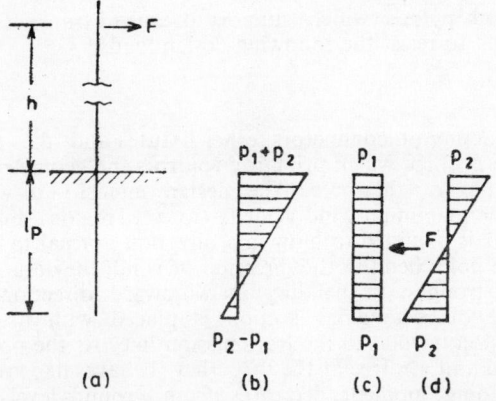

Fig. 15.2.1

the resultant of rectangular distribution of Fig. 15.2.1(c) should be equal to F. This resultant together with the force F near the top of the pole forms a clockwise couple equal to $F\left(h + \dfrac{l_p}{2}\right)$. Hence from the equilibrium of the pole, the forces p_1 and p_2 per unit length may be determined from

$$F = p_1\, l_p$$

or
$$p_1 = \frac{F}{l_p} \tag{15.2.1}$$

$$F\left(h + \frac{l_p}{2}\right) = \frac{p_2}{2} \times \frac{l_p}{2} \times \frac{2}{3}\, l_p$$

$$\tag{15.2.2}$$

or
$$p_2 = \frac{6F\left(h + \frac{l_p}{2}\right)}{l^2_p}$$

Knowing p_1 and p_2, the point of zero shear or maximum bending moment can be determined. This point of maximum bending moment occurs at a short distance below the ground level. When the pole is embedded in a block of concrete, most of the reactive forces are confined to the upper portion of the planting length, whereas the lower part of the planting length remains almost unstressed. If M is the maximum bending moment due to wind load F, the extreme fibre stresses are equal to (M/Z) in which the section modulus Z depends upon the type of cross-section,

In the case of a vierendeel pole shown in Fig. 15.2.2 (a), the section modulus Z may be based on the reduced cross-sectional area taking into account the web opening. In this pole, the most critical section xx is through the opening nearest to the ground level. Referring to Fig. 15.2.2(b), the section modulus at critical section $\ddot{X}X$ may be expressed as

$$Z = \frac{\dfrac{b}{12}(d^3 - d_0^3)}{\dfrac{d}{2}} \tag{15.2.3}$$

In the case of relatively large oblong openings, the vierendeel pole may be visualised as a rigid frame comprising two almost parallel stringers connected to each other by horizontal diaphragms or stiffeners between consecutive openings. The idealised vierendeel frame shown in Fig. 15.2.2 (c) may be analysed by the following approximate analysis which is similar to the cantilever method for a multistorey plane frame. The stringer AB on the windward side is in tension whereas stringer CD on the leeward side is in compression. From symmetry, the forces in the two stringers may be assumed to be equal in magnitude but opposite in sign. Assuming the inflection point to occur at midheight of the opening (section YY) and considering the equilibrium of the free body of the pole above the section YY shown in Fig. 15.2.2 (d),

$$2N_s \sin\theta + 2Q_s \cos\theta = F$$

or
$$2N_s\, \theta + 2Q_s \approx F \tag{15.2.4}$$

$$(N_s \cos\theta - Q_s \sin\theta)\left(\frac{d + d_0}{2}\right) = \frac{1}{2}F\, h'$$

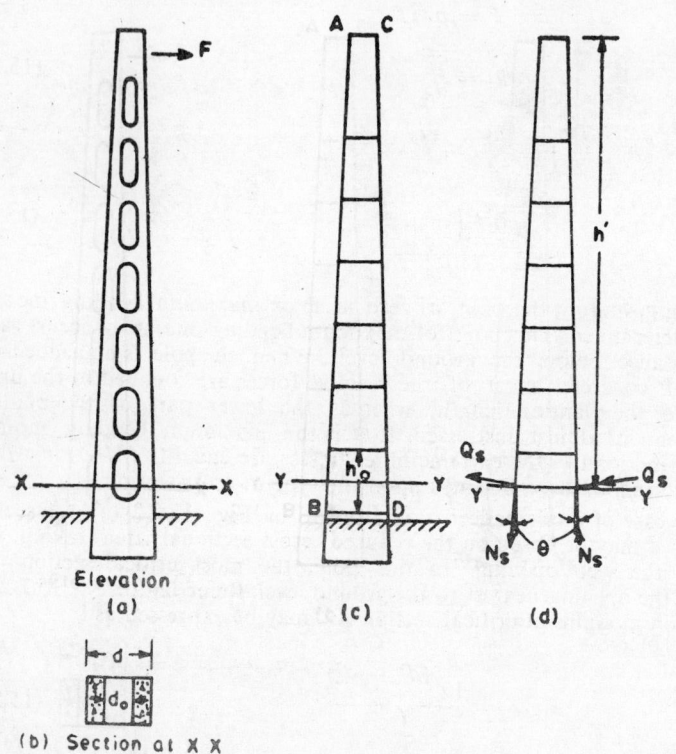

Fig. 15.2.2

or $(N_s - Q_s \theta) \left(\dfrac{d + d_0}{2} \right) \approx Fh'$ (15.2.5)

where $N_s =$ axial force in stringers

         $Q_s =$ shear force in stringers

         $\theta =$ inclination of stringers to vertical.

Ignoring the small term $Q_s \theta$ in Eq. (15.2.5) and solving for N_s and Q_s.

$$N_s = \frac{Fh'}{\left(\dfrac{d + d_0}{2} \right)}$$ (15.2.6)

$$Q_s = \frac{F}{2} - N_s \theta$$ (15.2.7)

At the critical section XX, each stringer carries axial force N_s, shear force Q_s and bending moment, $M_s = Q_s \dfrac{h_0}{2}$ in which h_0 is the height of the opening. Instead of the foregoing approximate analysis, the vierendeel frame may be analysed more rigorously using slope-deflection equations or stiffness approach. The geometry and proportions of a vierendeel pole are controlled by several considerations. The bending moment due to wind load is resisted primarily by the axial compressive and tensile forces in stringers. Hence the cross-section of stringers should be adequate for resisting these axial forces. The overall dimensions of the pole are guided by aesthetic and functional considerations. After selecting the overall depth of the cross-section and size of stringers, the horizontal dimensions of the opening are automatically decided. The vertical dimensions of the openings are controlled by the requirement regarding the spacing of stiffeners or diaphragms. The stiffeners should be spaced close enough to prevent the buckling of leeward stringer. In order to relieve stress concentration, the ends of the openings may be r)unded. The optimum shape of the openings with minimum stress concentration may be obtained by photoelastic analysis in which the vierendeel pole may be visualised as a thin plate with multiple openings. The openings serve to reduce the self weight of pole considerably. Some length of the pole near each end (say about a metre) should be without openings to reduce the chance of breakage during transportation.

15.2.7 Design Principles

The design of a pole supporting a transmission line is controlled primarily by the wind load. As the wind load is reversible, it is not possible to take advantage of eccentric prestressing. The poles are, therefore, prestressed concentrically. Concentric prestressing is appropriate also for resisting torsion caused by breakage of a conductor placed eccentrically. The design of poles, masts and pylons generally follow the following guidelines:

(i) The cross-section of a pole may be designed either as Type 1, 2 or 3. In other words, tension may or may not be permitted under full working load. The German National Railways permitted a maximum tensile stress of 4.8 MPa under frequent wind load and a fictitious tensile stress as high as 10.4 MPa under rarely occurring wind load. Poles designed as partially prestressed members develop cracks under the action of high wind but these cracks close under normal conditions.

(ii) Poles should have a high degree of resilience to enable them to withstand overloads and also impact or shock loads. A high degree of resilience may be ensured by adopting a relatively low intensity of prestress and by the provision of auxiliary untensioned reinforcement in the form of mild steel or high yield strength deformed bars. High intensity of prestress which tends to cause brittle failure has been abandoned in most countries.

(iii) The torsion caused by snapping of an eccentric cable is constant over the entire length of the pole. Hence from the stand point of torsion, the pole should have a uniform section. For the same reason, hollow or tubular sections are preferable in comparison to thin walled open sections.

(iv) Load factors for adequate serviceability and ultimate strength are specified in relevant codes. The Indian, Czechoslovakian and German codes prescribe load factors of 2.5, 2 and 1.75 against collapse. A lower value of load factor ranging from 1.1 to 1.5 is generally specified against collapse caused by combined bending and torsion due to snapping of cables.

(v) The poles should have a certain minimum flexural strength in the direction of transmission line to be able to resist the drag caused by broken cables. The codes usually specify that the flexural strength in the direction of transmission line should be at least equal to one-fourth of the flexural strength in the direction normal to the transmission line.

(vi) The prestressing tendons as well as the auxiliary untensioned reinforcement should be placed as near as possible to the outer faces in order to obtain maximum possible ultimate flexural strength.

(vii) In the case of poles having tapered sections, the intensity of prestress may increase to undesirable values towards the top of the pole unless the prestressing force is reduced progressively towards top. This may be achieved by debonding, dead ending or looping some of the tendons at suitable points along the length of the pole. Poles having slender sections tend to bend near the top when the prestressing force is constant throughout. The bending near the top of the slender section may be attributed to unintentional eccentricity of the prestressing tendons.

EXAMPLE 15.2.1

A prestressed concrete pole carries a wind load of 1.8 kN at a height of 8 m from the ground level. Determine the maximum bending moment in the pole assuming a linear distribution of passive soil pressure over the planting length of 1.2 m. Hence determine the maximum compressive and tensile stresses if the pole has a rectangular section 100 mm × 350 mm and is prestressed concentrically by means of 12 wires of 5 mm diameter carrying an effective stress of 950 MPa. Six wires are placed on either side of the major axis at an effective cover of 50 mm. Take modular ratio, $m = 6.6$.

Solution

From Eq. (15.2.1) and (15.2.2),

$$p_1 = \frac{1.8}{1.2} = 1.5 \text{ kN/m}$$

$$p_2 = \frac{1.8 (8 + 0.6)}{\frac{1}{6} \times 1.2^2} = 64.5 \text{ kN/m}$$

Let the point of zero shear force or maximum bending moment occur at a distance x below ground level. Referring to Fig. 15.2.1, the shear force at depth x from ground level,

$$Q = 1.8 - 1.5x - \frac{\frac{64.5}{0.6} (0.6 - x) + 64.5}{2} x$$

Putting $Q = 0$ and solving the resulting equation,

$$x = 0.0279 \text{ m}$$

Hence the maximum bending moment at 0.0279 m below the ground level, $M_{max} = 14.425$ kN.m. This bending moment is only slightly greater than the bending moment, $M = 1.8 \times 8 = 14.4$ kN.m at ground level. Effective prestressing force,

$$P = \frac{\pi}{4} \times 5^2 \times 10^{-6} \times 12 \times 950 = 0.224 \text{ MN}$$

Net area of concrete,

$$A_c = 0.1 \times 0.35 - \frac{\pi}{4} \times 5^2 \times 10^{-6} \times 12 = 0.03476 \text{ m}^2$$

moment of inertia of transformed section,

$$I_t = \frac{1}{12} \times 0.1 \times 0.35^3 + (6.6 - 1)\left(\frac{\pi}{4} \times 5^2 \times 10^{-6} \times 12\right)$$
$$\times 0.125^2 = 0.000378 \text{ m}^4$$

Hence maximum compressive and tensile stresses,

$$f_c = \frac{0.224}{0.03476} + \frac{0.014425 \times 0.175}{0.000378} = 13.12 \text{ MPa}$$

$$f_t = \frac{-0.224}{0.03476} + \frac{0.014425 \times 0.175}{0.000378} = 0.236 \text{ MPa}$$

EXAMPLE 15.2.2

A vierendeel pole carries a wind load of 1.6 kN at a height of 10 m from the ground level. The overall and planting lengths are 12 and 1.5 m respectively. The cross-section of the pole varies linearly from 120 mm × 120 mm at top to 120 mm × 360 mm at bottom. The first web opening with its centre located at 600 mm above the ground level has a height of 400 mm. The cross-section of each stringer is 120 mm × 90 mm. Determine the extreme fibre stresses due to wind load at the most critical section.

Solution

The overall depth of the cross-section increases from 120 mm at top to 360 mm at bottom. Hence,

$$\tan \theta \approx \theta = \frac{0.5 (360 - 120)}{12000} = 0.01$$

(i) Referring to Fig. 15.2.2, the most critical section XX is located at $10.5 - 0.4 = 10.1$ m from top. The bending moment at this section,

$$M = 1.6 (10.1 - 0.5) = 15.36 \text{ kN.m} = 0.01536 \text{ MN.m}$$

Overall depth at critical section,

$$d = 0.120 + 2 \times 0.01 \times 10.1 = 0.322 \text{ m}$$

Hence the width of opening,

$$d_0 = 0.322 - 2 \times 0.090 = 0.142 \text{ m}$$

From Eq. (15.2.3), the section modulus,

$$Z = \frac{\dfrac{1}{12} \times 0.12\,(0.322^3 - 0.142^3)}{\left(\dfrac{0.322}{2}\right)} = 0.001896 \text{ m}^3$$

Hence the extreme fibre stresses due to wind load,

$$f_{max} = \pm\,\frac{0.01536}{0.001896} = \pm\,8.1 \text{ MPa}$$

(ii) Alternatively, the stresses may be computed by treating the frame as a vierendeel frame. The overall depth at section YY (Fig. 15.2.2) through the centre of opening,

$$d = 0.12 + 2 \times 0.01\,(10.5 - 0.6) = 0.318 \text{ m}$$

Hence the width of opening,

$$d_0 = 0.318 - 2 \times 0.09 = 0.138 \text{ m}$$

From Eq. (15.2.6) and (15.2.7),

$$N_s = \frac{1.6\,(10 - 0.6)}{\left(\dfrac{0.318 + 0.138}{2}\right)} = 65.96 \text{ kN}$$

$$Q_s = \frac{1.6}{2} - 65.96 \times 0.01 = 0.14 \text{ kN}$$

Bending moment in stringer at section XX,

$$M_s = 0.14 \times 0.2 = 0.028 \text{ kN.m}$$

Hence the extreme fibre stresses,

$$f_{max} = \pm\left(\frac{65.96}{0.12 \times 0.09} + \frac{0.028}{\dfrac{1}{6} \times 0.12 \times 0.09^2}\right)$$

$$= \pm\,6280 \text{ kN/m}^2 \text{ or } \pm\,6.28 \text{ MPa}$$

EXAMPLE 15.2.3

A prestressed concrete pole of rectangular section has an overall length of 10 m. It has to carry a wind load of 1.9 kN at 0.5 m from top. The planting length is 1.5 m. Design a fully prestressed (Type 1) pole suitable for production by long line system of pretensioning. The grade of concrete is M-40 and the breaking stress of prestressing tendons is 1800 MPa. Check the adequacy of the design at the limit state of collapse in bending and also in torsion due to snapping of a conductor carrying a tension of 5 kN at an eccentricity of 300 mm.

Solution

Adopting a tapering rectangular section 120 mm × 120 mm at top and 120 mm × 360 mm at bottom, the cross-section of the pole at ground level is 120 mm × 324 mm. Hence the moment of inertia based on gross concrete section,

$$I_g = \frac{1}{12} \times 0.12 \times 0.324^3 = 0.00034 \text{ m}^4$$

Taking moment of inertia of transformed section I_t approximately equal to gross moment of inertia I_g,

$$I_t = I_g = 0.00034 \text{ m}^4$$

Bending moment at ground level due to wind load,

$$M = 1.9 \ (10 - 0.5 - 1.5) = 15.2 \text{ kN.m} = 0.0152 \text{ MN.m}$$

Extreme fibre stresses due to wind load,

$$f_w = \pm \frac{0.0152}{0.00034} \times \frac{0.324}{2} = 7.242 \text{ MPa}$$

Hence for no tension condition.

$$\frac{P}{A_c} = 7.242$$

Taking $\quad A_c = A_g = 0.12 \times 0.324 = 0.03888 \text{ m}^2,$

$$P = 7.242 \times 0.03888 = 0.2816 \text{ MN}$$

Hence 16 wires each of 5 mm diameter with an effective stress of 896 MPa may be provided. Eight wires may be provided on either side of the major axis in two layers, each layer having 4 wires with effective covers of 30 mm and 60 mm. The spacing of wires along the width of the section may be kept as 24 mm. The adequacy of the section may now be checked.

(i) *Initial stage*

$$A_c = 0.12 \times 0.324 - \frac{\pi}{4} \times 5^2 \times 10^{-6} \times 16 = 0.03857 \text{ m}^2$$

$$E_c = 5700 \sqrt{40} = 36050 \text{ MPa}$$

$$m = \frac{2 \times 10^5}{36050} = 5.55$$

$$I_t = \frac{1}{12} \times 0.12 \times 0.324^3 + (5.55-1) \times$$

$$\left(\frac{\pi}{4} \times 5^2 \times 10^{-6} \times 16\right) \left(\frac{0.324}{2} - 0.045\right)^2$$

$$= 0.00036 \text{ m}^4$$

Assuming the loss factor $\eta = 0.8$, the compressive stress at stress transfer,

$$f_c = \frac{0.2816}{0.8 \times 0.3857} = 9.13 \text{ MPa}$$

Hence according to the Indian code, $f_{cl} = 9.13/0.51 = 17.9$ MPa. However, as this value is less than $0.5 f_{ck}$ or 20 MPa, the stress transfer may occur when the strength of concrete is at least equal to 20 MPa.

Assuming a transmission length of 100 ϕ for transfer of stress by bond, full prestress develops at $100 \times 5 = 500$ mm from top. The intensity of initial prestress at this section is 22.2 MPa. As this stress is excessive and may cause bending of top slender portion, it is desirable to debond some of the tendons to reduce it to an acceptable value. For instance, one-half of the tendons may be debonded at midheight of the pole.

(ii) *Final stage*

At this stage, extreme fibre stresses are,

$$f_1 = - \frac{0.2816}{0.03857} - \frac{0.0152}{0.00036} \times \frac{0.324}{2} = -14.14 \text{ MPa}$$

$$f_2 = - \frac{0.2816}{0.03857} + \frac{0.0152}{0.00036} \times \frac{0.324}{2} = -0.46 \text{ MPa}$$

As the permissible compressive stress for M-40 concrete is 15.6 MPa, the above stresses are safe.

(iii) *Limit state of collapse in bending*

(a) **Bending about major axis**

$$b = 0.12 \text{ m}$$

$$d_p = 0.324 - (0.030 + 0.015) = 0.279 \text{ m}$$

Steel on tension side,

$$A_p = \frac{\pi}{4} \times 5^2 \times 10^{-6} \times 8 = 0.000157 \text{ m}^2$$

$$\frac{A_p f_{pu}}{bd_p f_{ck}} = \frac{0.000157 \times 1800}{0.12 \times 0.279 \times 40} = 0.211$$

Referring to Table 6.9.1,

$$\frac{f_p}{0.87 f_{pu}} = 1 \quad \text{or} \quad f_p = 0.87 \times 1800 = 1566 \text{ MPa}$$

$$\frac{x_u}{d_p} = 0.459 \quad \text{or} \quad x_u = 0.459 \times 0.279 = 0.128 \text{ m}$$

Hence from Eq. (6.9.1), the ultimate moment,

$$M_u = 0.000157 \times 1566 (0.279 - 0.42 \times 0.128)$$
$$= 0.0554 \text{ MN} \cdot \text{m}$$

The load factor against collapse is $0.0554/0.0152 = 3.64$, which being greater than 2.5, is adequate

(h) **Bending about minor axis**

$$b = 0.324 \text{ m}$$

$$d_p = 0.120 - (0.024 + 0.012) = 0.084 \text{ m}$$

$$A_p = 0.000157 \text{ m}^2$$

$$\frac{A_p f_{pu}}{bd_p f_{ck}} = \frac{0.000157 \times 1800}{0.324 \times 0.084 \times 40} = 0.26$$

$$\frac{f_p}{0.87 f_{pu}} = 1 \quad \text{or} \quad f_p = 0.87 \times 1800 = 1566 \text{ MPa}$$

$$\frac{x_u}{d_p} = 0.565 \quad \text{or} \quad x_u = 0.565 \times 0.084 = 0.0475 \text{ m}$$

Hence from Eq. (6.9.1), the ultimate moment,

$$M_u = 0.000157 \times 1566 (0.084 - 0.42 \times 0.0475)$$
$$= 0.0158 \text{ MN} \cdot \text{m}$$

As the ultimate moment capacity of the pole in the direction of the transmission line (minor axis bending) is greater than one-fourth of the ultimate moment capacity in the direction normal to the transmission line (major axis bending), the section is adequate.

(iv) *Limit state of collapse in torsion*

The most critical section for torsion is located at the level of conductors, i.e. at 0 5 m from top. At this section, the pole is subjected to torsion without bending due to snapping of the conductor. The torsional moment at critical section,

$$T = 5 \times 0.3 = 1.5 \text{ kN} \cdot \text{m} = 0.0015 \text{ MN} \cdot \text{m}$$

Ignoring the contribution of steel, the torque resisted by concrete alone may be computed from Eq. (7.4.10).

$$b = 0.12 \text{ m}$$

$$D = 0.132 \text{ m}$$

$$f_{pe} = \frac{0.2816}{0.12 \times 0.132} = 17.78 \text{ MPa}$$

$$k_p = \sqrt{1 + \frac{12 \times 17.78}{40}} = 2.517$$

$$T_c = 0.125 \times 0.12^2 \times 0.132 \left(1 - \frac{0.12}{3 \times 0.132}\right)$$

$$\times 2.517 \sqrt{40}$$

$$= 0.00264 \text{ MN} \cdot \text{m}$$

The load factor on collapse in torsion,

$$\frac{T_c}{T} = \frac{0.00264}{0.0015} = 1.76$$

Hence the section is adequate in torsion.

To satisfy the requirement of minimum shear reinforcement, closed rectangular stirrups of 6 mm diameter with a yield stress of 250 MPa may be provided at a spacing of 250 mm. In addition, two stirrups may be provided as close as possible to each end.

EXAMPLE 15.2.4

Determine the possible saving in prestressing steel in the pole of Example 15.2.3 if a tensile stress of 3 MPa is permitted under the service load.

Solution

Referring to the computations of Example 15.2.3, the effective prestrsssing force required to restrict the tensile stress in concrete to 3 MPa,

$$P = (7.242 - 3) \times 0.03888 = 0.1649 \text{ MN}$$

Hence only ten wires will suffice as against sixteen wires required in the case of fully prestressed (Type 1) design.

EXAMPLE 15.2.5

Design the pole of Example 15.2.3 as a partially prestressed (Type 3) member on the basis of maximum permissible fictitious tensile stress of 5

MPa. The modulus of rupture of concrete is 4.4 MPa. Use auxiliary reinforcement in the form of high yield strength deformed bars having a yield stress of 415 MPa.

Solution

Referring to the computations of Example 15.2.3, the effective prestressing force required to restrict the fictitious tensile stress to 5 MPa,

$$P = (7.242 - 5) \times 0.03888 = 0.0872 \text{ MN}$$

Hence six wires each of 5 mm diameter will suffice. Three wires may be placed on either side of the major axis at an effective cover of 60 mm.

For a load factor of 2.5, the design ultimate moment,

$$M_{ud} = 2.5 \times 1.9 \times 8 = 38 \text{ kN·m} = 0.038 \text{ MN·m}$$

Using the provisions of the Indian code, the ultimate moment of resistance of the section contributed by prestressing steel, $M_{up} = 0.0225$ MN·m. Hence the auxiliary untensioned reinforcement should be designed to provide ultimate moment,

$$M_{us} = M_{ud} - M_{up} = 0.038 - 0.0225 = 0.0155 \text{ MN·m}$$

Providing equal untensioned steel, $A_s = A_s'$ near each face at an effective cover of 25 mm and using the steel beam theory,

$$A_s = A_s' = \frac{0.0155}{0.87 \times 415 (0.324 - 0.050)} = 0.0001566 \text{ m}^2$$

Hence one high yield strength deformed bar of 10 mm diameter may be provided near each corner.

The partially prestressed pole in general enjoys all the advantages discussed in Sec. 10.2. In particular, it has the benefit of high resilience due to relatively low prestress (2.72 MPa) and moderate proportion of untensioned steel (0 8 percent). The slender portion near top has a much lower tendency to bend on account of unintentional eccentricity due to lower prestress (7.33 MPa) as compared to very high prestress (19.55 MPa) in fully prestressed pole of Example 15.2.3 unless it is reduced by debonding some of the tendons near top.

15.3 RAILWAY SLEEPERS

Next only to the transmission poles, the rail track sleeper or tie is perhaps the most suitable item for mass production as a precast prestressed element. The maintenance problems of wooden and metal sleepers are well known. In a sharp contrast, prestressed concrete sleepers require little maintenance and possess a long useful life. Besides, they are helpful in a better alignment of the track resulting in a more comfortable ride.

15.3.1 Scope

Prestressed concrete sleepers were first developed in France shortly before the second world war. Germany and Britain quickly followed suit. Progressively they have been adopted in all advanced and some of the developing countries. In their ambitious track modernisation plan, the Indian Railways have started using long welded rails (LWR), prestressed concrete sleepers and elastic fastenings on the trunk routes. The use of prestressed concrete sleepers is picking up fast and a demand of several million sleepers per year may be expected by the turn of the century

15.3.2 Advantages and Disadvantages

The relative advantages and disadvantages of prestressed concrete discussed in Sec. 1.14 apply in general, also to railway sleepers. In particular, the following points may be mentioned.

(a) Advantages

(i) Prestressed concrete sleepers are very appropriate for mass production using local materials.

(ii) With the progressive adoption of heavier rail section, LWR and the consequent faster traction, the use of prestressed concrete sleepers has become indispensable. Due to their greater weight and strength, these sleepers offer greater stability to the track and effectively prevent the buckling of LWR due to rise of temperature.

(iii) In conjunction with elastic fastenings, prestressed concrete sleepers are capable of maintaining the gauge, cross-level and alignment much better than other types of sleepers.

(iv) Prestressed concrete sleepers are very suitable for modern methods of track maintenance and mechanical packing of ballast. Due to their large weight and flat bottom, they retain the packing much better as compared to other sleepers.

(v) As concrete is poor conductor of electricity, prestressed concrete sleepers are better suited for track circuiting as compared to metal sleepers.

(vi) Prestressed concrete sleepers have high fire resistance. They are not attacked by insects and pests. Due to high grade concrete and freedom from cracks, they have a long useful life of the order of 40 to 50 years. It is, therefore, possible to synchronise the replacements of the rails and sleepers resulting in substantial economy.

(b) Disadvantages

(i) The damage to prestressed concrete sleepers in the event of derailment is much greater as compared to wooden or metal sleepers. Consequently, the restoration of track is costlier. Also these sleepers have no scrap value.

(ii) Prestressed concrete sleepers invariably require modern mechanised methods of alignment, ballast packing and track maintenance which entail high capital investment.

15.3.3 Classification and Manufacturing Techniques

The research and development of prestressed concrete sleepers have progressed concurrently in several countries. As a result, each country has adopted its own design appropriate for national track requirements, indigenous know-how and cost of materials and labour. As railway sleepers are mass produced in very large numbers, even a small saving in the unit cost results in a very substantial overall economy. Consequently, the sleepers are critically designed and produced with a high degree of standardisation. Although many types of prestressed concrete sleepers have been standardised in various countries, the most common type is the monoblock or beam type sleeper. This type of sleeper has been adopted in many countries including Britain, France, Germany, Japan and India. In Sweden, a prestressed concrete sleeper was developed which comprises two concrete blocks connected to each other by means of a pipe filled with concrete and containing high tensile steel bars for precompressing the concrete in the blocks. Another design comprises prestressed concrete longitudinal blocks under the rails connected to each other by means of flexible tie bars. Some advanced countries are working on the development of ballastfree permanent way comprising biaxially prestressed continuous slab for catering to future high speed and high density traffic.

In India prestressed concrete sleepers of the monoblock or beam type are being used currently on broad gauge trunk routes. In comparison to two block sleepers, the monoblock sleepers are superior in respect of maintenance of gauge and track alignment due to higher strength and rigidity against rotation at seatings and buckling of LWR on account of thermal changes. These sleepers which are like a beam of variable cross-section have an overall length of 2.75 m for broad gauge length of 1.676 m (5′ 6″). Both pretensioned and post-tensioned sleepers are being manufactured in several factories spread in different parts of the country.

(a) Pretensioned sleepers

Under the progressive development of pretensioned monoblock prestrested concrete sleepers, three types, viz. PCS-2, PCS-3 and PCS-12 have been evolved. These sleepers developed by Research Designs and Standards Organisation (RDSO) of Indian Railways at Lucknow are approximately trapezoidal in shape and weigh nearly 2.7 kN.

(i) PCS-2 sleeper

This later version of PCS-1 has three types, viz. PCS-2A, B and C. The three types differ only in respect of reinforcement. These sleepers have a trapezoidal cross-section 154 mm wide at top, 250 mm at bottom and a height of 210 mm at the rail seat. They have a cant of 1 in 20 on the top surface extending over 175 mm on either side of centre line of rail. The sleeper designed as PCS-2C sleeper is prestressed by means of 20 strands comprising 3 wires of 3 mm diameter carrying an initial stress of 1025 MPa. Each sleeper contains 12 stirrups of 6 mm diameter. The concrete has a minimum crushing strength of 52.5 MPa at 28 days.

(ii) PCS-3 sleeper

This later version has a width of 290 mm and height of 214 mm at ends and width of 230 mm and height of 180 mm at centre. It is prestressed by means of 22 wires of 5 mm diameter. The stirrups are the same as in PCS-2C sleeper.

(iii) PCS-12 sleeper

As a further development, PCS-12 sleeper was introduced. The widths at top and bottom and depth of the sleeper are 154 mm, 270 mm and 230 mm at ends; 154 mm, 220 mm and 180 mm at centre and 154 mm, 249 mm and 210 mm at rail seat. The sleeper is prestressed by means of 18 strands comprising 3 wires of 3 mm diameter with an initial stress of 1285 MPa. The 28-day strength of concrete is 52.5 MPa.

(b) Post-tensioned sleepers

In collaboration with M/s Dyckerhoff and Widmann of West Germany, Northern Railway established a plant at Allahabad for the manufacture of post-tensioned sleepers. The plant which went into production in 1981 has a planned capacity of 300 000 sleepers per year. The sleeper which weighs 2.95 kN has a width of 160 mm at top, 220 mm at bottom and depth of 180 mm at its centre. The initial and effective prestressing forces are 370 and 310 kN respectively. The strength of concrete at stress transfer and at 28 days are 45 MPa and 55 MPa respectively.

The relative advantages of pretensioning and post-tensioning techniques discussed in Chapter 3 apply in general also to prestressed concrete sleepers. Due to relatively short length of these units, the transfer of prestress to concrete in pretensioned sleepers may be achieved by a combination of bond and end anchors. The following three systems of pretensioning are being used for the manufacture of prestressed concrete sleepers.

(i) Long line system

In this system 30 to 40 moulds may be laid in a row on prestressing beds and stressed simultaneously.

(ii) Stress-bench system

This system is similar to long-line system except that the stressing of three or four units is performed on a stress-bench, which being mounted on wheels can be moved to steam chambers for accelerated curing.

(iii) Individual mould system

The stressing is carried out against the mould itself. Several moulds may be placed on a trolley and moved to steam chamber.

The post-tensioned sleepers being manufactured in the plant at Allahabad utilise U-shaped high tensile steel bars of 7.5 mm diameter known as 'hair pins' with slit nuts. The plant is highly mechanised equipped with automatic batching, mixing, transportation and placement of concrete. The compaction of concrete is achieved with the help of high frequency (9000 rpm) vibrators and simultaneous compression of 16 kN applied by a compression beam. The core bars secured to the moulds before placement of concrete are removed by means of core bar extractor immediately before demoulding. The units are then moved to heating stacks for steam curing. After concrete gains sufficient strength, the units are brought to prestressed line where U-shaped prestressing rods are inserted and slit nuts are tightened at the either end alongwith anchor washers and back rests at the other end. After the application of requisite prestressing force by means of hydraulic jacks, the grout mortar is injected and the ends are sealed.

15.3.4. Design Loads

The complexity of the dynamic loads imposed on a sleeper due to moving wheel loads arises from several factors including the following:

(i) the type of track (straight or curved), type of fittings and standard of contruction and maintenance

(ii) running characteristics, speed and standard of maintenance of rolling stock

(iii) magnitude of axle loads, their number and spacing

(iv) distribution of ballast reaction which is influenced by shape, rigidity, spacing and weight of sleepers; unit weight of rail, characteristics of the ballast and the standard of track maintenance.

On a straight track, the static load of 60 kN has been amplified to 150 kN by German and Indian Railways in order to include an impact factor of 1.5 due to dynamic effect of rolling loads. On a curved track, the vertical load on the outer rail is decreased and that on the inner rail is increased and at the same time a horizontal load acts in the outward direction on the outer rail due to centrifugal action. The Indian Railways have prescribed the equivalent static loads shown in Fig. 15.3.1, based on a sleeper spacing of 630 mm on a broad gauge track. The loads shown in Fig. 15.3.1(a) and (b) refer to a straight track and those in Fig. 15.3.1(c) refer to a curved track. For a straight track, two extreme cases of ballast pressure distribution have been considered. In Fig. 15.3.1(a) the ballast near the centre of sleeper is assumed to be loose and therefore offers no reaction. In Fig. 15.3.1(b), the ballast near the centre of sleeper has reached the condition of maximum compaction offering a reaction equal to one-half of that under the rails. In Fig. 15.3.1(c) which refers to a curved track, the outward horizontal load of 70 kN acts only on the outer rail and the vertical load W represents passive reaction offered by the other rail to maintain equilibrium of the sleeper. The loads for the metre gauge track are approximately two-thirds of those for broad gauge. These loads are shown in parentheses in the figure.

15.3.5. Analysis

The internal forces, viz. the bending moment and shear force due to service loads depend upon the type of sleeper. The most common monoblock sleeper behaves like a beam on elastic foundation. The complexity of the problem increases because the dynamic axle loads as well as the subgrade reaction of the ballast depend upon a large number of factors. The factors which influence the dynamic loads are enumerated in Sec. 15.3.4. The distribution of subgrade reaction depends upon the packing and compaction of the ballast. In a new track, the ballast under the rail seat is very well compacted whereas the ballast near the centre of the sleeper is kept loose. Consequently, subgrade reaction is confined in the regions near the rail seats. With passage of time, under service conditions the ballast near the rail seat works loose and that near the centre tends to get compacted. Consequently, the ratio of ballast reactions at midspan and rail seat increases progressively reaching a value as high as 0.5. In this case the hogging bending moment at midspan increases substantially and becomes comparable to the sagging bending moment at the rail seat. The actual distribution of ballast reaction has been idealised as shown in Fig. 15.3.1. for the sake of design. The bending moment and shear force in the sleeper may be computed readily using the equivalent static load and distribution of subgrade reaction shown in the figure. A bending moment of 13.3 kN.m at rail seat and a bending moment of 5.2 kN. m at midspan section have been prescribed for the design of a broad gauge sleeper.

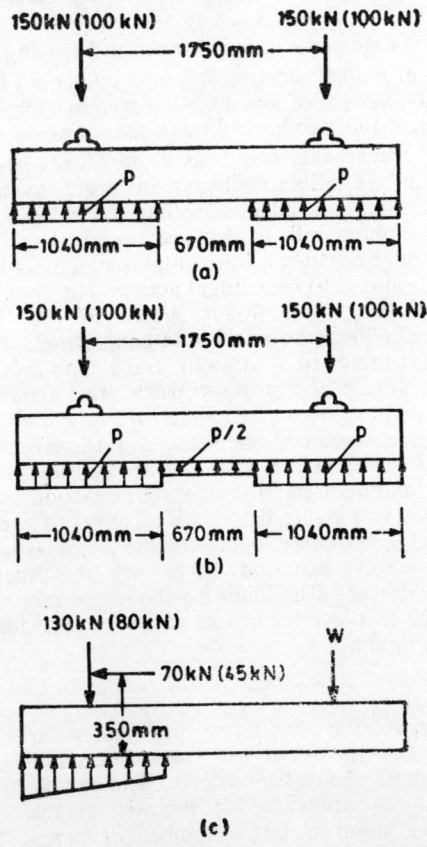

Fig. 15.3.1

15.3.6. Design Principles

The design of prestressed concrete sleepers is based on the following considerations:

 (i) resilience, ductility and fatigue resistance
 (ii) the characteristics of the external loads and reactive forces acting on the sleeper
 (iii) form and geometry of the sleeper including its shape, size and weight
 (iv) characteristics of the fastenings
 (v) impact load on sleeper in the event of derailment.

Due to uniformly high rigidity of monoblock or beam type sleeper, it is reasonable to assume the sleeper to be infinitely rigid. The German approach assumes the monoblock sleeper and also the supporting bed to be infinitely rigid and unyielding. In this concept, the design is based on the internal forces due to equivalent static load. A different approach was adopted in France where sleepers comprising two blocks connected by a tie rod were used. In this approach the sleeper is assumed to be supported by resilient bed capable of undergoing differential settlement. Due to the complexity of this approach, the design is evolved on empircal basis.

15.4 BEAMS, JOISTS AND GIRDERS

In view of the outstanding merits of prestressed concrete as a construction material enumerated in Sec. 1.14 coupled with the steep rise in the cost of production of structural steel, in many developing countries the prestressed concrete beams, joists and girders may compete favourably with rolled steel joists and girders. As beams, joists and girders are primarily flexural members, the advantage of eccentric prestressing can be taken to the fullest extent. Precast prestressed beams, joists and girders, either pretensioned or post-tensioned, find numerous applications in residential buildings, industrial structures, culverts and major bridges.

15.4.1 Residential Buildings

Most of the developing countries are faced with the problem of acute shortage of housing, particularly in fast growing urban areas. The urgent need for cheap and quick construction of houses is, therefore, evident. Precast prestressed beams and joists can offer a feasible solution to the problem. The spans in residential buildings usually vary in the range of 3 to 5 m. For these relatively small spans, precast pretensioned beams of approximately rectangular cross-section may be found very suitable. To facilitate demoulding and to minimise breakage during transportation, the sides of the cross-section may be made slightly sloping and the corners either rounded off or chamfered as shown in Fig. 15.4.1(a). These beams may be placed at a spacing ranging from 1 m to 2 m so as to keep the cross-sectional dimensions and consequently the weight within convenient limits keeping in view the problems associated with their handling and transportation. The beams may support either reinforced or prestressed concrete floor panels which may be either precast units suitable for mass production such as transmission line poles or railway sleepers. They may then be made available from ready stock in the same manner as rolled steel joists. Although this alternative method of construction for floors and roofs of residential buildings may prove to be both quicker and economical, it has not yet been used in a big way, perhaps due to general inertia and prejudices against new forms of construction.

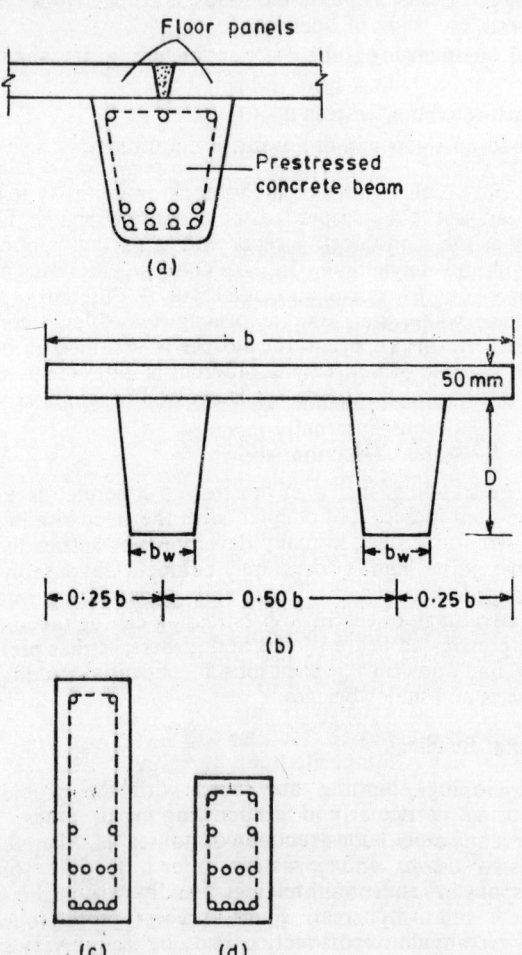

Fig. 15.4.

15.4.2. Public Buildings and Industrial Structures

Spans exceeding 5 m are common in public buildings and industrial structures. There are immense possibilities and varieties in framing systems for these structures. The following are the 'desirable features of floor and roof framing systems:

(i) high serviceability at working load and high ultimate strength
(ii) adaptability in span variations.
(iii) minimum total depth of construction
(iv) ease in the provision of openings for services
(v) aesthetically pleasing soffit
(vi) good insulation against heat and sound and high fire resistance.
(vii) minimum requirement of skilled labour
(viii) fast and convenient erection schedule
(ix) low capital investment and maintenance cost.

Precast prestressed joists and girders may form the main load bearing elements in flooring and roofing systems responsible for transferring the loads to supporting walls or columns. These girders usually have a thin walled open section such as symmetrical or unsymmetrical I, T or channel. In United States standardised sections in the form of single T and double T are extensively used. The Prestressed Concrete Institute (PCI) of United States has published a directory of precast prestressed concrete producers and their products which include beams, joists, girders, stemmed units, members with continuous internally formed voids, piles and a variety of other units. The double T-section shown in Fig. 15.4.1(b) has width b equal to 1.2 m, 1.8 m and 2.4 m. While the top flange has the same thickness (50 mm), the overall depth $D = 450, 650$ and 800 mm and minimum web thickness, $b_w = 64, 102$ and 121 mm. The single T-sections are of two types. While the first type of sections can be made from the same mould as for double T-section, the second type of sections is made in a mould which permits dimensional changes. The thicknesses of the flange and web of second type are greater than those of the first type and can be varied within specified limits. The section has a maximum flange width 3 m and the overall depth varies from 0.4 to 1.2 m. The double T and single T-sections have been used extensively up to 36 m in buildings and 18 m in bridges. The precast prestressed sections together with 50 mm thick cast-in-situ topping act like a composite floor. In the case of roofs, the appropriate insulation or roof topping may be applied directly to the precast prestressed units. The single T and double T sections have high structural efficiency due to relatively large section moduli for a given cross-sectional area. The main advantage of these sections is that a major portion of dead load acts at the time of stress transfer and can, therefore, be neutralised by additional eccentricity of prestress. Their main drawback is low torsional strength which is of much importance, particularly in buildings. The sections are usually manufactured as pretensioned units on 60 to 120 m long prestressing beds although they are sometimes manufactured by individual mould system.

A precast prestressed beam which is particularly suitable for asbestos or galvanised iron sheet roofing of industrial building, may have a non-prismatic section with horizontal soffit and sloping top. The cross-sections at midspan and end of such a beam are shown in Fig. 15.4.1(c) and (d) respectively. There are two main advantages of this roof beam.

The sloping top automatically provides for the drainage of rainwater. The eccentricity of prestress is maximum at midspan and decreases on and either side until it becomes zero at ends. The distribution of bending moment due to prestress is, therefore, more appropriate for simply supported beams. The beam is suitable for production by long line or individual mould method.

Precast prestressed joists of various shapes can be combined in several ways with plain or reinforced concrete elements to form a composite slab. They can also be used with cored or hollow units made from concrete or clay to serve as filler blocks. Composite slabs can also be formed by combining precast prestressed joists with poured gypsum and lightweight insulating material. Due to their excellent insulating property, lightweight filler blocks are particularly suitable for the construction of roof slabs in order to cut out the heat from direct sun. For small spans upto approximately 3 m, the joists may be placed directly on the supporting walls. In the case of longer spans, the joists may be supported on main girders. There are several advantages of composite slabs using precast prestressed joists. As their spacing is variable, the joists are adaptable to varying spans. The joist-filler block composite slab can accommodate openings of various sizes without difficulty and without weakening the structural system. Composite slabs using precast prestressed joists are generally found satisfactory in respect of appearance, fire resistance, stability during handling operations, stiffness and diaphragm action against horizontal loads. The precast pre-stressed units in a composite slab are placed wholly or partly in the flexural tension zone and are primarily responsible for resisting flexural tension. Among numerous possible arrangements for the construction of composite slabs for buildings and bridges, a few which are commonly adopted are shown in Fig 9.2.1.

15.4.3. Bridge Structures

The technique of precasting is eminently suited to bridge girders for quicker construction and minimum interference to traffic below the bridge. The girders may be cast in a central plant or in a temporary yard established near the work site. Girders up to about 40 m can be precast in a single piece and transported to site by means of truck cranes. In some western countries, particularly United States, the precast prestressed girders for bridges have been standardised to facilitate mass production under highly controlled and mechanised central plants to ensure high quality products. For instance, the American Association of State Highway and Transpor-tation Officials (AASHTO) has standardised precast prestressed bridge girders for spans ranging from 10.5 m to 42 m. These girders have a wide top flange for resisting flexural compression in combination with cast-in-place reinforced concrete deck slab. The web is relatively thin but the girder ends are flared and made solid to accommodate end anchorages and to resist safely the severe anchorage zone stresses. The bottom flange is also relatively wide to enable it to accommodate the prestressing tendons and to resist flexural compression at stress transfer. In order to ensure compo-site action by preventing longitudinal slip between the precat girders and cast-in-situ deck slab, the top surface of the precast girder is deliberately roughened to serve as a shear key. In addition, the stirrups in the precast girder are continued in the cast-in-place slab to serve as shear connectors by dowel action.

While precast prestressed girders cast in a single piece are feasible for simple spans upto about 40 m, larger spans upto about 50 m are possible by adopting continuous or balanced cantilever construction. For spans exceeding 50 m, box type section and segmental construction technique are eminently suitable. The box sections, which offer high strength and stiffness against bending, shear and torsion, may be precast in segments of suitable lengths either at place, central plant or temporary casting yard at work site. Each segment is prestressed transversely before erection. In order to ensure good alignment and close fitting of abutting surfaces in expoxy resin joints, each succeeding segment is cast against the leading face of the preceding segment. In the segmental technique of bridge construetion, the segments are lifted and launched one by one in the same manner as balanced cantilever construction. After each segment is launched to its final position, it is anchored by post-tensioning to the construction already existing. After the launching of all girders, the entire structure may be post-tensioned by continuous tendons to ensure continuity of the structure. The segmental technique of bridge construction is gaining popularity and many bridges using this technique have been constructed all over the world.

15.5. SLAB AND WALL PANELS

The acute shortage of housing and rapid pace of industrial development gave tremendous fillip to the technique of prefabrication of buildings and other structural systems. The major problem of cracking of large size flooring, roofing and wall panels due to shrinkage of concrete and temporary stresses during transportation and handling is eliminated by the technique of prestressing which induces permanent compression in concrete. By using large size panels, it is possible to construct a multistorey building without beams and columns. Precast prestressed solid and hollow panels are being used in numerous ways for flooring, roofing and wall.

15.5.1 Floor and Roof Slab Panels

Precast prestressed slab panels of solid section for flooring and roofing are being manufactured in thicknesses ranging from 50 mm to 300 mm. The main advantage of thinner panels is the ease of transportation and handling. However, they create problems associated with deflection and alignment. The difficulties due to unequal deflections of adjacent panels arise due to normal variations in quality of concrete, shrinkage of concrete and appreciable effect of even small errors in the eccentricity of prestress. Although these problems are minimised in thicker units, they suffer from the disadvantage of large self weight and the consequent difficulties in transportation and erection. In order to reduce self weight and yet maintain high strength and stifiness, hollow or cored slab panels were developed. These units ranging from 100 to 300 mm in thickness may have multiple longitudinal web openings or voids of rectangular, circular, elliptical, oval or other shapes. The size, shape and spacing of opening is chosen in such a way that top and soffit elements as well as the diaphragms have adequate strength and stiffness and there is minimum stress concentration. The voids may reduce the self weight from 25 to 50 per cent. The panels are usually manufactured by pretensioning technique either by long line or individual mould system. The units are often cast one over the other in several layers and moved to steam chamber for accelerated curing. The voids may be formed either by paper tubing or inflatable rubber tubes which are later deflated and withdrawn. While small slab units rest on beams, the large slab panels are those which span freely from wall to wall without the help of beams. These long panels may be manufactured on a continuous casting bed and later cut to required lengths after curing.

15.5.2 Lift Slabs

An ingenious method for the construction of multistorey buildings is the lift slab technique in which all slabs are cast one over the other at ground level and then lifted to their final position. The main advantages of this novel technique include almost total elimination of form work and scaffolding and placement of concrete only at ground level instead of high up in the air. The slabs are cast one by one and separated from each other by means of bond preventing material. The columns are cast to their full height and are provided with collars at appropriate places for the attachchment of slabs to them. When the top or roof slab has gained sufficient strength, it is post-tensioned by means of permanent prestressing tendons to resist service loads and also by temporary tendons, if necessary, to resist stresses caused during lifting. The roof slab is then lifted to its final position by means of a system of jacks coordinated by a central console. The slab is then attached to the columns permanently by moving the collars to their final appropriate position. A similar treatment is then given to the remaining slabs one by one until all slabs are secured to the columns. The columns have to be held plumb by means of temporary guys until the slabs are in position. The main disadvantage of lift slab construction is high equipment cost and expert supervision.

15.5.3. Wall Panels

Precast prestressed wall panel may be used for interior as well as exterior walls having finishes ranging from plane, exposed aggregate and a variety of intricate designs. A column free building is obtained when these wall panels are designed to serve as vertical supports. Wall panels of channel, single-T or double-T sections serve as excellent load bearing members. In order to provide good thermal insulation, particularly in the case of exterior walls, a sheet of insulating material may be sandwiched between layers of concrete on either side. The sandwiched panels of rectangular section may be made solid near both vertical edges for requisite strength and to accommodate the prestressing tendons. In the case of flanged panels, the insulation may be accommodated in the flange while the webs may be made solid.

15.5.4. Large Panel Construction

The techniques of precasting and prestressing may be employed with advantage for the construction of public buildings and industrial structures using large panels having flanged sections such as single or double-T for floors, roofs and walls. In western countries large panels having thin walled open sections have also been used extensively for the construction of multistorey apartment buildings. An alternative approach for the construction of apartment buildings is by means of precast prestressed room sized modules in the form of boxes which are erected one over the other fitted completely with plumbing, wiring, air-conditioning and other services. The main advantage of using room sized box modules is the ease of erection although it may require special facilities for handling and transportation.

15.6. COLUMNS AND OTHER UNITS

Precast columns are cast in horizontal position, thereby simplifying the formwork and greatly facilitating the placement and compaction of concrete. Columns of high quality concrete are obtained when they are precast in a central plant under strict supervision and control. As columns are essentially compression members, the question of prestressing them should be given careful consideration. In precast construction the beam-column connection is not as rigid as in cast-in-place construction. Consequently, the precast columns are subjected to relatively smaller bending moments. By proper design of connection details, it may even be possible to eliminate bending moments altogether thereby requiring the column to provide only vertical support. In such cases, the column does not benefit from prestressing in carrying the service loads. However, the prestress may still be beneficial in resisting the tensile stresses due to shrinkage, creep and thermal effects. The prestress may also be advantageous in resisting tensile stresses caused by self weight during transportation and handling, particularly when several storey high column is cast in a single-piece. These long columns usually have brackets or corbels at appropriate places for supporting the beams. When the object of prestressing is only to resist the temporary stresses during transportation and handling, the prestress may likewise be temporary which may be removed after erection. A beam-column connection, in which the beam transfers only a vertical reaction to the column, may be achieved by the provision of appropriate bearings made from tetrafluorethylene, elastomeric, neoprene or rubber pads which permit appreciable horizontal movements due to their relatively low shear resistance. Experience has shown that while these connections known as soft connections are helpful in relieving the columns from undesirable bending moments, they are not effective in resisting horizontal forces due to wind and earthquake. When the structure is required to possess high resistance to lateral forces, it is desirable to have hard beam-column connections which may be achieved in various ways. For instance, bearing plates rigidly connected to beam and column may be welded to each other. The hard connection can also be obtained by means of reinforcing bars and cast-in-place concrete. Precast columns in structures with hard connections may carry appreciable bending moments due to service loads. In this case the prestress may offer a real advantage.

A variety of other units such as fencing posts lampstands, goalposts and sheet piles may be mass produced as precast prestressed elements. Besides saving in the material cost, they possess a maintenance-free long useful life. The design aims at just the requisite strength and stiffness for the minimum weight. The prestressing force should be the minimum which is necessary to keep concrete in a state of perpetual compression. Substantial economy is possible through mass production, reuse of moulds, minimum materials and quality control.

REFERENCES

15.1 Abdul Rahman, P.M. and Streenath, G.H. *Precast prestressed concrete slab incorporating structural hollow clay blocks in floors and roofs*, The Indian Concrete Journal, Vol. 49, No. 5, May 1975.

15.2 Arvindan, N., *Mono-block prestressed concrete sleepers on Indian Railways—a performance review*, All India Seminar on Prestressed Concrete Structures, Proceedings Vol. 1, Madurai, 1979, pp. 239-54.

15.3 Felix, A., *Precast plate units for roof construction*, Proceedings of the Institution of Civil Engineers (London), Vol. 23, November 1962, pp. 321-6.

15.4 Fiorato, A.E., *Geometric variation in the columns of a precast concrete industrial building*, Journal of the Prestressed Concrete Institute, Vol. 18, No. 4. July-August 1973, pp. 50-60.

15.5 Gerwick Jr. D.C., *Long span prestressed concrete bridges utilising precast elements*, Journal of the Prestressed Concrete Institute, Vol. 9, No. 1, February 1964.

15.6 Holland, E.P. and Svab. L.E., *Design considerations for a precast prestressed apartment building—design problem*, Journal of the Prestressed Concrete Institute, Vol. 18. No. 6, November-December 1973, pp. 50-53.

15.7 Hoppe, *Mass production of prestressed concrete auxiliary members*, Journal of the Prestressed Concrete Institute. Vol. 8, No. 5, October 1963.

15.8 Korb, J.L., *Construction of post-tensioned roof panels*, Journal of the American Concrete Institute, Vol. 64, No. 7, August 1967, pp. 488-91.

15.9 Kunze Walter, E., *Survey of European concrete cross ties*, Proceedings of the Structural Division of ASCE. Vol. 88, April 1962, pp. 111-35.

15.10 Lague, D.J., *Load distribution test on precast prestressed hollow-core slab construction*, Journal of the Prestressed Concrete Institute, Vol. 16, No. 6, November-December 1971, pp. 10-8.

15.11 Mattock, A.H. and Kaar, P.H., *Continuous precast prestressed concrete bridges*, Journal of the PCA Research and Development Laboratories, Vol. 2, No. 1, January 1960, pp. 23-4.

15.12 Omsted, H., *Notes on post.tensioned lift slab*, Journal of the Prestressed Concrete Institute, Vol. 10, No. 6, December 1965, pp. 84-90.

15.13 PCI, *Precast prestressed concrete producers and products*, Prestressed Concrete Institute, Chicago, 1969.

15 14 PCI, *Manual on design of connections for precast prestressed concrete*, Prestressed Concrete Institute, Chicago, 1973, 99 p.

15.15 **Raths, C.H.**, *Design considerations for a precast prestressed apartment building — design of load bearing wall panels*, Journal of the Prestressed Concrete Institute, Vol. 19, No. 2, March-April 1974, pp. 68-92.

15.16 Sauter, S., *Tall precast towers*, Journal of the Prestressed Concrete Institute. Vol. 6, No. 3, September 1961, pp. 72-4.

15.17 Shah Navin, *Precast prestressed concrete components for a large span industrial building*, The Indian Concrete Journal, Vol. 50, No. 9, September 1976.

15.18 Shemic, M., *Bolted connections in large panel systems buildings'* Journal of the Prestressed Concrete Institute, Vol. 18, No. 1, January-February 1973, pp. 27-33.

15.19 SERC Report, *Investigations on prestressed concrete monoblock sleepers*, Project report of SERC, Madras, September 1975.

15.20 Symposium on *Prestressed concrete Pipes, Poles, Pressure Vessels and Sleepers*, Madras, 1972.

15.21 Varkay, I., *Load bearing wall panels*, Journal of the Prestressed Concrete Institute, Vol. 16, No. 4, July-August 1971, pp. 34-48.

PROBLEMS

15 1 A pole having an overall length of 10 m carries a resultant wind load of 1.6 kN at 0.6 m from top. The planting length is 1.4 m. The pole has a tapering rectangular section 100 mm × 100 mm at top which increases linearly to 100 mm × 300 mm at bottom. The pole is prestressed concentrically by means of 12 wires each of 5 mm diameter carrying an effective stress of 920 MPa. Using gross concrete section, determine the extreme fibre stresses in the pole at ground level.

15.2 A pole having an overall length of 12 m has to carry a resultant wind load of 2.5 kN at 0.6 m from top. The planting length is 1.5 m. Using concrete of grade M-45 and provisions of the Indian code, design the pole as (i) fully prestressed member (Type 1). (ii) moderately prestressed member (Type 2) restricting the tensile stress to 3 MPa and (iii) partially prestressed member (Type 3) restricting the fictitious tensile stress to 6 MPa. Compare the quantities of concrete and steel required in the three types.

15.3 A vierendeel pole having an overall length of 10.5 m carries a resultant wind load of 2.1 kN at 0.6 m from top. Design the pole adopting a planting length of 1.4 m. Use provisions of the Indian code. Check the design against torsion due to snapping of a conductor carrying a tension of 4 kN at an eccentricity of 400 mm.

15.4. Compare prestressed concrete sleepers with other types of sleepers in respect of strength. stiffness, resilience, durability. fire resistance, track aligment and maintenance, manufacture. handling, placement and over all cost.

15.5 Draw bending moment and shear force diagrams for a broad gauge sleeper for the loading and distribution of ballast reaction shown in

Fig. 15.3.1(a) and (b). Hence design the sleeper using concrete of grade M-50 and multi wire strands.

15.6 Design a prestressed concrete sleeper for metre gauge track taking the design load at each rail seat equal to 100 kN. Assume the ballast reaction to be of intensity p over end portions each of length 0.375 l and zero or 0.5 p in the central portion of length 0.25 l where l is the overall length of the sleeper. Use concrete of grade M-55 and long line system of pretensioning.

15.7 Design a precast prestressed concrete beam for a residential building for a clear span of 3 m. The beams spaced at 1 m to support precast concrete slab units of 50 mm thickness and cast-in-situ floor topping of 40 mm thickness. The design live load is 4 kN/m². Use long line system of pretensioning.

15.8 Design a prestressed concrete double-T section for the roof of an industrial structure. The width of each unit is 2 m and the clear span is 18 m. The precast units are topped by lime concrete terracing having an average thickness of 100 mm. The design live load is 1.5 kN/m². Adopt Lee McCall system of post-tensioning and concrete of grade M-45.

15.9 Design a prestressed concrete beam for A.C. sheet roof of an industrial building. The beams resting on walls have a clear span of 12 m and spacing of 3 m. The beam has a rectangular section with horizontal soffit and sloping top. The design live load is 0.75 kN/m². Use individual mould system of pretensioning and concrete of grade M-40.

15.10 With the help of suitable sketches, describe the applications of precast prestressed beams, joists and girders in residential, industrial and bridge structures. Point out the merits and demerits in each case.

List of Main Symbols

A	= cross-sectional area of member (effective)
A_{br}	= bearing area
A_c	= cross-sectional area of concrete
A_d	= total area of ducts
A_g	= gross cross-sectional area of member
A_p	= area of prestressing steel
$A_{p1}, A_{p2}, \ldots, A_{pn}$	= cross-sectional areas of tendons $1, 2, \ldots, n$
A_{pun}	= punching area
A_s	= area of longitudinal reinforcing steel in tension zone
A'_s	= area of longitudinal reinforcing steel in compressi cn zone
A_{sl}	= area of longitudinal reinforcing steel for torsion
A_t	= area of transformed section; also area cf stirrup (one leg) for resisting torsion
A_{tr}	= transformed cross-sectional area of member
A_v	= area of stirrup (both legs) for resisting shear
A_{vf}	= area of shear-friction reinforcement in a composite section
A_{vt}	= area of stirrup (both legs) for resisting shear and torsion
	= $A_v + 2A_t$
A_o	= area enclosed by shear flow in torsion
a	= depth of rectangular stress block
b	= breadth of beam
b'	= shorter centre line dimension of stirrup
b_f	= breadth of flange

b_l	= shorter diamension of rectangular section measured between centres of corner longitudinal bars
b_w	= breadth (thickness) of web or rib
b_1	= centre to centre distance between corner bars in the direction of the width
C	= total compression
C_c	= creep coefficient
C_{cs}	= coefficient for combined effect of creep and shrinkage
C_l	= coefficient for loss of prestress
C_{ep}	= coefficient for effective prestress
C_{sh}	= shrinkage coefficient
C_u	= ultimate compression
c	= concrete cover
c_1, c_2, c_3	= constants
D	= overall depth of beam
DL	= dead load
d	= effective depth of beam
d'	= longer centre line dimension of stirrup
d_d	= depth of centroid of duct area measured from top face
d_f	= depth (thickness) of flange
d_l	= longer dimension of rectangular section measured between centres of corner longitudinal bars
d_p	= distance of centroid of prestressing force from top face
d_s	= effective depth of reinforcing steel
d_v	= effective depth of beam in shear (larger of d_p and d_s)
d_1	= centre to centre distance between corner bars in the direction of depth.
EL	= earthquake load
E_c	= modulus of elasticity of concrete
E_{ci}	= initial tangent modulus of elasticity of concrete
E_{cs}	= secant modulus of elasticity of concrete
E_{ct}	= tangent modulus of elasticity of concrete

E_s	= modulus of elasticity of steel
e	= eccentricity
e_c	= eccentricity of centre of compression or line of thrust
e_{cl}, e_{cu}	= eccentricities of extreme lower and upper positions of centre of compression for stresses
e_k	= eccentricity of kern points
e_{cl}, e_{ku}	= eccentricity of lower and upper kern points
e_p	= eccentricity of prestressing force or cable line
$e_{p1}, e_{p2}, \ldots, e_{pn}$	= eccentricities of tendons 1, 2, . . ., n
F	= characteristic load
F_{bst}	= bursting tensile force
F_d	= design load
f	= characteristic strength
f_B	= stress in bottom fibre
f_c	= compressive strength
f'_c	= characteristic compressive (cylinder) strength of concrete
f_{ci}	= cube strength of concrete at transfer
f'_{ci}	= cylinder strength of concrete at transfer
f_{ck}	= characteristic compressive (cube) strength of concrete
f_{cr}	= modulus of rupture of concrete (flexural tensile strength)
f_{cu}	= cube compressive strength of concrete
f_{cy}	= cylinder compressive strength of concrete
f_d	= design strength
f_{dc}	= principal diagonal compressive stress
f_{dt}	= prinlipal diagonal tensile stress
f_e	= elastic stress in concrete
f_{ep}	= elastic stress in concrete at centroid of prestressing force
f_{fx}	= fictitious or hypothetical tensile stress in concrete at the tension face

f_m	= peak stress in concrete
f_p	= stress in prestressing steel
f_{pe}	= effective stress in prestressing steel after losses
f_{pi}	= initial stress in prestressing steel before losses
f_{pu}	= characteristic ultimate strength of prestressing steel
f_{py}	= yield strength of prestressing steel; also stress at 0.2 per cent offset.
f_{rc}	= residual compressive stress in concrete (in tension members).
f_s	= stress in reinforcing steel in tension zone
f_s'	= stress in reinforcing steel in compression zone
f_T	= stress in top fibre
f_t	= tensile strength of concrete
f_y, f_v', f_{sy}	= yield stresses of longitudinal tension steel, longitudinal compression steel and stirrup steel respectively
$f_1. f_2$	= principal stresses
G	= shear modulus
G_c	= shear modulus of elasticity of concrete
g	= acceleration due to gravity
h	= height
I	= moment of inertia (effective)
I_{cr}	= moment of inertia (effective)
I_g	= gross moment of inertia
I_t	= moment of inertia of transformed section
i	= Impact factor
J	= polar moment of inertia
j	= lever arm factor
K	= torsion constant
k	= radius of gyration; also coefficient for wave effect
k'	= stress factor in rectangular stress block
k_p	= prestress factor in torsion

k_1, k_2, k_3	= constants; also stress block parameters
LL	= live load
l_{bd}	= flexural bond length
l	= effective span
l_d	= development length
l_t	= transfer length or transmission length
l_x, l_y	= lengths of larger and shorter sides of slab.
$l_1, l_2, \ldots l_n$	= effective spans of continuous beam
M	= bending moment
M_{cr}	= cracking moment
$M_{cr,\,n}$	= cracking moment of non-prestressed beam
M_D	= bending moment due to dead loads
M_{dc}	= bending moment causing decompression at the extreme bottom fibre
$M_{e1}, M_{e2}\,M_{e3}$	= equivalent ultimate bending moments in modes 1, 2 and 3
M_L	= bending moment due to live loads
M	= bending moment due to prestress
M_{pi}	= bending moment due to prestress at initial stage
$\mathbf{M}_{pp}$	= primary bending moment due to prestress
M_{ps}	= secondary bending moment due to prestress
M_r	= resultant bending moment
	$= (M_p + M)$
M_t	= equivalent moment due to torque
M_u	= ultimate bending moment
M_{ud}	= design ultimate moment
$\mathbf{M}_w$	= bending moment due to working or serivce loads
M_y	= bending moment at yielding of tension steel
M_o	= bending moment causing decompression at the level of centroid of prestressing steel
m	= modular ratio

$$= E_s/E_c$$

$m_1, m_2, \ldots, m_n$	= modular ratio of layers, 1, 2, ..., n in a composite section
N	= axial tensile force
N_D	= axial tension due to dead loads
N_L	= axial tension due to live loads
N_{min}	= minimum axial tension
N_S	= axial tension due to service loads
n	= number
P	= prestressing force (effective)
P_i	= initial prestressing force before losses
P_k	= characteristic load in tendon
P_x	= prestressing force at a distance x from the jacking end
P_o	= prestressing force at the jacking end
$P_1, P_2, \ldots P_n$	= prestressing forces in tendons 1, 2, .., n respectively
p	= intensity of prestress in concrete
	$= P/A$
	= also, fluid pressure
p_i	= initial intensity of prestress at any point before losse
	$= P_i/A$
Q	= torsional shear flow
q	= intensity of torsional shear stress
R, r	= radius
S	= sag of cable profile
$S_1, S_2, \ldots, S_n$	= sags of cable profiles, 1, 2, ..., n in a continuous beam
s	= stirrup spacing
T	= torsional moment; also total tension in steel
T_c	= ultimate torsional moment carried by concrete
T_{c1}	= reduced ultimate torsional moment resisted by concrete
T_d, T_l, T_t	= ultimate torques controlled by compression in walls, longitudinal reinforcement and space truss respectively.

T_s	= ultimate torque resisted by steel
T_u	= ultimate torque
T_{up}	= ultimate torque of plain concrete section
t	= thickness; also time
t_d	= thickness of compression diagonals
u	= perimeter
u_t	= perimeter of tendon
u_o	= perimeter of shear flow in torsion
V	= shear force; also ultimate shear force
V_c	= ultimate shear force resisted by concrete
V_{cc}	= ultimate shear force resisted by concrete in a section cracked in flexure
V_{ci}	= nominal shear strength provided by concrete when diagonal cracking results from combined shear and moment
V_{cr}	= ultimate shear strength of concrete for the section cracked in flexure
V_{cw}	= nominal shear strength provided by concrete when diagonal cracking results from excessive principal tensile stress in web
V_{co}	= ultimate shear strength of concrete for the section uncracked in flexure.
V_D	= shear froce due to dead loads
V_L	= shear force due to live loads
V_p	= shear force induced by prestress
V_S	= shear force due to service loads
V_s	= ultimate shear force resisted by steel
V_u	= factored ultimate shear force
v	= flexural shear stress
v_c	= ultimate shear stress resisted by concrete
v_{ch}	= permissible horizontal shear stress at contact surface in composite seetion.
v_{chu}	= permissible ultimate horizontal shear stress at contact surface in composite section.
v_h	= horizontal shear stress along the contact surface in composite section

v_{hu}	= nominal ultimate horizontal shear stress at contact surface in composite section
v_{max}	= maximum permissible ultimate shear stress
W	= concentrated load
WL	= wind load
W_p	= concentrated force induced by prestress ˙
w	= intensity of distributed load; also crack width
w_b	= self weight of beam per unit length
w_c, w_w	= unit weight of concrete and water respectively
w_D	dead load per unit length
w_L	= live load per unit length
w_m, w_{max}, w_k	= mean, maximum and characteristic crack width respectively
w_p	= intensity of distributed force induced by prestres (effective)
w_{pi}	= intensity of distributed force induced by prestress at initial stage
w_{ub}	= intensity of unbalanced distributed load
w_{ubi}	= intensity of unbalanced distributed load at initial stage
x	= depth of neutral axis; also distance along x-axis
x_u	= depth of neutral axis at ultimate bending moment
y	= distance from centroidal axis; also distance along y-axis
y_B	= distance of bottom fibre from controidal axis
y_{po}	= side of loaded area in end block
y_T	= distance of top fibres from centroidal axis
y_o	= side of end block
Z	= section modulus
Z_B	= section modulus with respect to bottom fibre
Z_T	= section modulus with respect to top fibre
α	= angle change of cable
β	= flexural rotation or slope of beam
	$= \dfrac{d\Delta}{dx}$

	= also, inclination of principal stresses
γ	= shear strain
γ_f	= partial safety factor for load
γ_i	= factor of safety against lateral instability
γ_m	= partial safety factor for material
Δ	= deflection of beam
δ_m	= percentage reduction in moment due to redistribution of moments
ϵ	= normal strain in concrete
ϵ_c	= total creep strain of concrete
ϵ_{cc}	= unit creep strain or creep strain per unit stress
ϵ_e	= elastic strain in concrete
ϵ_p	= strain in prestressing steel
ϵ_{pe}	= effective strain in prestressing steel after losses
ϵ_{pi}	= initial strain in prestressing steel before losses
ϵ_{pu}	= strain in prestressing steel at ultimate or breaking stress f_{pu}
ϵ_{py}	= strain in prestressing steel at yield stress
ϵ_s	= strain in reinforcing steel
ϵ_{sh}	= shrinkage strain of concrete
ϵ_u	= ultimate or crushing strain of concrete
ϵ_y	= strain in reinforcing steel at onset of yielding
ϵ_o	= strain in concrete at peak compressive stress
η	= reduction factor for effective prestress or loss factor
θ	= slope of the cable line
μ	= coefficient of friction
ν	= Poisson's ratio
ν_c	= Poisson's ratio for concrete

ν_s	= Poisson's ratio for steel
ρ , ρ_{max}	= stress factor in torsion and its maximum value
ρ_p	= proportion of prestressing steel
σ_{cbc}	= permissible stress in concrete in bending compression at final stage
σ_{cbci}	= permissible stress in concrete in bending compression at initial stage
σ_{cc}	= permissible stress in concrete in direct compression at final stage
σ_{cci}	= permissible stress in concrete in direct compression at initial stage
σ_{ct}	= permissible tensile stress in concrete at final stage
σ_{cti}	= permissible tensile stress in concrete at initial stage
σ_{ft}	= maximum permissible value of hypothetical flexural tensile stress
τ	= torsional shear stress
τ_{bd}	= permissible bond stress
τ_c	= shear stress in concrete
$\tau_{c, max}$	= maximum permissible nominal shear stress
τ_{max}	= maximum torsional shear stress
τ_v	= nominal shear stress
	$= \dfrac{V}{bd}$
Φ	= curvature; also capacity or strength reduction factor
ϕ	= diameter of tendon; also torsional rotation per unit length
ψ	= torsional rotation
ψ_i , ψ_t	= slopes of stress-strain curves for concrete at origin and at any point
ψ_s	= slope of the chord joining origin to any point on the stress-strain curve for concrete

$\dfrac{d\psi}{dx}$ $=$ torsional slope or unit twist

Ω $=$ dimensionless constant

ω_p $=$ effective proportion of prestressing steel

$\quad = \rho_p \dfrac{f_{cu}}{f_{ck}}$

ω_p' $=$ effective reinforcement index

$\quad = \rho_p \dfrac{f_p}{f_c'}$

$\Omega_v,\ \Omega_t$ $=$ effectiveness factors for stirrup in shear and torsion respectively

Answers

1.1. (i) −0.4 m, (ii) zero. (iii) 0.6 m

1.2. (i) 200 kN.m, (ii) zero, (iii) −300 kN.m

1.4. 0.2 m, −120 kN.m

1.5. 0.202 m, −212.1 kN.m

1,6. 0.2 m, 38S kN.m

1.7. −0.2 m

1.8. −0.2 m, 0.3 m

1.9. $V_p = − 80$ kN, $M_p = − 200$ kN.m
$V_r = 70$ kN, $M_r = 175$ kN.m

1.10. $V_p = − 48$ kN, $M_p = − 120$ kN.m
$V_r = 32$ kN, $M_r = 80$ kN.m

1.11. Normal stresses due to prestressing. $f_T = 4\,375$ MPa,
$f_B = − 14.375$ MPa
Resultant stresses, $f_T = − 13.2$ MPa, $f_B = 3.2$ MPa

1.12. Normal stresses due to prestressing, $f_T = 4$ MPa,
$f_B = − 16$ MPa

1.13. Resultant stresses, $f_T = − 12.67$ MPa, $f_B = 0.67$

1.14. 0.1825 m, 0.365 m

1.15. (i) 564 kN.m (sagging), (ii) 1878 kN.m (sagging)

1.16. $−PS, 2/3\ PS, −1/3\ PS$

1.17. $\dfrac{wl^2}{24} − \dfrac{PS}{3}$

4.1. 45 MPa

4.2. 60 MPa

4.3. 25 MPa

4.4. 72 MPa

4.5. 67.5 MPa

4.6. 72 MPa

4.7. (i) 90 MPa, (ii) 79 MPa

4.8. 48.25 MPa

4.9. 77.32 MPa

4.10. 35.7 MPa

4.13. 40 MPa

4.14. (i) 242.6 kN, (ii) 237.3 kN, (iii) 194.7 kN

4.15. (i) 244.5 MPa, (ii) 905.5 MPa

5.1. Initial stage : $f_T = 0, f_B = -14$ MPa
Final stage : $f_T = -11$ MPa, $f_B = 0$

5.4. Initial stage : $f_T = 0, f_B = -14$ MPa
Final stage : $f_T = -12$ MPa, $f_B = 0$

5.7. 0.7425 MN.m

5.8. 50.16 kN/m

5.9. 0.9045 MN.m

5.10. 63.48 kN.m

5.11. 0.9481 MN.m

5.12. 1.53 MN.m

5.13. 2.765 MN.m

5.14. 5.239 MN.m

5.15. 0.879 MN.m

5.16. 5.416 MN.m

6.1. $e_{cl} = e_{kl} = 0.1$ m $\quad e_{cu} = e_{ku} = 0.1$ m

6.2. $e_{cl} = 0.1$ m $\quad e_{cu} = 0.15$ m

6.3. e_p (max) $= 0.2$ m $\quad e_p$ (min) $= 0.15$ m

6.4. e_p (max) $= 0.2$ m $\quad e_p$ (min) $= 0.1$ m

6.5. M_D (max) $= 0.168$ MN.m

6.6. M_D (max) $= 0.168$ MN.m

7.2. (i) 0.454 MPa, 106.8°,   (ii) 0.454 MPa, 73.2°

8.1. (i) 12.48 mm, (ii) 4.85 mm

8.2. (i) 0.0435 m, (ii) zero, (iii) 0.047 m

8.3. (i) 0.0395 m, (ii) 0.00874 m

8.4. (i) 0.1088 m, (ii) zero, (iii) 0.1175 m

9.1. 1.75 MN.m

9.2. 1.2 MN.m

9.3. 27.46 kN.m

9.4. 33.10 7N.m

9.5. 0.831 MN

9.6. 1.38 MN.m

9.7. 2.4 MN

9.8. 0.5 MPa

9.11. 0.6464 MN

9.12. 27.27 kN

9.13. 0.672 MN

10.1. (i) 0.108 MN.m (ii) 0.162 MN.m

10.2. (i) 0.706 (Chaikes), 0.643 (Ehurlimann)

      (ii) 0.661, (iii) 0.675

10.3. 13.75 MPa

10.4. -25.25 MPa

10.5. f_c=13.21 MPa, f_s=60.55 MPa, centre of compression=54.6mm below top face, $M = 160.3$ kN.m

10.6. 349.7 kN.m

10.7. 372.4 kN.m

10.9. 38.3 mm

10.10. 0.2036 mm

10.11. 0.163 mm

10.12.   186 kN.m

10.13.   Section is adequate

11.1. $P\left(\dfrac{Sb}{1} - e_A\right),\ P\left(\dfrac{Sa}{1} - e_B\right)$

11.3. $P\,(S - e_A)$

11.5. (d),

11.6   (a) $e_D = e_E = 0.15$ m, (b) $e_D = e_E = 0.1$ m,

   (c) $e_D = e_E = 0.1$ m

11.8. -78.3 kN.m

11.9.   0.6986 m

11.10. $f_T = 3.09$ MPa, $f_B = -13.09$ MPa.

11.11. -66.3 kN.m

11.12.   0.6542 m,

11.13.   166.67 kN

11.14.   29.63 kN.m

11.18. $M_B = 300$ kN.m, $M_F = 265$ kN.m, $M_C = 250$ kN.m

11.19.   187.5 kN.m

11.20.   Ductility conditions satisfied

12.1. $T_D = 57.76$ kN, $T_W = 300$ kN

12.2. $T_{dc} = 1.460$ MN, $T_{cr} = 1.787$ MN, $T_u = 2.334$ MN

12.3. $f_{rc} = 0.88$ MPa, $\dfrac{T_{dc}}{T_w} = 1.123$, $\dfrac{T_{cr}}{T_w} = 1.375$, $\dfrac{T_u}{T_w} = 1.795$

12.5. $f_{rc} = 2.25$ MPa, $p_{dc} = 1.11$ MPa, $p_{cr} = 1.40$ MPa,

   $p_{bst} = 1.57$ MPa

12.6.   0.594 MPa

12.7.   top $= -4.6$ MPa,   bottom $= 0.51$ MPa

12.9.   (i) 0.0713 MPa, (ii) 0.0923 MPa, (iii) 0.1418 MPa

12.10.   (i) $f_c = -4.712$ MPa, $f_1 = -5.784$ MPa and -0.216 MPa

   (ii) $f_c = -0.942$ MPa, $f_1 = -3.557$ MPa and -2.443 MPa

12.11. $T = 1.744$ MN/m, $M = -0.0444$ MN.m/m

12.12. $T = 0.323$ MN/m, $M = -0.0696$ MN.m/m

12.13. $T = 0$, $M = 0.00035$ MN.m/m

12.14. $T = -0.04$ MN/m, $M = -0.006$ MN.m/m

13.1. $f_{Ti} = f_{Bi} = -4.8$ MPa, $f_{si} = f'_{si} = -28.8$ MPa,
$f_T = -17.28$ MPa, $f_B = 0, f_S = -10.37$ MPa,
$f'_S = 93.31$ MPa, $f_{p1} = 939.3$ MPa, $f_{p2} = 877.1$ MPa

13.2. $\Phi_i = 0, \Phi = 0.00105$ radian/metre

13.3. $f_{c'cr} = 0.8316$ MPa. $f_{s'cr} = -59.756$ MPa,
Approx. values $f_{c'cr} = 0, f_{s'cr} = -64.75$ MPa

13.4. $f_T = -21.6$ MPa, $f_s = 250$ MPa, $f'_s = -250$ MPa,
$f_{p1} = 1200$ MPa, $f_{p2} = 720$ MPa

13.5. $f_T = -24$ MPa, $f_S = 250$ MPa, $f'_s = -250$ MPa,
$f_{p1} = 1464$ MPa, $f_{p2} = 660$ MPa

13.6. $W_C = 0.427$ MN, $W_t = 0.167$ MN, $W_b = 0.297$ MN,
$e_b = 0.941$ m.

13.7. 0.48 MN, 45 mm

13.8. 0.48 MN

13.9. 0.48 MN

13.10. 1.3044 MN, 1.2873 MN,

13.17. (i) 1.498 MN, (ii) 3.002 MN

13.18. (i) 28.86 m, (ii) 20.38 m

13.19. 1.596 MN

14.4. $f_T = -13.4$ MPa, $f_B = 1.4$ MPa

14.5. 12.74 kN/m²

14.6. 0.15 m

14.10. (i) $f_T = -0.012$ MPa, $f_B = -14.988$ MPa
(ii) $f_T = -8.844$ MPa. $f_B = -6.156$ MPa

14.12 12 kN/m²

14.14. 75 kN/m²

14.15. 1.988 MN

14.16. 83.33 kN/m²

14.17. 107 kN/m²

14.18. 39.7 kN/m²

15.1. -18.35 MPa and 2.41 MPa

Selected Book and Codes

1. ABBLES, P.W., *An Introduction to Prestressed Concrete, Vols. 1&2,* Concrete Publications Limited, London.

2. BELL, B.J., *Practical Prestressed Concrete Design,* Sir Issac Pitman and Sons, Limited, London.

3. COWAN, H.J., *The Theory of Prestressed Concrete Design, Statically Determinate Structures,* McMillan and Company Limited, London.

4. DAYARATNAM, P., *Prestressed Concrete Structures,* Oxford and IBH Publishing Company, New Delhi.

5. EVANS, R.H. AND BENNETT, E.W., *Prestressed Concrete Theory and Design,* Chapman, and Hall, London.

6. GUYON, Y., *Prestressed Concrete, Vols. 1&2,* C.R. Books Limited, London.

7. GUYON, Y., *Limit State Design of Prestressed Concrete, Vols. 1&2,* Applied Science Publishers, London.

8. KOMENDANT, A.E., *Prestressed Concrete Structures,* McGraw-Hill Book Company, New York.

9. KRISHNA RAJU, N., *Prestressed Concrete,* Tata McGraw-Hill Publishing Company Limited, New Delhi.

10. LEONHARODT, F., *Prestressed Concrete, Design and Construction,* Wilhelm Ernst and Sohn, Berlin.

11. LIBBY, J.R. *Modern Prestressed Concrete, Design and Construction Methods,* Van Nostrand Reinhold Company, U.S.A.

12. LIN, T.Y. AND BURNS, N.H., *Design of Prestressed Concrete Structures,* John Wiley and Sons, New York.

13. MAGNEL, G., *Prestressed Concrete,* Concrete Publications Limited, London.

14. MALLICK, S.K. AND GUPTA, A.P. *Prestressed Concrete,* Oxford and IBH Publishing Company, Calcutta.

15. NAWY, E.G., *Prestressed Concrete, A fundamental Approach,* Prentice Hall International, Inc. Englewood Cliffs.

16. PRESTON, H.K. AND SOLLENBERGER, *Modern Prestressed Concrete,* McGraw-Hill Book Company, New York.

17 RAMASWAMY, G.S., *Modern Prestressed Concrete Design*, Sir Issac Pitman Limited, London.

18. I.S. 456-1978, *Indian Standard Code of Practice for Plain and Reinforced Concrete*, Bureau of Indian Standards, New Delhi.

19. I.S. 1343-1980, *Indian Standard Code of Practice for Prestressed Concrate*, Bureau of Indian Standards, New Delhi.

20. B.S. 8110-1985, *Code of Practice for the Structural Use of Concrete*, British Standards Institution, Linford Wood, Milton Keynes.

21. ACI: 318-1989, *Building Code Requirements for Reinforced Concrete*, American Concrete Institute, Detroit.

22. CEB-FIP, *Model Code for Concrete Structures*, CEB-FIP International Recommendations, Comite Euro-International du Beton (CEB), 1978.

Index